W0227524

Flora Australiensis

A Description of the Plants of the
Australian Territory

VOLUME 5: MYOPORINEAE TO PROTEACEAE

GEORGE BENTHAM
FERDINAND VON MUELLER

CAMBRIDGE
UNIVERSITY PRESS

CAMBRIDGE UNIVERSITY PRESS

Cambridge, New York, Melbourne, Madrid, Cape Town,
Singapore, São Paolo, Delhi, Tokyo, Mexico City

Published in the United States of America by Cambridge University Press, New York

www.cambridge.org
Information on this title: www.cambridge.org/9781108037426

This edition first published 1870
This digitally printed version 2011

ISBN 978-1-108-03742-6 Paperback

CAMBRIDGE LIBRARY COLLECTION

Books of enduring scholarly value

Life Sciences

Until the nineteenth century, the various subjects now known as the life sciences were regarded either as arcane studies which had little impact on ordinary daily life, or as a genteel hobby for the leisured classes. The increasing academic rigour and systematisation brought to the study of botany, zoology and other disciplines, and their adoption in university curricula, are reflected in the books reissued in this series.

Flora Australiensis

George Bentham (1800–84) was one of Britain's most influential botanists, whose own collection of plant specimens numbered more than 100,000. Although he donated his herbarium to the Royal Botanic Gardens, Kew in 1854, he continued to make significant contributions to the field, including this exhaustive, seven-volume work detailing the plant life of Australia, which was published from 1863 to 1878. It was part of a series of works commissioned by the British government to document the flora in its colonies. Using the extensive numbers of specimens at Kew – and with the help of Ferdinand Mueller (1825–96), a German botanist in Australia – Bentham was able to compile descriptions of more than 8,000 species of Australian plants, making these volumes the first completed compendium of the flora of any large continental area. Volume 5, published in 1870, gives the details of 14 orders of monopetalae and monochlamydeae dicotyledon flora.

FLORA AUSTRALIENSIS.

FLORA AUSTRALIENSIS:

A DESCRIPTION

OF THE

PLANTS OF THE AUSTRALIAN TERRITORY.

BY

GEORGE BENTHAM, F.R.S., P.L.S.,

ASSISTED BY

FERDINAND MUELLER, M.D., C.M.G., F.R.S. & L.S.,
GOVERNMENT BOTANIST, MELBOURNE, VICTORIA.

VOL. V.

MYOPORINEÆ TO PROTEACEÆ.

PUBLISHED UNDER THE AUTHORITY OF THE SEVERAL GOVERNMENTS
OF THE AUSTRALIAN COLONIES.

LONDON:
L. REEVE & CO., 5, HENRIETTA STREET, COVENT GARDEN.

1870.

LONDON :
SAVILL, EDWARDS AND CO., PRINTERS, CHANDOS STREET,
COVENT GARDEN.

CONTENTS.

———

CONSPECTUS OF THE ORDERS CONTAINED IN THE FIFTH VOLUME.

Class. I. DICOTYLEDONS.

Subclass II. MONOPETALÆ.

(Continued from Vol. IV.)

(Ovary in the following Orders superior, usually 2-celled or the cells divided so as to be apparently 4-celled, with 1 pair of ovules or rarely 1 ovule or 2 superposed pairs of ovules to each true cell.)

XC. Myoporineæ. Shrubs. Leaves alternate. Stamens usually 4, in pairs anthers when open reniform and 1-celled by confluence. Ovary not lobed, the style terminal. Micropyle and radicle superior.

XCI. Selagineæ. Herbs or small undershrubs. Leaves alternate or the lower ones opposite. Stamens usually 4, in pairs; anthers straight, 1-celled. Ovary not lobed, the style terminal. Micropyle and radicle superior.

XCII. Verbenaceæ. Herbs shrubs or trees. Leaves opposite or rarely alternate. Stamens 2 or 4, in pairs, or rarely equal and isomerous with the corolla-lobes; anthers 2-celled. Ovary not at all or scarcely lobed, the style terminal. Micropyle and radicle inferior.

XCIII. Labiatæ. Herbs or shrubs. Leaves opposite. Stamens 2 or 4, in pairs; anthers 2-celled or 1-celled by abortion or by confluence. Ovary deeply lobed, the style nearly basal between the lobes. Micropyle and radicle inferior.

XCIV. Plantagineæ. Herbs. Leaves radical or tufted, rarely scattered. Flowers regular, *the corolla with 4 spreading scarious lobes.* Stamens 4 or fewer, equal; anthers 2-celled. Ovary not lobed, with a terminal style. *Seeds peltate.*

Subclass III. MONOCHLAMYDEÆ.

Perianth really or apparently simple, the lobes or segments all calycine or herbaceous, or all petal-like or scarious, or entirely wanting (rarely petals or petal-like staminodia in a few *Euphorbiaceæ* or *Phytolaccaceæ*).

* *Ovary (except in* Nyctagineæ?) *formed of several carpels, but 1-celled and usually 1-ovulate (except in a very few* Phytolaccaceæ *and* Amarantaceæ). *Embryo excentrical or curved; albumen mealy, rarely wanting* (Curvembryæ).

XCV. Phytolaccaceæ. Herbs undershrubs or rarely shrubs. Leaves alternate, without stipules. Ovules 1 to each carpel.

XCVI. Chenopodiaceæ. Herbs or undershrubs, often succulent or scaly tomentose. Leaves alternate or rarely opposite, without stipules. Perianth usually herbaceous. Ovary (of 2 or 3 carpels) 1-celled, with 2 or 3 styles or style-branches and only 1 ovule.

XCVII. Amarantaceæ. Herbs or undershrubs, rarely shrubs. Leaves alternate or opposite, without stipules. Perianth usually more or less scarious or coloured. Ovary (of 2 or 3 carpels) 1-celled, with 2 or 3 styles or style-branches and only 1 ovule or rarely a cluster of ovules, bearing no relation in number to that of the carpels.

XCVIII. PARONYCHIACEÆ. Herbs with the character of *Amarantaceæ*, except that the leaves (usually opposite) are accompanied by small scarious stipules or connected by a raised line or narrow membrane.

XCIX. POLYGONACEÆ. Herbs or shrubs. Leaves alternate ; stipules usually thin or scarious, forming a sheath or ring round the stem. Ovary (of 2 or 3 carpels) 1-celled, with 2 or 3 styles or style-branches and only 1 ovule.

C. NYCTAGINEÆ. Herbs shrubs or trees. Leaves usually opposite, without stipules. Lower portion of the perianth persistent and enclosing the ovary and fruit, the upper portion deciduous or withering. Ovary 1-celled with 1 ovule and an undivided style.

** *Ovary apocarpous or more frequently reduced to a single more or less oblique carpel, 1-celled with a single one or a pair of ovules and a single excentrical or oblique style or stigma. Embryo small or amygdaloid; albumen fleshy or none.*

CI. MYRISTICEÆ. Trees. Leaves alternate. Flowers diœcious. Perianth-lobes 1-seriate valvate. Stamens united in a central column. Carpel 1. Embryo very small at the base of a ruminate albumen.

CII. MONIMIACEÆ. Trees or shrubs. Leaves opposite. Perianth-lobes in 2 or more rows. Stamens opposite the perianth-lobes or indefinite. Carpels usually several. Embryo very small in a fleshy albumen.

CIII. LAURINEÆ. Trees or shrubs with alternate or rarely opposite leaves or (in *Cassytha*) leafless parasitical twiners. Perianth-segments usually in 2 rows. Stamens opposite the perianth-segments ; *anther-cells opening in deciduous valves.* Carpel 1. Fruit succulent. Albumen none. Radicle superior.

CIV. PROTEACEÆ. Trees or shrubs, rarely undershrubs. Leaves alternate or rarely opposite. Perianth-segments 4, valvate. Stamens opposite the perianth-segments and inserted on them. Carpel 1. Albumen none. Radicle inferior.

FLORA AUSTRALIENSIS.

ORDER XC. MYOPORINEÆ.

Flowers irregular or rarely nearly regular. Calyx persistent, more or less deeply toothed or divided into 5, rarely 4, lobes or segments. Corolla with 4 or 5, rarely more, lobes more or less 2-lipped or nearly equal, imbricate in the bud, the upper lip or lobes outside (or rarely inside ?). Stamens usually 4, in pairs, inserted in the tube of the corolla and alternating with its lower lobes, rarely nearly equal and as many as corolla-lobes ; anther-cells opening longitudinally, at first nearly parallel, confluent at the apex, and usually when open forming a single reniform cell. Ovary free, not lobed, normally 2-celled, with 2 collateral ovules, or 2 or 3 superposed pairs of collateral ovules in each cell, attached to the incurved margins of a placenta projecting from the dissepiment so as to divide each cell more or less perfectly into 2, with one of the ovules of each pair in each half-cell, or sometimes the ovary divided from the first into 2 to 4, or in extra-Australian species more, cells with one ovule in each cell. Ovules pendulous, anatropous, with a superior micropyle. Style simple, undivided, or obscurely notched at the apex. Fruit a dry or succulent drupe, the endocarp 2- to 4- or rarely more-celled or 1-celled by abortion, or separating into as many pyrenes. Seeds usually solitary in each cell or half-cell, very rarely 2 or 3 superposed, albuminous in the species where they have been seen ripe, but the albumen sometimes thin ; embryo straight, with a superior radicle.— Shrubs or rarely trees. Leaves alternate, scattered, or rarely strictly opposite, undivided. Flowers axillary, solitary, or in clusters of 2, 3, or more. Bracts at the base of the pedicels very small or none, no bracteoles on the pedicels.

The Order is chiefly Australian, and two genera entirely so ; a third extends sparingly over the Indian Archipelago, the Pacific Islands, and tropical Africa ; and there is a fourth monotypic tropical American genus. The affinity of this Order with Verbenaceæ is so striking that some have proposed uniting the two, but the superior radicle has, on examination, proved so constant a distinction, that I have followed Brown and others in maintaining the two as separate Orders. The three genera, very distinct as to the majority of their species, run so much into one another, through intermediate species with the characters differently combined, that it is impossible to ascribe to them any absolute limits.

Corolla usually campanulate, nearly regular, rarely shortly
  cylindrical at the base.   Ovary usually 2- to 4- or more
  celled, with 1 ovule in each cell, rarely 2-celled with 2 ovules
  in each cell .   .   .   .   1. MYOPORUM.
Corolla usually tubular at the base, with a more or less irre-
  gular limb.   Ovary 2-celled, with 2 or rarely 1 ovule in each
  cell .   .   .   .   2. PHOLIDIA.
Corolla usually tubular at the base, with a more or less irre-
  gular limb.   Ovary 2-celled, with 2 or 3 superposed pairs, or
  rarely 1 pair only of ovules in each cell .   .   .   .   3. EREMOPHILA.

1. MYOPORUM, Banks and Soland.

(Polycœlium *and* Disoon, *DC.*)

Calyx divided to the middle or nearly to the base into 5 lobes or segments not enlarged after flowering.   Corolla-tube usually short and
almost campanulate or shortly cylindrical at the base, lobes 5, nearly
equal and regular, or the lowest rather larger.   Stamens 4, alternating
with the lower lobes, or rarely 5, all nearly equal, and scarcely protruding or shortly exserted.   Ovary 2- to 4-celled, or in species not
Australian 5- or 6-celled, with 1 ovule in each cell, or rarely 2-celled
with 2 ovules in each cell.   Drupe usually small, but more or less
succulent. — Shrubs (or undershrubs?).   Leaves alternate or rarely
opposite, entire or toothed.   Pedicels axillary, usually clustered.   Flowers
small, mostly white.

The genus is represented by a few species in the Indian Archipelago and the Pacific
islands, and by one species in tropical Africa.   Of the thirteen Australian species here
enumerated, one may be the same as a New Caledonian one, the others appear to be all
endemic.

SECT. I. **Eumyoporum.**—*Calyx small, narrow.   Ovary 2- to 4-celled, with 1 ovule
in each cell.   Fruit globular or ovoid, not compressed.*

Erect or divaricate shrubs.   Leaves from lanceolate to obovate.
  Corolla more or less bearded inside, or rarely quite
  glabrous.   Perfect stamens 4.
Leaves acute or acuminate, entire or very rarely slightly
  serrate.   Corolla-lobes usually shorter than the tube . .   1. *M. acuminatum.*
Leaves obtuse, acute, or acuminate, some usually serrate.
  Corolla-lobes usually as long as the tube.   (Southern or
  Western seacoast or salt plant) .   .   .   .   2. *M. serratum.*
Erect shrubs.   Leaves lanceolate or linear, entire.   Corolla not
  bearded at the throat.   Stamens, 5.
Corolla 2 to 3 lines long .   .   .   .   3. *M. deserti.*
Corolla 4 to 5 lines long .   .   .   .   4. *M. laxiflorum.*
Diffuse or procumbent shrubs.   Leaves linear or cuneate, thick.
  Corolla-lobes bearded at the base, as long as the tube, or
  nearly so.   Fruit globular .   .   .   .   5. *M. parvifolium.*
Corolla-lobes glabrous, much shorter than the tube.   Fruit
  ovoid .   .   .   .   6. *M. brevipes.*
Diffuse or weak shrub.   Leaves opposite .   .   .   .   .   .   7. *M. oppositifolium.*

SECT. II. **Disoon.**—*Calyx small, narrow.   Ovary 2-celled, with 1 ovule in each
cell.   Fruit compressed.*

Fruit very flat, acute (about 3 lines long).
  Leaves linear-lanceolate, acute, entire or scarcely toothed,
  1½ to 3 in. long .   .   .   .   8. *M. platycarpum.*

Leaves oblong or lanceolate, obtuse, serrate, ½ to ¾ in. long.
  Plant very glutinous . . . . 9. *M. Beckeri.*
Fruit small and very obtuse, somewhat flattened (about 1 line
  long).
Leaves narrow, linear, 2 to 4 in. long . . . . 10. *M. floribundum.*
Leaves minute, crowded, cordate . . . . 11. *M. salsoloides.*

SECT. III. **Chamæpogonia.**—*Calyx-segments herbaceous. Ovary 2-celled, with 2 ovules in each cell. Fruit somewhat compressed.*

Calyx-segments 2 to 4 lines long . . . . 12. *M. debile.*

Species insufficiently known.
Calyx and foliage of some forms of *M. serratum,* but ovary
  and fruit said to be 2-celled, with 2 ovules or seeds in each
  cell . . . . 13. *M. mucronulatum.*

SECT. 1. EUMYOPORUM.—Calyx small, narrow. Ovary 2- 3- or 4-celled, with 1 ovule in each cell. Fruit globular or ovoid, not compressed.

The first six species of this Section, however different in extreme cases, run so much into each other that they might almost be reduced to varieties of a single one.

1. **M. acuminatum,** *R. Br. Prod.* 515. An erect glabrous shrub, exceedingly variable in stature, breadth of leaves, and size of flowers. Leaves alternate, in the common forms varying from elliptical-oblong to lanceolate or linear, more or less acuminate, much contracted towards the base, quite entire, and 1½ to 3 in. long, but sometimes the broader ones almost obovate and rather obtuse, or all smaller, or very rarely a few of the leaves marked here and there with a few distant teeth. Pedicels 2 to 4 lines long, in axillary clusters of 2 to 4 or rather more, or rarely solitary. Calyx-tube very short, segments narrow, acute, rather rigid, rarely above 1 line long. Corolla almost campanulate, about 3 to 4 lines long, the lobes nearly equal, spreading, rather shorter, or sometimes much shorter than the tube, more or less bearded inside as well as the tube, the hairs sometimes almost disappearing from the lobes, but on a close examination I have very rarely found them quite absent as in *M. deserti.* Stamens 4 without any rudiment of the fifth in the numerous flowers examined, although such a rudiment has been observed by others; anthers very shortly protruding. Ovary most frequently 4-celled, but occasionally with only 3 cells and ovules. Drupe nearly globular, 2 to 3 lines diameter, or rarely larger.

N. Australia. Dampier's Archipelago and Cygnet Bay, N.W. coast, *A. Cunningham;* Nichol Bay, N.W. coast, *Ridley's Expedition.*

Queensland. Common along the coast from Cape Upstart, *M'Gillivray,* to Moreton Bay, *A. Cunningham* and others.

N.S. Wales. Very common from Port Jackson to the northern frontier and in the desert interior to the Murray and the Barrier range.

Victoria. On the Murray and adjoining deserts, but apparently replaced on the south coast by *M. serratum.*

W. Australia. Murchison river, *Oldfield, Drummond,* 6th coll. n. 137.

This truly polymorphous species, usually distinguished from *M. serratum* by its acute or acuminate entire leaves, cannot, however, be separated from it by any positive characters; and on the other hand has been subdivided into several races, or supposed species, of which the following are the most marked :—

1. *ellipticum.* Leaves rather broad and scarcely acuminate. Flowers moderate-

sized.—*M. ellipticum*, R. Br. Prod. 515; A. DC. Prod. xi. 707.—About Port Jackson, *R. Brown, Sieber*, n. 223, and others.

2. *acuminatum.* Leaves rather broad, acuminate, mostly 3 to 4 in. long. Flowers larger than in *M. ellipticum.*—*M. acuminatum*, R. Br. Prod. 515; A. DC. Prod. xi. 707.—Barnard and Frankland islands, *M'Gillivray*; Brisbane river, Moreton Bay, *F. Mueller* and others: Port Jackson, *R. Brown, Sieber*, n. 222 and others: Hastings river, *Beckler. Pogonia glabra*, Andr. Bot. Rep. t. 283; *Andreusia glabra*, Vent. Jard. Malm. t. 108, although figured with 5 equal stamens, is probably this form.

3. *parviflorum.* Leaves of the typical form, or rather smaller and narrow. Flowers smaller, the beard of the corolla copious or rare, or sometimes none at all.—*M. tenuifolium*, R. Br. Prod. 515; A. DC. Prod. xi. 711, au Forst.?—Queensland coast, Keppel and Shoalwater Bays, etc. *R. Brown;* islands off the coast, *M'Gillivray, F. Mueller*, and others; Rockingham Bay and Rockhampton, *Dallachy* and others; Moreton Bay and other parts of the coast, *A. Cunningham.* Some of the N.W. specimens appear also to belong to this form.

4. *angustifolium.* Leaves narrow-lanceolate or almost linear, but on longer petioles and more acute than in *M. deserti*, the lobes of the corolla sometimes almost, or even quite, glabrous, but often much bearded, and the upper stamen wanting (or small and abortive?)—*M. montanum*, R. Br. Prod. 515; A. DC. Prod. xi. 708; *M. Cunninghamii*, Benth, in Hueg. Enum. 78; A. DC. l. c. 707; *M. cyanantherum* and *M. Dampieri*, A. Cunn. in A. DC. l. c. 708.—Port Jackson and Mount Hunter, *R. Brown*, but chiefly in the interior of Queensland and N. S. Wales, extending to the Murray, the Barrier Range, and to Cooper's Creek. To this form also belong most of the specimens from the N.W. coast as well as those from Murchison river.

The species is closely allied to, and perhaps should include, the New Caledonian *M. tenuifolium*, Forst., a name which in that case would claim the priority over Brown's. But on examining our New Caledonian specimens (Viellard n. 1091 and Deplanche n. 356), I find that, although they resemble some Queensland ones of the var. *parviflorum* yet the corolla is more perfectly glabrous inside, and the fifth stamen is present, although with a narrow barren anther. I have great doubts, however, whether this character will prove constant.

2. **M. serratum,** *R. Br. Prod.* 516. An erect or somewhat diffuse shrub, attaining several feet and usually glabrous, still more variable in foliage and flowers than *M. acuminatum*, and sometimes very difficult to distinguish from that species, whilst some of the forms enumerated below may be thought by many to be specifically distinct. Generally speaking the leaves are elliptical-oblong or lanceolate, obtuse or acute, and more or less serrate, but in a few maritime specimens they are all or nearly all entire, cuneate-oblong and obtuse. Flowers usually smaller than in *M. acuminatum*, and several in each axil, but sometimes quite as large as in that species. Calyx-lobes or segments varying fron lanceolate and under ¾ line long, to subulate and 1½ lines. Corolla-lobes usually as long as the tube, abundantly or sparingly bearded inside, at least at the base. Stamens 4, not exceeding the corolla-lobes, and sometimes scarcely protruding from the tube. Ovary-cells 2 to 4, but more frequently 3 than 2 or 4, with 1 ovule in each cell. Fruit globular or ovoid, not compressed, from 1 to 2 lines diameter, or even larger, but all the large ones appear to be deformed by the puncture of some insect.—A. DC. Prod. xi. 709; Bartl. in Pl. Preiss. i. 350; *M. insulare*, R. Br. Prod. 516; A. DC. l. c. 708; Bartl. in Pl. Preiss. i. 349; *M. tasmanicum*, A. DC. Prod. x. 709; Hook. f. Fl. Tasm. i. 287.

N. S. Wales. Apparently rare and only towards the Victorian frontier.

Victoria. Along the whole coast from Gipps Land to the Glenelg, *F. Mueller* and many others; Wimmera, *Dallachy.*

Tasmania. Kent's group, *R. Brown;* common along the N. coast, *J. D. Hooker.*
S. Australia. Spencer's gulf and other parts of the coast, *R. Brown* and others.
W. Australia. From the Great Bight, *Maxwell,* and King George's Sound, *R.
Brown* and many others, to Murchison River, *Oldfield, Drummond;* Dirk Hartog's
Island, *A. Cunningham, Milne,* and the Abrolhos, *Bynoe.*
The principal forms are the following :—
1. *obovatum.* Leaves obovate, oblong, rather broad, obtuse. Ovary very frequently
2-celled only, as figured Bot. Reg. 1845, t. 15.—*M. adscendens,* R. Br. Prod. 516.
A. DC. Prod. xi. 710.—Chiefly in Tasmania and the sea-coast of Victoria, S. Australia,
and King George's Sound.
2. *apiculatum.* Leaves linear-cuneate, ½ to 1 in. long, obtuse or mucronate, thick
and often entire.—*M. apiculatum,* A. DC. Prod. xi. 707.—Station uncertain, probably
W. coast, *Baudin's Expedition.* The above quoted specimens from Dirk Hartog's
Island and the Abrolhos, are very near it, but the leaves are broader. They are still
thick and entire, but some of the Murchison river specimens have both the narrow
and entire, and large and serrate leaves (all thick) on the same specimen.
3. *tuberculatum.* Leaves narrow, mostly serrate, rather thick and obtuse, sprinkled
or covered with raised glandular dots.—*M. tuberculatum,* R. Br. Prod. 516; A. DC.
Prod. xi. 710. Bartl. in Pl. Preiss. i. 349.—King George's Sound, *R. Brown* and others ;
Swan river, *Preiss. n.* 1351 (the latter approaching the var. *apiculatum*).
4. *subserratum.* Leaves mostly oblong or lanceolate, serrate, not very thick. Calyx-
segments short, as in all the preceding forms.—*M. subserratum,* Nees in Pl. Preiss. i.
350.—S. coast of W. Australia, King George's Sound, *R. Brown* and others, extending
to Swan river, *Drummond, Preiss. n* 1247, and eastward to the Great Bight, *Maxwell.*
This may be considered as the typical *M. serratum,* the typical *M. insulare* only
differing in the leaves being altogether larger.
5. *pubescens.* Like the preceding variety, but the branches, leaves and calyxes
copiously pubescent, with short spreading hairs. Ovary 2-celled only in the flowers
examined.—Gale's Brook, W. Australia, *Maxwell.*
6. *glandulosum.* Leaves small in some specimens, 1 to 2 in. long in others; almost
ovate, very tuberculate-glandular (more so than in the var. *tuberculatum*). Flowers
small.—*M. viscosum,* R. Br. Prod. 516 ; A. DC. Prod. xi. 710 ; *M. glandulosum,* A.
DC. l. c. 709, and (according to A. DC.) *Bertolonia glandulosa,* Spin. Jard. S. Sebast.
25. f. 2.—Coast of S Australia, *R. Brown* and others.
7. *gracile.* Leaves usually narrow, always acute and mostly serrate, thinner than
in most of the preceding forms. Pedicels slender. Ovary cells usually 3.—*M. caprarioides,* Benth. in Hueg.
Enum. 77 ; A. DC. Prod. xi. 707, *M. gracile,* Bartl. in Pl. Preiss. i. 350, A. DC. l.c. 708.
—Common in W. Australia, *Preiss. n.* 1350, *Drummond, Oldfield.* Some of Brown's
specimens are very nearly, if not quite, identical with this form.
8. *parviflorum.* Leaves small and narrow, sessile or nearly so, often tuberculate.
Flowers and fruits very small. Calyx-segments slender, as in the last variety. Ovary
cells 2 or 3.—Murchison river, *Oldfield.*
All the above varieties appear to be connected by several intermediate forms.

3. **M. deserti,** *A. Cunn.* ; *Benth. in Hueg. Enum.* 78. An erect,
glabrous shrub, nearly resembling the narrow-leaved varieties of *M.
acuminatum,* but the leaves still narrower, linear or linear-lanceolate,
acute or almost obtuse, entire, rather thick, 1 to 2 in. long, and narrowed into a very short petiole. Pedicels often several together, rather
thick, and almost always remarkably recurved. Calyx and corolla
about the size of the smaller-flowered varieties of *M. acuminatum,* but
the corolla very regular, without any or with scarcely any hairs in the
throat. Stamens 5, all equal in every one of the numerous flowers
examined, the anthers not exserted. Ovary 2-celled or very rarely 3-
celled, with 1 ovule in each cell. Fruit ovoid, "yellowish," 2 to 3 lines
long, not compressed, usually with 2 cells and seeds.—A. DC. Prod. xi.

707; *M. strictum* and *M. patens*, A. Cunn. in A. DC. Prod. xi. 708; *M. dulce*, Benth. in Mitch. Trop. Austr. 384; *M. rugulosum*, F. Muell. in Linnæa, xxv. 427.

Queensland. Burdekin river, *F. Mueller;* Belyando and Balonne rivers, *Mitchell;* Nerkool Creek, *Bowman;* Darling Downs, *Lau.*

N. S. Wales. Lachlan river, *A. Cunningham;* from the Murray and Darling to the Barrier Range, *Victorian and other Expeditions;* Mudgee, *Woolls;* New England, *C. Stuart.*

Victoria. Murray river and Bacchus Marsh, *F. Mueller.*

S. Australia. From the Murray to St. Vincent's Gulf, Flinders Range, &c., *F. Mueller;* in the interior, *M'Douall Stuart's Expedition.*

W. Australia. Æstuary of the Murchison, *Oldfield;* Shark's Bay, *Milne.*

4. **M. laxiflorum,** *Benth.* An erect shrub, closely allied to *M. deserti,* with the foliage and habit of the broader leaved forms of that species, differing chiefly in the larger flowers. Leaves narrow-lanceolate, acute, contracted into a short petiole. Pedicels solitary or 2 or 3 together, often ½ in. long. Calyx-lobes rather longer than in *M. deserti.* Corolla fully 5 lines long, the lobes bearded inside at the base or nearly glabrous. Stamens 5, all equal, included in the tube. Ovary 2-celled, with one ovule in each cell. Drupe succulent, with a hard putamen, not compressed.—*Eremophila myoporoides,* F. Muell. Fragm. v. 23.

Queensland. Cape river, *Bowman ;* Rockhampton, *Thozet.*

I examined 4 ovaries and found them all 2-celled, with 1 ovule in each cell, and the drupe I cut across had also only 2 seeds; but in one drupe dissected by F. Mueller there were 4 cells and seeds. The ripe drupes were, however, all loose in the sheets with the specimens, and this one may have got mixed among them from some other species.

5. **M. parvifolium,** *R. Br. Prod.* 516. Stems procumbent, extending sometimes to 2 ft. or more, the whole plant glabrous. Leaves scattered, rather crowded, linear or linear-spathulate, obtuse, or rarely almost acute, entire, thick, and sometimes succulent, contracted at the base, and sometimes shortly petiolate, all under ½ in. in some specimens, above 1 in. in others. Flowers solitary or 2 or 3 together, on slender pedicels, sometimes very short, but often ½ in. long, or even more. Calyx-segments rather acute, about 1¼ to 1½ line long. Corolla campanulate, glabrous inside or nearly so, about 4 lines long, the lobes at least as long as the tube. Stamens 4, often exceeding the lobes. Ovary 3- or 4-celled, with 1 ovule in each cell. Drupe ovoid-globular, attaining about 2 lines, the putamen with 3 or 4 cells and seeds, or fewer by abortion.—A. DC. Prod. xi. 710, Bot. Mag. t. 1693.

Victoria. Murray river, *F. Mueller, Herrgott.*

Tasmania. Flinders Island, *Milligan.*

S. Australia. Memory Cove, Spencer's Gulf, *R. Brown;* W. of Mount Sturgeon, *Robertson;* lagoons near Rivoli Bay and Holdfast Bay, *F. Mueller;* Port Lincoln, *Wilhelmi;* Spencer's Gulf, *Warburton.*

W. Australia. Goose Island Bay, *R. Brown.*

M. humile, R. Br. Prod. 516; A. DC. Prod. xi. 710, is founded upon specimens of what appears to me to be a slight variety of *M. parvifolium,* with rather shorter and broader leaves.

6. **M. brevipes,** *Benth.* The specimens have the aspect of some of the short, thick-leaved ones of *M. parvifolium,* but the stems may be erect. Leaves linear, obtuse, very thick, all under ½ in. long. Flowers

of the size of those of *M. parvifolium*, but different in shape. Pedicels mostly solitary, not above 2 lines long. Calyx-segments acute, about 1 line long. Corolla glabrous inside, the tube about 2 lines long, not much dilated upwards, the lobes nearly equal, scarcely 1 line long. Stamens 4, rather shorter than the lobes. Ovary 2-celled, with 1 ovule in each cell. Fruit oblong, not compressed, about 2 lines long.

S. Australia. From M'Douall Stuart's journey into the interior; only known from very few specimens.

7. **M. oppositifolium,** *R. Br. Prod.* 516. A weak shrub, ascending to 3 or 4 feet, usually bearing numerous resinous tubercular glands. Leaves opposite, sessile, and usually stem-clasping, lanceolate or oblong-lanceolate, acute, serrate, ¾ to 1½ in. long, the margins often recurved. Pedicels rather slender. Calyx-segments rather narrow, acute, rather long. Corolla very open, the lobes somewhat longer than the tube, very slightly bearded inside at the base. Stamens 4, the anthers short. Ovary 2- or 3-celled, with 1 ovule in each cell. Fruit small, globular, not compressed.—A. DC. Prod. xi. 710.

W. Australia. King George's Sound, *R. Brown, A. Cunningham,* and many others.

SECT. 2. DISOON. Calyx small, narrow. Ovary 2-celled, with 1 ovule in each cell. Fruit compressed.

8. **M. platycarpum,** *R. Br. Prod.* 516. A tall shrub, or small tree, quite glabrous. Leaves linear-lanceolate, acute, entire, or with a few small distant teeth in the upper part, 1⅓ to 3 in. long, rather thick, contracted into a short petiole. Pedicels often 6 or more in the axils, 1 to 2 lines long. Calyx not ¾ line long, acutely lobed. Corolla more or less bearded inside at the throat, sometimes scarcely 2 lines long with the stamens included, in other specimens twice as large with exserted stamens. Ovary 2-celled, with 1 ovule in each cell. Fruit ovate or ovate-oblong, acute, much flattened, about 3 lines long.— A. DC. Prod. xi. 711.

N. S. Wales. Murray and Darling rivers, *Victorian and other Expeditions.*
Victoria. Wimmera, *Dallachy.*
S. Australia. Spencer's Gulf, *R. Brown;* Encounter Bay, *Whittaker;* Murray Scrub, *Behr.;* Elders and Flinders Range, Lakes Hindmarsh and Gairdner, *F. Mueller.*

9. **M. Beckeri,** *F. Muell.* An erect, much-branched shrub of several feet, strongly scented, and very viscous. Leaves alternate, oblong or lanceolate, rarely almost ovate, serrate, about ½ to ¾ in. long, contracted into a petiole. Flowers shortly pedicellate, often 2 or 3 in the axil. Calyx 1 to 1¼ lines long, deeply divided. Corolla-tube cylindrical to about 2 lines, then expanded into a small campanulate throat, the lobes spreading to 5 or 6 lines diameter, slightly bearded inside towards the base, the middle lower one larger and broader than the others, and slightly notched. Anthers 4, shortly protruding from the tube. Ovary 2-celled, with 1 ovule in each cell. Fruit ovate, acute, much flattened, exceeding the calyx, but not seen quite ripe.—*Disoon Beckeri,* F. Muell. Fragm. iv. 48; vi. 150; *Eremophila Beckeri,* F. Muell. Fragm. i. 156.

W. Australia. *Drummond, n.* 338 ; Phillips river and sand hummocks, Eyre's Relief, *Maxwell.*

Notwithstanding some approach in the form of the corolla to that of *Pholidia*, this species agrees in other respects much more with *Myoporum*, approaching very near *M. platycarpum* in the ovary and fruit.

10. **M. floribundum,** *A. Cunn. ; Benth. in Hueg. Enum.* 78. A glabrous, strong-scented shrub of 5 or 6 ft. Leaves very narrow, linear, acute, entire, 2 to 4 in. long on the main branches, often much smaller on the lateral ones. Flowers small, often numerous, in axillary clusters of 3 to 6, on filiform pedicels of 1 to 2 lines. Calyx-segments rather thick, acute, $\frac{1}{2}$ to $\frac{3}{4}$ lines long. Corolla-tube about 1 line long, lobes rather longer, almost acute, not much imbricate in the bud, glabrous inside. Stamens 4, rather longer or shorter than the corolla-lobes. Ovary compressed with a nerve-like border, 2-celled, with 1 ovule in each cell. Style filiform, the stigma obtuse. Fruit compressed, or at length somewhat turgid, very obtuse, almost truncate, 1 to 1¼ lines long. — *Disoon floribundus*, A. DC. Prod. xi. 703 ; F. Muell. Fragm. i. 126.

N. S. Wales. Rocky banks of the Nepean river, *A. Cunningham.*
Victoria. Snowy river, *F. Mueller.*

11. **M. salsoloides,** *Turczan. in Bull. Soc. Imp. Nat. Mosc.* 1863, ii. 226. An erect, very much branched shrub of several feet, quite glabrous, but often glandular-tuberculate. Leaves very small (under 1 line long), but numerous, sessile, cordate, often broader than long, very thick, spreading. Flowers solitary in the axils, on short pedicels. Calyx not 1 line long, deeply divided into lanceolate lobes. Corolla 2½ to 3 lines long, the lobes longer than the tube, spreading, nearly equal. Stamens 4, exserted. Ovary 2-celled, with 1 ovule in each cell. Fruit small, obtuse, somewhat compressed, like that of *M. floribundum.* — *Disoon cordifolius*, F. Muell. Fragm. i. 126 ; vi. 150.

W. Australia. *Drummond, 5th coll. n.* 339 ; Gordon, Phillips, and Salt rivers, *Maxwell.*

SECT. 3. CHAMÆPOGONIA. Calyx-segments herbaceous. Ovary 2-celled, with 2 ovules in each cell. Fruit somewhat compressed.

12. **M. debile,** *R. Br. Prod.* 516. A low glabrous shrub, with a thick stock and decumbent or ascending stems, attaining sometimes 2 ft. or more, the branches often glandular-tuberculate. Leaves alternate, very shortly petiolate, or nearly sessile, elliptical oblong or lanceolate, entire or with a few small distant acute teeth, and often one or two larger ones on each side near the base, 1½ to 3 or even 4 in. long. Pedicels solitary in the axils or in pairs, rarely so long as the calyx, Calyx-segments linear but leaf-like, acute, 3 to 4 lines long. Corolla pink or purplish, the tube about as long as the calyx, the lobes not half so long, more or less bearded inside at the base. Stamens included in the tube. Ovary 2-celled, with 2 collateral ovules in each cell. Fruit ovoid, somewhat compressed, 3 to 4 lines long, often furrowed on each side, 2-celled. Seeds either 2 in each cell more or less separated by an

imperfect dissepiment, or more frequently solitary by the abortion of the other ovule.—Bot. Mag. t. 1830, A. DC. Prod. xi. 711; *Pogonia debilis*, Andr. Bot. Rep. t. 212; *Andreusia debilis*, Vent. Jard. Malm. under n. 108; *Myoporum diffusum*, R. Br. Prod. 516, A. DC. Prod. xi. 711; *Capraria calycina*, A. Gray, in Proc. Amer. Acad. vi. 49; Benth. Fl. Austral. above, iv. 503.

Queensland. Keppel and Shoalwater Bays and Broad Sound, *R. Brown ;* Dawson river, Brisbane river, Moreton Bay, *F. Mueller ;* Rockhampton, *Dallachy* and others; Nerkool Creek, Connor's river, *Bowman ;* Darling Downs, *Lau.*

N. S. Wales. Port Jackson to the Blue Mountains, *R. Brown* and others ; Hunter's river, *A. Cunningham* and others ; Clarence river, *Beckler ;* Richmond river, *Fawcett.*

Since the publication of the last volume, Dr. Torrey has kindly re-examined and sent me a flower from the specimen inadvertently described by A. Gray as a *Capraria*, of which it had so much the aspect, and to which species I had probably myself referred it on a first hasty sorting.

Doubtful Species.

13. **M. mucronulatum,** *A. DC.* Prod. xi. 706. A glabrous shrub with the aspect foliage and flowers nearly of the var. *apiculatum* of *M. serratum*, but said to have a very different ovary and fruit. Leaves oblong or oblanceolate, obtuse, mucronate, much contracted into a petiole, rather thick, quite entire, 1 to 2 in. long. Pedicels 2 to 4 together, 2 to 4 lines long. Calyx-lobes short. Corolla-lobes slightly bearded inside. Fruit ovoid-globular, 2-celled, with 2 seeds in each cell, not separated by any spurious dissepiment.

N. S. Wales. "East Coast" *Herb. Mus. Par.* I have seen the specimen described by De Candolle, but have not had the opportunity of examining the ovary or fruit. The stations given for Australian plants from the collections of Baudin and other early navigators are not to be depended upon, the "côte occidentale" or "côte orientale" being sometimes attached to plants from the Recherche Archipelago or from the north coast.

2. PHOLIDIA, R. Br.

(Pseudopholidia, *A. DC.;* Sentis, Duttonia and Pholidiopsis, *F. Muell.*)

Calyx divided to the base, with 5 or rarely 4 segments, often unequal, somewhat dilated and much imbricate at the base, acuminate, not enlarged after flowering. Corolla-tube shortly cylindrical at the base, expanded into an obliquely campanulate throat, the limb of five spreading lobes, not very unequal, the 2 upper ones usually rather more united. Stamens 4, didynamous, usually exserted from the corolla-tube, but shorter than the lobes. Ovary 2-celled, with 2 ovules, or very rarely only 1 in each cell; style usually longer than in *Myoporum*, and hooked at the end. Fruit a dry or rarely succulent drupe, 2-celled, or more or less perfectly 4-celled, with 1 seed in each cell. —Shrubs. Leaves alternate scattered or irregularly opposite, entire or toothed. Flowers axillary, solitary and sessile, or on very short pedicels (except in *P. santalina*).

The genus is limited to Australia. In the typical forms it differs from *Myoporum* in the more perfectly divided calyx, the shape of the corolla, the more didynamous stamens, as well as in the ovary and fruit and inflorescence, but *Myoporum Beckeri* has almost the corolla and *M. debile* the ovary of *Pholidia*, whilst *Pholidia brevifolia* and perhaps

P. Delisserii have only 1 ovule in each of the ovary-cells as in the majority of *Myopora.* From *Eremophila, Pholidia* differs in the more regular corolla, the calyx-segments never enlarged after flowering, the stamens not exserted, the ovules never superposed in each half cell, and the fruit not separating into 1 seeded nuts as in the section *Eremocosmos,* nor so succulent as in *Stenochilus,* but none of the latter characters are constant through all the species of *Eremophila,* and the fruits of some species are as yet unknown. The distinction between *Pholidia* and *Eremophila* is not, therefore, more definite than that between *Pholidia* and *Myoporum.*

Leaves mostly opposite, hoary or white, almost scaly, usually re-
    curved at the end. Flowers sessile or nearly so.
    Leaves narrow-linear, ¾ to 1 in. long . . . . 1. *P. Dalyana.*
    Leaves narrow-linear, rarely above ¼ in. long . . . . 2. *P. scoparia.*
    Leaves obovate or oblong, 3 to 4 lines long . . . . 3. *P. Delisserii.*
Leaves alternate, obovate, or ovate, 3 to 5 lines long. Flowers
    sessile or nearly so.
    Leaves very thick, complicated and recurved, glaucous or hoary 4. *P. crassifolia.*
    Leaves rather thick, white on both sides, resinous . . . . . 5. *P. resinosa.*
    Leaves rather thin, green, often toothed.
        Leaves obovate or cuneate, acute, mostly toothed.   Ovules 2
            in each cell . . . . 6. *P. Behriana.*
        Leaves ovate or elliptical-oblong, entire.   Ovules 2 in each
            cell . . . . 7. *P. Woollsiana.*
        Leaves ovate, entire or toothed.   Ovules 1 in each cell . . 8. *P. brevifolia.*
Leaves alternate, entire, crowded or imbricate, ½ to 1 in. long.
    Flowers sessile or nearly so.
    Leaves oval or oblong, white-tomentose. Ovary glabrous . . 9. *P. imbricata.*
    Leaves linear, acute, glabrous. Ovary woolly . . . . . . 10. *P. densifolia.*
Leaves alternate, narrow or small, not crowded. Flowers sessile
    or nearly so.
    Leaves small, erect, with a few large tubercles. Fruit com-
        pressed, obtuse, not exceeding the calyx . . . . 11. *P. gibbifolia.*
    Leaves linear, entire. Branches divaricate, often spinescent.
        Fruit beaked . . . . 12. *P. divaricata.*
    Leaves linear, not gibbous, erect. Branches erect, hoary-pube-
        scent or nearly glabrous. Leaves 2 to 4 lines long . . . . 13. *P. microtheca.*
        Very viscid-pubescent. Leaves 1 in. long. . . . . . . 14. *P. adenotricha.*
    Leaves narrow-lanceolate, above 1 in. long. Pedicels about ½ in.
        long . . . . 15. *P. santalina.*

1. **P. Dalyana,** *F. Muell.* Very closely allied to *P. scoparia,* and perhaps a variety only, differing in the leaves longer and more slender, usually ¾ to 1 in. long, the corolla-tube not so much contracted at the base, and the ovary densely villous instead of being scaly only.—*Eremophila Dalyana,* F. Muell. Fragm. v. 22.

S. Australia. Between Cooper's Creek and Stoke's Range, *Howitt's Expedition.* There is but a single specimen (*Herb. F. Muell.*), and I could only examine one ovary, which was 2-celled as in *P. scoparia,* but one ovule of each cell was very small and probably abortive.

2. **P. scoparia,** *R. Br. Prod.* 517. An erect shrub, hoary or almost silvery, with a close more or less scaly indumentum, the branches rigid, but not thick, with prominent angles decurrent from the leaves. Leaves mostly opposite or nearly so, narrow-linear with hooked points, rather thick, keeled underneath, channelled above, rarely exceeding ½ in. in length. Flowers of a pale violet blue, solitary on short axillary pedicels, without bracts. Calyx 1¼ to 1½ line long,

deeply divided into 5 acute keeled segments. Corolla 8 to 9 lines long, the narrow part of the tube twice as long as the calyx, hairy inside at the top, the broad, almost campanulate, upper part or throat at least as long as the narrow part, the lobes much shorter, and nearly equal. Stamens inserted at the top of the narrow part of the tube, and about as long as the broad part. Ovary 2-celled, with 2 ovules in each cell. Fruit ovoid or ovoid-oblong, nearly 3 lines long, the putamen completely 4-celled, with 1 seed in each cell.—Endl. Iconogr. t. 66; A. DC. Prod. xi. 713; *Eremophila scoparia,* F. Muell. in Proc. R. S. Tasm. iii. 296, Fragm. v. 22.

N. S. Wales. Nandirooga Creek and towards the Barrier Range, *Victorian Expedition.*

S. Australia. Head of Spencer's Gulf, *R. Brown;* in the scrub from the Murray river to St. Vincent's Gulf, *F. Mueller;* Gawler ranges, *Sullivan;* Lake Gairdner, *Babbage;* head of the Great Australian Bight, *Delisser.*

3. **P. Delisserii,** *F. Muell.* A shrub of 2 to 3 ft., the branches and foliage hoary or white with a close tomentum, and sometimes glandular-tubercular. Leaves mostly opposite, from obovate to oblong, obtuse, recurved, contracted into a short petiole, 3 to 4 lines long. Flowers sessile, only seen in very young buds; but according to F. Mueller's description and figure, the calyx-segments narrow, woolly, imbricate, about 2½ lines long, not enlarged after flowering, the corolla nearly 8 lines long, tomentose outside, the cylindrical portion of the tube nearly as long as the calyx, with a ring of wool inside at the top, the broad part campanulate, the lobes ovate-oblong, obtuse, nearly equal. Ovary densely woolly with white plumose hairs, and, as far as I could ascertain in the very young one examined, with only 1 ovule in each cell.—*Eremophila Delisserii,* F. Muell. Fragm. v. 108, t. 41.

W. Australia. N.W. of the head of the Great Australian Bight, *Delisser;* a single fragment in *Herb. F. Mueller.*

4. **P. crassifolia,** *F. Muell. in Linnæa,* xxv. 430. An erect shrub of several feet, quite glabrous, but often glaucous and minutely scaly. Leaves broadly ovate, obtuse, thickly coriaceous, folded longitudinally, and often recurved, narrowed into a short petiole, 3 to 5 lines long. Flowers small, almost sessile, and solitary in the axils. Calyx-segments ovate-lanceolate, acutely acuminate, keeled, with thin and sometimes slightly ciliate margins, about 2 lines long, the inner ones rather smaller. Corolla not twice as long as the calyx, the lobes nearly equal and rather longer than the tube, the narrow part of the tube very short. Stamens didynamous. Ovary 2-celled, with 2 ovules in each cell. Fruit small, rugose, slightly compressed, obtuse, shorter than the calyx, the putamen more or less completely 4-celled when ripening all the seed.—*Eremophila crassifolia,* F. Muell. in Proc. R. Soc. Tasm. iii. 297.

S. Australia. About Mount Greenly, Dombey Bay, Spencer's Gulf, *Wilhelmi;* Venus Bay, *Warburton.*

5. **P. resinosa,** *Endl. Nov. Stirp. Dec.* 50. Branches densely covered with a white tomentum and sprinkled with resinous tubercles. Leaves

alternate, obovate, obtuse with a minute point, rather thick, flat, 4 lines long, hoary with stellate hairs on both sides. Peduncles axillary, solitary, exceedingly short. Calyx-segments linear-lanceolate, acute, 3 lines long. Corolla funnel-shaped, 6½ lines long, the tube scarcely exceeding the calyx, the throat dilated, the lobes nearly equal, scarcely more than ½ line long, the upper ones recurved, the lower ones spreading; all sprinkled outside with stellate hairs. Stamens included in the tube. Ovary densely tomentose, 4-celled, with 1 ovule in each cell (or 2-celled with 2 ovules in each cell ?)—A. DC. Prod. xi. 713; *Eremophila resinosa, F. Muell. in Proc. R. Soc. Tasm.* iii. 296.

W. Australia. Between King George's Sound and Swan river, *T. S. Roe.* The specimen is a very poor one. The above description is taken chiefly from Endlicher's.

6. **P. Behriana,** *F. Muell. in Linnæa,* xxv. 430. A low shrub, with erect, rather slender glabrous or pubescent branches. Leaves obovate or cuneate, mucronate-acute, often toothed in the upper part, contracted at the base but scarcely petiolate, rather rigid and occasionally veined, 3 to 4 lines long. Flowers nearly sessile, solitary in the axils. Calyx-segments narrow-lanceolate, acute, the outer ones 2½ lines long, the inner ones rather smaller. Corolla about 4 to 5 lines long, the tube gradually enlarged from near the base, the lobes short and broad, the middle lower one broader than the others, all bearded inside at the base. Stamens included in the tube. Ovary oblong, 2-celled, with 2 ovules in each cell (or one occasionally abortive ?).—*Eremophila Behriana* or *Behrii.* F. Muell. in Proc. R. Soc. Tasm. iii. 296, Rep. Babb. Exped. 18.

S. Australia. In the scrub near Gawler river, *Behr.;* Tumby Bay and hills near Port Lincoln, *F. Mueller:* Kangaroo Island, *Waterhouse.*

7. **P. Woollsiana,** *F. Muell.* An erect shrub of 3 to 4 ft., the branches hoary with a very short minutely plumose pubescence. Leaves alternate, ovate or oblong, elliptical, obtuse or almost acute, 1-nerved, rather thick, glabrous, entire (or rarely slightly toothed ?), narrowed at the base but sessile, 2 to 4 lines long. Flowers " puce-coloured," axillary, on very short pedicels. Calyx-segments narrow-lanceolate, acute or rather obtuse, not 2 lines long. Corolla 5 to 6 lines long, the cylindrical part of the tube about as long as the calyx, the obliquely campanulate broad part rather longer, the lobes short and broad, the 2 upper ones more united, and the middle lower one broader than the others. Stamens not exserted. Ovary 2-celled, with 2 ovules in each cell. Fruit scarcely as long as the calyx, slightly compressed, broadly ovate, rugose, slightly furrowed on each side, more or less perfectly 4-celled, or with fewer cells and seeds by abortion.—*Eremophila Woollsiana,* F. Muell. Fragm. i. 125, t. 7.

W. Australia. Salt river, Stokes Inlet, Oldfield river, *Maxwell.*

8. **P. brevifolia,** *Benth.* A tall, erect, much-branched shrub, quite glabrous. Leaves alternate, sessile and more or less stem-clasping, broadly ovate, obtuse, entire or coarsely toothed, usually concave, 3 to 4 or sometimes 5 lines long. Flowers " white," solitary, on very short axillary pedicels. Calyx-segments very narrow, acute, about 2 lines

long. Corolla-tube nearly 5 lines long, broad, hairy inside, very shortly contracted at the base, the lobes scarcely 1 line long, the middle lower one broader than the others. Ovary quite glabrous, 2-celled, with only 1 ovule in each cell in all the flowers examined. Fruit not known. —*Myoporum brevifolium,* Bartl. in Pl. Preiss. i. 350; *Pseudopholidia brevifolia,* A. DC. Prod. xi. 704.

W. Australia. Swan river, *Drummond,* 1st coll. n. 445, *Preiss.* n. 2335 and 2382.

9. **P. imbricata,** *Benth.* An erect shrub, densely clothed with a hoary or white tomentum. Leaves crowded and imbricate, ovate or oblong, obtuse, sessile and very shortly contracted at the base, thick and soft, entire, ½ to nearly 1 in. long. Flowers sessile and shorter than the leaves. Calyx-segments narrow, softly tomentose, 2 to 2½ lines long, not enlarging after flowering. Corolla glabrous, apparently of the shape of other *Pholidiæ,* but not seen perfect. Drupe glabrous, ovoid, as long as the calyx, slightly succulent, the endocarp hard, completely 4-celled, with one seed in each cell.

W. Australia. Between Moore and Murchison rivers, *Drummond,* 6th coll. n. 147. I have seen the ovary only in a far advanced state, but could find no trace of the lower abortive ovules of most species of *Eremophila.*

10. **P. densifolia,** *F. Muell.* Branches virgate, glabrous or sprinkled with a small minutely plumose pubescence, which is also sometimes on the margins of the leaves, and more abundant on the margins of the sepals and on the ovary. Leaves rather crowded, linear or narrowly linear-lanceolate, very acute, sessile, thick, convex underneath, mostly about ½ in. long on the main branches, much smaller on the lateral ones. Flowers almost sessile and nearly exceeding the leaves. Calyx-segments narrow, acute, the outer ones about 2 lines long, the inner ones smaller. Corolla "blue," about 5 lines long, the narrow part of the tube very short, the lobes scarcely as long as the tube, the 4 upper ones almost acute, contorted in the bud, the lowest broader and obtuse. Stamens didynamous. Ovary 2-celled, with 2 ovules in each cell, densely tomentose-villous. Drupe tomentose, slightly compressed, obtuse, shorter than the calyx, the endocarp completely divided into 4 cells (or fewer by abortion) with 1 seed in each cell.—*Eremophila densifolia,* F. Muell. Fragm. ii. 160.

W. Australia. E. Mount Barren and Stokes Inlet, *Maxwell.*

11. **P. gibbifolia,** *F. Muell.* An erect shrub of 1 to 2 ft., with numerous virgate branches, glabrous or minutely scaly-pubescent. Leaves linear or linear-oblong, sessile, erect, 1 to 3 lines long, remarkable for several large tubercular swellings on the back. Flowers nearly sessile. Calyx-segments subulate-acuminate, often above 2 lines long. Corolla fully ½ in. long, the tube shortly cylindrical at the base, the upper part much dilated and oblique, the lobes short. Stamens didynamous. Ovary oblong, 2-celled, with 2 ovules in each cell. Fruit oblong, compressed, not exceeding the calyx, completely 4-celled, with 1 seed in each cell, or with fewer cells and seeds by abortion.—*Duttonia gibbifolia,* F. Muell. in Hook. Kew Journ. viii. 73, t. 1., and in Trans.

Vict. Inst. 1855, 41; *Eremophila gibbosifolia*, F. Muell. Rep. Babb. Exped. 18.

Victoria. Wimmera, *Dallachy.*
S. Australia. Rocky hills between the Murray river and St. Vincent's Gulf, *F. Mueller;* Tattiara Country, *Wood.*

12. **P. divaricata,** *F. Muell. in Hook. Kew Journ.* viii. 201, and in *Trans. Phil. Soc. Vict.* i. 47. An erect shrub of several feet, with divaricate branches, sometimes spinescent and quite glabrous, or with a line of short hairs above each leaf. Leaves linear or linear-cuneate, obtuse, narrowed at the base, rarely above ½ inch long and often smaller, usually drying black. Flowers "purple or white, often spotted," solitary and nearly sessile in the axils. Calyx-segments 4 or 5, broad at the base, acuminate, slightly pubescent or ciliate, about 2 to 2¼ lines long. Corolla densely stellate-pubescent outside, under ½ in. long, the cylindrical base shorter than the calyx, the throat obliquely campanulate, the middle lower lobe rather larger and broader than the others and very hairy inside, the hairs continued to the base of the tube as in most *Pholidias,* the 2 upper lobes shortly united. Filaments hairy at the base. Ovary glabrous, 2-celled, with 2 ovules in each cell. Ripe fruit ovoid at the base, tapering into a beak, the whole about 4 lines long, more or less perfectly 4-celled, with 1 seed in each cell.—*Sentis rhynchocarpa,* F. Muell. Fragm. iv. 48, vi. 150; *Eremophila divaricata,* F. Muell. in Trans. R. Soc. Tasm. iii. 293.

N. S. Wales. Murray river, *F. Mueller;* Darling river, *Victorian Expedition, Mrs. Ford;* tributaries of the upper Darling, *Bowman.*
S. Australia. Murray desert, *F. Mueller;* Lake Alexandrina, *Hildebrand.*

13. **P. microtheca,** *F. Muell.* An erect almost heath-like shrub, the branches and young leaves hoary with a very short minutely plumose, almost farinaceous pubescence, the older foliage glabrous or nearly so. Leaves rather crowded, linear, somewhat obtuse, slightly contracted at the base, 2 to 4 lines long. Flowers "lilac," on very short axillary pedicels. Calyx-segments plumose-pubescent, about 2 lines long. Corolla like that of *P. Woollsiana,* but rather smaller. Ovary rugose, 2-celled, with 2 ovules in each cell. Fruit rugose like that of *E. Woollsiana,* but smaller and not compressed, the only one seen was, however, not quite perfect.—*Eremophila Woollsiana* var. *angustifolia,* F. Muell. Fragm. ii. 160 ; *E. microtheca,* F. Muell. Herb.

W. Australia. Port Gregory, Murchison river, *Oldfield.*

14. **P. adenotricha,** *F. Muell.* Densely clothed with a glandular-ferruginous or dingy viscid pubescence. Leaves crowded, linear, obtuse, rather thick and soft, viscid-pubescent on both sides, about 1 in. long. Flowers solitary and sessile. Calyx-segments linear, nearly equal, and not so imbricate as in the other species, ciliate and viscid-pubescent, 3 to 3½ lines long, not enlarged after flowering. Corolla glabrous outside, about ¾ in. long, the cylindrical part of the tube nearly as long as the calyx, the upper part much enlarged; the lobes broad, short and nearly equal. Stamens not examined Drupe slightly succulent, ovoid, shorter

than the calyx, the putamen thick and long, 4-celled, with 1 seed in each cell.—*Eremophila adenotricha*, F. Muell. Herb.

W. Australia. *Herb. F. Mueller.*

15. **P. santalina,** *F. Muell.* An erect glabrous shrub of several feet, slightly glandular-verrucose. Leaves narrow lanceolate, acuminate, entire, narrowed into a rather long petiole, rather thick, 1½ to 2 in. long. Flowers "white," solitary in the axils, on pedicels usually of about ½ in., thickened under the flower. Calyx-segments narrow, acuminate, not 2 lines long, imbricate at the base. Corolla-tube with the cylindrical part nearly as long as the calyx, the upper part broad, about 3 lines long, glabrous inside or nearly so, the lobes scarcely 2 lines long, the 4 upper ones ovate, spreading, with short recurved points or almost obtuse, the 2 uppermost of them ascending, the middle lower lobe twice as broad as the others. Stamens included, didynamous. Ovary glabrous, 2-celled, with 2 ovules in each cell. Drupe succulent, the putamen more or less perfectly 4-celled, with 1 seed in each cell, or more frequently reduced by abortion to 1 or 2 cells and seeds.— *Pholidiopsis santalina*, F. Muell. in Linnæa, xxv. 429; *Eremophila santalina*, F. Muell. in Proc. R. Soc. Tasm. iii. 295.

S. Australia. Rocky hills near Cudnaka, *F. Mueller.*

This species differs from all the others of the genus in its elongated pedicels, like those of *Myoporum;* they are, however, solitary, and the calyx, corolla, and ovary are those of *Pholidia* rather than of *Myoporum* or *Eremophila.*

3. EREMOPHILA, R. Br.

(Stenochilus, *R. Br.* Eremodendron, *DC.*)

Calyx divided to the base into 5 segments or rarely 5-lobed, often but not always enlarged after flowering. Corolla-tube usually broad from the base or constricted above the ovary, more or less elongated and incurved, very rarely with the cylindrical base of *Pholidia*, the limb oblique or 2-lipped, 5-lobed. Stamens 4, didynamous, often exserted. Ovary 2-celled, with 2 or 3 superposed pairs of ovules in each cell, of which, however, the lower pairs remain usually unfecundated, or in a very few species only one pair in each cell at the time of flowering. Style filiform. Fruit, where known, a dry or succulent drupe, the putamens separating into 4 1-seeded pyrenes, or 4-celled with one seed in each cell, or fewer cells and seeds by abortion.—Shrubs. Leaves alternate or scattered. Flowers solitary, or in a few species several together in the axils, usually pedicellate, without bracts.

The genus is limited to Australia. As will be seen by the above character, there is no positive combination of characters to separate it from *Pholidia*, being connected with that genus, as *Pholidia* is with *Myoporum*, by exceptional species. The habit is, however, different, and there are always either the superposed ovules or the enlarged fruiting calyx, and often the succulent fruit, to distinguish *Eremophila*. On the other hand, the five sections into which I have divided *Eremophila* may perhaps one day be admitted as genera, which I have been unwilling to do whilst the ripe fruit of so many species is unknown, especially as there appears to be a greater proportion of intermediate species between them than between the three Myoporineous genera here adopted. F. Mueller

(Fragm. vi.) unites *Pholidia* with *Eremophila*, retaining *Myoporum*, *Disoon*, and *Sentis* as distinct, but has not published the definite distinctive character he relies upon.

SECT. I. **Eriocalyx.**—*Calyx-segments not overlapping, thick and soft, densely tomentose, not becoming scarious after flowering. Ovary with two pairs of ovules in each cell. Fruit unknown.*

Flower nearly sessile. Leaves obovate to lanceolate, short, densely
  tomentose.
  Corolla and ovary tomentose. Corolla lobes all broad and obtuse.
    Stamens included . . . . 1. *E. Mackinlayi*.
  Corolla and ovary glabrous.
    Corolla lobes obtuse. Stamens included . . . . 2. *E. Bowmanni*.
    Corolla upper lobes small and acute. Stamens exserted . . 3. *E. leucophylla*.
Flowers distinctly pedunculate. Leaves obovate, oblong, densely
  tomentose, ½ to ¾ in. long. Corolla tomentose. Ovary glabrous 4. *E. Forrestii*.
Flowers distinctly pedunculate. Leaves linear or linear-lanceolate,
  mostly above 1 in. long.
  Calyx clothed with a long loose plumose wool. Leaves woolly
    when young, at length nearly glabrous . . . . 5. *E. eriocalyx*.
  Calyx shortly stellate-tomentose. Leaves closely tomentose,
    not becoming glabrous . . . . 6. *E. Maitlandi*.

SECT. II. **Eremocosmos.**—*Calyx-segments not at all or scarcely overlapping at the base, more or less enlarged, veined, and scarious after flowering. Ovary with 2 to 4 pairs of ovules in each cell (except in E. oppositifolia and E. Paisleyi). Fruit (where known) dry, the endocarp separating into distinct pyrenes.*

Enlarged calyx-segments more or less cuneate and obtuse.
  Leaves small, short, broad, thick, and hoary . . . . 7. *E. rotundifolia*.
  Leaves linear or linear-lanceolate. Corolla lobes obtuse. Stamens
    included or shortly exserted.
    Ovary shortly hairy, with 1 pair of ovules to each cell.
      Corolla above 1 in. long. Stamens often exserted. Leaves
        1 to 2 in. long . . . . 8. *E. oppositifolia*.
      Corolla ½ in. long, stamens included . . . . 9. *E. Paisleyi*.
    Ovary very woolly, with 2 to 4 pair of ovules to each cell.
      Leaves narrow, linear . . . . 10. *E. Sturtii*.
      Leaves linear-lanceolate . . . . 11. *E. Mitchelli*.
Enlarged calyx-segments oblong or lanceolate, acute. Leaves
  linear or linear-lanceolate.
  Corolla-lobes all broad. Stamens included . . . . 12. *E. Clarkii*.
  Corolla upper lobes rather acute. Stamens exserted . . . . 13. *E. Latrobii*.

SECT. III. **Platycalyx.**—*Calyx campanulate, 5-lobed. Flowers and fruit of* Platychilus.

Single species . . . . 14. *E. Macdonellii*.

SECT. IV. **Platychilus.**—*Calyx-segments much imbricate at the base (except in the first species), the outer ones usually broader. Corolla-lobes all broad and obtuse, or the upper ones scarcely acute. Stamens included or scarcely exserted. Ovules in 2 or 3 pairs in each cell. Fruit of* Stenochilus.

Calyx-segments small or narrow and acute, not enlarged after
  flowering. Leaves long, linear or lanceolate.
  Corolla tube not much enlarged upwards. Calyx-segments nearly
    linear . . . . 15. *E. graciliflora*.
  Corolla tube much enlarged upwards. Calyx-segments small,
    very acute, from a broad base . . . . 16. *E. longifolia*.
Calyx-segments ovate or lanceolate, acute, not exceeding 3 lines in
  flower, nor much enlarged afterwards. Plants very glabrous,
  often drying blue.
  Erect, virgate and very glutinous. Leaves narrow-linear.
    Corolla-tube cylindrical at the base, as in *Pholidia* . . . . 17. *E. Drummondii*.

Very divaricately branched. Leaves narrow-linear. Corolla-
tube very broad and enlarged from the base . . . . . . 18. *E. polyclada.*
Moderately spreading. Leaves lanceolate or linear-lanceolate,
usually long. Corolla-tube broad and enlarged from the
base . . . . 19. *E. bignoniæflora.*
Calyx-segments lanceolate or the outer ones ovate, 3 to 6 lines
long. Plant hoary-tomentose or at length glabrous . . . 20. *E. Freelingii.*
Calyx-segments broad-lanceolate, 4 to 6 lines long, more or less
hirsute.
Leaves linear or lanceolate . . . . 21. *E. Goodwinii.*
Leaves obovate or oblong, serrulate . . . . 22. *E. Willsii.*
Calyx-segments obtuse, very much enlarged coloured and sca-
rious after flowering. Leaves linear-lanceolate.
Stamens included. Plant hoary-tomentose or glabrous . . . 23. *E. platycalyx.*
Stamens exserted. Plant glabrous . . . . 24. *E. viscida.*

SECT. V. **Stenochilus.**—*Calyx-segments imbricate at the base, usually enlarged
after flowering. Corolla 4 upper lobes short and acute, the fifth lowest more deeply
separated and sometimes narrow. Stamens exserted (except* E. alternifolia). *Ovules
2 or 3 pairs, or rarely only 1 pair in each cell. Drupe (except* E. alternifolia) *succulent, with a thick bony putamen not separating into nuts.*

Peduncles usually shorter than the calyx, not flexuose.
Calyx-segments lanceolate, small in flower and not much en-
larged afterwards.
Leaves linear or lanceolate, hoary-tomentose or at length
glabrous . . . . 25. *E. Brownii.*
Leaves ovate-oblong, crowded, tomentose, more or less floccose 26. *E. subfloccosa.*
Calyx-segments oblong, rather obtuse, enlarged after flowering 27. *E. Oldfieldii.*
Peduncles longer than the calyx, very spreading, usually
flexuose.
Leaves narrow-lanceolate, entire. Ovules, 2 pairs in each
cell.
Lowest corolla-lobe obtuse. Calyx much enlarged after
flowering . . . . 28. *E. Duttonii.*
Lowest corolla-lobe acute. Calyx-segments acute scarcely
enlarged after flowering . . . . 29. *E. maculata.*
Leaves lanceolate or ovate, often denticulate. Ovules 1 pair
in each cell.
Leaves mostly lanceolate. Calyx-segments lanceolate,
scarcely enlarged after flowering . . . . 30. *E. denticulata.*
Leaves mostly ovate. Calyx-segments ovate, much enlarged
after flowering . . . . 31. *E. latifolia.*
Leaves narrow-linear. Calyx-segments much enlarged after
flowering. Ovules 1 pair in each cell. Stamens included . 32. *E. alternifolia.*

SECT. 1. ERIOCALYX.—Calyx-segments not overlapping at the base,
thick and soft, densely tomentose, sometimes enlarged but not scarious
after flowering. Ovary with 2 pairs of ovules in each cell.

The species of this section have rather more the habit of *Pholidia* than of *Eremophila*,
but the shape of the corolla and the superposed ovules are those of the latter genus.
Most of the species have however been described from very imperfect specimens.

1. **E. Mackinlayi,** *F. Muell. Fragm.* iv. 80. A shrub of several feet,
densely clothed with a hoary or yellowish soft and almost woolly
tomentum. Leaves obovate broadly ovate or almost orbicular, contracted below the middle but broadly sessile, and sometimes dilated at
the very base, thick, 4 to 8 lines long in the only specimens seen.

Flowers " purple," nearly sessile and solitary. Calyx-segments narrow-lanceolate, rather obtuse, densely tomentose, about 4 lines long, scarcely imbricate, apparently becoming enlarged after flowering. Corolla rather above 1 in. long, slightly constricted above the ovary, then almost campanulate, tomentose outside, partially woolly inside, the lobes all broad, obtuse, or with a very short point in the centre, the middle lowest one rather broader than the others. Stamens included. Ovary woolly-tomentose, with 2 pairs of ovules in each cell. Fruit unknown.

W. Australia. Sharks Bay, *Maitland Brown* (a single specimen in herb. F. Muell.).

2. **E. Bowmanni,** *F. Muell. Fragm.* ii. 139. Densely clothed with a white or hoary tomentum, either short and close or looser and plumose. Leaves oblong or lanceolate, obtuse, rather thick, entire, contracted at the base but scarcely petiolate, tomentose on both sides, the midrib prominent underneath, under $\frac{1}{2}$ in. long when broad, nearly 1 in. when narrow. Flowers " blue," solitary, on pedicels of 3 to 6 lines. Calyx-segments 5 to 7 lines long, oblong-linear, rather obtuse, tomentose on both sides, rather unequal, but scarcely imbricate. Corolla glabrous outside, about 1 in. long, the tube slightly contracted above the ovary, then broad, the lobes broad and obtuse, the middle lower one rather narrower than the others. Stamens included. Ovary glabrous, narrow, with 2 pairs of ovules to each cell.

N. S. Wales. Darling desert, *Neilson* (with short leaves and a close tomentum); tributaries of the upper Darling, *Bowman* (with long leaves and a loose tomentum)—both mere fragments in herb. F. Mueller.

3. **E. leucophylla,** *Benth.* Densely clothed with a white or hoary tomentum either close and short or looser and plumose. Leaves obovate or elliptical-oblong, obtuse, distinctly petiolate, under $\frac{1}{2}$ in. long, thick, tomentose on both sides, the midrib prominent underneath. Pedicels solitary, 1 to 2 lines long. Calyx-segments lanceolate or linear, scarcely acute, about 4 lines long, tomentose on both sides, not imbricate. Corolla glabrous, about $\frac{3}{4}$ in. long, scarcely constricted above the ovary, the tube broad and slightly incurved, the upper lobes small and acute, the lowest not seen perfect. Stamens exserted. Ovary glabrous, rather short, with 2 pairs of ovules in each cell, all very near the base.

W. Australia. Sharks Bay, *Milne.*

4. **E. Forrestii,** *F. Muell. Fragm.* vii. 49. A shrub densely covered with a white or yellowish almost floccose tomentum. Leaves opposite or alternate, shortly petiolate, oblong or obovate-oblong, very obtuse, contracted at the base, thick and soft, $\frac{1}{2}$ to $\frac{3}{4}$ in. long in the specimen. Flowers axillary, on peduncles of 2 to 4 lines. Calyx-segments narrow, not overlapping, almost obtuse, thick and soft, stellate-tomentose, 6 to 7 lines long, not enlarging after flowering. Corolla rather longer than the calyx, minutely tomentose outside, the lobes ovate, mucronate, rather more than 2 lines long, the 2 (upper?) ones rather narrower

than the others and shortly united in an (upper ?) lip of the same length as the others and inside in the bud. Stamens included. Ovary glabrous, with 2 superposed pairs of ovules in each cell. Drupe small with a hard putamen, but the only one seen not yet ripe.

W. Australia. Lake Barlee, *Forrest* (*Herb. F. Mueller*). Described from a single specimen in which what appeared to be the upper lip of the corolla (but possibly the middle bifid lobe of the lower lip) was certainly inside in the bud, whilst in all Myoporineæ which I have been able to examine in bud I have uniformly found the upper lip outside.

5. **E. eriocalyx,** *F. Muell. Fragm.* i. 236. A shrub of 3 or 4 feet, hoary with a close stellate tomentum, the pedicels and calyx thickly covered with a much looser wool consisting of branched hairs. Leaves linear-lanceolate, obtuse, entire, flat or with revolute margins, contracted into a very short petiole, 1 to 2 in. long. Flowers "red," solitary on pedicels shorter than the calyx. Calyx-segments narrow, scarcely imbricate, ½ in. long. Corolla woolly-pubescent outside, only seen very imperfect. Ovary narrow, glabrous.

W. Australia. Murchison river, *Oldfield.* The specimens are very bad, and I have been unable to ascertain the true structure of the ovary or the form of the corolla.

6. **E. Maitlandi,** *F. Muell.* A tall erect shrub, hoary or white all over with a soft dense but close or scarcely floccose tomentum. Leaves linear-lanceolate, entire, contracted at the base but scarcely petiolate, 1 to 2 in. long, coriaceous, hoary-tomentose even when old. Pedicels solitary, about ½ in. long. Calyx-segments oblong-lanceolate, rather obtuse, not overlapping, 8 to 9 lines long, tomentose outside, more glabrous inside. Corolla broad, above 1 in. long, slightly pubescent outside, the lobes not seen perfect. Ovary acuminate, glabrous, with 2 pairs of ovules in each cell.

W. Australia. Sharks Bay, *Maitland Brown, Milne,* the specimens all very imperfect.

Sect. II. Eremocosmos.—Calyx-segments not at all or scarcely overlapping at the base, more or less enlarged veined and scarious after flowering. Ovary with 2 to 4 pairs or in two species with only 1 pair of ovules in each cell. Fruit where known dry, the endocarp separating into distinct pyrenes.

This might be considered as the typical *Eremophila,* and in most species the flowers and fruit are so different from those of *Stenochilus,* that it seems difficult to unite them in one genus, were it not that in other species the several characters are very differently combined. Both the original *Eremophilas* of Brown are exceptional in having but one pair of ovules to each cell of the ovary, and one of them, *E. alternifolia,* in the over-lapping calyx-segments and deeply separated lowest corolla-lobe, is closely connected with *Stenochilus,* under which I have classed it, although it has the fruit of *Eremocosmos.* There are too many species in which the ripe fruit is unknown, to admit of its structure being taken at present as an absolute sectional character.

7. **E. rotundifolia,** *F. Muell. Fragm.* i. 207. Hoary with a close tomentum and glandular-tuberculate. Leaves nearly orbicular, or broader than long, 3 to 4 lines diameter, thick, often complicate, with a short recurved, obtuse point, abruptly contracted at the base into a short

broad petiole. Calyx-segments oblong-spathulate or almost obovate, about 5 lines long. Corolla not seen. Unripe fruit hoary-pubescent, oblong, almost perfectly 4-celled, and apparently separable into distinct pyrenes.

S. Australia. N.W. interior, *M'Douall Stuart's Expedition*, described from mere fragments in herb. F. Mueller, and of very doubtful affinity.

8. **E. oppositifolia,** *R. Br. Prod.* 518. A small elegant spreading tree of 20 to 30 ft. (*A. Cunn.*), or a tall shrub, quite glabrous or the young shoots hoary or yellowish with a close minute tomentum. Leaves scattered or here and there opposite, linear-lanceolate, acuminate and often ending in a hooked point, contracted into a short petiole, 1 to 2 in. long. Flowers solitary in the axils, on pedicels of 2 to 3 lines. Calyx-segments oblong-spathulate, 6 to 8 lines long, much contracted below the middle. Corolla nearly 1 in. long, glabrous outside and in, the tube incurved, the lobes all short and obtuse, the 2 upper ones more united and the lowest broader than the others. Stamens as long as the corolla, or the longest pair shortly exserted. Ovary shortly villous, with one pair of ovules only in each cell. Fruit not seen quite ripe, but appears either to open in 4 valves leaving the central placenta with 4 pendulous seeds, or to separate into 4 dry cocci, the central placenta at length splitting into 4.—A. DC. Prod. xi. 712; F. Muell. in Proc. R. Soc. Tasm. iii. 294, and Rep. Babb. Exped. 16; *E. arborescens*, A. Cunn.; F. Muell. in Proc. R. Soc. Tasm. iii. 293; *E. Cunninghamii*, R. Br. App. Sturt. Exped. 21; *Eremodendron Cunninghamii*, A. DC. Prod. xi. 713.

N. S. Wales. Barren wastes near the termination of the Lachlan river, *A. Cunningham;* deserts of the Murray and Darling, *Victorian Expedition;* Mount Murchison, *Bonney.*

Victoria. Murray river towards the junction with the Murrumbidgee, *F. Mueller.*

S. Australia. Head of Spencer's Gulf, *R. Brown;* Elder's and Flinders' Ranges, *F. Mueller.* Lakes Gregory, Hart, Campbell, &c., *Babbage's Expedition.*

The specific name is unfortunately chosen, for the leaves are usually alternate, and rarely as opposite as in the original imperfect specimens.

9. **E. Paisleyi,** *F. Muell. Rep. Babb. Exped.* 17. Very similar to the narrow-leaved specimens of *E. Mitchelli*, with the same habit and glabrous linear-lanceolate leaves, but the flowers are smaller on very short pedicels and usually several together in axillary clusters as in *Myoporum*. Calyx glandular-pubescent, the segments shortly united at the base and not overlapping, narrow-oblong, obtuse, and 2 lines long at the time of flowering, afterwards enlarged, obovate-oblong, very obtuse and 3 lines long, thin and veined. Corolla about ½ in. long, pubescent outside, scarcely hairy inside, the tube cylindrical at the base, dilated upwards, the 4 upper lobes ovate, obtuse, about half as long as the tube, the lowest rather longer and twice as broad as the others. Stamens included. Ovary slightly hairy, tapering upwards, with only one pair of ovules in each cell at the top of the rather long cavity. Fruit not seen ripe, but apparently that of the section *Eremocosmos*.

S. Australia. Mayerte, Lake Gairdner, *Babbage.* The clustered pedicels are quite exceptional in the genus.

10. E. Sturtii, *R. Br. App. Sturt. Exped.* 22. An erect very much branched strong-scented and viscid shrub of several ft., glabrous or very minutely hoary-pubescent. Leaves narrow linear, usually ending in a hooked point, entire, contracted at the base and often petiolate, rarely above 1 in. long. Flowers " purplish," numerous but solitary in each axil, on pedicels of 3 to 4 lines. Calyx-segments obovate or oblong, membranous and rather rigid, obtuse, coloured and veined, rather variable in shape and size but usually attaining 4 or 5 lines when the flowering is over. Corolla pubescent, about ½ in. long, the narrow base of the tube short, the upper part broadly campanulate, bearded inside, the 4 upper lobes short broad and obtuse, the 2 uppermost more united than the others, the middle lowest lobe larger and broader than the others, notched or 2-lobed and woolly inside. Stamens included. Ovary very villous with 2 or 3 pairs of ovules to each cell. Fruit when young like that of *E. Mitchelli* but not acuminate, not seen quite ripe.—F. Muell. in Proc. R. Soc. Tasm. iii. 294 and Rep. Babb. Exp. 17.

N. S. Wales. Deserts of the Lachlan and Darling to the Barrier Range, *Victorian and other Expeditions.*

S. Australia. *Sturt;* between Stoke's Range and Cooper's Creek, *Howitt's Expedition.*

The species scarcely differs from *E. Mitchelli*, except in its smaller flower and narrower leaves.

11. E. Mitchelli, *Benth. in Mitch. Trop. Austr.* 31. A tall shrub or small tree of 10 to 30 ft., glabrous viscid and strongly-scented. Leaves linear-lanceolate, obtuse or with a hooked point, entire, contracted into a petiole, 1-nerved, 1 to 2 in. long. Flowers solitary in the axils, on pedicels of 3 to 4 lines. Calyx-segments oblong or cuneate-oblong, obtuse, membranous, veined, glabrous or pubescent on the edges, 4 to 5 lines long. Corolla about ¾ in. long, the cylindrical part of the tube about 2 lines, the broad part above twice as long, the middle lower lobe broader than the others, shortly 2-lobed, woolly inside. Stamens shorter than the corolla. Ovary very woolly, with 3 or 4 superposed pairs of ovules in each cell. Fruit ovoid, almost acuminate, half as long as the calyx, the exocarp thin and membranous, the endocarp separating into 4 nuts each with 1 or with 2 superposed seeds.—F. Muell. in Proc. R. Soc. Tasm. iii. 294, and Rep. Babb. Exped. 17.

Queensland. Elevated stony lands on the Bogan, Narran, Maranoa, Belyando, &c., *Mitchell;* Port Denison, *Fitzalan, Dallachy;* Rockhampton, *Thozet* and others ; Suttor river, *Fitzalan, Sutherland;* Armadilla, *Barton;* Darling Downs, *Lau.*

N. S. Wales. Lachlan river, *A. Cunningham;* between the Bogan and Lachlan, *L. Morton;* Castlereagh, *Woolls.*

12. E. Clarkii, *F. Muell. Fragm.* i. 208. An erect shrub, attaining 6 to 8 ft., usually glabrous and often glaucous. Leaves linear or linear-lanceolate, entire or with rather distant serratures, contracted into a short petiole, 1 to 2 in. long. Pedicels solitary or 2 together, ½ to 1 in. long, spreading and usually incurved and dilated at the end. Calyx-segments broadly lanceolate, acute, ½ in. long when in flower,

¾ in. or more when in fruit, not at all or scarcely overlapping at the base. Corolla "pale purple," slightly pubescent, above 1 in. long, the tube scarcely constricted above the ovary, broad and slightly incurved, the lobes all broad, the 2 uppermost more united and very obtuse, the lateral ones more acute, the middle lower one broader than the others. Stamens included. Ovary glandular-dotted and very hirsute, with 2 pairs of ovules in each cell. Fruit hirsute with long hairs, ovate, rather acute, much shorter than the calyx, dry and like that of *E. alternifolia*, but not seen quite ripe.

W. Australia. Murchison river, *Oldfield*; Sharks Bay, *Maitland Brown.*

13. **E. Latrobei,** *F. Muell. in Proc. R. Soc. Tasm.* iii. 294, *Rep. Babb. Exped.* 17, *and Fragm.* i. 125, *t.* 8. An erect much-branched shrub, attaining 10 ft., more or less clothed with a close hoary minutely stellate tomentum, rarely almost glabrous, usually also glandular-tuberculate. Leaves linear or narrow-lanceolate, flat or with recurved margins, obtuse or rarely acute, entire, contracted into a short petiole, ½ to 1 in. long or rather more. Pedicels solitary, rarely exceeding ¼ in. Calyx-segments lanceolate, acute, not at all or scarcely overlapping at the base, under ½ in. long when in flower, attaining sometimes ¾ in. in fruit. Corolla about 1 in. long "spotted," glabrous outside, the tube broad incurved not contracted above the ovary, the lobes almost acute, the 4 upper ones erect and nearly equal, the lowest often shorter and separated to near the middle of the corolla. Stamens exserted. Ovary glabrous, narrow-conical, with 2 pairs of ovules to each cell. Drupe nearly dry, ovoid-conical, 4 to 5 lines long, readily splitting into 2 or 4 at the top, the endocarp sometimes readily separating into 4 nuts. —*E. tuberculata,* F. Muell. in Proc. R. Soc. Tasm. iii. 294.

N. Australia. Sturt's Creek and Newcastle Range, *F. Mueller;* between Strangways river and Rupert's Range, *M'Douall Stuart.*
Queensland. Suttor, Burdekin, and Mackenzie rivers, *F. Mueller.*
N. S. Wales. Darling desert, *Barton,* and thence to the Barrier Range, *Victorian and other Expeditions.*
S. Australia. Cooper's Creek, *Howitt's Expedition;* Thomson river, *A. C. Gregory;* head of the Great Australian Bight, *Delisser.* The corolla approaches that of *Stenochilus,* but the calyx and fruit are those of *Eremocosmos.*

SECT. 3. PLATYCALYX.—Calyx campanulate, divided to the middle only into 5 lobes. Flowers and fruits of *Platychilus.*

14. **E. Macdonellii,** *F. Muell. Rep. Babb. Exped.* 18. A branching shrub, apparently diffuse, sometimes quite glabrous, more frequently clothed with a hoary or white close stellate or plumose tomentum, or with long spreading hairs, or with both. Leaves oblong-linear or lanceolate, obtuse or acute, contracted at the base but scarcely petiolate, rarely above ½ in. long. Pedicels solitary, ¼ to above ½ in. long. Calyx campanulate, with acute or acuminate lobes about as long as the tube, which is sometimes prominently angled and membranous, sometimes more herbaceous and tomentose without prominent ribs, the whole calyx varying from 4 to 8 or even 9 lines long. Corolla "blueish," glabrous outside, 1 to 1½ in. long, the tube not constricted

above the ovary and much dilated upwards, the lobes all broad and nearly ½ in. long, the upper ones more united, the three lower more spreading, the lowest rather broader than the others but not notched. Stamens included. Ovary narrow, glandular-dotted and tipped with a few hairs, with 2 pairs of ovules in each cell. Fruit very succulent, ovoid-globose, acute, not seen ripe but already as long as the somewhat enlarged calyx.

S. Australia. Cooper's Creek, *Wright;* Wills Creek, *Howitt's Expedition;* Lake Gregory and other parts of the interior, *Babbage's and M'Douall Stuart's Expeditions;* towards Spencer's Gulf, *Warburton.*

The calyx of this species is exceptional in the whole Order of Myoporineæ.

SECT. 4. PLATYCHILUS.—Calyx-segments much imbricate at the base (except in the first two species) the outer ones usually broader. Corolla-lobes all broad and obtuse or the upper ones scarcely acute. Stamens included or scarcely exserted. Ovules in two or three superposed pairs in each cell of the ovary. Fruit (of *Stenochilus*) succulent, with a thick bony putamen, not separating into pyrenes.

15. **E. graciliflora,** *F. Muell. Fragm.* i. 208. A shrub of several feet, the young shoots more or less hoary-tomentose, the adult leaves usually glabrous. Leaves linear-lanceolate, acutely acuminate and the points sometimes incurved, quite entire, contracted into a petiole, 1 to 2½ in. long. Pedicels solitary, under ½ in. long. Calyx-segments linear-lanceolate, not overlapping at the base, scarcely exceeding 3 lines when in flower, broader 6 to 8 lines long and acute when in fruit. Corolla "red," more slender than in most species, scarcely incurved, under 1 in. long, sprinkled with short spreading hairs, the tube not contracted above the ovary, gradually but not much enlarged upwards, the lobes nearly equal, oval-oblong, obtuse, the 2 upper ones rather more united than the others. Stamens included; anther-cells narrow. Ovary oblong, glandular-dotted, glabrous, with 2 pairs of ovules in each cell. Drupe small, dry, depressed-globular, the putamen hard.

W. Australia. Murchison river, *Oldfield.*

16. **E. longifolia,** *F. Muell. in Proc. R. Soc. Tasm.* iii. 295. A tall erect shrub, the young shoots minutely hoary-tomentose, the older foliage nearly glabrous and often drying black. Leaves scattered, linear or almost linear-lanceolate, obtuse or tapering into a recurved point, rather thick but flat, 2 to 4 or even 5 in. long, contracted into a short petiole. Pedicels solitary or 2 together, varying in length from 2 or 3 lines to ½ in., stout or slender, erect or spreading. Calyx-segments triangular or lanceolate, acute or acuminate, rarely 2 lines long, united at the base and scarcely overlapping, usually woolly-ciliate on the margins. Corolla velvety-pubescent outside, ¾ to 1 in. long, the tube gibbous at the base, contracted over the ovary, the remainder much dilated and slightly incurved, the lobes all ovate and obtuse the two uppermost rather smaller and the lowest often but not always more deeply separated than the others. Stamens shortly exserted. Ovary thick and fleshy, with 2 pairs of ovules in each cell. Fruit ovoid or

globular, very succulent, with a thick hard putamen, completely 4-celled and not separating into pyrenes.—*Stenochilus longifolius,* R. Br. Prod. 517 and App. Sturt. Exped. 23; A. DC. Prod. xi. 714; *S. salicinus,* Benth. in Mitch. Trop. Austr. 251 and *S. pubiflorus,* Benth. l. c. 273.

N. Australia. Sturt's Creek, *F. Mueller;* in the interior, *M'Douall Stuart's Expedition.*

Queensland. Suttor river, *F. Mueller, Bowman* (the latter with smaller flowers and fruit), *Sutherland;* Belyando river, *Mitchell;* Armadilla, *Barton;* Darling downs, *Lau.*

N. S. Wales. Lachlan river to Liverpool Plains and all the brushes of the interior, *A. Cunningham, Fraser;* from the Murray and Darling to the Barrier Range, *Victorian and other Expeditions.*

Victoria. Murray and Avoca rivers, and Lake Hindmarsh, *F. Mueller;* Wimmera, *Dallachy.*

S. Australia. Spencer's Gulf, *R. Brown;* S. coast, *Sturt;* N. of Adelaide, *Whittaker.*

W. Australia. Swan river, *Drummond;* Murchison river, *Oldfield.*

17. **E. Drummondii,** *F. Muell. Fragm.* vi. 147. An erect, virgate, much-branched shrub, glabrous and more or less glutinous. Leaves alternate, linear-filiform, obtuse or with a short straight or incurved point, usually about 1 in. but sometimes nearly 2 in. long. Pedicels solitary or 2 together, often above ½ in. long. Calyx-segments very much imbricate, lanceolate or ovate-lanceolate, acute, nearly 3 lines long. Corolla glabrous outside, incurved, 7 to 9 lines long, the cylindrical base short, the broad part of the tube much longer, the four upper lobes almost acute, the middle lower one broader than the others, very obtuse, pubescent inside at the base. Stamens included. Ovary glabrous, 2-celled, with 2 pairs of ovules in each cell. Young fruit oblong-conical, nearly as long as the calyx, the putamen almost perfectly 4-celled, with 1 seed in each cell.

W. Australia. *Drummond, n.* 64, with rather broad calyx-segments, and *n.* 74 with the segments still broader, almost ovate.

18. **E. polyclada,** *F. Muell. in Proc. R. Soc. Tasm.* iii. 294. A glabrous shrub of 4 to 8 feet, with very divaricate rigid intricate branches, the smaller ones almost spinescent, the specimens usually drying black. Leaves mostly very spreading, distant, linear or narrowly linear-lanceolate, acute, entire, narrowed at the base, 1 to 2 in. long. Pedicels solitary, often recurved, 2 to 4 lines long. Calyx-segments much imbricate, broad, obtuse or acuminate, with spreading or recurved points. Corolla glabrous outside, ¾ to 1 in. long, the tube broad, almost campanulate, gradually enlarged from the base and not contracted above the ovary, the lobes all very broad, the 2 upper ones more united and the middle lower one twice as broad as the others and emarginate, the whole corolla bearded inside especially under the upper lobes. Stamens scarcely exserted from the tube, shorter than the lobes. Ovary oblong, glabrous, with 2 pairs of ovules to each cell. Fruit tapering into a beak exceeding the calyx, but not seen quite ripe.— *Pholidia polyclada,* F. Muell. in Hook. Kew Journ. viii. 201, and in Trans. Phil. Soc. Vict. i. 47.

Queensland. Desert on the Suttor, *F. Mueller, Sutherland;* Cape river, *Bowman;* Curriwillighie, *Dalton.*
N. S. Wales. Darling and Murray desert, *F. Mueller, Victorian and other Expeditions.*
S. Australia, Great marsh of the interior, *Sturt.*
The species is nearly allied to *E. bignoniæflora,* differing chiefly in the narrow leaves and tapering ovary and fruit.

19. **E. bignoniæflora,** *F. Muell. in Proc. R. Soc. Tasm.* iii. 294 and *Pl. Vict.* ii. t. 55. A strong-scented tall shrub or small tree, quite glabrous and often glutinous. Leaves lanceolate or linear-lanceolate, acuminate, entire, contracted into a short petiole, 2 to 6 in. long. Pedicels solitary, ¼ to ½ in. long, more or less flattened, often recurved, but not turned up again. Calyx-segments imbricate at the base, ovate, obtuse or rarely acute, thickened in the middle, 2 to 3 lines long. Corolla glabrous outside, about 1 in. long, scarcely contracted above the ovary, the tube gradually enlarged from the base, the lobes all broad and short, the 2 uppermost more united, the lowest twice as broad as the others and 2-lobed. Stamens shortly exserted from the tube but shorter than the corolla-lobes. Ovary 2-celled with 2 pairs of ovules to each cell. Drupe ovate, acute, ½ in. long or more, succulent, the putamen hard and bony, more or less completely 4-celled.—*Stenochilus bignoniæflorus,* Benth. in Mitch. Trop. Austr. 386.

N. Australia. Sturt's Creek and Gilbert river, *F. Mueller.*
Queensland. Balonne river, *Mitchell;* Suttor river, *Bowman, Sutherland;* Rockhampton, *Herb. F. Mueller.*
N. S. Wales. Murray and Darling desert, *Dallachy and Goodwin.*
Victoria. Murray desert, *Irvine.*

20. **E. Freelingii,** *F. Muell. in Proc. R. Soc. Tasm.* iii. 295. An erect shrub, more or less hoary-tomentose and glutinous or the foliage at length glabrous. Leaves crowded, lanceolate, acute, entire, contracted into a rather long petiole, ¾ to 1½ in. long. Peduncles solitary, mostly 2 to 3 lines long. Calyx-segments much imbricate, ovate or lanceolate, rather acute or acuminate, not dilated upwards, the outermost usually much broader and larger than the inner ones. Corolla above 1 in. long, pubescent outside, the tube constricted above the ovary, then enlarged, the 4 upper lobes rather broad and acute, the 2 uppermost more united than the others, the middle lower lobe broader and obtuse. Stamens included. Ovary ovoid, with 2 or 3 pairs of ovules in each cell, suspended from short broad flat erect superposed funicles. Drupe not seen perfect, apparently nearly dry, with a 4-celled putamen.

S. Australia. Lake Torrens, *Howitt's Expedition;* between Stoke's Range and Cooper's Creek, *Wheeler* (both with the calyx 2 to 3 lines long); near Lake Torrens, *Hawker in Freeling's Expedition* (with the outer calyx-segments above ½ in. long, and the corolla also large).

21. **E. Goodwinii,** *F. Muell. Rep. Babb. Exped.* 17. A shrub of several feet, more or less glandular and viscid and often hirsute with spreading hairs which are rarely wanting on the calyxes and pedicels. Leaves linear or linear-lanceolate, acutely acuminate, entire, scarcely

contracted or even dilated at the base and sessile, the midrib often very prominent underneath, mostly 1 to nearly 2 in. long. Pedicels solitary, ½ to 1 in. long or even more. Calyx-segments much imbricate, lanceolate, very acute, 4 to 8 lines long, the outer one usually broader and the 2 innermost smaller than the others. Corolla more or less pubescent outside, ¾ to above 1 in. long, the lobes broad, obtuse or shortly acute, the 2 uppermost more united, the middle lowest lobe scarcely broader than the others. Stamens included. Ovary shortly ovoid, very hairy, with 2 pairs of ovules in each cell. Fruit very obtuse or retuse, hairy, 4 to 5 lines long, very thick with a thick bony 4-celled putamen.

N. S. Wales. Darling river and Mount Murchison, *Dallachy and Goodwin.*
S. Australia. N.W. interior, Mount Freeling. &c., *M'Douall Stuart's Expedition;* between Stoke's Range and Cooper's Creek, *Wheeler.*

22. **E. Willsii,** *F. Muell. Fragm.* iii. 21, t. 20. Branches and foliage more or. less covered with a glandular rust-coloured pubescence and somewhat glutinous. Leaves obovate-oblong, obtuse or almost acute, entire or serrulate, contracted at the base but not petiolate, about 1 in. long. Pedicels solitary, hispid, short. Calyx-segments much imbricate, ½ to ¾ in. long and enlarging after flowering, the outermost almost ovate, the innermost narrow, all acute. Corolla glabrous, or slightly pubescent outside, "blue," the lobes all broad obtuse or very shortly acute, the middle lower one rather broader than the others. Stamens included. Ovary narrow, densely tomentose, with 2 pairs of ovules in each cell. Fruit not seen.

N. Australia. Finke river, *M'Douall Stuart's Expedition.*

23. **E. platycalyx,** *F. Muell. Fragm.* v. 109. A shrub of about 10 feet, more or less hoary-tomentose or almost glabrous, the branches often glandular-verrucose. Leaves lanceolate, broad or narrow, tapering into a short petiole, entire and rather thick, above 1 in. long. Calyx-segments much imbricate, almost like those of an *Ipomæa,* oval-oblong in the bud rather thick and very obtuse, but as the flower expands very soon enlarging, almost orbicular, thin, coloured and veined, attaining ½ in. diameter. Corolla glabrous outside, above 1 in. long, the tube broad, slightly constricted above the ovary, the lobes all broad. Stamens included. Ovary oblong, tapering upwards, slightly glandular-tomentose or glabrous, with 2 pairs of ovules in each cell.

W. Australia. *Drummond;* Sharks Bay and 300 miles up Murchison river, *Maitland Brown.*

24. **E. viscida,** *Endl. Nov. Stirp. Dec.* 51. Glabrous and glutinous. Leaves elliptical-lanceolate, entire, 1½ to 2 in. long, 4 to 5 lines broad. Peduncles solitary or 2 together, 5 lines long, dilated under the flowers. Calyx-segments obovate, obtuse, enlarged scarious and veined when in fruit and then 5 lines long and 4 broad. Corolla-tube broad, 5 lines long, the lobes all broad obtuse 3 lines long, the 2 uppermost more united in an upper lip. Stamens much exserted. Drupe small.—DC. Prod. xi. 712; F. Muell. in Proc. R. Soc. Tasm. iii. 294.

W. Australia. *Roe.* I have not examined this species; from the above description abridged from Endlicher's it appears to be near *E. platycalyx,* but with long exserted stamens. The only specimen I have seen (in the Herbarium of the Imperial Botanic Garden at Vienna) has no corolla.

SECT. 5. STENOCHILUS.—Calyx-segments imbricate at the base, usually enlarged after flowering. Corolla 4 upper lobes short and acute, the fifth lowest more deeply separated and sometimes narrow. Stamens exserted (except in *E. alternifolia*). Ovules 2 or 3 pairs or rarely only 1 pair in each cell of the ovary. Drupe usually succulent, with a thick bony putamen not separating into pyrenes.

25. **E. Brownii,** *F. Muell. in Proc. R. Soc. Tasm.* iii. 297. A shrub attaining sometimes several feet, rarely quite glabrous, more frequently with the branches and young shoots and sometimes the adult foliage hoary or white with a close almost mealy tomentum. Leaves lanceolate or rarely elliptical oblong or cuneate, obtuse or acute, entire or very rarely marked with a few serratures, contracted into a petiole, very variable in size, most frequently ¾ to 1 in. long, but in some specimens all under ½ in. and crowded. Pedicels solitary, usually shorter than the calyx. Flowers "yellow, red, or with these colours variously mixed." Calyx-segments imbricate, broadly or narrowly lanceolate, acuminate or almost obtuse, varying from scarcely above 1 line to above 3 lines long, the outer ones usually larger than the inner. Corolla glabrous or slightly pubescent outside, usually about 1 in. but in some specimens only 8 or 9 lines long, the tube constricted above the ovary, then dilated and incurved, the 4 upper lobes short narrow and acute, with sometimes an accessory one between the 2 uppermost, the lowest lobe narrow, rolled back, separated to about the middle of the corolla. Stamens exserted, usually long. Ovary with 2 pairs of ovules to each cell. Fruit ovoid or almost globular, succulent, 4 to 5 lines diameter when perfect, the putamen hard, almost perfectly 4-celled with 1 seed in each cell.—*Stenochilus glaber,* R. Br. Prod. 517; A. DC. Prod. xi. 714; Endl. Iconogr. t. 92; Bot. Mag. t. 1942; Bot. Reg. t. 572; *S. viscosus,* Grah. in Edinb. Phil. Journ. vi. 387 and in Bot. Mag. t. 2930; A. DC. Prod. xi. 715; *Eremophila Grahami,* F. Muell. in Proc. R. Soc. Tasm. iii. 297; *S. ochroleucus* A. Cunn. (*S. maculati* var. A. DC.), A. DC. Prod. xi. 715; *S. albicans* and *S. subcanescens* Bartl. in Pl. Preiss. i. 351; A. DC. l. c.; *Eremophila albicans,* F. Muell. in Proc. R. Soc. Tasm. iii. 297; *S. incanus,* Lindl. Bot. Reg. 1839, Misc. 70; *Eremophila incana,* F. Muell. in Proc. R. Soc. Tasm. iii. 297.

Queensland. Between Warrego and the Maranoa, *Barton.*
N. S. Wales. Lachlan river, Peel's Range, &c., *A. Cunningham;* from the Lachlan, Murray, and Darling to the Barrier Range, *Victorian and other Expeditions.*
Victoria. Murray desert, *F. Mueller;* Wimmera, *Dallachy.*
S. Australia. Fowler's Bay, and head of Spencer's Gulf, *R. Brown,* from the Murray to St. Vincent's and Spencer's Gulf, and Lake Torrens *F. Mueller* and others; Kangaroo island, *Waterhouse;* Lake Gairdner, *Babbage.*
W. Australia. From Swan river to the northward, *Drummond, 1st coll. n.* 441, 442, *Preiss. n.* 2303, 2304, 2318, *Fraser* and others; Murchison river, *Oldfield;* Sharks Bay, *Milne, Maitland Brown;* Phillips and Fitzgerald river, *Maxwell.*

This the typical *Stenochilus* from which the following species diverge more or less, connecting it with the other sections of *Eremophila*, is itself exceedingly variable, from cottony white all over to perfectly glabrous, as well as in the shape of the leaves and size of the leaves and flowers. The following appears almost deserving to be reckoned a distinct species :—

Var. *viridiflora*, F. Muell. Diffuse, with small crowded leaves and small flowers, viscid, pubescent and green. The lower lobe smaller than in the typical form. W. Australia, *Drummond*, (*2nd coll. ?*) *n.* 162 ; Upper Kalgan river, *F. Mueller.*

26. **E. subfloccosa,** *Benth.* Young shoots thickly covered with a loose plumose almost floccose tomentum, wearing off from the older leaves. Leaves crowded, elliptical-oblong, obtuse, entire, contracted at the base but sessile or nearly so, ½ in. long or rather more, rather thick, the older ones apparently glutinous. Flower solitary, sessile or nearly so, longer than the leaves. Calyx-segments much imbricate, narrow, acute, loosely tomentose, 3 lines long or rather more, the outer ones linear-lanceolate, the inner narrow-linear. Corolla glabrous outside or sprinkled with short hairs, 7 to 8 lines long, slightly constricted above the ovary, then incurved and enlarged, the 4 upper lobes small and acute, the lowest one much shorter, also acute. Stamens exserted. Ovary glabrous, with only one pair of ovules in each cell.

W. Australia. In the interior, *Roe*, also *Drummond* (*in herb. F. Muell.*)

27. **E. Oldfieldii,** *F. Muell. Fragm.* i. 208. An erect shrub of several feet, or small tree of 10 feet (*Oldfield*), glabrous and probably glutinous, or the branchlets and young shoots minutely hoary. Leaves linear or lanceolate, acute or almost obtuse, entire, contracted into a petiole, flat but rather thick, 1 to 2 in. long or rarely more. Flowers " red with a yellow base," solitary, on pedicels of ¼ to ½ in. Calyxsegments much imbricate, oblong, from almost cuneate to lanceolate, obtuse or acute, 3 to 4 lines long at the time of flowering, enlarging to ½ in. or more. Corolla glabrous outside or nearly so, about ¾ in. long, the tube broad from the base and scarcely constricted above the ovary, the lobes all obtuse or scarcely acute, the 4 upper ones short, the lowest broadly oblong and separate to near the middle of the corolla. Stamens more or less exserted. Ovary short, obtuse, glabrous, with 2 or 3 pairs of ovules in each cell.

W. Australia. Murchison river, *Oldfield;* Sharks Bay, *Milne.*

28. **E. Duttoni,** *F. Muell. Rep. Babb. Exped.* 16. An erect glutinous shrub, glabrous or the young shoots slightly tomentose. Leaves narrow-lanceolate, entire, tapering into a long acute point, contracted at the base but scarcely petiolate, 1 to 2 in. long. Pedicels solitary, ½ in. long or more, very spreading and turned up towards the end. Flowers " orange-red." Calyx-segments ovate, acute or acuminate, and 4 to 6 lines long at the time of flowering, afterwards often enlarged, broad, coloured, almost scarious and veined. Corolla usually glabrous outside, slightly bearded inside, 1 to 1¼ in. long, the tube constricted above the ovary, then enlarged and slightly curved, the 4 upper lobes short and acute, the lowest oblong, obtuse, separate to about ⅓ of the corolla. Stamens exserted. Ovary glabrous or slightly glandular-pubescent, with a pair of ovules to each cell. Fruit suc-

culent, shining, shorter than the enlarged calyx, the putamen hard and bony, usually 4-celled, with one seed in each cell.

N. S. Wales. Near the Barrier Range, *Victorian Expedition;* Mount Murchison, *Bonney.*
S. Australia. Cooper's Creek, *Wright;* Northern interior, *M'Douall Stuart.*

29. **E. maculata,** *F. Muell. in Proc. R. Soc. Tasm.* iii. 297. A tall shrub, with rigid divaricate branches, more or less hoary-tomentose or pubescent, the adult foliage usually glabrous. Leaves mostly lanceolate, varying however from elliptical-oblong to linear, acute or obtuse, entire, contracted into a petiole, rarely above 1 in. long, flat and green on both sides or hairy when young. Pedicels solitary, often above ½ in. long, very spreading or reflexed but turned up again under the flowers. Calyx-segments much imbricate and ovate at the base, acuminate, 2 to 3 lines long or more. Corolla glabrous outside, "red, more or less variegated with yellow or quite yellow," 1 in. long or more, the broad tube constricted above the ovary, the upper part slightly incurved and not much dilated, the 4 upper lobes short and acute, the lowest one narrow, recurved, separated to below the middle of the corolla. Stamens usually but perhaps not always exserted. Ovary glabrous, with 2 or 3 pairs of ovules to each cell. Fruit ovoid-globular, shortly acuminate, above ½ in. diameter, very succulent, with a hard bony putamen, completely 2-celled and less perfectly 4-celled. Seeds small, without so much albumen as in some species.—*Stenochilus maculatus,* Ker. in Bot. Reg. t. 647; R. Br. App. Sturt. Exped. 23; *S. racemosus* Endl. Nov. Stirp. Dec. 50; A. DC. Prod. xi. 715; *S. curvipes,* Benth. in Mitch. Trop. Austr. 221.

N. Australia. Attack Creek, *M'Douall Stuart's Expedition.*
Queensland. Warrego river, *Mitchell;* Isaacs and Fitzroy rivers, *Bowman and others;* Curriewillighie, *Dalton;* Darling Downs, *Lau.*
N. S. Wales. Lachlan river, *A. Cunningham;* Murray, Darling, and Lachlan rivers to the Barrier Range, *Victorian and other Expeditions;* Junction of the Murray and Murrumbidgee, *F. Mueller.*
Victoria. Murray river, *F. Mueller.*
S. Australia. Murray river towards Moriunda, *F. Mueller.*
Var. *brevifolia.* Leaves oblong or obovate-oblong, very obtuse, mostly about ½ in. long.
N. Australia. Hammersley Ranges, N.W. coast, *Maitland Brown.*
N. Australia. Murchison river, 300 miles above the Geraldine, *Oldfield;* 100 miles E. of York, *Roe.*

30. **E. denticulata,** *F. Muell. Fragm.* i. 125. A shrub of several feet, glabrous or nearly so and glutinous. Leaves lanceolate ovate-lanceolate or oblong-elliptical, acute or acuminate, entire or serrulate, contracted into a rather long petiole, 1 to 2 in. long. Pedicels solitary, ½ to 1 in. long, very spreading and incurved under the flowers. Calyx-segments much imbricate at the base, lanceolate or ovate-lanceolate, acute or acuminate, 2 to 4 lines long. Corolla "red," glabrous outside, about 1 in. long, slightly constricted above the ovary, but broad even there, and enlarged and incurved upwards, the 4 upper lobes very small and acute, the lowest lobe narrow, recurved, separated to below the middle of the corolla. Stamens exserted. Ovary rather short,

ovoid, glabrous, with only one pair of ovules to each cell. Drupe succulent, but not seen ripe.

W. Australia, *Drummond;* Phillips river and sand hummocks, Eyre's Relief,
Maxwell.

31. **E. latifolia,** *F. Muell. in Linnæa,* xxv. 428, *and in Proc. R. Soc.
Tasm.* iii. 293. A spreading shrub of 2 to 3 ft., the young shoots
slightly hoary-pubescent, otherwise glabrous and usually glutinous.
Leaves ovate obovate or ovate-lanceolate, obtuse, mostly denticulate
and often undulate, contracted into a rather long petiole, ½ to 1 in.
long. Pedicels solitary, slender, above ½ in. long, very spreading and
curved upwards at the end. Calyx-segments much imbricate, broadly
ovate or obovate, obtuse, herbaceous, 3 to 4 lines long when in flower,
enlarging sometimes to 4 or 5 lines, very broad rigid and veined when
in fruit, the outermost one often smaller than the others. Corolla
glandular-pubescent outside, ¾ to 1 in. long, slightly constricted above
the ovary, then broad and incurved; the 4 upper lobes short and acute,
the lowest narrow, reflexed, separated to below the middle of the
corolla. Stamens exserted. Ovary depressed, 2-celled, with only one
pair of ovules to each cell. Fruit depressed-globular, half as long as
the calyx, succulent, with a hard bony almost completely 4-celled
putamen.—*Stenochilus serrulatus,* A. Cunn. in DC. Prod. xi. 715.

N. S. Wales. Peel's Range, *A. Cunningham;* Lachlan and Darling rivers to the
Barrier Range, *Victorian and other Expeditions.*
S. Australia. Near Cudnaka, *F. Mueller;* Lake Gillies, *Burkitt;* N. interior,
M'Douall Stuart's Expedition.
W. Australia, *Drummond, Harper.*

32. **E. alternifolia,** *R. Br. Prod.* 518 *and App. Sturt. Exped.* 22.
A tall erect much-branched shrub, the young shoots minutely hoary,
otherwise glabrous. Leaves scattered, linear-terete, usually ending in a
recurved point, entire, contracted into a short petiole, rarely above 1 in.
long. Pedicels solitary, very spreading or reflexed but turned up at
the end. Calyx-segments much imbricate, ovate or almost orbicular,
scarious, veined, coloured, the inner ones 3 to 4 lines long, the 2 outer
ones smaller. Corolla " red, spotted with purple," glabrous outside,
¾ to 1 in. long, the short base of the tube almost globular, constricted
above the ovary, then dilated and somewhat incurved; the 4 upper
lobes short and acute, the lowest lobe broader, obtuse, very spreading
and separated to the middle of the tube. Stamens included. Ovary
glabrous, with only one pair of ovules to each cell. Fruit ovoid or
ovoid-conical, the exocarp very thin, the endocarp readily separating
into 4 acuminate pyrenes, with 1 seed in each.—A. DC. Prod. xi. 712;
F. Muell. in Proc. R. Soc. Tasm. iii. 294.

N. S. Wales. Darling river, *Giles.*
S. Australia. Spencer's Gulf, *R. Brown, Warburton;* Murray Scrub, *Behr.;*
Flinders' Range and Lake Torrens, *F. Mueller;* Lake Gillies, *Burkitt;* Lake Gairdner,
Babbage's Expedition.
This species has the calyx and corolla of *Stenochilus,* with the included stamens and
the fruit of *Eremocosmos.*
Var. *latifolia,* F. Muel. Leaves thick and nerveless, but flat, and 1 to 1¼ lines broad.
—Head of the Great Bight, *Delisser.*

Order XCI. SELAGINEÆ.

Flowers irregular. Calyx persistent, more or less deeply toothed or divided into 3 to 5 lobes, or into 2 or 3 distinct sepals. Corolla with 4 or 5 lobes more or less obliquely declinate or rarely 2-lipped. Stamens usually 4, in pairs, inserted at the summit of the tube of the corolla and alternating with its lower lobes; anthers 1-celled (by the confluence of the 2 cells?) Ovary free, not lobed, 2-celled with 1 pendulous ovule in each cell. Style simple, undivided at the apex. Fruit small, dry, readily separating into two 1-seeded nuts or reduced to a single one by abortion. Seeds pendulous, albuminous; embryo straight, with a superior radicle.—Herbs or undershrubs usually small. Leaves alternate or rarely opposite, the floral ones often dissimilar and reduced to bracts. Flowers solitary within each floral leaf, usually sessile, forming dense or interrupted terminal spikes.

A small extratropical Order, chiefly South African, with one northern genus, the only Australian species having been probably introduced from the Cape. The order is closely allied to and forms as it were the S. African representative of the Australian Myoporineæ, differing more in habit than in any positive character except the reduction of the ovules to one only or two in the whole ovary, which appears to be constant in Selagineæ, and only occurs in a very few species of Myoporineæ. The irregularity of the corolla assumes also a somewhat different type in the two Orders.

*1. DISCHISMA, Chois.

Calyx divided to the base into 2 sepals. Corolla with a cylindrical tube, the limb obliquely declinate (1-lipped). Stamens 4.

*1 **D. capitatum,** *Chois. in Mem. Soc. Gen. and in DC. Prod.* xii. 7. A diffuse small but hard branching annual, more or less hirsute or sprinkled with crisped hairs. Lower leaves opposite, the upper ones alternate, linear with a few distant teeth or entire, mostly about or under ½ in. long. Flowers small (blue?), in terminal spikes which after flowering become very dense, ovoid or oblong, and from ½ to above 1 in. long, each flower sessile within a floral leaf or bract, broad at the base with a linear herbaceous point longer than the flower, and after flowering the broad bases of the bracts are closely imbricated concealing the fruit. Sepals small very thin and ciliate (not ½ line long). Corolla tube about 1 line long, the upper lip almost obsolete, the lower as long as the tube, with small lateral lobes and a larger lower one. Fruit oblong, about 1 line long, separating into 2 narrow nuts.

W. Australia. *Drummond, 2nd coll. n.* 150, *3rd. coll. n.* 292; apparently abundant, as numerous specimens were gathered each time, but most likely introduced from the Cape where the species is not uncommon.

Order XCII. VERBENACEÆ.

Flowers irregular or rarely regular. Calyx persistent, truncate toothed or lobed. Corolla with 4 or 5, rarely 6 to 8, lobes or rarely truncate, the lobes more or less 2-lipped or nearly or quite equal, imbricate in the bud, the upper lip or uppermost lobe or sometimes the lateral ones outside. Stamens inserted in the tube of the corolla, usually 4 in pairs or nearly equal and alternating with its lower lobes, or when the corolla is regular 4 to 8 alternating with its lobes. Anthers 2-celled, the cells opening longitudinally and usually parallel. Ovary not lobed or only shortly 4-lobed, usually more or less perfectly divided into 2 or 4 cells or half-cells, with 1 ovule in each cell or half-cell, either anatropous and erect from the base, or more or less amphitropous and attached laterally or near the top so as to appear pendulous. Style terminal, simple, entire or more frequently with 2 short stigmatic lobes. Fruit dry or more or less drupaceous, the whole fruit or the endocarp separating into 2 or 4 nuts or pyrenes or quite indehiscent and 2- or 4-celled, and sometimes with an additional central cavity between the carpels having the appearance of a third or fifth empty cell. Seeds solitary in each cell half-cell or pyrene, erect, with or without albumen, the testa usually membranous. Embryo straight, with thick cotyledons and an inferior radicle.—Herbs shrubs trees or woody climbers. Leaves opposite whorled or rarely alternate, without stipules, entire toothed or divided. Inflorescence various.

A large Order, ranging over both the New and the Old World, most abundant within the tropics, but with several extratropical species, both in the northern and southern hemispheres. Of the twenty Australian genera, three are large American ones of which a very few species, including those found in Australia, have become more or less generally diffused over the Old World, five are most numerously represented in tropical Asia and Africa, but three of them are also American, and one of these extends in a single species beyond the tropics as far as southern Europe, one, consisting chiefly of maritime plants, is as common in the New as in the Old World, and the remaining eleven genera are purely endemic, with the exception of *Faradaya*, which is represented by a second species in the South Pacific Islands.

The structure of the flower in Verbenaceæ has been considerably elucidated in several points by the observations of H. Bocquillon (Revue du groupe des Verbénacées, Paris, 1861-1863), but his arrangement of the genera appears to me to be neither natural nor practical, removing as it does for instance *Clerodendron* far from *Premna*, to place it next to *Priva*, which again is placed at some distance from *Verbena*, and in a totally different group from *Stachytarpheta*. The regularity or irregularity of the flower is in some cases not well marked, nor sufficiently in accord with other characters to serve as a basis for the primary division of the Order, and the degree in which the placentary margins of the carpellary leaves protrude into the cavity of the ovary, meeting only or cohering in the centre, thus more or less completely dividing it into cells, is, in this Order, a difference of degree only, the placentation never having the truly parietal character of that of Gesneriaceæ. I have therefore returned to the old division of the Order into the main tribes adopted by Schauer in the Prodromus, with some minor modifications as to a few genera which had been imperfectly known or incorrectly described.

TRIBE 1. **Verbeneæ.**—*Ovules ascending from the base of the cells. Flowers in racemes or spikes sometimes contracted into heads, solitary within each bract, without bracteoles.*

Ovary 2-celled, with 1 ovule in each cell. Spikes dense, usually
 contracted into heads.
 Fruit a drupe . . . . 1. LANTANA.
 Fruit dry, separating into small nuts . . . . 2. LIPPIA.
Ovary 4-celled, with one ovule in each cell. Spikes elongated . 3. VERBENA.

Tribe 2. **Viticeæ.**—*Ovules laterally attached above the base or near the top. Flowers in cymes or if solitary or in spikes usually accompanied by 2 bracteoles besides the subtending bract or leaf.*

Subtribe 1. **Chloantheæ.**—*Ovary not lobed. Fruit small, dry, or rarely drupaceous. Shrubs or herbs usually very cottony or woolly, rarely nearly glabrous, glutinous or resinous. Seeds where known always albuminous.*

Corolla small, regular or nearly so. Stamens isomerous. Style entire or minutely 2-lobed. Flowers in dense woolly spikes.

Flowers 5–8-merous. Corolla truncate or very shortly lobed 4. Lachnostachys.
Flowers 5-merous. Corolla distinctly lobed . . . . . 5. Newcastlia.
Flowers 4-merous. Corolla distinctly lobed . . . . . 6. Physopsis.
Style rather shortly 2-lobed. Flowers 4-merous in heads or corymbs . . . . 7. Mallophora.
Style deeply 2-lobed. Flowers 5-merous in heads cymes or panicles. . . . . 8. Dicrastyles.
Corolla more or less 2-lipped, or unequally 5-lobed. Stamens 4.
Leaves decurrent. Corolla-tube elongated. Anthers without appendages . . . . 9. Chloanthes.
Leaves not decurrent. Corolla-tube broad. Anther-cells usually with small appendages at the lower end . . . . 10. Pityrodia.
Calyx-lobes much enlarged spreading and veined after flowering . . . . 11. Cyanostegia.
Corolla small, the tube narrow, the upper lip erect. Leaves mucronately toothed . . . . 12. Denisonia.
Corolla small, the tube broad. Fruit a succulent drupe (dry in the preceding genera) . . . . 13. Spartothamnus.

Subtribe 2. **Euviticeæ.**—*Ovary not at all or scarcely lobed. Fruit a drupe. Shrubs or trees. Seeds without albumen (or rarely in* Vitex *with a scanty albumen?).*

Corolla small, usually 4-lobed, with a short tube. Stamens included or not much exserted.
Cymes axillary. Style dilated at the top or truncate . . . 14. Callicarpa.
Cymes or panicles terminal. Style 2-lobed . . . . . 15. Premna.
Corolla 5-lobed, with a slender tube. Stamens long. Fruiting calyx enlarged and spreading (except in *C. hemiderma*) . . 16. Clerodendron.
Corolla-tube broad, limb 4- or 5-lobed. Stamens not exceeding the upper lip. Drupe with a bony 4-celled putamen. Leaves simple . . . . 17. Gmelina.
Corolla-tube short, limb 5-lobed. Stamens often exserted. Drupe with 2 or 4 distinct pyrenes. Leaves often digitately compound (sometimes simple) . . . . 18. Vitex.

Subtribe. 3. **Oxereæ.**—*Ovary distinctly 2- or 4-lobed.*

Calyx 2-cleft. Tall climber with large flowers and fruits. Stamens exserted . . . . 19. Faradaya.

Tribe 3. **Avicennieæ.**—*Fruit a 2-valved capsule. Seed solitary, without integuments; embryo with large folded cotyledons.*

Single genus . . . . 20. Avicennia.

Tribe 1. Verbeneæ.—Ovules ascending from the base of the cells. Flowers in racemes or spikes, sometimes contracted into heads, solitary within each bract, without bracteoles.

1. LANTANA. Linn.

Calyx small and thin, truncate or sinuately toothed. Corolla-tube slender; the limb spreading, 4- or 5-lobed, nearly regular or slightly 2-lipped. Stamens 4, included in the tube. Ovary 2-celled, with one ovule in each cell erect from the base. Fruit a more or less succulent drupe, the putamen 2-celled or dividing into two 1-celled pyrenes.— Shrubs or rarely herbs. Leaves opposite. Flowers in pedunculate axillary heads, rarely lengthening into spikes, each one sessile or nearly so within a small bract without bracteoles.

A considerable genus, chiefly from tropical or subtropical America, with two or three Asiatic or African species, which however may also have been of American origin. The ovary in this and the following genus, as shown by Bocquillon, although con-taining only 2 cells corresponding to the half-cells of other genera, is yet bicarpellary, one half only of each carpel being developed.

1. **L. Camara,** *Linn.* ; *Schau. in DC. Prod.* xi. 598. A tall shrub with long weak branches, often armed with short recurved prickles, and more or less hairy. Leaves petiolate, ovate or slightly cordate, crenate, 2 to 3 in. long, wrinkled and very rough with short stiff hairs. Flowers yellow or orange, turning to a deep red ; the heads not lengthening into spikes. Bracts linear-lanceolate, shorter than the corolla. Corolla-tube 3 to 4 lines long, lobes of the limb short and broad.

A common species in tropical America, frequently cultivated for ornament, and, escaping from gardens, now naturalised on the Hastings and Clarence rivers, *Beckler*, and probably in other parts of **N. S. Wales** and **Queensland.** As already observed in my " Flora Hongkongensis," the species should probably include as varieties several of those described by Schauer, in DC. Prod. xi. 597 and 598, as distinct.

2. LIPPIA. Linn.

(Zapania, *Scop.*)

Calyx membranous, either flattened with 2 keels or wings and 2-lobed, each lobe either entire or 2-toothed, or the whole calyx more equally tubular or globular and 2- or 4-toothed. Corolla-tube cylin-drical or dilated upwards, the limb more or less distinctly 2-lipped, the upper lip entire or 2-lobed, the lower 3-lobed, all the lobes flat and spreading. Stamens 4, included in the tube or scarcely protruding. Ovary 2-celled, with 1 ovule in each cell erect from the base. Fruit not succulent, separating more or less readily into two indehiscent nuts.—Herbs or shrubs often glandular and aromatic or strong-scented. Leaves opposite or whorled, undivided. Flowers small, in simple spikes or heads, each one sessile in the axil of a single bract, without bracteoles, the bracts often closely imbricate.

A considerable American genus, a few species of which, including the two Australian ones, are also more or less widely spread over the warmer regions of the Old World. Bocquillon's character of the genus (Revue, p. 147), taken probably from the examina-tion of a single species, will not apply to a large portion of the genus, including the commonest species of all, *L. nodiflora.*

Prostrate or creeping perennial. Leaves obovate or cuneate.
  Peduncles in one axil of each pair. Calyx flat . . . . 1. *L. nodiflora.*
Shrub with straggling branches. Leaves ovate. Peduncles oppo-
  site. Calyx globular . . . . 2. *L. geminata.*

1. **L. nodiflora,** *Rich.; Schau. in DC. Prod.* xi. 585. A prostrate
or creeping perennial, with shortly ascending flowering branches, hoary
with closely appressed hairs or nearly glabrous. Leaves from obovate
to linear-cuneate, coarsely toothed at the apex, $\frac{1}{2}$ to 1 in. long, narrowed into a petiole. Peduncles axillary but only one to each pair of
leaves and much longer than them, each one bearing a spike at first
short and ovoid, and sometimes very small, at length cylindrical, and
when luxuriant attaining $\frac{1}{2}$ to $\frac{3}{4}$ in. or even more. Bracts closely imbricate, broadly spathulate, more or less fringed or toothed at the end,
nearly $1\frac{1}{2}$ lines long. Calyx shorter than the bract, membranous, flat,
divided in front nearly to the base, at the back to about the middle,
into two keeled lobes, entire or 2-toothed at the apex. Corolla-tube
scarcely exceeding the calyx, the lower lip twice as long as the upper
one and about half as long as the tube. Fruit not one line long,
readily separating into two nuts, with one half of the calyx adhering to
each.—*Zapania nodiflora,* Lam.; R. Br. Prod. 514.

N. Australia. Victoria river, *F. Mueller.*
Queensland. Shoalwater Bay and Broad Sound, *R. Brown;* Port Denison,
Fitzalan; Fitzroy river, *Thozet;* Bowen river, *Bowman;* Moreton island, *M'Gillivray.*
W. Australia. Murchison river, *Oldfield.*
  The species is very common in waste lands on banks and in sandy places, &c., all
over the warmer parts of the world. It is very variable in the breadth of the leaves,
the size of the spikes and flowers, the points and teeth of the bracts, &c.

2. **L. geminata,** *Kunth; Schau. in DC. Prod.* xi. 582. A strongly
scented shrub, with long straggling branches, more or less hirsute, the
young shoots often hoary. Leaves opposite or rarely in whorls of
three, from broadly ovate to ovate-oblong, obtuse, crenate, very rugose,
$\frac{3}{4}$ to $1\frac{1}{2}$ in. long when broad, longer when narrow. Peduncles much
shorter than the leaves, and mostly in both axils, each one bearing a
small head of pink flowers becoming ovoid as the flowering advances.
Bracts very broad, herbaceous, hispid, 1 to $1\frac{1}{2}$ lines long. Calyx
shorter than the bract, membranous, nearly globular, neither flattened
nor ribbed, 2-lobed, the lobes broad and obscurely 2-toothed. Corolla-
tube $1\frac{1}{2}$ lines long, dilated upwards, the lobes short broad and nearly
equal. Fruit separating into two nuts.

Queensland. About Rockhampton, *Dallachy* and others. Probably introduced
from South America, where it is often common, ranging from Buenos Ayres to Mexico.

3. VERBENA. Linn.

Calyx 5-toothed. Corolla with a distinct tube, the limb spreading,
rather unequally 5-lobed. Stamens 4 or rarely 2, included in the
tube. Ovary 4-celled, with one ovule in each cell erect from the base.
Fruit not succulent, enclosed in the calyx, separating more or less
readily into 4 1-seeded nuts. — Herbs or rarely shrubs. Leaves

opposite, entire or divided. Flowers small, alternate, in simple or branched spikes, each one in the axil of a small bract without bracteoles.

The genus comprises a large number of American species, with only two natives of the warmer regions of the Old World, including one of the Australian ones; the other species here enumerated is an introduced one from America.

Leaves mostly deeply lobed or divided. Spikes long and slender, with distant flowers . . . . 1. *V. officinalis.*
Leaves narrow, toothed. Spikes rather close, in a terminal cluster or panicle . . . . 2. *V. bonariensis.*

1. **V. officinalis,** *Linn.; Schau. in DC. Prod.* xi. 547. An erect perennial, 1 to 2 ft. high, with long spreading wiry branches, sometimes nearly glabrous, usually with closely appressed hairs, sometimes more coarsely hirsute, or the inflorescence rough with glandular hairs. Lower leaves petiolate, obovate or oblong, coarsely toothed or cut; upper ones either deeply pinnatifid and lobed or toothed, or the uppermost small and lanceolate. Flowers usually very small, in slender spikes lengthening often to 8 or 10 in., the lower ones becoming distant as the spike lengthens, the whole corolla sometimes not 2 lines long, but in the larger-flowered forms the tube about 1½ lines, and the lower lip about as long.—*R. Br. Prod.* 514.

Queensland. Broad Sound and Shoalwater Bay, *R. Brown;* Rockingham Bay, *Dallachy;* Suttor river, *Bowman;* Armadilla, *Barton.*

N. S. Wales. Port Jackson, *R. Brown;* Blue mountains, *Miss Atkinson;* Clarence river, *Beckler;* Richmond river, *Fawcett;* Darling river, *Victorian and other Expeditions.*

Victoria. Port Phillip, *F. Mueller;* Melbourne, *Robertson;* Portland, *Allitt;* Skipton, *Whan.*

S. Australia. Near Adelaide, *Blandowski;* towards Spencer's Gulf, *Warburton.*
The species is common in a great part of Europe and temperate Asia, in waste places on roadsides, &c.; more rare and perhaps introduced into North America, South Africa, and within the tropics. It may also be introduced only into some of the Australian stations.

Var. *macrostachya.* Flowers rather larger, and the spikes very glandular, hirsute.— *V. macrostachya,* F. Muell. Fragm. i. 60.—Peak Downs, *F. Mueller;* Rockhampton, *Bowman.*

*2. **V. bonariensis,** *Linn.; Schau. in DC. Prod.* xi. 541. An erect coarse rigid herb of 2 to 4 ft., the stems scarcely branched, acutely 4-angled and roughly hispid especially on the angles. Leaves sessile, lanceolate or the lower ones ovate-lanceolate, 1½ to nearly 3 in. long, coarsely toothed, hirsute, the upper ones distant small and narrow. Flowers in rather close spikes of ½ to ¾ in., which are usually clustered at the end of the branches of a rigid corymbose trichotomous panicle, and generally assume a blueish purple hue. Bracts acute, ciliate, hirsute, 1 to 1½ lines long. Calyx shorter than the bract. Corolla-tube shortly exceeding the calyx, the lobes broad and spreading.

Queensland. Brisbane river, Moreton Bay, *F. Mueller.*
N. S. Wales. Near Sydney, *F. Mueller;* New England, *C. Stuart;* Hastings river, *Beckler.*

The species is common in waste places and pastures in extratropical South America, and has spread as a weed of cultivation over South Africa, the Mauritius, and some other countries, and is evidently introduced only into Australia.

TRIBE 2. VITICEÆ.—Ovules amphitropous, laterally attached above the base or sometimes so near the top as to appear pendulous, but the micropyle always inferior, the funicle either very short or more or less lengthened and then arising from the base of the placenta. Flowers in axillary or terminal cymes or heads, or, if solitary, on axillary or spicate pedicels, usually accompanied by two bracteoles besides the subtending bract.

In many of the genera of this tribe the ovary is not perfectly divided into cells, the incurved ovuliferous margins of the carpellary leaves not meeting in the centre at the time of flowering, and the ovary has in these cases been described by Bocquillon as one-celled with parietal placentas. The ovules are, however, never placed on the inner face of the expanded placentas as in Gesneriaceæ and other Orders with a normally parietal placentation, and usually, after flowering, the placentas meet in the centre and unite, or the endocarp grows and hardens so as completely to enclose each seed, forming a fruit perfectly divided into cells, or only leaving a small central cavity, described by earlier authors as a third or a fifth empty cell.

SUBTRIBE 1. CHLOANTHEÆ.—Ovary not lobed. Fruit small, dry, the mesocarp not succulent, the endocarp thin or hard, 4-celled or more frequently separating into two 2-celled or four 1-celled nuts, sometimes reduced by abortion to a single 1-seeded nut. Shrubs undershrubs or rarely herbs, usually very cottony or woolly, with branched hairs, rarely more glabrous and glutinous or resinous. Flowers often solitary within each bract or floral leaf, but sometimes in cymes as in Euviticeæ.

The ripe seeds have been observed in a few species only, and these have shown a rather copious albumen. This character may not, however, be constant in the sub-tribe. The ten following genera—perhaps all that strictly belong to the subtribe—are all endemic in Australia.

4. LACHNOSTACHYS, Hook.

(Walcottia, *F. Muell.* Pycnolachne, *Turcz.*)

Calyx broadly campanulate, 5- to 8-lobed, valvate in the bud, densely woolly outside, glabrous inside. Corolla shorter than the calyx, broadly campanulate, truncate or very shortly and equally 5- to 8-lobed. Stamens 5 to 8, exserted, opposite to the calyx-lobes, inserted on the margin of the corolla or between its lobes; anthers without appendages. Ovary 2-celled, with 2 ovules in each cell laterally attached below the top. Style slender, entire or minutely 2-lobed. Fruit enclosed in the calyx, hard, usually 1-celled and 1-seeded by abortion.—Erect shrubs clothed with a dense cotton or wool consisting of intricate branched hairs. Leaves opposite, sessile, undivided. Flowers opposite and sessile in dense terminal woolly spikes. Bracts often imbricate in 4 rows in the young spikes, but very deciduous; bracteoles minute or none.

The genus is endemic in W. Australia. In the two species first published the disse-piment of the ovary is very thin, and readily breaks off from the walls of the cavity, and as only one ovule enlarges, the whole on a hasty examination has the appearance of a 1-celled uniovulate ovary; this with the short membranous corolla with marginal stamens, suggested the idea of a staminal cup, and induced the referring the plant to Amarantaceæ, a view which Moquin adopted without re-examining the ovary. A more careful scrutiny shows however an ovary characteristic of a considerable group of Verbenaceæ, and scarcely to be found in any other Order; and this affinity is fully confirmed by the since described *Lachnostachys Walcottii*, which has the corolla-lobes shortly developed between the stamens. The habit and peculiar rudimentum are entirely those of *Newcastlia*, which has also regular flowers with isomerous stamens, and of which one species, *N. spodiotricha*, only differs from *Lachnostachys* in the greater development of the corolla-lobes, and in the parts of the flowers being in fives only, whilst in one species of *Lachnostachys* they are in sixes or rarely in fives, and in the three others in eights or rarely in sevens. F. Mueller has, however, placed *Lachnostachys* in Buettneriaceæ, from which it appears to me to differ essentially in the position of the stamens, opposite to not alternating with the calyx-lobes, and in the structure of the ovary, independently of the habit, the pistil, and the supposition that the cup supporting the stamens is a corolla and not the united base of the filaments only.

Spikes simple. Flowers 6-merous, rarely 5-merous . . . . . 1. *L. albicans.*
Spikes paniculate. Flowers 8-merous, rarely 7-merous.
  Leaves ovate or oblong with revolute margins, 1 to 3 in. long.
    Wool very long and dense . . . . 2. *L. verbascifolia.*
  Leaves ovate or oblong, nearly flat, ¾ to 1 in. long. Tomentum
    thick but close and short . . . . 3. *L. ferruginea.*
  Leaves oblong-linear, with revolute margins. Panicle much
    branched. Tomentum close and short. Corolla distinctly
    lobed . . . . 4. *L. Walcottii.*

1. **L. albicans,** *Hook. Ic. Pl. t.* 414.—A shrub of several ft., with rather thick erect branches, hoary or white as well as the foliage with a close but dense cottony wool. Leaves oblong-lanceolate, obtuse or almost acute, erect, decussate, thick, the margins often narrowly recurved and slightly rugose-crenulate, ¾ to 1½ in. long. Spikes terminal, simple, the flowers concealed in a dense silky-woolly mass, 1 to 2 in. long, and ½ to ¾ in. diameter. Calyx about 1½ lines long and opening to 2 lines diameter, divided to about ⅓ into 5 or more, frequently 6, broad triangular lobes, the external wool at least twice as long as the whole calyx. Corolla rather shorter than the calyx, glabrous outside, hirsute inside with long hairs, truncate, the filaments quite marginal, without lobes between them, and only to be traced down the tube by a darker vein. Ovary densely villous. Young fruit shorter than the calyx, thick and hard, with only one ovule enlarged, but not seen ripe.—Moq. in DC. Prod. xiii. ii. 298; Nees in Pl. Preiss. i. 631.

W. Australia, *Drummond, n.* 13, 434; Wellington district, *Preiss, n.* 1377.

2. **L. verbascifolia,** *F. Muell. Fragm.* vi. 158. A tall stout shrub, very densely clothed in every part with long silky-woolly hairs, more or less ferruginous. Leaves crowded, decussate, ovate or oblong, obtuse, very thick and soft, with revolute margins, 1 to 3 in. long. Spikes apparently few, in a short dense terminal panicle, but the inflorescence imperfect in our specimens. Calyx 1½ to nearly 2 lines long, divided like that of *L. ferruginea* into 8 finely pointed lobes, but the external wool longer and more silky. Corolla short, the filaments quite marginal, without intervening lobes. Ovary certainly 2-celled,

with 2 ovules in each cell, but as in *L. ferruginea* and *L. albicans,* only 1 ovule enlarges.

W. Australia, *Drummond,* 5th coll. n. 237.

I have not succeeded in finding ripe seeds in our specimens, but F. Mueller has observed them to be albuminous with a straight embryo as in the allied genera.

3. **L. ferruginea,** *Hook. Ic. Pl. t.* 415. A tall shrub, covered with a soft dense cottony-wool, thicker than in *L. Walcottii,* almost floccose, and of a more rusty colour. Leaves oblong · ovate or almost orbicular, obtuse, soft and very thick, the margins thickened · underneath but scarcely recurved, ¾ to 1 in. long. Spikes rather dense, 1 to 2 in. long, and ½ to ¾ in. diameter, several in a terminal spreading panicle. Bracts (or floral leaves) orbicular, thick and brown, imbricate in 4 rows in the young spike, but soon falling off, leaving each flower a globular woolly mass of 2 to 3 lines diameter. Calyx nearly 2 lines long, divided to below the middle into 8 or sometimes 7 narrow finely pointed lobes. Corolla rather shorter than the calyx, glabrous outside, bearded with long hairs inside, but not so densely so as in *L. albicans.* Stamens quite marginal, without lobes between them. Young fruit as in *L. albicans,* but not seen ripe.—Moq. in DC. Prod, xiii. ii. 298.

W. Australia, *Drummond,* n. 14, 202, 438.

4. **L. Walcottii,** *F. Muell. Fragm.* ii. 140. A tall shrub, covered with a close but soft cottony-wool sometimes almost floccose. Leaves oblong-linear, obtuse, thick, with closely revolute margins, ¾ to 1½ in. long. Spikes not very dense but many-flowered, 1 to 2 in. long, numerous in dense corymbose or pyramidal panicles of several inches diameter, each flower a woolly ball of about 2 lines diameter, showing in the centre a small glabrous corolla only when fully expanded. Calyx about 1½ lines diameter when spread open, divided to the middle into 8 acute lobes. Corolla rather shorter than the calyx, with 8 very short rounded reticulate lobes. Stamens inserted in the notches and prominently decurrent in the tube to the base of the corolla, where there are a few hairs inside. Ovary glabrous or minutely pubescent. Fruit already somewhat enlarged, apparently ripening 2 or 3 seeds and divided into as many cells by spurious dissepiments, but not seen ripe. —*Walcottia eriobotrya,* F. Muell. Fragm. i. 241; *Pycnolachne ledifolia,* Turcz. in Bull. Soc. Imp. Nat. Mosc. 1863, ii. 215.

W. Australia. Murchison river, *Walcott and Oldfield, Drummond,* 6th coll. n. 219, 220.

5. NEWCASTLIA, F. Muell.

Calyx campanulate, 5-lobed, valvate in the bud. Corolla-tube campanulate, the limb of 5 nearly equal lobes. Stamens 5 ; anthers without appendages. Ovary small, completely 2-celled with 2 ovules in each cell laterally attached above the middle. Style slender, entire, or minutely 2-lobed. Fruit not exceeding the calyx, not succulent, apparently separating into 4 nuts.—Densely woolly or cottony shrubs.

Leaves opposite, undivided. Flowers opposite and sessile, in dense
terminal woolly spikes. Bracts and bracteoles small and deciduous.

The genus is endemic in Australia, and closely allied to *Physopsis* and *Mallophora*,
differing from them chiefly in the 5-merous flowers.

Wool loose. Corolla-lobes short. Stamens shorter than the
  corolla . . . . 1. *N. cladotricha.*
Tomentum close. Corolla-lobes ending in a fine point. Stamens exserted . . . . 2. *N. spodiotricha.*

1. **N. cladotricha,** *F. Muell. in Hook. Kew Journ.* ix. 22, *Fragm.* i.
184, *t.* i. *and* iii. 21. An erect shrub, attaining 2 to 3 ft., densely
clothed with white or rust-coloured woolly branching hairs, and
strongly scented. Leaves sessile, narrow-oblong to ovate-lanceolate,
obtuse, rounded at the base, the margins slightly recurved, ¾ to above
1 in. long, thick, very rugose, reticulate underneath, loosely hirsute
or tomentose on both sides with branched hairs. Flowering spikes at
first short and dense but lengthening sometimes to 2 in. and interrupted. Bracts ovate or ovate-lanceolate, imbricate in the very young
spike, but falling off early. Calyx about 1½ lines long, very woolly, the
lobes rather shorter than the tube. Corolla glabrous outside, the tube
broad, about as long as the calyx, lobes short and spreading. Stamens
about as long as the tube, inserted above a woolly ring near the base.
Ovary glabrous. Ovules attached near the top, but the seed enlarges
upwards so as to be attached near the base. The fruit not seen however quite ripe.

N. Australia. Sturt's Creek, near Mount Mueller, *F. Mueller.*

2. **N. spodiotricha,** *F. Muell. Fragm.* iii. 21, *t.* 21. A shrub or
undershrub, clothed with a rather shorter and closer tomentum than
that of *N. cladotricha.* Leaves very shortly petiolate, ovate-oblong,
obtuse, 1 to 2 in. long, rugose above, reticulate underneath, tomentose
on both sides. Flowers in terminal cottony spikes, with 1 or 2 pairs of
lateral ones at the base, forming a pyramidal panicle. Bracts not seen.
Calyx nearly sessile, about 2 lines long, the lobes much shorter than
the tube. Corolla-tube as long as the calyx, very hairy inside near the
top, the lobes narrow, ending in a point nearly as long as the tube, and
longer than the lobes themselves. Stamens inserted at the top of the
corolla-tube between the lobes, and longer than them, the upper
anthers sometimes abortive. Ovary glabrous, the ovules attached at or
near the top.

N. Australia. Between lat. 17° 30′ and 18° 30′, *M'Douall Stuart.*

6. PHYSOPSIS, Turczan.

Calyx tubular, 4-toothed. Corolla-tube short, cylindrical, the limb
of 4 nearly equal spreading lobes. Stamens 4, included in the tube;
anthers without appendages. Ovary 2-celled, with 2 ovules in each cell,
laterally attached above the middle, but usually only 1 ovule perfect.
Style slender, very shortly 2-lobed. Fruit dry, enclosed in the calyx,

often reduced to a single cell and seed.—Woolly shrub. Leaves scattered, undivided. Flowers small, opposite and sessile, in a dense woolly spike, each one within a small bract.

The genus consists of a single species endemic in Australia, differing from *Mallophora* chiefly in inflorescence.

1. **P. spicata,** *Turcz. in Bull. Soc. Imp. Nat. Mosc.* 1849, ii. 35. An erect shrub, with rather stout woolly-tomentose virgate branches. Leaves scattered or irregularly opposite, sessile, oblong, obtuse, with recurved margins, narrowed at the base, rarely exceeding ½ in., glabrous or slightly scabrous and nerveless on the upper side, cottony-white underneath. Spikes dense, either solitary or clustered at the ends of the branches, usually 1 to 1½ in. long, each flower sessile within a linear bract, which is glabrous inside, woolly outside, and very deciduous. Calyx enveloped in cottony-wool forming an ovoid mass about 3 lines long, the calyx itself, when stripped of its wool nearly tubular and very shortly 4-toothed. Corolla-tube scarcely exceeding the calyx, slightly thickened inside at the throat, the lobes broad and obtuse. Stamens inserted above the middle of the tube, the filaments very short. Ovary glabrous, inserted on a disk, in the very young bud completely 2-celled with 2 ovules in each cell, but at the time of flowering usually very oblique with only one perfect ovule.

W. Australia, *Drummond, 4th coll. n.* 234.

7. MALLOPHORA, Endl.

(Lachnocephalus, *Turcz.*)

Calyx deeply divided into 4 lobes. Corolla-tube short, cylindrical, the limb of 4 equal spreading lobes. Stamens 4, shortly exserted; anthers without appendages. Ovary 2-celled, with 2 ovules in each cell laterally attached above the middle. Style filiform with 2 linear lobes. Fruit dry, 4-celled, with 1 seed in each cell.—Cottony or woolly undershrub. Leaves opposite or scattered, undivided. Flowers small, sessile, in dense cottony-woolly heads which are either solitary or corymbose at the ends of the branches.

The genus is limited to a single species endemic in Australia, closely allied to the two preceding genera, but with a more divided style and the inflorescence nearer to that of *Dicrastyles.*

1. **M. globiflora,** *Endl. Nov. Stirp. Dec.* 64. Stems from a woody base rather slender, apparently ascending or erect, branching, 1 to 1½ ft. high, covered as well as the foliage with a close white intricate tomentum. Leaves sessile or nearly so, linear or oblong, obtuse, ¼ to nearly ½ in. long, narrowed at the base, rather thick, flat, cottony-white on both sides or becoming at length nearly glabrous above and then rugose. Flower-heads dense, either solitary or more frequently several in terminal corymbs, each flower sessile within a woolly bract, the outer bracts of each head rather larger than the others, but none of them exceeding the calyx. Calyx enveloped in a long dense woo'

forming globular masses of 2 to 3 lines diameter, within the wool the calyx is deeply divided into linear membranous lobes. Corolla-tube scarcely so long as the calyx, the lobes small, obtuse, woolly outside. Stamens inserted within the tube. Ovary cottony, the 4 ovules usually perfect. Style-lobes linear, but much shorter than in *Dicrastyles*. Fruit, according to Endlicher, tomentose and 4-celled.—*Lachnocephalus lepidotus*, Turcz. in Bull. Soc. Imp. Nat. Mosc. 1849, ii. 36.

W. Australia, *Drummond, n. 72, 555, and 4th coll. n. 235.*

Bocquillon (Rev. Verb. p. 138) places *Lachnocephalus* (*Mallophora*) in his section with irregular flowers, but the corolla appears to me to be as nearly regular as in *Dicrastyles* and other so-called regular-flowered Verbenaceæ.

8. DICRASTYLES, Drumm.

Calyx more or less deeply divided into 5 lobes. Corolla-tube short, the limb of 5 nearly equal short lobes. Stamens 5, exserted; anthers without appendages. Ovary 2-celled, with 2 ovules in each cell, laterally attached at or above the middle. Style deeply divided into 2 slender branches or lobes. Fruit small, dry, 4-celled, with 1 seed in each cell.—Cottony or woolly undershrubs or small shrubs. Leaves opposite or scattered, undivided. Flowers small, in cymes collected into corymbose panicles, more rarely contracted into dense solitary or corymbose heads. Bracts and bracteoles very deciduous.

The genus is limited to Australia. It is nearly related to *Mallophora*, but the inflorescence is usually looser, the flowers pentamerous, and the style much more deeply divided.

Leaves petiolate, lanceolate, very rugose but flat, 1½ to 3 in. Cymes in a pyramidal panicle . . . . 1. *D. ochrotricha.*
Leaves nearly sessile, oblong, rugose but flat, ½ to 1 in. Cymes in loose corymbose panicles . . . . 2. *D. fulva.*
Leaves sessile, oblong, rugose, with revolute margins, under 1 in. Cymes very dense, in corymbose panicles . . . . . . 3. *D. reticulata.*
Leaves linear, with revolute margins, ¼ to ½ in. Flowers small, in corymbose panicles . . . . 4. *D. parvifolia.*
Leaves sessile, very rugose, with revolute margins. Flowers in dense heads. Diffuse, *Filago*-like plant . . . . 5. *D. stœchas.*

1. D. ochrotricha, *F. Muell. Fragm.* iv. 161. An erect shrub of 1 to 2 ft., densely clothed with a rather close woolly tomentum, often assuming a golden yellow colour. Leaves opposite, lanceolate, obtuse, contracted into a rather long petiole, thick soft and woolly on both sides when young, scabrous and rugose above when old, reticulate and tomentose underneath, 1½ to nearly 3 in. long, the upper floral ones reduced to small bracts. Cymes opposite, pedunculate, forming a short pyramidal terminal panicle, the pedicels and calyxes very woolly-hirsute with short golden-yellow branching hairs. Pedicels ½ to 3½ lines long, thicker than in the other species. Calyx about 2 lines long, divided to rather below the middle into rather broad obtuse lobes. Corolla scarcely exceeding the calyx, the tube woolly outside, the lobes small. Stamens less exserted than in the other species. Ovary very woolly as well as the entire part of the style; ovules attached very near the top.

Style-branches glabrous, about as long as the entire part. Fruit small, depressed globular, not seen quite ripe.—*Pityrodia exsuccosa*, F. Muell. Fragm. i. 60.

N. Australia. Sturt's Creek, near Mount Wilford, *F. Mueller.*

2. **D. fulva,** *Drumm. in Hook. Kew Journ.* vii. 56. A perennial or undershrub with erect simple or branched stems of 1 to 2 feet, densely clothed with a whitish, or more frequently brownish cottony wool, sometimes almost floccose in the lower part, looser and longer towards the inflorescence. Leaves mostly opposite, narrow-ovate or oblong, obtuse, narrowed at the base, $\frac{1}{2}$ to above 1 in. long, very thick soft and reticulate-rugose, but the wrinkles concealed by the wool till it wears off with age. Flowers very numerous, in broad trichotomous corymbose panicles, the clusters when young forming globular woolly heads surrounded by woolly bracts, but much looser when fully out, when the bracts fall away and each flower is on a little filiform pedicel of $\frac{1}{2}$ to 1 line. Calyx about 1 line long, hirsute with branched hairs, not so woolly as in *Mallophora*, divided to the middle or more deeply into rather unequal lobes. Corolla almost campanulate, about $1\frac{1}{2}$ lines long, the lobes as long as the tube and equal. Stamens inserted a little below the lobes and as long as or rather longer than them. Ovary tomentose. Style hairy with glabrous branches about as long as the entire part.—*Pityrodia myriantha*, F. Muell. Fragm. i. 236.

W. Australia. Northern districts, *Drummond;* Murchison river, *Oldfield;* Dirk Hartog's island, *Martin.*

3. **D. reticulata,** *Drumm. in Hook. Kew Journ.* vii. 57. An undershrub or shrub with the general habit of *D. fulva,* the stems erect or ascending covered with the same dense cottony wool of a white or brownish hue. Leaves opposite or scattered, sessile, oblong or ovate-lanceolate, obtuse, usually smaller than in *D. fulva,* less narrowed at the base, the margins revolute, much wrinkled above and reticulate underneath, pubescent above and hoary-tomentose underneath, but the cotton not dense enough to conceal the reticulations. Flowers rather larger than in *D. fulva,* in dense heads of $\frac{1}{2}$ to 1 in. diameter, which are either several together in terminal corymbs or solitary on the side branches, the outer bracts often leaflike, the inner ones small. Pedicels about 1 line long. Calyx rather more than 1 line long, divided to the base into linear segments. Corolla 2 to $2\frac{1}{2}$ lines long, the lobes shorter than the tube. Stamens longer than the corolla-lobes. Ovary tomentose. Style rather longer than in *D. fulva,* hairy with glabrous branches.

W. Australia, *Drummond,* 4th coll. n. 94.

Mallophora corymbosa, Endl. Nov. Stirp. Dec. 64, appears to me from a cursory inspection without examination of the specimen in the Vienna herbarium, to be a very woolly variety of *Dicrastyles reticulata.*

4. **D. parvifolia,** *F. Muell. Fragm.* ii. 160. An erect undershrub or shrub of 1 to 2 ft., with numerous rather slender branches, the whole plant hoary or white with a close tomentum. Leaves linear, obtuse,

with revolute margins, from $\frac{1}{4}$ to rather above $\frac{1}{2}$ in. long, losing the tomentum on the upper side when old, and then somewhat rugose. Flowers in compact head-like cymes, forming trichotomous corymbose panicles as in *D. fulva*, but very much smaller and the bracts very small. Calyx almost sessile, scarcely above $\frac{1}{2}$ line long, divided almost to the base into oblong segments. Corolla about 1 line long, very broad and open, the lobes much longer than the tube and one larger than the others. Stamens 5 as in the other species, longer than the corolla. Ovary tomentose. Style-branches longer than the entire part.—*D. rosmarinifolia*, Turcz. in Bull. Soc. Imp. Nat. Mosc. 1863; ii. 226.

W. Australia, *Drummond, n.* 176, 236; Young river, East river, and Oldfield river, *Maxwell.*

5. **D. stœchas,** *Drumm. in Hook. Kew Journ.* vii. 57. A diffuse, much branched, low undershrub, with the aspect of a *Gnaphalium* or *Filago*, densely clothed in every part with white wool. Leaves opposite or scattered, sessile, oblong, obtuse, with revolute margins, 2 to 4 lines long, thick and soft, very rugose under the white wool. Flowers numerous in dense terminal woolly heads of $\frac{1}{2}$ to $\frac{3}{4}$ in. diameter, not paniculate. Bracts small except the outer ones which are sometimes leafy and 1 to 2 lines long. Calyx about 1 line long, divided to the base into linear lobes. Corolla $1\frac{1}{2}$ lines long, the lobes rather shorter than the tube. Stamens longer than the corolla-lobes. Ovary as well as the entire part of the style tomentose or woolly, containing but a single cell and ovule in all the flowers examined, but they were all somewhat enlarged after fecundation, probably as in the rest of the genus 2-celled at an earlier stage, but in the rather numerous speci-mens in the collections before me I have not succeeded in finding either buds or far advanced fruits.

W. Australia, *Drummond, 5th coll. suppl. n.* 95.

9. CHLOANTHES, R. Br.

Calyx more or less deeply divided into 5 narrow herbaceous lobes. Corolla-tube elongated, usually incurved and dilated upwards; limb 2-lipped, the upper lip erect at the base with two spreading lobes, the lower lip three-lobed, spreading. Stamens 4, somewhat didynamous, inserted below the middle of the tube above a ring of cottony hairs, shorter than the upper lip; anthers without any or with very obscure appendages. Ovary imperfectly or almost perfectly 2-celled with 2 ovules in each cell laterally attached. Style very shortly 2-lobed. Fruit a dry 4-celled drupe, the endocarp separating into 2 hard 2-celled nuts, leaving between them a central cavity reaching halfway up. Seeds solitary in each cell, ascending, with a thin testa and copious albumen.—Perennials undershrubs or shrubs, more or less cottony woolly or glandular-hirsute. Leaves opposite or in whorls of three, narrow, bullate-rugose and decurrent along the stem. Flowers axillary, solitary, shortly pedicellate, with a pair of bracteoles below the calyx, the upper flowers sometimes forming a leafy spike.

The genus is limited to Australia. The transformation of the imperfectly 2-celled ovary into a completely 4-celled fruit in this and the following genera is effected by the growth of the endocarp round the seeds, filling up the cavity in the upper part, but usually leaving in the lower part a vacuity or so-called fifth empty cell.

Flowers mostly axillary and distant. Eastern species.
Leaves lanceolate, the margins scarcely revolute, shortly hispid on both sides, not cottony underneath. Flowers large, the tube narrow . . . . 1. *C. glandulosa.*
Leaves narrow with very recurved margins, white underneath.
Corolla tube narrow (variable in size), glabrous inside except the ring of hairs near the base . . . . 2. *C. stœchadis.*
Corolla-tube short and broad, hairy inside under the upper lip 3. *C. parviflora.*
Flowers in short leafy spikes at the ends of the branches. Western species . . . . 4. *C. coccinea.*

Pityrodia uncinata and *P. Bartlingii* are usually placed in *Chloanthes*, of which they have the anthers, but the corollas as well as scattered non-decurrent leaves are those of *Pityrodia.*

1. **C. glandulosa,** *R. Br. Prod.* 514. An erect perennial or undershrub, nearly resembling *C. stœchadis*, but coarser and taller. Leaves lanceolate or linear-lanceolate, bullate-rugose and decurrent as in that species, but mostly 1½ to 3 in. long, the margins less revolute, and both sides muricate or hispid with short rigid glandular hairs, not white or woolly underneath. Flowers axillary, 1½ in. long, the calyx fully ½ in., the peduncles 3 to 4 lines long, with short linear bracts below the middle or near the base, the shape and structure of the flowers and fruits otherwise as in *C. stœchadis.*—Schau. in DC. Prod. xi. 531.

N. S. Wales. Grose river, *R. Brown;* Blue Mountains, *A. and R. Cunningham.*
Further observations may possibly show this to be a luxuriant variety of *C. stœchadis.*

2. **C. stœchadis,** *R. Br. Prod.* 514. A perennial or undershrub, with erect simple or branched stems of 1 to 2 ft. Leaves opposite, linear or linear-lanceolate, but often almost terete owing to the revolute margins, obtuse, rarely above 1 in. long, exceedingly bullate-rugose and scabrous-muricate on the upper or outer surface, which is decurrent along the stem to the next pair of leaves, the under-surface woolly-white but often quite concealed by the revolute margins. Flowers "yellowish," on very short axillary pedicels, with a pair of linear bracteoles rugose like the leaves but shorter than the calyx, inserted about the middle of the pedicel. Calyx 4 to 5 lines long, more or less clothed with woolly hairs inside and out, divided to the middle or lower down into 5 lanceolate or linear herbaceous lobes, bullate like the leaves. Corolla in the typical form above 1 in. long, pubescent outside, the tube gradually dilated upwards, and slightly curved, glabrous inside except a ring of woolly hairs above the ovary, the upper lip erect, somewhat concave, with two short spreading lobes, the lower lip divided into three acute very spreading lobes, the middle one rather longer and more reflexed than the others. Ovary densely

villous. Fruit separating into two hemispherical reticulate hairy cocci, the exocarp membranous, the endocarp and placenta forming in each a bony 2-celled nut with 1 seed in each nut.—Schau. in DC. Prod. xi. 532; *C. lavandulifolia,* Sieb. in Spreng. Syst. ii. 756.

N. S. Wales. Heaths about Port Jackson, *R. Brown, Sieber, n.* 185 *and* 186, and many others.

Var. *parviflora.* Flowers smaller, but shaped like those of *C. stœchadis,* the corolla about ¾ in. long.—Waverley hills, Sydney, *Mossman;* Wooloomooloo, *A. Cunningham,* and in some other N. S. Wales collections.

3. **C. parviflora,** *Walp. Rep.* iv. 58. An erect perennial or undershrub, with the habit and foliage of *C. stœchadis,* but the calyx is more deeply divided, and the corolla, not above ½ in. long, has the throat or upper portion of the tube broader and very hairy inside below the upper lip with long whitish hairs, the lobes are also much shorter and all obtuse. Stamens usually shorter than in *C. stœchadis,* but variable. Fruit as in *C. stœchadis.* The colour of the flower is uncertain, being variously described as purple, light blue, yellow, or yellowish.— Schau. in DC. Prod. xi. 532.

Queensland. Rockingham Bay, *Dallachy.*

N. S. Wales. In the interior, *Lhotzky (Schauer).* I have not seen the original specimens, but Walpers's character applies rather to this than to the small-flowered variety of *C. stœchadis.*

4. **C. coccinea,** *Bartl. in Pl. Preiss.* i. 352. An erect slightly branched undershrub or shrub of about 1 to 2 ft., the stems usually clothed with a white cottony wool, concealed however by the decurrent leaves. Leaves opposite or in whorls of three, narrow and nearly terete owing to the revolute margins, obtuse, ½ to 1 in. long, bullate-rugose with the tubercles very regular in three or four longitudinal rows, coriaceous, shining, slightly tuberculate or muricate and decurrent along the stem, the white cottony under surface usually completely concealed. Flowers scarlet according to Preiss, nearly sessile and axillary, but collected into short leafy spikes or heads at or near the summits of the branches. Pedicels not 1 line long, the linear bracteoles near the base. Calyx 4 to 5 lines long, deeply divided, hirsute with long white woolly hairs. Corolla-tube about ½ in. long, gradually dilated upwards, the lobes about 3 lines long, almost acute. Stamens exserted from the tube.—Schau. in DC. Prod. ix. 831.

W. Australia, *Drummond, n.* 97, 142; Hay district, *Preiss, n.* 2339.

Chiefly distinguished from the eastern *C. stœchadis* by the more rigid regularly bullate leaves, and by the inflorescence.

There is apparently a fifth species with decurrent leaves, of which the specimens from Depuech island, N.W. Coast, *Bynoe,* are in a state of very young bud, insufficient for description.

10. PITYRODIA. R. Br.

(Quoya, *Gaudich.* Dasymalla, *Endl.*)

Calyx more or less deeply 5-lobed. Corolla-tube broad, usually short, more rarely elongated; limb of 5 spreading lobes more or less distinctly 2-lipped, or oblique with the lowest lobe much larger than the others. Stamens 4, usually didynamous; included or shortly exserted; anther-cells all, or one of each anther, or those of one pair of anthers, tipped at the lower end by a small or very prominent appendage rarely entirely deficient. Ovary imperfectly or almost perfectly 2-celled, with 2 ovules in each cell laterally attached to a short or very long funicle. Style very shortly 2-lobed, and often dilated at the base of the lobes. Fruit a dry drupe, the endocarp separating into two 2-celled nuts with one seed in each cell, or reduced by abortion to one or two single-seeded nuts. Seeds ascending, with a thin testa and copious albumen.—Shrubs or undershrubs, more or less clothed with cottony wool. Leaves scattered or irregularly opposite, not decurrent. Flowers solitary, or in cymes or clusters, axillary or collected in terminal cymes or leafy spikes.

The genus is limited to Australia, differing from *Chloanthes* in foliage, in the corolla-tube usually shorter and much broader, and in the appendages to the anthers which are usually very distinct, although in a few species very much reduced or obsolete. The corolla also of some species is very nearly that of *Chloanthes parviflora*, but the peculiar foliage of the latter genus, which is constant, may be sufficient to maintain it as distinct.

Leaves oblong or lanceolate, the margins slightly recurved. Calyx (not exceeding 3 lines), with lanceolate lobes. Flowers axillary.
  Leaves petiolate, lanceolate, very rugose, 1 to 2 in. long. Tomentum ferruginous . . . . 1. *P. salvifolia.*
  Leaves nearly sessile, mostly obtuse, under ¾ in. long. Tomentum hoary or white . . . . 2. *P. hemigenioides.*
Leaves linear or lanceolate with revolute margins, very rugose (as in *Chloanthes*) but not decurrent. Calyx (4 to 6 lines) with linear plumose lobes. Flowers axillary or in spikes.
  Leaves mostly linear, the floral ones exceeding the flowers . 3. *P. uncinata.*
  Leaves mostly lanceolate, the floral ones not exceeding the flowers . . . . 4. *P. Bartlingii.*
Leaves ovate oblong or lanceolate, flat thick and soft. Calyx (3 to 6 lines) with narrow or lanceolate lobes. Flowers usually in dense or interrupted spikes.
  Calyx about 5 lines long, woolly-tomentose.
    Corolla-tube shorter than the calyx; middle lobe twice as broad as the others . . . . 5. *P. verbascina.*
    Corolla-tube longer than the calyx, the three lower lobes broad . . . . 6. *P. racemosa.*
  Calyx scarcely 3 lines long, densely plumose-hairy. Corolla-tube not exceeding the calyx; middle lobe twice as broad as the others . . . . 7. *P. Drummondii.*
Leaves without revolute margins. Calyx-lobes ovate oblong or spathulate, very obtuse.
  Leaves very rugose, contracated below the middle but sessile and dilated at the base.
    Corolla-tube much longer than the calyx, gradually dilated upwards . . . . 8. *P. dilatata.*
    Corolla-tube broadly campanulate, not much exceeding the calyx . . . . 9. *P. cuneata.*
  Leaves petiolate, broadly ovate or orbicular. Calyx-lobes enlarged and thin after flowering.
    Very thickly woolly-tomentose. Calyx 6 to 8 lines long. Corolla-lobes nearly of equal breadth . . . . 10. *P. Oldfieldii.*

Hoary or white with a close tomentum. Calyx 3 to 5 lines
long. Corolla with the lowest lobe twice as broad as the
others . . . . 11. *P. atriplicina.*
Like *P. atriplicina,* but smaller, with a looser inflorescence
and smaller flowers . . . . 12. *P. paniculata.*

1. **P. salvifolia,** *R. Br. Prod.* 513. A shrub of spreading growth, attaining 6 to 8 ft., with a strong sage-like scent, the branches densely
clothed with a woolly tomentum usually rust-coloured, but sometimes
whitish. Leaves opposite, shortly petiolate, lanceolate or almost linear,
obtuse or rather acute, 2 to 3 or even 4 in. long, very rugose, pubescent
above, cottony and rusty or whitish underneath. Flowers nearly sessile,
in axillary clusters of 2, 3, or rarely more. Bracts very narrow, the
outer ones shorter than the calyx, the inner ones much smaller. Calyx
turbinate-campanulate, nearly 3 lines long, prominently ribbed,
tomentose, divided to rather below the middle into lanceolate acuminate
lobes. Corolla white, scarcely exceeding the calyx, the tube broadly
campanulate, with a dense ring of hairs inside below the stamens ; lobes
shorter than the tube, the 2 upper ones rather more united, the middle
lower one rather broader than the others. Filaments very short;
anthers almost exserted, the two lower ones with prominent appendages,
the two upper ones with shorter ones, and in one bud I found a fifth
rudimentary stamen. Ovary glabrous ; ovules attached close to the
top with a short funicle. Fruit almost completely 4-celled.—Schau.
in DC. Prod. xi. 628 ; *Premna salvifolia,* Spreng. Syst. ii. 755.

Queensland. Northumberland islands, *R. Brown*; barren rocky hills, Cleveland
bay, *A. Cunningham;* near Rockhampton, *O'Shanesy ;* near Mount Hedlow, *Dallachy.*

2. **P. hemigenioides,** *F. Muell.* A rigid divaricate shrub of 1 to
3 ft., the branches cottony-white or woolly-tomentose. Leaves sessile
or the larger ones shortly petiolate, narrow-ovate or oblong, rather
crowded on the branches, obtuse, the margins recurved, rounded or
cordate at the base, from about ¼ to above ½ in. long, rather rigid, at
first cottony-white but becoming glabrous and minutely rugose above,
reticulate underneath. Flowers solitary, shortly pedicellate, or nearly
sessile in the axils of the upper smaller leaves. Bracteoles linear, leafy.
Calyx about 3 lines long, turbinate-campanulate, strongly ribbed, divided to much below the middle into lanceolate lobes. Corolla 5 to
6 lines long, the tube short and much dilated, the lobes about as long
as the tube, the 2 upper ones shortly united, the middle lower one
rather broader than the others. Stamens slightly exceeding the tube ;
anther-cells with minute appendages, one cell of each of the upper
ones occasionally abortive. Ovary tomentose, ovules attached at or
near the top by a very short funicle.—*Chloanthes hemigenioides* or
Quoya hemigenioides, F. Muell. Fragm. vi. 156.

W. Australia. Northern districts, *Drummond ;* Dirk Hartog's island, *Milne.*
N. Australia. A single specimen without flowers from *M'Douall Stuart's
Expedition* appears to belong to this species.

3. **P. uncinata,** *Benth.* An erect branching shrub of 1 to 2 ft., the
branches covered with white cottony wool. Leaves crowded but not

decurrent, scattered or in irregular whorls of three, linear or linear-lanceolate, usually tapering towards the end and often but not always terminating in a hooked blunt point, the margins recurved or revolute, more or less bullate-rugose, the half-concealed under surface woolly-white or nearly glabrous. Flowers solitary in the axils of the upper leaves, forming long leafy spikes, usually very woolly-hairy, the floral leaves mostly exceeding the flowers. Pedicels very short. Calyx about 4 lines long, deeply divided into narrow membranous hairy lobes, sometimes slightly bullate and muricate at the end. Corolla-tube scarcely ½ in. long, much dilated upwards and slightly incurved, the limb 2-lipped, half as long as the tube, the upper lip very shortly 2-lobed, the lower of 3 very spreading lobes. Stamens included in the tube or the lower ones shortly exserted. Anther-cells (all ?) without any appendages. Ovary tomentose, without any hypogynous disk; ovules attached at or near the top, with a very short or scarcely any funicle.—*Chloanthes uncinata,* Turcz. in Bull. Soc. Imp. Nat. Mosc. 1863. ii. 194; *C. bullata,* F. Muell. Fragm. vi. 156.

W. Australia, *Drummond, 4th coll. n.* 160, *J. S. Roe ;* Oldfield river, *Maxwell.*

Var. *exserta.* Coarser, the leaves mostly cordate and stem-clasping. Flowers rather larger, the stamens exserted.—Cape Arid, *Maxwell.*

This and the following species are usually placed in *Chloanthes,* and the anther-appendages, conspicuous in the majority of *Pityrodia,* are here very obscure, but the shape of the flowers as well as the scattered non-decurrent leaves appear to me to place them much better in the latter than in the former genus.

4. **P. Bartlingii,** *Benth.* Stems, from a woody base, 1 to 2 ft. high, densely clothed with white wool often intermixed with long soft hairs and sometimes turning to a reddish brown. Leaves scattered or in whorls of three, not much crowded, lanceolate or linear, the margins more or less revolute, bullate and hairy or nearly glabrous outside, more or less woolly-white underneath. Flowers solitary or 3 together on short peduncles in the axils of the upper leaves, which are always smaller and sometimes reduced to bracts, forming terminal spikes of 3 to 4 in., very hairy with long soft hairs, and either continuous or interrupted and more leafy at the base. Calyx divided to the base into linear or linear-lanceolate membranous hairy segments of 4 to 6 lines. Corolla usually about ½ in. long, very much dilated above the inner ring of woolly hairs into a broad campanulate throat but oblique and somewhat incurved, the lobes all short and broad. Stamens slightly exserted; anthers with minute obtuse appendages sometimes almost obsolete, the upper pair usually smaller than the lower. Ovary tomentose, the ovules attached near the top to exceedingly long flexuose filiform funicles.—*Chloanthes Bartlingii,* Lehm. Ind. Sem. Hort. Hamb. 1844; Bartl. in Pl. Preiss. i. 352; Schau. in DC. Prod. xi. 531.

W. Australia. Swan river and Darling range, *Drummond, 1st coll. n.* 447, *Preiss, n.* 2340, *Oldfield* and others.

The leaves are exceedingly variable, sometimes all narrow and under ½ in. long; in some large-flowered specimens lanceolate, 1½ in. long, not much revolute and very densely woolly underneath ; in other specimens narrower and so much revolute as completely to conceal the wool. Some of Oldfield's specimens have very small leaves, the floral ones broad and almost ovate, with rather smaller flowers.

Some **Victorian** specimens, from near Swan Hill on the Murray, *W. Ross*, may belong to some *Pityrodia* allied to *P. uncinata* or *P. Bartlingii*, but being without flower or fruit they cannot be determined.

5. **P. verbascina,** *F. Muell.* A stout erect shrub, densely clothed as in *P. Oldfieldii* with cottony wool, often floccose and sometimes assuming a golden or orange-red hue in the upper part of the plant. Leaves opposite or in whorls of three, ovate or oblong, obtuse, sessile or contracted into a petiole, 1 to 2 in. long, very thick and soft, the veins concealed by the wool, the floral ones smaller, the upper ones shorter than the calyx. Flowers very shortly pedunculate, usually several together in the upper axils, forming a dense or interrupted more or less leafy spike. Bracts small or none. Calyx about 5 lines long, very woolly, very deeply divided into narrow 3-nerved segments. Corolla about ½ in. long, tomentose, the tube much dilated, the 2 upper lobes short and broad, the 2 lateral ones smaller and triangular, the lowest one very much larger and more than twice as broad as any of the others. Stamens included or shortly exserted; anther-appendages variable, usually one large one to one cell of each anther, the other cell without any or with a smaller one. Ovules attached at or near the top by a very short funicle. Fruit obovoid or depressed, often oblique, about 1½ lines long, usually ripening only 2 seeds (one to each carpel) but occasionally all 4 are enlarged.— *Chloanthes verbascina,* F. Muell. Fragm. i. 233; *Quoya verbascina,* F. Muell. Fragm. iv. 80.

W. Australia. Murchison river, *Oldfield, Drummond, 6th coll. n.* 140.

6. **P. racemosa,** *Benth.* An erect shrub or undershrub of about 2 ft., densely covered with white wool, sometimes loose and floccose, sometimes shorter and closer. Leaves opposite, oblong or oval, very obtuse, sessile and sometimes stem-clasping, mostly ¾ to above 1 in. long, thick and soft, reticulate-rugose but the wrinkles concealed by the tomentum, and otherwise flat. Flowers "resembling in colour and in shape those of the garden sage," solitary or more frequently in cymes of 3 to 5, the peduncles very variable in length but always shorter than the leaves, forming an interrupted terminal leafy raceme. Bracts usually short. Calyx about 5 lines long, divided almost or quite to the base into 5 narrow membranous 3-nerved segments. Corolla nearly 1 in. long, the tube much dilated upwards, the 4 upper lobes nearly equal in length, broad, spreading, pubescent outside, the lowest twice as broad as the others and glabrous. Stamens as long as the tube or shortly exserted, the anther-cells linear, at length diverging, with short appendages. Ovules attached near the top, with short funicles. Fruit ripening, occasionally at least, all the four seeds.—*Quoya (?) racemosa,* Turcz. in Bull. Soc. Imp. Nat. Mosc. 1863. ii. 194; *Chloanthes stachyodes,* or *Quoya stachyodes,* F. Muell. Fragm. v. 50, vi. 158; *Dasymalla axillaris* and *D. terminalis,* Endl. Nov. Stirp. Dec. 11, 12.

W. Australia, *Drummond,. 3rd coll. n.* 141, *5th coll. n.* 73; near Mount Walter, *Herb. F. Mueller,* collector not named.

7. **P. Drummondii,** *Turcz. in Bull. Soc. Imp. Nat. Mosc.* 1863. ii. 213. An undershrub with long erect slightly branched stems attaining 3 or 4 ft., the lower part of the stem and leaves usually loosely tomentose and more or less floccose, the upper part of the stem and sometimes the whole plant except the inflorescence and calyxes quite glabrous. Leaves opposite or in whorls of three, oval-oblong or lanceolate, obtuse, often irregularly crenate, contracted at the base, mostly ¾ to 1½ in. long, rugose but otherwise flat, the floral ones very small or entirely wanting. Flowers small, whitish, in dichotomous cymes sometimes pedunculate and many-flowered, sometimes condensed into sessile opposite heads in distant pairs forming a long interrupted terminal raceme-like panicle. Calyx scarcely 2 lines long, divided to about ¾ into linear membranous slightly unequal lobes, clothed outside with hairs sometimes very short, more frequently very long, repeatedly forked, soft and often assuming a purple hue. Corolla about 5 lines long, the tube very broad and oblique, the 2 upper lobes short and erect, the 2 lateral ones rather larger, the lowest twice as broad and longer than the others. Ovary glabrous or slightly woolly, usually with only 2 perfect ovules, each one attached to an exceedingly long filiform and several times folded funicle. Fruit not seen.—*Chloanthes loxocarpa,* F. Muell. Fragm. ii. 22; *Quoya loxocarpa,* F. Muell. Fragm. iv. 80.

W. Australia. Murchison river, *Oldfield, Drummond, 6th coll. n.* 141; Flinders' Bay, *Collie.*

The indumentum of the calyx and sometimes of the whole plant is exceedingly variable.

8. **P. dilatata,** *F. Muell.* A branching shrub, densely clothed with a white cottony wool, more or less floccose on the branches and calyxes, shorter on the leaves and sometimes disappearing from the old ones. Leaves obovate or oblong-spathulate, narrowed below the middle, dilated and stem-clasping at the base, thick and much bullate-rugose on the upper surface, reticulate underneath, otherwise flat or nearly so. Flowers all axillary, mostly solitary on short pedicels, forming sometimes a long interrupted leafy spike. Bracteoles short, linear. Calyx 4 to 5 lines long, divided to the base into 5 narrow somewhat spathulate membranous segments, very thickly woolly outside. Corolla 9 to 10 lines long, the tube gradually dilated upwards but scarcely more so than in some species of *Chloanthes,* upper lobes of the limb erect and acute, lateral ones reflexed, the lowest one rather larger than the others and reflexed. Upper pair of stamens about as long as the corolla-tube, lower pair longer; anther-cells with minute appendages. Ovary very densely villous on a small glabrous disk; ovules attached above the middle to rather long erect funicles or almost sessile and attached at or near the top. Fruit not seen.—*Chloanthes dilatata* or *Quoya dialata,* F. Muell. Fragm. vi. 157.

W. Australia, *Drummond, n.* 210.

9. **P. cuneata,** *Benth.* A rigid divaricate shrub, densely clothed with cottony stellate or branched white or yellowish hairs, more woolly

and sometimes floccose on the branches, shorter and more scabrous on the leaves. Leaves opposite, obovate or cuneate, very obtuse, contracted below the middle, sessile and usually dilated or almost auriculate at the base, $\frac{1}{2}$ to $\frac{3}{4}$ in. long, very thick and reticulate on both sides, quite flat or the margins recurved only at the narrow base. Flowers "blue," solitary or in pedunculate cymes of three, all axillary. Calyx-lobes ovate, obtuse, more or less distinctly arranged in two lips and about as long as the tube at the time of flowering, when in fruit more distinctly and deeply 2-lipped, membranous, reticulate and very woolly outside. Corolla exceeding the calyx, very broadly campanulate above the inner ring of hairs, the lowest lobe much broader than the others. Stamens shortly exceeding the corolla; anther-cells with short appendages at the base. Fruit woolly-hairy, rather above 1 line long, and broader than long, dividing into two 2-celled nuts, the cavity of their inner faces broad short and only slightly excavated.—*Quoya cuneata,* Gaud. in Freyc. Voy. Bot. 454, t. 66.

W. Australia. Sharks Bay, *Gaudichaud;* waste places, Sharks Bay, rare, *Milne.* The specimens examined being far advanced, the details of the flower are chiefly taken from Gaudichaud's figure and description.

10. **P. Oldfieldii,** *F. Muell.* An erect shrub of 2 to 3 ft., the branches and leaves thicker and more densely tomentose-woolly or scabrous-hispid with branched hairs than in any other species except *P. verbascina.* Leaves opposite, broadly ovate obovate or almost rhomboidal, very obtuse, flat but the margins minutely undulate, $\frac{3}{4}$ to $1\frac{1}{2}$ in. long, narrowed into a short petiole or almost sessile but not dilated at the base. Flowers "pink," solitary, or 3 together on very short axillary peduncles, rarely exceeding the leaves. Calyx 6 to 8 lines long, very thickly woolly, divided to nearly the middle, the lobes oblong, very obtuse, membranous, 3-nerved, reticulate, the two upper ones rather higher connate than the others. Corolla about $\frac{3}{4}$ in. long, tomentose outside, much dilated, the lobes all broad and nearly equal. Anther-appendages very distinct. Ovary very woolly-hirsute ; ovules attached at or near the top by very short funicles. Fruit not seen ripe.— *Chloanthes Oldfieldii,* F. Muell. Fragm. i. 234; *Quoya Oldfieldii,* F. Muell. Fragm. iv. 80, but not *Q. cuneata* Gaud., to which it is referred by F. Muell. Fragm. vi. 157.

W. Australia. Murchison river, *Oldfield; Drummond, 6th coll. n.* 139.

11. **P. atriplicina,** *F. Muell.* A tall much-branched shrub, white or hoary with a dense but close and short tomentum, sometimes looser and almost floccose on the branches. Leaves opposite, broadly ovate obovate or orbicular, $\frac{1}{2}$ to above 1 in. diameter, contracted into a short petiole, the venation usually concealed by the tomentum. Flowers in the upper axils, in pedunculate cymes or rarely solitary, forming often a short broad leafy panicle. Bracts small. Calyx varying from 3 to 5 lines in length, the tube ovoid or turbinate, the lobes ovate or oblong, obtuse, shorter than the tube. Corolla 6 to 8 lines long, tomentose outside, much dilated, the lobes short and broad, the lowest

twice as large as the others and very broad. Anthers of the lower stamens with appendages to one or to both of the cells, of the upper stamens usually without appendages. Ovary densely tomentose, ovules attached at or near the top with very short funicles. Fruit not seen.—*Chloanthes atriplicina,* F. Muell. Fragm. i. 235 ; *Quoya atriplicina,* F. Muell. Fragm. iv. 80.

W. Australia. Murchison river, *Oldfield; Drummond, 6th coll. n.* 138 ; Sharks Bay, *Maitland Brown.*

12. **P. paniculata,** *F. Muell.* Evidently nearly allied to *P. atriplicina,* with the same close white indumentum, and perhaps a variety or even a different state only of the same species, but only known from a small fragment, showing looser cymes with very small oblong floral leaves and smaller flowers, the calyx rather more deeply divided into oblong-spathulate or almost obovate lobes. The structure of the flowers appears to be the same as that of *P. atriplicina.*—*Chloanthes paniculata,* or *Quoya paniculata,* F. Muell. Fragm. iv. 80.

W. Australia. Sharks Bay, *Maitland Brown (Herb. F. Muell.).*

11. CYANOSTEGIA. Turczan.

(Bunnya, *F. Muell.*)

Calyx broadly campanulate, expanding after flowering, opening very flat, membranous, reticulate, the margin sinuate-toothed or 5-lobed. Corolla broadly campanulate, glabrous inside, the limb 5-lobed, the 2 upper lobes rather longer than the 3 lower. Stamens 4, inserted near the base of the corolla ; anthers large, oblong, without appendages. Ovary small, depressed, 2-celled, with 2 ovules in each cell laterally attached ; style elongated, minutely and often unequally 2-lobed. Fruit in the centre of the enlarged calyx, small, hard, usually more or less oblique, and ripening 1 or 2 ascending or oblique seeds, resembling those of *Chloanthes* and *Pityrodia.*—Glabrous and apparently glutinous shrubs. Leaves opposite, undivided, not decurrent, the upper floral ones reduced to small bracts. Peduncles axillary, 1- or 3-flowered, forming a loose terminal panicle. Bracteoles small.

The genus is limited to Australia, and nearly allied to the two preceding ones although without any cottony wool.

Leaves linear, often folded lengthwise . . . . 1. *C. angustifolia.*
Leaves lanceolate, flat . . . . 2. *C. lanceolata.*
Leaves cuneate-oblong . . . . 3. *C. Bunnyana.*

1. **C. angustifolia,** *Turcz. in Bull. Soc. Imp. Nat. Mosc.* 1849, ii. 36. An erect glabrous shrub, the upper part often appearing glutinous. Leaves linear, obtuse or with a short hooked point, flat or concave with the margins folded inwards, often bordered by small distant teeth, 1 to 2 in. long, rather thick and nerveless, the floral ones much smaller, the upper ones reduced to small bracts. Peduncles opposite, the upper ones usually short, 1-flowered, with a pair of small bracteoles under the calyx, the lower ones often 3-flowered or growing out into a flowering branch, the whole forming a loose pyramidal panicle. Calyx

at the time of flowering not much longer than the corolla, when expanded under the fruit as much as ¾ in. diameter, pubescent outside, of a blueish hue, irregularly sinuate-toothed or more deeply and broadly 5-lobed. Corolla nearly 3 lines long, apparently purple, pubescent outside, the lobes shorter than the tube. Anthers shortly protruding. Fruit not seen perfect.—*C. intermedia*, Turcz. l. c.

W. Australia, *Drummond*, 3rd coll. n. 140, 4th coll. n. 161.

2. **C. lanceolata,** *Turcz. in Bull. Soc. Imp. Nat.. Mosc.* 1849, ii. 35. An erect glabrous shrub closely allied to *C. angustifolia*,. and united with it by F. Muell. Fragm. vi. 154, under the name of *C. Turczaninovii*, but the leaves, in the numerous specimens seen, are constantly lanceolate and flat, varying from 1 in. to above 2 in. in length. Flowers rather larger and more numerous than in *C. angustifolia*. Fruit small, hard, depressed, hirsute with long branched hairs, most frequently 1-seeded.

W. Australia, *Drummond*, (1st coll. ?) n 440, 3rd coll. n. 139.

3. **C. Bunnyana,** *F. Muell. Fragm.* v. 36. An erect glabrous and apparently glutinous shrub of 3 to 4 ft. closely resembling the two other species, except that the leaves are cuneate-oblong and slightly toothed towards the end, and the flowers form a very long narrow panicle in the few specimens seen. Calyx quite glabrous outside, and usually more entire than in the other species, but some calyxes of *C. angustifolia* are in like manner only slightly sinuate-toothed. Fruit hirsute with branched hairs as in *C. lanceolata*.—*Bunnya cyanocalyx*, F. Muell. Fragm. v. 36, t. 39.

N. Australia. Roebuck Bay, N.W. Coast, *Martin*.

The three species, as suggested by F. Mueller, may prove to be forms of one very variable species, but as yet we have no intermediates.

12. DENISONIA, F. Muell.

Calyx 10-ribbed, divided to the middle into 5 narrow lobes, not much enlarged after flowering. Corolla-tube not much dilated, limb 2-lipped, the upper lip erect with 2 spreading lobes, the lower lip 3-lobed, spreading. Stamens 4, inserted above the middle of the tube, shortly exserted; anther-cells divergent, without appendages. Ovary 2-celled, with 2 ovules in each cell laterally attached near the top. Style filiform, very shortly 2-lobed. Fruit dry, 4-celled, the endocarp separating into two 2-celled nuts. Seed solitary in each cell, albuminous. —Aromatic shrub, scarcely tomentose. Leaves in whorls of 3 or scattered, not decurrent. Flowers solitary in the axils, on short pedicels, with a pair of bracteoles under the calyx.

The genus is limited to a single species, endemic in Australia and scarcely differing from *Chloanthes*, except in the non-decurrent leaves.

1. **D. ternifolia,** *F. Muell. Fragm.* i. 124, t. 2. An erect shrub of several ft. with a strong aromatic odour, the branches virgate, clothed with a short glandular tomentum intermixed with long slightly

branched spreading hairs.   Leaves sessile, ovate to lanceolate, acute and mucronate, bordered by acute mucronate teeth, $\frac{3}{4}$ to $1\frac{1}{2}$ in. long, rigid but not thick, sprinkled with short hairs, nearly smooth above, with very strong raised veins and reticulations underneath.   Flowers shorter than the leaves, the bracteoles linear, acute.   Calyx narrow campanulate, about 4 or at length 5 lines long, the lobes lanceolate, acute. Corolla-tube rather shorter. than the calyx, hairy inside below the insertion of the stamens; the upper lip usually shorter than the lower, and the middle lower lobe usually but not always elongated.   Ovary tomentose at the top.   Fruit oblong, attenuate at the base, nearly 2 lines long, tomentose.   Seeds narrow, tapering at the base.

N. Australia. Towards the sources of the Seven-Emu and M'Arthur rivers, Gulf of Carpentaria, *F. Mueller.*

13. SPARTOTHAMNUS, A. Cunn.

Calyx very open, 5-lobed.   Corolla-tube short and broad; limb spreading, 5-lobed, the middle lower lobe rather larger than the others. Stamens 4, exserted; anther-cells with minute tips at the lower end. Style filiform, with 2 rather long branches.   Ovary imperfectly 2-celled, with 2 ovules in each cell laterally attached at or above the middle.   Fruit a small globular succulent drupe, the endocarp separating into 4 1-seeded pyrenes, usually separated at the base by a central cavity.   Seeds albuminous.—Shrub or undershrub, with few small distant leaves, all opposite and not decurrent.   Flowers very small, solitary in the axils, with small bracteoles.

The genus is limited to a single species, endemic in Australia, very nearly allied to *Pityrodia*, but with a different habit, and differs from all the preceding genera by its succulent drupe, and from the following ones by its albuminous seeds and solitary flowers.

1. **S. junceus,** *A. Cunn. in Loud. Hort. Brit.* 600, *and in Walp. Rep.* vi. 694.   An undershrub or shrub of several ft., glabrous or pubescent with branching hairs, the branches divaricate, rigid, acutely 4-angled, broomlike and appearing almost leafless, the smaller branchlets often 2 or 3 together at the nodes.   Leaves small and distant, often reduced to small scales, all opposite, when more developed especially on young shoots $\frac{1}{4}$ to $\frac{1}{2}$ in. long, lanceolate or ovate-lanceolate, entire or with a few coarse teeth, the margins recurved.   Flowers very small, on short pedicels with 2 small bracteoles about the middle.   Calyx about 1 line long.   Corolla shortly exceeding the calyx, the tube broad, with a ring of short hairs inside near the top, the lobes rather longer than the tube. Ovary glabrous.   Fruit quite smooth, (orange-red?), 1 to 2 lines diameter.—A. DC. Prod. xi. 705.

Queensland. Brisbane river, Moreton Bay, *A. Cunningham;* Suttor range and Burnet river, *F. Mueller;* Wide Bay, *Leichhardt;* Cape and Isaacs rivers, *Bowman;* sandstone ridges of the interior, *Mitchell;* Armadilla, *Barton.*

N. S. Wales. Sterile country, Mount Aiton, Peel's Range, *A. Cunningham;* Macleay and Clarence rivers, *Beckler;* New England, *C. Stuart.*

This plant had been placed in Myoporineæ and retained there by A. De Candolle (who had no specimens in an examinable state) owing to Walpers having erroneously described the radicle as superior. F. Mueller, in referring it correctly to Verbenaceæ (Fragm. vi. 153) adduces *Teucridium*, Hook. f. from New Zealand, as a second species, which however can scarcely be admitted—the anthers and lobed ovary and fruit of the latter plant showing a nearer relation to *Oxera* and a few other genera which connect Verbenaceæ with the tribe Ajugoideæ of Labiatæ. The albuminous seeds branching hairs and other characters of *Spartothamnus* are quite those of Chloantheæ.

SUBTRIBE 2. EUVITICEÆ.—Ovary not at all or scarcely lobed. Fruit a drupe. Shrubs or trees. Flowers in cymes or clusters, axillary or in terminal corymbose or racemose panicles. Seeds without albumen, (or with a scanty albumen in some species of *Vitex?*).

14. CALLICARPA, Linn.

Calyx truncate or 4- rarely 5-toothed. Corolla with a short tube, the limb spreading, of 4, rarely 5, lobes, nearly equal. Stamens 4, rarely 5, shortly exserted. Ovary 4-celled, with 1 ovule in each cell laterally attached at or above the middle; style filiform, dilated and truncate or very shortly 2-lobed. Fruit a small succulent drupe, the endocarp of 4 distinct 1-seeded nuts or pyrenes. Seed without albumen. —Shrubs, rarely undershrubs, more or less cottony or woolly with stellate hairs or rarely glabrous, and often with numerous resinous glandular dots especially on the under side of the leaves. Leaves opposite, undivided. Flowers small, in axillary cymes, with very small bracts.

A considerable tropical and subtropical genus, chiefly Asiatic, with a few African and American species. Of the three Australian species two are widely dispersed over the Indian Archipelago, one of them extending to the Khasia mountains and the other at least to the Malayan Peninsula; the third, supposed to be endemic in Australia, is also perhaps not sufficiently distinct from another Asiatic species. The genus is most readily distinguished from *Premna* by the inflorescence, and by the flowers more regular with isomerous stamens. The differences in the fruit may not be constant.

Leaves acute at the base, glabrous above, white-tomentose underneath. Cymes dense. Corolla glabrous . . . . . 1. *C. cana.*
Leaves rounded at the base, pubescent or velvety above, somewhat floccose underneath. Cymes dense. Corolla glabrous or nearly so, lilac or purple . . . . 2. *C. pedunculata.*
Leaves acute at the base, green on both sides. Cymes very loose. Corolla densely pubescent, white . . . . 3. *C. longifolia.*

1. **C. cana,** *Linn. Mant.* 198. A "small shrub," the tomentum close and short, usually white, slightly floccose on the older branches. Leaves petiolate, ovate, very shortly acuminate; bordered by rather broad more or less mucronate teeth, acutely narrowed at the base, 2 to 3 in. long in most of the Australian specimens, much larger and more acuminate in those from the Indian Archipelago, glabrous above when full-grown except a slight tomentum on the principal veins which is rather more abundant in R. Brown's specimens, membranous when large, rather firmer and more rugose when small, white-tomentose underneath. Flowers small, in rather dense cymes, the common peduncle usually about as long as the petiole. Calyx about ¾ line long,

minutely 4- or 5-toothed. Corolla glabrous, twice as long as the calyx.
Stamens exserted; anthers with small glands along the connectivum.
Drupe depressed-globular, not above 1 line diameter, slightly succulent,
the endocarp of 4 hard nuts.—Schau. in DC. Prod. xi. 643; *C. adenan-*
thera, R. Br. Prod. 513.

N. Australia. Victoria river, *F. Mueller;* Groote Island, Gulf of Carpentaria,
R. Brown.

Queensland, *Bowman;* Edgecombe Bay and Port Denison, *Dallachy;* Gilbert
river, *Daintree.*

The species extends over the Indian Archipelago to the Malayan peninsula and the
Philippine islands. The Timor and Javanese specimens, correctly referred here by
Schauer, differ but slightly from the Australian ones in their larger more acuminate
leaves. The figure in Bot. Mag. t. 2107, represents a much more woolly plant, but is
perhaps a variety only. *C. bicolor*, Juss., Schau. in DC. Prod. xi. 642, and *C. erioclona*,
Schau. l. c. 643, appear to me both to be precisely the common Archipelago form of
C. cana. The *C. adenanthera* referred by Schauer with doubt to *C. longifolia* Lam.
appears to me to be the true *C. cana.*

2. **C. pedunculata**, *R. Br. Prod.* 513. A shrub of 3 or 4 ft., with the
tomentum rather loose and more or less floccose on the branches.
Leaves petiolate, ovate or ovate-lanceolate, acuminate with the point
often rather long, rounded or scarcely contracted at the base, usually
3 or 4 in. long in the Australian specimens, soft green and velvety with
scattered hairs above, more tomentose underneath but not very white.
Cymes rather loose, the peduncle often shortly exceeding the petiole.
Flowers of *C. cana*, the corolla purplish or lilac. Fruit usually much
larger than in that species, the succulent drupes often at least 1½ lines
diameter, purple when fresh.—*C. dentata*, Roth? in Wall. Cat. n. 6319,
but probably not the true plant of Roth; *C. lanata*, Schau. in DC.
Prod. xi. 644, not of Lam.

Queensland. Northumberland islands, Broad Sound and Shoalwater Bay, *R.*
Brown; Endeavour river, *A. Cunningham;* Dawson river, *F. Mueller;* Rockingham
Bay and Rockhampton, *Dallachy* and others; Brisbane river, Moreton Bay, *W. Hill*,
F. Mueller and others.

N. S. Wales. Clarence river, *Beckler;* Tweed river, *C. Moore.*

The species is also in the Archipelago, and is closely allied to the widely diffused *C.*
macrophylla, Vahl. Schauer refers it to " *C. lanata*, Vahl. Symb. iii. 13," but, if he had
turned to the page he quotes, he would have seen that the name is Linnæus', not
Vahl's, and relates to the very different Ceylon species which Schauer has published
as new under the name of *C. Wallichiana.*

3. **C. longifolia,** *Lam. ex. Schau. in DC. Prod.* xi. 645, *but not* C.
adenanthera, *Br.* A shrub of several ft., the tomentum short, not so
abundant as in most species and rarely whitish, although sometimes
reddish on the young shoots. Leaves petiolate, oblong or oblong-
lanceolate, acuminate with a long point, serrate, much narrowed at the
base, 4 to 6 in. or even longer, membranous, green and nearly glabrous
or sprinkled with very short hairs above, more copiously tomentose and
glandular underneath but usually green or very slightly rusty or
whitish. Flowers "white," smaller than in the other species, in very
loose repeatedly forked cymes, the common peduncle scarcely exceeding
the petiole, but the branches elongated and slender. Calyx about ½ line

long. Corolla about 1 line, pubescent outside. Fruit globular, white, about 1 line diameter.

Queensland. Rockingham Bay, *Dallachy.*

The species is widely spread over the Indian Archipelago, extending into India to Khasia and East Bengal.

15. PREMNA, Linn.

Calyx truncate or sinuately toothed. Corolla-tube short, the limb of 3, 4 or rarely 5 short teeth or lobes, nearly equal or slightly 2-lipped. Stamens 4, shorter than the corolla or rarely exserted. Ovary 4-celled, with 1 ovule in each cell laterally attached at or above the middle. Style filiform, with 2 short acute stigmatic lobes. Fruit a small succulent drupe, with a hard 4-celled undivided kernel. Seeds solitary in each cell, without albumen.—Shrubs or trees. Leaves opposite, undivided. Flowers small, in terminal trichotomous panicles, or in opposite cymes or clusters forming a terminal spike-like thyrsus.

A considerable genus, limited to the tropical and subtropical regions of the Old World. Although some, especially among the Asiatic species, are very well marked, there are a number of forms including the *P. integrifolia* and *P. serratifolia* of Linnæus which seem to pass into each other by numerous intermediates, and it would require a much more detailed study of good specimens from different localities than can now be devoted to them, to determine whether or not they can be classed into tolerably distinct races. To these would belong the first two of the following Australian forms here regarded as species, the next two are rather more distinct and apparently endemic ; the fifth, also endemic, is a much more marked one.

Calyx very shortly and obtusely 3 lobed or obscurely 2-lipped.
  Leaves very obtuse. Corolla-tube scarcely exceeding the calyx 1. *P. obtusifolia.*
  Leaves shortly acuminate. Corolla-tube nearly twice as long
    as the calyx . . . . 2. *P. integrifolia.*
Calyx with a somewhat expanded obscurely toothed margin.
  Leaves acuminate, thin, glabrous, the petiole short . . . . 3. *P. limbata.*
Calyx 5-toothed, the margin not dilated.
  Leaves ovate, acuminate, on rather short petioles . . . . 4. *P. Dallachiana.*
  Leaves deltoid, much acuminate (Poplar-like), on long petioles 5. *P. acuminata.*

1. **P. obtusifolia,** *R. Br. Prod.* 512. A shrub of 3 to 6 feet, glabrous in the typical form except a minute pubescence on the inflorescence and sometimes a row of hairs along the principal veins on the underside of the leaves. Leaves broadly ovate obovate or almost orbicular, usually broadly obtuse, very rarely with a short obtuse point, cordate or very obtuse at the base, mostly 3 to 6 in. long and sometimes nearly as broad, the petiole varying from ¼ to 1 in. in length. Flowers white or greenish, often very numerous, in terminal trichotomous corymbose panicles, sometimes shorter than the leaves, sometimes 6 to 8 in. diameter. Bracts very small and narrow. Calyx rarely above 1 line long and usually rather shorter, obscurely and irregularly 2-lipped, or rather very shortly and broadly 3-lobed, the upper lobe broader than the others and entire or obscurely 3-toothed, the two lower lobes entire, the whole calyx spreading open under the fruit but not otherwise enlarged. Corolla-tube about as long as the calyx, very hairy inside at the throat, the limb as long as the tube,

4-lobed, the upper inner lobe rather larger and less spreading than the others. Stamens inserted in the throat and nearly as long as the lobes. Style with very short stigmatic lobes. Drupe 2 to 2½ lines diameter. —Schau. in DC. Prod. xi. 637 ; *P. glycycocca,* F. Muell. Fragm. iii. 36.

N. Australia. Islands of the Gulf of Carpentaria and shores of the mainland, *R. Brown, Henne, Hulse.*

Queensland. Rockingham Bay, *Dallachy;* islands off the coast from Cape York to Cape Flattery, *F. Mueller, W. Hill, Henne.*

Very closely allied to some maritime forms of *P. integrifolia* but the leaves more obtuse, of a firmer consistence, and the corolla-tube shorter. These characters may not however prove constant.

P. attenuata, R. Br. Prod. 512, Schau. in DC. Prod. xi. 637, may possibly be a variety of *P. obtusifolia,* with leaves of the same consistence and equally obtuse and a similar inflorescence, but the leaves are narrow-obovate or oblong and cuneate or narrowed at the base. There is however in Brown's herbarium only a single specimen in fruit from the N. Coast, insufficient for determining whether it be really distinct.

Var. ? *velutina.* Leaves softly tomentose on both sides, otherwise the foliage and flowers quite those of *P. obtusifolia*—Rockingham Bay, *Dallachy*, who notes that it is a small tree with a spreading head.

2. **P. integrifolia,** *Linn.? var.* A tree or shrub, either quite glabrous or with a slight pubescence on the foliage and inflorescence. Leaves ovate, shortly acuminate, broad or rather narrow, usually rather smaller than in *P. obtusifolia.* Inflorescence and calyx entirely as in that species, but the corolla (in the Australian form) larger, the tube nearly twice as long as the calyx.—*P. ovata* and *P. media,* R. Br. Prod. 512; Schau. in DC. Prod. xi. 637.

N. Australia. Islands of the Gulf of Carpentaria, *R. Brown.*

The specimens I saw in Brown's herbarium did not appear to me to be at all different from some Asiatic ones of *P. integrifolia,* a very common sea-coast plant united by Schauer in DC. Prod. xi. 632 with *P. serratifolia,* Linn., under the latter name, which however appears to be the least appropriate of the two Linnæan ones for the consolidated species. The whole question however of the species of *Premna* requires a thorough revision.

3. **P. limbata,** *Benth.* A climbing shrub (*Dallachy*), the young branches and inflorescence minutely tomentose. Leaves ovate, mostly acuminate, rounded or broadly cordate at the base, 4 to 6 in. long, membranous, glabrous or pubescent along the veins underneath, the petioles 1 in. long or more. Panicles trichotomous, but not so spreading as in *P. obtusifolia* and the flowers larger, the pedicels often above 1 line long. Calyx fully 1 line long, the margin slightly expanded into broad very short obtuse or retuse teeth. Corolla-tube twice as long as the calyx, the upper inner lobe not much larger than the others.

Queensland. Rockingham Bay, *Dallachy.*

4. **P. Dallachyana,** *Benth.* A fine spreading shrub (*Dallachy*), the younger branches and inflorescence minutely tomentose, the older branches glabrous with a light-coloured bark, the foliage usually drying black. Leaves ovate, acuminate, entire, obtuse or narrowed at the base, mostly 2 to 3 in. long, glabrous or scarcely pubescent along the principal veins underneath. Panicles trichotomous, corymbose, not

large. Calyx scarcely 1 line long, more or less distinctly 5-toothed, the teeth very short, acute or obtuse but not dilated. Corolla-tube twice as long as the calyx and considerably dilated upwards, hairy inside at the throat, the lobes broad, the upper inner one larger than the others but entire. Stamens usually longer than the corolla. Drupe depressed-globular, not usually exceeding the calyx.

Queensland. Port Denison and in the scrub about Edgecombe Bay, *Dallachy;* Fort Cooper, *Bowman.* The calyx-lobes are more obtuse and irregular in the Edge- combe Bay specimens than in the others.

5. **P. acuminata,** *R. Br. Prod.* 512. A small tree of spreading habit, the inflorescence and foliage more or less hoary with a short close tomentum. Leaves broadly cordate-ovate, deltoid or almost rhomboidal, acuminate, entire or coarsely and irregularly toothed, 2 to 4 in. long, more or less pubescent or tomentose on both sides, 3- or 5- nerved (the first and often the second pair of primary veins starting from the base of the midrib), the petioles usually more than half as long as the leaves. Panicles very loose, the primary branches tricho- tomous, the ulterior ones dichotomous, the whole panicle sometimes 8 to 10 in. broad. Flowers nearly sessile. Calyx tomentose, nearly 1½ lines long, shortly and obtusely 5-toothed but the 3 upper teeth usually smaller and sometimes united as in *P. obtusifolia.* Corolla-tube shortly exserted, the lobes ovate, shorter than the tube, the upper inner one not very different from the others. Drupe depressed- globular, about 2 lines diameter.—Schau. in DC. Prod. xi. 637; F. Muell. Fragm. iii. 36; *P. cordata,* R. Br. l. c.; Schau. l. c.

N. Australia. Islands of the Gulf of Carpentaria and adjoining mainland, *R. Brown, Henne;* Point Cunningham, Cygnet Bay, N. W. coast, and Goulburn islands, *A Cunningham;* Victoria river, *F. Mueller;* Escape Cliffs, *Hulse;* also from *M'Douall Stuart's Expedition.*

The species is readily distinguished by its Poplar-like foliage and loose inflorescence.

16. CLERODENDRON, Linn.

Calyx campanulate or inflated, 5-toothed or 5-lobed, enlarged and spreading under the fruit (except in *C. hemiderma* and in some species not Australian). Corolla-tube slender, often very long; limb spreading, nearly equally 5-lobed. Stamens 4, exserted and often very long. Ovary 4-celled, with 1 ovule in each cell laterally attached at or above the middle. Style filiform, with 2 short acute stigmatic lobes. Fruit a more or less succulent or almost dry drupe, the endocarp separating into 4 one-celled or rarely into 2 two-celled pyrenes.—Trees or shrubs or rarely herbs or woody climbers. Leaves opposite or in whorls. Flowers in loose heads or cymes, usually forming terminal corymbose or thyrsoid panicles or rarely axillary.

A considerable tropical genus, chiefly Asiatic, with a few African or American species. Of the Australian species—which are here limited to eight, but might almost equally well be raised to ten or eleven, or reduced to four or five—one is a sea-coast plant widely spread over tropical Asia, the others appear to be endemic.

Calyx minutely toothed.
　Woody climber. Flowers small in compact cymes. Corolla-
　　tube 3 lines long. Fruit oblong, pubescent, small . . . . 1. *C. hemiderma.*
　Erect shrub. Flowers few in axillary loose cymes. Corolla-tube
　　1 in. long. Fruit obovoid, glabrous, rather large . . . . 2. *C. inerme.*
Calyx 5-lobed to the middle.
　Corolla-tube about 3 lines long. Stamens shortly exserted.
　　Leaves large and broad . . . . 3. *C. Tracyanum.*
　Corolla-tube ¾ to nearly 1 in. long. Stamens long. Leaves
　　usually tomentose.
　Leaves usually broad. Flowers mostly in dense terminal
　　corymbs . . . . 4. *C. tomentosum.*
　Leaves usually narrow. Flowers mostly in axillary cymes . 5. *C. lanceolatum.*
　Corolla tube 1 to 1¼ in. long. Stamens long. Leaves usually
　　glabrous or tomentose only when young . . . . 6. *C. floribundum.*
　Corolla-tube 2 in. long or more. Stamens long. Leaves of
　　C. floribundum . . . . 7. *C. Cunninghamii.*
Species insufficiently known, with broad tomentose, very rugose
　leaves . . . . 8. *C. costatum.*

1. **C. hemiderma,** *F. Muell.* A tall woody climber, the young
parts and inflorescence more or less hoary-pubescent, the leaves becoming glabrous when full-grown. Leaves shortly petiolate, broadly
ovate, obtuse or shortly and obtusely acuminate, mostly 2 to 3 in. long,
green on both sides. Flowers small for the genus, numerous, in rather
compact trichotomous cymes either terminal or on short branches or
leafless divaricate peduncles in the upper axils. Primary bracts sometimes oblong-lanceolate and contracted into a petiole, but most of them
very small and narrow. Calyx shortly pedicellate, narrow-campanu-
late or obovoid, about 1½ lines long, with 5 minute teeth. Corolla-
tube slender, shortly exserted but not exceeding 3 lines, glabrous
inside; lobes about half as long as the tube, more or less silky-pubescent outside. Stamens about twice as long as the corolla-lobes.
Fruiting calyx often above 2 lines long but remaining narrow. Fruit
oblong, obtuse, pubescent or hirsute, 3 to 4 lines long, 4-celled in the
upper portion where the endocarp closes round the seeds and separates
into 4 narrow nuts, the lower seedless portion assuming the appearance
of a wing to each nut, whilst the lower portion of the dissepiment
remains attached to the receptacle after the nuts have fallen in a
cuneate-oblong shape three-toothed at the top and nearly as long as
the calyx.—*C. (Hemiderma) Linnæi*, F. Muell. Fragm. vi. 151, not of
Thwaites.

Queensland. Cape York, *Daemel;* Rockingham Bay, *Dallachy;* Selheim river,
Bowman; Rockhampton, *Thozet.*

This plant has a singular resemblance with the Cingalese *C. Linnæi*, Thw. which has
the same climbing habit, foliage, and inflorescence, but rather larger flowers, the outer
bracts much larger, broader, and foliaceous, and the fruit, although nearly similar in
shape, is much more normal, without the flat winglike bases of the nuts or the persistent
axis upon which F. Mueller has founded his sectional character of *Hemiderma.*

2. **C. inerme,** *R. Br. Prod.* 511. A shrub attaining 6 ft. or more,
glabrous or the young shoots slightly pubescent, the branches sometimes dilated and hardened at the base of the leaves, but not spinescent.
Leaves on rather long petioles, ovate or elliptical, obtuse or shortly

acuminate, entire, mostly 2 to 3 in. long. Peduncles axillary, often
nearly as long as the leaves, bearing usually 3, but sometimes a cyme
of 7 or even more pedicellate flowers. Bracts minute. Calyx campanulate, slightly dilated on the margin, truncate and minutely toothed,
more open and 3 lines long when in fruit. Corolla-tube about 1 in.
long, the lobes about 4 lines. Stamens protruding about 1 in. beyond
the throat. Drupe obovoid, usually about $\frac{1}{2}$ in. long, but sometimes
much larger, the exocarp thick and spongy or almost corky, with a
crustaceous endocarp. Seeds with thick cotyledons and a very short
radicle.—Schau. in DC. Prod. xi. 660.

N. Australia. Islands of the Gulf of Carpentaria, *R. Brown;* Albert river, *F.
Mueller;* Port Essington, *Armstrong;* Adams Bay, *Hulls.*
Queensland. Bay of Inlets and Endeavour river, *Banks and Solander;* frequent
along the coast from Cape York to Rockhampton, *A. Cunningham, M'Gillivray, F.
Mueller, Dallachy,* and many others.
N. S. Wales. Richmond river, *Henderson.*

3. **C. Tracyanum,** *F. Muell.* A tall shrub or small tree, rather
bare of foliage, the young branches foliage and inflorescence more or
less velvety-pubescent or hirsute. Leaves broadly ovate, shortly acuminate, rounded or broadly cordate at the base, 4 to 8 in. long or more,
on petioles of 1 to 4 in. Flowers in terminal trichotomous cymes, very
dense at the time of flowering, 3 to 4 in. diameter when in fruit. Bracts
very small and deciduous. Calyx shortly pedicellate, villous outside
glabrous inside, turbinate-campanulate and about 2$\frac{1}{4}$ lines long at the
time of flowering, divided to near the middle into acute lobes, enlarged
and coloured after flowering, in some specimens with the fruit apparently ripe broadly campanulate and about 4 lines diameter, in others
still more enlarged and opening almost flat to about 5 lines diameter.
Corolla-tube above 3 lines long, slender and the 5 lobes of the limb
nearly equal and spreading as in the rest of the genus. Stamens rather
longer than the corolla-lobes. Drupes succulent, 3 to 4 lines diameter,
enclosing 4 distinct 1-seeded pyrenes. — *Premna Tracyana* or *Vitex
Tracyana,* F. Muell. Fragm. v. 61.

Queensland. Rockingham Bay, *Dallachy.* Although the flowers are small, their
structure and that of the fruit appears to me to be entirely those of *Clerodendron,* and
not at all those of *Premna* or *Vitex.*

4. **C. tomentosum,** *R. Br. Prod.* 510. A tall shrub or small tree,
the foliage and inflorescence usually velvety-pubescent, the older leaves
rarely quite glabrous. Leaves on rather long petioles, ovate elliptical
or almost lanceolate, shortly acuminate, acute or rarely rounded at the
base, 2 to 4 in. long. Flowers in the normal state numerous, in compact terminal corymbs, with rarely a few peduncles bearing small cymes
in the upper axils. Calyx campanulate, softly pubescent, about 3 lines
long when in flower, divided to about the middle into acute or rather
obtuse lobes. Corolla-tube under 1 in. long, the lobes 3 to 4 lines.
Stamens protruding by from $\frac{1}{2}$ to 1 in. Fruiting calyx expanding to
$\frac{3}{4}$ in. diameter, the drupe black and shining.—Andr. Bot. Rep. t. 607;
Bot. Mag. t. 1518; Schau. in DC. Prod. xi. 662.

Queensland. Brisbane river *F. Mueller.*

N. S. Wales. Port Jackson to the Blue Mountains, *R. Brown, Sieber, n.* 267, *Macarthur,* (Sydney woods, Paris Exhibition, 1855, n. 104), and many others; northward to Hastings, Macleay, and Clarence rivers, *Beckler* and others; southward to Kiama, *Harvey.*

Var.? *mollissima,* very softly villous, referrible perhaps to *C. lanceolatum,* but with the broader leaves and terminal inflorescence of *C. tomentosum.*

N. Australia. Roebuck Bay, N.W. Coast, *Martin ;* near Caledon Bay, Gulf of Carpentaria, *R. Gull.*

Ventenat's plate of *Volkameria tomentosa,* Jard. Malm., t. 84, represents an abnormal garden state, flowering very sparingly in the upper axils, connecting this species as well with the following *C. lanceolatum* as with some pubescent small-flowered forms of *C. floribundum.*

5. **C. lanceolatum,** *F. Muell. Fragm.* iii. 145. A tall shrub or small tree, the foliage and inflorescence softly velvety-pubescent or glabrous. Leaves on rather long petioles, lanceolate or ovate-lanceolate, acute or rather obtuse, 2 to 3 in. long. Cymes in the upper axils several-flowered, shorter than the leaves, on short peduncles. Bracts narrow, acute, or the outer ones more leafy. Calyx broadly campanulate, about $2\frac{1}{2}$ lines long when in flower, divided to near the middle into obtuse lobes. Corolla usually pubescent outside, the tube about $\frac{3}{4}$ in. long, or rather more, the lobes scarcely 3 lines. Stamens rather long. Fruiting calyx enlarged, coloured, very open, 4 to 5 lines diameter in the specimens seen, the lobes recurved. Drupe depressed-globular.

N. Australia. N. W. coast, Nickol Bay, *Gregory, Walcott;* Depuech Island, *Bynoe.*

The leaves are narrower in most of the glabrous specimens than in the pubescent ones, but some of Bynoe's pubescent ones have them also very narrow. The species is not very definitely separated from *C. tomentosum.*

6. **C. floribundum,** *R. Br. Prod.* 511. A tall shrub or small tree, usually quite glabrous or the young parts tomentose. Leaves on rather long petioles, usually ovate or elliptical, but varying from broadly ovate and cordate to lanceolate, obtuse acute or acuminate; acute rounded or cordate at the base, usually 2 to 3 in. long but sometimes twice as large. Cymes sometimes all loose and few-flowered in the upper axils, sometimes forming a broad terminal corymbose panicle. Flowers to the naked eye apparently glabrous, but often sprinkled with a minute pubescence visible under a lens. Calyx $2\frac{1}{2}$ to 3 lines long, more acute at the base than in *C. tomentosum,* the lobes acute, about as long as the tube or rather shorter. Corolla-tube usually rather above 1 in. long, but not exceeding $1\frac{1}{4}$ in., in other specimens rather longer, the lobes short in proportion. Stamens exserted by about 1 in. Fruiting-calyx expanding to above $\frac{1}{2}$ in. diameter, with a narrow base of 1 to 2 lines.—Schau. in DC. Prod. xi. 671.

N. Australia. N. Coast, *R. Brown;* N. W. Coast, *Bynoe;* Upper Victoria river, *F. Mueller ;* islands of the gulf of Carpentaria, *Henne.*

Queensland. Endeavour river, *Banks and Solánder,* Harvey's islands, Keppel Bay and Northumberland islands, *R. Brown;* Percy island and Port Curtis, *M Gillivray,* Port Denison, *Fitzalan ;* Rockingham Bay, *Dallachy ;* Rockhampton, *Dallachy* and others ; in the interior, *Mitchell.*

South Australia. Daly waters, *Waterhouse.*

The species is exceedingly variable. In general it is to be distinguished from *C. tomentosum* by the absence of pubescence, more acute calyxes, longer flowers and looser inflorescence. Sometimes the inflorescence approaches that of *C. inermis*, more frequently it is at least on the main branches almost as abundant as in *C. Cunninghamii*. An apparently common abnormal state, produced evidently by the puncture of insects, has axillary cymes with few flowers, in most of which the corolla assumes a campanulate ovoid or globular form of considerable thickness and much regularity.

C. attenuatum and *C. medium*, R. Br. Prod. 510, 511, Schau. in DC. Prod. xi. 671, are very slight modifications, which I am quite unable to distinguish from the common forms.

C. ovatum, R. Br. Prod. 511, Schau. in DC. Prod. xi. 671, or *C. cardiophyllum*, F. Muell. Fragm. iii. 144, is a form with large, very broad, somewhat coriaceous leaves and loose inflorescence, which may at first sight appear very distinct, but there are quite as many specimens which might equally well be referred to this or to some of the commoner varieties.

C. coriaceum, R. Br. Prod. 511, Schau. in DC. Prod. xi. 671, of which the specimens are very indifferent, is much the same as *C. ovatum*, but with pubescent leaves, which connect it with *C. tomentosum*. I have not seen the flowers, and without the corolla there appears to be no positive character to distinguish *C. floribundum* from the two preceding and the following species. *C. ovatum* and *C. tomentosum* R. Br. are both much like the garden specimens figured by Ventenat as *Volkameria tomentosa*.

7. **C. Cunninghamii,** *Benth.* A tall shrub, either quite glabrous or the under side of the leaves and inflorescence more or less tomentose. Leaves ovate, scarcely acuminate, often narrowed at the base, sometimes above 6 in. long and membranous, sometimes much smaller and firmer, the petiole varying from under 1 in. to above 2 in. Flowers numerous in a broad terminal corymb sometimes dense sometimes loose, with the calyx and structure nearly of *C. floribundum*, but remarkable for the long slender corolla-tube, usually exceeding 2 in., the lobes broad, not above 3 lines long. Stamens rather long. Fruiting calyx more or less funnel-shaped, shortly contracted at the base, the margins very spreading or recurved. Drupe 4 or 5 lines diameter, ripening 2 to 4 distinct pyrenes.

N. Australia. S. Goulburn Island, *A. Cunningham;* Escape Cliffs, *Hulls.*
Queensland. Cape York, *Daemel, Jardine;* Endeavour river, *A. Cunningham.*

Some of Brown's specimens, as well as others seen only in fruit and referred to *C. floribundum*, may perhaps belong to *C. Cunninghamii*. Some of *F. Mueller's* from Gilbert river, with more pubescent leaves, are very doubtful.

8? **C. costatum,** *R. Br. Prod.* 511. A tall shrub. Leaves very broadly ovate, obtuse, 4 to 5 lines long, reticulate-rugose and velvety-tomentose underneath. Inflorescence a terminal corymbose panicle, not exceeding the leaves but looser than in *C. tomentosum.* Flowers not seen. Fruiting calyx enlarged and drupe of *C. floribundum.*—Schau. in DC. Prod. xi. 671.

Queensland. Endeavour river, *Banks and Solander.*
The foliage is that of *Gmelina Leichhardtii*, but the fruit undoubtedly that of *Clerodendron*, and not of *Gmelina*.

17. GMELINA, Linn.

Calyx 4- or 5-toothed or sinuate-lobed. Corolla-tube much dilated upwards or almost campanulate; limb oblique, with 4 or 5 spreading lobes, the two upper ones sometimes united in an upper lip. Stamens 4, in pairs, shorter than the corolla. Ovary 4-celled with 1 ovule in each cell laterally attached at or above the middle; style filiform, unequally 2-lobed at the top. Fruit a succulent drupe, the putamen hard or bony, 4-celled or rarely 2-celled. Seeds solitary in each cell, without albumen. Trees or tall shrubs. Leaves opposite, undivided. Flowers often rather large, pale purplish pink or blue, or in species not Australian, yellow, in cymes arranged in irregular terminal panicles, sometimes almost reduced to simple racemes. Bracts small.

The genus extends over tropical Asia and the Indian Archipelago. The Australian species, although with the aspect of some Asiatic ones, appear to be all endemic.

Leaves glabrous (above 6 in.). Panicle long and narrow. Cymes
pedunculate . . . . 1. *G. macrophylla.*
Leaves glabrous (under 6 in.). Cymes forming sessile clusters
along the rhachis of the panicle or of its branches . . . . 2. *G. fasciculiflora.*
Leaves tomentose underneath. Cymes pedunculate in a loosely
pyramidal panicle . . . . 3. *G. Leichhardtii.*

1. **G. macrophylla,** *Benth.* A tall tree, glabrous except the inflorescence, which is slightly tomentose. Leaves ovate or ovate-oblong, obtuse, broad and sometimes almost cordate at the base, 8 to 10 in. long, coriaceous and glabrous but not shining, on petioles of 1 to 1½ in. Flowers "pale blue," in a long terminal thyrsoid panicle, the cymes on opposite peduncles, the pedicels very short. Calyx campanulate, about 2 lines long, truncate or sinuate-lobed, scarcely enlarged or sometimes slightly expanded under the fruit. Corolla villous outside with appressed hairs, the tube declinate and much dilated upwards, about ½ in. long, lobes 5, broad, about ¼ in. long, the middle lower one rather larger than the others. Stamens ascending under the upper lobes, anther-cells diverging. Ovary glabrous, 4-celled. Upper lobe of the style minute. Drupe obovoid-truncate, ½ to ¾ in. long, closely resembling that of *G. arborea.*—*Vitex macrophylla,* R. Br. Prod. 512, Schau. in DC. Prod. xi. 695 ; *Vitex Dalrympleana,* F. Muell. Fragm. iv. 128 ; v. 72 ; *Ephielis simplicifolia,* Seem. Fl. Vit. 189.

N. Australia. Port Essington, *Armstrong.*
Queensland. Cape Grafton, *Banks and Solander ;* Cape York, *Daemel ;* Rockingham Bay, *Dallachy.*

2. **G. fasciculiflora,** *Benth.* A tall tree, nearly glabrous except the inflorescence which is densely ferruginous-tomentose. Leaves ovate, obtuse or obtusely acuminate; usually broad, 3 to 5 in. long, somewhat coriaceous, shining above, the primary veins much raised underneath but not nearly so reticulate as in *G. Leichhardtii.* Cymes reduced to dense opposite clusters sessile along the branches of a terminal panicle, the floral leaves at the base of the clusters reduced to broad bracts shorter than the calyxes. Pedicels very short. Calyx broadly campanulate, ferruginous-villous, about 2 lines long, truncate or more or less distinctly toothed. Corolla "pale purple," villous, the tube short but much more dilated and oblique than in *G. Leichhardtii,* the lower lip fully ½ in. long, with a large broad middle lobe, the upper lobes all broad but much shorter than the lowest. Fruit not seen.— *Vitex Leichhardtii,* var. *glabrata,* F. Muell.

Queensland. Rockingham Bay, *Dallachy.*

3. **G. Leichhardtii,** *F. Muell.* A fine timber tree, attaining a great height, the young branches and inflorescence tomentose. Leaves ovate, scarcely acuminate but rather acute, rounded or cuneate at the base, 3 to 6 in. long, somewhat coriaceous, quite glabrous and almost rugose on the upper side, much reticulate with raised veins and densely and softly tomentose underneath, the petiole often above 1 in long. Flowers " white with purple markings," numerous in opposite pedunculate cymes forming loose ovoid or shortly pyramidal terminal panicles. Calyx broadly turbinate-campanulate, truncate, tomentose and not 2 lines long at the time of flowering, enlarged and spreading under the fruit. Corolla villous outside, the tube very broad and dilated upwards, twice as long as the calyx, the lobes ovate, above 2 lines long, the 2 upper ones rather larger and shortly united in an upper lip. Stamens incurved, the longer pair about as long as the upper lip; anther-cells diverging. Fruits in the specimens seen all deformed by insects, the calyx opening out horizontally to a diameter of 6 to 8 lines and obscurely sinuate-toothed.—*Vitex Leichhardtii,* F. Muell. Fragm. iii. 58.

Queensland. Myall Creek, *Leichhardt ;* Moreton Bay, *W. Hill, Queensland woods, London Exhibition,* 1862, *n.* 30 ; Pine river, *Fitzalan.*

N. S. Wales. Clarence river, *Beckler, W. Moore, N. S. Wales woods, London Exhibition,* 1862, *n.* 68 *and* 171, "White Beach," also *Macarthur, Paris Exhibition,* 1855, *n.* 193.

18. **VITEX,** Linn.

Calyx 5-toothed or lobed. Corolla-tube short; limb spreading, 5-lobed, the lowest lobe larger and longer than the others and sometimes notched. Stamens 4, in pairs, ascending and exserted beyond the upper corolla-lobes. Ovary 2-celled or more or less perfectly 4-celled, with 1 ovule to each half-cell or cell, laterally attached at or above the middle. Style filiform, shortly and acutely 2-lobed. Fruit a succulent drupe, the putamen separating into 4 hard one-seeded pyrenes (or fewer by abortion). Seeds without albumen.—Trees or shrubs. Leaves opposite, usually of 3 or 5 digitate leaflets, very rarely single (or of a single leaflet). Flowers in cymes, sometimes axillary but usually in terminal panicles either simple and spike-like or branched. Bracts very small.

A considerable tropical and subtropical genus, chiefly Asiatic or African with a few American species, and one species extending to S. Europe. Of the four Australian species, one is widely spread over the Old World within the tropics, the three others are endemic.

Leaves white underneath, undivided or of 3 or 5 leaflets . . . . 1. *V. trifolia.*
Leaves green on both sides, undivided . . . . 2. *V. lignum vitœ.*
Leaves green on both sides, of 3 or 5 leaflets.
   Flowers in loose thyrsoid panicles, mostly terminal   . . . . 3. *V. acuminata.*
   Flowers in very loose dichotomous cymes on axillary peduncles 4. *V. glabrata.*

1. **V. trifolia,** *Linn; Schau. in DC. Prod.* xi. 683. A shrub sometimes decumbent and low, in some varieties tall and erect, the branches, under side of the leaves and inflorescence mealy-white. Leaves very variable, simple or of 3 or 5 leaflets often white on both sides, but

usually becoming nearly glabrous on the upper side at least when old. Flowers nearly white or pale blue, in small nearly sessile opposite cymes, forming short terminal panicles, either simple and spike-like or slightly branched, the floral leaves reduced to short bracts. Calyx in the typical forms about 2 lines long, very shortly 5-toothed, the corolla-tube nearly twice as long as the calyx, the 4 upper lobes short, the lowest twice as large and often as long as the tube and both calyx and corolla more or less mealy outside. Ovary 2-celled, with 2 ovules in each cell. Drupe globular.

N. Australia. Victoria river, *F. Mueller;* islands of the Gulf of Carpentaria, *Henne,* and mainland, *F. Mueller, Landsborough.*

Queensland. Along the coast from Cape York to Moreton Bay, *R. Brown, A. Cunningham, F. Mueller, Dallachy,* and many others.

The species is a very common Asiatic one, chiefly maritime, and varying very much as to foliage, the three following principal Australian forms agreeing more or less with Asiatic varieties, but some of them passing into species which in Asia are considered as perfectly distinct.

a obovata. Decumbent. Leaflets (or simple leaves) mostly solitary, obovate or rounded, 1 to 1½ in. long, rarely especially on flowerless branches 3-foliolate and less obtuse. A strictly maritime variety in N. Australia and Queensland as in tropical Asia.— *V. ovata,* Thunb., Hook. and Arn. Bot. Beech. 206, t. 47, R. Br. Prod. 511.

β acutifolia. Decumbent or erect. Leaflets 3 or sometimes 5 or only 1, ovate or ovate-lanceolate, acute or acuminate, the middle one often above 2 in. long.— *V. trifolia,* R. Br. Prod. 511.—Common along the coast of Queensland and appears to be not so strictly maritime as the obovate-leaved form.

γ parviflora. Erect. Leaflets 5 or sometimes 3, ovate or ovate-lanceolate, acute. Flowers much smaller than in the two preceding forms, and resembling those of the Asiatic *V. Negundo,* from which this variety is scarcely to be distinguished.—In N. Australia on the Gulf of Carpentaria, and Moreton Bay in Queensland.

There are numerous intermediate specimens connecting the above three principal forms.

2. **V. lignum-vitæ,** *A. Cunn., Schau. in DC. Prod.* xi. 692. A tall handsome tree, the young branches petioles and inflorescence rusty-tomentose or pubescent. Leaves all simple (or unifoliolate ?), oblong or oval-elliptical, shortly acuminate, narrowed at the base, 1½ to 4 in. long on a petiole of ¼ to ½ in., somewhat coriaceous, shining on the upper side, paler underneath, conspicuously veined, quite glabrous or with a slight pubescence on the midrib underneath, those of barren branches sometimes broadly and unequally lobed, those of flowering branches usually entire but occasionally showing a few prominent angles or short lobes. Flowers few, in small loose axillary cymes. Calyx truncate, about 1 line long or rather more, rarely tomentose. Corolla tomentose outside, the tube 3 to 4 lines long, broad and incurved, the middle lower lobe not exceeding the others so much as in some species. Stamens shortly exserted beyond the upper lobes.—F. Muell. Fragm. iii. 58.

Queensland. Brisbane river, Moreton Bay, *A. Cunningham, W. Hill, F. Mueller, C. Moore, Queensland woods, London Exhibition,* 1862, *n.* 29.

N. S. Wales ? "Sydney woods," *Paris Exhibition,* 1855, *n.* 3.

3. **V. acuminata,** *R. Br. Prod.* 512. A small or large tree, the young shoots and inflorescence more or less hoary-pubescent, the adult leaves

glabrous or nearly so. Leaflets 3 or rarely 5, ovate-oblong, elliptical or almost lanceolate, mostly acuminate, contracted at the base into a petiolule sometimes very short, sometimes ¼ in. long, the terminal leaflets often 3 to 4 in. long or even more, the lateral ones usually shorter, all membranous, green on both sides, paler and usually glandular-dotted underneath, the common peduncle usually shorter than the leaflets. Flowers small, in loose thyrsoid panicles of 2 to 4 in., terminal or in the upper axils. Pedicels very short. Calyx 1 to 1¼ in. long, truncate or minutely toothed. Corolla pubescent outside, villous inside in the throat and the base of the limb, the tube about twice as long as the calyx, the lower lip nearly as long as the tube. Stamens shortly exserted beyond the upper lobes. Ovary 2-celled, with 2 ovules in each cell. Drupe in Dallachy's specimen nearly globular, about ½ in. diameter, the putamen bony and 4-celled as in *Gmelina*, bearing outside a number of radiating riblike excrescences, and perhaps therefore diseased, in R. Brown's smaller and apparently more perfect. Seed not seen perfect.—Schau. in DC. Prod. xi. 695; F. Muell. Fragm. v. 34; *V. melieopea*, F. Muell. Fragm. v. 35.

N. Australia. Vansittart and Careening Bays, N. W. Coast, *A. Cunningham;* Arnhem Bay and Islands of the Gulf of Carpentaria, *R. Brown.*

Queensland. Cape York, *M'Gillivray, W. Hill, Daemel;* Wide Bay, *Bidwill;* Rockingham Bay, *Dallachy;* Rockhampton, *Dallachy* and others.

V. Timoriensis, Walp.; Schau. in DC. Prod. xi. 686, from Timor, may be the same as *V. acuminata.*

4. **V. glabrata,** *R. Br. Prod.* 512. A tree or shrub, quite glabrous or rarely with a minute tomentum on the young shoots and inflorescence. Leaflets 3 or rarely 5, from broadly ovate and obtuse to elliptical-oblong and acuminate, 2 to 4 in. long, but usually much broader than in *V. acuminata* and less narrowed at the base, the petiolules about ½ to ¾ in. long, and the common petiole above 2 in. Flowers white, rather small, in very loose dichotomous cymes on axillary peduncles sometimes nearly as long as the petiole, but the whole inflorescence almost always shorter than the leaf. Pedicels shorter or rarely longer than the calyx. Calyx about 1¼ lines long, truncate or minutely toothed. Corolla-tube at least twice as long as the calyx, and the lower lip nearly as long as the tube. Stamens shortly exserted beyond the upper lobes. Fruit, in the specimens seen, 1-seeded by abortion, obovoid, about ½ in. long.—Schau. in DC. Prod. xi. 695; *V. Cunninghamii,* Schau. l. c. 691.

N. Australia. Careening Bay, N. W. Coast, *A. Cunningham;* Victoria and Fitzmaurice rivers and Macadam range, *F. Mueller;* Groote island, Gulf of Carpentaria, *R. Brown;* Port Essington, *Armstrong;* Adams Bay, *Hulls.*

Queensland. Cape York, *Daemel;* Gilbert river, *Daintree.*

The cultivated specimen described by F. Muell. Fragm. v. 35 as *V. glabrata,* is the New Zealand *V. littoralis,* A. Cunn., remarkable for its large differently shaped corolla. C. Moore's George-river specimen may be the same, but if so, it is probably cultivated also.

SUBTRIBE 3. OXEREÆ.—Ovary distinctly 2- or 4-lobed.

19. FARADAYA, F. Muell.

Calyx closed before flowering, then dividing into 2 valvate segments. Corolla-tube dilated upwards, limb 4-lobed, the upper lobe broad and emarginate, the three lower nearly equal. Stamens 4, didynamous, exserted. Ovary shortly 4-lobed, 4-celled in the upper portion, with one ovule in each cell laterally attached. Fruit a drupe, the putamen 1-celled and 1-seeded by abortion (or 4-lobed with 4 distinct pyrenes?).—Woody climbers. Leaves opposite, undivided. Flowers rather large, in terminal panicles.

Besides the Australian species, there are three from the S. Pacific islands. The nearest affinity of the genus appears to be with the New Caledonian Oxera.

1. **F. splendida,** *F. Muell. Fragm.* v. 21, 212. A tall woody climber, quite glabrous. Leaves ovate, acuminate, rounded or cordate at the base, 6 in. to nearly 1 ft. long, prominently penniveined, the petiole 1 to 2 in. long. Flowers large, white, in a terminal corymbose panicle. Bracts small, subulate. Pedicels $\frac{1}{4}$ to $\frac{1}{2}$ in. long. Calyx before expanding obovoid and acuminate, dividing into 2 acuminate segments 8 to 10 lines long. Corolla-tube above 1 in. long, the lobes flat, nearly $\frac{3}{4}$ in. long. Filaments sprinkled with hairs below the middle, inserted in the throat of the corolla; anthers with 2 parallel cells. Ovary tomentose. Drupe only one seen which was 1-seeded, about 2 in. long, contracted at the base and apparently proceeding from one lobe of the ovary, the other lobes remaining abortive at the base, as there is no scar of the style at the upper end. Seed not seen.

Queensland. Rockingham Bay, *Dallachy.*

TRIBE 3. AVICENNIEÆ.—Fruit a 2-valved capsule. Seed solitary, without integuments; embryo with large folded cotyledons.

20. AVICENNIA, Linn.

Calyx divided to the base into 5 distinct segments or sepals. Corolla-tube short and broad; limb of 4 nearly equal spreading lobes or the upper one rather larger. Stamens 4, inserted in the throat, with the anthers slightly protruding. Ovary 1-celled, with 4 ovules collaterally suspended from a central column, which has 4 angles between the ovules, imperfectly dividing the ovary into 4 cells. Fruit a compressed capsule, the pericarp opening in 2 valves. Seed solitary, erect, without integuments, (the integuments of the ovule not developed); embryo with 2 very large cotyledons folded longitudinally, a very hairy radicle, and a prominent plumula, which germinates before the fruit drops off as in *Rhizophora*, &c.—Shrubs. Leaves opposite, undivided. Flowers in small cymes in the upper axils or in terminal panicles.

The genus consists of very few species, widely distributed over the warmer maritime regions of the New and the Old World, and very nearly related to each other. The Australian species is the typical and most common form.

1. **A. officinalis,** *Linn.; Schau. in DC. Prod.* xi. 700. An erect shrub, varying much in height, the branches inflorescence and under-

side of the leaves white or silvery with a very close tomentum, more silky on the flowers, the upper side of the leaves usually glabrous when full grown, black and shining when dry. Leaves coriaceous, usually lanceolate or ovate-lanceolate, 2 to 3 in. long, acute and contracted into a petiole, but varying to elliptical or obovate, and very obtuse. Cymes contracted into small heads on rigid angular peduncles, which are often 2 together in the upper axils or several in a small terminal leafy thyrsus. Bracts shorter than the sepals. Sepals orbicular or broadly ovate, concave, hirsute and ciliate, about 1¼ line long. Corolla-tube shorter than the sepals, lobes ovate, rather longer than the tube, the upper inner one rather larger than the others. Ovary very hairy.— *A. tomentosa,* Jacq.; R. Br. Prod. 518; Schau. in DC. Prod. xi. 699; Wall. Pl. As. Rar. t. 271; Wight, Ic. t. 1481.

N. Australia, Queensland, N. S. Wales, Victoria, S. Australia, and **W. Australia,** extending along the sea-coast all round the Australian continent, *R. Brown* and many others, but no specimens seen from Tasmania. It is a common sea-coast shrub in tropical Asia, Africa, and America.

ORDER XCIII. **LABIATÆ.**

Flowers irregular or rarely nearly regular. Calyx persistent, 5-toothed or 2-lipped. Corolla with a distinct tube and 4 or 5 lobes more or less 2-lipped or nearly equal, imbricate in the bud, the upper lip or lobes usually and perhaps always outside. Stamens 2 or 4, in pairs, inserted in the tube of the corolla and alternating with its lower lobes. Anthers either 2-celled, or 1-celled by the confluence of the 2 cells or by the abortion of one of them. Ovary 4-lobed, with one erect ovule in each lobe. Style single, arising from the centre of the ovary, terminating in 2 short stigmatic lobes. Fruit enclosed in the calyx, consisting of 4 small seed-like nuts. Seeds solitary in each nut, without albumen. Embryo straight or slightly incurved (except in *Scutellaria*); radicle short, inferior, cotyledons thick.—Herbs or shrubs, very rarely arborescent, the stem and branches usually 4-angled. Leaves always opposite or whorled, without stipules, toothed or more rarely entire or divided. Flowers in opposite cymes or rarely solitary, forming frequently (by the extreme shortness or abortion of the common peduncle and branches of each cyme, the pedicels alone being developed) clusters called *false-whorls* or *verticillasters,* consisting of 3, 5, or more pedicels on each side of the stem; these false-whorls either in the axils of the stem-leaves or more frequently forming terminal racemes or panicles, the floral leaves subtending the clusters being reduced to small bracts. Real bracts, subtending the pedicels, usually abortive or reduced to bristles, rarely more prominent and leaflike. Foliage and green parts often studded with glandular dots filled with resinous oil rendering the plants highly aromatic. Nuts when soaked in water frequently emitting a thick mucilaginous coating.

A large Order generally distributed over every part of the globe. Of the twenty Australian genera (excluding *Hyptis*) four are extratropical genera of the northern hemisphere, very sparingly extending into or represented in the Southern hemisphere or in mountainous regions of the tropics, three are also extratropical and chiefly northern, but also more or less numerous within the tropics, especially in móuntain ranges, two are tropical, chiefly Asiatic but also represented in America, six are confined to the tropical regions of the Old World and chiefly or entirely Asiatic, the remaining five, constituting the distinct tribe *Prostanthereæ*, are endemic in Australia and chiefly extratropical.

TRIBE 1. **Ocimoideæ.**—*Stamens 4, declinate. Anthers 1-celled by the confluence of the 2 cells into one. Nuts dry, smooth or minutely granular.*

Corolla lower lobe flat or nearly so, not longer than the upper lobes.
 Corolla-tube shorter than or scarcely exceeding the calyx.
  Style with 2 short stigmatic lobes.
   Fruiting calyx deflexed with a broad decurrent upper lobe.
    False-whorls equal, 6-flowered . . . . 1. OCIMUM.
   Fruiting calyx erect spreading or deflexed, the upper lobe scarcely decurrent. Flowers very small in one-sided racemes . . . . 2. MOSCHOSMA.
  Corolla-tube exserted. Style capitate or clavate at the end. Calyx of *Ocimum* . . . . 3. ORTHOSIPHON.
 Corolla lower lobe concave or boat-shaped, longer than the upper lobes (except in *Plectranthus longicornis*).
  Filaments free. Corolla sometimes spurred . . . . 4. PLECTRANTHUS.
  Filaments united at the base in a sheath round the style. Corolla never spurred . . . . 5. COLEUS.
 Corolla lower-lobe saccate, contracted at the base and abruptly deflexed . . . . 6. HYPTIS.

TRIBE 2. **Satureieæ.**—*Stamens 4 or 2, erect and equal or ascending in pairs, the upper pair the shortest. Anthers 2-celled or 1-celled by the confluence of the 2 cells into one. Nuts dry, smooth or minutely granular. Corolla-lobes all equally spreading, or if 2-lipped the upper lip scarcely concave and not hoodshaped.*

Corolla-lobes nearly equal.
 Anthers terminal 1-celled. Filaments bearded . . . . . . 7. DYSOPHYLLA.
 Anthers 2-celled. Filaments glabrous.
  Perfect stamens 4 . . . . 8. MENTHA.
  Perfect stamens 2 . . . . 9. LYCOPUS.
Corolla-lobes forming 2 lips.
 Stamens 4, erect, diverging. Flowers in heads or short spikes with imbricate bract-like floral leaves . . . . * ORIGANUM.
 Stamens 4, ascending in pairs. Calyx 2-lipped. Corolla-tube curved upwards above the base . . . . * MELISSA.

TRIBE 3. **Monardeæ.**—*Stamens 2 ; anthers with one oblong or linear perfect cell, the other either quite abortive or deformed or sterile or separated from the upper one by a filiform connective. Nuts dry, smooth or minutely granular. Calyx and corolla usually 2-lipped.*

Connective of the anthers filament-like, transversely fixed on the short real filament . . . . 10. SALVIA.
Connective of the anthers short, continuous with the end of the filament, the junction marked by a minute tooth . . . . . * ROSMARINUS.

TRIBE * **Nepeteæ.**—*Stamens 4, in pairs, the upper ones the longest. Anthers 2-celled. Nuts dry, smooth or granular-tuberculate. Calyx 15-nerved. Corolla upper lip concave or hoodshaped.*

Calyx straight or incurved, scarcely 2-lipped . . . . * NEPETA.

TRIBE 4. **Stachydeæ.**—*Stamens 4, ascending in pairs, the lower ones the longest.*

Anthers, at least the upper ones, 2-celled. Nuts dry, smooth or granular-tuberculate. Calyx usually 5- or 10-nerved. Corolla upper lip concave or hoodshaped.

Calyx 2-lipped, the lips closed after flowering.

Calyx upper lip flat, 3-toothed, lower 2-lobed. False-whorls
   6-flowered, in dense terminal spikes . . . . 11. PRUNELLA.
Calyx-lips entire, the upper one with a hollow scale-like protuberance on the back. Flowers opposite in pairs . . . . 12. SCUTELLARIA.
Calyx 5- or 10-toothed, not 2-lipped.

Stamens included in the corolla-tube   Upper corolla-lip narrow.

Calyx often 10-toothed . . . . * MARRUBIUM.
Stamens protruding from the short upper lip . . . . . . 13. ANISOMELES.
Stamens ascending under the upper lip.

Calyx 5-toothed, not much enlarged after flowering. . . . * STACHYS.
Calyx 5-toothed, very large open and membranous, at least
   after flowering . . . . * MOLUCCELLA.
Calyx 8- or 10-toothed.

Corolla upper lip short very hairy . . . . 14. LEUCAS.
Corolla upper lip very long. (Flowers large, scarlet) . . * LEONOTIS.

TRIBE 5. **Prostanthereæ.**—*Stamens 4, in pairs; anthers all with two perfect cells, or one cell of all the anthers or also both cells of the lower pair sterile or abortive. Nuts prominently reticulate-rugose. Seeds albuminous. Calyx various. Corolla upper lip erect, usually short, throat broad.*

Calyx 2-lipped, the lips entire or nearly so. Anthers with 2 perfect cells, the connective not elongated . . . . 15. PROSTANTHERA.
Calyx 2-lipped or 5-toothed. Anthers with an elongated con-
   nective bearing at the upper end one perfect cell.
Connective with the lower end reduced to a small tooth.
   Leaves pungent-pointed . . . . 16. HEMIANDRA.
Connective with the lower end dilated linear or clavate or
   bearing an imperfect cell, usually bearded in the upper
   anthers. Leaves obtuse or rarely acute, not pungent . . 17. HEMIGENIA.
Calyx nearly equally 5-toothed. Lower anthers sterile and re-
   duced to 2 linear or clavate lobes.
Connective of the upper anthers elongated, the lower end di-
   lated and bearded. Corolla upper lip very concave or hood-
   shaped . . . . 18. MICROCORYS.
Upper anthers of one cell almost sessile on the filament.
   Corolla upper lip flat, 2-lobed . . . . 19. WESTRINGIA.

TRIBE 6. **Ajugoideæ.**—*Stamens 4, in pairs exserted from the very short trun-cate or deeply slit upper lip of the corolla (except very rarely in genera not Australian). Nuts prominently reticulate-rugose, not succulent. Seeds without albumen.*

Corolla with the four upper lobes lateral, equal or the upper ones
   longer, the stamens exserted from between the 2 upper ones . 20. TEUCRIUM.
Corolla with the upper lip exceedingly short and truncate, the
   stamens exserted beyond it, the lateral lobes oblong, forming
   part of the lower lip . . . . 21. AJUGA.

The introduced plants belonging to the genera marked above with the asterisk * are the following :—

Origanum vulgare, Linn., Benth. in DC. Prod. xii. 193. A perennial with a shortly creeping rootstock and erect stems of 1 to 2 ft. Leaves petiolate, ovate or ovate-lanceo-late, slightly toothed. Flowers purple or rarely white, in globular compact heads, forming a terminal trichotomous rather compact panicle. Bracts or bract-like floral leaves imbricate, as long as the calyx. Calyx very hairy inside, with short nearly equal teeth. Corolla-lobes nearly equal in length, the upper one broad and nearly erect. Stamens 4, erect, diverging, the two lower longer ones always, and often all 4 exserted.—Very common in the temperate regions of the northern hemisphere, and now introduced at Plenty Creek, Victoria, *F. Mueller.*

Melissa officinalis, Linn.; Benth. in DC. Prod. xii. 240. A rather coarse erect branching perennial of 1 to 3 feet, usually hairy. Leaves broadly ovate, crenate. Flowers few together in loose axillary false-whorls, all turned to one side. Calyx 2-lipped, the upper lip rather flat, 3-toothed, the lower 2-lobed. Corolla whitish, twice as long as the calyx, the tube curved upwards above the base, the upper lip erect, emarginate, the lower spreading, 3-lobed. Stamens 4, in pairs, ascending under the upper lip, the lower ones the longest.—A native of southern Europe and western Asia, long since cultivated (in England under the name of *Balm*) and readily naturalizing itself in the vicinity of gardens, in which localities it is also established in Victoria, *F. Mueller*.

Rosmarinus officinalis, Linn.; Benth. in DC. Prod. xii. 360. An erect shrub of 2 to 4 ft. Leaves sessile, linear, entire, with revolute margins. Flowers white or pale blue, opposite in pairs, in very short axillary racemes. Calyx 2-lipped, the upper lip entire, the lower 2-lobed. Corolla upper lip erect, emarginate, lower lip 3 lobed, the lateral lobes oblong and erect, the lowest large, broad, and spreading. Stamens 2, ascending, the filaments with a small tooth below the middle (indicating the junction of the real filament with the filament-like connective). Anthers 1-celled.—This the well-known *Rosemary*, a native of southern Europe, and of early cultivation among sweet herbs, has been introduced into Hope Valley, Victoria, *F. Mueller*.

Nepeta Cataria, Linn.; Benth. in DC. Prod. xii. 383. An erect herbaceous hoary-pubescent branching perennial, attaining 2 or 3 feet. Leaves petiolate, ovate-cordate, acute, coarsely toothed, often whitish underneath. Flowers pale blue or nearly white, numerous in compact false-whorls, forming short oblong terminal spikes, with frequently one or more false-whorls lower down. Calyx 15-ribbed, the orifice oblique, 5-toothed. Corolla upper lip erect, concave, lower lip longer, spreading, 3-lobed. Stamens 4, ascending in pairs under the upper lip, the upper or inner pair the longest. Anthers 2-celled. —Common on roadsides, in hedges, &c. in many parts of Europe and Asia, and readily naturalized in other temperate regions; occurs in the neighbourhood of Adelaide.

Marrubium vulgare, Linn.; Benth. in DC. Prod. xii. 453. An erect hard branching perennial, the stems thickly covered with white cottony wool. Leaves petiolate, orbicular, soft, whitish, and much wrinkled. Flowers small, of a dirty white, in dense axillary false-whorls. Calyx with 10 small hooked teeth. Corolla upper lip narrow, erect, 2-cleft, lower lip spreading, 3-lobed. Stamens 4, included in the corolla-tube. Anthers 2-celled.—A roadside weed of European or Asiatic origin, now naturalized in many parts of the world, and gathered by various collectors in N. S. Wales, Victoria, and S. Australia.

Stachys arvensis, Linn.; Benth. in DC. Prod. xii. 477. A decumbent or slightly ascending slender hairy annual, from an inch or two to nearly a foot long. Leaves small, petiolate, ovate. Flowers small, of a pale purple, in false-whorls of 2 to 6 or rarely 8, forming loose leafy spikes. Calyx with 5 nearly equal teeth as long as the tube. Corolla scarcely longer than the calyx, the upper lip erect, concave and entire, the lower lip spreading, 3-lobed. Stamens 4, in pairs, ascending under the upper lip. Anthers 2-celled.—A common weed of cultivation in Europe and temperate Asia, carried out with European crops to various parts of the world, and well established even in tropical countries, now abundant in some parts of Queensland, N S. Wales, and Victoria. —*S. palustris*, Linn.; Benth. in DC. Prod. xii. 470, a tall erect perennial, with nearly sessile, oblong or lanceolate leaves and rather large pale purple flowers in false-whorls of 6 to 8, forming terminal spikes, a common northern plant in ditches, moist banks, &c., has been gathered on Richmond river in N. S. Wales by *Fawcett*.

Moluccella lævis, Linn.; Benth. in DC. Prod. xii. 513. A glabrous erect or ascending annual of 1 to 2 ft. Leaves on long petioles, broadly ovate or almost orbicular. Flowers in distant false-whorls of about 6, the floral leaves gradually smaller but all on long petioles, the bracts connate at the base. Calyx very large, campanulate, oblique, and membranous, the margin 5-angled with a small point at each angle, attaining sometimes nearly 2 in. diameter. Corolla shorter than the calyx, whitish, the upper lip erect, concave, entire, the lower spreading, 3-lobed. Stamens 4, ascending in pairs under the upper lip. Anthers 2-celled.—A native of the Mediterranean region, now established in New England, *C. Stuart*, and on the Murrumbidgee, *F. Mueller*.

Leonotis leonurus, R. Br.; Benth. in DC. Prod. xii. 536. A shrub attaining several feet. Leaves oblong-lanceolate, contracted into a short petiole. Flowers of a rich scarlet, few or many, in axillary false-whorls. Calyx 10-toothed. Corolla often fully 2 in.

long, hairy, the upper lip long erect and concave, the lower lip short and spreading. Stamens 4, ascending in pairs under the upper lip. Anthers 2-celled.—A native of the Cape of Good Hope, long since cultivated in gardens for its showy flaming flowers, now established with other South African plants in West Australia.

TRIBE 1. OCIMOIDEÆ.—Stamens 4, declinate towards the lower lobe of the corolla. Anthers when fully out 1-celled by the confluence of the two cells. Nuts dry, smooth or minutely granular. Calyx with the upper tooth often much broader than the others. Corolla with the four upper lobes flat, the lowest alone forming the lower lip, and flat concave or saccate.

1. OCIMUM, Linn.

Fruiting-calyx enlarged and reflexed, the upper tooth orbicular or ovate with the margins decurrent, forming an upper lip, the 4 lower teeth small, pointed, equal or the two lowest with longer points. Corolla-tube straight, rarely exceeding the calyx, 4 upper lobes united in a broad shortly 4-lobed upper lip, the fifth lower lobe entire, flat or slightly concave, about as long as the upper lip. Stamens 4, declinate, the 2 upper ones usually with a tooth or tuft of hairs near the base; anther-cells confluent. Style-lobes subulate or somewhat flattened. Nuts smooth or minutely granular.—Herbs undershrubs or rarely small shrubs. Foliage usually densely dotted and highly scented. Flowers in false-whorls of 6, rarely 10, arranged in terminal racemes, the floral leaves reduced to small deciduous bracts.

The genus extends over the tropical and subtropical regions of the New as well as the Old World, two or three species having been very long in cultivation amongst aromatic herbs. The only Australian species is a slight, almost endemic variety of one of the common Asiatic ones.

1. **O. sanctum,** Linn.; Benth. in DC. Prod. xii. 38, var. angustifolium. A branching perennial or undershrub, usually forming a thick woody base, but sometimes flowering the first year so as to appear annual, rarely exceeding 1 ft. in height, more or less hirsute with spreading or reflexed hairs. Leaves on rather long petioles, mostly oblong-lanceolate, ½ to 1½ in. long, bordered by a few coarse teeth or nearly entire, more rarely ovate. Flowers small, purple, (F. Mueller) or pure white (Bowman), in loose whorls of 6, forming terminal racemes, the bract-like floral leaves very small. Pedicels slender, often as long as the calyx. Calyx scarcely 1 line long at the time of flowering, when in fruit 2 to 2½ lines long, the upper lobe orbicular, the lateral ones small and acute, the 2 lowest with long subulate points. Corolla-tube nearly as long as the calyx, the lips as long as the tube, the lower lobe or lip broadly ovate, or nearly orbicular. Stamens shortly exserted, the 2 upper ones with tufts of hairs near the base of the filaments. Style-lobes slightly flattened.—O. anisodorum and O. caryophyllinum, F. Muell. Fragm. iv. 46.

N. Australia. Victoria river, Hooker's and Sturt's Creeks, F. Mueller; Gulf of Carpentaria, F. Mueller, Landsborough; N. Kennedy district, Daintree; in the interior, lat. 18° 30', M'Douall Stuart's Expedition.

Queensland. Burdekin and Suttor rivers and Peak Downs, *F. Mueller;* Bowen, Isaacs and Suttor rivers, *Bowman.*

Widely diffused over tropical Asia, extending into Africa, but usually with broader leaves than in Australia. It is frequently cultivated in East Indian gardens, about temples, &c., for its strong aromatic properties.

F. Mueller observes that his *O. anisodorum* and *O. caryophyllinum* (which I am quite unable to distinguish from each other) are closely allied to *O. basilicum;* they differ however not only in the small flowers and long pedicels, but more essentially in the tooth of the upper (not the lower) stamens being replaced by the tuft of hairs of *O. sanctum.* From the common Indian form of *O. sanctum* (which by some mistake I stated in the Prodromus to be inodorous) I can perceive no difference, except in the narrow leaves and usually less hispid stems, but some Australian specimens have ovate leaves, and the hairs of the stem are very variable, whilst a very few Indian ones have narrow leaves. The calyx-lobes are more ciliate in the margin in Australia than in India, but I have not seen the hairs closing the orifice as in the Mascarene *O. gracile.*

2. MOSCHOSMA, Reichb.

Fruiting-calyx somewhat enlarged, erect spreading or reflexed, the upper tooth broad, with the margins scarcely decurrent, forming an upper lip, the 4 lower teeth small, acute, nearly equal. Corolla-tube straight, rarely exceeding the calyx, the 4 upper lobes united in a broad 3- or 4-lobed upper lip, the fifth lower lobe entire, flat or slightly concave, about as long as the upper lip. Stamens 4, declinate, the filaments without any appendage; anther-cells confluent. Style shortly clavate at the end, entire or emarginate. Nuts smooth or minutely granular.—Herbs, usually annual and much branched. Flowers minute, in false whorls of 6 to 10, all turned to one side and numerous in slender one-sided racemes, the floral leaves reduced to very small deciduous bracts. Foliage said to have a musky smell.

The genus comprises but very few species extending over tropical Asia and Africa. Of the two Australian species, one is common in East India, the other appears to be endemic.

Fruiting-calyx reflexed or very spreading, companulate. Racemes
loose . . . . 1. *M. polystachyum.*
Fruiting-calyx erect or slightly spreading, tubular-campanulate.
Racemes dense . . . . 2. *M. australe.*

1. **M. polystachyum,** *Benth.; DC. Prod.* xii. 48. An erect slender much branched annual of 1 to 2 ft., slightly pubescent or nearly glabrous, the stems acutely 4-angled. Leaves on long petioles, ovate or ovate-lanceolate, acuminate, toothed, 1 to 2 in. long, the upper ones small. Racemes numerous, terminating the main branches as well as short axillary branchlets, slender and loose. Flowers white or very pale blue, about ¾ line long. Calyx minute when in flower, when in fruit 1 to 1½ lines long, rather broadly campanulate, very spreading or reflexed, the upper lobe very broad and slightly decurrent. Corolla-tube about as long as the calyx, the lips about as long as the tube.—*Ocimum polystachyon,* Linn. Mant. 567; *Plectranthus parviflorus,* R. Br. Prod. 506, not of Willd.

N. Australia. Upper Victoria river, *F. Mueller* (a single specimen in herb. Hooker).

Queensland. Broad Sound, *R. Brown;* Wide Bay, *Bidwill;* Port Denison, *Fitzalan;* Rockingham Bay, *Dallachy;* Amity and Nerkool Creeks, *Bowman;* Rockhampton, *O'Shanesy;* Balonne river, *Mitchell.*
The species is a common weed in tropical Asia, extending into Africa.

2. **M. australe,** *Benth.; DC. Prod.* xii. 48. An erect herb of 2 or 3 ft., closely allied to *M. polystachyum,* but coarser, the foliage and calyxes usually more pubescent and the racemes more dense and spike-like. Fruiting-calyx more erect, longer and not so broad in proportion as in *M. polystachyum,* the upper lobe not so broad, and the lower ones broader than in that species; the corolla-tube more slender, slightly exceeding the calyx, the lips shorter than the tube. Middle-lobe of the upper lip emarginate in *M. australe* (lip 4-lobed), entire in *M. polystachyum* (lip 3-lobed), according to R. Brown, a difference, however, which I have failed to perceive.—*Plectranthus moschatus,* R. Br. Prod. 506.

N. Australia. Islands of the Gulf of Carpentaria, *R. Brown;* Upper Victoria and Fitzmaurice rivers, *F. Mueller;* Liverpool river, N. coast, *A. Cunningham.*

3. ORTHOSIPHON, Benth.

Fruiting-calyx enlarged and reflexed, the upper tooth orbicular or ovate with the margins decurrent, forming an upper lip, the 4 lower teeth small and pointed. Corolla-tube straight or somewhat curved, longer than the calyx, 4 upper lobes united in a broad 3- or 4-lobed upper lip, lower lobe entire, flat or slightly concave, as long as or (in a species not Australian) longer than the upper lip. Stamens 4, declinate, without appendages; anther-cells confluent. Style clavate or capitate at the end, entire or slightly notched. Nuts small, granular-punctate.—Perennial herbs. Flowers in false-whorls of 6 or rarely fewer, in long interrupted or short and dense racemes, the floral leaves reduced to bracts.

The genus is chiefly Asiatic, with two or three African species and two slightly anomalous American ones. The only Australian species is a widely spread Asiatic one. The genus has the calyx and habit of *Ocimum,* from which it differs in the elongated corolla-tube and capitate stigma.

1. **O. stamineus,** *Benth.; DC. Prod.* xii. 52. A loosely branched perennial of 1 to 3 ft., slightly hoary-pubescent or nearly glabrous. Leaves petiolate, ovate or ovate-lanceolate, often acuminate, 1 to 2 in. long, irregularly and coarsely toothed, or rarely regularly crenate or almost entire. Flowers white or pale blue, in whorls of 4 to 6, in loose but rather short terminal racemes, and by their long stamens resembling those of a *Clerodendron.* Pedicels about as long as the flowering calyx, which scarcely exceeds 1½ lines. Corolla-tube slender, 4 to 5 lines long or even more, the lips very spreading, shorter than the tube. Stamens filiform, two or three times as long as the corolla-tube, with very small anthers. Style still longer, the stigmatic end small and clavate. Fruiting-calyx attaining ½ in., the upper lobe ovate, obtuse and decurrent, the lateral ones nearly as long, acute, with short points,

the lower ones connate to the middle, and produced into long incurved subulate points.

Queensland. Cape York, *M'Gillivray, Veitch;* Port Denison, *Fitzalan;* Burdekin river, *Bowman;* Rockingham Bay, *Dallachy.* Frequent in the Indian Archipelago, extending on the Asiatic continent to Assam and Silhet.

4. PLECTRANTHUS, Lher.

Fruiting-calyx in the Australian species reflexed, the upper tooth broad and sometimes decurrent, the 2 lowest long and pointed, the lateral ones shorter, in some other species the teeth all nearly equal. Corolla-tube longer than the calyx, gibbous or produced into a spur on the upper side; upper lip 3- or 4-lobed, lower lip entire, concave, longer than or rarely rather shorter than the upper one. Stamens declinate, free, without any appendage; anther-cells confluent. Style shortly bifid. Nuts smooth or slightly granular.—Herbs, undershrubs or in species not Australian, shrubs. Flowers usually numerous, rarely only 6, in false-whorls, often developed into loose opposite cymes forming terminal panicles, in the Australian species more compact forming interrupted racemes, the floral leaves reduced to small deciduous bracts.

The genus is widely spread over tropical and subtropical Asia and Africa, one of the African species having been also found in Brazil (probably introduced). Of the three Australian species, one extends to the Pacific Islands, the two others appear to be endemic. The common Indian forms have none of them been as yet detected in Australia.

Flowers in false-whorls of 6, lower calyx-lobes obtuse. Corolla with
a long spur . . . . 1. *P. longicornis.*
Flowers in false-whorls of 10 or more. Lower calyx-lobes very
obtuse or aristate. Corolla not spurred.
False-whorls loose, of about 10 flowers. Fruiting-calyx 2 to 3
lines long . . . . 2. *P. parviflorus.*
False-whorls dense, of about 20 flowers. Fruiting calyx about 1
line long . . . . 3. *P. congestus.*

1. P. longicornis, *F. Muell. Fragm.* v. 51. A perennial, more or less pubescent or villous, with short rather rigid hairs. Leaves usually 2 or 3 pairs near the base of the stem, obovate or oblong, obtuse, sinuate or coarsely toothed especially near the base, 1½ to 3 in. long, contracted into a short petiole; below the lowest pair there are also usually 2 or 3 pairs of very small nearly sessile orbicular leaves. Flowering stems erect, often above 1 ft. high, simple or slightly branched, leafless except at the base, the floral leaves being all reduced to ovate reflexed bracts; rarely the stock emits also long weak decumbent stems with a few distant pairs of narrow leaves. Flowers deep purple (*Dallachy*) or blue (*F. Mueller*), in false-whorls of 6, forming long slender racemes. Pedicels 1 to 2 lines long. Calyx at first very small and open, when in fruit about 2 lines long, very much incurved, deeply 2-lipped, the upper lip formed of the broad obtuse slightly decurrent and recurved upper tooth with the 2 very small lateral lobes at its base, the lower lip as long as the upper, obtuse and emarginate, incurved and concave. Corolla-tube slender, nearly 2 lines long, produced at the base into a long narrow conical spur; upper lip erect, broad, shortly 4-lobed, shorter than the tube; lower lip oblong-obovate, very concave, rather shorter than the upper. Stamens shorter than the corolla, the two upper ones inserted much lower down than the lower ones.

Queensland. Cape York, *M'Gillivray;* Rockingham Bay, *Dallachy.*

This species has precisely the calyx and something of the habit of *Coleus Africanus,* Benth. (which probably includes *Plectranthus Palisoti,* Benth.) In the latter plant however the foliage is different, the flowers much more numerous in the false-whorls, and as far as I can ascertain, the corolla and stamens are truly those of a *Coleus.*

2. **P. parviflorus,** *Willd.; Benth. in DC. Prod.* xii. 67, *not of R. Br.* An erect herb or undershrub very variable in size and indumentum, sometimes under 1 ft. and much branched at the base, sometimes attaining 2 or 3 ft.; the stems rather fleshy, the whole plant pubescent villous or tomentose or rarely nearly glabrous. Leaves in the lower part of the stem petiolate, ovate to orbicular, coarsely crenate, rounded or cordate at the base, from under 1 in. to 2 or even 3 in. diameter, usually rather thick soft rugose and tomentose or villous, but sometimes nearly glabrous and smooth. Flowers small, of a bluish purple, in false-whorls of about 10, forming long terminal leafless racemes either simple or slightly branched, the floral leaves reduced to minute bracts falling off from the very early buds. Pedicels usually longer than the very small flowering calyx; the fruiting calyx reflexed, much curved, striate, 2 to nearly 3 lines long, the upper tooth or lobe broad obtuse and slightly decurrent, the 4 lower ones incurved, very acute or subulate-acuminate, the 2 lateral ones as long as, the 2 lowest longer than, the upper one. Corolla-tube about twice as long as the calyx, declinate and slightly gibbous at the base but not spurred; upper lip short and erect, the 2 upper lobes rather larger than the lateral ones, the lower lip twice as long; very concave. Stamens nearly as long as the lower lip.—*P. graveolens,* R. Br. Prod. 506; *P. australis,* R. Br. Prod. 506, Bot. Reg. t. 1098, Benth. in DC. Prod. xii. 67.

N. Australia. Port Essington, *Armstrong;* Gilbert river, *F. Mueller;* Kennedy district, *Daintree.*

Queensland. Broad Sound, Shoalwater Bay, Northumberland island, *R. Brown;* Burdekin river, *Fitzalan;* Lizard island, *M'Gillivray;* Whitsunday island, *Henne;* Rockhampton, *Dallachy, Bowman;* Brisbane river, Moreton Bay, *F. Mueller;* top of Mount Faraday, *Mitchell.*

N. S. Wales. Port Jackson to the Blue Mountains, *R. Brown* and others; northward to Hastings, Macleay and Clarence rivers, *Beckler* and others; New England, *C. Stuart;* southward to Twofold Bay, *F. Mueller.*

Victoria. Snowy river, *F. Mueller.*

The species extends to New Caledonia and other islands of the South Pacific and to the Sandwich Islands, the differences formerly observed between the Australian and Sandwich Island plants disappearing in the larger series of specimens now before us.

3. **P. congestus,** *R. Br. Prod.* 506. A tall herb, attaining according to Dallachy 10 to 12 ft., usually hoary-tomentose. Leaves ovate, shortly acuminate or obtuse, coarsely crenate, narrowed into a petiole, soft and rugose, tomentose on both sides, about 2 or 3 in. long in the specimens seen but the lower ones probably longer. Flowers small,

blue, numerous (often above 20), in dense false-whorls, forming long almost spike-like interrupted racemes of which several are collected in a large terminal leafless panicle, the floral leaves reduced to minute very deciduous bracts. Pedicels very short. Calyx villous and copiously dotted and scarcely above 1 line long when in fruit, declinate or reflexed, the upper lobe broadly ovate, obtuse, not decurrent, the lateral ones acute but not much narrower, the 2 lowest narrower and rather longer, incurved and very acute. Corolla-tube about 2 lines long, abruptly declinate and slightly gibbous on the upper side below the middle but not spurred; upper lip about half as long as the tube, the 2 upper lobes broadly obovate, the 2 lateral ones very much smaller; the lower lobe or lip twice as long as the upper one and very concave. —Benth. in DC. Prod. xii. 66.

Queensland. Endeavour river, *Banks and Solander;* Rockingham Bay, *Dallachy.*

5. COLEUS, Lour.

Fruiting-calyx usually declinate or reflexed, the upper tooth broad, scarcely decurrent, the lateral ones truncate or acute, the two lower usually longer, more connate and acute. Corolla-tube longer than the calyx, declinate or bent down, not spurred; upper lip short, 3- or 4-lobed, the lower much longer, entire, very concave or boat-shaped. Stamens 4, more or less connate in a tube round the style; anther-cells confluent. Style shortly bifid at the top. Nuts small, smooth.—Herbs rarely shrubs. Flowers in false-whorls of 6 or more, sometimes very dense, sometimes growing out into opposite variously branched cymes, forming terminal leafless racemes or panicles, the floral leaves reduced to small deciduous bracts.

The genus extends over tropical Asia and Africa, the only Australian species being apparently the same as a common one in the Archipelago, although represented by endemic forms or varieties, which however require further investigation. The genus differs from *Plectranthus* chiefly in the monadelphous stamens.

1. **C. scutellarioides,** *Benth.; DC. Prod.* xii. 73. A tall herb or undershrub, the typical form pubescent or nearly glabrous, with slender branches. Leaves petiolate, ovate, acuminate or obtuse, slightly crenate-toothed and more or less purple underneath in the typical form, but varying much in the Australian varieties, mostly 1½ to 3 in. long. Flowers rather small and numerous at first, in rather compact false-whorls forming long slender terminal racemes, but in most varieties as the flowering advances the two primary branches on each side lengthen considerably, converting the false-whorl into two opposite sessile once forked cymes, with the pedicels arranged along each branch. Calyx very small when in flower, enlarged afterwards, deflexed, the tube striate, the broad upper lobe slightly decurrent, the lateral ones rather shorter and very obtuse, the 2 lowest much longer, connate to near the end where they form two small points. Corolla-tube slender, slightly gibbous at the base, then abruptly bent down, the throat dilated espe-

cially in the Australian varieties, of a pale bluish white as well as the upper lip, the lower boat-shaped lip or lobe of a deeper blue. Stamens not exceeding the lower lobe.—*Ocimum scutellarioides*, Linn.; Bot. Mag. t. 1446; *Plectranthus scutellarioides*, R. Br. Prod. 506.

N. Australia, *R. Brown* and others. The species appears to be widely distributed over the Indian Archipelago, and ought perhaps to include *C. atropurpureus*, Benth. and some others. The flowers are in the typical form rather smaller, and the fruiting cymes usually less developed than in Australia, where moreover the following varieties appear very distinct in the few specimens we possess.

Var. *angustifolia*. Leaves lanceolate, acuminate, pubescent underneath, coarsely toothed. False-whorls in some specimens compact, in others the cymes more developed.—Table Hill, Victoria river and Macadam Range, *F. Mueller;* Port Essington, *Armstrong.*

Var. *laxa*. Glabrous. Leaves broad, not acuminate, coarsely toothed. Inflorescence loose, the 2 branches on each side of the false-whorl at length much elongated, flowers large —Roe river, York Sound, N.W. coast, *A. Cunningham;* Roebuck Bay, N.W. coast, *Martin;* Arnhem's Land, *R. Brown, Mackinlay.*

Var. *limnophila*, F. Muell Loosely villous, almost woolly. Leaves ovate, acuminate, coarsely toothed, rather large. Inflorescence at least as loose and flowers as large as in the preceding variety.—Nicholson and Macarthur rivers, *F. Mueller.*

* 6. HYPTIS, Jacq.

Fruiting-calyx erect, with 5 subulate or acute teeth. Corolla-tube cylindrical, the 4 upper lobes flat and erect or spreading, the lowest lobe saccate, contracted at the base and abruptly deflexed. Stamens 4, declinate, without appendages; anther-cells confluent. Nuts various. —Herbs, undershrubs or shrubs, exceedingly diversified in habit and inflorescence, but always known by the saccate lower lobe of the corolla.

A very large genus, entirely American, tropical or subtropical, a few of the species are however now spread over various tropical regions of the Old World, especially Africa, and amongst the most weedy of these is the only one hitherto found in Australia.

* 1. **H. suaveolens,** *Poit.; Benth. in DC. Prod.* xii. 126. A coarse erect herb of 2 to 6 ft., more or less hirsute. Leaves petiolate, broadly ovate, irregularly toothed, often cordate, 1 to 3 in. long, the upper floral ones gradually smaller and passing into small bracts. Peduncles in the upper axils 2 to 4 lines long or almost none, bearing little heads of 3 to 5 small flowers, sometimes all in irregular axillary false-whorls, sometimes forming irregular almost leafless panicles or spikes. Fruiting calyx 3 to 4 lines long, obliquely campanulate, truncate and ciliate on the margin, with 5 subulate teeth. Corolla about 3 lines long, the lips shorter than the tube.

N. Australia. Garden Bay, Port Essington, *Leichhardt.* A common tropical American weed, now found in many parts of the Old World, and probably introduced into Australia from the Indian Archipelago.

TRIBE II. SATUREIEÆ.—Corolla-lobes spreading, not hood-shaped, equal or more or less two-lipped. Stamens 4 or 2, distant, erect, divaricate or connivent in pairs under the upper lip, the upper pair shorter or abortive. Anthers 2-celled, or rarely as in Ocimoideæ

1-celled by the confluence of the cells.　Nuts dry, smooth or minutely granular.

7. DYSOPHYLLA, Blume.

Calyx equally 5-toothed.　Corolla with a short tube; the limb of 4 equal or nearly equal lobes.　Stamens 4, nearly equal, exserted, the filaments bearded; anthers small, terminal, nearly globular, 1-celled. —Herbs.　Leaves opposite or whorled.　Flowers in dense false-whorls, forming close cylindrical terminal spikes.

The genus comprises several tropical Asiatic species, including among them the only Australian one.　The commonest and widest spread of them, *D. auricularia*, Blume, has however, not yet been detected in Australia.　Hasskarl has proposed the uniting the genus with *Pogostemon*, of which it would at any rate form a very marked section, characterised by the inflorescence and habit as well as by the more regular corolla.

1. **D. verticillata,** *Benth.; DC. Prod.* xii. 157.　A glabrous or somewhat pubescent herb, decumbent at the base or floating in water, ascending to 1 ft. or more.　Leaves in whorls of 4 to 6 or rarely more, sessile, linear or linear-lanceolate, entire, with recurved margins, 1 to 2 in. long or even more.　Flowers very small and numerous, the false-whorls crowded in dense cylindrical spikes of ¾ to 2 in., the tips of the bracts subtending the false-whorls sometimes appearing in the young spikes in 4, 6 or 8 rows.　Calyx softly villous, scarcely one line long when in fruit, and much smaller in the flower.　Corolla-tube shorter than the calyx, the lobes as long as the tube.　Stamens shortly exceeding the corolla-lobes, the filaments bearded with reddish purple hairs.

N. Australia. Boggy situations, S. Goulburn island, *A. Cunningham;* Lagoons, Mitchell's river, Gulf of Carpentaria, *Leichhardt,* in both cases the common form with glabrous stems and foliage.

Queensland.—Edges of waterholes, Rockingham Bay, *Dallachy,* a pubescent variety also found in India, but not so common as the glabrous one.

The species is widely spread over East India and the Archipelago.　Thwaites is probably right in uniting with it as varieties *D. crassicaulis,* Benth., and *D. tetraphylla,* Wight, Ic. t. 1444.

8. MENTHA, Linn.

Calyx regular or slightly 2-lipped, 5-toothed.　Corolla-tube not at all or scarcely exceeding the calyx; limb 4-lobed, the lobes all equal and spreading or the upper one broader and notched.　Stamens 4, equal, erect, distant; filaments glabrous; anthers with 2 parallel cells.　Style shortly bifid.　Nuts smooth.—Herbs, usually copiously dotted and strongly scented.　Flowers small, in false-whorls usually dense rarely few-flowered, all axillary or in species not Australian, forming terminal spikes, with the floral leaves reduced to bracts.　Bracts within the false-whorls minute, or rarely subulate and as long as the calyx.

The genus is chiefly extratropical and most abundant in Europe and Northern Asia, with one or two North American species scarcely different from Asiatic ones, so also the very few tropical Asiatic forms are but slight modifications of common northern ones. The six following Australian species are all endemic. They appear as difficult to define by positive characters, and to pass into each other as gradually as the species allied to *M. arvensis* in the northern hemisphere. They all belong to the group with axillary false-whorls, and one of them comes very near indeed to the almost ubiquitous *M. arvensis*, the genuine forms of which have not however as yet appeared in Australia.

Leaves all toothed. Flowers numerous in the false whorls. Pedicels usually as long as the calyx . . . . 1. *M. laxiflora.*
Leaves mostly toothed. Flowers less numerous, the corolla fully 5 lines long. Pedicels short . . . . 2. *M. grandiflora.*
Leaves all or mostly entire. Pedicels very short (except where the flowers are very few).
Flowers numerous, rather large. Calyx-teeth subulate. Corolla under 4 lines long. Leaves lanceolate . . . . 3. *M. australis.*
Flowers less numerous and smaller. Calyx-teeth lanceolate-subulate. Leaves small ovate . . . . 4. *M. gracilis.*
Flowers few in the false-whorls and small. Calyx-teeth lanceolate, somewhat hairy inside. Leaves nearly sessile . . . 5. *M. serpyllifolia.*
Flowers few in the false-whorls and small. Calyx-teeth short, densely hairy inside. Leaves ovate to oblong-linear . . 6. *M. satureioides.*

Besides the above, the two following European species have been sent amongst the plants introduced into Australia.

M. viridis, Linn.; Benth. in DC. Prod. xii. 168. Stems erect, 1 to 2 feet high, glabrous as well as the foliage. Leaves sessile or nearly so, ovate-lanceolate, irregularly serrate. False-whorls in a loose cylindrical terminal leafless spike, the floral leaves reduced to small narrow bracts. Calyx usually hairy.—Borders of streams, Mount Lofty Range, *F. Mueller.*

M. aquatica, Linn.; Benth. in DC. Prod. xii. 170. Erect, 1 to 2 ft. high, more or less hirsute. Leaves petiolate, ovate or ovate-lanceolate, serrate, the floral ones similar except the upper ones, which are reduced to small bracts. Flowers numerous in a rather large terminal nearly globular false-whorl, with the addition frequently of one or more similar false-whorls a little lower down or in the upper axils.—Borders of streams, &c., Karrie Dale and throughout the Warree country, W. Australia, *Walcott;* Tone river, *Maxwell.*

1. **M. laxiflora,** *Benth. in DC. Prod.* xii. 174. Stems weak, procumbent, from under 1 ft. to near 2 ft. high, more or less hirsute on the angles as well as the foliage. Leaves petiolate, ovate or ovate-oblong, obtuse, toothed, the floral ones similar. Flowers in loose axillary false-whorls, the pedicels usually as long as or even longer than the calyx. Calyx hispid, about 1¼ lines long, the lobes lanceolate, as long as the tube, scarcely villous inside. Corolla twice as long as the calyx, the limb 4-lobed, the upper lobe scarcely notched.

Victoria.—Port Philip, *Gunn;* Buffalo Range, Mount William, Mount Disappointment, Ballarat, &c., *F. Mueller;* Ararat, *Green.*

This is not easily to be distinguished from some rather anomalous loose-flowered European forms of *M. arvensis*, the foliage is quite different from that of the more common Australian species.

2. **M. grandiflora,** *Benth. in Mitch. Trop. Austr.* 362 *and in DC. Prod.* xii. 698. Softly pubescent. Leaves petiolate, ovate, mostly toothed, ½ to ¾ in. long. False-whorls all axillary, loose and not many-flowered, but the pedicels all shorter than the calyx. Calyx about 2½ lines long, the teeth lanceolate or lanceolate-subulate, scarcely villous inside. Corolla fully 5 lines long, the tube rather longer than

the calyx, the upper lobe deeply notched or shortly bifid.   Stamens and style much longer than the corolla.

Queensland. On the Maranoa, *Mitchell.* The specimens are but few, and may hereafter prove to be exceptional, but the flowers are much larger than those of any *Mentha* known to me, and the foliage is nearer that of *M. laxiflora* than of *M. australis*, with which F. Mueller unites the species, without however having seen the specimens.

3. **M. australis,** *R. Br. Prod.* 505.   Stems erect or ascending, branched, 1 to 2 feet high, usually scabrous on the angles.   Leaves lanceolate or rarely ovate-lanceolate, quite entire or here and there sparingly toothed, contracted into a short petiole or almost sessile, glabrous or hoary-pubescent especially underneath, often above 1 in. long.   Flowers in axillary false-whorls, usually numerous, on very short pedicels or quite sessile.   Calyx narrow, hoary-pubescent or villous, 2 to nearly 3 lines long, the teeth subulate or narrow-lanceolate, very acute, slightly villous inside.   Corolla-tube not exceeding the calyx, the lobes shorter than the tube, the upper one more or less deeply 2-lobed, the whole corolla under 4 lines long.—Benth. in DC. Prod. xii. 174; Hook. f. Fl. Tasm. i. 281; *Micromeria australis*, Benth. Lab. Gen. et Sp. 380.

Queensland. In the interior, *Mitchell.*
N. S. Wales. Hawkesbury river, *R. Brown;* Darling river, *Mrs. Ford;* on the Murrumbidgee, *F. Mueller.*
Victoria. *Mitchell;* Wendu vale, *Robertson,* Yarra river and Sandy Creek, *F. Mueller.*
Tasmania. Port Dalrymple, *R. Brown;* common in marshy places, *J. D. Hooker.*
S. Australia. From the Murray river, *Behr.,* to St. Vincent's Gulf, *F. Mueller* and others; towards Spencer's Gulf, *Warburton;* Cooper's Creek, *Howitt's Expedition.*

The differences observed in the upper lobe of the corolla of this and the allied species, deeply lobed, notched only, or entire, require further observation to ascertain whether they are really of specific constancy.

4. **M. gracilis,** *R. Br. Prod.* 505. Very nearly allied to *M. australis*, of which Archer considers it as a variety, but a smaller more diffuse and branching plant, rarely attaining 1 ft., hoary-pubescent or glabrous like that species, but the stems much smoother.   Leaves petiolate, ovate or rarely ovate-lanceolate, entire or scarcely toothed, under 1 in. and often not ½ in. long.   Flowers much smaller than in *M. australis* and rather fewer in the false-whorls, the pedicels exceedingly short or scarcely any.   Calyx 1½ to about 2 lines long, the teeth lanceolate or lanceolate-subulate, not very hairy inside.   Corolla-tube shorter than the calyx, the lobes much shorter than the tube, the upper one scarcely notched.—Benth. in DC. Prod. xii. 174; Hook. f. Fl. Tasm. i. 281; *M. diemenica,* Spreng. Syst. ii. 724; *Micromeria gracilis,* Benth. Lab. Gen. et Sp. 380.

Victoria. Mouth of the Glenelg river, *Allitt;* near Melbourne, *Adamson* (with larger narrower leaves, but small flowers); Creswick, *Whan;* Gipps Land, *F. Mueller.*
Tasmania. Port Dalrymple, *R. Brown;* in stony places not unfrequent, *J. D. Hooker.*
S. Australia. Onkaparinga river, *F. Mueller;* Rapide Bay, *Malpas.* The northern specimens referred to this species by F. Mueller (Fragm. vi. 109) appear to me all to belong to *M. satureioides.*

5. **M. serpyllifolia,** *Benth. in DC. Prod.* xii. 174. A slender diffuse perennial, often creeping at the base, the filiform ascending stems rarely above 6 in. long, glabrous or very slightly pubescent. Leaves very shortly petiolate or almost sessile, ovate, entire or rarely obscurely toothed, under ½ in. and often not ¼ in. long. Flowers small, few, in axillary false-whorls of 4 to 6 or sometimes reduced to 2 opposite flowers. Calyx about 1½ lines long, the teeth lanceolate, acute, villous inside but much less so than in *M. satureioides.* Corolla not twice as long as the calyx, the upper lobe usually shortly bifid.— Hook. f. Fl. Tasm. i. 281; *Micromeria sessilis* and *M. affinis,* Hook. f. in Hook. Lond. Journ. vi. 274.

Victoria. King river, Guichen Bay, Mount Disappointment, Wilson's Promontory, &c., *F. Mueller.*

Tasmania. Not uncommon in marshes in various parts of the island, *Gunn.*

S. Australia. Torrens river, *F. Mueller.*

Possibly a variety of *M. gracilis.* The Victorian and S. Australian specimens are less characteristic than the Tasmanian ones.

6. **M. satureioides,** *R. Br. Prod.* 505. A small much-branched erect or diffuse perennial, under 1 ft. and often only a few inches high, glabrous or minutely hoary-pubescent. Leaves petiolate or sometimes almost sessile, usually oblong or oblong-lanceolate, but varying from ovate to almost linear, obtuse, entire, under ½ in. long when broad, sometimes nearly 1 in. when narrow. Flowers usually 6 or fewer rarely rather more, in axillary false-whorls. Pedicels shorter than the calyx and often very short. Calyx 1½ to nearly 2 lines long, the teeth shortly lanceolate or triangular, always densely villous inside with white hairs, readily distinguishing this species from all others. Corolla small, the upper lobe usually shortly bifid.—Benth. in DC. Prod. xii. 174; *Micromeria satureioides,* Benth. Lab. Gen. et Sp. 380; Bartl. in Pl. Preiss. i. 354.

Queensland. Burnett river and Moreton Bay, *F. Mueller*; Rockhampton, *Thozet;* Darling Downs, *Lau.*

N. S. Wales. Port Jackson to the Blue Mountains, *R. Brown, Sieber, n.* 491, and others; northward to Hastings, Macleay and Clarence rivers, *Beckler;* New England, *C. Stuart;* Castlereagh river, *C. Moore;* in the interior to Lachlan river, *A. Cunningham;* on the Murrumbidgee, *F. Mueller.*

Victoria. Bacchus Marsh, Loddon river, Creswick Creek, Macalister river, &c., *F. Mueller;* Wimmera, *Dallachy.*

S. Australia. Near Bethanie, *Behr.;* Onkaparinga river, St. Vincent's Gulf, Port Lincoln, &c., *F. Mueller.*

W. Australia. King George's Sound, *Huegel;* Blackwood river, *Walcott;* Swan river, *Drummond, 1st coll. n.* 458, *Preiss, n.* 2322 a, 2323, 2324.

9. LYCOPUS, Linn.

Calyx equally 4- or 5-toothed. Corolla-tube short, limb nearly equally 4-lobed. Two upper stamens reduced to small filiform staminodia sometimes capitellate at the top; the 2 lower ones perfect, distant, with 2-celled anthers. Style shortly bifid. Nuts smooth, with three callous angles and truncate at the top.—Perennial herbs, usually emitting stolones. Flowers small, usually numerous, in dense axillary false-whorls. Bracts within the false-whorls minute, or the outer ones as long as the calyx.

The genus consists of very few species, or varieties, dispersed over the temperate regions of the northern hemisphere, the only Australian species scarcely differing from some of the northern forms.

1. **L. australis,** *R. Br. Prod.* 500. An erect herb, attaining sometimes 2 or 3 ft., glabrous or sprinkled with a few minute hairs. Leaves lanceolate, acuminate, bordered by a few rather coarse acute teeth, contracted into a short petiole or nearly sessile, often 3 or 4 in. long, usually scabrous with very short rigid hairs or small tubercles. Flowers in dense axillary false-whorls, intermixed with subulate or linear-lanceolate bracts, of which the outer ones often exceed the calyx. Calyx 1½ lines long or rather more, with 5 acute lanceolate teeth longer than the tube. Corolla scarcely exceeding the calyx, the lobes shorter than the tube. Staminodia small, usually clavate at the end, but apparently variable; perfect stamens longer or shorter than the corolla.—Benth. in DC. Prod. xii. 178; Hook. f. Fl. Tasm. i. 282.

Queensland. Burnett river, *Daly.*
N. S. Wales. Port Jackson, *R. Brown* and others; New England, *C. Stuart.*
Victoria. Port Phillip, *R. Brown;* Wendu river, *Robertson;* near Melbourne, *F. Mueller.*
Tasmania. Derwent river, *R. Brown;* not uncommon in moist shady places, *Gunn.*
S. Australia. Third Creek, *F. Mueller.*

TRIBE III. MONARDEÆ.—Stamens 2; anthers with one linear or oblong-linear perfect cell, the other cell either entirely abortive or barren and deformed, or rarely perfect in species or genera not Australian, the connective usually elongated and filiform. Corolla usually 2-lipped. Nuts smooth or minutely granular.

10. SALVIA, Linn.

Calyx 2-lipped, the upper lip entire or with 3 minute teeth, the lower lip 2-cleft. Corolla with the upper lip erect, concave or arched, entire or scarcely notched, the lower lip spreading, 3-lobed, the middle lobe often notched or divided. Stamens 2, but easily mistaken for 4, the real filaments very short, the filiform connective appearing like branches of the filament, with a single oblong-linear anther-cell at the upper end, and at the lower end a small empty cell, usually much deformed or quite rudimentary, rarely almost perfect.—Herbs, shrubs or trees, exceedingly diversified in habit and inflorescence.

A very large genus, widely distributed over the temperate and warmer regions of the globe, although within the tropics the majority of species are mountain plants. In Australia however it is exceptionally limited to a single species, and that a common one in tropical Asia, and belonging, moreover, to an Asiatic section sparingly represented in Africa.

1. **S. plebeia,** *R. Br., Prod.* 501. An erect branching pubescent or hairy coarse annual, 1 to 2 or even 3 ft. high, the inflorescence some-

times glandular-viscid. Leaves petiolate, oblong or lanceolate, obtuse or acute, rugose, 1½ to 3 in. long. Flowers exceedingly small, in false-whorls of 6, forming branched paniculate racemes. Calyx pubescent, ovoid and 1 line long when in flower, reflexed campanulate and 2 lines long when in fruit, the upper lip entire, recurved, obtuse or obscurely pointed. Corolla scarcely longer than the calyx, the upper lip short erect and concave. Connectives of the anthers free at the lower end with a small barren anther-cell.—Benth. in DC. Prod. xii. 355.

Queensland. Burdekin river, *F. Mueller;* Rockhampton, *Thozet, O'Shanesy;* Kennedy District, *Daintree;* Darling Downs, *Lau.*

N. S. Wales. Hawkesbury, Nepean and Paterson rivers, *R. Brown;* Nepean river, *Woolls;* Argyle county, *A. Cunningham;* Hastings river, *Beckler;* New England, *C. Stuart.*

Victoria. Tambo, Snowy and Broadribbe rivers, *F. Mueller.*

The species is common in E. India, extending from Cabul to the Philippines and northward to Pekin and Japan.

Amongst the introduced plants in F. Mueller's and others herbaria are the following species of *Salvia:*—

S. verbenaca, Linn.; Benth. in DC. Prod. xii. 294, a coarse erect slightly branched perennial of 1 to 2 ft. Lower leaves petiolate, ovate, coarsely toothed or lobed, rugose, the upper ones sessile broader and shorter. Flowers small, blue, in false-whorls of about 6, forming terminal slightly branched racemes, the floral leaves reduced to heart-shaped bracts, the upper lip of the calyx with minute connivent teeth, the corolla not twice the length of the calyx, with a somewhat arched upper lip. Connectives of the anthers dilated and cohering at the lower end. A common European weed said to be naturalised near Bathurst, at Swan Hill, &c.

S. pratensis, Linn.; Benth. l.c. 289. Near *S. verbenaca,* with a similar floral structure, but the stems less leafy, the leaves being chiefly radical, and the flowers very much larger, of a deep blue, with a long arched upper lip. Also a common European plant of which there is an Australian fragment in Herb. F. Mueller, with the following.

S. sclarea, Linn.; Benth. l.c. 281. A coarse herb more branched than the preceding. Leaves mostly radical, broad, soft, very rugose and often woolly white, the bract-like floral ones broad and more or less coloured. Calyx much larger than in the preceding, more open, with the upper lip deeply 3-toothed. Corolla pale blue, rather large, the upper lip arched. A native of the Mediterranean region, established (in Victoria?) as an escape from gardens.

S. coccinea, Linn.; Benth. l.c. 343. Erect branching and almost shrubby, hoary-pubescent. Leaves rather small. Corolla scarlet, the upper lip short and erect, the lower large and broad. A North American plant, a very old inmate of gardens, said to have established itself about Rockhampton.

TRIBE IV. STACHYDEÆ.—Stamens 4, didynamous, ascending under the upper lip of the corolla, the lowest pair the longest. Anthers 2-celled. Corolla with the upper lip concave or keeled. Nuts dry, smooth or granular-tuberculate.

11. PRUNELLA, Linn.

(Brunella *of older authors.*)

Calyx 2-lipped, the upper lip flat, truncate, shortly 3-toothed, the lower with 2 lanceolate lobes. Corolla-tube as long as or longer than the calyx, the upper lip erect, short, broad, concave, nearly entire, the lower one spreading, 3-lobed. Stamens 4, in pairs, ascending under the upper lip, each filament with a small tooth under the anthers.

Anther-cells distinct, divaricate. Style bifid at the top. Nuts oblong, smooth.—Perennial herbs, usually decumbent at the base. Flowers in false-whorls of 6, forming dense terminal spikes, with bract-like floral leaves.

A genus of very few species, very widely dispersed over the temperate regions and tropical mountains of both the New and the Old World. The only Australian species is the common one over the whole range of the genus.

1. **P. vulgaris,** *Linn. ; Benth. in DC. Prod.* xii. 410. Stems procumbent or shortly creeping at the base, the flowering branches ascending sometimes to above 1 ft., more or less sprinkled as well as the foliage with short rigid hairs, rarely glabrous. Leaves petiolate, lanceolate or ovate-lanceolate, acute or obtuse, entire or somewhat toothed, 1 to 3 in. long. Flowers purplish blue or rarely white, in false-whorls of 6, forming a dense terminal spike, the bract-like floral leaves broad, shortly pointed, often coloured, about as long as the calyx. Fruiting-calyx usually about 4 lines long, the upper lip broad and flat, the lobes of the lower lip linear-lanceolate, acute, as long as the upper lip. Corolla varying from a little longer than the calyx to twice as long.—R. Br. Prod. 507 ; Hook. f. Fl. Tasm. i. 282.

N. S. Wales. Port Jackson to the Blue Mountains, *R. Brown* and others ; Hastings, Macleay, and Clarence rivers, *Beckler ;* New England, *C. Stuart.*

Victoria. Towards the Glenelg, *Robertson ;* moist shady places, Creswick Creek, Ballan, &c., *F. Mueller ;* Emu Creek, *Whan.*

Tasmania. Port Dalrymple, *R. Brown ;* common throughout the island by waysides, in pastures, &c., *J. D. Hooker.*

S. Australia. Torrens river, *F. Mueller ;* Rapide Bay, *Malpas.*

The species is common in Europe, northern Asia and North America, extending within the tropics into the mountainous regions of Asia and South America.

12. SCUTELLARIA, Linn.

Calyx divided into 2 entire lips, the upper one bearing on its back a hollow scale-like protuberance. Corolla with a rather long tube, and small nearly closed lips, the upper one concave, emarginate, the lower lip convex, spreading, emarginate at the end, the lateral lobes more frequently connate with the upper lip than with the lower. Stamens 4, in pairs, ascending under the upper lip ; anthers ciliate, those of the upper pair 2-celled, those of the lower 1-celled by the abortion of the second cell. Style with the upper stigmatic lobe exceedingly short. Nuts granular-tuberculate, raised on a short oblique stalk.—Herbs or rarely shrubs. Flowers solitary within each floral leaf, either opposite and axillary or in terminal racemes or spikes.

The genus is widely distributed over the temperate and some of the warmer regions both of the New and the Old World. The Australian species are both endemic, although one of them bears considerable resemblance to a common northern one.

Pubescent. Leaves often above 1 in. long. Corolla about 5 lines long, the lower lip much longer than the upper . . . . 1. *S. mollis.*
Glabrous or nearly so. Leaves rarely above ½ in. Corolla about 3 lines long, the lower lip rather longer than the upper . . . . 2. *S. humilis.*

1. **S. mollis,** *R. Br. Prod.* 507. A perennial with a slender creeping rootstock and weak slightly branched ascending stems of 1 ft. or more, the angles acute and pubescent. Leaves petiolate, ovate or the upper ones ovate-lanceolate, the lower ones sometimes cordate, coarsely toothed, pubescent with short hairs or more densely villous when young, the larger ones above 1 in. long, the upper floral ones gradually smaller and narrower, almost passing into bracts. Pedicels axillary, turned both to one side, 2 to 3 lines long. Calyx hirsute, scarcely as long as the pedicel. Corolla pale blue, about 5 lines long, the tube shortly exserted, the lower lip considerably longer than the upper one. —Benth. in DC. Prod. xii. 428.

N. S. Wales. Port Jackson to the Blue Mountains, *R. Brown, A. and R. Cunningham,* and others.

Victoria. Nangatta mountains, *F. Mueller.*

2. **S. humilis,** *R. Br. Prod.* 507. A perennial with a slender creeping rootstock and ascending stems, like *S. mollis,* but much smaller in all its parts and nearly glabrous or only minutely pubescent. Stems usually under 6 in. and rarely when very luxuriant nearly a foot long and then not so weak as in *S. mollis.* Lower leaves petiolate, broadly ovate or almost orbicular, usually cordate, marked with a few deep crenatures or almost lobed, rarely above ½ in. long, the lower floral ones often the largest on long petioles and almost deltoid, the upper ones gradually smaller, narrower and with shorter petioles, but none quite sessile. Pedicels axillary, both turned to one side, 1 to 3 lines long. Calyx minutely pubescent. Corolla about 3 lines long, the lower lip rather longer than the upper one.—Benth. in DC. Prod. xii. 427; Hook. f. Fl. Tasm. i. 283.

N. S. Wales. Port Jackson, *R. Brown;* Mudgee, *Woolls;* Liverpool plains, *A. Cunningham;* New England, *C. Stuart.*

Victoria. Yarra river, Darebin Creek, Bacchus marsh, &c., *F. Mueller* and others.

Tasmania. Port Dalrymple, *R. Brown;* very common in rich soil in the northern parts of the island, *J. D. Hooker.* Some of *Story's* specimens remarkably luxuriant and nearly 1 ft. high.

S. Australia. Kangaroo island, *R. Brown, Sealy.*

The species is nearly allied to the European and Asiatic *S. minor* and to the N. American *S. parvula.*

13. ANISOMELES, R. Br.

Calyx 5-nerved, 5-toothed. Corolla-tube about as long as the calyx, the upper lip erect, entire, short and somewhat concave; lower lip longer, spreading, 3-lobed, the middle lobe larger than the others, emarginate or 2-lobed. Stamens 4, in pairs, projecting from the upper lip of the corolla; anthers of the upper pair 1-celled, of the lower pair 2-celled, all the cells parallel and transverse. Nuts smooth.—Coarse herbs. Flowers in false-whorls either dense or developed into opposite cymes, all axillary or forming terminal racemes.

The genus consists of very few but very variable species, common in tropical Asia, scarcely extending into E. Africa. The Australian forms, whether regarded as one or as four or five species, are supposed to be endemic, but they approach very near to a few of the narrower-leaved E. Indian varieties of *A. ovata* and *A. Heyneana*.

1. **A. salvifolia,** *R. Br. Prod.* 503. A coarse erect herb, attaining 2 to 3 ft. or even more, very variable in indumentum and in the developement of the inflorescence, frequently hoary-tomentose or almost woolly without spreading hairs, or when the plant is greener often hispid with spreading hairs especially on the angles of the stem, the calyxes and inflorescence more or less glandular-viscid in the hispid forms, the glands less conspicuous or entirely concealed in the tomentose ones. Leaves lanceolate or ovate-lanceolate, or rarely almost ovate in the small flowered forms, coarsely toothed, the larger ones 2 to 4 in. long contracted into a petiole, the floral ones passing into small bracts, all from thick soft and densely woolly tomentose on both sides to green and pubescent only or almost glabrous. Flowers variously described as pink, blue or pale purple, in loose false-whorls generally turned to one side, the lower ones sometimes axillary, the upper ones forming interrupted or dense irregular terminal racemes or thyrsoid panicles. As the flowering advances the primary branches of each cyme often lengthen out to above ½ in., with the flowers all erect on the upper side. Pedicels very short, each in the axil of a small linear bract. Calyx usually about 2 lines when the flower first opens, but lengthening to 3, 4 or even 5 lines, the tube 5-angled, the teeth lanceolate or more or less subulate, shorter than the tube. Corolla-tube scarcely so long as the calyx and the upper lip still shorter; lower lip at least twice as long as the upper. Nuts shining.—Benth. in DC. Prod. xii. 455.

N. Australia. From the N.W. coast to the Gulf of Carpentaria.

Queensland. From Cape York to Moreton Bay, but not seen from far into the interior.

The very different aspects given to the specimens by the diversities in the indumentum must at first suggest the existence of several species, and, had we only Brown's specimens, we should without hesitation have adopted the three he has proposed; but with the very large number from various localities now before me, I am unable to assign any positive limits to any of the following :—

1. Covered with a very dense soft white or hoary tomentum, sometimes thick and almost woolly, sometimes close and cottony—the typical *A. salvifolia*, Br.—Islands of the Gulf of Carpentaria, *R. Brown;* Arnhem's Land, *M'Kinlay*; Victoria river and Sturt's Creek, *F. Mueller;* Escape Cliffs, *Halls;* Kennedy district, Queensland, *Daintree;* Nerkool Creek, *Bowman;* Rockingham Bay, *Dallachy*.

2. More loosely tomentose, the hairs scarcely spreading, the tomentum less white, passing through some specimens into the next form.—N.W. coast and Port Essington, *A. Cunningham;* Sweers island, *Henne :* and the commonest form throughout Queensland, not far from the coast, and in the adjoining islands, *A. Cunningham, Fraser, M'Gillivray, Bidwill, F. Mueller, Henne, Dallachy, Bowman,* and many others.

3. Greener, pubescent, or nearly glabrous, scarcely tomentose, but without spreading hairs, the inflorescence usually looser and the flowers smaller than in the densely tomentose forms.—Rockingham Bay, *Dallachy ;* Trinity island, *M'Gillivray*. To this form I should refer *A. moschata* and *A. inodora*, R. Br. Prod. 503, Benth. in DC. Prod. xii. 455, the former from Keppel and Shoalwater Bays, leading towards the common Queensland form (n. 2), the latter from Prince of Wales island and Arnhem N. Bay, more nearly glabrous than any other, all very near to the E. Indian *A. Heyneana*.

4 Scarcely tomentose, green, more or less hirsute with spreading hairs.—Rockingham Bay, *Dallachy ;* Gilbert river, *F. Mueller ;* the latter specimens more hispid still than Dallachy's.

14. LEUCAS, R. Br.

Calyx erect, straight or curved and oblique at the top, 8- to 10-ribbed, 8- to 10-toothed. Corolla-tube not longer than the calyx, the upper lip erect, concave, entire or rarely notched, very villous outside; lower lip spreading, 3-lobed. Stamens 4, ascending in pairs under the upper lip; anthers 2-celled. Style with the upper stigmatic lobe much shorter than the lower. Nuts dry, smooth, triangular, obtuse.—Herbs or undershrubs. Flowers in axillary false-whorls, white or rarely purplish.

A considerable genus, spread over tropical and subtropical Asia and Africa. The only Australian species is closely allied to, if not identical with, one of the Asiatic forms.

1. **L. flaccida,** *R. Br. Prod.* 505. An annual with a hard branching decumbent base and ascending or erect flowering branches, often virgate or wiry and above a foot long, the whole plant pubescent or tomentose with soft appressed hairs. Leaves shortly petiolate, ovate, crenate, ½ to 1 in. long, the upper floral ones small. Flowers 6 to 20 or even more together in axillary false-whorls, the pedicels exceedingly short, the subtending floral leaves usually exceeding the flowers, the bracts within the false whorls very small. Calyx about 3 lines long when in flower and not much enlarged afterwards, straight, 10-ribbed, with 10 short softly subulate teeth, all equal or the alternate ones rather smaller. Corolla white (or blue according to Dallachy), not half as long again as the calyx.—Benth. in DC. Prod. xii. 526.

Queensland. Endeavour river, *Banks and Solander;* Rockingham Bay, *Dallachy;* Rockhampton, *Bowman, O'Shanesy.*

A few specimens of apparently the same species have been received from the Eastern Archipelago and from Burmah, and the whole may not be specifically distinct from the common Pacific island *L. decemdentata*, Sm. (*Stachys decemdentata*, Forst.), which however has much smaller calyxes with shorter teeth.

Var.? *petiolaris.* Petioles longer than the calyx, as in *L. decemdentata*, but the fragmentary specimens appear to be in an abnormal state, and the calyxes are those of *L. flaccida.*—Cape York, *M'Gillivray.*

TRIBE V. PROSTANTHEREÆ. — Stamens 4, in pairs, all bearing anthers, but sometimes one cell of all the anthers or both cells of the lower pair sterile or abortive. Nuts prominently reticulate-rugose or rarely nearly smooth, the pericarp rather thick. Seeds albuminous. Calyx usually 10- or 13-nerved, very rarely 15-nerved. Corolla with a broad throat, the upper lip erect, concave or nearly flat, broad and emarginate or 2-lobed; lower lip spreading, 3-lobed, longer than or very rarely shorter than the upper lip.

R. Brown describes the albumen of *Prostanthera* as "parcum aut nullum." I have found it present in the seeds of a considerable number of species of the different genera of the tribe which I have now examined, sometimes scanty, often rather copious, although never perhaps so thick as is figured by Labillardière. F. Mueller has also represented the albumen of *Prostanthera spinosa* rather thicker than I observed it, but the exact proportion of the embryo to the albumen may vary in different seeds of the same plant, according to the condition under which they ripened.

15. PROSTANTHERA, R. Br.

(Chilodia and Cryphia, *R. Br.* Klanderia, *F. Muell.*)

Calyx-tube usually striate, the limb 2-lipped, the lips entire or the lower one slightly emarginate. Corolla-tube short, dilated into a broad campanulate throat, the upper lip erect, usually short, slightly concave, broadly 2-lobed; the lower lip spreading, 3-lobed, longer or in one section shorter than the upper lip, the middle lobe larger and usually emarginate or 2-lobed. Stamens 4, in pairs; anthers 2-celled, the connective prominent at the back, sometimes cristate and usually produced into one or two linear appendages adnate at the base or in their whole length to the back of the cells, most frequently one of them rarely both free at the end or in a great part of their length, sometimes produced beyond the cell and tipped with a crest or tuft of short points or hairs, but sometimes both appendages very short or obsolete, the anther-cells themselves usually tipped by short points, not crested, and distinct from, although sometimes mistaken for, the ends of the connective-appendages. Style shortly bifid at the end. Nuts reticulate-rugose, attached either obliquely at the base or adnate higher up. Seeds albuminous.—Shrubs or undershrubs studded with resinous glands and usually strongly scented. Leaves opposite. Flowers solitary in the axils of the stem-leaves, or opposite in terminal racemes, the floral leaves reduced more or less to deciduous bracts. Pedicels with a pair of bracts usually close under the calyx.

The genus is limited to Australia, the greater number of the species are extratropical and two only are natives of West Australia.

SECT. I. **Euprostanthera.**—*Corolla-throat short and broad, upper lip short, very broad, erect, lower lip much longer with a large spreading middle lobe. Calyx-lips usually closed over the fruit.*

SERIES 1. **Racemosæ.**—*Flowers in terminal racemes, the floral ones all or mostly reduced to membranous or broad acuminate or very deciduous bracts.*

Leaves mostly above 1 in. long, on rather long petioles, flat or nearly so.
  One anther-appendage much longer than the cell. Corolla pubescent.

Leaves mostly 1 to 2 lines long, not so thick, crenate, with
slightly revolute margins. Plant pubescent or hirsute . . 9. *P. violacea.*
Leaves on short petioles, with revolute margins.
  Leaves mostly under ¼ in., ovate, crenate, very rugose. Plant
hirsute. Anther-appendages shorter than the cells . . . . 10. *P. incana.*
  Leaves ½ to 1 in., ovate lanceolate or linear, entire, scarcely
rugose. Plant hirsute. Both anther-appendages shortly exceeding the cells . . . . 11. *P. hirtula.*
  Leaves from ¼ in. and ovate to 1 in. and linear entire, sometimes
echinate. Plant pubescent or nearly glabrous. One anther-
appendage exceeding the cell . . . . 12. *P. denticulata.*

Series 2. **Convexæ.**—*Leaves small or narrow, sessile or shortly petiolate, convex
or with revolute margins when dry (nearly flat in* P. cuneata*). Flowers axillary, the
floral leaves similar to the stem-ones or rather smaller.*

Leaves obovate ovate or orbicular, mostly under ¼ in. long.
  Pubescent-hirsute. Leaves ovate.
    Leaves very rugose, crenate. Anther-appendage not ex-
ceeding the cell . . . . 13. *P. rugosa.*
    Leaves not rugose, entire. Anther-appendage half as long
again as the cell . . . . 14. *P. marifolia.*
  Slightly pubescent. Leaves orbicular or rhomboidal. Anther-
appendage not exceeding the cell . . . . 15. *P. rhombea.*
  Branches pubescent. Margins of the leaves very slightly recurved. Anther-appendage twice as long as the cell.
    Spinescent with opposite slender spines of ½ in. Leaves ovate
often complicate and recurved . . . . 16. *P. spinosa.*
    Unarmed. Leaves obovate or cuneate, nearly flat . . . . 17. *P. cuneata.*
Leaves linear, ¼ to 1 in. long.
  Calyx glabrous inside or nearly so. Anther-appendage twice as
long as the cell. Plant glabrous or nearly so . . . . . . 18. *P. linearis.*
  Calyx with a raised transverse pubescent line inside at the base
of the upper lip. Anther-appendage short or none.
    Glabrous or nearly so. Leaves smooth, mostly ½ in. or more.
Anther-appendage nearly as long as the cell . . . . . 19. *P. phylicifolia.*
    Pubescent or hirsute. Leaves about ¼ in. long, scabrous, de-
cussate. Anther-appendage nearly as long as the cell . . 20. *P. decussata.*
    Glabrous or nearly so. Leaves smooth, about ⅛ in. long.
Anther-appendage quite obsolete . . . . 21. *P. empetrifolia.*

Series 3. **Subconcavæ.**—*Leaves (small or narrow), sessile or very shortly petiolate, concave, or with incurved margins or flat, the margins never recurved. Flowers
axillary, the floral leaves similar to the stem ones. Anther-appendage about twice as
long as the cell.*

Leaves narrow, quite entire (above ¼ in.). Plant glabrous or
hoary with white appressed hairs.
  Leaves oblong-linear or lanceolate, mostly ½ to 1 in. long.
Branches white.
    Calyx lips nearly equal . . . . 22. *P. lithospermoides.*
    Calyx upper lip much longer than the lower one . . . . 23. *P. Behriana.*
  Leaves linear-terete, channelled above. Western species.
    Calyx upper lip much larger than the lower . . . . . 24. *P. Baxteri.*
    Calyx-lips nearly equal . . . . 25. *P. canaliculata.*
  Leaves very narrow-linear and nearly 1 in. long, or linear-
oblong and much shorter. Eastern species.
    Calyx upper lip large and membranous, lower much smaller.
Leaves mostly very narrow.
      Calyx upper lip broad, under 5 lines long . . . . . 26. *P. nivea.*
      Calyx upper lip ovate, ½ in. long . . . . 27. *P. striatiflora.*
    Calyx-lips small, nearly equal. .Leaves mostly linear-oblong 28. *P. saxicola.*

Leaves entire or toothed, usually small and narrow. Plant more
   or less glandular-pubescent or viscid.
Calyx upper lip larger than the lower. Plant very viscid-
   pubescent. Leaves mostly about 3 lines long.
   Leaves entire . . . . 29. *P. odoratissima.*
   Leaves prominently toothed or pinnatifid . . . . 30. *P. euphrasioides.*
Calyx-lips nearly equal. Plant slightly viscid-pubescent.
   Leaves about 2 lines, entire or slightly toothed . . . . 31. *P. cryptandroides.*
Leaves under 2 lines long, ovate or oblong. Plant nearly glabrous.
   Calyx-lips nearly equal . . . . 32. *P. eurybioides.*

SECT. II. **Klanderia.**—*Corolla-tube incurved, dilated upwards, the upper lip erect and concave, the lower lip spreading, shorter or not longer. Calyx-lips usually equally open. Anther-appendage small or none.*

Leaves petiolate, flat, rather thick, above ¼ in. and often ½ in.
   long. Corolla twice as long as the calyx.
   Leaves orbicular . . . . 33. *P. ringens.*
   Leaves oblong or scarcely obovate . . . . 34. *P. Leichhardtii.*
Leaves scarcely petiolate, small, the margins recurved.
   Pedicels very short. Corolla not half as long again as the
      large calyx . . . . 35. *P. calycina.*
   Pedicels longer than the calyx. Corolla twice as long as the
      calyx . . . . 36. *P. chlorantha.*
Leaves under 2 lines long, oblong, recurved, thick, the margins
   not recurved . . . . 37. *P. microphylla.*
Leaves linear-terete, with incurved margins, 1 to 4 lines long . 38. *P. aspalathoides.*

SECT. 1. EUPROSTANTHERA.—Calyx with a distinctly striate tube, the upper lip after flowering usually slightly turned back, the lower lip turned up against it, closing the orifice of the tube. Corolla with the tube very shortly narrow at the base, the throat very broadly campanulate and oblique, the upper lip short, broad, erect, emarginate or broadly 2-lobed, the lower lip larger, with 3 broad spreading lobes, the middle one larger, notched or 2-lobed. Stamens usually concealed in the tube (short or incurved) or not very prominent.

The corolla in this section, although varying in size and in a slight degree in the proportion of the lobes, appears, as far as can be judged from dried specimens, remarkably uniform in general shape. The degree of development of the anther-appendages, although generally constant in species, does not appear to agree with other characters sufficiently to be available for subsectional groups.

SERIES 1. RACEMOSÆ.—Flowers in terminal racemes, the floral leaves, or at least the upper ones reduced to bracts, either small and membranous, or ovate and acuminate and very different from the stem-leaves, or in most cases so deciduous as to be rarely observable when the plant is in flower.

1. **P. lasianthos,** *Labill. Pl. Nov. Holl.* ii. 18, t. 157. A tall shrub, sometimes attaining the dimensions of a moderate sized tree, glabrous except the flowers. Leaves petiolate, usually oblong-lanceolate, rather acute, serrate, flat or the margins recurved, of a firm consistence, not rugose, dark green above, pale or glaucous and minutely dotted underneath, 2 to 3 in. long; in some Victorian specimens shorter broader more entire and almost coriaceous. Flowers opposite in pairs,

in short leafless racemes, forming a terminal panicle often leafy at the base, but the leaves under the upper racemes very much reduced and those under the pairs of flowers entirely abortive. Pedicels short. Bracts linear, shorter than the calyx and sometimes very minute. Calyx slightly pubescent, attaining 3 lines when in fruit, the tube obscurely striate, the upper lip broad, the lower rather smaller. Corolla "white tinged or spotted with pink" or "pale blue" hairy inside and out, often ½ in. long, the lobes very broad. Anthers with the longer appendage about twice as long as the cell, the other short and adnate. —Br. Prod. 508; Benth. in DC. Prod. xii. 559; Hook. f. Fl. Tasm. i. 283; Andr. Bot. Rep. t. 641; Bot. Reg. t. 143; Bot. Mag. t. 2434.

N. S. Wales.—Blue Mountains, *R. Brown* and others; New England, *C. Stuart;* southward to Illawarra, *Shepherd,* and Twofold Bay, *Mossman, F. Mueller,* and others.

Victoria.—Banks of streams, &c., near Melbourne, *Adamson;* Dandenong ranges, Mount Disappointment, and various parts of Gipps Land, *F. Mueller.*

Tasmania.—Derwent river and Port Dalrymple, *R. Brown;* common by the margins of forests, banks of streams, &c., *J. D. Hooker.*

Var. *subcoriacea,* F. Muell., leaves smaller and firmer, flowers rather smaller.—Grampians, *Wilhelmi.*

2. **P. prunelloides,** *R. Br. Prod.* 508. A tall shrub, quite glabrous except the flowers or minutely scaly-pubescent, the angles of the branches often prominent, and sometimes crisped or denticulate. Leaves on rather long petioles, ovate, obtuse, entire or with a few coarse irregular teeth, of the firm consistence of those of *P. lasianthos,* 1 to 2 in. long. Flowers in simple terminal racemes, or with one pair of branches at the base. Floral leaves reduced to broad membranous obtuse concave ciliate bracts, about as long as the calyx, and enclosing it in the young bud, but falling off long before the flowering. Pedicels short, with a pair of very deciduous linear-lanceolate bracts under the calyx. Calyx and corolla of *P. lasianthos,* or the latter rather less hairy. Anthers with one appendage about twice as long as the cell, the other very short or obsolete.—Benth. in DC. Prod. xii. 559.

N. S. Wales. Blue Mountains, *R. Brown, A. Cunningham* and others.

3. **P. cœrulea,** *R. Br. Prod.* 508. A tall shrub, glabrous or minutely scaly-pubescent, intermediate in foliage between *P. lasianthos* and *P. prunelloides,* differing from both in the glabrous corolla and short antherappendages, the angles of the branches sometimes prominent. Leaves petiolate, in the typical form ovate-lanceolate or lanceolate, slightly serrate, 1 to 2 in. long, and much like those of *P. lasianthos,* in the more northern specimens nearer to those of *P. prunelloides.* Flowers in simple terminal racemes or with one pair of branches at the base, the floral leaves reduced to ovate concave acuminate membranous bracts, falling off from the very young bud. Bracts under the calyx very small or obsolete. Calyx of *P. lasianthos;* corolla rather smaller than in that species, "blue" and quite glabrous. Anthers with one appendage very shortly free at the end, and about as long as the cell, the other much shorter and adnate.—Benth. in DC. Prod. xii. 559.

N. S. Wales. Grose river, *R. Brown;* shaded ravines, Wollondelly river. *A. Cunningham,* (both with narrow leaves) ; New England, *C. Stuart* with the ovate leaves of *P. prunelloides.*

4. **P. ovalifolia,** *R. Br. Prod.* 509. A densely bushy strong-scented shrub, more or less hoary with a minute appressed pubescence. Leaves petiolate, ovate to oblong, obtuse, entire, rather thick and flat, rarely exceeding ½ in. and mostly smaller. Flowers rather small, in short loose terminal racemes, the floral leaves small, bract-like and deciduous, or the lower pair more leaf-like. Calyx not above 2 lines long, the lips nearly equal and both entire. Corolla "purple," about 4 or 5 lines long, slightly pubescent outside. Anthers with both appendages adnate nearly to the end, and not projecting beyond the cells. —*P. atriplicifolia,* A. Cunn. in Benth. Lab. 451, and in DC. Prod. xii. 560.

Queensland. Shoal bay passage, *R. Brown;* Wide Bay, *Bidwill.*
N. S. Wales. Mount Lindsay, *W. Hill;* barren hills S.W. of Lachlan river, *A. Cunningham.* Some specimens from Mudgee, *Woolls,* belong also probably to this species, unless indeed they represent an entire-leaved variety of *P. incisa.*

Var.? *latifolia.* Leaves broadly ovate or almost orbicular and occasionally with one or two slight crenatures, very much larger than in *P. rotundifolia,* and as much smaller than in *P. cærulea*—*P. ovalifolia,* Benth. in DC. Prod. xii. 560.—N.W. interior of N. S. Wales, *Fraser;* head of Hastings river, *C. Moore.*

5. **P. melissifolia,** *F. Muell. Fragm.* i. 19. A slender loosely branched shrub, with the habit and short hoary pubescence of *P. incisa* and *P. Sieberi,* of which it may be a variety as suggested by F. Mueller. Leaves much larger than in those species, usually ovate, 1 to 1½ or even 2 in. long, rather thin, coarsely toothed. Racemes longer and looser than in *P. incisa,* all leafless. Calyx-lobes rather larger than in that species, nearly equal, the upper one entire, the lower one emarginate or 2-lobed. Corolla about twice as long as the calyx. Anthers with the appendages adnate almost to the end and not exceeding the cells.

Victoria. Ranges near Cape Otway and Port Phillip, *F. Mueller.*

6. **P. incisa,** *R. Br. Prod.* 509. A slender much branched shrub, quite glabrous or more frequently slightly hoary with a minute pubescence. Leaves from ovate-lanceolate to oblong, obtuse, usually bordered by a few coarse teeth, contracted into a rather long petiole, rather thick and flat in some specimens, thinner with the margins slightly recurved in others, green above, pale underneath, mostly ½ to 1 in. long. Flowers rather small, in short but slender terminal racemes, the lowest pair of floral leaves sometimes like the stem leaves but smaller, the others reduced to small bracts falling off before the flowering. Pedicels slender, but shorter than the calyx ; bracts small. Calyx about 2 lines long or scarcely 3 lines when in fruit, the tube rather broad, striate, pubescent, the upper lip very broad, entire; the lower lip longer, narrower, obtuse, and usually emarginate. Corolla expanding to about 5 lines diameter, the lobes all broad. Anthers with both appendages adnate nearly or the shorter one quite to the end, and neither of them exceeding the cell.—Benth. in DC. Prod. xii. 559.

N. S. Wales. Grose river, *R. Brown ;* Blue Mountains, *Caley, Miss Atkinson ;* Avoca valley, Blue Mountains, *Wilhelmi ;* Lansdown river and Port Macquarrie, *C. Moore ;* Hastings river, *Beckler.*

There are two common forms of this species, arising perhaps from the degree of exposure of their stations, the one with rather thick less-toothed leaves seems almost to pass into *P. ovalifolia,* the other with thinner paler more cut leaves. Both are in Brown's herbarium, but chiefly the former. The comparative size and shape of the two calyx-lips appears to be variable.

Var. ? *pubescens,* F. Mueller. The whole plant very pubescent. Leaves more obovate or cuneate. Anthers with one appendage rather more prominent (in the only flower examined) the calyx and other characters those of the typical form. Possibly however a distinct species.—Forest rivulets near Twofold Bay, *F. Mueller ;* Port Macquarrie, *C. Moore.*

7. **P. Sieberi,** *Benth. Lab. Gen. et Sp.* 451, *and in DC. Prod.* xii. 559. A tall slender much branched shrub, slightly pubescent and closely resembling the thin-leaved forms of *P. incisa,* and perhaps a variety. Leaves usually more deeply toothed and more contracted at the base, the racemes shorter and more leafy, and the calyx-lips both broad and very nearly equal.—*P. incisa,* Sieb. Pl. Exs. not of R. Br.

N. S. Wales. Port Jackson, *Sieber. n.* 189 ; shady woods on the coast from Port Jackson to the Illawarra, *A. Cunningham.* It remains to be ascertained how far the characters separating this from *P. incisa,* derived chiefly from the calyx, may prove constant.

8. **P. rotundifolia,** *R. Br. Prod.* 509. A tall very bushy shrub, attaining sometimes 6 or 7 ft., the branches very shortly hoary-pubescent, the foliage nearly glabrous. Leaves broadly ovate orbicular or spathulate, on rather long petioles, very obtuse, entire or with a few large crenatures, all under ½ in. long and often all under ¼ in. Flowers in short close terminal racemes, the lower ones sometimes in the axils of leaves like the stem ones, but the upper floral leaves always reduced to small deciduous bracts. Pedicels shorter than the calyx, with linear deciduous bracts. Calyx about 2 lines long when in flower, somewhat enlarged afterwards, the tube striate, the lips broad and nearly equal. Corolla rather larger than in *P. ovalifolia.* Anthers with both appendages adnate nearly to the end, and not protruding beyond the cells.— Benth. in DC. Prod. xii. 560, Hook. f. Fl. Tasm. i. 284, t. 89 ; *P. retusa,* R. Br. Prod. 509 ; Benth. in DC. Prod. xii. 560 ; *P. cotinifolia,* A. Cunn. in Benth. Lab. Gen. et Sp. 452, and in DC. Prod xii. 560.

N. S. Wales. Barren rocky country W. of Wellington valley, *A. Cunningham ;* Lower Macquarrie river, *Bowman.*

Victoria. Buffalo Range, Bacchus Marsh, Mount Zero, Avon and Genoa rivers, *F. Mueller ;* Mount Arepiles, *Dallachy.*

Tasmania. Port Dalrymple, *R. Brown ;* abundant on N. and S. Esk rivers, *J. D. Hooker.*

S. Australia. S. E. part of the colony, *J. E. Woods.*

Brown's specimens of *P. retusa* differ from the typical *P. rotundifolia* but very slightly, in the leaves more constantly crenate.

9. **P. violacea,** *R. Br. Prod.* 509. A slender divaricately branched twiggy shrub, pubescent, with very short but rigid hairs. Leaves very small, shortly but distinctly petiolate, broadly ovate or orbicular, more or less crenate, with revolute margins, rarely exceeding 2 lines and often not above one line long. Flowers usually blueish purple, in 2 or 3 pairs,

forming little compact terminal racemes, the small bract-like floral leaves very deciduous. Calyx shortly pubescent, 1½ to 2 lines long, the tube striate, the upper lip very broad, the lower rather longer and narrower, both usually quite entire. Corolla not twice as long as the calyx, and sometimes scarcely exceeding it, the throat very broad. Anthers with both appendages adnate, one shortly free at the end, but shorter than the cell.—Benth. in DC. Prod. xii. 563; Bot. Reg. t. 1072; *P. retusa,* Sieb. Pl. Exs. not of R. Br.

N. S. Wales. Port Jackson to the Blue Mountains, *R. Brown, Sieber, n.* 199 and others.

Var. *albiflora.* Corolla white, but no other difference.—*P. thymifolia,* A. Cunn. in Benth. Lab. Gen. et Sp. 455, and in DC. Prod. xii. 563.—Springwood, Blue Mountains, growing with the typical form, *A. Cunningham.*

10. **P. incana,** *A. Cunn. in Benth. Lab. Gen. et Sp.* 455, *and in DC. Prod.* xii. 563. A handsome shrub of 5 or 6 ft., more densely hirsute and more robust in all its parts than *P. violacea* and *P. rugosa,* some varieties of which it sometimes resembles. Leaves on very short petioles, ovate, prominently crenate, bullate-rugose, with recurved margins, 4 to 6 lines long, the lower floral ones similar but smaller. Flowers small, in several pairs crowded together at the ends of the branches into short racemes leafy at the base, the upper floral leaves reduced to small very deciduous bracts. Pedicels short. Calyx hirsute, about 2 lines long, both the lips broad and entire, the lower one scarcely longer than the upper. Corolla scarcely twice as long as the calyx. Anthers with both appendages adnate, one shortly free, but shorter than the cell, the other still shorter.

N. S. Wales. Rocky ridges, Nepean river, *A. Cunningham.* Some imperfect specimens from Bent's river, *Woolls,* probably belong to the same species.

11. **P. hirtula,** *F. Muell.* A shrub of 3 to 5 ft., pubescent or hirsute with the rigid hairs of *P. marifolia,* to which this species is nearly allied, differing in inflorescence and in the larger leaves. Leaves very shortly petiolate, ovate-lanceolate oblong or almost linear, obtuse, entire, with revolute margins, usually hirsute above and whitish underneath, scarcely or not at all rugose, rarely under ½ in. and often nearly 1 inch long, the floral ones smaller, the upper ones reduced to small deciduous bracts. Flowers larger than in *P. marifolia,* in the upper axils, forming an interrupted terminal more or less leafy raceme, or sometimes nearly all axillary. Pedicels short; bracts small and setaceous. Calyx hirsute, 2½ to 3 lines long, both the lips broad and nearly equal, entire or the lower one retuse. Corolla nearly twice as long as the calyx, glabrous or slightly hairy. Anthers with both the appendages shortly exceeding the cells.

Victoria. Buffalo Range, Mount Disappointment, *F. Mueller;* Grampians, *Wilhelmi.*

Var. *angustifolia.* Leaves narrow and rather less hirsute.—Genoa Peak, *F. Mueller*

12. **P. denticulata,** *R. Br. Prod.* 509. A robust shrub, with virgate or long and loose sometimes slender but rigid branches, pubescent

with short crisped hairs. Leaves sessile or nearly so, from $\frac{1}{4}$ in. or under and broadly lanceolate to $\frac{1}{2}$ in. long or more and narrow-linear, obtuse or almost acute, with recurved margins, not toothed but often bearing on the upper surface near the margin a few short rigid hairs on raised tubercles resembling minute prickles, all very spreading, the floral ones passing into ovate acuminate coloured bracts. Flowers usually in distant pairs, forming interrupted terminal racemes often leafy at the base. Pedicels short. Bracts narrow-linear, close under the calyx. Calyx more or less hirsute, 2 to 3 lines long, the lips broad, entire, nearly equal or the lower one rather smaller. Corolla glabrous or nearly so. Anthers with one appendage nearly half as long again as the cell, the other short and adnate.—Benth. in DC. Prod. xii. 561.

N. S. Wales. Port Jackson, *R. Brown, Woolls,* and others; granitic ranges W. of Bathurst, *A. Cunningham.*

Victoria. Buffalo Range, Futter's Range, Bendigo, Grampians, *F. Mueller;* Wimmera, *Dallachy.*

The name is unfortunate as the leaves are not toothed, and it is only occasionally that the asperities on the upper surface give them a denticulate appearance. Most of the Port Jackson specimens have rather broad lanceolate leaves, Cunningham's have narrow and long linear leaves, Brown's are intermediate. The Victorian and a few of the Port Jackson specimens have the leaves mostly under $\frac{1}{4}$ in., with few or none of the tubercular prickles. The species is remarkable for the acuminate bract-like floral leaves, and like *P. hirtula* forms almost a passage from the racemose to the axillary inflorescences.

SERIES 2. CONVEXÆ.—Leaves small or narrow, sessile or shortly petiolate, convex or with revolute margins at least when dry (in *P. cuneata* and *P. spinosa,* the margins usually flat but occasionally narrowly recurved). Flowers axillary, the floral leaves all similar to the stem ones or rather smaller, and the flowering branch usually growing out beyond the flowers.

13. **P. rugosa,** *A. Cunn. in Benth. Lab. Gen. et Sp.* 456, *and in DC. Prod.* xii. 563. A robust divaricately-branched shrub, pubescent or hirsute with short rigid hairs. Leaves small, sessile or nearly so, ovate, crenate, with revolute margins, very rugose, from $1\frac{1}{2}$ to 4 lines long, the floral ones all similar though sometimes smaller. Flowers small, not numerous, all axillary, scattered along the branches or rarely 2 or 3 pairs together near the ends. Pedicels very short, the bracts very small. Calyx not exceeding 2 lines, hispid, the upper lip broad, short, distinctly or obscurely 3-toothed, the lower lip longer, usually emarginate. Corolla not twice as long as the calyx. Anthers with both appendages adnate, one shortly free at the end but shorter than the cell, the other still shorter.

N. S. Wales. Mountainous country bordering on Hunter's River, *A. Cunningham;* Monkey Creek, *Woolls.*

14. **P. marifolia,** *R. Br. Prod.* 509. An undershrub with twiggy branches, pubescent or hirsute with short rigid hairs. Leaves sessile or very shortly petiolate, ovate or ovate-lanceolate, obtuse, entire, with revolute margins, scabrous-hispid above but not rugose, whitish under-

neath, 2 to 4 lines long. Flowers all axillary, but sometimes forming interrupted leafy racemes, the floral leaves all like those of the stem. Pedicels very short; bracts subulate. Calyx more or less hirsute, 2 to 2½ lines long, the lips broad, nearly equal and usually entire, often assuming a blueish tint. Corolla not twice as long as the calyx, glabrous or .sparingly hirsute. Anthers with one appendage about half as long again as the cell, the other short and adnate.—Benth. in DC. Prod. xii. 562.

N. S. Wales. Port Jackson to the Blue Mountains, *R. Brown, A. Cunningham, Woolls* and others.

15. **P. rhombea,** *R. Br. Prod.* 509. A shrub or undershrub, with long divaricate almost terete branches more or less pubescent. Leaves nearly sessile, orbicular or almost rhomboidal, glabrous or sparingly ciliate, entire, with revolute margins and almost bullate, 2 to 3 lines diameter. Flowers small, in the upper axils, on very short pedicels. Calyx usually not 2 lines long, shortly pubescent and very glandular, the lips nearly equal, the upper one very broad, the lower one narrower. Corolla not twice as long as the calyx, glabrous. Anthers with the appendages adnate, the longer one very shortly free at the end but shorter than the cell, the other still shorter.—Benth. in DC. Prod. xii. 563.

N. S. Wales. Blue Mountains, *R. Brown, Woolls;* Illawarra, *Shepherd.*

16. **P. spinosa,** *F. Muell. in Hook. Kew Journ.* viii. 168, *in Trans. Phil. Soc. Vict.* i. 48 *and Pl. Vict.* ii. t. 56. A rigid but slender divaricate shrub with hirsute branches and remarkable for the numerous small branchlets reduced to opposite divaricate spines, about ½ in. long, either leafless or with a pair of small leaflets at their base. Leaves very small, shortly petiolate, ovate, obtuse, entire, complicate and recurved, the margins usually slightly revolute, rarely above 2 lines long. Pedicels axillary, 4 to 8 lines long, with minute setaceous bracts above the middle. Calyx more or less hirsute, 2 to 3 lines long, the lips not very broad, entire and nearly equal. Corolla "lilac," slightly hairy outside, not twice as long as the calyx. Anthers with one appendage about twice as long as the cell, the other short and adnate.

Victoria. About springs and on irrigated rocks in the Grampians, *F. Mueller, Wilhelmi;* summit of Mount Arapiles (with very hirsute leaves), *Dallachy.*
S. Australia. Tattiara country, *J. E. Woods;* scrub near Wallan's Hut and Cygnet Bay, Kangaroo Island, *Waterhouse.*

17. **P. cuneata,** *Benth. in DC. Prod.* xii. 560. A much branched spreading shrub of 2 or 3 ft., more or less pubescent or villous with short crisped hairs. Leaves sessile or nearly so, often crowded on the short branchlets, obovate-cuneate or almost orbicular, obtuse, entire or crenate at the end, flat or recurved and complicate, the margins often slightly revolute, rather thick, glabrous or pubescent, mostly about 2 lines, rarely above 3 lines long, the floral ones scarcely smaller. Flowers all axillary, but sometimes crowded into terminal leafy racemes.

Pedicels very short. Bracts linear, ciliate, often as long as the calyx-tube. Calyx 2½ to 3 lines long or even longer when in fruit, the tube prominently striate, the lips nearly equal, at least as long as the tube, broad and entire, or the lower one emarginate. Corolla white with purple spots, nearly glabrous, twice as long as the calyx. Anthers with one appendage about twice as long as the cell, the other short and adnate.—Hook. f. Fl. Tasm. i. 284. t. 90.

Victoria. Summit of many of the Australian Alps, Haidinger and Munyong ranges, Mount Kosciusko and others, at an elevation of 4000 to 7000 ft., *F. Mueller.*

Tasmania. Sterile gravelly soil on the S. Esk river, abundant, *Gunn, Archer* and others.

18. **P. linearis,** *R. Br. Prod.* 509. A tall erect shrub, glabrous or slightly pubescent. Leaves nearly sessile, linear, obtuse, entire, the margins more or less revolute in drying, ½ to above 1 in. long, the upper floral ones similar but smaller. Flowers all axillary, but the upper ones sometimes forming terminal interrupted leafy racemes. Pedicels short. Bracts small, filiform. Calyx glabrous or slightly pubescent, about 2 lines long or longer when in fruit, the lips not very broad, nearly equal, entire, without any or only a very slight trace of the transverse downy line of the three following species. Corolla sprinkled with a few hairs or hairy all over, about twice as long as the calyx. Stamens longer than in most species of this section, and the anther-cells more divergent; the longest appendage nearly twice as long as the cell, the other short and adnate.—Benth. in DC. Prod. xii. 561.

N. S. Wales. Port Jackson to the Blue Mountains, *R. Brown, A. Cunningham* and others.

19. **P. phylicifolia,** *F. Muell. Fragm.* i. 19. A robust bushy shrub sometimes small but attaining often several ft., glabrous or hoary-pubescent with very short somewhat crisped hairs. Leaves sessile or nearly so, oblong-linear, obtuse, entire, with revolute margins, usually thicker and broader than in *P. linearis,* in some specimens all under ½ in., in others ½ to ¾ in. long. Flowers all axillary. Pedicels shorter than the calyx, with linear-setaceous bracts close under the calyx, or at some distance from it. Calyx 2 to 2½ or rarely 3 lines long, the tube prominently striate, the lips ovate, the upper one with a transverse rather broad cottony line inside at the base, the lower one at first nearly equal to, at length much smaller than, the upper one. Corolla "whitish," glabrous or very sparingly pubescent, nearly twice as long as the calyx. Anther-appendages short and adnate or one of them with a small point not cristate and scarcely exceeding the cell.

Queensland. Glass-houses, *F. Mueller,* a single specimen in the Hookerian herbarium.

N. S. Wales. New England, *C. Moore.*

Victoria. Mount M'Farlane, Mitta-Mitta mountains, rocks at Maneroo, *F. Mueller.*

20. **P. decussata,** *F. Muell. Fragm.* i. 126. A robust shrub of few feet, with numerous short leafy branches, pubescent or hirsute with crisped or spreading hairs. Leaves sessile or nearly so, crowded and

decussate on the smaller branches but not clustered in the axils, linear, obtuse, with revolute margins, somewhat coriaceous, scabrous-hirsute with minute almost prickle-like hairs like those of *P. denticulata*, mostly about ¼ in. long. Flowers all axillary, on very short pedicels. Bracts linear, half as long as the calyx. Calyx short and broad, strongly ribbed, glandular-hirsute, scarcely above 2 lines long when in fruit, the lips nearly orbicular and equal, with a transverse downy curved line inside at the base of the upper one. Corolla shortly exceeding the calyx (perhaps not fully developed), glabrous or nearly so. Anther-appendages adnate, the longer one very shortly free at the end, and not at all or scarcely exceeding the cell.

Victoria. Rocky summits of the M'Alister range and Mount Mueller, *F. Mueller*.

21. **P. empetrifolia,** *Sieb. in Spreng. Syst. Cur. Post.* 226. An erect much branched but rather slender shrub, glabrous or sprinkled with a few short appressed hairs. Leaves sessile, linear, acute, entire, with revolute margins, rarely above ½ in. long. Flowers all axillary. Pedicels short, with a pair of linear bracts close under the calyx. Calyx about 2 lines long, the tube prominently striate, the lips broad and about equal, the upper one with a prominent transverse downy line inside at the base, the lower one often emarginate. Corolla fully twice as long as the calyx. Anthers with the connective prominent at the back, but without any or only very rudimentary appendages. Upper lobe of the style short.— *Chilodia scutellarioides*, R. Br. Prod. 507; Benth. in DC. Prod. xii. 558; Bot. Mag. t. 3405.

N. S. Wales. Port Jackson to the Blue Mountains, *R. Brown, Sieber, n.* 187 and many others.

The genus *Chilodia* was founded mainly upon the absence of the anther-appendages, the degree of development of which is very different in different species of *Prostanthera;* they are quite obsolete also in some species of the section *Klanderia,* and scarcely distinguishable in some other *Prostantheræ,* where they are completely adnate to the back cell. The second character, the transverse rib inside the calyx, appears to be no more than the transverse pubescent zone of the two preceding species, rather narrower and more raised in this one, but quite disconnected with the venation of the calyx.

SERIES 3. SUBCONCAVÆ.—Leaves (small or narrow) sessile or very shortly petiolate, concave or with incurved margins or flat, the margins never recurved. Flowers all axillary, the floral leaves similar to the stem-ones. Anthers with one appendage about twice as long as the cell.

The species of this group in most cases appear to pass into each other so as to make it very difficult to draw any definite lines between them when the specimens are numerous.

22. **P. lithospermoides,** *F. Muell. Fragm.* vi. 107. A shrub of 6 to 8 ft., hoary with appressed hairs, the young shoots silky. Leaves very shortly petiolate or nearly sessile, oblong-lanceolate or almost linear, obtuse or acute, entire, flat or concave, the margins not recurved, the larger ones 1 to 2 in. long, the floral ones smaller but similar. Flowers axillary, on very short pedicels. Bracts nearly as long as the

calyx. Calyx silky-pubescent, about 2½ lines long, the lips nearly
equal, entire or sinuate-toothed. Corolla white (*F. Mueller*) or blue
(*O'Shanesy*), softly pubescent, not twice as long as the calyx. Anthers
with one appendage about twice as long as the cell, the other short and
adnate.

Queensland. Table Mount, Rockhampton, *O'Shanesy ;* Armadilla, *Barton.*

23. **P. Behriana,** *Schlecht. in Linnæa*, xx. 610. A shrub of 5 or 6 ft.
with erect branches, hoary or white with short closely appressed hairs.
Leaves sessile, linear or oblong, mostly obtuse, entire, thick, flat or
concave, the margins never recurved, often above 1 in. long when
narrow, much shorter when broad, or in some specimens the narrow
ones also under ½ in. Flowers axillary, sessile or very shortly pedicellate, the linear bracts very deciduous. Calyx pubescent or villous, about
4 lines long, the tube sulcate-striate, the upper lip erect, entire, about
as long as the tube, the lower lip scarcely half so long, entire or notched.
Corolla " white," villous outside, twice as long as the calyx. Anthers
with one appendage about twice as long as the cell, the other short and
adnate.—Benth. in DC. Prod. xii. 700.

S. Australia. Rocky valley of the Tonunda, *Behr.;* Murray river, *F. Mueller ;*
near Adelaide, *Blandowski.*

24. **P. Baxteri,** *A. Cunn. in Benth. Lab.* 452, *and in DC. Prod.* xii.
561. An erect heath-like shrub of 1 to 3 or 4 ft., the branches white
with a close appressed tomentum. Leaves sessile, linear-terete, obtuse,
entire, with involute margins, slender and mostly about ½ in. long in the
typical form, or rarely nearly ¾ in., quite glabrous. Flowers all axillary,
on very short pedicels, with setaceous bracts. Calyx when in flower
2 to 2½ lines long, often 4 lines when in fruit, more or less hoary
or white with appressed hairs or rarely glabrous, the tube striate, the
upper lip large and often shortly and obtusely acuminate, the lower one
much smaller. Corolla pubescent, not twice as long as the calyx.
Anthers with one appendage nearly twice as long as the cell, the other
short and adnate.

W. Australia. King George's Sound or to the eastward, *Baxter ;* Thomas river,
Maxwell.

Var. *crassifolia.* Leaves under ½ in. long, thicker and broader than in the type.—
Phillips river and Eyres Range, *Maxwell.*

25. **P. canaliculata,** *F. Muell. Fragm.* vi. 105. An erect heath-
like shrub of 1 to 2 ft., the branches hoary with minute appressed hairs
or nearly glabrous. Leaves sessile or nearly so, linear-terete, obtuse,.
entire, with involute margins, under ½ in. and often scarcely ¼ in. long.
Flowers all axillary, on short pedicels, the bracts apparently wanting.
Calyx about two lines long, prominently striate, glabrous, the lips
shorter than the tube and nearly equal. Corolla about twice as long as
the calyx, more or less hairy or nearly glabrous. Anthers with one
appendage about twice as long as the cell, the other short and adnate.

W. Australia. Kalgan river, *Oldfield, F. Mueller ;* towards Cape Riche, *Drummond, 4th coll. n.* 166, *5th coll. n.* 343 ; Fitzgerald river, *Maxwell.*

The species is near *P. Baxteri*, but readily distinguished by the calyx, very near also to the eastern *P. saxicola*, but the leaves much narrower.

Var. *? canosericea.* Leaves small, rather flatter, and sometimes distinctly petiolate, silvery-white on both sides, with the same tomentum as the branches.—*Drummond, 4th coll. n.* 164.

26. **P. nivea,** *A. Cunn. in Benth. Lab. Gen. et Sp.* 452, *and in DC. Prod.* xii. 561. A beautiful shrub of 3 to 6 ft., glabrous except the corolla or sprinkled with a few appressed hairs, especially on the young shoots, and usually of a pale green. Leaves sessile, linear-terete with incurved or involute margins, or flat when fresh, acute or obtuse, rather slender, mostly ½ to 1 in. long, the upper floral ones smaller. Flowers rather large, of a snow white or tinged with pale blue, all axillary, the upper ones forming interrupted leafy racemes. Pedicels much shorter than the calyx, with short setaceous bracts. Calyx 2½ to 3 lines long when in flower and not much enlarged afterwards, slightly pubescent, the tube prominently ribbed, the lips broad and ciliate, the upper one much larger than the lower. Corolla pubescent or villous, twice as long as the calyx. Anthers with one appendage about twice as long as the cell, the other short and adnate.—Bot. Mag. t. 5658.

N. S. Wales. Barren rocky hills on the Lachlan river, *A. Cunningham ;* Castlereagh river, *C. Moore ;* New England, *C. Stuart.*

Victoria. Mountains of Bacchus Marsh, Mount Korong, Mount Hope, Station Peak, &c., *F. Mueller.*

Var. *induta.* Branches and foliage hoary or white with appressed hairs, the young shoots silky. Flowers rather larger.—Castlereagh river, *C. Moore.*

27. **P. striatiflora,** *F. Muell. in Linnæa,* xxv. 425. A rigid much branched shrub, sometimes quite low, sometimes attaining 5 or 6 ft., glabrous or the young shoots hoary or silky with short appressed hairs, the smaller branches sometimes almost spinescent. Leaves sessile or nearly so, oblong-lanceolate or linear, mostly obtuse, entire, flat or with incurved margins, rather thick, usually under ½ in. long, but when narrow often ¾ in., the floral ones gradually smaller, the upper ones shorter than the calyx. Flowers all axillary, but the upper ones crowded into terminal leafy racemes or spikes. Pedicels very short. Calyx-tube not 2 lines long, prominently striate, the upper lip ovate, obtuse, often ½ in. long when in fruit, the lower lip not half so large. Corolla nearly twice as long as the calyx, glabrous or sprinkled with a few hairs, white streaked with red and tinged with yellow at the base of the broad middle lobe of the lower lip. Anthers with one appendage about twice as long as the cell, the other short and adnate.

Queensland. Newcastle range, *Sutherland.*

N. S. Wales. From the Lachlan and Darling to the Barrier Range, *Victorian and other Expeditions.*

S. Australia. Cudnaka, Arkaba, and Lake Torrens, *F. Mueller ;* towards Cooper's Creek, *Wheeler ;* Lake Gillies, *Burkitt* (a short-leaved form, and another with narrow white or hoary leaves), Mount Morphett, *M'Douall Stuart's Expedition.*

The species is very nearly allied to *P. nivea,* and sometimes difficult to distinguish from it. In general it has shorter broader leaves, a more dense inflorescence, the upper lobe of the calyx longer and not so broad, and the corolla more glabrous.

Var. ? *sericea.* The whole plant white with soft silky hairs. Leaves rather short and broad.—Gawler ranges, S. Australia, *Mrs. Sullivan ;* a small fragment in Herb. F. Mueller, under the name of *P. Sullivaniæ,* but quite insufficient to judge whether it be any more than a variety of *P. striatiflora,* corresponding with the white silky varieties of *P. canaliculata* and others of this series.

28. **P. saxicola,** *R. Br. Prod.* 509. A slender shrub or undershrub, sometimes having the appearance of a rigid annual; the stems much branched at the base and ascending to from 6 in. to 1 ft. in the typical form, more erect and attaining 2 ft. in others, hoary with minute appressed hairs or nearly glabrous. Leaves very shortly petiolate or nearly sessile, oblong or linear-oblong, obtuse, entire, flat, rather thick, 2 to 4 lines long in the typical form, the young ones sprinkled with appressed hairs, the older ones usually glabrous. Flowers small, in few pairs in the upper axils. Pedicels short, with setaceous bracts. Calyx about 2 lines long, more or less hispid with spreading hairs especially in the small typical specimens, the lips nearly equal. Corolla nearly twice as long as the calyx, glabrous or slightly pubescent. Anthers with one appendage about twice as long as the cell, the other short and adnate.

N. S. Wales. George's river, *R. Brown* (the specimens all under 1 ft. and the calyx very hispid).

Var. *major.* Taller, more shrubby, leaves longer and narrower, calyx less hispid and sometimes almost glabrous.—*P. saxicola,* A. Cunn.; Benth. in DC. Prod. xii. 562 ; *P. pimeleoides,* F. Muell. Fragm. vi. 107.—Rocky ranges near Bathurst, *A. Cunningham ;* New England, *C. Stuart.*

29. **P. odoratissima,** *Benth. in Mitch. Trop. Austr.* 291, *and in DC. Prod.* xii. 700. A small erect bushy shrub or undershrub, more or less pubescent with glandular hairs, the branches and foliage apparently viscid and very strongly scented. Leaves sessile and often clustered in the axils, linear or almost lanceolate, obtuse, entire, thick, flat or concave, the margins never recurved, mostly about ¼ in. long. Flowers axillary, on short pedicels, the bracts linear, obtuse, thick and often as long as the calyx. Calyx 2½ lines or at length 3 lines long, rather narrow, prominently striate, the lips ovate, the upper one considerably longer than the lower. Corolla not seen perfectly open, glabrous outside when in bud. Anthers with one appendage about twice as long as the cell, the other short and adnate.

Queensland. Mantuan Downs, *Mitchell.*

30. **P. euphrasioides,** *Benth. in Mitch. Trop. Austr.* 360, *and in DC. Prod.* xii. 700. A small bushy shrub, villous with white spreading hairs intermixed with glandular ones and often viscid. Leaves on very short petioles or almost sessile, often clustered in the axils, linear-oblong, obtuse, with 2 or 3 prominent obtuse teeth on each side or almost pinnatifid, mostly about 3 lines long, rather thick, flat, the margins never recurved. Flowers all axillary, rather large. Pedicels short but slender. Bracts linear, obtuse, rather long. Calyx pubescent or hirsute, about 3 lines or at length sometimes 4 lines long, the

lips entire, broad, the upper one usually larger than the lower. Corolla more than twice as long as the calyx, nearly glabrous outside, hairy inside. Anthers with one appendage about twice as long as the cell, the other short and adnate.

Queensland. On the Maranoa, *Mitchell ;* Hodgson's Creek and Dogwood Creek, *Leichhardt* ; Cape river and Broad Sound, *Bowman.*

31. **P. cryptandroides,** *A. Cunn. in Benth. Lab. Gen. et Sp.* 453, *and in D C. Prod.* xii. 561. A heath-like shrub, with slender virgate branches, glabrous or slightly glandular-pubescent and viscid. Leaves sessile or very shortly petiolate and sometimes clustered in the axils, linear or linear-lanceolate, obtuse, entire or with 2 or 3 short obtuse teeth on each side, flat or concave, the margins not recurved, 2 to 3 lines long. Flowers all axillary, on very short pedicels. Bracts close under the calyx, linear-lanceolate and almost as long as the calyx. Calyx about 2 lines long, slightly hairy, prominently ribbed but rather thin, the lips ovate, obtuse, nearly equal. Corolla glabrous outside, not twice as long as the calyx. Anthers with one appendage about twice as long as the cell, the other short and adnate.

N. S. Wales. Sandstone Hills, N.W. branch of Hunter's river, *A. Cunningham.* Nearly allied to *P. euphrasioides,* although the slender stems, small leaves, &c., give it a very different aspect.

32. **P. eurybioides,** *F. Muell. in Hook. Kew Journ.* viii. 168, *and in Trans. Phil. Soc. Vict.* i. 48. A shrub with slightly hoary-pubescent branches. Leaves small, sessile, often clustered in the axils, from ovate to oblong-linear, obtuse, entire, thick, concave, usually glabrous, 1 to 2 lines long, the floral ones similar or passing into broader bracts. Flowers axillary, but usually 3 or 4 pairs crowded at the ends of the branches so as to form short leafy racemes, the floral leaves shorter than the calyx. Pedicels short; bracts short and obtuse. Calyx nearly 3 lines long, coloured, glabrous or nearly so, prominently ribbed, the lips nearly equal, the lower one sometimes retuse. Corolla glabrous, more than twice as long as the calyx. Anthers with one appendage about twice as long as the cell, the other short and adnate.

S. Australia. Mallee scrub, near Mount Barker, *F. Mueller.*
The specimens are few and small, and the habit may be different when more fully developed. The affinities of the species appear, however, to be rather with the present series than with the *Racemosæ.*

SECT. 2. KLANDERIA.—Calyx with the tube less prominently striate than in *Euprostanthera,* the lips nearly equal, and usually equally open after flowering. Corolla-tube narrow at the base, usually incurved and dilated upwards, the upper lip erect concave or arched, the lower lip shorter or at any rate not longer and spreading. Anther-appendages very short and adnate or quite obsolete, or rarely one very delicate one about as long as the cell.

The shape of the corolla is so different from that which is so nearly uniform in *Eu-prostanthera,* that this section might well be considered as a distinct genus, were it not that in other respects some species come so near to different typical species of *Prostanthera*, as to prevent their having any distinguishing habit. As a sectional name I have preferred F. Mueller's generic name *Klanderia*, to that of *Cryphia* previously established by Brown, the latter being derived from a character probably abnormal in the particular flower examined.

33. **P. ringens,** *Benth. in Mitch. Trop. Austr.* 363, *and in DC. Prod.* xii. 700. A much branched shrub, glabrous or the branches slightly pubescent. Leaves on rather long petioles, broadly ovate or orbicular, obtuse, entire, rather thick, flat, under ½ in. diameter. Flowers all axillary, on very short pedicels. Calyx glabrous, herbaceous, scarcely striate, fully 3 lines long, the lips broad, equal, entire, not half so long as the tube. Corolla-tube shortly exserted, the upper lip concave, 2-lobed, longer than the lower one. Stamens nearly as long as the corolla; anthers without any prominent appendages to the connective.

Queensland. On the Maranoa, *Mitchell*. The leaves resemble those of the larger varieties of *P. rotundifolia*, but are larger and entire, the flowers are totally different.

34. **P. Leichhardtii,** *Benth.* A bushy shrub, the branches slightly pubescent, the foliage glabrous or nearly so. Leaves distinctly petiolate, oblong or obovate-oblong, obtuse, entire, rather thick, flat, rarely exceeding ½ in. Flowers apparently all axillary, on very short pedicels, without bracts on the specimens seen. Calyx about 3 lines long when in flower, somewhat enlarged afterwards, the tube broad, scarcely striate, the lips equal, entire, scarcely half as long as the tube. Corolla slightly hairy outside, the tube shortly exceeding the calyx and scarcely dilated, the upper lip 4 to 5 lines long, concave, emarginate, the lobes of the lower lip much shorter, all fringed with rather long hairs. Stamens nearly as long as the upper lip of the corolla; anthers without any prominent appendages to the connective.

Queensland. Bottletree Creek, *Leichhardt*. F. Mueller (Fragm. vi. 106) includes this in *P. ovalifolia* of which it has nearly the foliage, but the flowers are totally different. It is very near to *P. ringens*, but with differently shaped leaves, and perhaps the corollas may not be quite the same. The two will require further comparison on better specimens.

35. **P. microphylla,** *A. Cunn. in Benth. Lab. Gen. et Sp.* 454 *and in DC. Prod.* xii. 562. A low bushy or scrubby shrub rarely above 1 ft. high, more or less scabrous-pubescent. Leaves very shortly petiolate, oblong or rarely oval-oblong, obtuse, thick, recurved from the base to the end, but without recurved or revolute margins, often all under 1 line long and very rarely exceeding 2 lines. Pedicels axillary, much shorter than the calyx and often not above ½ line long. Bracts small, close under the calyx. Calyx obscurely striate, pubescent or nearly glabrous, usually about 3 lines or rarely 4 lines long, the lips much shorter than the tube, equal and obtuse. Corolla scarlet, slightly pubescent, fully twice as long as the calyx, the tube exserted and slightly incurved, enlarged upwards, the upper lip erect, concave, emarginate, the lower lip much shorter. Stamens exserted but shorter than the upper lip; anthers with one appendage about as long as or shortly exceeding the cell, but very delicate and easily overlooked.—Benth. in DC. Prod. xii.

562; *P. coccinea*, F. Muell. in Hook. Kew Journ. viii. 168, and in Trans.
Phil. Soc. Vict. i. 48.

N. S. Wales. Euryalean scrub, S.W. of Lachlan river, *A. Cunningham.*
Victoria. Dry arid places, Avoca and Murray Desert, *F. Mueller*; Lake Waringra,
Dallachy.
S. Australia. Tumby Bay, *Wilhelmi ;* Venus Bay, *Warburton.*
W. Australia. Towards Cape Riche, *Drummond*, 5th *coll. n.* 341; Phillips,
Oldfield and Salt rivers, Eyre's Range, E. Mount Barren, &c., *Maxwell.*

Cryphia serpyllifolia, R. Br. Prod. 508, Benth. in DC. Prod. xii. 558, from Memory
Cove, appears to me from the inspection of the original specimens to be identical with
this plant. Brown, it is true, describes the corolla as shorter than calyx and concealed
within it and derives from that circumstance his generic name. But he probably had
only an imperfectly developed flower to examine. His specimens now show only
calyxes past flower, and a few very young buds. *C. microphylla*, R. Br. l.c., from the
same locality is evidently, as suggested by Brown, a minute-leaved variety of the same
plant. The specimens have no flowers.

36. **P. aspalathoides,** *A. Cunn. in Benth. Lab. Gen. et Sp.* 453 *and in
DC. Prod.* xii. 562. A low rigid bushy shrub, slightly scabrous-pubescent and sometimes perhaps viscid. Leaves linear-terete, very obtuse,
channelled above, rather thick, contracted into a very short petiole,
crowded on the smaller branchlets, rarely above 3 lines long. Pedicels
axillary, usually very short and always much shorter than the calyx,
the bracts close under the calyx. Calyx more or less pubescent, 4 or
rarely 5 lines long, somewhat striate at the base, the lips obtuse or
shortly acuminate, nearly equal and much shorter than the tube.
Corolla twice as long as the calyx, slightly pubescent or nearly glabrous,
the tube exserted, much enlarged upwards and incurved, the lips
short, the upper one erect, emarginate and very broad, the lower one
short, with 3 ovate lobes. Stamens exserted but not exceeding the
upper lip; anthers without any appendage to the connective which is
only slightly fringed.

N. S. Wales. Barren wastes S. W. of the Lachlan river, *A. Cunningham, Fraser,
Mitchell.*
Victoria. Murray Desert, *F. Mueller ;* Wimmera, *Dallachy.*
S. Australia. Sandy scrub, Kangaroo island, *Waterhouse.*

F. Mueller unites this and the two following species with *P. microphylla* under the
name of *P. coccinea*, but the very marked differences in the foliage as well as in the
shape of the corolla appear to me to be constant in all the specimens seen.

37. **P. calycina,** *F. Muell.* A rigid bushy shrub, more or less
hoary-pubescent with short rigid hairs. Leaves very shortly petiolate,
ovate or ovate-oblong, obtuse, entire, with recurved margins but the
whole leaf rather incurved than recurved, contracted at the base, rarely
under 2 lines and sometimes 4 lines long. Flowers axillary, nearly
sessile or on pedicels not exceeding 1 line, the bracts short, linear, close
under the calyx. Calyx 5 to 7 lines long, not striate, but with a prominent rib on each side decurrent from the junction of the lips, which
are broad, obtuse, nearly equal and much shorter than the tube.
Corolla slightly pubescent outside, with a broad straight tube slightly
enlarged upwards and not exceeding the calyx, the lips very short, the

upper one erect and broad, the lower one broadly 3-lobed, and rather shorter than the upper one, the whole corolla not exceeding the calyx by more than a quarter of its length. Anthers without any or only with an exceedingly short appendage to the connective, but one cell tipped with a minute point.

S. Australia. Port Lincoln, *Wilhelmi;* Venus Bay, *Warburton.*

This species has the leaves nearly of *P. chlorantha,* but larger, and the large nearly sessile calyx and the corolla are very different from those of that species. The specimens seen are but very few.

38. **P. chlorantha,** *F. Muell. Herb.* A rigid divaricate shrub, with numerous small sometimes almost leafless branches, more or less sprinkled or scabrous with short crisped or almost stellate hairs. Leaves very small and shortly petiolate or almost sessile, broadly ovate or rhomboidal, obtuse, with revolute margins, all under 2 lines and mostly not 1 line long. Pedicels axillary, slender, 3 to 5 lines long, with a pair of bracts near the calyx or at a little distance below. Calyx pubescent, often reddish when dry, about 5 lines long, the lips nearly equal, more or less acuminate, rather shorter than the tube. Corolla " green," about twice as long as the calyx, slightly pubescent, the tube exserted, incurved, gradually enlarged, the limb very oblique, the upper lip erect concave emarginate, the lower one rather shorter, the lateral lobes ovate, the middle one broader. Anthers shortly exserted from the tube, without any appendage to the connective.—*Klanderia chlorantha,* F. Muell. in Linnæa, xxv. 426.

S. Australia. Mount Barker Creek, *L. Fischer;* Encounter Bay, *Whittaker;* Cygnet Bay, Kangaroo island, *Waterhouse.*

P. Caleyi, Benth. Lab. Gen. et Sp. 454, and in DC. Prod. xii. 562, from N. S. Wales, which, owing to the dispersion of the Lambertian herbarium, I am unable now to re-examine, must be very near *P. chlorantha,* with the same foliage and long pedicels; but, if the character I gave proves correct, it differs in the shorter corolla, and the presence of a short appendage to the connective. The form of the corolla having however not been specially described, I am unwilling formally to admit the species without further confirmation.

16. **HEMIANDRA,** R. Br.

Calyx 2-lipped or 5-toothed. Corolla with a broad campanulate throat, the upper lip short, erect, broadly 2-lobed, the lower longer, spreading, 3-lobed, the middle lobe often 2-lobed. Stamens 4, in pairs; anthers 1-celled, the connective elongated and produced beyond its insertion on the filament in a small tooth-like or shortly linear appendage. Style shortly bifid at the end. Nuts reticulate-rugose, attached to above the middle. Seeds albuminous.—Shrubs or undershrubs, usually diffuse but rigid. Leaves opposite, entire, narrow, rigid, pungent-pointed. Flowers axillary, solitary, with a pair of bracts under the calyx.

The genus is limited to West Australia. It only differs from *Hemigenia* in the pungent-pointed leaves and the shortness of the posterior end of the connectivum of

the anthers, and might be considered as a section of that genus were there any advan-
tage in doing so compensating for the inconvenience of the change in nomenclature.

Calyx 2-lipped, the upper lip entire or with small lateral lobes, the
    lower 2-lobed.
    Calyx-lobes very acute or pungent-pointed . . . . 1. *H. pungens.*
    Calyx-lobes, at least the lower ones, very obtuse . . . . . . 2. *H. leiantha.*
Calyx with 5 rigid subulate nearly equal teeth . . . . 3. *H. incana.*

1. **H. pungens,** *R. Br. Prod.* 502. A diffuse or spreading (rarely
erect ?) rigid shrub, sometimes quite dwarf, sometimes ascending to 1
or 2 ft., quite glabrous or the branches only or also the leaves and
calyxes hispid with rigid spreading hairs, often intermixed on the
branches with a minute pubescence. Leaves sessile, linear or linear-
lanceolate, rigid, acute with pungent points, flat or concave, with 1 to
5 parallel nerves very prominent underneath. Pedicels shorter than
the leaves and sometimes very short, the bracts linear or lanceolate,
rigid and pungent-pointed, close under the calyx. Flowers very
variable in size, white or pink with darker spots. Calyx 2-lipped, the
middle lobe of the upper lip broad, tapering into a pungent point, the
lateral ones small and rounded or quite obsolete, the lower lip smaller,
with 2 acute usually pungent-pointed lobes. Corolla-tube exserted
and dilated into a broad throat, lobes of the lower lip usually all emarginate or crenate, the middle one much larger and 2-lobed, rolled over
the anthers in the bud. Connective of the anthers forming usually a
very small tooth below its insertion on the filament. Disk cup-shaped
and thick, enclosing the base of the ovary. Style shortly and equally
2-lobed. Nuts attached by their inner face to above the middle.—
Benth. in DC. Prod. xii. 564; Lemaire, Jard. Fleur. t. 126.

W. Australia. Very abundant from King George's Sound, *R. Brown* and many
others, to Swan river, *Drummond, Preiss, Oldfield,* and others, and everywhere very
variable as to the size of the flowers and the hairs. The following are the most marked
forms :—

    a. *grandiflora.* Glabrous or hispid. Leaves usually narrow, 1 to 1½ in. long. Calyx
about ½ in., corolla 1¼ to 1½ in. long.—*H. linearis,* Benth. in Hueg. Enum. 79, and in
DC. Prod. i. 564; *H. longifolia,* Bartl. in Pl. Preiss. i. 356.—Chiefly from Swan river,
Drummond, 1st coll. also n. 10, 139, 449; *Preiss, n.* 2305, 2317.

    b. *glabra.* Glabrous or scarcely hispid. Leaves spreading, broad and about ½ in.
long, or narrow, more erect, and longer. Calyx about ¼ in. long, with the lobes of the
lower lip shorter, acute but not pungent, and much inflexed after flowering. Corolla
nearly ¾ in. long.—*H. glabra.* Benth. in Hueg. Enum. 79; Bartl. in Pl. Preiss. i. 356;
H. juniperina, Bartl. in Pl. Preiss. i. 355.—Chiefly Swan river, *Drummond, 1st coll.
n.* 450, *2nd coll. n.* 144, 145, *3rd coll. n.* 191, 192, *Preiss, n.* 2307, 2308.

    c. *diffusa.* Usually dwarf decumbent and hispid. Leaves spreading, linear-lanceo-
late, mostly about ½ in. long. Calyx ¼ in.; corolla a little more than ½ in. long.—*H.
brevifolia* and *H. hirsuta,* Benth. in Hueg. Enum. 79, and in DC. Prod. xii. 564; Bartl.
in Pl. Preiss. i. 355.—The commonest form about King George's Sound, especially on
the sandy shores, *R. Brown, Baxter, A. Cunningham,* &c.

    d. *hispida.* The same as the var. c, but more erect and more hispid, the upper lip
of the calyx after flowering rather more enlarged.—*H. rupestris,* Hueg. Bot. Arch. t. 4;
Benth. in Hueg. Enum. 78, and in DC. Prod. xii. 564; Bartl. in Pl. Preiss. i. 354; *H.
emarginata,* Lindl. Bot. Reg. 1841, Misc. 72 (from the character given)—Rocky hills,
chiefly about King George's Sound, *Huegel* and others, *Drummond, n.* 12, 183, 193,
448.

e. incana. Pubescence short and more or less hoary, giving the plant the aspect of *H. incana*, but with the calyx of *H. pungens*. Corolla small, minutely pubescent.— *Drummond, 3rd coll. n.* 171, and some specimens of *1st coll. n.* 450, *Preiss, n.* 2306.

It is possible that the observation of flowers in the recent state may supply characters to distinguish amongst the above at least two more definite varieties or species, but in the great majority of·dried specimens the corollas are too much injured to ascertain their precise form and size.

2. **H. leiantha,** *Benth.* An erect bushy rigid shrub of 1 to 4 ft., our specimens all entirely glabrous. Leaves of the glabrous varieties of *H. pungens* and varying like them from lanceolate to linear, recurved or nearly straight, ½ in. to above 1 in. long, pungent-pointed, rigid, 3- to 5-nerved, smooth and shining. Flowers sessile or shortly pedicellate, usually about ¾ in. long. Calyx with the upper lip broad, obtuse or scarcely acute in the centre and not pungent-pointed, the lower lip smaller 2-lobed, the lobes always very obtuse. Corolla of *H. pungens*, but quite glabrous. Anthers slender, the lower end of the connective much more prominent than in any of the flowers examined of *H. pungens*, although less so than in *H. incana*.

W. Australia. Murchison river, *Oldfield* (several forms differing chiefly in the length and breadth of the leaves), *Drummond*, (a single specimen in herb. F. Muell.)

3. **H. incana,** *Bartl. in Pl. Preiss.* i. 357. A shrub probably low and bushy, much branched, hoary-pubescent or shortly hispid. Leaves rather crowded, linear or linear-lanceolate, rigid, pungent-pointed, mostly 5-nerved, ½ to 1 in. long. Pedicels very short. Calyx 3 to 4 lines long, narrower than in *H. pungens*, rigid, striate, with 5 rigid pungent linear-subulate teeth, as long as the calyx, and nearly equal or more or less arranged in 2 lips. Corolla like that of *H. pungens* or perhaps with the lower lip not so long, but not seen very perfect. Connective produced below its insertion on the filament into a tooth usually longer than in the two preceding species, but perhaps variable. —*Benth. in DC. Prod.* xii. 565.

W. Australia. Swan river, *Drummond, n.* 75; *Preiss, n.* 2316. Resembles at first sight the var. *incana* of *H. pungens*, but the calyx is very different.

All the above species and varieties of *Hemiandra* require further illustration from the examination of fresh flowers, for in dried specimens the rigidity of the foliage has interfered very much with the proper desiccation of the corollas, which are usually withered up or destroyed.

17. HEMIGENIA, R. Br.

(*Colobandra, Bartl.* Atelandra, *Lindl.*)

Calyx 2-lipped or 5-toothed. Corolla with a dilated throat; the upper lip erect, more or less concave, emarginate or 2-lobed, the lower lip longer, spreading, 3-lobed, the middle lobe usually larger and often 2-lobed. Stamens 4, in pairs, anthers 1-celled, the connective elongated, produced beyond the insertion into an appendage or sterile branch, which in the upper pair is usually short dilated and bearded or crested

at the end with short hairs, in the lower pair or rarely in both pairs
glabrous and attenuate or bearing an imperfect cell at the end. Style
shortly bifid at the end. Nuts reticulate-rugose, attached to the middle
or higher up. Seeds albuminous.—Shrubs or undershrubs. Leaves
opposite or in whorls of 3, entire, obtuse or rarely acute and never
pungent-pointed. Flowers all axillary, solitary or rarely clustered, with
a pair of bracts under the calyx. Corolla hairy inside at the insertion
of the stamens and usually at the base of the lower lip.

The genus is limited to Australia, and, with the exception of two species, to West
Australia.

Sect. I. **Homalochilus.**—*Calyx 2-lipped, the lips broad, the upper one entire or
broadly and shortly 3-lobed, the lower one entire or shortly 2-lobed and closed over the
orifice of the tube, as in* Prostanthera. *Lower end of the connective of the lower
anthers attenuate or slightly clavate.*

Leaves opposite.
  Leaves obovate or oblong-cuneate. Pedicels very short.
    Flowers 1½ in. long, the corolla twice as long as the calyx . 1. *H. macrantha.*
  Leaves narrow. Pedicels as long as the calyx. Corolla not
    much exceeding the calyx . . . . 2. *H. rigida.*
Leaves in whorls of 3. Flowers small.
  Pedicels slender, longer than the calyx . . . . 3. *H. ramosissima.*
  Pedicels very short . . . . 4. *H. microphylla.*

Sect. II. **Atelandra.**—*Calyx 2-lipped, the upper lip 3-lobed, the lower deeply 2-lobed,
all the lobes acuminate (in the last two species the bilabiation less distinct). Lower end
of the connective of the lower anthers attenuate. Leaves obtuse, contracted into a short
petiole.*

Plant softly hoary or silky-villous.
  Leaves mostly oblong and above 1 in. long on the main stems,
    shorter and more obovate on the branches . . . . . 5. *H. incana.*
  Leaves mostly obovate and rarely exceeding ½ in. . . . . 6. *H. canescens.*
Plant closely hoary or silvery. Leaves mostly obovate or orbicu-
  lar, rarely exceeding ¼ in. . . . . 7. *H. podalyrina.*
Plant minutely glandular-pubescent, not hoary. Leaves mostly
  obovate or oval-oblong, ⅓ to 1 in. . . . . 8. *H. platyphylla.*
Plant glabrous or minutely hoary-pubescent. Calyx irregularly
  2-lipped.
  Leaves oblong or obovate-oblong . . . . 9. *H. glabrescens.*
  Leaves narrow-oblong or linear-cuneate . . . . 10. *H. obtusa.*

Sect. III. **Hemigenia.**—*Calyx-teeth 5, nearly equal, subulate-acuminate or acute.
Lower end of the connective of the lower anthers attenuate. Leaves sessile (except in
H. humilis).*

Branches silky-villous or woolly. Flowers on the main branches
  clustered in the axils, rarely solitary on the smaller branch-
  lets.
  Leaves flat or concave, erect or spreading.
    Leaves linear-lanceolate, cuneate or narrow-oblong, usually
      1 in. or longer . . . . 11. *H. sericea.*
    Leaves broadly oblong, mostly about ½ in. . . . . . . 12. *H. barbata.*
  Leaves complicate and recurved, oblong or ovate-lanceolate,
    mostly about ½ in. long . . . . 13. *H. curvifolia.*
Glabrous hoary-pubescent or hirsute. Flowers all solitary.
  Leaves mostly opposite. Western species.
    Leaves oblong-cuneate, scabrous-pubescent or hirsute . . 14. *H. scabra.*

Leaves linear-oblong or cuneate. Plant hoary-pubescent . 15. *H. humilis.*
Leaves very narrow-linear or terete.
 Pedicels 2 to 3 lines long. Calyx-teeth shorter than the
  tube . . . . 16. *H. westringioides.*
 Pedicels not 1 line long. Calyx-teeth longer than the tube 17. *H. teretiuscula.*
Leaves in whorls of 3 or 4. Eastern species.
 Leaves narrow-linear or terete . . . . 18. *H. purpurea.*
 Leaves oblong-cuneate . . . . 19. *H. cuneifolia.*

Sect. IV. **Diplanthera.**—*Calyx-teeth 5, nearly equal, subulate-acuminate or acute.
Lower end of the connective of the lower anthers bearing an imperfect cell at the end.
Leaves sessile, opposite.*

Leaves oblong or oblong-cuneate, ¼ to nearly ½ in. long. Plant
 glabrous or minutely pubescent . . . . 20. *H. Drummondii.*
Leaves oblong, 2 to 3 lines long. Plant hirsute. Flowers very
 small . . . . 21. *H. pimelifolia.*
Leaves linear, ¼ to ½ in. long. Plant glabrous . . . . . 22. *H. diplanthera.*

Sect. 1. Homalochilus.—Calyx 2-lipped, the lips broad, the
upper one entire or broadly and shortly 3-lobed, the lower one entire
or shortly 2-lobed and closed over the orifice of the tube as in *Prostanthera.* Lower end of the connective of the lower anthers attenuate
or slightly clavate.

The species of this section differ considerably from each other in habit, but yet are
not closely connected with any of those of other sections, and are all remarkable for their
Prostanthera-like calyx.

1. **H. macrantha,** *F. Muell. Fragm.* i. 210. A shrub with erect
virgate rather stout branches, hoary or white as well as the foliage with
a close stellate tomentum, which disappears from the older leaves.
Leaves opposite, erect or scarcely spreading, obovate or oblong-cuneate,
very obtuse and sometimes minutely mucronate, contracted into a short
petiole, rigidly coriaceous, with few rather prominent primary veins, ½
to 1 in. long. Flowers all axillary, on short pedicels. Bracts linear.
Calyx attaining ½ to ¾ in. after flowering, the upper lip ovate, contracted upwards but obtuse, entire, the lower one much shorter, with 2
short obtuse or almost acute lobes. Corolla glabrous, nearly 1½ in.
long, the tube exserted and not much dilated at the throat, the upper
lip long and narrow, arcuate, concave, emarginate, the sides spreading ;
lower lip shorter, with 3 rather narrow lobes. Stamens ascending
under the upper lip and nearly as long, all the anthers with the lower
end of the connective long linear and glabrous. Nuts very prominently
reticulate.

W. Australia. Murchison river, *Oldfield, Drummond, 6th coll. n.* 142 ; Lagrange
Bay, *Martin.*

The stellate pubescence of this species appears to be exceptional in the tribe and
almost in the Order.

2. **H. rigida,** *Benth. in DC. Prod.* xii. 565. A glabrous shrub,
apparently diffuse or loosely spreading as in *Hemiandra,* but the foliage
not at all pungent. Leaves opposite, linear-oblong or linear-cuneate,
obtuse, entire, contracted into a short petiole, coriaceous, concave,
nerveless except the scarcely prominent midrib, mostly ¾ to 1 in. long.

Pedicels axillary, slender, longer than the calyx but shorter than the leaves. Bracts from a broad base acutely acuminate, nearly as long as the calyx. Calyx 3 to 4 lines long, the upper lip broad, acute or acutely acuminate, the lower one shorter, with 2 acute lobes. Corolla not much longer than the calyx, glabrous outside, the upper lip short with 2 broad lobes, the lower lip much longer, with a large middle lobe emarginate or 2-lobed, all the lobes crenulate. Connective of the upper anthers clavate at the lower end and minutely bearded, of the lower anthers glabrous.

W. Australia, *Drummond, 4th coll. n.* 146.

3. **H. ramosissima,** *Benth. in DC. Prod.* xii. 565. A slender shrub, apparently diffuse, glabrous or with opposite lines of minute hairs decurrent on the branches. Leaves in whorls of 3, nearly sessile, linear, obtuse or acute, entire, rather rigid, 1-nerved, rarely above ½ in. long. Pedicels axillary, filiform, about as long as the leaves, very spreading, with a pair of linear-subulate bracts under the calyx. Calyx about 1½ lines long, broadly campanulate, glabrous, the upper lip broad, recurved, shortly and broadly 3-lobed, the lower one more or less distinctly 2-lobed with obtuse or scarcely acute lobes, and curved over the tube as in *Prostanthera.* Corolla not seen open. Anthers in the young bud similar to those of *H. rigida.*

W. Australia. Between Swan river and King George's Sound, a single specimen in the Hookerian herbarium, with numerous perfect calyxes, but all past flower.

4. **H. microphylla,** *Benth. in DC. Prod.* xii. 565. A much-branched erect somewhat virgate shrub, with numerous small leaves and flowers, and quite glabrous. Leaves mostly in whorls of three, oblong lanceolate or almost linear, obtuse, entire, rather thick, under ¼ in. long and sometimes not ¼ in. Flowers all axillary, on very short pedicels. Bracts linear, acute, shorter than the calyx. Calyx 1½ to 2 lines long, like that of a *Prostanthera,* with 2 broad lips closed after flowering, the upper one rounded, obtuse and entire, the lower one rather smaller, entire or retuse. Corolla not twice as long as the calyx, glabrous outside, the tube about as long as the calyx, the upper lip short, erect, 2-lobed, the lower one longer, spreading, with undulate emarginate lobes, the middle one 2-lobed. Connective of the upper anthers dilated and slightly bearded at the lower end, that of the lower ones attenuate and glabrous.

W. Australia, *Drummond, 3rd coll. n.* 191 (*and* 151?); Harvey and Gordon rivers, *Oldfield.*

SECT. 2. ATELANDRA.—Calyx more or less distinctly 2-lipped, the upper lip 3-toothed or 3-lobed, the middle lobe usually larger than the lateral ones, the lower lip deeply 2-lobed, all the lobes acuminate. Lower end of the connective of the lower anthers attenuate or rarely clavate and glabrous.

5. **H. incana,** *Benth. in DC. Prod.* xii. 566. An undershrub or shrub attaining 2 or 3 ft., covered in every part with hoary or silky short

hairs, appressed and rather short on the stems and leaves, longer and more spreading on the calyxes and inflorescence. Leaves on the main stems oblong, obtuse, contracted into a very short petiole and 1 to 2 in. long, shorter and more petiolate on the side branches, the floral ones gradually smaller, and the upper ones scarcely exceeding the flowers. Flowers " pink " or " purple," shortly pedicellate or almost sessile, all axillary, but sometimes crowded into short axillary or terminal leafy racemes. Bracts linear or setaceous. Calyx very villous, nearly 3 lines long, the teeth lanceolate, very acute, in 2 lips, the upper one 3-lobed with the middle lobe larger, the lower deeply 2-lobed. Corolla about ½ in. long, shortly villous outside, the tube about as long as the calyx, the upper lip erect concave and emarginate, the lower one spreading and twice as long, with a large 2-lobed middle lobe. Lower end of the connective of the lower anthers clavate.—*Atelandra incana*, Lindl. Swan Riv. App. 40. t. 5 (the corolla reversed in the figure); *A. polystachya*, Lindl. l.c.; *Hemigenia polystachya*, Benth. in DC. Prod. xii. 566; *Colobandra robusta*, Bartl. in Pl. Preiss. i. 357.

W. Australia. Swan river, *Drummond*, 1st coll. n. 451, *Preiss, n.* 2313 ; Harvey river, *Oldfield.*

6. **H. canescens,** *Benth. in DC. Prod.* xii. 566. A shrub of 1 to 2 ft., much less robust than *H. incana*, clothed with a hoary or silky-white pubescence, sometimes short and appressed, sometimes dense, long, and loose, and often wearing off from the upper surface of the leaves. Leaves opposite or very rarely in whorls of three, obovate from almost orbicular to oblong, obtuse, very spreading or recurved, contracted into a short petiole, thick and soft, rarely exceeding ½ in., sometimes very silky-white, sometimes quite green. Flowers small, solitary in the axils, on pedicels usually very short and rarely above 1 line. Calyx rather broad, 2 lines long when in flower, often 3 lines in fruit, the teeth rather broad, acute or almost obtuse, scarcely so long as the tube, the 3 upper ones nearly equal or the middle one larger, the 2 lower ones united in a shortly 2-lobed lip. Corolla about twice as long as the calyx, pubescent outside, apparently like that of *H. incana*, but not seen very perfect. Anthers of *H. incana—Colobandra canescens*, Bartl. in Pl. Preiss. i. 358 (from the character given).

W. Australia, *Drummond*, 1st coll., 3rd coll n. 149 ; Hay district, *Preiss, n.* 2314 (*Bartling*) ; Salt river, *Maxwell.*

Var. *mollis.* More hirsute, the hairs more spreading, white or dark colour d, often mixed with a few glandular hairs.—*Colobandra mollis,* Bartl. in Pl. Preiss. i. 358 ; *Hemigenia mollis,* Benth. in DC. Prod. xii. 566.—York district, *Preiss, n.* 2310, also some specimens among *Drummond's, n.* 149.

Colobandra lanata, Bartl. l.c. 359 (*Hemigenia lanata,* Benth. l.c.) of which the flowers are unknown, is probably a more woolly form of the same species.

7. **H. podalyrina,** *F. Muell. Fragm.* vi. 112. A spreading shrub of 1 to 2 ft., with the general aspect and foliage of *H. canescens*, of which it may possibly be a variety, but the indumentum is very close and short, hoary or silvery or reddish at the ends of the branches. Leaves obovate

or orbicular, contracted into a very short petiole, rarely above ½ in.
diameter. Flowers in the upper axils, on very short pedicels. Bracts
small, linear. Calyx open, 2 to 3 lines long, covered with the same
appressed tomentum as the rest of the plant, the teeth short and broad,
almost obtuse and more or less distinctly forming 2 lips, the middle
upper tooth usually the largest. Corolla about ½ in. long, slightly pubescent outside, densely bearded inside at the throat. Anthers of *H.
incana* and *H. canescens.*

W. Australia. Rocks on the Kalgan river, *Oldfield, Maxwell.*

8. **H. platyphylla,** *Benth. in DC. Prod.* xii. 566. A slender shrub
or undershrub of about 2 ft., the branches and foliage pubescent with
minute glandular hairs intermixed sometimes with a few longer ones,
not glandular on the young shoots and pedicels, but not hoary or silky
like the three preceding species. Leaves opposite, obovate to oval-oblong,
obtuse, narrowed into a short petiole, 1-nerved or obscurely triplinerved,
½ to 1 in. long or smaller on the side branches. Flowers "lilac," all
axillary on short pedicels. Calyx after flowering about 3 lines long,
2-lipped, the upper lip 3-lobed, the lower rather shorter and 2-lobed, all
the lobes rather broad, acute or shortly acuminate but irregular. Corolla (which I have not seen) 5 lines long, the limb pubescent outside,
the upper lip concave, bifid, the lower twice as long with undulate crenate lobes. Lower end of the connective of the upper anthers scarcely
bearded.—*Colobandra platyphylla,* Bartl. in Pl. Preiss. i. 358.

W. Australia. Mount Bakewell, York district, *Preiss, n.* 2319 (*Herb. DC. and
F. Muell.*).

Colobandra subvillosa, Bartl. in Pl. Preiss. i. 359 (*Hemigenia subvillosa,* Benth. in
DC. Prod. xii. 566), from the same York district, is said to be very similar to *H. platyphylla,* but with villous branches, and is probably a variety only, as there are occasionally some non-glandular hairs on the typical *H. platyphylla.*

9. **H. glabrescens,** *Benth. in DC. Prod.* xii. 566. A shrub with
slender branches, slightly hoary as well as the young foliage with short
appressed hairs, at length nearly glabrous and without glandular hairs.
Leaves opposite, oblong or obovate-oblong, obtuse, contracted into a
short petiole, mostly ½ to ¾ in. long, or on the side shoots under ½ in.,
green and glabrous or nearly so when full-grown. Flowers small, axillary, on short pedicels. Bracts linear. Calyx rather broadly campanulate, villous with spreading hairs, 2 lines, or after flowering 3 lines long,
the teeth subulate-acuminate, irregularly 2-lipped. Corolla about twice
as long as the calyx, pubescent outside, the upper lip short, broad and
concave, the lower lip longer. Connective of the upper anthers bearded
at the lower end, of the lower anthers glabrous and clavate at the tip.

W. Australia, *Drummond, 1st coll. n.* 452.

10. **H. obtusa,** *Benth. in DC. Prod.* xii. 567. A slender, apparently
diffuse or spreading shrub, glabrous or more or less hoary with short
appressed hairs. Leaves opposite, oblong or cuneate, usually narrow
and sometimes almost linear, rarely almost obovate, obtuse, contracted

into a petiole, under ½ in. long. Flowers small, axillary, on pedicels sometimes above 1 line long, but usually short; bracts subulate. Calyx 2 lines, or after flowering nearly 3 lines long, shortly pubescent, the tube turbinate, the teeth not longer than the tube, acute or almost obtuse, more or less distinctly 2-lipped or almost equal. Corolla not twice as long as the calyx, pubescent outside. Connective of the upper anthers shortly bearded at the lower end, that of the lower anthers glabrous.

W. Australia, *Drummond, 2nd coll. n.* 147 ; plains near Observatory hill, Salt Lagoons, *Maxwell.*

SECT. 3. HEMIGENIA.—Calyx-teeth nearly equal, subulate-acuminate or acute. Lower end of the connective of the lower anthers attenuate. Leaves sessile, except in *H. humilis.*

11. **H. sericea,** *Benth. in Hueg. Enum.* 80, *and in DC. Prod.* xii. 567. A stout shrub of several feet, with erect branches more or less silky-villous or at length glabrous. Leaves opposite, sessile, erect or spreading, lanceolate or oblong, obtuse or mucronate, often contracted at the base, coriaceous, flat or concave, more or less silky-villous or silvery white when young, becoming glabrous when old, with few veins besides the midrib, mostly above 1 in. long. Flowers sessile or very shortly pedicellate, usually clustered in the axils with linear or linear-lanceolate bracts, rarely solitary on young side branches. Calyx usually about 3 lines long, the tube turbinate, the teeth lanceolate-subulate, nearly equal, longer than the tube. Corolla usually about ½ in. long, glabrous or nearly so outside. Connective of the upper stamens with the lower end broad and bearded, that of the lower stamens attenuate and glabrous. —Bartl. in Pl. Preiss. i. 360.

W. Australia. Swan river, *Fraser, Huegel, Drummond,* 1*st coll., Preiss, n.* 2333, *Oldfield* and others.

Var. *parviflora.* Leaves usually but not always narrow, and more contracted at the base, sometimes narrow-linear, silvery-white or nearly glabrous. Flowers smaller but variable in size, the calyx sometimes scarcely above 2 lines long.—*H. parviflora,* Bartl. in Pl. Preiss. i. 359 ; Benth. in DC. Prod. xii. 567.—With the typical form, *Drummond, Preiss, n.* 2321.

Drummond, n. 148 of the 3rd coll. with nearly glabrous very narrow or rather broad leaves, and n. 453 of the 1st coll. with silvery-white leaves, are in many respects intermediate forms, and to these varieties of *H. sericea* should probably be added *H. argentea,* Bartl. l.c. 360 ; Benth. in DC. l.c. 567, which I have not seen.

Var. *lanosa,* F. Muell. Leaves and flowers of the typical form, but the whole plant, especially the young parts densely woolly with long soft silky hairs.—*Drummond's last coll.*

12. **H. barbata,** *Bartl. in Pl. Preiss.* i. 360 ? Very closely allied to *H. sericea,* but the short broad leaves and loose indumentum give it a very different aspect. Young branches densely clothed with long loose spreading but silky hairs, which wear off with age. Leaves oblong, mostly rather above ½ in. long, and nearly ¼ in. broad, loosely silky-villous on both sides. Flowers small, usually 2 in each axil, nearly sessile, each with 2 linear membranous hairy bracts. Calyx 2 to 2½ lines long, silky-villous, the teeth nearly equal, soft, acutely acuminate,

rather shorter or longer than the tube. Corolla glabrous, at least in the bud. Anthers of *H. sericea*, the lower end of the connective of the lower ones slightly clavate or attenuate.

W. Australia, *Drummond, n.* 77 (*Preiss, n.* 2320 ?). I have not seen Preiss's specimens, but Drummond's agree much better with Bartling's description than the plant I referred to *H. barbata* in DC. Prod. xii. 566, which is but one of the small-flowered varieties of *H. sericea.*

13. **H. curvifolia,** *F. Muell. Fragm.* i. 210. A shrub of 2 or 3 ft., the branches woolly-hirsute, the young shoots silky-villous, the older foliage becoming glabrous. Leaves opposite, sessile, ovate-lanceolate or oblong, mostly acute, rigid, complicate, recurved, ½ in. long or rather more. Flowers rather small, clustered in the axils and in every respect like those of the small-flowered varieties of *H. sericea.*

W. Australia. Rocky hills, Hill river, *Oldfield.* There are but very few small specimens, more complete ones may possibly show this to be an extreme form of *H. sericea.*

14. **H. scabra,** *Benth.* Apparently an undershrub with slightly-branched erect stems of ½ to ¾ ft., scabrous-pubescent or shortly hirsute as well as the foliage. Leaves opposite, or very rarely in whorls of 3, oblong-cuneate, very obtuse or truncate at the end, contracted at the base but scarcely petiolate, 4 to 8 lines long. Flowers small, solitary in the axils, on short pedicels. Bracts linear-lanceolate, often as long as the calyx. Calyx nearly 3 lines long, the teeth acute, rather broad, all equal, as long as the tube. Corolla only seen in bud. Connective of the upper anthers with the lower end dilated and bearded, that of the lower anthers attenuate and glabrous.

W. Australia, *Drummond.*

15. **H. humilis,** *Benth. in DC. Prod.* xii. 567. A low shrub or undershrub, much branched at the base, usually under 6 in. high, but growing out sometimes to near 1 ft., hoary with a minute velvety pubescence wearing off from the older foliage, and with a few rigid spreading hairs about the inflorescence. Leaves opposite, linear linear-cuneate or oblong, obtuse, contracted into a short petiole, rather thick, flat, rarely above ½ in. long, and mostly shorter. Flowers small, solitary in the axils, on very short pedicels. Bracts linear-subulate, usually ciliate with rigid hairs. Calyx 2 lines, or after flowering 3 lines long, often hirsute and sometimes with a few glandular hairs, the tube turbinate, the teeth subulate-acuminate, nearly equal, longer than the tube. Corolla scarcely twice as long as the calyx, pubescent outside. Connective of the upper anthers broad and bearded at the lower end, that of the lower anthers narrow and glabrous or scarcely minutely bearded.

W. Australia, *Drummond, 2nd coll. suppl. n.* 49.

16. **H. westringioides,** *Benth. in DC. Prod.* xii. 568. A slender shrub with virgate branches, glabrous or minutely hoary-pubescent. Leaves opposite, sessile, very narrow linear or terete and channelled above, obtuse or mucronate-acute, contracted at the base, ½ to 1 in. long.

Flowers solitary in the axils, on pedicels of 2 to 3 or rarely 4 lines, with small subulate bracts near the calyx. Calyx usually minutely hoary-pubescent, about 3 lines long, the teeth broad, acute, nearly equal, much shorter than the tube. Corolla above twice as long as the calyx, glabrous outside. Connective of the upper anthers broad and bearded at the lower end, that of the lower anthers narrow and glabrous.

W. Australia, *Drummond, 3rd coll. n.* 152.

17. **H. teretiuscula,** *F. Muell. Fragm.* vi. 111. A slender branching shrub, with the habit and foliage of *H. westringioides,* but quite glabrous, the pedicels very short, the calyx-teeth narrow, acute and about as long as the tube, and the corolla scarcely so large as in *H. westringioides,* of which it is probably a variety.

W. Australia. Stokes Inlet and Kydenup Range, *Maxwell.*

18. **H. purpurea,** *R. Br. Prod.* 502. A slender twiggy heath-like shrub or undershrub, glabrous or with longitudinal rows of a minute pubescence on the branches. Leaves in whorls of 3 or 4, linear-terete, mucronate-acute or obtuse, channelled above, contracted at the base, and sometimes shortly petiolate, rarely above ½ in. long. Flowers " purple " or " blue," solitary and pedicellate or almost sessile in the upper axils. Bracts linear, shorter than the calyx. Calyx 2 to 2½ lines long, the tube turbinate, the teeth linear or linear-lanceolate, longer than the tube. Corolla not twice as long as the calyx, slightly pubescent outside, the lower lip twice as long as the upper. Connective of the upper anthers broad and bearded at the lower end, that of the lower anthers glabrous at the lower end, the cell at the upper end apparently perfect as usual in the genus.—Benth. in DC. Prod. xii. 568 ; *H. Sieberi,* Benth. Lab. Gen. et Sp. 457, and in DC. Prod. l.c.

N. S. Wales. Port Jackson to the Blue Mountains, *R. Brown, Sieber, n.* 191, *A. Cunningham,* and many others. On comparing a large number of specimens I am now persuaded that those with four leaves in a whorl (*H. Sieberi,* Benth.) do not otherwise differ from those which have only three.

19. **H. cuneifolia,** *Benth.* A shrub probably of 2 or 3 ft., glabrous except the corolla. Leaves in whorls of 3, oblong-cuneate, obtuse or mucronate-acute, contracted into a rather long petiole, flat, green on both sides, about ½ in. long. Flowers small, solitary in the axils, shortly pedicellate. Bracts small, acute. Calyx 1½ to 2 lines long, quite glabrous, striate, the teeth all equal, acute, shorter than the tube. Corolla pubescent outside, not twice as long as the calyx, the upper lip broad erect concave and emarginate as in the preceding species, the lower lip longer and spreading. Upper stamens as in *H. purpurea,* with the connective dilated and bearded at the lower end, the lower stamens with the connective short, the cell at the upper end ovate but perhaps not perfect, the lower end linear and glabrous.

N. S. Wales. George river, very rare, *Woolls;* Macleay river, *Beckler.*

The above specimens are referred by F. Mueller, Fragm. vi. 110, to *Westringia glabra,* which has something of the general aspect of this plant, but differently shaped

leaves and very different corolla and anthers. The New England plant there mentioned is the true *W. glabra.* In *H. cuneifolia,* the cell of the lower stamens in the
two flowers examined appeared to be not quite so perfect as is usual in *Hemigenia,*
showing thus a passage to the genus *Microcorys,* although still nearer to *Hemigenia,*
of which it has the corolla.

SECT. 4. DIPLANTHERA.—Calyx-teeth 5, nearly equal, subulate-
acuminate or acute. Lower end of the connective of the lower anthers
and sometimes of all the anthers bearing a second imperfect cell. Leaves
sessile, opposite.

20. **H. Drummondii,** *Benth.* A perennial or undershrub (sometimes shrubby ?) all the specimens showing several simple or slightly
branched stems erect from the rootstock, glabrous or with opposite decurrent lines of short hairs. Leaves opposite, sessile, oblong or oblong-
cuneate, obtuse, entire, 1-nerved, minutely ciliate or quite glabrous,
under ½ in. long. Pedicels short, with linear or linear-lanceolate ciliate
bracts. Calyx about 3 lines long, sprinkled or ciliate with a few long
spreading hairs, the teeth lanceolate, longer than the tube. Corolla
twice as long as the calyx, the tube rather long, the upper lip concave
below the lobes. Anthers rather large, the connective of the upper
ones dilated and bearded at the lower end, that of the lower anthers
bearing a second smaller cell probably sterile.

W. Australia, *Drummond, last coll.*

21. **H. pimelifolia,** *F. Muell. Fragm.* vi. 112. A shrub with slender
divaricate branches, the young ones and foliage hirsute with long spreading hairs and opposite rows of shorter ones on the branches. Leaves
opposite, sessile, obovate or oblong, obtuse, concave, rather thick, 2 to
3 lines long. Flowers solitary in the axils, nearly sessile. Bracts linear
or linear-lanceolate. Calyx hirsute with long spreading hairs, under 2
lines long, the teeth narrow-lanceolate, nearly equal, rather obtuse,
longer than the tube. Corolla scarcely exceeding the calyx, the lobes
less unequal than in the other species. Connective of the upper anthers
with the lower end dilated and bearded, that of the lower anthers with
the lower branch elongated and terminating in a second cell nearly as
large as the perfect one, but perhaps sterile.

W. Australia. Murchison river, *Oldfield.*

22. **H. diplanthera,** *F. Muell. Fragm.* vi. 111. A heath-like erect
glabrous shrub or undershrub, sometimes bushy and under 6 in., sometimes more straggling, and attaining 1 to 3 ft., the smaller branches
slender. Leaves opposite, linear or rarely linear-oblong, acute or obtuse, concave, contracted at the base, and sometimes shortly petiolate.
Pedicels solitary in the axils, shorter than the calyx. Bracts subulate.
Calyx 2 to 2½ lines long, the tube turbinate, the teeth nearly equal,
broad or narrow, very acute, rarely as long as the tube. Corolla
" white," the upper lip rather longer than the calyx, 2-lobed and
scarcely concave below the lobes, the lower lip nearly twice as long as
the calyx, 3-lobed, all the lobes nearly equally 2-lobed. Anthers all

with one perfect cell, the lower lobe of the connective ending in a second smaller and perhaps sterile slender cell in the lower stamens, and sometimes also in the upper ones, but sometimes the lower end of the latter dilated and bearded as in the other species of the genus, and in one flower I found a fifth imperfect stamen.

W. Australia, *Drummond, last coll.*

18. MICROCORYS, R. Br.

(Anisandra, *Bartl.*)

Calyx campanulate, 5-toothed. Corolla with a dilated throat, the upper lip very concave or hood-shaped, with the addition sometimes of 2 flat spreading lobes; lower lip spreading, 3-lobed. Stamens 2 (the upper ones) perfect, the connective elongated, produced beyond the insertion into a short lower branch usually dilated and bearded at the end, the 2 lower stamens (or staminodia) sterile and short, the anthers reduced to a small connective, with 2 linear or linear-clavate parallel branches. Style shortly bifid at the end. Nuts reticulate-rugose, attached to the middle or higher up. Seeds albuminous.—Shrubs or undershrubs. Leaves opposite or more frequently in whorls of 3 or 4, all entire. Flowers all axillary or rarely in terminal leafy heads, solitary within each floral leaf, with a pair of bracts under or below the calyx. Corolla usually hairy inside the tube, especially at the insertion of the stamens.

The genus is limited to Western Australia.

SECT. 1. **Hemigenioides.**—*Leaves opposite. Corolla-tube exserted, the upper lip concave, shortly lobed, not much shorter than or as long as the lowest lobe.*
Corolla narrow, above 1 in. long.
     Leaves linear, with revolute margins. Pedicels rather long,
         upper corolla-lip longer than the lower . . . . 1. *M. longifolia.*
     Leaves ovate or elliptical-oblong with recurved margins. Pedi-
         cels short. Upper corolla-lip nearly as long as the lower . 2. *M. longiflora.*
Corolla broad, ½ in. long or less, the upper lip nearly as long as
     the lower.
     Leaves narrow linear . . . . 3. *M. tenuifolia.*
     Leaves ovate or oblong, flat . . . . 4. *M. loganiacea.*

SECT. 2. **Anisandra.**—*Leaves in whorls of 3 or 4, flat or concave. Corolla-tube included or rarely exserted, the upper lip very concave or hood-shaped with 2 anterior spreading usually large lobes.*
Flowers in terminal leafy heads or spikes, the floral leaves much
     broader than the stem ones.
     Leaves ovate . . . . 5. *M. capitata.*
     Leaves oblong-lanceolate . . . . 6. *M. pimelecides.*
Flowers all in the axils of leaves similar to the stem ones.
     Leaves linear-cuneate, 3 to 5 lines long. Calyx silky or hoary 7. *M. subcanescens.*
     Leaves under 3 lines, linear or oblong.
         Calyx densely hirsute, with long spreading hairs.
             Corolla-tube as long as the calyx . . . . 8. *M. ericifolia.*
         Calyx glabrous or slightly pubescent.
             Corolla-tube twice as long as the calyx . . . . 9. *M. exserta.*
             Corolla-tube not exceeding the calyx . . . . 10. *M. glabra.*

SECT 3. **Microcorys.**—*Leaves in whorls of 3 or 4. Corolla-tube included, the upper lip very short, concave or hood-shaped, without large anterior lobes, the lower lip much longer.*

Leaves linear, rarely above 3 lines long. Flowers very small.
   Calyx glabrous or scarcely pubescent . . . . 11. *M. virgata.*
   Calyx very densely hirsute with white hairs . . . . . . 12. *M. barbata.*
Leaves orbicular, flat, 1 to 2 lines diameter. Flowers very small.
   Calyx very densely hirsute . . . . 13. *M. lenticularis.*
Leaves broad, 3 to 6 lines long. Flowers large. Calyx glabrous
   or slightly hoary.
   Leaves flat, obovate . . . . 14. *M. obovata.*
   Leaves ovate, with revolute margins . . . . 15. *M. purpurea.*

SECT. 1. HEMIGENIOIDES.—Leaves opposite. Corolla-tube exserted, the upper lip concave, shortly lobed, not much shorter than or as long as the lowest.

1. **M. longifolia,** *Benth. in DC. Prod.* xii. 568. Apparently a tall shrub, the branches and young shoots hoary with a minute tomentum. Leaves opposite, sessile, linear, obtuse, with revolute margins, becoming glabrous above when old, hoary or white underneath, above 1 in. and often nearly 2 in. long. Pedicels axillary, slender, spreading, often ½ in. long, with a pair of small linear-subulate bracts a little distance below the calyx, and there usually bent. Calyx rather narrow, slightly hoary, the tube nearly 3 lines long, the teeth narrow, acute, rather shorter than the tube, the uppermost usually larger than the others. Corolla slightly pubescent outside, above 1 in. long, the tube exserted and somewhat dilated upwards, the upper lip narrow, erect, concave, slightly emarginate, the lower one shorter, spreading, with 3 ovate entire lobes. Upper anthers with one large fertile cell, the lower end of the connective rather long, dilated at the end but scarcely bearded. Nuts less prominently reticulate than in most species.—*Hemigenia longifolia,* Benth. in Hueg. Enum. 80.

W. Australia. Swan river, *Huegel, Drummond, 2nd coll. n.* 214.

The long corollas of this and the following species give them more the aspect of *Hemigenia* than of *Microcorys,* but the lower anthers in both have the two linear sterile lobes of *Microcorys* without any perfect cell.

2. **M. longiflora,** *F. Muell. Fragm.* vi. 113. A divaricately branched rather slender shrub, apparently 1 to 2 ft. high, the branches and young shoots minutely pubescent. Leaves opposite, petiolate, ovate to elliptical-oblong, obtuse, with recurved margins, glabrous above when old, pale or hoary underneath, mostly about ½ in. long, but varying from ¼ to ¾ in. Pedicels short but slender, often recurved, the bracts very short, linear-setaceous. Calyx 3 to 4 lines or rarely at length 5 lines long, the teeth lanceolate, acute, the upper one often as long as the tube, the others smaller. Corolla nearly 1 in. long, the tube slightly dilated upwards and much longer than the calyx, the lips short, the upper one erect and concave, the lower one scarcely longer, with three ovate lobes. Connective of the upper anthers dilated and bearded at the lower end.

W. Australia. Between Swan river and Cape Riche, *Drummond, 5th coll. n.* 340.

3. **M. tenuifolia,** *Benth.* A slender shrub, with much of the aspect of *Hemigenia westringioides,* but the branches and young shoots hoary or white with minute appressed hairs, and the stamens those of *Microcorys.* Leaves opposite, very narrow linear, obtuse or mucronate, thick, with slightly recurved margins, mostly ½ to ¾ in. and sometimes nearly 1 in. long. Pedicels filiform, 2 to 3 lines long, with short setaceous bracts at some distance from the calyx. Calyx narrow, hoary with a minute pubescence, about 3 lines long, the teeth narrow-lanceolate, acute, rather shorter than the tube. Corolla pubescent, the tube not exceeding the calyx, the upper lip short, concave, broadly 2-lobed, the lower. lip (rather longer?) 3-lobed. Connective of the upper stamens long, the lower end short, dilated and bearded.

W. Australia, *Drummond, 4th coll. n.* 172. The foliage and inflorescence are like those of *M. longifolia* but more slender, and the short broad corolla is very different.

4. **M. loganiacea,** *F. Muell. Fragm.* vi. 113. A shrub or undershrub, the stems not much branched, ½ to 1 ft. high or rather more, minutely scabrous-pubescent. Leaves opposite, ovate or oval-oblong, obtuse, flat, rather thick, contracted into a short petiole, nerveless except the midrib, minutely scabrous-pubescent or glabrous. Flowers nearly sessile or on pedicels of about 1 line. Bracts linear-lanceolate, sometimes as long as the calyx. Calyx more or less hirsute, about 3 lines long or at length nearly 4 lines, the teeth linear-lanceolate, longer than the tube and nearly equal. Corolla pubescent outside, scarcely twice as long as the calyx, the throat broad and open, the upper lip concave, with 2 short broad lobes, the lower one but little longer, with 3 broad fringed or crenate lobes. Connective of the upper anthers dilated and slightly bearded at the lower end. Nuts glabrous, slightly reticulate.

W. Australia. Towards Cape Riche, *Harvey, Drummond, 4th coll. n.* 168, *Maxwell.*

SECT. 2. ANISANDRA.—Leaves in whorls of three or rarely four, flat or concave. Corolla-tube included or rarely exserted, the upper lip very concave or hood-shaped, with 2 anterior spreading usually large lobes.

5. **M. capitata,** *Benth. in DC. Prod.* xii. 568. An erect rigid shrub, the branches often clustered and divaricate under the old inflorescences, the whole plant usually glabrous except the inflorescence. Leaves in whorls of 3, on exceedingly short petioles, ovate, acute or rather obtuse, coriaceous, flat, 1-nerved, under ½ in. long, the floral ones much broader not so thick and ciliate. Flowers solitary under each floral leaf, but collected in terminal globular or ovoid heads, the floral leaves imbricate and about as long as the calyx. Bracts very small, setaceous and ciliate or quite abortive. Calyx 3½ to 5 lines long, glabrous or slightly glandular-pubescent, the teeth rather broad, very acute or acu-

minate, much shorter than the tube. Corolla not much longer than the calyx, the upper lip erect, conspicuously helmet-shaped, with 2 anterior spreading lobes ; lower lip spreading, 3-lobed. Filaments of the upper anthers short, the connective long and slender, with the lower end broadly oblong and shortly bearded.— *Westringia capitata,* Bartl. in Pl. Preiss. i. 362.

W. Australia, *Drummond, n.* 98, 4*th coll. n.* 143 ; Mount Baldhead, *Preiss, n.* 2334.

6. **M. pimeleoides,** *F. Muell. Fragm.* i. 156. A robust shrub, quite glabrous or with minute opposite lines of pubescence along the branches. Leaves crowded, in whorls of 3, oblong-lanceolate, obtuse or almost acute, minutely petiolate, thick, concave, 3 to 4 lines long, the floral ones broader, almost ovate. Flowers almost sessile and crowded at the ends of the branches, but not so distinctly capitate as in *M. capitata,* the floral leaves about as long as the calyxes. Bracts small, linear or oblong. Calyx nearly 3 lines long, the teeth ovate or oblong-ovate, mucronate, as long as the tube, rather spreading. Corolla shortly pubescent outside, nearly twice as long as the calyx, the tube not exserted, the upper lip hood-shaped, with two large broad anterior spreading lobes, the lower lip of three broad undulate emarginate lobes. Connective of the upper anthers long, with a large acuminate lower lobe slightly bearded at the end.

W. Australia. Phillips range, *Maxwell.*

7. **M. subcanescens,** *Benth.* An apparently small slender shrub, the branches and young shoots hoary or white with a close almost silky pubescence, the adult foliage glabrous. Leaves mostly in rather distant whorls of 3, linear-oblong or slightly cuneate, very obtuse, contracted into a very short petiole, coriaceous, 1-nerved, flat or concave, 3 to 5 lines long. Flowers axillary, nearly sessile. Bracts linear, at least as long as the calyx-tube. Calyx about 3 lines long, silky-pubescent, the teeth narrow-lanceolate, rather obtuse, about as long as the tube. Corolla pubescent outside, the tube shorter than the calyx, the upper lip scarcely exceeding the calyx-teeth, erect, very concave with 2 anterior spreading lobes, the lower lip longer, the lateral lobes obovate, the middle one broad and emarginate. Connective of the upper anthers with the lower lobe broad and ciliate.

W. Australia, *Maxwell.*

8. **M. ericifolia,** *Benth. in DC. Prod.* xii. 569. A heath-like shrub of several ft., with virgate branches minutely pubescent in decurrent lines or glabrous. Leaves in whorls of 3, or rarely of 4, linear or oblong, obtuse, contracted at the base, thick and flat or concave, glabrous, sometimes all scarcely above 1 line long, in other specimens 2 to 3 lines, the floral ones similar but usually shorter than the calyx. Flowers axillary but usually crowded in leafy racemes towards the ends of the branches. Pedicels very short or nearly 1 line long. Bracts linear, often as long as the calyx-tube. Calyx 2 to 2½ lines long or after

flowering nearly 3 lines, narrower and more contracted at the base than that of *M. barbata*, densely hispid with spreading hairs longer and more rigid than in that species. Corolla glabrous or slightly pubescent outside, the tube as long as the calyx, the upper lip very concave or almost hood-shaped at the base with 2 large obovate-oblong erect but laterally spreading lobes, lower lip spreading, with a large and broad middle lobe. Connective of the upper anthers rather long, the lower lobe short and bearded. Nuts glandular-pubescent.

W. Australia, *Drummond,* 1*st coll. n.* 453 *or* 455, 2*nd coll. suppl. n.* 70. Notwithstanding some general resemblance to *M. barbata* this species is readily distinguished by the calyx twice as long, narrower and with much longer and more spreading hairs, and by the large lobes of the upper lip of the corolla.

M. parvifolia, Benth. in DC. Prod. xii. 569 (*Drummond,* 1*st coll. n.* 456) has small leaves and rather smaller flowers, but I have now seen too many ambiguous specimens to admit of distinguishing it even as a marked variety. A specimen of Roe's has the leaves all small and much broader than usual, some of them almost ovate.

9. **M. glabra,** *Benth. in DC. Prod.* xii. 569.—A shrub of 1 to 3 ft. usually much branched, but less virgate and looser than *M. virgata*, the typical form glabrous in every part. Leaves in whorls of 3 or very rarely of 4, erect or spreading, scarcely petiolate, linear, obtuse, thick, flat or concave, rarely under 3 lines and often 4 lines long. Flowers axillary, on short pedicels, with very small linear obtuse bracts. Calyx 1½ lines long when in flower, afterwards attaining 2 lines, glabrous or slightly pubescent, the teeth lanceolate, usually obtuse and rather shorter than the tube, sometimes very short. Corolla quite glabrous outside, the tube shorter than the calyx, the upper lip very concave with 2 broad open anterior lobes, the lower lip much longer. Connective of the upper anthers longer than in *M. virgata*, the lower lobe dilated and bearded.—*Anisandra glabra*, Bartl. in Pl. Preiss. i. 361; *Microcorys brevidens*, Benth. in DC. Prod. xii. 569.

W. Australia, *Drummond,* 3*rd coll. n.* 150, 4*th coll. n.* 104; Konkoberup hills, *Preiss, n.* 2328; Salt river, Plantagenet and Stirling ranges, *Maxwell.*

Var. *gracilis.* Very slender in all its parts and heath-like. Leaves very narrow often almost terete, 2 to 3 lines long. Flowers smaller but the essential characters entirely those of *M. glabra.*—" Poor soil," no precise station, *Maxwell.*

Var.? *pubescens.* Branches foliage and calyx pubescent with very short spreading hairs, the corolla and stamens as in the typical form.—Salt river ranges (with the leaves as in the typical form), Phillips river (with shorter and broader leaves), *Maxwell.*

10. **M. exserta,** *Benth.* A shrub, probably of 2 or 3 ft., glabrous except the corolla, with numerous virgate branches. Leaves rather crowded, in whorls of 3 or 4, linear, obtuse or almost acute, contracted into a short petiole, thick, flat or concave, under 3 lines and often not 2 lines long. Flowers nearly sessile in the upper axils. Bracts very small, linear, deciduous. Calyx glabrous or slightly pubescent, under 2 lines long when in flower, above that when in fruit, the teeth ovate, slightly mucronate, shorter than the tube. Corolla pubescent outside, the tube slender, exserted (about half as long again as the calyx), the upper lip short, broad, very concave, with spreading anterior lobes, the

lower lip much longer, with broad lobes.   Connective of the upper anthers rather long, with a broad lower lobe bearded at the end.

W. Australia.   East river, Stokes Inlet, *Maxwell*, and some specimens in *Drummond's 3rd coll. n.* 151.   With the habit and foliage of *M. ericifolia* and *M. virgata,*
this is at once distinguished by the corolla-tube, which, when perfect, is fully 3 lines
long.

SECT. 3. MICROCORYS.—Leaves in whorls of three or rarely four. Corolla-tube not exceeding the calyx-teeth, the upper lip very short, concave or hood-shaped, slightly emarginate, without the two large spreading lobes of *Anisandra,* the lower lip much longer, spreading.

11. **M. virgata,** *R. Br. Prod.* 502.   An erect shrub with erect or
spreading slender virgate branches, the whole plant glabrous except the
corolla.   Leaves in whorls of 3, linear, obtuse, contracted at the base,
but scarcely petiolate, thick, flat or concave, rarely exceeding 3 lines,
the floral ones similar or rather smaller.   Flowers all axillary and nearly
sessile, but frequently forming terminal leafy racemes.   Bracts very
small and falling off early so as to be rarely seen.   Calyx 1¼ or rarely
1½ lines long, glabrous, the teeth ovate, obtuse or shortly mucronate,
shorter than the tube, sometimes shortly ciliate.   Corolla hirsute outside
with long stiff hairs (except the upper part of the lower lobes) the upper
lip not exceeding the calyx, very broad, concave, very shortly emarginate or sinuate-lobed, lower lip more than twice as long, with broadly
obovate spreading lobes.   Connective of the upper anthers short, the
lower dilated and bearded lobe nearly as long as the perfect cell.—Benth.
in DC. Prod. xii. 569; Bartl. in Pl. Preiss. i. 362.

W. Australia.   Boggy ground, King George's Sound, *R. Brown, A. Cunningham ;*
rocks of Mount Wulgenup, *Preiss, n.* 2330 ; also *Drummond, 4th coll. n.* 169.

M. selaginoides, Bartl. in Pl. Preiss. i. 363, Benth. in DC. Prod. xii. 569, from moist
shady bogs, Twopeopled Bay, *Preiss, n.* 2332, from the single not very good specimens
seen, appears to be a slight variety of *M. virgata,* with the branches minutely hoary-
pubescent, and the leaves slightly scabrous with minute hairs.

12. **M. barbata,** *R. Br. Prod.* 502.   A shrub with slender virgate
branches, glabrous or with minutely pubescent decurrent lines, the
foliage quite glabrous.   Leaves in whorls of 3, linear, obtuse, thick,
concave or almost terete, contracted into a short petiole, 2 to 3 or very
rarely 4 lines long, the floral ones similar but often smaller.   Flowers
very small, all axillary but forming long leafy racemes and very conspicuous from the white silky hairs of the calyx.   Calyx scarcely 1¼
lines long, densely hirsute with white spreading hairs, the teeth broadly
oblong, almost obtuse, about as long as the tube.   Corolla pubescent
outside, the upper lip scarcely exceeding the calyx, broad, concave,
shortly sinuate-lobed, the lower lip much longer and spreading.   Connective of the upper anthers short, the lower dilated and bearded lobe
nearly as long as the perfect cell.   Nuts hirsute.—Benth. in DC. Prod.
xii. 569.

W. Australia.   Lucky Bay, *R. Brown ;* to the eastward of King George's Sound ?
Baxter, Drummond, 4th coll. n. 167 ; Kojonerup valley, Oldfield river and Esperance
Bay, *Maxwell.*

13. **M. lenticularis,** *F. Muell. Fragm.* vi. 113. A shrub with the slender virgate branches and hispid calyxes of *M. barbata*, quite glabrous or with minutely pubescent decurrent lines. Leaves in whorls of three, very broadly ovate or orbicular, obtuse, thick, flat or concave, shortly but distinctly petiolate, 1 to nearly 2 lines diameter. Flowers in the upper axils nearly sessile. Calyx after flowering ovoid-globular, densely hirsute with whitish spreading hairs, rather above 1 line long, the teeth rather broad and usually shorter than the tube. Corolla and stamens not seen. Nuts pubescent.

W. Australia, *Drummond, 3rd coll. n.* 196.

14. **M. obovata,** *Benth. in DC. Prod.* xii. 569. An erect bushy shrub, the branches and young shoots hoary with minute appressed hairs, the adult foliage glabrous. Leaves mostly in whorls of three, obovate, obtuse, contracted into a very short petiole or almost sessile, flat, coriaceous, nerveless, 3 to 4 lines long. Flowers axillary, on very short pedicels, with small linear-setaceous ciliolate bracts. Calyx about 2 lines long or at length rather longer, glabrous or nearly so, the teeth acute, rather broad, about as long as the tube. Corolla nearly ½ in. long, slightly pubescent outside, the tube longer than the calyx, the upper lip short, broad, concave, without the spreading anterior lobes of *Anisandra*, the lower lip three times as long, with broad lobes. Connective of the upper anthers dilated and bearded at the lower end.

W. Australia, *Drummond, n.* 69 *and 3rd coll. n.* 195.

15. **M. purpurea,** *R. Br. Prod.* 502. A bushy or spreading shrub of 2 to 3 ft., the branches and young shoots hoary-pubescent with short appressed hairs, the adult foliage often glabrous. Leaves in whorls of 3, on very short petioles, ovate, obtuse or scarcely acute, with recurved margins, green above, pale or hoary and long retaining their pubescence underneath, rarely above ½ in. long. Flowers "purple" or "puce-coloured," all axillary and distant, on short pedicels with minute bracts. Calyx hoary-pubescent, 2½ to nearly 3 lines long, the teeth narrow, much longer than the tube. Corolla-tube shorter than the calyx, the upper lip not at all or scarcely exceeding the calyx-teeth, broad, concave, very shortly lobed, the lower lip much longer and spreading. Connective of the upper anthers about as long as the cell, the lower lobe very short, broad and bearded at the end. Nuts glabrous.—Benth. in DC. Prod. xii. 569.

W. Australia. Lucky Bay, *R. Brown ;* overhanging rocks, Cape Arid, *Maxwell.* Some very bad specimens of *Drummond's* in herb. F. Mueller, with shorter and more oblong leaves and a more compact inflorescence, may nevertheless possibly belong to this species.

Westringia serpyllifolia, Bartl. in Pl. Preiss. i. 362 ; Benth. in DC. Prod. xii. 571, from near Mount Manypeak ; *Preiss, n.* 2312, of which I formerly saw a very imperfect specimen in Herb DC., and of which the corolla and stamens are unknown, is most probably the *Microcorys purpurea*.

19. WESTRINGIA, Sm.

Calyx campanulate, 5-toothed. Corolla with a short tube and dilated throat; the upper lip erect but flat and broadly 2-lobed, the lower spreading, 3-lobed. Stamens 2 (the upper ones) perfect, the anthers 1-celled with a short slightly prominent connective not produced below its insertion on the filament, the 2 lower stamens (or staminodia) sterile and short, the anthers reduced to a small connective with 2 linear or linear-clavate parallel branches. Style shortly bifid at the end. Nuts reticulate-rugose, attached to the middle or higher up. Seeds albuminous.—Shrubs. Leaves in whorls of three, four or rarely more, all entire. Flowers all axillary or rarely in terminal leafy heads, with a pair of bracts under the calyx usually very small and sometimes almost obsolete. Corolla usually hairy inside the tube, especially at the insertion of the stamens.

The genus is limited to Australia. With the exception of *W. cephalantha*, the species are so closely allied, and run so much into each other as to render it exceedingly difficult to assign to them any tangible characters. The chief differences observed are in the number of leaves in the whorl, in the relative abundance or absence of the hoary tomentum, and in the length of the teeth of the calyx, none of which are quite constant in any one species. The corolla might perhaps in some instances supply better characters, but they can only be ascertained by the observation of living specimens; the want of any concavity in the upper lip at once distinguishes the genus from *Microcorys*.

Flowers in globular terminal heads with bract-like floral leaves
  not exceeding the calyx . . . . 1. *W. cephalantha.*
Flowers all in the axils of leaves not differing from the stem
  ones (distant or crowded at the ends of the branches).
  Leaves very white underneath, often not much revolute.
    Calyx-teeth above half as long as the tube.
     Leaves in threes, oblong-elliptical, ½ to 1 in. long . . 2. *W. grandifolia.*
     Leaves in fours, oblong-lanceolate or linear, ⅓ to 1 in.
      long . . . . 3. *W. rosmariniformis.*
     Leaves in fours, 3 to 4 lines long . . . . 4. *W. brevifolia.*
  Leaves very much revolute, rigid. Calyx-teeth very short.
    Leaves mostly in fours . . . . 5. *W. Dampieri.*
    Leaves mostly in threes . . . . 6. *W. rigida.*
  Leaves much revolute, rigid, in fives or sixes. Calyx-teeth
    as long as the tube . . . . 7. *W. senifolia.*
  Leaves narrow-linear, slender, much revolute, mostly in
    threes. Calyx-teeth as long as the tube . . . . . 8. *W. eremicola.*
  Leaves not much revolute or flat, green on both sides as
    well as the calyx.
    Leaves in threes, linear, often above 1 in. long. . . 9. *W. longifolia.*
    Leaves in threes, oblong-elliptical or lanceolate, under 1
     in. long . . . . 10. *W. glabra.*
    Leaves in fours, oblong-elliptical or lanceolate, under ½
     in. long . . . . 11 *W. rubiæfolia.*

1. **W. cephalantha,** *F. Muell. Fragm.* vi. 110. An erect bushy shrub, glabrous except the inflorescence. Leaves in whorls of 3, 4 or rarely 5, linear, obtuse or almost acute, with revolute margins, shining and black when dry, rarely exceeding ¼ in., the floral ones passing into linear concave rather thin ciliolate bracts. Flowers sessile and

solitary within each floral leaf, collected into dense globular terminal
heads of 3 or 4 lines diameter, the bract-like floral leaves not exceeding
the calyxes and the real bracts usually wanting. Calyx $1\frac{1}{4}$ to $1\frac{1}{2}$ lines
long, the teeth very obtuse, irregularly separating to below the middle.
Corolla pubescent outside, the tube about as long as the calyx, the
upper lobes erect and obovate, the lower ones spreading, all flat obovate
and entire, or the middle lower one emarginate. Stamens of *Westringia*,
except that, at least in the flowers examined, the anthers of the lower
pair (or staminodia) are entirely abortive.

W. **Australia,** *Drummond,* 4th coll. n. 170 and 5th coll. suppl. n. 76. Although
the inflorescence is so nearly that of *Microcorys capitata,* the foliage and flowers are
very different.

2. **W. grandifolia,** *F. Muell. Herb.* A tall shrub with the habit
and white indumentum of *W. rosmariniformis.* Leaves in whorls of 3,
oblong-elliptical, $\frac{1}{2}$ to 1 in. long, the margins slightly revolute, green
above, white underneath. Flowers all axillary. Calyx very white,
about 4 lines long, the teeth nearly as long as the tube. Corolla
pubescent, nearly that of *W. rosmariniformis,* but the lobes appear to be
shorter and broader, and the upper lip rather shorter than the lower.

Queensland. Glasshouse mountains, *F. Mueller,* and probably the same species
but the specimens not in flower, Biroa, *Leichhardt.*

F. Mueller now proposes to reduce this to *W. rosmariniformis,* but the characters
appear as distinct as those of most *Westringiæ.*

3. **W. rosmariniformis,** *Sm. Tracts,* 282, t. 3. A robust bushy
shrub of several ft., the branches underside of the leaves and calyxes
hoary or silvery-white with densely appressed hairs. Leaves in whorls
of 4, oblong-lanceolate lanceolate or linear, acute or obtuse, $\frac{1}{2}$ to 1 in.
long, coriaceous, glabrous and shining on the upper side, the margins
more or less recurved or revolute. Flowers almost sessile, all axillary,
with short linear bracts. Calyx about 3 lines long, the teeth acute,
varying from half the length to nearly the length of the tube. Corolla
pubescent outside, not twice as long as the calyx, the upper lip deeply
2-lobed, equal to or longer than the tube, the lower lip scarcely so
long. Anthers of the staminodia with 2 linear clavate parallel lobes.
Nuts glabrous, reticulate-rugose. — Benth in DC. Prod. xii. 570;
R. Br. Prod. 501; *Cunila fruticosa,* Willd. Spec. Pl. i. 122; *W. rosma-
rinacea,* Andr. Bot. Rep. t. 214.

N. S. Wales. Sandy hills near the sea coast, Port Jackson, *R. Brown, Sieber,*
n. 266, and others; northward to Port Macquarrie, *Backhouse,* and southward to Cape
Howe, *Mossman.*

4. **W. brevifolia,** *Benth. Lab. Gen. et Sp.* 459, *and in DC. Prod.* xii.
570. Very near *W. rosmariniformis,* with which F. Mueller proposes to
unite it, but it is a much smaller plant and its short leaves and small
flowers give it a very different aspect. Leaves in whorls of 4, oblong
or elliptical-lanceolate, hoary or white underneath, usually 3 to 4 lines
long, or in Hannaford's very luxuriant specimens about $\frac{1}{2}$ in. Flowers
like those of *W. rosmariniformis* but much smaller, the stamens usually

more exserted. Calyx-teeth as in that species about half the length of the tube or rather more.—Hook. f. Fl. Tasm. i. 285, t. 91.

W. Australia. South of Launceston, *Gunn, Hannaford.*

5. **W. Dampieri,** *R. Br. Prod.* 501. A rigid bushy shrub usually more scrubby than *W. rosmariniformis* but attaining several feet, the young shoots and under-side of the leaves often hoary, but less white than in that species, sometimes the whole plant slightly but equally hoary or altogether nearly green. Leaves in whorls of 4, or very rarely of 3 on the side-branches, linear, much revolute, the upper surface smooth or scabrous, varying in length but usually about ½ in. Flowers axillary, nearly sessile, about the size of those of *W. rosmariniformis*, but the corolla more hirsute and the calyx with a striate tube of about 2 lines and the teeth whether narrow or broad always very short and acute, usually not ¼ the length of the tube.—Benth. in DC. Prod. xii. 570; Bartl. in Pl. Preiss. i. 361; Bot. Mag. t. 3308.

Tasmania. Peaks of Flinders island and between Huon river and Oyster Bay, *Milligan;* Port Esperance, *Oldfield;* South Port, *C. Stuart.* The Tasmanian specimens, however, although undistinguishable from several western ones of *W. Dampieri,* may be only a 4-leaved variety of *W. angustifolia,* Br., the common 3-leaved form of which I am unable to distinguish from *W. rigida.*

W. Australia. King George's Sound and to the eastward, *R. Brown, Baxter, Drummond, n. 47 and 5th coll. n. 342, Preiss, n. 2325, Maxwell.*

6. **W. rigida,** *R. Br. Prod.* 501. A rigid bushy scrubby shrub, scarcely to be distinguished from *W. Dampieri* by any constant characters. It varies also like that species in the indumentum, sometimes hoary all over or green all over, usually the young shoots and under-side of the leaves only hoary and never so white as in *W. rosmariniformis.* Leaves mostly in whorls of 3, but here and there of 4, linear, obtuse or mucronate-acute, rigid with much revolute margins, usually glabrous above when full-grown and either smooth and shining or scabrous with minute tubercles, varying from scarcely 2 lines long and thick and broad in some specimens, to above ½ in. and more slender in others, with every intermediate size. Flowers of *W. Dampieri,* with the same very short teeth to the calyx, which varies from very hoary to quite green.— Benth. in DC. Prod. xii. 570; Bartl. in Pl. Preiss. i. 361; *W. grevillina,* F. Muell. in Hook. Kew Journ. viii. 169 and in Trans. Phil. Soc. Vict. i. 49.

Victoria. Avoca and Murray rivers, *F. Mueller;* Wimmera, *Dallachy.*
Tasmania. See below, *W. angustifolia.*
S. Australia. Fowler and Petrel Bay, Waldegrave and Flinders islands, *R. Brown;* Salt Creek, *Behr.;* Murray Creek to St. Vincent's Gulf, *F. Mueller;* Port Lincoln, *Wilhelmi.*
W. Australia, *Drummond, n. 194, Harvey;* and rocky shores of Rottenest Island, *A. Cunningham, Preiss, n. 2309;* Sharks Bay, *Milne;* Murchison river, *Oldfield.*

This species differs generally from *W. Dampieri* in the shorter more rigid leaves in whorls of 3 only; but none of these characters are constant, the primary branches even of the most characteristic short-leaved forms having occasionally the leaves in fours, the short calycine teeth and the corollas are the same in both species which might well be united as suggested by F. Mueller, in which case the name of *W. Dampieri* might be given to the whole.

W. cinerea, R. Br. Prod. 501, Benth. in DC. Prod. xii. 570, only differs from the longer-leaved form of *W. rigida* in being more hoary than usual, the plant figured Bot. Mag. t. 3307 is still more hoary with the leaves longer than in any of the Continental wild specimens.

W. angustifolia, R. Br. Prod. 501, Benth. in DC. Prod. xii. 571 ; Hook. f. Fl. Tasm. i. 285, from the central and southern parts of Tasmania, *R. Brown* and others, does not appear to me to differ in the slightest degree from the longer-leaved continental specimens of *W. rigida;* in S. and W. Australia, as in Tasmania, the leaves are sometimes very scabrous sometimes quite smooth. The specimens from between Oyster Bay and Huon river, *Milligan*, and from the Mersey river, *C. Stuart*, have the leaves sometimes in threes sometimes in fours, thus still further connecting *W. rigida* with *W. Dampieri.* F. Mueller refers them to *W. rosmariniformis*, but they have all the very short calyx-teeth of *W. Dampieri.* A few specimens of C. Stuart's have remarkably long leaves, attaining almost 2 inches.

7. **W. senifolia,** *F. Muell. in Hook. Kew Journ.* viii. 169, *and in Trans. Phil. Soc. Vict.* i. 49. An erect robust bushy shrub of 2 to 4 ft., more or less hirsute with white hairs, sometimes silky but looser and longer than in the other species, occasionally wearing off from the older leaves. Leaves crowded, in whorls of 5 or 6, linear, acute or almost obtuse, rigid, the margins much revolute, mostly about ½ in. long. Flowers axillary, sessile. Calyx-tube 1½ or at length 2 lines long, the teeth subulate-acuminate, nearly or quite as long as the tube. Corolla slightly pubescent outside, the lobes all nearly equal or the middle lower one rather longer, and not very much exceeding the calyx-teeth.

Victoria. Mount Aberdeen, Buffalo Range, *F. Mueller.*

Var. *canescens.* Shortly hoary, like *W. rigida*, but with the crowded whorls of 5 or 6 leaves, the long calyx-teeth and small corollas of *W. senifolia.*

W. Australia. Phillips river, *Maxwell.*

8. **W. eremicola,** *A. Cunn. in Benth. Lab. Gen. et Sp.* 459, *and in DC. Prod.* xii. 571. A shrub of several ft. with erect often virgate rather slender branches more or less hoary or silky-pubescent as well as the leaves with appressed hairs, the older foliage becoming glabrous. Leaves usually in whorls of 3, narrow-linear, acute or mucronate, with revolute margins, more slender than in the preceding species and rarely above ½ in. long. Flowers rather small, usually distant. Calyx hoary, the tube about 1½ lines long, the teeth subulate or rarely lanceolate-subulate, as long as or sometimes longer than the tube. Corolla pubescent outside, smaller than in any of the preceding species except *W. senifolia*, the upper lip shorter than the middle lobe of the lower lip and not deeply lobed, the lobes all rather narrow and emarginate.—Bot. Mag. t. 3438 ; *W. longifolia*, Lindl. Bot. Reg. t. 1481, not of R. Br.

Queensland. Brisbane river, Moreton Bay, *F. Mueller, C. Stuart.*

N. S. Wales. Arid wastes on the Lachlan. *A. Cunningham, Fraser.*

Victoria. Genoa and Towamba rivers, *F. Mueller* (included by him in Fragm. vi. 110, in *W. longifolia*).

Var. ? *quaterna.* Leaves in whorls of 4.—Shoalhaven gullies, near Glenroch, *Herb. F. Mueller* (the collector not named).

9. **W. longifolia,** *R. Br. Prod.* 501. An erect shrub of several
ft., without any of the hoary tomentum or hairs of the preceding species
either on the foliage or calyxes and very rarely and only in a very
slight degree on the young branches. Leaves in whorls of 3, narrow-
linear, the margins somewhat revolute or nearly flat, above ½ in. and
mostly above 1 in. long. Flowers rather small, axillary. Calyx-tube
1 to 1¼ lines long, 5-ribbed, green and smooth, the teeth lanceolate-
subulate, usually about as long as the tube and often ciliate. Corolla
pubescent outside, the tube exserted and usually exceeding the calyx-
teeth, dilated upwards, the upper lip much shorter than the tube,
broadly 2-lobed, the middle lobe of the lower lip much longer. Connective of the perfect anthers very prominent and (in some specimens
at least) almost winged at the back.—Benth. in DC. Prod. xii. 571;
Prostanthera linearis, Sieb. Pl. Exs., not of R. Br.

N. S. Wales. Port Jackson to the Blue Mountains, *R. Brown, Sieber, n.* 180,
and many others.

10. **W. glabra,** *R. Br. Prod.* 501. A bushy shrub of 2 or 3 ft.,
quite glabrous or the young branches slightly silky-pubescent, the
leaves on both sides and the calyxes quite green. Leaves in whorls of
3, shortly petiolate, from oblong-elliptical to lanceolate, from under
½ in. to nearly 1 in. long, acute or obtuse, flat or the margins slightly
recurved, smooth and often shining on the upper surface. Flowers of
W. longifolia, the calyx-teeth usually lanceolate, about as long as the
tube.—Benth. in DC. Prod. xii. 571; *W. violacea,* F. Muell. in Hook.
Kew Journ. viii. 169, and in Trans. Phil. Soc. Vict. i. 49.

Queensland. Shoalwater Bay, *R. Brown.*
N. S. Wales. New England, *C. Stuart.*
Victoria. Near the Goulburn river, *F. Mueller.*
For the other stations mentioned by F. Mueller, Fragm. vi. 110, see *Hemigenia
cuneifolia.*

11. **W. rubiæfolia,** *R. Br. Prod.* 501. A dense bushy shrub of 1
to 3 ft., quite glabrous or the young branches pubescent. Leaves in
whorls of 4, oblong-elliptical or lanceolate, mucronate-acute or
almost obtuse, the margins usually slightly recurved, glabrous, smooth
and shining above, paler but not hoary underneath. Flowers rather
small, all axillary but usually crowded towards the ends of the
branches. Calyx green, the tube ribbed, scarcely above 1 line long,
the teeth acute, nearly or quite as long as the tube. Corolla pubescent
like those of *W. longifolia* and *W. glabra.*—Benth. in DC. Prod. xii. 571;
Hook. f. Fl. Tasm. i. 285.

Tasmania. Derwent river, *R. Brown;* abundant throughout the colony, ascending
to 3000 feet, *J. D. Hooker.*

Var.? *subsericea.* Upper leaves and inflorescence silky-pubescent. Leaves thick,
not at all revolute, 3 to 4 lines long. Corolla nearly glabrous, the lobes apparently narrower than in the typical form.—Head of the Douglas river, *Milligan.*

This variety in some measure connects *W. rubiæfolia* through *W. brevifolia* with
W. rosmariniformis, whilst, if the number of leaves in the whorl be neglected, *W. rubiæfolia* passes through *W. glabra* and *W. longifolia* into *W. eremicola* and *W. senifolia,*

leaving only *W. rigida* and *W. Dampieri* to be distinguished by the shortness of the calyx-teeth.

TRIBE VI. AJUGOIDEÆ.—Stamens 4, in pairs, exserted from the very short truncate or deeply slit upper lip of the corolla (except very rarely in genera not Australian). Nuts prominently reticulate-rugose, not succulent, usually attached to near the middle. Seeds without albumen.

Some genera of this tribe approach very nearly to *Vitex* and its allies in Verbenaceæ.

20. TEUCRIUM, Linn.

Calyx-teeth 5, equal or the upper one more frequently larger than the others. Corolla-tube short, the 4 upper lobes nearly equal or the 2 uppermost larger, all 4 lateral, erect or declinate, the middle lower lobe larger, obovate or oblong, spreading and usually concave. Stamens 4, in pairs, exserted from between the upper corolla-lobes and arched over the corolla. Anthers reniform, 1-celled by confluence of the cells. Style shortly bifid at the end. Nuts laterally attached to near or to above the middle, reticulate-rugose or rarely nearly smooth.—Herbs undershrubs or shrubs, showing considerable diversity in habit and inflorescence. Leaves entire, toothed or variously divided.

The genus is widely distributed over the temperate regions of the globe, chiefly in the northern hemisphere, with a few tropical chiefly mountain species. The Australian species are all endemic, although in some measure allied, in some instances to S. African, in others to Himalayan or European species.

Peduncles longer than the calyx, 1- or more-flowered.
  Plant hoary or white, rigid. Leaves mostly entire. Peduncles
    rigid, all 1-flowered . . . . .1. *T. racemosum.*
  Plant green, nearly glabrous. Leaves mostly entire. Peduncles
    slender, all 1-flowered or the lower ones 3- or 5-flowered . . 2. *T. integrifolium.*
  Plant green, pubescent or villous. Leaves toothed or cut or the
    upper ones entire. Peduncles slender, 3- or more flowered . 3. *T. corymbosum.*
Flowers sessile or nearly so.
  Flowers all axillary. Leaves narrow, 3-lobed. Corolla two upper
    lobes twice as large as the next pair . . . . 4. *T. fililobum.*
  Flowers in more or less leafy spikes. Leaves mostly 3- or 5-lobed.
    Corolla lateral lobes as large as the upper pair . . . . . 5. *T. sessiliflorum.*
  Flowers in terminal spikes. Leaves toothed or rarely lobed. Co-
    rolla with all 4 upper lobes small and distant . . . . . . 6. *T. argutum.*

1. **T. racemosum,** *R. Br. Prod.* 504. A perennial or undershrub with a woody rootstock and erect rigid more or less branched stems, from 6 in. to above 1 ft. high, hoary or white as well as the foliage and inflorescence, with a close minute tomentum scarcely wearing off from the upper surface of the older leaves. Stem-leaves linear-lanceolate or oblong-linear, obtuse, entire or very rarely 3-lobed, contracted into a short petiole, from under ½ in. to above 1 in. long, the margins sometimes recurved and occasionally undulate-crisped; the lower leaves in some specimens 3 together on each side of the stem; the upper and floral ones gradually smaller, more sessile, broader at the base, the

uppermost very small. Peduncles all 1-flowered, rigid, spreading, as long as or longer than the floral leaf, forming a stiff terminal more or less leafy raceme. Calyx 2 to 2½ lines long, the teeth nearly equal, as long as or longer than the tube. Corolla-limb sparingly hirsute outside, the 4 upper lobes in lateral pairs, all nearly equal oblong and erect, the middle lower one twice as long. Nuts more or less pubescent, the adnate part of the inner face very hard.—Benth. in DC. Prod. xii. 576.

Queensland. In the interior, *Mitchell;* Armadilla, *Barton;* Curriwillighie, *Dalton.*

N. S. Wales. Swampy flats on the Lachlan river, *A. Cunningham;* Upper Hunter river, *Woolls;* from the Lachlan and Darling to the Barrier Range, *Victorian and other Expeditions.*

Victoria. On the Murray, *F. Mueller.*

S. Australia. Murray river and St. Vincent's Gulf, *F. Mueller* and others; Spencer's Gulf, *R. Brown, Wilhelmi;* in the interior, *M'Douall Stuart's Expedition;* towards Cooper's Creek, *Howitt's Expedition.*

W. Australia. Murchison river, *Oldfield, Drummond, 6th coll. n.* 143.

Var. *tripartitum,* F. Muell. Flowers and leaves very small, the lower leaves three together on each side of the stem, perhaps single tripartite leaves, for here and there there is a trifid one, but mostly the three are quite distinct from the base.—Murray river, *F. Mueller.*

2. **T. integrifolium,** *F. Muell. Herb.* An erect perennial of 6 in. to 1 ft., with the habit of some forms of *T. racemosum,* with which F. Mueller now unites it, but not so rigid and without any of the white tomentum so constant in that species, the calyx and young shoots very rarely slightly pubescent or hirsute and the corolla usually hirsute outside, the rest of the plant glabrous. Leaves lanceolate oblong or almost linear, the lower ones petiolate and ½ to 1 in. long, the upper floral ones smaller and more sessile, all entire. Peduncles much more slender than in *T. racemosum,* but mostly shorter than the leaves, the upper ones and sometimes all 1-flowered, the lower ones often 3- or 5-flowered as in *T. corymbosum.* Nuts pubescent.

N. Australia. Dry lagoons, Arnhem's Land, Hooker's Creek, *F. Mueller.*

Queensland. In the interior, *Mitchell;* Bowen river, *Bowman;* Flinders river, *Sutherland;* Suttor river, *Dorsay;* Armadilla, *Barton;* Curriwillighie, *Dalton.*

This species closely connects *T. racemosum* and *T. corymbosum,* being as near to the one as to the other, and, as appears to me, cannot well be referred to either without uniting all three into one.

3. **T. corymbosum,** *R. Br. Prod.* 504. An erect perennial, not usually much branched, from under 1 ft. to 3 ft. high, pubescent with very short hairs passing sometimes into a hoary tomentum on the under side of the leaves; thin leaves mostly ovate or ovate-lanceolate, irregularly and deeply toothed or lobed, contracted into a distinct petiole, more or less rugose, green on both sides or hoary underneath, 1 to 1½ in. long, passing into smaller, narrower and less cut floral leaves, which are sometimes as well as the upper stem ones all entire or slightly toothed, or all the leaves are oblong or broadly lanceolate and more regularly toothed, or all are rather broad and deeply lobed. Peduncles

slender, shorter or longer than the leaves, bearing a loose cyme of 3 to 7 or rarely more flowers, usually rather smaller than in *T. racemosum*, but sometimes at least as large, otherwise the same as in that species, the calyx-teeth nearly equal and as long as or longer than the tube, the 4 upper lobes of the corolla nearly equal and oblong, the middle lower one twice as long. Nuts pubescent.—Benth. in DC. Prod. xii. 577, Hook. f. Fl. Tasm. i. 285 ; *Scoparia australis*, Sieb. in Schult. Syst. iii. Mant. 66 ; *Anisomeles australis*, Spreng. Syst. Cur. Post. 226.

N. S. Wales. Port Jackson to the Blue Mountains, *R. Brown, Sieber, n.* 184, and others; northward to Clarence and Hastings rivers, *Beckler ;* New England, *C. Stuart.*

Victoria. Yarra-Yarra, *Robertson ;* Latrobe, Avon, Snowy and Macalister rivers, *F. Mueller.*

Tasmania. Not uncommon in dry places in various parts of the colony, *J. D. Hooker.*

S. Australia. Beds of creeks and rocky hills, Wulpena, Barula, Crystal Brook, *F. Mueller ;* Lake Gillies, *Burkitt ;* Mount Searle, *Warburton.*

Var. ? *hirsutum.* The whole plant viscid and hirsute with rather long spreading hairs. Leaves rather large, contracted into a very short petiole or quite sessile. Cymes loose, several flowered. Perhaps a distinct species.

Queensland. Mountain tops near Rockhampton, *Bowman, O'Shanesy ;* Liverpool Range, *Leichhardt.*

T. lanceolatum, Benth, Lab. Gen. et Sp. 666 and in DC. Prod. xii. 576, from Bathurst, *A. Cunningham*, is founded on luxuriant flowering branches of *T. corymbosum*, of which the lower leaves are wanting and the upper ones are all lanceolate and entire, or slightly toothed.

T. petrophilum, F. Muell. in Linnæa, xxv. 426 from South Australia is founded on stout specimens of *T. corymbosum*, rather more hoary than usual.

4. **T. fililobum,** *F. Muell. Herb.* Apparently shrubby, with diffuse or erect branched stems of 6 in. to 1 ft., pubescent with very short spreading hairs. Leaves rather crowded, deeply divided into 3 or rarely 5 narrow-linear lobes with revolute margins, the whole leaf rarely above ½ in. long, the floral ones similar. Flowers rather large, solitary in the axils, nearly sessile or on pedicels rarely exceeding 1 line. Calyx broadly campanulate, green, slightly pubescent, about 3 lines long, the teeth subulate-acuminate or very acute, as long as the tube. Corolla pubescent outside, the 2 upper lobes broad and very obtuse, the 2 lateral ones much smaller, the middle lower lobe again larger. Nuts glabrous or scarcely pubescent.

W. Australia, *Drummond, n.* 65, *2nd coll. n.* 213, *4th coll. n.* 169; Puttingup, *Maxwell.*

Drummond's specimens, n. 65, were referred by De Vriese (Goodenovieæ, p. 183) to *Leschenaultia tubiflora*, and so named by him in Herb. Hooker.

5. **T. sessiliflorum,** *Benth. in DC. Prod.* xii. 580. A perennial, with ascending or erect slightly branched stems under 6 in. high, glabrous or slightly pubescent as well as the foliage, the inflorescence usually hirsute. Leaves oblong-cuneate or linear-cuneate, 3- or 5-lobed, the lobes either short and broad or long and linear, the margins usually recurved and the primary veins prominent underneath, usually contracted into a broad petiole and varying in length from ½ in. to above

1 in., the floral ones short, broad and sessile, distant or close and imbricate, usually toothed only at the end. Flowers small, in dense or
interrupted terminal leafy spikes, each one solitary in the axil of the
floral leaf and sessile. Calyx hirsute, about 2 lines long, the teeth
equal short and broad. Corolla lobes only shortly exceeding the calyx.
—*T. trifidum*, Schlecht. Linnæa, xx. 609, not of Retz.

Victoria. Murray river, *F. Mueller;* Wimmera, *Dallachy.*
S. Australia, *Behr. ;* Rocky Creek and head of Spencer's Gulf, *F. Mueller ;* Venus
Bay, *Warburton.*
W. Australia, *Drummond, 2nd coll. n.* 211.

6. **T. argutum,** *R. Br. Prod.* 504. A perennial, with erect simple
or slightly branched stems of 6 in. to 1 ft., the whole plant pubescent
or hirsute but green. Leaves on rather slender petioles, ovate-lanceolate
or ovate, sometimes hastate, regularly serrate or crenate, or rarely less
regularly toothed or lobed, more or less rugose, mostly from ¾ to 1½ in.
long, the floral ones reduced to small lanceolate or rarely ovate persistent bracts. Flowers not very constantly opposite, sessile or shortly
pedicellate, in loose or dense terminal spikes. Calyx 2 to 3 lines long,
the teeth acute, shorter than the tube, the upper ones especially the
uppermost one rather broad, the 2 lowest narrow. Corolla with the 4
upper (lateral) lobes very small acute and distant, the middle lower
lobe much larger and obovate. Nuts glabrous.—Benth. in DC. Prod.
xii. 584.

Queensland. Gilbert river, *F. Mueller;* Port Denison, *Fitzalan, Dallachy;* Rock-
ingham Bay, *Dallachy;* Rockhampton, *O'Shanesy, Dallachy;* Nerkool Creek and
Bowen river, *Bowman ;* Moreton Bay, *Backhouse;* near Mount Owen, *Mitchell.*
N. S. Wales. Hawkesbury river, *R. Brown ;* Nepean river, *Woolls ;* Tweed and
Richmond rivers, *C. Moore;* New England, *C. Stuart;* M'Leay and Clarence rivers,
Beckler· (the latter with smaller leaves and flowers).
Var. *incisa.* Leaves small, mostly lobed.—Dawson river, *F. Mueller;* Darling
Downs, *Lau;* Armadilla, *Barton.*

21. AJUGA, Linn.

Calyx-teeth 5, equal. Corolla-tube short or long, the upper lip very
short, truncate or emarginate, the lower lip long and spreading, the
lateral lobes oblong, small, the middle lobe much larger, emarginate or
bifid. Stamens 4, in pairs, exserted from the upper lip and arched
over the corolla; anthers reniform, 1-celled by the confluence of the
cells. Style shortly bifid at the end. Nuts laterally attached to near
or above the middle, reticulate-rugose.—Herbs, usually diffuse or ascending or with spreading radical leaves and shortly erect stems. Flowers
in false-whorls in the axils of floral leaves gradually smaller than the
stem-leaves, the upper ones sometimes forming terminal leafy spikes.
Bracts linear, or very small or none.

The genus is widely dispersed over the extratropical regions of the Old World, and
chiefly in the mountain districts within the tropics, but wanting in America. The two
Australian species, as usually defined, are endemic, one of them however scarcely to be
distinguished from a common northern one.

Floral leaves entire or with very few coarse teeth, and smaller or
  narrower than the flowerless ones. Flowers 5 lines to 1 in. long .   1. *A. australis*.
Floral leaves like the stem ones, ovate, deeply sinuate-toothed and rugose. Flowers not exceeding 3 lines . . . .   2. *A. sinuata*.

1. **A. australis,** *R. Br. Prod.* 503. A perennial, more or less pubescent or villous, without stolones, with erect or ascending simple
stems from 2 or 3 in. to above 1 ft. long, flowering nearly from the base.
Leaves chiefly radical, obovate or oblong, coarsely toothed, contracted
into a long petiole, often 3 to 4 in. and sometimes still longer ; lower
floral leaves nearly similar or smaller and narrower, passing into sessile
oblong or lanceolate entire ones, all longer than the flowers or the
upper ones very small. Flowers blue, nearly sessile, in false whorls of
from about 6 to above 20, exceedingly variable in size. Bracts linear,
the outer ones sometimes as long as the calyx, the inner ones or nearly
all very small or obsolete. Calyx villous or nearly glabrous, from about
2 lines to nearly 4 lines long, the teeth acute, shorter than the tube.
Corolla-tube from the length of the calyx to twice as long, always
with a transverse ring of hairs inside above the ovary, the upper lip
truncate or emarginate, sometimes exceedingly short, in some of the
larger flowers nearly 1 line long, but always twice as broad as long,
and never longer than the space between the base of the lower lip and
the lateral lobes ; the middle lobe of the lower lip usually longer than
the tube. Nuts glabrous.—Benth. in DC. Prod. xii. 597 ; Hook. f.
Fl. Tasm. i. 286 ; *A. diemenica*, Benth. Lab. Gen. et Sp. 695 and in DC.
Prod. xii. 597 ; *A. virgata* and *A. tridentata*, Benth. Lab. Gen. et Sp.
700, 701 ; and in DC. Prod. xii. 601, 602.

Queensland. Keppel Bay, *R. Brown;* Percy isles, *A. Cunningham;* Port Curtis,
M'Gillivray; Rockingham Bay and Rockhampton, *Dallachy* and others ; Moreton Bay,
C. Stuart; Mount Faraday, *Mitchell.*

N. S. Wales. Port Jackson to the Blue Mountains, *R. Brown* and others ;
Lachlan river, *A. Cunningham;* from thence and the Darling to the Barrier
Range, *Victorian and other Expeditions;* New England, *C. Stuart;* Hastings, Macleay
and Clarence rivers, *Beckler* and others.

Victoria. Near Melbourne, *Adamson;* Yarra, Broken and Murray rivers, *F. Mueller;*
Wimmera, *Dallachy;* mouth of the Glenelg, *Allitt;* Ballarook Forest, Creswick,
Whan.

Tasmania. Common in damp meadows, &c. throughout the colony, *J. D. Hooker.*

S. Australia. From the Murray to St. Vincent's Gulf, *F. Mueller* and others ;
Mount Searle, *Warburton;* Kangaroo island, *Waterhouse.*

The characters on which I had formerly, from the examination of few and some of
them very imperfect specimens, distinguished four species distributed into two
sections, have entirely broken down by the comparison of the numerous specimens now
before me, comprised in above eighty sheets, from a great variety of stations, and all
must evidently be referred to a single species, not separable from the northern *A.
genevensis* by any marked characters, but differing chiefly in the greater development
of the floral leaves, which are narrower more herbaceous and less toothed than in *A.
genevensis*, and never broad and imbricate as in *A. pyramidalis*. In some of the
Australian va ieties the flowerless leaves are almost entirely radical, large and on long
petioles, and the floral leaves lanceolate or oblong, scarcely toothed and not twice as
long as the flowers ; in others there are a few flowerless leaves at the base of the stems
forming a gradual passage from the radical to the floral ones which are all several times
as long as the flowers. Then as to size and indumentum, some specimens from the

interior of N. S. Wales and from S. Australia are very hoary-villous all over, with radical leaves 6 in. long and robust stems of above 1 ft., the flowers themselves nearly 1 in. long ; others both from the North and the South have the flowers scarcely 5 lines long, the common size being between these two extremes, with every degree of villosity from almost glabrous in some Tasmanian ones to the above-mentioned exceedingly villous ones. The very indifferent specimen which I had published as *A. virgula* appears to be a long drawn-up flowering stem of an old plant not otherwise differing from a form which now proves to be frequent. *A. tridentata*, with a similar habit but still more drawn out and less villous, has all the floral leaves broadly sessile, ovate and deeply 3-toothed or 3-lobed, which give a very different aspect to the plant, but these characters are not as yet confirmed by any more perfect specimens.

2. **A. sinuata,** *R. Br. Prod.* 503. A low diffuse much-branched hirsute perennial, not exceeding 6 in. Leaves ovate or oblong, deeply and irregularly sinuate-toothed, the radical ones on rather long petioles, 1 to 3 in. long, the floral ones on shorter petioles, $\frac{1}{2}$ to $\frac{3}{4}$ in. long, but all rugose and prominently and obtusely toothed like the radical ones. Flowers not 3 lines long, about 6 in the whorl. Calyx hirsute. Corolla upper lip not exceeding the calyx-teeth.—Benth. in DC. Prod. xii. 598.

N. S. Wales. Hunter's river, *R. Brown;* Macleay river, *Herb. F. Mueller.*

The species is nearer to the Himalayan *A. parviflora* than to any other, the habit the floral leaves and the very small flowers prevent its union with *A. australis*, unless the whole of the section *Bugula* from Europe, Asia, Africa and Australia, with the exception of *A. lobata, A. reptans*, and *A. orientalis*, be joined together under the Linnæan name of *A. genevensis.*

Order XCIV. **PLANTAGINEÆ.**

Flowers regular. Sepals 4. Corolla small, scarious, with an ovate or cylindrical tube and 4 spreading lobes, imbricate in the bud. Stamens 4, or rarely fewer, inserted in the tube of the corolla and alternate with its lobes, usually long; anthers 2-celled, the cells parallel, opening longitudinally. Ovary free, 1- 2- or 4-celled, with one or more ovules in each cell. Style simple, terminal, entire, with 2 opposite longitudinal stigmatic lines. Capsule opening transversely or indehiscent. Seeds peltate, laterally attached, albuminous. Embryo straight or slightly curved, parallel to the hilum.—Herbs with radical tufted or spreading leaves, rarely branched and leafy. Flowers in heads or spikes or rarely solitary, on leafless axillary peduncles, each one sessile within a small bract.

A small *Order*, widely spread over the globe, but chiefly in the temperate regions of the Old World. The only Australian genus is the principal one of the Order, which, besides that one, only contains two others, both monotypic, one from the mountains of S. America, the other European and aquatic, both of them very anomalous.

1. **PLANTAGO,** Linn.

Flowers hermaphrodite, in heads or spikes. Stamens 4. Capsule 2- or 4-celled; the other characters those of the Order.

The geographical range of the genus is the same as that of the Order. Among the Australian ones, besides those that are introduced, one extends to New Zealand and the Antarctic islands, and possibly to the extratropical mountains of South America; the others, as far as hitherto ascertained, are all endemic, but the discrimination of some of the very variable species of the genus is as yet very far from being satisfactorily carried out. The characters derived from the exserted or included styles or stamens have been shown by A. Gray to be dimorphic or subsexual and not specific, and there remains often little to be relied upon. but the shape of the leaves, the density of inflorescence, the size of the flowers and similar eminently variable differences.

Ovary 2-celled with 1 ovule in each cell. Flowers in very dense
  ovate or broadly oblong spikes or heads . . . . 1. *P. lanceolata.*
Ovary apparently 4-celled, with 1 ovule in each cell (2-celled, with
  2 ovules in each cell separated by spurious dissepiments).
  Flowers closely appressed in narrow cylindrical spikes . . . . 2. *P. coronopus.*
Ovary 2-celled, with 2 ovules in each cell without spurious disse-
  piments.
  Spikes very slender, interrupted. Flowers small. Leaves usually
    rather broad. Calyx not exceeding ¾ line . . . . 3. *P. debilis.*
  Spikes rigid, not very close, above 1 in. long (excepting depaupe-
    rated specimens with few flowers). Leaves usually narrow.
    Calyx 1 to 1½ lines long . . . . 4. *P. varia.*
  Spikes dense, cylindrical, ½ to 1 in. long, the buds closely imbri-
    cate. Leaves rather broad, villous on both sides . . . . . 5. *P. antarctica.*
  Spikes dense, ovoid-oblong or cylindrical, under ¾ in. long.
    Leaves usually broad and rather thick. Flowers glabrous.
    Corolla-lobes narrow . . . . 6. *P. tasmanica.*
Ovary 2-celled, with 2 superposed pairs of ovules in each cell.
  Dwarf alpine plants with thick leaves, the flowers in heads of
  2 to 6 or solitary.
  Leaves ⅓ to 1½ in. long., rosulate on a short thick stock . . . 7. *P. Brownii.*
  Leaves not above ¼ in. long, the fresh ones rosulate at the ends
    of the slender branches of a densely tufted stock covered with
    the remains of old leaves . . . . 8. *P. Gunnii.*

Besides the above, *P. major*, Linn.; Dcne. in DC. Prod. xiii. i. 694, a common European weed, with broadly-ovate large leaves, long cylindrical spikes, and the ovary 2-celled, with usually more than 2 superposed pairs of ovules in each cell, has established itself between Bridgewater and New Norfolk in Tasmania (*Herb. F. Mueller*).

1. **P. lanceolata,** *Linn.; Dcne. in DC. Prod.* xiii. i. 714. Stock usually more or less woolly or silky-hairy. Leaves radical, lanceolate or oblong-lanceolate, acute, entire or minutely and obscurely toothed, contracted at the base, more or less hairy or glabrous. Scapes long. Flowers in a close dense ovate spike or head, often becoming oblong when old. Lower sepals usually united almost to the top. Ovary 2-celled with only one ovule in each cell.

Very common in Europe and temperate Asia and now established in many parts of **Victoria** and **Tasmania.**

Var. *eriophylla*, Dcne. l.c. 715. Leaves and sometimes the inflorescence more or less densely covered with long silky-woolly hairs.

Victoria. Yarra-Yarra and Forest Creek, *F. Mueller;* heath ground near Portland, *Barclay.*

2. **P. coronopus,** *Linn.; Dcne. in DC. Prod.* xiii. i. 732. Plant more or less hirsute. Leaves radical, linear, acute, entire or pinnatifid with linear lobes, which are themselves sometimes deeply toothed or pinnatifid. Scapes usually under 6 in. long. Spikes dense and cylindrical but narrow, ½ to 2 in. long, the flowers closely imbricate when

dry, the 2 upper sepals with prominent scarious usually ciliate keels or
wings. Ovary apparently 4-celled with one ovule in each cell, but
really 2-celled only, with a spurious dissepiment between the two collateral ovules reaching to but not cohering with the wall of the cell,
the capsule often only 1- or 2-seeded by abortion.

Victoria. Portland, *Allitt.*
Tasmania. Roadsides, George Town, Perth, &c., *Gunn* and others.
S. Australia. Holdfast Bay, *F. Mueller.*
W. Australia. Swan river, *Drummond, n.* 225.
Very common in temperate regions of the northern hemisphere in the Old World,
especially in maritime or in sandy districts, and thence probably introduced into
Australia.

3. **P. debilis,** *R. Br. Prod.* 425. Very near *P. varia,* and according
to F. Mueller only a variety of that species. Stock without any or with
only a few long woolly hairs between the leaves, the foliage and inflorescence hirsute or nearly glabrous. Leaves radical, oblong or
lanceolate, entire or toothed, usually broader and shorter than in *P.
varia.* Scapes very slender, from 2 or 3 in. to 1 ft. high, the flowers
all distant when fully out, and much smaller than in *P. varia,* forming
a slender interrupted spike often occupying above half the scape.
Calyx not exceeding ¾ line when in fruit and still smaller at the time
of flowering, the opaque centre of the sepals with a few appressed hairs
or more frequently glabrous. Ovary 2-celled, with 2 collateral ovules
in each cell. Capsule with 4 or fewer seeds.—Dcne. in DC. Prod. xiii. i.
701.

Queensland. Brisbane river, Moreton Bay, *F. Mueller, C. Stuart*; Armadilla,
Barton.
N. S. Wales. Blue Mountains, *R. Brown, Woolls;* New England, *C. Stuart;*
Hastings, Clarence, and Macleay rivers, *Beckler.*
Victoria. Taralgin Creek and Hobson's Creek, *F. Mueller.*

P. Cunninghamii, Dcne. in DC. Prod. xiii. i. 702, from a single leaf preserved in the
Hookerian herbarium, is probably this species ; Cunningham's original specimen appears
to have remained in the herbarium of the Paris Museum.

4. **P. varia,** *R. Br. Prod.* 424. A perennial often flowering the
first year so as to appear annual, but forming ultimately a thick stock
with the membranous dilated imbricate bases of the leaves enveloped
in long reddish brown woolly or silky hairs sometimes very copious,
in other specimens very few or scarcely any. Leaves all radical, erect
or rosulate, usually lanceolate or linear-lanceolate, entire or bordered
by a few teeth, with 1, 3 or 5 nerves prominent underneath, contracted
into a long petiole, more or less hirsute, the whole leaf under 2 in. in
some specimens, 6 in. to 1 ft. long in others and varying occasionally
but rarely to ovate-lanceolate, more frequently to linear. Scapes longer
than the leaves, bearing in the upper portion a rather dense or more
or less interrupted spike from 1 to 3 or 4 in. long, more or less
hirsute, with short or long hairs more appressed than on the leaves ;
in starved specimens grown in very dry places, the spikes are sometimes
reduced to very few or even only 2 or 3 flowers clustered at the end of
a short scape. Calyx sessile within a bract rather shorter and narrower

than the sepals and not at all or very narrowly scarious on the margins. Sepals all free, 1¼ to 1½ lines long, obtuse with broad scarious margins, the centre opaque, from copiously hirsute to quite glabrous. Corolla-tube about as long as the calyx, lobes ovate, usually broad, acute or almost obtuse, much imbricate in the bud, one entirely outside and the opposite one entirely inside. Ovary 2-celled with 2 collateral ovules in each cell. Capsule shortly conical at the top or obtuse, circumsciss, ripening all four seeds or sometimes only one or two of them.—Dcne. in DC. Prod. xiii. i. 701; Nees in Pl. Preiss. i. 490; Hook. f. Fl. Tasm. i. 302; *P. debilis,* Nees in Pl. Preiss. i. 491, not of R. Br.

Queensland. In the interior, *Mitchell.*

N. S. Wales. Chiefly in the interior, Morley's plains, *A. Cunningham;* Darling river, *Woolls;* Lachlan and Darling rivers to the Barrier Range, *Victorian and other Expeditions;* northward to Clarence river and Mount Mitchell, *Beckler;* New England, *C. Stuart.*

Victoria. Port Phillip, *R. Brown;* Wendu Vale, *Robertson;* near Melbourne, *Adamson;* thence to the Avoca, Murray, and Ovens rivers, *F. Mueller;* Wimmera, *Dallachy.*

Tasmania. Port Dalrymple and Derwent river, *R. Brown;* abundant everywhere in the colony, *J. D. Hooker.*

S. Australia. Murray river to St. Vincent's and Spencer's Gulfs, *F. Mueller* and others; in the interior to Cooper's Creek, *Wheeler;* Kangaroo island, *Waterhouse.*

W. Australia. King George's Sound to Swan river, *Drummond, n.* 224, 393, 714, 738, *Preiss, n.* 1968, 1970, *Oldfield* and others; Murchison river, *Oldfield.*

The variations of this polymorphous species are so complicated that I have been un-able to assign them any definite limits as to characters or to geographical range, and it would appear that no less than ten of the supposed species enumerated by Decaisne should be included in it, the characters derived from supposed duration, from minutiæ in the form and hairiness of the sepals and bracts, and from the breadth and acuteness of the corolla-lobes having entirely broken down. The typical *P. varia* has the woolly hairs at the base of the leaves copious, the sepals not very obtuse and hispid on the opaque centre and extends over the whole range of the species. It would include Nees's *P. debilis* or *P. exilis,* Dcne. in DC. Prod. xiii. i. 702, *P. runcinata,* Dcne. l.c. 702 and *P. consanguinea* Dcne. l.c. 703.

P. hispida of most authors, but scarcely of Brown, has the long hairs or so-called beard at the base of the leaves few or none, the sepals broad and obtuse, quite glabrous or slightly hispid on the opaque centre. It is also found at almost all the habitats of the typical form and would include *P. Mitchelli* and *P. Drummondii,* Dcne. l.c. 701, and also, from the character given, *P. Gaudichaudii,* Barn. Monogr. Plantag. 15; Dcne. l.c. 702.

P. struthionis, A. Cunn.; Dcne. l.c. 702, and *P. sericophylla,* Dcne. l.c. 702, both founded on indifferent N. S. Wales specimens of Cunningham's, appear to be luxuriant states of the species, with the glabrous or slightly hispid sepals of the preceding form, but with rather more woolly hair at the base of the leaves and the leaves and scapes drawn out to nearly 1 ft. in length, the leaves of *P. struthionis* being rather broader than in *P. sericophylla.*

Starved specimens of both the above principal forms occur, especially in Victorian and Tasmanian collections, with the spikes reduced to very few flowers almost collected into heads, but not so compact as in *P. tasmanica* and usually with the appearance of annuals. It is probably to one of these that belongs the *P. bellidioides,* Dcne. l.c. 701, described from a Tasmanian specimen of Gunn's, but which I have not precisely iden-tified.

P. hispida, R. Br. Prod. 425, from the seacoast, Port Dalrymple, is a small very hispid form with the narrow hispid calyxes of the typical form but with rather smaller flowers and without the long woolly hairs at the base of the leaves. It passes into the following :—

Var.? *parviflora.* Smaller in all its parts than *P. varia,* but otherwise closely resembling the typical form with narrow leaves and hirsute flowers. Spikes cylindrical, rather dense or loose, ½ to 1 in. long or rather more. Sepals rather above 1 line long, the opaque centre usually hirsute. Capsule with a conical end exceeding the calyx and longer than in other varieties of this species.—N. S. Wales and more frequently in Victoria, also Kangaroo island, *R. Brown.*—Perhaps a distinct species and in some measure connecting *P. varia* with *P. debilis.*

It is very possible that the study of the different forms thus included in *P. varia* on the living plants by local botanists, who would bestow on them the time and patience that has been devoted to European *Rubi, Batrachian Ranunculi,* &c., might point out several permanent races, of which dried specimens without indication of the circumstances of their growth give no indication.

5. **P. antarctica,** *Dcne. in DC. Prod.* xiii. i. 703? *Hook. f. Fl. Tasm.* i. 303. Stock usually woolly-hairy at the base of the leaves. Leaves radical, broadly lanceolate or ovate-lanceolate, acute, entire, with broad petioles, villous on both sides like the villous specimens of *P. tasmanica,* but usually larger. Spikes dense and cylindrical although narrower than in *P. tasmanica,* ½ to above 1 in. long, the buds very closely imbricate in several rows. Bracts and sepals quite glabrous or rarely with a few marginal cilia.

Victoria. Cobra mountains, *F. Mueller* (rather doubtful).
Tasmania. Marshes of St. Patrick's river, *Gunn;* South Esk river, *C. Stuart.*

I have seen but very few specimens, and with J. D. Hooker do not feel certain of their correct identification with Verreaux's Tasmanian plant described by Decaisne. The flowers are smaller and the spikes more dense than in *P. varia,* the foliage is nearer that of *P. tasmanica,* and it appears to differ from both in the close imbrication of the buds before expanding.

6. **P. tasmanica,** *Hook. f. in Hook. Lond. Journ.* vi. 276, *and Fl. Tasm.* i. 303. A small alpine species. Leaves radical, rosulate, oblong or oblong-lanceolate or sometimes almost obovate, more rarely narrow-lanceolate, contracted into a short pétiole, entire or rarely coarsely toothed, usually of a thick consistence, from densely hoary-hirsute on both sides to quite glabrous. Scapes short, rarely attaining 6 in., with a dense ovoid oblong or cylindrical spike, rarely ¾ in. long, and sometimes reduced to 2 or 3 flowers, glabrous or intermixed with a few hairs. Sepals glabrous, about 1 line long, with less of the scarious margin than in *P. varia.* Corolla-lobes narrower than in that species and more acute. Ovary 2-celled with two collateral ovules in each cell.—Dcne. in DC. Prod. xiii. i. 703; *P. glabrata,* Hook. f. in Hook. Lond. Journ. vi. 276, Dcne. l.c. 703; *P. leptostachys,* Hook. f. l.c. (*P. Daltoni,* Dcne. l.c.)

Victoria. Munyong mountains, sources of the Yarra, Baw-baw mountains, *F. Mueller.*
Tasmania. Abundant in wet marshy places on the mountains at an elevation of 3000 to 4000 ft., *J. D. Hooker, Gunn, Milligan* and others.

P. Archeri, Hook. f. Fl. Tasm i. 303, from the Western mountains, Tasmania, *Archer,* and from the summit of Mount Wellington, *F. Mueller,* seems to be a small state of the same species with more obovate leaves. F. Mueller thinks that *P. tasmanica* itself is only an alpine form of *P. varia.*

7. **P. Brownii,** *Rapin in Mem. Soc. Linn. Par.* vi. 484. A small plant, with a densely tufted stock, woolly amongst the leaves or quite glabrous. Leaves radical, rosulate, thick, somewhat fleshy, and of a

bright shining green when fresh, oblong-lanceolate or spathulate, entire or with a few teeth, quite glabrous. Scapes sometimes scarcely any, often shorter than the leaves but sometimes longer, glabrous as well as the flowers. Flowers only 2 to 4 rarely as many as 6 in small terminal spikes or heads, and sometimes reduced to a single one, each flower about the size of those of *P. varia*. Sepals obtuse, without much of the scarious margin, 1¼ lines long or rather more. Corolla-lobes usually rather narrow. Ovary 2-celled with 2 superposed pairs of ovules in each cell, but often few of them only ripening into seeds.—Dcne. in DC. Prod. xiii. i. 727; Hook. f. Fl. Tasm. i. 304; *P. carnosa*, R. Br. Prod. 425; not of Lam.; *P. stellaris*, F. Muell. Fragm. ii. 23, Pl. Vict. ii. t. 66.

Victoria. Summits of the Munyong Mountains, *F. Mueller.*

Tasmania. Port de l'Esperance, *R. Brown;* South Cape, *Gunn;* South Port, *C. Stuart;* Macquarrie Harbour, *Milligan.*

The species is also in New Zealand and in the Auckland islands, and may be the same as one or more of the antarctic American ones referred to the same section by Decaisne.

P. paradoxa, Hook. f. in Hook. Lond. Journ. vi. 277 and Fl. Tasm. i. 304, from Lake St. Clair, *Gunn*, Mersey river, *Archer*, Kermandee river, *Oldfield*, Mount Lapeyrouse, *C. Stuart*, seems to me to be a small state of *P. Brownii* reduced to 1 or 2 flowers on the scape as in *P. Gunnii*, but with the foliage and stock of the small specimens of *P. Brownii*. The specimens before me from C. Stuart's collection show a gradual series from the small 1-flowered to the largest 4- to 6-flowered states. Brown's own specimens have mostly only 1 or 2 flowers.

8. **P. Gunnii,** *Hook. f. in Hook. Lond. Journ.* v. 446, *t.* 13, *and Fl. Tasm.* i. 304. A small densely tufted almost moss-like plant, the stems closely packed but slender, branching and growing to 1 in. or more, covered with thin persistent remains of old leaves and ending in a spreading tuft of fresh ones. Leaves from narrow ovate-oblong to linear-lanceolate, acute or obtuse, contracted into a short petiole, rather thick, with ciliate margins and sometimes a few hairs on the upper surface, otherwise glabrous and smooth, rarely exceeding 3 lines. Peduncles shortly exceeding the leaves, more or less hairy, bearing 1 or rarely 2 small flowers. Sepals glabrous. Ovary 2-celled with 2 superposed pairs of ovules in each cell, but usually only one appears to ripen.— Dcne. in DC. Prod. xiii. i. 728.

Tasmania. Loftiest parts of the Western Mountains, *Gunn, Archer.*

ORDER XCV. **PHYTOLACCACEÆ.**

Perianth of 5 rarely 4 divisions or lobes, either all herbaceous, or scarious or petal-like on the margins or coloured inside, imbricate in the bud. Staminodia (or petals?) in a few genera not Australian 5 or fewer, minute and stipitate. Stamens as many as perianth-divisions and alternate with them or more numerous, inserted on the torus or at the base of the perianth; anthers 2-celled, the cells parallel, opening by longitudinal slits. Ovary either of a single somewhat excentrical carpel, or of several carpels either distinct or united in a ring round the centre of the torus or, in a genus not Australian, forming a single 1-celled ovary with the ovules in a ring round a central column. Ovules solitary in each carpel, ascending, amphitropous or anatropous. Styles as many as carpels, proceeding from their upper inner angle, free or united at the base, stigmatic along their inner edge. Carpels of the fruit variously enlarged, free or united, dry or succulent, indehiscent or dehiscent along their inner or outer edge or both. Seed ascending, sometimes accompanied by a small arillus or strophiole; testa membranous or crustaceous. Albumen mealy or somewhat fleshy, copious, scanty, or sometimes none. Embryo usually much curved, rarely folded or straight; cotyledons narrow or broad and convolute; radicle inferior.—Herbs, undershrubs, or rarely shrubs or trees, usually glabrous. Leaves alternate, usually entire. Flowers hermaphrodite or unisexual, in terminal axillary or leaf-opposed spikes racemes or clusters, rarely solitary, usually accompanied by a subtending bract and 2 bracteoles.

The Order is chiefly American and African, a very few species extending into Asia, as weeds of cultivation or otherwise introduced. The Australian genera are all endemic, one of them nearly allied to, but quite distinct from, an American genus, the others belonging to a series (or genus in an extended sense) exclusively Australian. All have unisexual flowers, whilst those of the American and African genera are, with the exception of a single species, hermaphrodite.

Ovary of a single carpel. Fruit a burr with hooked bristles. Flowers unisexual. Filaments filiform . . . . 1. MONOCOCCUS.
Ovary of several carpels united in a ring. Fruit succulent. Flowers hermaphrodite . . . . * PHYTOLACCA.
Ovary of 2 or more carpels united round a central column. Flowers unisexual. Anthers sessile or nearly so.
Perianth deeply 4-lobed. Carpels 2. Flowers axillary . . . 2. DIDYMOTHECA.
Perianth sinuate-toothed. Carpels several.
Carpels 4 to 20, separating at their maturity, and opening at the outer edge or at both outer and inner edges. Flowers axillary . . . . 3. GYROSTEMON.
Carpels from above 20 to 50, separating at their maturity and opening on the inner edge only. Flowers in spikes or racemes . . . . 4. CODONOCARPUS.
Carpels about 20, connate in a globular almost woody indehiscent fruit. Male flowers in terminal spikes; females solitary and axillary . . . . 5. TERSONIA.

The male plants in some species of each of the last four genera are undistinguishable from each other except by slight differences in the inflorescence.

The genus *Phytolacca*, marked above with an asterisk* is American, but one species, *P. octandra*, Linn.; Moq. in-DC. Prod. xiii. ii. 32, long since cultivated in various parts of the Old World, has established itself in some parts of N. S. Wales and Victoria near the towns. It is erect and herbaceous, attaining 6 to 10 ft. and not much branched, with large ovate-lanceolate acute leaves. Flowers hermaphrodite, almost sessile in pedunculate racemes, either terminal or almost leaf-opposed. Perianth small, of 5 divisions. Stamens usually 8. Carpels usually 8, united in the fruit in a depressed succulent almost black berry more or less prominently 8 ribbed. The flowers are occasionally 9- or 10-merous and then they only differ from *P. decandra*, Linn., which is more frequently cultivated in Southern Europe, in the sessile or exceedingly shortly pedicellate flowers.

1. MONOCOCCUS, F. Muell.

Flowers unisexual, monœcious or diœcious. Perianth of 4 distinct divisions. Stamens in the males 10 to 20, filaments filiform; anthers oblong-linear. Ovary in the females of a single carpel; style very short, hooked, and decurrent in a bearded line along the inner edge of the carpel. Fruit dry, indehiscent, covered with hooked bristles. Seed with a thin testa; albumen unilateral; embryo transversely folded, the cotyledons broad and convolute.—Shrub. Leaves membranous. Flowers in racemes either terminal or in the upper axils, the males usually in separate racemes or on separate individuals, but sometimes a few females at the base of the male racemes or a few males at the summit of the female racemes.

The single species known is endemic in Australia. It is in habit and most of the characters nearly allied to the American genus *Petiveria*, differing in the unisexual flowers, the more numerous stamens and the glochidiate bristles of the ovary and fruit numerous and spreading, instead of being 2 or 3 only and closely reflexed.

1. **M. echinophorus,** *F. Muell. Fragm.* i. 47. A straggling shrub, sometimes more erect and attaining 5 or 6 ft. Leaves petiolate, from ovate to lanceolate, obtusely acuminate, contracted at the base, membranous, 2 to 4 in. long. Racemes slender, often 5 or 6 in. long. Flowers rather distant, shortly pedicellate, each within a lanceolate acute bract shorter than the calyx and often shortly adnate to the base of the pedicel. Bracteoles 2, small, close under the perianth. Divisions of the perianth membranous, very thin, about 1 line long, obtuse. Filaments rather shorter than the anthers, often connate at the base in a short column when there is no rudiment of the ovary, rarely free round a rudimentary or imperfect ovary. Female flowers usually without stamina or staminodia. Ovary of a single oblique carpel, the straight inner (stigmatic?) edge densely bearded and terminating in a small hooked style, the back and sides of the carpel echinate with rigid hooked bristles. Fruiting carpel about 2 lines long without the long hooked bristles with which it is covered, forming an adhesive burr.

Queensland. Port Denison, *Fitzalan;* Edgecombe Bay, *Dallachy;* Broad Sound, *Bowman;* Rockhampton, *Thozet, O'Shanesy;* Brisbane river, Moreton Bay, *W. Hill, F. Mueller.*

N. S. Wales. Clarence river, *Beckler;* Richmond river, *C. Moore.*

2. DIDYMOTHECA, Hook. f.

Flowers diœcious. Perianth small, deeply 4-lobed. Stamens in the males about 8 or 9, the anthers oblong, nearly sessile, radiating in a single series round a flat central disk. Ovary in the females of 2 carpels adnate along the inner edge to a centre column, not at all or scarcely dilated at the top. Styles or stigmas linear, proceeding from the summit of the central column. Fruiting carpels scarcely separating from the axis, dry, opening in 2 valves along the outer edge. Seeds with a crustaceous testa and a small membranous arillus or strophiole. Embryo curved round a central albumen.—Erect herb. Leaves linear. Flowers small, all axillary and nearly sessile, with 2 bracteoles under the perianth.

The single species is endemic in Australia. The male individuals would be generically undistinguishable from *Gyrostemon*, but for the more deeply lobed calyx, which in the females is accompanied by the reduction of the carpels to two, the herbaceous habit giving the plant moreover a sufficiently distinct facies to justify its retention as a separate genus.

1. **D. thesioides,** *Hook. f. in Hook. Lond. Journ.* vi. 279, *and Fl. Tasm.* i. 309, t. 93. Stems erect, often hard at the base with rather slender virgate branches, from under 1 ft. to nearly 2 ft. high, glabrous and smooth or slightly scabrous with minute papillæ. Leaves linear, the lower ones sometimes rather broad and flat, 1 to 2 in. long, and narrowed into a short petiole, the upper ones narrower smaller and sessile, and sometimes nearly all very narrow and almost terete or filiform, the floral ones often not ½ in. long. Flowers very small and almost sessile in the upper axils, all turned to one side and slightly nodding, with a pair of minute bracteoles under the perianth. Perianth not ½ line long, unequally divided nearly to the base into 4 segments varying much in breadth. Anthers nearly 1 line long. Fruit 1½ to 2 lines broad, the central axis not 1 line long, emarginate between the carpels, the styles often persistent in the notch and arched over the carpels.—Moq. in DC. Prod. xiii. ii. 37; *D. Drummondii,* Moq. l.c.; *D. veroniciformis,* F. Muell. in Linnæa, xxv. 438.

Tasmania. Near Launceston, *Lawrence;* Flinders island, *Gunn, Milligan.*
S. Australia. Port Lincoln, Dombey Bay, *Wilhelmi.*
W. Australia, *Drummond, n.* 216, *Preiss, n.* 1226 (referred in Pl. Preiss. ii. 397 doubtfully to *Euphorbiaceæ*); King George's Sound, *F. Mueller.*

3. GYROSTEMON, Desf.

(Cyclotheca, *Moq.*)

Flowers diœcious. Perianth very open under the fruit, very shortly and obtusely or obscurely sinuate-toothed. Stamens in the males either 8 to 12 radiating in a single series round a central disk or numerous covering the whole disk, the anthers oblong, nearly sessile. Ovary in the females of 4 to 20 carpels, more or less connate in a ring round a central column slightly or scarcely dilated at the top. Styles or stigmas from ovate to linear, free or slightly connected in a ring round the summit of the central column, persistent or more frequently deciduous. Fruiting carpels separating from each other and from the axis, opening usually in 2 valves both along the outer and the inner edge.—Shrubs (except *G. brachystigma?*). Leaves linear. Flowers axillary, sessile or on short recurved pedicels, with 2 bracteoles usually very small under the perianth.

The genus is limited to Australia.

Flowers sessile. Ovary of 4 to 10 carpels. Styles short ovate.
　Fruit small, much depressed. Leaves few, distant . . . . .　1. *G. brachystigma.*
Flowers sessile. Ovary of 8 to 15 carpels. Styles linear. Fruit
　nearly globular. Stamens 8 to 12 in a single ring . . . .　2. *G. cyclotheca.*
Flowers shortly pedicellate. Ovary of 15 to above 20 carpels.
　Styles shortly linear. Fruit obovoid or pear-shaped. Sta-
　mens 30 to 50 covering the whole disk . . . .　3. *G. ramulosus.*

1. **G. brachystigma,** *F. Muell.* Lower part of the stem not seen,
but probably shrubby, the branches elongated, almost rush-like, with
few distant linear-terete rather thick leaves, mostly small, the lower
ones sometimes 1 in. long. Male flowers not seen. Females sessile in
the axils, very small, the open perianth scarcely 1 line diameter.
Ovary of 4 to 10, usually about 6 carpels, the angles prominent.
Stigmas ovate, flat, round a somewhat dilated central disk or summit
of the column. Fruits broadly turbinate or almost hemispherical, under
2 lines diameter, much depressed in the centre, the angles prominent
before the dehiscence.—*Amperea?* subnuda, Nees in Pl. Preiss. ii. 229.

W. Australia, *Drummond;* York district, *Preiss, n.* 1233.

2. **G. cyclotheca,** *Benth.* A much branched shrub, sometimes low
and straggling sometimes erect and attaining 4 or 5 ft. Leaves narrow-
linear and flat, or linear-terete slender and almost filiform, acute or
with hooked points, from ½ to 1 in. long. Flowers small, on very short
axillary recurved pedicels or quite sessile, the open perianth about 1½
lines diameter. Stamens in the males about 8 to 12, in a single ring
round the flat central disk. Carpels in the females 8 to 12 or rarely
more; styles or stigmas linear, rather long, in a ring round the small
scarcely dilated summit of the column. Fruit nearly globular, scarcely
or not at all depressed in the centre, about 3 lines diameter but rather
variable in size, the dorsal angles of the carpels very prominent, separating at maturity and opening both at the inner and outer edge as in
G. ramulosus.—*Cyclotheca Australasica,* Moq. in DC. Prod. xiii. ii. 38;
Gyrostemon ramulosus, Schlecht. in Linnæa xx. 632, not of Desf.; *Didy-*
motheca pleiococca, F. Muell. Pl. Vict. i. 198, t. suppl. 9.

N. Australia? Sturt's Creek near Mount Mueller, *F. Mueller;* between Bonney
river and Mount Morphett, *M'Douall Stuart's Expedition* (in both cases only male
specimens and the identification uncertain).
　Victoria. In the Mallee Scrub, from the Wimmera through the north-western part
of the colony, growing especially on sandhills, *F. Mueller.*
　S. Australia. Encounter Bay, *Whittaker;* Murray river to St. Vincent's and
Spencer's Gulfs, *Behr., F. Mueller* and others; Kangaroo island, *Waterhouse.*
　W. Australia. Swan river, *Drummond, 1st coll.;* Swan and Murchison rivers,
Oldfield.

　Notwithstanding the difference in the stamens which is so often of no more than specific value in diœcious plants, this appears to me to be much nearer to *Gyrostemon ramulosus*
than to *Didymotheca.* The size and and in some respects the shape of the fruit as well
as the number of carpels are often so similar that there are some fruiting specimens
which can scarcely be distinguished from those of *G. ramulosus* but by the shortness of
the pedicels.

3. **G. ramulosus,** *Desf. in Mem. Mus.* vi. 17, t. 6. An erect bushy much branched shrub of 3 to 8 ft., often of a somewhat fleshy habit. Leaves linear-terete, thick or slender, 1 to 3 in. long. Flowers small, on axillary reflexed pedicels of 2 to 4 lines. Calyx varying from 1 to 2 lines diameter when open. Stamens in the males from 30 to 50, crowded in several series covering the whole centre of the flower. Ovary in the females small, nearly hemispherical, the carpels varying from about 15 to above 20, the styles linear and rather long when perfect. Fruit obovoid turbinate or more or less pear-shaped, slightly depressed on the top, with a small central disk or summit of the column, round which the styles sometimes persist to the maturity of the fruit, but often wear off early, .the dorsal angles of the carpels prominent from an early stage, the carpels at maturity separating and opening more or less at the inner as well as at the dorsal angle.—Moq. in DC. Prod. xiii. ii. 38.

N. Australia. Bay of Rest, N. W. coast, *A. Cunningham;* Fincke river, *M'Douall Stuart's Expedition.*

W. Australia. King George's Sound, *Maxwell* (a single specimen in Herb. F. Mueller) ; near Stirling Range, *F. Mueller* ; Swan river, *Drummond,* 1*st coll.;* Point Henry and Murchison river, *Oldfield;* Sharks Bay and Dirk Hartog's island, *Milne* (originally described by Desfontaines from Sharks Bay specimens).

I have presumed that all the male specimens with the anthers covering the centre of the flower belong to this species, but they vary considerably, and some lead me into doubt whether there may not be another species with the male flowers of *G. ramulosus* and the females of *G. cyclotheca;* the foliage shows no constant difference and the pedicel of the flower, both male and female, is somewhat variable in length.

4. CODONOCARPUS, A. Cunn.

(Hymenotheca, *F. Muell.*)

Flowers diœcious or monœcious. Perianth very open under the fruit, very shortly and obtusely or obscurely sinuate-toothed. Stamens in the males 10 to 20 radiating in a single series round a central disk, the anthers oblong, nearly sessile. Ovary in the females of 20 to 50 carpels connate in a ring round a central column dilated into a flat disk at the top. Styles or stigmas short or linear, free or slightly connected in a ring round the terminal disk. Fruiting carpels closely connected till near their maturity, separating when ripe from each other and from the central column and opening only along their inner edge. Seeds of adjoining carpels alternately placed near the top and below the top of the carpel, each with a small membranous aril or strophiole.—Tall shrubs or trees. Leaves linear or broad. Flowers in leafless racemes, axillary or terminal or the females on the leafless bases of the year's shoot. Bracteoles usually very small under the perianth.

The genus is limited to Australia.

Leaves narrow-linear. Styles rather long. Carpels 30 to 40 . . 1. *C. pyramidalis.*
Leaves lanceolate, tapering into a long point. Carpels 40 to 50 . 2. *C. australis.*
Leaves obovate to broadly lanceolate, obtuse or shortly pointed.
   Styles short, conical. Carpels about 20 to 30 . . . . . . 3. *C. cotinifolius.*

1. **C. pyramidalis,** *F. Muell.* A tall arborescent shrub of pyramidal growth. Leaves narrow-linear, acute, contracted into a short petiole or almost sessile, 2 to 3 in. long. Flowers diœcious, in lateral or axillary racemes, the males on very short pedicels, the females on rather long ones. Perianth about 1½ lines diameter. Ovary almost hemispherical, nearly 3 lines diameter, the central disk or dilated summit of the short column depressed; carpels from 30 to 40; styles or stigmas linear, rather long and recurved over the margins of the carpels. Fruit somewhat pear-shaped, 7 to 8 lines long, 6 to 7 lines diameter at the top and rounded and not so much expanded as in *C. australis.—Gyrostemon pyramidalis*, F. Muell. in Linnæa, xxv. 438; *Hymenotheca pyramidalis*, F. Muell. Fragm. i. 202.

S. Australia. Elder's Range, *F. Mueller;* in the interior, *M'Douall Stuart.*

2. **C. australis,** *A. Cunn. Herb.; Moq. in DC. Prod.* xiii. ii. 39. A tree of 30 ft. with numerous slender flexuose branches. Leaves lanceolate, tapering into a long narrow point and contracted into a long petiole, 1½ to 3 in. long. Flowers not seen. Fruits on long pedicels along the leafless bases of axillary branches (racemes of which the axis has grown out into a leafy branch?). Perianth 2 to 2½ lines diameter. Fruit turbinate, almost campanulate, 7 to 8 lines long, very broad at the apex, and rather deeply depressed in the centre, the disk or dilated summit of the central column 3 to 4 lines diameter; carpels 40 to 50, quite connate when young, their dorsal edges forming prominent ribs when approaching maturity, and finally separating completely, the sides then thin and transparent, empty and indehiscent in the lower part, broader and opening at the inner edge in the upper seed-bearing portion, but in alternate carpels close to the top or a little above the middle.—*Gyrostemon attenuatus*, Hook. Bot. Misc. i. 244, t. 53.

Queensland. Brisbane river, Moreton Bay, *A. Cunningham, Fraser.*
N. S. Wales. Richmond river, *Fawcett.*

It must have been owing to some mistake that the carpels have been described as having 3 or 4 ovules in each, of which only one comes to maturity. I can find no trace of any more than one in the youngest fruits on the specimens. The second seed figured in the *Botanical Miscellany* evidently belongs to the adjoining carpel. Two carpels sometimes remain unseparated and the valves are so thin that without careful examination the two seeds appear to be in one carpel, one a little above and overlaying the other as represented by the artist.

3. **C. cotinifolius,** *F. Muell. Pl. Vict.* i. 200. Usually a tall shrub or small tree, but attaining sometimes 40 ft., of a pale or glaucous green. Leaves from broadly obovate or ovate to elliptical-oblong or almost lanceolate, obtuse or shortly pointed, contracted into a rather long petiole, 1 to 2 in. long. Flowers diœcious or monœcious but usually the two sexes in separate racemes in the upper axils, sometimes forming a terminal panicle, the males on very short, the females on rather long pedicels. Perianth about 2 lines diameter in the females, rather smaller in the males. Stamens 15 to 20. Ovary about twice as long as the calyx, broadly turbinate, depressed in the centre, consisting of

20 to 30 or rather more carpels, the ovules in alternate carpels inserted near the top or about the middle of the cavity so as to give the appearance of biseriate cells or carpels. Styles or stigmas shortly conical and soon wearing off or falling off in a ring. Fruit obconical or obovoid, much less expanded .at the top and much less depressed in the centre than in *C. australis*, about 5 lines diameter, the carpels less distinctly biseriate than when young, separating and dehiscent on the inner edge as in *C. australis.—Gyrostemon cotinifolius*, Desf. in Mem. Mus. viii. 116, t. 10; Moq. in DC. Prod. xiii. ii. 39; *G. pungens*, Lindl. in Mitch. Three Exped. ii. 121; *G. acaciæformis*, F. Muell. in Linnæa, xxv. 439.

N. Australia. In the interior lat. 20°, *M'Douall Stuart's Expedition.*

N. S. Wales. Lachlan, Darling, and Murray deserts, *Mitchell, Victorian and other Expeditions;* Mount Murchison, *Bonney.*

Victoria. Mallee scrub on the Murray, *Hughan;* Wimmera, *Dallachy.*

S. Australia. Flinders range, *F. Mueller;* Cooper's river, *A. C. Gregory.*

W. Australia, *Drummond, n.* 40; Murchison river, *Oldfield;* Sharks Bay *Milne.*

5. TERSONIA, Moq.

Flowers diœcious. Perianth very open under the fruit, very shortly and obtusely or obscurely sinuate-toothed. Stamens in the males 8 to 12 or rather more, radiating in a single series round a central disk; anthers oblong, nearly sessile. Ovary of the females (in *T. brevipes*) of 15 to 20 carpels completely connate into a plurilocular fleshy ovary not ribbed but marked with horizontal raised concentric zones. Styles or stigmas linear, radiating round a broad central disk or dilated summit of the central column. Fruit (where known) depressed globular, indehiscent, almost woody. Seeds oblong, all on the same level, with a small membranous aril or strophiole. Embryo folded. Albumen scanty or none.—Diffuse shrubs or herbs. Leaves linear. Flowers sessile or nearly so, the males in terminal racemes or interrupted spikes, the females (where known) solitary in the axils of the stem-leaves. Bracteoles under the perianth small.

The genus is endemic in W. Australia.

Leaves long, narrow linear . . . . 1. *T. brevipes.*
Leaves short, distant. Branches flexuose or twining . . . . . 2. *T. subvolubilis.*

1. **T. brevipes,** *Moq. in DC. Prod.* xiii. ii. 40 (*as to the female plant*). A decumbent shrub of 2 or 3 ft. with ascending flowering branches. Leaves linear or linear-terete, rather thick, 1 to 2 in. long or even more. Male flowers all in terminal leafless spikes or racemes, the floral leaves all reduced to small bracts or the lower ones rarely exceeding the flowers. Anthers radiating to a diameter of nearly 3 lines when fully out. Female flowers all sessile and solitary in the axils of distant stem-leaves. Perianth scarcely 1½ lines diameter. Styles or stigmas nearly 2 lines long, the central disk bordered within them by a raised undulate margin. Fruit 6 to 8 lines diameter, armed with raised almost scale-like hard tubercles more or less arranged in

irregular horizontal zones. Seeds small, oblong.—*Gyrostemon ramu-losus*, Lehm. Pl. Preiss. i. 243, not of Desf.; *G. angustifolius*, Schnitzl. Ic. Fam. Nat. t. 208.*

W. Australia. Swan river, *Preiss. n.* 1234; Swan and Murchison rivers, *Oldfield*.

The male specimens described by Moquin belong to *Gyrostemon ramulosus*, Desf.

2. **T. ? subvolubilis,** *Benth.* Stems numerous, herbaceous diffuse and flexuose, many of them apparently twining, attaining 1 ft. and more, the flowering branches ascending, slightly scabrous, perhaps viscid when fresh. Leaves small, linear or linear-lanceolate, acute, contracted at the base, rarely exceeding ½ in. and usually few and distant. Male flowers in terminal interrupted spikes, precisely like those of *T. brevipes;* females unknown.

W. Australia. Oldfield river and Phillips Ranges, rare, *Maxwell*.

The female flowers and fruits of this species being as yet unknown, the genus must be in some measure uncertain, but the inflorescence of the males and their general habit agree with none but *Tersonia*.

ORDER XCVI. **CHENOPODIACEÆ.**

(Salsolaceæ, *Moq. in DC. Prod.* xiii. ii.)

Perianth small, with 5 or fewer segments or lobes, herbaceous or rarely thin and transparent or somewhat scarious, imbricate in the bud. Stamens 5 or rarely fewer, opposite the perianth-segments and usually inserted at or near their base; anthers 2-celled, the cells opening longi-tudinally. Ovary free, 1-celled, with a single ovule erect or suspended from an erect funicle. Styles or style-branches 2 or 3, stigmatic along their whole inner edge or rarely towards the end only. Fruit 1-celled and indehiscent, membranous or succulent, enclosed in or resting on the persistent perianth which is sometimes enlarged or altered in form. Seed solitary, erect or horizontal, usually orbicular and flattened. Embryo coiled round a mealy albumen, or spirally twisted without any or with scarcely any albumen.—Herbs or undershrubs, often succulent and very frequently hoary or white, especially the young parts, with a minute and mealy or more dense and scaly tomentum, or in some genera villous or woolly. Leaves alternate or very rarely opposite, sometimes none. Stipules none. Flowers small, usually sessile and clustered, either axillary or in axillary or terminal dense or interrupted spikes or panicles, and often unisexual. Bracts inconspicuous, or, in some genera, 1 bract and 2 bracteoles more or less conspicuous.

A considerable Order, spread over the greater part of the world, but most abundant in maritime or saline situations in the Old World, a few species, in identical or closely allied forms, being quite cosmopolitan. Of the fifteen Australian genera seven have the general distribution of the Order, the remaining eight appear to be endemic.

TRIBE 1. **Chenopodieæ.**—*Branches continuous. Leaves flat, glabrous, mealy, scaly or glandular. Testa crustaceous. Embryo curved round a mealy albumen.*

Perianths equally 5- or 4-lobed, herbaceous, not much enlarged in fruit.

Fruit a small succulent berry . . . . 1. RHAGODIA.

Fruit a dry nut enclosed in the perianth . . . . 2. CHENOPODIUM.
Perianths minute white and dry, with 3 or 1 clavate segments . . 3. DYSPHANIA.
Male perianths small, equally 5- or 4-lobed, females much enlarged
in fruit with 2 broad appressed segments enclosing the fruit . . 4. ATRIPLEX.

TRIBE 2. **Camphorosmeæ.**—*Branches continuous. Leaves narrrow, entire, flat or terete, glabrous, villous-tomentose or woolly. Testa membranous. Embryo curved round a mealy albumen.*

Fruiting-perianths globular or depressed, membranous, herbaceous
or succulent, the lobes horizontally (or rarely conically) closing
over the fruit. Seed horizontal or oblique.
Fruiting-perianth without appendages, succulent or coriaceous,
glabrous or the lobes slightly pubescent . . . . 5. ENCHYLÆNA.
Fruiting-perianth surrounded by 3 to 5 distinct or by 1 continuous
annular horizontal wings . . . . 6. KOCHIA.
Fruiting-perianth membranous or herbaceous, enveloped in wool or
long hairs, without any or with horn-like or spinescent dorsal
appendages . . . . 7. CHENOLEA.
Fruiting-perianth hard, at least at the base, the lobes usually membranous and withering.
Fruiting-perianth with dorsal wings.
Flowers solitary. Wings 2 or 3, nearly vertical. Seed horizontal . . . . 8. BABBAGIA.
Flowers 2 together divaricate and connate at the base. Wings
5, horizontal. Seed vertical . . . · . . . . 9. DIDYMANTHUS.
Fruiting-perianth without any or with spinescent dorsal appendages.
Fruiting-perianth tomentose or woolly, with 2 opposite diverging dorsal spines rarely wanting. Seed horizontal or
oblique . . . . 10. SCLEROLÆNA.
Fruiting-perianth glabrous, without appendages or rarely with
5 small erect spines. Seed horizontal or oblique . . . . 11. THRELKELDIA.
Fruiting-perianth glabrous or slightly hairy, with 3 to 5 divaricate dorsal spines. Seed vertical . . . . 12. ANISACANTHA.

TRIBE 3. **Salicornieæ.**—*Branches articulate, fleshy. Leaves none. Flowers more or less immersed. Testa various. Embryo curved or folded with little or no albumen.*
Single genus . . . . 13. SALICORNIA.

TRIBE 4. **Salsoleæ.**—*Branches continuous. Leaves narrow, flat or terete, entire. Testa various. Embryo spirally coiled, without albumen.*

Perianth small, herbaceous, without appendages. Testa crustaceous.
Embryo coiled in a flat spire . . . . 14. SUÆDA.
Perianth rigid, with dorsal horizontal wings. Testa membranous.
Embryo coiled in a conical or biconvex spire. (Leaves and bracts
pungent) . . . . 15. SALSOLA.

TRIBE 1. CHENOPODIEÆ.—Branches continuous (not articulate).
Leaves flat, often triangular or hastate, glabrous mealy scaly or glandular. Testa of the seed crustaceous. Embryo curved round a mealy
albumen.

1. RHAGODIA, R. Br.

Flowers polygamous, mostly hermaphrodite or female, but sometimes
almost diœcious. Perianth deeply 5-cleft, the lobes or segments obtuse,

concave, scarcely enlarged in fruit, and either closing over the fruit or expanded under it. Stamens 5 or fewer, filaments more or less flattened. Ovary globular or nearly so. Styles 2 or very rarely 3, shortly subulate, very shortly united at the base. Fruit a small depressed-globular berry. Seed flattened, horizontal ; testa crustaceous. Embryo circular, enclosing a mealy albumen.—Shrubs undershrubs or rarely herbs. Leaves alternate or some or nearly all opposite, flat, entire. Flowers small, sessile or very rarely pedicellate, in clusters or rarely solitary, in interrupted terminal spikes or panicles, without bracts. Perianth tomentose outside, glabrous inside.

The genus is exclusively Australian, differing from *Chenopodium* in the succulent pericarp and usually in the more shrubby habit. The species are often very variable in stature and foliage and very difficult to mark out by positive characters. They are moreover often represented in herbaria by specimens so imperfect as to leave a large proportion of the determinations doubtful. The succulent pericarps or berries appear to vary in colour, even in the same species, from red or purple to yellow, but perhaps the collectors' notes in this respect are not all to be trusted.

Panicle usually much branched.
  Flowers polygamous.
    Leaves almost all alternate, mostly narrow and green
      above, paler or whiter underneath than above, the mar-
      gins often recurved . . . . 1. *R. Billardieri.*
    Leaves alternate or opposite, mostly broad, flat, green or
      white on both sides . . . . 2. *R. parabolica.*
  Flowers diœcious, very small. Leaves oblong or broad, flat,
    pale or white on both sides . . . . 3. *R. dioica.*
Inflorescence nearly single or panicle not much branched.
  Leaves rather thick and fleshy, flat or concave, mostly alter-
    nate. Plant not spinescent.
    Fruiting perianth 2 to 3 lines diameter when open and
      much larger than the fruit. Leaves mostly hastate   4. *R. Gaudichaudiana.*
    Fruiting perianth 1 to 1½ lines diameter and usually not
      broader than the fruit.
      Leaves rarely above ¼ in. long, linear cuneate or rarely
        obovate . . . . 5. *R. crassifolia.*
      Leaves linear, ½ to 1 in. long . . . . 6. *R. Preissii.*
      Leaves broadly obovate to oblong, ½ to 1 in. long . . 7. *R. obovata.*
  Leaves flat, rather thin, mostly alternate broad and small.
    Plant spinescent . . . . 8. *R. spinescens.*
  Leaves thin, green, opposite or alternate. Plant usually
    slender or weak.
    Leaves mostly opposite, ovate or hastate, very obtuse . . 9. *R. hastata.*
    Leaves mostly opposite, lanceolate or broad, all acute . . 10. *R. nutans.*
    Leaves alternate, linear . . . . , ... 11. *R. linifolia.*

R. chenopodioides, Moq. Chenopod. Enum. 11, and in DC. Prod. xiii. ii. 52, from Port Jackson, *Gaudichaud*, is unknown to me, the character given agrees with that of *R. Billardieri*, except that the flowers are said to be pedicellate, which may have been an accidental anomaly in the specimens described.

1. R. Billardieri, *R. Br. Prod.* 408. A shrub either straggling or diffuse, or erect and attaining 5 or 6 ft., the foliage and young shoots somewhat fleshy, rarely quite green, but usually with less of the mealy tomentum than most species, except on the inflorescence. Leaves alternate or rarely here and there opposite or nearly so, usually oblong-lanceolate or linear-oblong, but varying to quite linear or when small to ovate or broadly hastate, always obtuse, contracted into a short petiole, from scarcely ½ in. to above 1 in. long, usually green above when full grown and pale or whitish underneath, the margins often recurved and never incurved. Panicle terminal, usually much branched and 1½ to above 2 in. long, the branches rather slender and divaricate, the flowers small, polygamous, in distinct clusters. Perianth about 1 line diameter, or when fully expanded under the fruit attaining 1½ lines, lobed to about the middle. Berry when fully ripe as broad as or broader than the expanded perianth.—*Chenopodium baccatum*, Labill. Pl. Nov. Holl. i. 71, t. 96; *Rhagodia baccata*, Moq. in DC. Prod. xiii. ii. 50; Hook. f. Fl. Tasm. i. 312; *R. Candolleana*, Moq. Chenop. Enum. 10 (with small broad leaves approaching the var. *congesta*).

N. S. Wales. Port Jackson, *R. Brown, Woolls;* Macleay river and Ash island, *Beckler* (the latter specimens not good and rather doubtful); Twofold Bay, *A. Cunningham* and others.

Victoria. Abundant along the sea-coast from the Glenelg to the eastern frontier, *F. Mueller* and others.

Tasmania. Port Dalrymple, *R. Brown;* common in salt marshes especially on the north coast, *J. D. Hooker, Labillardière* and others.

S. Australia. Lower Murray river and round St. Vincent's Gulf, *F. Mueller* and others.

W. Australia. Goose-island and Lucky Bays, *R. Brown;* common along the sea-coast from King George's Sound round to Swan and Murchison rivers, *Fraser, Drummond, n.* 209, *Oldfield* and others, and eastward to the Great Bight, *Maxwell.*

Among the numerous forms which this species assumes, many of them no doubt owing to local influences, the following are the most distinct in the dry state:—

Var. *congesta,* Hook. f. Fl. Tasm. i. 312. Low and much branched, densely crowded with leaves and inflorescences; leaves mostly ovate small and here and there hastate.—*Chenopodium congestum,* Hook. f. in Hook. Lond. Journ. vii. 280; *Rhagodia congesta,* Moq. in DC. Prod. xiii. ii. 51.—Tasmania and Victoria.

Var. *linearis.* Leaves all narrow-linear.—*R. radiata,* Nees in Pl. Preiss. i. 637, Moq. in DC. Prod. xiii. ii. 50.—Chiefly in W. Australia where it passes very gradually into the oblong or linear-oblong leaved form there very abundant, and readily distinguished from the linear-leaved form of *R. crassifolia* by the recurved or revolute margins of the leaves and the more branched panicle.

2. **R. parabolica,** *R. Br. Prod.* 408. An erect shrub, attaining sometimes 8 or 10 ft., but often much lower, mealy-white all over or in the more slender specimens the adult leaves green. Leaves opposite or alternate, on rather long petioles, broadly ovate obovate or almost rhomboidal, usually rounded at the end and obtuse or shortly mucronate, contracted or rarely obtusely hastate at the base, rarely exceeding 1 in. and often under ½ in. long. Flowers, as in *R. Billardieri,* polygamous, in distinct clusters along the divaricate branches of a terminal panicle, sometimes very dense and crowded, sometimes loose and slender. Perianth mealy-tomentose and sometimes densely so, the lobes ovate, obtuse, more united at the base than in most species, expanding to above 1 line diameter, but the succulent ripe fruit still larger. Seed about 1 line diameter.—Schlecht. in Linnæa, xx. 574; Moq. in DC. Prod. xiii. ii. 51; *R. reclinata,* A. Cunn. Herb.; Moq. l. c. 51 (with the leaves less white than in the typical form).

Queensland. In the interior, *Mitchell, Bowman;* Armadilla, *Barton;* Curri-willighie, *Dalton;* between Burnett and Dawson rivers, *F. Mueller.*

N. S. Wales, *Leichhardt;* Liverpool plains, *A. Cunningham;* Richmond river, *Henderson.*

S. Australia. Spencer's Gulf, *R. Brown;* Murray Scrub, *Behr.;* Murray river and Salt Creek, *F. Mueller.*

This may possibly prove to be an inland variety of *R. Billardieri.*

3. **R. dioica,** *Nees in Pl. Preiss.* i. 636. A tall shrub, with the habit of *R. parabolica,* usually hoary or white. Leaves alternate or nearly opposite, on rather long petioles, oblong obovate ovate or rarely broadly hastate, very obtuse, flat, rather thin, pale or hoary on both sides, often above 1 in. long. Flowers smaller than in either of the preceding species, diœcious in all the specimens seen, very numerous in a dense much branched ovate or pyramidal panicle of two or three inches. Fruiting perianth small, the fruit rather larger, but not exceeding 1 line diameter.—Moq. in DC. Prod. xii. ii. 50.

W. Australia, *Drummond;* Swan river, *Preiss, n.* 1253, Murchison river, *Oldfield.*

4. **R. Gaudichaudiana,** *Moq. Chenop. Enum.* 11, *and in DC. Prod.* xiii. ii. 53. A very divaricate or prostrate shrub, covered with a dense soft white tomentum which usually persists even on the old leaves, the branchlets sometimes but very rarely almost spinescent. Leaves mostly alternate, from broadly orbicular or deltoid to lanceolate, often hastate at the base, rather thick, flat or concave, ½ to 1 in. long. Flowers polygamous (or sometimes diœcious?) larger than in the allied species, solitary or clustered in nearly simple interrupted spikes or along the divaricate branches of short terminal panicles. Fruiting perianth 2 to 3 lines diameter, lobed to the middle, very cottony outside, glabrous and coloured or green inside, much larger than the fruit and either closed over it or expanded under it.

N. S. Wales. Darling river, *Victorian Expedition.*

S. Australia. Gawler Range, *Sullivan.*

W. Australia. Murchison river, *Oldfield;* Sharks Bay, *Gaudichaud* (I have not seen the latter specimens).

The specimens from N. S. Wales and from S. Australia have the perianth not quite so large as those from Murchison river, but they are not so far advanced. There are many other specimens in Herb. F. Mueller from the desert interior of N. S. Wales and S. Australia, mostly mere scraps barely in flower, which may belong to *R. Gaudichaud-iana,* but which cannot be safely distinguished from some of the following species.

5. **R. crassifolia,** *R. Br. Prod.* 408. A dwarf or diffuse much branched scrubby shrub, or the branches somewhat elongated in narrow-leaved forms, nearly green or more or less hoary-tomentose. Leaves mostly alternate, linear or scarcely oblong in the typical form, rarely cuneate or almost obovate in some varieties, obtuse, contracted into a short petiole, rather thick, flat or concave, rarely ½ in. long. Flowers and fruits small, clustered or rarely solitary, in short terminal nearly simple interrupted spikes or slightly branched panicles. Fruiting perianth not exceeding the fruit.

N. Australia. Sturt's Creek, *F. Mueller.*

Victoria. Wimmera, *Dallachy.*

S. Australia. Kangaroo island, *R. Brown ;* towards Spencer's Gulf, *Warburton.*
W. Australia. Scattered over the treeless plains N. W. of the Great Bight, *De-lisser ;* Point Henry and Murchison river, *Oldfield.*

Var. *latifolia.* Leaves obovate ovate hastate or broadly oblong, all very obtuse and under ½ in. long.—*R. crassifolia,* Moq. in DC. Prod. xiii. ii. 52.—Dirk Hartog's island, *A. Cunningham.*

There are several imperfect specimens from the desert interior, in Herb. F. Mueller and others which may belong to this species but cannot be determined with any certainty. *R. Drummondii,* Moq. in DC. Prod. xiii. ii. 52, from W. Australia, *Drummond, n.* 133, seems to be a half-starved small and narrow-leaved state of this species. The flowers are said to be diœcious, our specimens are not sufficient to ascertain whether they are really so. I have not seen *Drummond's n.* 715, described by Moq. l.c. as *R. parvifolia,* but the character entirely agrees with that of the small broad-leaved forms of *R. crassifolia.*

6. **R. Preissii,** *Moq. in DC. Prod.* xiii. ii. 49. A much branched slender divaricate undershrub, usually green or slightly hoary except the inflorescence which is whiter. Leaves mostly alternate, linear or linear-oblong, obtuse, contracted into a petiole, thick and fleshy, green or scarcely hoary, flat or with incurved margins, ½ to 1 in. long. Flowers and fruits of *R. crassifolia,* of which this may prove to be a variety with longer leaves and with the inflorescence usually more elongated and slender.—*R. linifolia.* Nees in Pl. Preiss. i. 637, not of R. Br.

W. Australia. Swan river, *Drummond, n.* 716, *Preiss, n.* 125 ; eastern interior of W. Australia, *Harper.*

7. **R. obovata,** *Moq. Chenop. Enum.* 10, *and in DC. Prod.* xiii. ii. 51. An erect scrubby shrub, more or less hoary or white. Leaves alternate or here and there opposite, from broadly ovate obovate or ovate-hastate to oblong, obtuse, contracted into a petiole, thick and fleshy, flat or slightly concave, equally hoary or mealy-white on both sides when young, the older ones often turning black or lead colour when dry, ½ to 1 in. long. Flowers as in the two preceding species in distinct clusters in a simple interrupted spike or slightly branched panicle, about the size of those of *R. Billardieri.* Fruiting perianth not exceeding the perfectly ripe fruit.

W. Australia. Sharks Bay, *Milne, Denham.* The specimens described by Moquin are said to be from the south and west coast of Australia, but were probably all from Sharks Bay, the Australian stations in the herbarium of the Paris Museum being very frequently erroneous.

8. **R. spinescens,** *R. Br. Prod.* 408. A divaricately branched rather slender shrub, usually low and straggling or prostrate, sometimes more erect and bushy and attaining several ft., mealy-white or at length nearly glabrous, the smaller branchlets often (but not always) terminating in slender spines. Leaves mostly alternate, obovate ovate orbicular or deltoid, in some specimens nearly ½ in. long, rather narrow, thin and but slightly mealy, in others all under ¼ in. broad, rather thick and very mealy-white, with many intermediate states, always flat or concave. Flowers small, polygamous, in small clusters or almost solitary in short terminal interrupted spikes or panicles, or almost

solitary in the upper axils. Fruit small ($1\frac{1}{4}$ lines diameter when dry), the perianth expanded under it but not projecting beyond.—Moq. in DC. Prod. xiii. ii. 53.

Queensland. Burdekin river, *F. Mueller ;* near Rockhampton, rare, *O'Shanesy ;* Armadilla, *Barton.*

N. S. Wales. Liverpool plains and Lachlan river, *A. Cunningham;* Darling river, *Mitchell;* and thence to the Barrier Range, *Victorian Expedition;* Castlereagh river, *C. Moore.*

Victoria. Murray and Avoca rivers, *F. Mueller.*

S. Australia. Spencer's Gulf, *R. Brown* ; Cudnaka and Murray river, *F. Mueller ;* Gawler Ranges, *Sullivan;* N.E. of Lake Gairdner, *Babbage.*

R. prostrata, A. Cunn. Herb.; Moq in DC Prod. xiii. ii. 52, is described from specimens which have no spines, but which in all other respects agree with *R. spinescens,* and may be either an unarmed variety or a mere state in which the spines are not yet developed.

9. **R. hastata,** *R. Br. Prod.* 408. A procumbent or divaricately branched undershrub, spreading to 2 or 3 ft., green or slightly mealy-white when young. Leaves opposite or rarely alternate, petiolate, ovate-hastate or almost rhomboidal, very obtuse or emarginate, the basal lobes short obtuse or rarely acute, under 1 in. and often under $\frac{1}{2}$ in. long. Flowers small, clustered, usually in compact simple or slightly branched spikes, either terminal or in the upper axils and shorter than the leaves, rarely more slender and elongate. Perianth-divisions oblong, not contracted at the base or stipitate as in *Chenopodium triangulare,* which this species sometimes resembles. Fruit $\frac{1}{2}$ to $\frac{3}{4}$ line diameter, usually red.—Moq. in DC. Prod. xiii. ii. 53.

Queensland. Rockhampton, *Dallachy* and others; Nerkool Creek, *Bowman;* Moreton Bay, *Leichhardt, C. Stuart.*

N. S. Wales. Port Jackson, *R. Brown* and others; Liverpool plains, *A. Cunningham;* New England, *C. Stuart;* Hastings river, *Beckler.*

Victoria. Bacchus marsh, *F. Mueller.*

10. **R. nutans,** *R. Br. Prod.* 408. Herbaceous, prostrate or procumbent and slender, often extending to 1 to 2 ft., green or the young foliage more or less mealy-white, the stems rarely almost woody at the base. Leaves opposite or here and there alternate, on rather slender petioles, from broadly hastate with very prominent basal lobes to lanceolate and angular only at the base, always acute, rather thin and green, rarely 1 in. long and often all under $\frac{1}{2}$ in., the upper ones gradually smaller. Inflorescence simple or with a few short branches, terminal or in the upper axils, under 1 in. long or rarely elongated, sometimes nodding at the end, sometimes slightly leafy at the base, with one or two solitary flowers or small clusters in the axils of the upper leaves. Flowers very small, the males with 2 or 3 stamens and a rudimentary pistil, the females without any or with only 1 stamen. Fruit about $\frac{1}{2}$ line diameter, the pericarp red and succulent when fresh, thin when dry.—Moq. in DC. Prod. xiii. ii. 53; Hook. f. Fl. Tasm. i. 312.

Queensland. Brisbane river, *F. Mueller ;* Curriwillighie, *Dalton.*

N. S. Wales. Lachlan river, *A. Cunningham;* Lachlan and Darling rivers to the

Barrier Range, *Victorian and other Expeditions;* Hastings river, *Beckler;* Ballandool river, *Locker.*

Victoria. Wendu vale, *Robertson;* Tambo river, *F. Mueller;* Creswick, *Whan.*

Tasmania. Derwent river, *R. Brown, J. D. Hooker;* abundant in plains near Ross, *Gunn;* S. Esk river, *C. Stuart.*

S. Australia. Kangaroo island, *R. Brown;* Murray river, Salt Creek, Port Adelaide, *F. Mueller;* towards Cooper's Creek, *Wheeler, Howitt's Expedition.*

12. **R. linifolia,** *R. Br. Prod.* 408. A diffuse or procumbent herb or undershrub, more slender even than *R. nutans* and like that species green or the young shoots very slightly mealy. Leaves alternate, linear or linear-lanceolate, rather acute, contracted into a short petiole, thin and green on both sides, from under $\frac{1}{2}$ in. to nearly 2 in. long. Inflorescence almost filiform, rarely above 1 in. long, simple or slightly branched or forming a slender divaricate leafy panicle. Flowers very small, solitary or in small clusters, the females mostly pedicellate. Perianth glabrous. Fruit smaller than in any other species, the pericarp red and pulpy when fresh, thin when dry.—*Moq. in DC. Prod.* xiii. ii. 49.

Queensland. Broad Sound, *R. Brown;* Bay of Inlets, *Banks and Solander;* in the interior, *Mitchell;* Curriwillighie, *Dalton;* Darling Downs, *Lau;* Rockhampton, *O'Shanesy.*

N. S. Wales. Hunter's river, *A. Cunningham;* in the interior, *Leichhardt;* Camden district, *Woolls;* New England, *C. Stuart;* Ballandool river, *Locker.*

2. CHENOPODIUM, Linn.

(Ambrina, *Moq.*, Blitum, *Moq.* (*partly*)).

Flowers hermaphrodite or rarely polygamous. Perianth herbaceous, deeply divided into 5 or rarely 4 or 3 lobes or segments which are obtuse and concave or rarely acute and erect, scarcely altered or slightly enlarged after flowering. Stamens 5 or fewer, filaments filiform or flattened. Ovary globular or ovoid; styles 2 or rarely 3, usually united at the base. Fruit depressed or ovoid, partially or completely covered by the persistent perianth, pericarp dry, membranous, distinct from or inseparable from the seed. Seed horizontally flattened, or vertical and less compressed; testa crustaceous; embryo circular, enclosing a mealy albumen.—Herbs or rarely shrubs or undershrubs. Leaves alternate, flat, entire toothed or divided. Flowers small, sessile in clusters, either axillary or in interrupted terminal spikes or panicles.

The genus is widely distributed over the globe, but appears to be really indigenous chiefly in temperate and subtropical regions, some species, including four of the Australian ones, probably of European origin, are amongst the most generally dispersed weeds of cultivation. Of the remaining eight Australian species one is also in New Zealand and New Caledonia, the other seven appear to be endemic although one of them is perhaps too closely connected with an East Asiatic one.

The precise limits to be assigned to the genus are as yet very uncertain. The last four species here included, with the seeds all erect and the inflorescence axillary, are certainly nearly allied to the European *Blita* originally characterized by the succulent perianth, but recently extended to the majority of *Chenopodia* with erect seeds. The adoption of the latter character entails however the assigning *C. nitrariacea* and *C.*

Bonus-henricus to *Blitum*, a most unnatural combination, and leaves *C. glaucum* and *C. rubrum*, in which the seeds of some of the flowers are often erect, ambiguous between the two genera. I have therefore followed F. Mueller in reuniting them, at least as to the Australian species, and the very variable consistence of the fruiting perianth in *C. carinatum* and *C. rubrum*, leaves it very doubtful whether even the Linnean *Blita*, with their berry-like fruits, can be distinctly separated from *Chenopodium*.

SECT. 1. **Rhagodioides.**—*Spinescent shrub. Flower-clusters in terminal spikes. Seeds vertical.*

Plant hoary or mealy-white. Leaves entire . . . . 1. *C. nitrariacea.*

SECT. 2. **Chenopodiastrum.**—*Herbs mealy-white or glabrous. Flower-clusters in terminal or axillary spikes or panicles. Seeds all or mostly horizontal.*

Erect and mealy-white or almost glabrous. Spikes terminal, often
   paniculate.
   Leaves (usually very hoary or white) entire or very rarely hastate 2. *C. auricomum.*
   Leaves (green or mealy-white underneath or on both sides) at
    least the lower ones coarsely sinuate-toothed . . . . . 3. *C. album.*
Erect green annual. Cymes or panicles axillary and loose.
   Leaves broad, coarsely-toothed . . . . 4. *C. murale.*
Decumbent or prostrate herbs (mostly annual).
   Leaves entire or hastate. Stamen usually 1.
   Stems weak and elongated. Leaves green or scarcely mealy.
    Flower-clusters in terminal interrupted spikes . . . . 5. *C. triangulare.*
   Stems short, much-branched. Leaves small, mealy. Flower-
    clusters small in the upper axils or scarcely spicate . . 6. *C. microphyllum.*
Lower leaves sinuate-toothed, mostly green above, white under-
   neath. Stamen usually 1 . . . . 7. *C. glaucum.*

SECT. 3. **Botryois.**—*Erect glandular aromatic herbs or undershrubs not mealy. Seeds all or mostly horizontal.*

Leaves narrow, mostly sinuate-toothed. Flower-clusters axillary,
   forming a leafy panicle . . . . 8. *C. ambrosioides.*

SECT. 4. **Orthosporum.**—*Decumbent glandular herbs not mealy. Seeds all vertical. Flower-clusters all axillary.*

Perianth-segments broad, concave with a thickened keel . . . 9. *C. carinatum.*
Perianth-segments narrow, nearly erect, with a thickened keel.
   Minute filiform plant . . . . 10. *C. pumilio.*
Perianth-segments linear, erect, the keel dilated into a broad
   fringed wing or crest . . . . 11. *C. cristatum.*
Perianth-segments lanceolate, erect, the keel much thickened and
   angular at the base . . . . 12. *C. atriplicinum.*

SECT. 1. RHAGODIOIDES.—Spinescent shrub. Flower-clusters in ter-minal spikes. Seeds all vertical.

1. **C. nitrariacea,** *F. Muell.* A rigid divaricately branched or prostrate shrub or undershrub, hoary or mealy-white all over with a minute tomentum, the smaller branchlets often spinescent but not nearly so slender as in *Rhagodia spinescens.* Leaves alternate, sometimes clustered at the base of the flowering branchlets, linear oblong or linear-spathulate, very obtuse, entire, contracted into a short petiole, from under ½ in. to nearly 1 in. long. Flowers sessile, usually clustered in interrupted or dense spikes, either simple and terminal or forming short divaricate branches to a terminal panicle, mostly hermaphrodite

with a few males intermixed.   Perianth-segments broad, thick, concave, slightly imbricate in the bud.   Stamens 5, shortly exserted, the
filaments flat and glabrous.   Ovary ovoid, erect, the styles short, rather
thick, united at the base.   Fruit enclosed in the unaltered perianth.
Pericarp membranous.   Seed erect, flat; embryo circinate, the radicle
usually inferior.—*Rhagodia nitrariacea,* F. Muell. in Trans. Phil. Inst.
Vict. ii. 73.

N. Australia.   N.W. coast, *Bynoe,* the specimens in bud and in some measure
doubtful.
N. S. Wales.   Darling river, *Victorian Expedition, Mrs. Ford.*
Victoria.   Murray and Avoca rivers, *F. Mueller.*
W. Australia.   Swan river, *Drummond.*
Some other specimens referred to this species by F. Mueller appear to me to belong
to *Rhagodia spinescens,* but are too young to determine.   In all those which 1 have
quoted as typical, I have uniformly found the seed, either already enlarged after flowering or quite ripe, erect and enclosed in a thin dry pericarp.

SECT. 2. CHENOPODIASTRUM, Moq.—Herbs, mealy-white or gla-
brous.   Flower-clusters in terminal or axillary spikes or panicles.   Seeds
all or mostly horizontal.

2. **C. auricomum,** *Lindl. in Mitch. Trop. Austr.* 94.   Erect and
probably tall, more or less white or hoary all over, apparently herbaceous and not spinescent.   Leaves on rather long petioles, ovate or
oblong, very obtuse. entire or rarely hastate with prominent basal
lobes, mostly ¾ to 1½ in. long.   Flowers in little dense globular clusters
along the branches of a terminal panicle, sometimes distinct and rather
distant, sometimes crowded into dense spikes.   Perianth-segments
broad, concave, closing over the fruit.   Stamens 5, shortly exserted.
Ovary small, globular, contracted into a long neck or united base of
the styles.   Pericarp depressed-globose, membranous.   Seed very flat,
horizontal.   Embryo annular.—Moq. in DC. Prod. xiii. ii. 460.

N. Australia.   Upper Victoria river and Sturt's Creek, *F. Mueller ;* Gulf of Carpentaria, *Landsborough ;* in the interior, *M'Douall Stuart's Expedition.*
Queensland.   Narran river, *Mitchell ;* Curriwillighie, *Dalton ;* Suttor and Bowen
rivers, *Bowman.*
N. S. Wales.   Darling river and Duroodoo, *Victorian Expedition.*
This species undoubtedly comes near to some forms of *C. album,* differing in its
entire more tomentose leaves and larger flowers.   It appears to be still more closely
allied to and perhaps not really distinct from the East Asiatic *C. acuminatum,* Willd.
C. furfuraceum, Moq. in DC. Prod. xiii. ii. 64, from the Straits of Entrecasteaux,
Tasmania, is unknown to me.   The character given agrees with that of *C. auricomum,* of which however I have seen no specimen from Tasmania, nor from the south
coast of the continent of Australia.

3. **C. album,** *Linn. ; Moq. in DC. Prod.* xiii. ii. 70.   A tough annual
usually erect, 1 to 2 ft. high, of a pale green or more or less mealy-
white, especially the flowers and the under side of the leaves.   Leaves
petiolate, the lower. ones ovate or rhomboidal, more or less sinuate-
toothed or angular, the upper ones usually narrow and entire.   Clusters
of flowers in short dense or interrupted spikes, simple or slightly
branched, the lower ones axillary, the upper ones or sometimes nearly

all in a long terminal panicle leafy at the base. Segments of the fruiting perianth broad, concave, somewhat thicker in the centre or keeled, contracted and united at the base, completely closing over the fruit. Stamens usually 5. Seeds all horizontally flattened, smooth and shining, the pericarp exceedingly thin.—*C. lanceolatum,* R. Br. Prod. 407; Moq. in DC. Prod. xiii. ii. 62; *C. Browneanum,* Roem. and Schult. Syst. vi. 275.

Queensland. Nerkool Creek, *Bowman;* Armadilla, *Barton;* Warwick, *Beckler* (the specimen bad and somewhat doubtful).

N. S. Wales. Paterson's river, *R. Brown;* Liverpool plains, *Leichhardt;* Paramatta, *Woolls.*

Victoria. Melbourne, *Adamson;* Bacchus marsh and Snowy river, *F. Mueller;* Skipton, *Whan.*

W. Australia. *Drummond, n.* 224.

The species is a very common weed in Europe and temperate Asia, and has spread as such over many other parts of the world. Whether it be really indigenous or introduced only into Australia is uncertain. In N. S. Wales and Queensland it is said to be known under the name of *Fat-Hen.*

C. biforme, Nees in Pl. Preiss. i. 636, from Swan river, *Preiss, n.* 1256, described from a single specimen which I have not seen, may be one of the numerous forms of *C. album.* It is described as having the inflorescence flowers and indumentum of *C. album,* but with the leaves, especially in their dentation, more like those of *C. murale,* to which Moquin refers it in DC. Prod. xiii. ii. 69.

4. **C. murale,** *Linn.; Moq. in DC. Prod.* xiii. ii. 69. A rather stout erect or decumbent much branched annual, from under 1 ft. to nearly 2 ft. high, usually green, but sometimes with a slight whitish meal on the young shoots. Leaves on long petioles, broadly ovate triangular or rhomboidal, deeply and irregularly toothed, 1 to above 2 in. long. Flowers small, green or slightly mealy, the clusters in much-branched rather slender spikes, forming loose leafless cymes or panicles usually much shorter than or rarely as long as the leaves, almost all axillary, rarely lateral or terminal. Segments of the fruiting perianth broad, concave, somewhat keeled, closing over the fruit or nearly so. Stamens usually 5. Seeds all horizontally flattened, opaque or somewhat rugose, the margins thick and obtuse or thin and acute. Pericarp not readily separable from the seed.—*C. erosum,* R. Br. Prod. 407; Moq. in DC. Prod. xiii. ii. 68; Hook. f. Fl. Tasm. i. 313.

Queensland. Rockhampton, rare, *O'Shanesy.*

Victoria. Near Melbourne, Murray river, and Gipps Land, *F. Mueller.*

Tasmania. Kent's Group, Bass's Straits, *R. Brown.*

This is another European weed now widely dispersed over various temperate and warm regions of the globe. The Australian specimens I have seen are mostly single ones, and it is therefore probably introduced only. Brown's specimens have the inflorescences more compact, but they are still in young bud and some European ones are precisely similar to them.

5. **C. triangulare,** *R. Br. Prod.* 407. Stems weak procumbent or straggling, extending sometimes to 2 ft. or more, the whole plant green or with but little of white meal on the young shoots. Leaves on rather long petioles, from ovate to oblong or to broadly hastate in the typical form, obtuse or shortly mucronate, under 1 inch long, the upper ones often and sometimes all lanceolate. Flowers very small, in clusters or

little cymes in a terminal interrupted spike or along the short distant branches of a slender terminal panicle, or the lower ones in the axils of the upper leaves. Perianth-segments broad and concave, sometimes closing over and covering the fruit, sometimes smaller and much contracted at the base. Stamen usually 1. Styles short. Seed flat, horizontal, about ½ line diameter, in a very thin membranous pericarp. —*C. trigonon,* Roem. and Schult. Syst. vi. 275 ; Moq. in DC. Prod. xiii. ii. 65.

Queensland. Armadilla, *Barton.*
N. S. Wales. Paramatta, *Woolls;* Namoi river, *Leichhardt;* New England, *C. Stuart.*

The habit is nearly that of *Rhagodia hastata,* but the fruit is never succulent, and the inflorescence rather different. The *C. triangulare* of Forskähl being reduced by Moquin to *C. murale,* there seems no reason to suppress Brown's name of *C. triangulare* for the present species.

Var. *stellulatum.* Perianth-segments with a rather small concave lamina contracted at the base into a linear stipes (reduced to the somewhat prominent midrib). Leaves of the typical form.—New England, *C. Stuart.*

Var. *angustifolium.* Leaves linear-lanceolate or the lower ones lanceolate-hastate. Perianth of the typical form.—To this belong the Queensland specimens and some from New England.

6. **C. microphyllum,** *F. Muell. in Trans. Phil. Inst. Vict.* ii. 74. A small much-branched prostrate or diffuse plant apparently perennial and more or less mealy-white. Leaves numerous, small, petiolate, ovate rhomboidal triangular or broadly lanceolate, entire, 2 to 3 lines long, hoary or white on both sides or becoming nearly green above. Flowers few together in small rather loose clusters in the upper axils, scarcely forming very short terminal spikes. Perianth very scaly, mealy, the segments concave, shortly united and keeled but not contracted at the base. Stamen usually 1. Seed flat, horizontal.

N. S. Wales. On the Billabong, *W. Bissett.*
Victoria. Bacchus marsh, *F. Mueller;* Wimmera, *Dallachy.*
S. Australia. Near the Barossa Range, *Behr.;* Enfield, *F. Mueller.*

7. **C. glaucum,** *Linn.; Moq. in DC. Prod.* xiii. ii. 72. An annual, much-branched diffuse and prostrate or decumbent at the base, the stems ascending to 1 ft. or more, glabrous striate and furrowed. Leaves petiolate, the lower ones broadly lanceolate or almost rhomboidal or hastate, coarsely sinuate-toothed, often above 1 in. long, the upper ones gradually smaller narrower and more entire, the uppermost passing into small bracts, all green above and more or less white underneath. Flowers small, nearly glabrous, in clusters or short leafless irregular spikes, the lower clusters or spikes axillary and much shorter than the leaves, the upper ones forming terminal interrupted spikes leafy at the base only. Perianth-segments rather thin, or the keel somewhat thickened, closely appressed on the fruit but not completely covering it. Stamen usually 1 only. Fruits about ½ line diameter, mostly depressed with a horizontally flat seed, but some of the lateral ones occasionally with a vertical seed and the perianth-segments reduced to 4 or 3.—

Hook. f. Fl. Tasm. i. 313 ; *C. ambiguum,* R. Br. Prod. 407 ; Moq. in DC. Prod. xiii. ii. 67.

N. S. Wales. Paramatta, *Woolls;* Ash island, *Beckler.*
Victoria. Along the coast from the Glenelg, *Robertson* and others, to Gipps Land, *F. Mueller* and others.
Tasmania. Port Dalrymple, *R. Brown;* common on the seacoast near high-water mark, *J. D. Hooker.*
S. Australia. Kangaroo island, *R. Brown;* Bethanie, *F. Mueller.*
W. Australia. *Drummond, n.* 225 (in some herbaria 235) ; Port Gregory, *Oldfield.*
The species is common in many parts of Europe and temperate Asia, and occurs here and there in other parts of the globe.
C. littorale, Moq. in DC. Prod. xiii. ii. 65, described from a specimen of Caley's in the Paris Herbarium, which I have not seen, may, from the character given, be a form either of this species or of *C. album.*

SECT. 3. BOTRYOIS, Moq.—Glandular aromatic herbs or undershrubs, not mealy. Seeds all or nearly all horizontal.

*8. **C. ambrosioides,** Linn.; Moq. in DC. Prod.* xiii. ii. 72. An erect much-branched annual of 1 to 2 ft., not mealy but more or less glandular-dotted and strongly aromatic. Leaves lanceolate or oblong, acute or obtuse, the lower ones irregularly toothed or sinuate, contracted into a short petiole, from under 1 in. to above 2 in. long, the upper ones smaller and entire, passing into small linear or linear-lanceolate acute petiolate bracts, all green on both sides, glandular underneath. Flowers very small and numerous, solitary or clustered in the axils of bracts which are either minute or leafy and longer than the clusters, the clusters forming more or less leafy slender interrupted spikes, arranged in a large leafy panicle occupying the greater part of the plant. Fruiting perianth about ⅓ line diameter, the lobes short, completely or almost completely covering the fruit. Seeds smooth and shining, all or mostly horizontal.

Queensland. Moreton Bay, *F. Mueller;* Rockhampton, *O'Shanesy.*
N. S. Wales. Port Jackson, *R. Brown* and others ; New England, *C. Stuart.*
W. Australia, *Drummond, n.* 207.
A common weed in southern Europe, northern Africa, and western Asia, and spread with cultivation over many parts of the world. It is probably introduced only into Australia as suggested in R. Brown's notes, and on that account omitted in his Prodromus.

SECT. 4. ORTHOSPORUM, R. Br.—Decumbent glandular herbs not mealy. Seeds all vertical. Flower-clusters all axillary.

9. **C. carinatum,** *R. Br. Prod.* 407. Stems much-branched and procumbent or prostrate at the base, ascending to from ½ to 1 ft. or more, the whole plant more or less glandular-pubescent. Leaves on long petioles, ovate or oblong, obtuse, coarsely sinuate-toothed, usually rather thick and rugose, glandular-scabrous on both sides, ¼ to 1 in. long, the upper floral ones often much reduced, and sometimes all the leaves almost orbicular and small. Flowers small, in dense globular clusters in almost all the axils, the upper ones sometimes forming interrupted more or less leafy spikes. Perianth-segments erect, incurved,

broadly oblong, concave and almost boat-shaped, with a thickened broad obtuse keel, more or less pubescent or hirsute. Stamen usually 1. Fruit small, ovoid, erect, the pericarp inseparable from the seed.— *Salsola carinata*, Spreng. Syst. i. 923; *Ambrina carinata*, Moq. Chenop. Enum. 41; *Blitum carinatum* and *B. glandulosum*, Moq. in DC. Prod. xiii. ii. 81, 82; *Chenopodium glandulosum*, F. Muell. Fragm. vii. 11.

Queensland. Moreton Bay, *W. Hill*, *F. Mueller*, and others; Peak Downs, *F. Mueller;* Rockhampton, *O'Shanesy;* Armadilla, *Barton.*

N. S. Wales. Port Jackson, *R. Brown, J. D. Hooker;* Bengalla, *Leichhardt;* Clarence river, *Beckler;* Murray and Darling rivers, *Victorian and other Expeditions.*

Victoria. Yarra-Yarra, *F. Mueller;* Skipton and Creswick, *Whan;* Lockwood, *Bissil.*

S. Australia. Bethanie, *Behr.;* Mount Barker, Lofty Ranges, Lake Torrens, *F. Mueller.*

W. Australia, *Drummond, n.* 165, 715.

The species is also in New Zealand and New Caledonia. In most of Drummond's specimens and in some others, the fruiting perianth has a tendency to dry black and become rather thick, showing an approach to the European typical *Blita.*

10. **C. pumilio,** *R. Br. Prod.* 407. A branching decumbent filiform annual of about 1 in., more or less hoary with crisped or glandular hairs. Leaves on slender petioles, ovate or oblong, entire, 1 to 2 lines long. Flowers minute, axillary, solitary or 2 or 3 together on very short pedicels. Perianth-segments 4 or 5, linear, erect, concave, slightly incurved, nearly ½ line long when in fruit, hirsute with a few crisped hairs.—*Blitum pumilio*, Moq. in DC. Prod. xiii. ii. 82; *Ambrina pumilio*, Moq. Chenop. Enum. 42.

S. Australia. Kangaroo island, *R. Brown.* Possibly a diminutive form of *C. carinatum.*

11. **C. cristatum,** *F. Muell. Fragm.* vii. 11. Diffuse or procumbent, with ascending flowering branches of 1 ft. or more, the whole plant slightly glandular-pubescent. Leaves on long petioles, from ovate to oblong-lanceolate, obtuse, coarsely toothed, narrowed at the base, ½ to 1 in. long, green and glandular-scabrous on both sides. Flowers in dense globular clusters, all axillary. Perianth-segments linear, erect, not incurved, acute, about 1 line long when in fruit, the keel dilated into a broad fringed crest or wing. Fruit ovoid, erect, enclosed in the perianth. Styles very slender.—*Blitum cristatum*, F. Muell. in Trans. Phil. Inst. Vict. ii. 73.

N. S. Wales. Darling river, *Victorian Expedition.*
Victoria. Murray river, *F. Mueller.*
S. Australia. Flinders Range, *F. Mueller.*

12. **C. atriplicinum,** *F. Muell. Fragm.* vii. 11. Apparently perennial, branching at the base only, with numerous ascending or erect stems under 1 ft. and often under 6 in., flowering from near the base, of a pale green and slightly glandular-pubescent. Lower leaves on long slender petioles, from lanceolate to broadly hastate, otherwise entire, rather thick, ¾ to 1 in. long, the upper ones smaller lanceolate and entire, but all petiolate. Flowers in dense sessile axillary clusters

shorter than the petioles. Perianth-segments 4 or 5, erect, lanceolate, rather above 1 line long, the points somewhat spreading, the keel much thickened and irregularly angular at the base. Stamen 1. Seed erect, rugose, enclosed in the perianth.—*Blitum atriplicinum*, F. Muell. in Trans. Vict. Inst. 1855, 133, and in Hook. Kew Journ. viii. 204.

N. S. Wales. Darling desert, *Victorian Expedition.*
Victoria. Wimmera, *Dallachy.*
S. Australia. Flinders Range, *F. Mueller.*

3. DYSPHANIA, R. Br.

Flowers polygamous. Perianth of 1 to 3 minute segments, which when in fruit are clavate, concave or hood-shaped, white and almost transparent. Stems 1 to 3. Ovary ovoid; styles 1 or 2, very finely filiform. Fruit ovoid, the pericarp inseparable from the seed. Seed erect, testa crustaceous with a very thin membranous inner integument. Embryo circular enclosing a mealy albumen ; radicle inferior.—Small annuals. Leaves alternate, flat, entire. Flowers minute, in clusters either all axillary or in terminal spikes, the females numerous, the hermaphrodite ones few in each cluster.

The genus is limited to Australia. It is nearly allied to the section *Orthosporum* of *Chenopodium,* but readily distinguished by the remarkable perianth.

Fruiting perianth of 3 (rarely 2) segments falling off with the
  fruit. Style 1. Plant of 1 to 3 in.
Flower-clusters forming a dense terminal leafless spike . . ʻ. 1. *D. plantaginella.*
Flower-clusters closely contiguous but axillary, forming a leafy
  spike . . . . 2. *D. littoralis.*
Fruiting perianth usually of a single segment. Styles 2. Plant
  of 3 to 6 in. Flower-clusters all axillary and distinct . . . 3. *D. myriocephala.*

1. **D. plantaginella,** *F. Muell. Fragm.* i. 61. An erect branching annual of 1 to 3 in., slightly glandular-hairy. Stem leaves in the lower part of the plant petiolate, ovate or obovate, obtuse, entire, 2 to 4 lines long. Flowers resembling those of *D. littoralis,* but the clusters crowded in dense terminal cylindrical leafless spikes of 1 to 2 in., and consequently occupying the greater portion of the plant. Perianth of 3 obovate-clavate concave segments, about ¼ line long, and falling off with the fruit. Style 1, very deciduous.

N. Australia. Sturt's Creek, *F. Mueller.*

2. **D. littoralis,** *R. Br. Prod.* 411. A small plant apparently annual, although sometimes hard and perhaps fleshy at the base, with ascending branching stems of 2 to 3 in., glabrous or nearly so. Leaves all petiolate, ovate or oblong, obtuse, entire, rather thick and sometimes fleshy, not above 2 lines long. Flower-clusters all axillary, but nearly all close together, forming a terminal leafy spike occupying the greater part of the plant, the lower clusters sometimes rather more distant. Flowers numerous in the cluster, chiefly females. Perianth of 3 or rarely 2 segments falling off together and enclosing the fruit, the segments all equal, obovate, clavate, concave, contracted at the base, about ¼

line long.   Fruit still shorter, obovoid, somewhat oblique; style 1, very finely filiform and very deciduous.   Stamens 1 or 2 but difficult to find, the anthers falling off early from the very minute flowers.

N. Australia.   Moist salt places on the N. coast (snatched up in the hurry of escape from an armed native in close pursuit, and never seen again), *R. Brown.*
S. Australia.   Flooded ground S. of Wills Creek, *Howitt's Expedition.*

3. **D. myriocephala,** *Benth.*   A diffuse or procumbent glabrous or slightly glandular-pubescent annual, much larger than the two preceding species, although the ascending branching stems rarely exceed 6 in.   Leaves petiolate, oblong or lanceolate, obtuse or scarcely acute, rarely above ¼ in. long.   Flower-clusters all axillary and distinct, very numerous, occupying the greater part of the plant, globular and scarcely exceeding 1 line in diameter when in fruit, and often much smaller, although containing 10 to 20 or even more flowers, chiefly females, with a very few hermaphrodite or male ones.   Segments of the fruiting perianth single and falling off separately, about ¼ line long, obovoid-clavate and as it were inflated, shortly contracted at the base.   Seed ovoid like that of *D. littoralis,* but more regular and slightly flattened; styles 2, very fine, but shorter than the single one of *D. littoralis.*   Stamens 1 or 2, with very short broad filaments and comparatively large anthers.—*D. littoralis,* Moq. in DC. Prod. xiii. ii. 86, not of R. Br.

Victoria.   Sandy occasionally flooded banks of the Murray near the junction of the Golgol, *F. Mueller.*
W. Australia, *Drummond, n.* 206.

Moquin's description is taken from a specimen of Drummond's in which he had correctly observed the two styles, but in which I have always found in every cluster as many or nearly as many fruits as enlarged perianth-segments, but as these fall off separately, it is difficult to ascertain whether there may not sometimes be two to one fruit. Moquin in describing three has probably followed Brown's character founded on the true *D. littoralis.*

4. ATRIPLEX, Linn.

(Obione and Theleophyton, *Moq.*)

Flowers unequal.   Male perianth nearly globular, deeply divided into 5, rarely fewer segments.   Stamens 5 or fewer.   Female perianth very small at the time of flowering, 2-toothed or 2-lobed, enclosing the ovary.   Styles 2, free or united at the base.   Fruiting perianth much enlarged and variously shaped, the tube very small or large, flat or variously thickened, the limb of 2 variously shaped segments or *valves* closely appressed, at least at the margin (except in *A. campanulata*), entire or toothed.   Fruit entirely enclosed in the tube or between the valves.   Pericarp membranous, very thin.   Seed compressed, vertical; testa crustaceous, often thin with a very thin inner integument sometimes scarcely distinct.   Embryo surrounding a mealy albumen, the radicle superior lateral or inferior.—Herbs or shrubs, more or less mealy or scaly-tomentose.   Leaves alternate or the lower ones rarely opposite, flat, entire hastate or sinuate-toothed.   Male flowers in globular clusters, either detached from the females in close or interrupted simple or paniculate spikes, or axillary and then each cluster usually surrounded by females; female flowers usually in axillary clusters, rarely solitary or the clusters in terminal leafless panicles. Bracts subtending the male as well as the female flowers usually minute, or quite obsolete.

The genus is widely distributed over most parts of the globe, chiefly in maritime or subsaline districts, some species also frequenting rich cultivated grounds. Of the thirty Australian species, one is a common European weed of cultivation possibly of modern introduction into Australia, two are also in New Zealand, the others appear all to be endemic, for although one is nearly allied to a New Caledonian species, another to a South African one, and others may be more or less compared with other exotic ones, there are none which I have been able precisely to identify. The specific characters are in many instances taken chiefly from the fruiting perianths, which are so extraordinarily diversified in the genus, and which evidently vary also to a certain degree even on the same individual. It may therefore possibly be shown hereafter that in some instances the distinctions here relied upon may not prove sufficiently constant to retain their specific value.

Moquin, relying apparently on observations communicated by Fengl, considers that the lobes or valves of the female perianth of *Atriplex* are really bracts (bracteoles), for, he says, in monstrous female flowers of *Atriplex* and normally in *Exomis*, minute perianth-segments occur within these bracts. Trusting implicitly to his observations I should, with most recent botanists, have adopted his views, but that, on a careful examination of the various forms assumed by this perianth in Australian species and of its structure at the time of flowering, I could by no means reconcile its insertion and development with any other view than that of its being the homologue of the male perianth. This induced me to examine a considerable number of flowers and fruits of both species of *Exomis*. In *E. albicans* I find the structure quite that of *Atriplex*, nor can I discover anything that might be taken for minute perianth-segments unless it be sometimes some rudimentary stamens. A small additional perianth-lobe occurred once, not inside of the two valves, but in their sinus on one side, and I have occasionally but very rarely seen three valves to the perianth of *Atriplex*. In *Exomis axyrioides* the case is quite different. I find the female perianth abortive or reduced to minute scales and the quasi-petiolate bracts described by Moquin appear to me to be real subtending bracts or floral leaves, one only to each flower, although when in fruit, owing to the abortion of some of the ovaries, there may be 2 or 3 bracts to one fruit. but never two opposite ones united at the base and enclosing the fruit, as in *Atriplex* and in *Exomis albicans*, which latter species ought surely to be restored to *Atriplex*.

A few modern botanists have. after Pliny, treated the name *Atriplex* as of the neuter instead of the feminine gender. As there is classical authority for both, I have preferred following Linnæus, De Candolle and the great majority of botanists in treating it as feminine.

In the arrangement of the Australian species I have been unable to retain Moquin's distinction between *Atriplex* and *Obione* even as sectional The thickening of the perianth over the fruit may be observed in every degree from flat and membranous to hard and terete, and in species so closely allied as *A. inflata* and *A. holocarpa*, or as *A. Drummondi* and *A. isatidea*, the radicle is superior in the one and inferior or lateral in the other. The deviation from the normal position of the seed, transverse instead of parallel to the valves, in *A. Billardieri* is remarkable, but is scarcely sufficient for separating, on that character alone a single species from a large genus otherwise so natural and so well defined.

SERIES 1. **Paniculatæ.**—*Diœcious or semidiœcious scaly tomentose shrubs. the male clusters in more or less branched or paniculate dense or interrupted leafless spikes.*

Diœcious, both sexes paniculate.
    Fruiting perianths flat.
        Fruiting perianths reniform. Spikes in male panicles interrupted.
            Fruiting perianths on a slender stipes . . . . . . 1. *A. stipitata*.

Fruiting-perianths sessile . . . . 2. *A. Moquiniana.*
Fruiting perianths ovate or broadly cordate, with a short solid
base or stipes. Leaves narrow (female inflorescence more
simple and leafy) . . . . 3. *A. paludosa.*
Fruiting perianths ovate or slightly cordate, quite sessile.
Leaves mostly obovate. Female panicles more branched ˉ 4. *A. Drummondii.*
Fruiting perianths with thick convex valves.
Leaves elliptical or oblong, 1½ to 3 in. Female panicles
branched . . . . 5. *A. isatidea.*
Leaves mostly orbicular, ½ to 1 in. Female inflorescence
more simple and leafy . . . . 6. *A. nummularia.*
Semi-diœcious. Female flowers solitary or very few in the axils
of the stem-leaves of the male plants, more clustered but all
axillary in the females.
Leaves mostly oblong. Fruiting perianths broadly triangular
or rhomboid, flat or thickened over the fruit, with a turbinate
solid base . . . . 7. *A. cinerea.*
Leaves ovate or lanceolate sometimes hastate. Fruiting perianths thickened to the margin.
Fruiting perianths 1½ to 3 lines diameter′ . . . . . . 8. *A. rhagodioides.*
Fruiting perianths 4 to 5 lines diameter . . . . . . 9. *A. incrassata.*

SERIES 2. **Vesicariæ.**—*Small bushy or decumbent more or less scaly tomentose
shrubs, diœcious or monœcious. Male flowers in short terminal dense spikes, females
axillary. Fruiting perianth orbicular, membranous, with large membranous appendages
on each face.*
Leaves obovate or oblong-lanceolate, usually white, 4 to 8 lines
long. Eastern species mostly monœcious . . . . 10. *A. vesicaria.*
Leaves narrow, less white, mostly 2 to 4 lines long. Western
species mostly diœcious . . . . 11. *A. hymenotheca.*

SERIES 3. **Oleraceæ.**—*Monœcious annuals usually tall or spreading, green or
slightly mealy. Flowers clustered in the axils and in terminal panicles. Valves of the
fruiting perianth flat or muricate. (Introduced species.)*
Erect plant of 4 or 5 ft. Leaves broad. Flowers crowded in a
long panicle. Fruiting perianth broad, thin and flat, the valves
free to the base . . . . * *A. hortensis.*
Erect and 2 or 3 ft., or spreading or procumbent. Leaves narrow
except the lower ones. Flowers usually in distant clusters.
Fruiting perianth thickened at the base, the valves united to
near the middle . . . . 12. *A. patula.*

SERIES 4. **Glomeratæ.**—*Monœcious decumbent procumbent or spreading herbs,
scaly-tomentose or very rarely green. Male flowers in globular clusters surrounded by
a few females in the upper axils or rarely forming a short terminal spike, females clustered in the lower axils without males. Fruiting perianths more or less compressed,
conspicuously 2-valved.*
Fruiting perianth flat, rhomboidal, the valves free almost or quite
to the base, closing over the fruit.
Leaves narrow, entire. Male flowers in short terminal compact spikes. Fruiting perianth with a small turbinate solid
base . . . . 13. *A. humilis.*
Leaves broad, mostly sinuate. Male flowers axillary or the
upper ones spicate. Fruiting perianth triangular with a
broad base . . . . 14. *A. velutinella.*
Fruiting perianth with a compressed turbinate base half enclosing
the fruit, shorter than or not longer than the valves
Leaves nearly orbicular, about 1 in. diameter. Fruiting perianth stipitate, the valves more than twice as broad as the
tube . . . . 15. *A. angulata.*

Leaves narrow, usually green, ½ to 1 in. long. Fruiting perianth
sessile, rhomboidal, the valves not broader than the tube .   . 16. *A. semibaccata.*
Leaves 2 to 3 lines long. Fruiting perianth rhomboidal, about
1 line diameter, with a short solid base . . . . 17. *A. exilifolia.* ·
Fruiting perianth with a globular ovoid or slightly compressed
tube enclosing the fruit, the valves shorter than the tube.
Spreading (or erect ?). Leaves broad, mostly toothed and 2 in.
long or more. Fruiting perianth 1 to 1¼ lines diameter .   . 18. *A. Muelleri.*
Diffuse. Leaves obovate or oblong, rarely above 2 lines long.
Fruiting perianth 1 to 1¼ lines diameter . . . . . . 19. *A. elachophylla.*
Procumbent. Leaves narrow, ¼ to ½ in. long. Fruiting perianths
clustered, nearly 1 line diameter . . . . 20. *A. microcarpa.*
Prostrate. Leaves narrow, 1 to 2 lines long. Fruiting perianths
clustered, scarcely ½ line diameter . . . . 21. *A. prostrata.*
Procumbent. Leaves ovate, 1 to 1½ lines long. Fruiting peri-
anths 1 or 2 in the axils, scarcely ½ line diameter . . . . 22. *A. pumilio.*
Fruiting perianth minute, the valves free, spreading. Small erect
plant glabrous and green. Leaves 2 lines long . . . . . 23. *A. glomulifera.*
Fruiting perianth with an obliquely campanulate slightly compressed tube with appendages on the shorter face. Valves
toothed, unequal and scarcely appressed . . . . 24. *A campanulata.*

SERIES 5. **Parviloba.**—*Monœcious spreading or procumbent herbs or undershrubs
scaly-tomentose or mealy. Inflorescence of the Glomeratæ Fruiting perianths not
compressed, enclosing the fruit, the orifice small closed by small erect appressed valves.*
Fruiting perianth cylindrical, narrow.
Valves of the fruiting perianth minute, entire, without appen-
dages . . . . 25. *A. leptocarpa.*
Valves 2-horned with a minute central lobe and a dorsal appen-
dage between the horns . . . . 26. *A. limbata.*
Fruiting perianth inflated and spongy.
Fruiting perianth hemispherical or turbinate with an almost
flat top and acute or winged margin. Radicle lateral or
almost inferior . . . . 27. *A. halimoides.*
Fruiting perianth turbinate-globular, 4 to 6 lines diameter.
Radicle superior . . . . 28. *A. holocarpa.*
Fruiting perianth depressed-globular, not 2 lines diameter.
Radicle superior . . . . 29. *A. spongiosa.*

SERIES 6 (or SECTION 2). **Theleophyton.**—*Monœcious prostrate crystalline herb.
Flowers axillary. Perianth obovoid with short valves. Seed compressed, at right
angles with, not parallel to the valves.*
Single species . . . . 30. *A. Billardieri.*
* *A. hortensis,* Linn. (*Atriplex* sect. *Dichospermum* Moq. in DC. Prod. xiii. ii. 90,
91), an erect green annual of 4 or 5 ft., with large broad leaves, and numerous flowers
crowded in a long terminal panicle, the fruiting perianths broad, thin, flat and entire,
intermixed with a few small regular 5-cleft perianths with horizontal seeds, a plant of
east European or west Asiatic origin, very long cultivated as a vegetable under the
name of *Orache,* has been sent from N. S. Wales and from Victoria as an escape from
gardens.

SERIES 1. PANICULATÆ.—Diœcious or semidiœcious scaly-tomentose shrubs, the male clusters in more or less branched or paniculate
dense or interrupted leafless spikes, the females either also paniculate or
spicate or in axillary clusters.

1. **A. stipitata,** *Benth.* An erect bushy rather slender shrub,
scaly white or somewhat fulvous all over. Leaves from obovate to

narrow-oblong, very obtuse, entire, contracted into a short petiole, mostly rather thick, ½ to ¾ in. long. Flowers diœcious, the males numerous in little globular clusters scarcely 2 lines diameter, all distinct and somewhat distant, in slightly branched terminal panicles or almost simple spikes; females in smaller clusters, the lower ones often solitary in the axils, the upper ones in a somewhat leafy spike or panicle, some of them sessile ovoid or globular shortly 2-lobed, enclosing an apparently perfect ovary but soon falling off, the greater number more or less stipitate at a very early age with broad flat valves. Fruiting perianth on a slender stipes of 2 to 4 lines, with a small campanulate tube half enclosing the fruit; valves flat, reniform, entire, 4 to 5 lines broad. Seed orbicular, flat, the radicle superior.—*A. reniformis*, F. Muell. Fragm. vii. 9, as to the eastern stations, not of R. Br.

N. S. Wales. Desert of the Darling, *Victorian Expedition, Mrs. Ford;* also in *Leichhardt's* collection.

Victoria. In the N. W. portion of the colony, *L. Morton* (the specimens not in fruit and therefore doubtful).

S. Australia. Murray scrub, *Behr., F. Mueller;* Gawler's range, *Sullivan;* towards Spencer's Gulf, *Warburton.*

2. **A. Moquiniana,** *Webb.; Moq. in DC. Prod.* xiii. ii. 97 (from the character given). A small much-branched scaly-tomentose shrub, with the habit foliage and inflorescence of *A. stipitata* but with a different fruiting perianth. Leaves obovate or broadly oblong, entire, ½ to ¾ in. long. Flowers diœcious, the males in distinct globular clusters usually smaller than in *A. stipitata* but forming similar panicles; females in panicles or spikes leafy at the base as in *A. stipitata,* but I have not observed in them any dimorphism. Fruiting perianth sessile, orbicular-cordate or almost reniform but usually as long as broad, flat with scarcely any tube. Seed orbicular; radicle lateral.—*A. reniformis,* F. Muell. Fragm. vii. 9, as to the Western specimens.

N. Australia. Bay of Rest, N. W. Coast, *A. Cunningham.*

W. Australia. Murchison river, *Oldfield;* Sharks Bay, *Milne;* Abrolhos islands, *Bynoe.*

3. **A. paludosa,** *R. Br. Prod.* 406. An erect spreading or diffuse shrub (or undershrub?) covered with a white or fulvous scaly tomentum. Leaves lanceolate or oblong, usually narrow, obtuse, entire, contracted into a short petiole, ½ to 1 in. long. Flowers diœcious or nearly so, the males in little globular distinct or distant clusters in terminal panicles rarely reduced to interrupted spikes; the female inflorescence more simple and leafy, the lower clusters all axillary. Fruiting perianth on a thick stipes sometimes very short, sometimes above 1 line long, the valves in the typical form broadly ovate-triangular, flat, mostly acute, often toothed towards the base, 2 to 3 lines long and broad, truncate at the base and almost entirely free. Seed enclosed in the valves, with a lateral radicle.—Moq. in DC. Prod. xiii. ii. 102 (partly).

Victoria. In maritime wet sandy places, Port Albert, *F. Mueller.*

Tasmania. Port Dalrymple, *R. Brown.*

W. Australia. King George's Sound, *R. Brown.*

Var. *cordata*. Fruiting perianth nearly orbicular and broadly cordate.—*A. reniformis*, R. Br. Prod. 406; Moq. in DC. Prod. xiii. ii. 101.

S. Australia. Kangaroo Island, *R. Brown, Waterhouse;* Port Adelaide, *F. Mueller.*

Var.? *appendiculata*. Fruiting perianth of the shape of that of the var. *cordata*, but longer, at least 5 lines diameter, with a small foliaceous appendage at the base of the disk on one side, showing an approach to the perianth of *A. vesicaria*.—N. W. of the head of the Great Bight, *Delisser*. A small fragment in fruit only and the affinity uncertain. There are also in Herb. F. Mueller specimens from various other localities which may belong to *A. paludosa*, but being in leaf only they cannot be determined.

4. A. Drummondii, *Moq. in DC. Prod.* xiii. ii. 102. An erect bushy shrub attaining 3 or 4 ft., white or fulvous with a scaly tomentum. Leaves obovate or oblong, obtuse, entire, contracted into a short petiole, mostly ½ to 1 inch long. Flowers dioecious, both sexes in terminal panicles, the male clusters rather small, in numerous short dense spikes; the female panicles rather more leafy at the base with numerous flowers, the fruiting panicles dense. Fruiting perianth flat, broadly ovate-triangular, slightly cordate, 2½ to 3 lines long, the valves free, membranous, entire or slightly toothed at the base. Fruit flat, raised on a small thickened base within the valves. Radicle lateral.—*A. paludosa*, Nees in Pl. Preiss. i. 633, not of R. Br.; *A. paludosa* var. *obovata*, Moq. in DC. Prod. xiii. ii. 102.

W. Australia, *Drummond, n.* 134 (♂) and 135 (♀); Rottenest Island, *Preiss, n.* 1255; Port Gregory, *Oldfield;* Fitzgerald flats, *Maxwell.*

Although allied to *A. paludosa*, this species appears to be sufficiently distinct in foliage and inflorescence probably at least, as well as in the quite sessile fruiting perianth. Moquin describes the valves as elliptical, but it is evident that in the specimens he saw, as in all Drummond's, they were not yet fully formed.

5. A. isatidea, *Moq. Chenop. Enum. 63, and in DC. Prod.* xiii. ii. 101. An erect robust shrub, attaining 16 ft. (*Oldfield*), densely scaly-tomentose. Leaves elliptical obovate or oblong, obtuse, entire or slightly sinuate, contracted into a rather long petiole, thick and soft, 1½ to 3 in. long. Flowers dioecious, both sexes in dense much-branched terminal panicles, the males without the axillary female flowers of *A. cinerea*, the females sometimes with a very few male flowers intermixed. Fruiting perianths with a thick solid turbinate base, the valves semicircular or almost rhomboidal, thick, entire, 3 to 4 lines diameter, shortly united at the base, the disk smooth, tuberculate or muricate with soft appendages; intermixed with these are other perianths not half so large and more orbicular, in which however I have not found perfect seeds. Fruit half-included in the closed base of the perianth, covered by the appressed valves. Radicle superior.—*A. halimus* var. *erecta*, Nees in Pl. Preiss. i. 633 (Moquin).

W. Australia. Swan river, *Drummond, n.* 226 (or 223?), *Fraser, Preiss, n.* 1259; Murchison river and S. W. Bay, *Oldfield;* Sharks Bay, *Milne;* Abrolhos islands, *Bynoe.*

6. A. nummularia, *Lindl. in Mitch. Trop. Austr.* 64. An erect shrub attaining several ft., with spreading branches, the whole plant covered with a scaly tomentum. Leaves on rather long petioles, mostly orbicular, rather thick, entire or scarcely sinuate-toothed, more rarely bordered by numerous small teeth, ½ to 1 in. diameter, or on some luxuriant branches nearly 2 in. Flowers diœcious, the male clusters forming dense oblong or shortly cylindrical spikes, in more or less branched terminal panicles, either leafless or sparingly leafy at the base, the females also clustered in dense terminal rather more leafy spikes or panicles, with a few flowers also in the axils of the upper stem-leaves. Fruiting perianth sessile, from ovate to orbicular, 2 to 3 lines long, rounded or truncate not cordate at the base, the valves free nearly to the base, thickened and hardened over the fruit at the base, the remainder flat and herbaceous, entire or toothed on each side towards the base. Radicle superior.—Moq. in DC. Prod. xiii. ii. 460.

Queensland. Darling Downs, *Lau.*
N. S. Wales. Macquarrie river, *Mitchell;* Castlereagh river, *Woolls;* Darling river, *Mrs. Ford.*
Victoria. Murray scrub, *F. Mueller, Herrgott.*
S. Australia. In the interior, *Howitt's Expedition,* also *M'Douall Stuart's Expedition.*

Moquin in Herb. Hook. had referred this plant to *A. capensis,* to which it bears some resemblance, but appears to me sufficiently distinct. The *A. halimus,* Br., quoted by Moquin with doubt under *A. capensis* (in DC. Prod. xiii. ii. 100) is *A. cinerea.*

7. **A. cinerea,** *Poir. Dict. Suppl.* i. 471. A branching shrub, sometimes low or slightly decumbent, more frequently erect and attaining several feet, white or ashy grey all over with a scaly tomentum. Leaves oblong or lanceolate, rarely almost ovate, obtuse, entire, contracted into a short petiole, mostly 1 to 2 in. long, but in some specimens scarcely exceeding 1 in. or smaller. Flowers semidiœcious, the males in dense globular clusters of 2 to 4 lines diameter collected into a terminal spike either short and interrupted at the base or 2 or 3 in. long with a few short densely oblong or cylindrical branches, the flowers often not quite sessile in the clusters. Female flowers in axillary clusters on the female plants, and also 1 or 2 female flowers in the axils of the upper stem-leaves of the male plants. Fruiting perianths with an obovoid or turbinate solid base, 1 to nearly 2 lines long, the valves broadly triangular or rhomboidal, from under 3 lines to above 4 lines diameter, free almost from the base, entire, flat or thickened over the fruit, smooth or rarely with 1 or 2 tubercles on the disk. Fruit at the base of the valves. Radicle ascending.—Moq. in DC. Prod. xiii. ii. 101; Hook. f. Fl. Tasm. i. 314; *A. halimus,* R. Br. Prod. 406, not of Linn.; *A. halimus β. ascendens* Nees in Pl. Preiss. i. 633; *A. elæagnoides,* Moq. Enum. Chenop. 65.

Queensland. Moreton Bay, *A. Cunningham.*
N. S. Wales. Botany Bay, *Banks and Solander;* Ash Island, *Herb. F. Mueller;* Lord Howe's Island, *Milne, M'Gillivray.*
Victoria. Seashore, Portland, *Robertson, Allitt;* Port Phillip, Brighton and Station Peak, *F. Mueller.*
Tasmania. Abundant upon all the coasts near high-water mark, *J. D. Hooker.*
S. Australia. Spencer's Gulf, *Warburton;* Kangaroo Island, *F. Mueller, Waterhouse.*

W. Australia. Swan river, *Drummond, n.* 230; Champion Bay and Murchison river, *Oldfield.*

The western specimens are mostly males with rather small leaves, but they can be readily distinguished from the preceding ones by the dense compact male inflorescence and by the female flowers (very minute in most specimens in which the males are scarcely expanded) always present in the axils of the upper stem-leaves. *A. hypoleuca,* Nees in Pl. Preiss. i. 633, or *A. prostrata,* Moq. in DC. Prod. xiii. ii. 99, not of R. Br., appears to belong to this species, but the specimens I have seen are not in flower. *A. prostrata,* Br., is a totally different plant.

8. **A. rhagodioides,** *F. Muell. in Trans. Phil. Inst. Vict.* ii. 74. A divaricately branched scaly-tomentose scrubby shrub, closely allied to and perhaps a variety of *A. cinerea.* Leaves shortly petiolate, ovate or lanceolate, often angular or almost hastate at the base, otherwise entire, acute and under $\frac{1}{2}$ in. long when narrow, more obtuse and larger when broad. Flowers semidiœcious, the male clusters forming an interrupted terminal spike or slightly branched panicle, the females solitary or nearly so in the axils of the upper stem-leaves in the male plant, more numerous and often clustered when there are no males. Fruiting perianth sessile, orbicular or nearly rhomboid, $1\frac{1}{2}$ to nearly 3 lines diameter, the valves thick convex and corky, united about half way up, the margins entire. Seed orbicular, with the radicle ascending or nearly superior, but in many perianths the seed is abortive.

Victoria, or **S. Australia.** Murray scrub, *F. Mueller.*
W. Australia. Murchison river, *Oldfield.*

The specimens are insufficient to determine whether this is more than a variety or state of *A. cinerea.* The foliage however is rather different.

9. **A. incrassata,** *F. Muell. Rep. Babb. Exped.* 20. Shrubby. Leaves ovate, hastate, nearly sessile, entire or indistinctly toothed, scaly-tomentose. Male flowers unknown, females clustered in the axils (*F. Mueller*). Fruiting perianth sessile, orbicular, 4 to 5 lines diameter, the valves much thickened with a narrow flat margin, quite smooth outside. Fruit raised to the centre of the perianth on a very broad flattened solid base. Seed not seen ripe.

S. Australia. Emu springs, *Babbage's Expedition.*

This may prove not to be distinct from *A. rhagodioides,* it is however but very imperfectly known, and I have only seen a few detached enlarged but unripe perianths in Herb. F. Mueller.

SERIES 2. VESICARIÆ.—Small bushy or decumbent shrubs, more or less scaly-tomentose, diœcious or monœcious. Male flowers in short dense terminal spikes, females axillary. Fruiting perianth orbicular, membranous, with large membranous appendages on each face.

10. **A. vesicaria,** *Heward, MS.* A bushy shrub, apparently erect, covered with a scaly tomentum. Leaves oblong oblong-lanceolate or rarely almost obovate, obtuse or almost acute, entire, contracted into a short petiole, from under $\frac{1}{2}$ in. to about $\frac{3}{4}$ in. long. Flowers monœcious (or sometimes diœcious ?), the males in small clusters forming rather dense terminal leafless spikes of $\frac{1}{2}$ to 1 in.; females few together

in axillary clusters. Fruiting perianth nearly orbicular, 3 to 5 lines diameter, the valves membranous, very shortly connate, very obtuse or obscurely acuminate, the margins entire, flat but each with a membranous inflated appendage on the disk nearly as large as the valve itself. Seed rather large, compressed; radicle lateral.

Queensland. In the interior, *Mitchell.*

N. S. Wales. Molle's plains, *A. Cunningham;* Murray and Darling desert, *Victorian Expedition, F. Mueller.*

S. Australia. Crystal Brook, *F. Mueller;* Gawler Ranges, *Sullivan* (with more obovate leaves).

11. **A. hymenotheca,** *Moq. in DC. Prod.* xiii. ii. 101. An erect and bushy or procumbent shrub, the branches and foliage minutely scaly but not so white as most species. Leaves lanceolate or oblong, entire, contracted into a petiole, rather thick, ¼ to ½ in. long with smaller ones often clustered in the axils. Flowers diœcious, the male clusters forming cylindrical terminal leafless spikes rarely above ½ in. long, the perianth-segments thicker and darker coloured than in most species, the female flowers axillary, solitary (or 2 or 3 together?). Fruiting perianth nearly orbicular, ¼ to ½ in. diameter, the valves membranous, free almost from the base, the margins entire, flat but each with a large membranous inflated appendage on the disk. Seed not seen ripe.

W. Australia, *Drummond, n.* 128 (♂) *and* 129 (♀). The specimens although numerous are not good, and the fruiting perianths are few, but all have the membranous appendages of *A. vesicaria,* which appears to have escaped Moquin's attention.

SERIES 3. OLERACEÆ.—Monœcious annuals, usually tall or spreading, green or slightly mealy. Flowers clustered in the axils and in terminal panicles. Valves of the fruiting perianth flat or muricate.

12. **A. patula,** *Linn.; Moq. in DC. Prod.* xiii. ii. 95. An erect spreading or prostrate annual, usually 1 to 2 ft. long, either quite green or somewhat mealy-white, never so thickly scaly as most species. Leaves petiolate, the lower ones usually lanceolate-hastate, coarsely toothed or somewhat lobed, often 3 in. long or more, the upper ones lanceolate and entire. Flowers clustered in slender interrupted spikes forming narrow terminal panicles leafy at the base, the upper floral leaves reduced to small bracts, the female flowers mixed with the males or a few in separate axillary clusters. Fruiting perianths ovate or rhomboidal, usually acute, the valves united to near the middle, entire or toothed, smooth or muricate on the disk, very variable in size and shape but usually under 2 lines diameter. Radicle lateral or ascending.—Hook. f. Fl. Tasm. i. 314; *A. australasica,* Moq. in DC. Prod. xiii. ii. 96.

Queensland. Islands of Moreton Bay, *F. Mueller.*

N. S. Wales. Paterson's river, *R. Brown;* Ash Island, *Herb. F. Mueller.*

Victoria. Abundant in gardens about Melbourne, *Adamson, F. Mueller.*

Tasmania. Abundant in saline marshes near Launceston, *Gunn.*

S. Australia. Holdfast Bay and Gawler ranges, *F. Mueller.*

W. Australia. Port Gregory, *Oldfield.*

The species is very common in Europe and a great part of Asia, including several described as distinct by Moquin, and is probably only of modern introduction in Australia.

SERIES 4. GLOMERATÆ.—Monœcious decumbent procumbent or spreading herbs, scaly-tomentose or mealy-white. Male flowers in globular clusters surrounded by a few females in the upper axils or rarely forming short terminal spikes. Females clustered in the lower axils without males. Fruiting perianths more or less compressed, conspicuously 2-valved.

13. **A. humilis,** *F. Muell. Fragm.* iv. 48. Stems hard and more or less decumbent at the base, ascending to about 1 ft. in our specimens, the branches and foliage mealy or minutely scaly-tomentose. Leaves nearly sessile, mostly lanceolate or oblong-linear, obtuse, entire, contracted at the base, rarely above ½ in. long, but the lower ones not seen. Flowers monœcious, the males in compact sessile terminal spikes of ¼ to ½ in., the females all axillary and densely clustered. Fruiting perianths broadly rhomboid, 1½ to 2½ lines diameter, with a small thick turbinate solid base, the valves entire, free almost to the base, herbaceous and reticulate. Seed orbicular; radicle superior.

N. Australia. Subsaline banks of Flinders river, Gulf of Carpentaria, *F. Mueller.*

14. **A. velutinella,** *F. Muell. Rep. Babb. Exped.* 20. Apparently herbaceous and procumbent, with elongated branching stems, the whole plant scaly-tomentose. Leaves sessile or rarely the lower ones contracted into a short broad petiole, ovate or rhomboid, coarsely sinuate-toothed, mostly above 1 in. long, or the upper ones oblong and nearly entire, much smaller, and passing into small floral bracts. Flowers monœcious, in sessile clusters, the upper ones with numerous males surrounded by several females, either in the uper axils or 2 to 6 of the last clusters forming an interrupted terminal spike; the lower axillary clusters all female. Fruiting perianths flat, triangular ovate, about 3 lines long, the hardened base very short and broad; the valves almost acute, entire or with a few short teeth on each side at the base, herbaceous, scaly-tomentose, free almost to the base. Seed broadly orbicular; the radicle inferior or lateral.

N. S. Wales. Darling Desert, *Victorian Expedition.*
S. Australia. Stuart's Creek, *Babbage's Expedition.*

15. **A. angulata,** *Benth.* Mealy or almost scaly-tomentose and probably herbaceous. Leaves on long petioles which are winged below the lamina, orbicular or broadly rhomboid, very obtuse, angular or sinuate, ¾ to 1½ in. diameter. Flowers monœcious, the male clusters in the upper axils accompanied by a few females or 2 or 3 forming a short terminal spike, the lower axillary clusters small and all female. Fruiting perianth not yet quite ripe, raised on a stipes of about 1 line, with a turbinate compressed tube of a little more than 1 line, and large green toothed valves, 3 lines broad or more. Fruit enclosed in the tube, but raised on a short solid base. Radicle superior.

S. Australia. Subsaline plains near Cudnaka, *F. Mueller;* Murray river, *W. Ross.*

This plant has the foliage almost of *A. nummularia*, but the fruiting perianths cannot be referred to any of those of allied species. I have only seen three small specimens.

16. **A. semibaccata,** *R. Br. Prod.* 406, *not of Moquin.* Stems herbaceous, procumbent or prostrate, much branched and slender, spreading to 1 or 2 ft., the whole plant green and nearly glabrous or mealy-white. Leaves petiolate, oblong oblanceolate or cuneate and ½ to 1 in. long or shorter and obovate, obtuse, entire or sinuate-toothed, rather thin. Flowers monœcious, the males in little globular clusters in the upper axils surrounded by a few females, and a few females alone in the lower axils. Fruiting perianth more or less rhomboidal, 1½ to 2½ lines long, and nearly as broad in the centre, the lower half a flattened triangular tube closed at the base, usually thickened (or fleshy when fresh?) and prominently 3-nerved, the upper half consisting of the flat appressed triangular valves, entire or toothed at the base, herbaceous at least at the margins. Fruit half-enclosed in the tube. Radicle lateral.— *A. denticulata,* Moq. in DC. Prod. xiii. ii. 97.

Queensland. Keppel Bay, *R. Brown;* Rockhampton, *O'Shanesy;* Darling Downs, *Lau;* Armadilla, *Barton;* Curriwillighie, *Dalton.*

N. S. Wales. Port Jackson, *R. Brown;* Liverpool plains, *A. Cunningham, Leichhardt;* Ballandool river, *Locker.*

Victoria. Murray river, *Herrgott;* Wimmera, *Dallachy* (in leaf only and doubtful); Little river, *Fullagar.*

S. Australia. Holdfast Bay, Gawler river, Port Adelaide, *F. Mueller.*

W. Australia, *Drummond, n.* 222, (or 228?).

17. **A. exilifolia,** *F. Muell. Fragm.* vii. 9. A prostrate herb, with a hard almost woody base, but apparently annual, the stems about 1 ft. long, with numerous shortly ascending branches, the whole plant minutely scaly-hoary. Leaves shortly petiolate, obovate orbicular or rhomboidal, the larger ones on the main stems scarcely ¼ in. long, but most of them smaller. Flowers monœcious, in axillary clusters, small and not numerous, the males and females mixed. Fruiting perianths with a short solid turbinate base, broadly deltoid or rhomboid, scarcely above 1 line diameter, the valves flat. entire or 3-toothed.

W. Australia, *Drummond, n.* 249. The habit is that of *A. prostrata,* but on a very much larger scale, and the fruiting perianth much larger and flatter.

18. **A. Muelleri,** *Benth.* An erect or spreading annual of 1 or 2 ft., with a hard base, more or less mealy-white, but not so densely scaly as the shrubby species. Leaves petiolate, broadly obovate ovate or rhomboidal, coarsely and irregularly sinuate-toothed or lobed, mostly from under 1 in. to about 2 in. long and rather thin. Flowers small, monœcious, all axillary, the males in the upper axils in little globular heads surrounded by a few females, the females alone clustered in the lower axils. Fruiting perianths sessile, 1 to 1½ lines diameter, with a hard compressed globular smooth tube, the valves short broad appressed, shortly toothed. Fruit enclosed in the tube. Radicle superior.

Queensland. Peak Downs, *F. Mueller;* Armadilla, *Barton.*

N. S. Wales. Liverpool plains, *Leichhardt.*

Victoria. Lagoons on the Murray, *F. Mueller.*

S. Australia. In the interior, *Howitt's Expedition.*

This is referred by F. Mueller, Fragm. vii. 9, to the European *A. rosea,* but that is a coarser, much more scaly-tomentose species, the fruiting perianth is larger, broader and flatter, the disk reticulate and sometimes muricate, and the radicle of the seed lateral. *A. Muelleri* is, as observed by F. Mueller, allied to *A. semibaccata*, but more mealy and otherwise different both in foliage and fruiting perianth. The true *A. rosea* is said by Moquin in DC. Prod. xiii. ii. 92, to be found in "New Holland," but the special authority is not given, and I have seen no Australian specimen of it.

19. **A. elachophylla,** *F. Muell. Fragm.* vii. 8. A small slender much-branched plant, hard and almost woody, diffuse or procumbent, the specimens not exceeding 6 in., more or less scaly-tomentose. Leaves shortly petiolate, from obovate to oblong or almost lanceolate, obtuse, rather thick, rarely exceeding 2 lines. Flowers monœcious, the males in globular clusters of less than 1 line diameter, sessile within a floral leaf, either terminal or on a short axillary peduncle-like branchlet; the females axillary, solitary or 2 or 3 together, Fruiting perianth rhomboid-globular, but slightly compressed, 1 to 1¼ line diameter, hard, scaly-tomentose, with 2 very short broad green valves, usually toothed on the margin. Fruit enclosed in the tube. Seed compressed; radicle superior.

N. Australia. Desert of Sturt's Creek, *F. Mueller.*

20. **A. microcarpa,** *Benth.* A small diffuse or procumbent herb, with a hard stem and numerous ascending branches not exceeding 6 in. clothed with a scaly tomentum. Leaves very shortly petiolate, oblong or lanceolate, obtuse, entire, under ½ in. long. Flowers very numerous and small, in axillary clusters, a small head of males surrounded by females in the upper axils, all females in the lower ones. Fruiting perianth rhomboidal, compressed, scarcely 1 line long and broad, membranous, densely and softly tomentose, the lower half closed, the upper moiety consisting of 2 entire valves. Radicle of the seed superior.

N. ·S. Wales. Clay flats, Banaroo (Darling desert), *Victorian Expedition.*
It is possible that this and the following species may prove to be varieties of *A. pumilio*, but at present they appear to me to be quite distinct.

21. **A. prostrata,** *R. Br. Prod.* 406, *not of Moquin.* A prostrate scaly-tomentose annual, with slender much branched leafy stems extending from an inch or two to half a foot or rather more. Leaves very shortly petiolate, oblong or rarely ovate-lanceolate, obtuse, mostly entire, 1 to 2 lines long. Flowers monœcious, the male clusters in the upper axils surrounded by a few females, the females alone several together in the lower axils. Fruiting perianth tomentose, obovoid-rhomboidal, slightly compressed, scarcely ½ line diameter, closed to above the broad middle, the valves short and entire. Seed parallel to the valves; radicle superior.—*A. decumbens,* Roem. and Schult. Syst. vi. 289.

S. Australia. Kangaroo Island, *R. Brown.*

22. **A. pumilio,** *R. Br. Prod.* 406. A small scaly-tomentose herb, with a short hard decumbent stem and numerous branches ascending to 1 or 2 in. Leaves sessile or very shortly petiolate, ovate obovate or oblong, entire or sinuate-toothed, 1 to 1½ lines long. Flowers monœcious, the males in 2 or three little clusters in the upper axils (with 1

or 2 females ?) scarcely forming very short leafy spikes, the females in the lower axils solitary or two together without males. Fruiting perianth ovate, tomentose, not very flat, about ½ line long, the valves entire or toothed, shorter than the tube. Fruit enclosed in the tube, but bursting it irregularly when ripe. Radicle superior.—Moq. in DC. Prod. xiii. ii. 92.

S. Australia. St. Peter's isles, *R. Brown.*

The testa of the seed is thin, as observed by Brown, but it is of the brown colour of other species, and appears to me to be rather thinly crustaceous than truly membranous.

23. ? **A. glomulifera,** *Nees in Pl. Preiss.* i. 634. A much-branched glabrous erect herb of a finger's length. Leaves oblong or lanceolate-spathulate, obtuse, entire, contracted into a petiole, about 2 lines long, green and fleshy. Flowers (monœcious ?) very minute, the female clusters in almost all the axils, the upper ones about the size of a poppy seed, the lower ones smaller, all very dense, the individual flowers scarcely conspicuous to the naked eye. Fruiting perianth pedicellate, the valves free, longer than the fruit but spreading, obovate nearly orbicular, entire, thin, sprinkled with a few stipitate glands.—Moq. in DC. Prod. xiii. ii. 103.

W. Australia. Cultivated grounds, head of Swan river, Preiss, n. 1257. I have not seen this species. It is said to be probably allied to *A. prostrata* and *A. pumilio,* but the perianth valves ("leaflets of the involucre") are differently described, and the plant is said to be glabrous and green.

24. **A. campanulata,** *Benth.* A perennial, with a hard almost woody stock and rather slender procumbent branching stems extending to 1 or 2 ft., the whole plant nearly glabrous or mealy-white. Leaves shortly petiolate, obovate or oblong, entire or coarsely angular-toothed, mostly under ½ in. or rarely nearly 1 in. long. Flowers monœcious, all axillary, the males forming little globular heads or clusters of little more than 1 line diameter surrounded by several females, or all the flowers female in the lower axils. Fruiting perianth very shortly stipitate, the tube obliquely campanulate, slightly compressed, about 1 line long in front, longer at the back ; the limb much dilated, very oblique, the valves unequal and scarcely appressed, each one 3-lobed ; with 2 small herbaceous appendages on the front or shorter face of the tube. Fruit enclosed in the tube. Radicle superior.

N. S. Wales. Darling river, *Victorian Expedition.* Included by *F. Mueller* in *A. leptocarpa,* to which it approaches in habit, foliage and inflorescence, but the fruiting perianth is totally different.

Series 5. PARVILOBÆ.—Monœcious spreading or procumbent herbs or undershrubs, scaly-tomentose or mealy. Flowers axillary, the males in globular clusters in the upper axils usually surrounded by females, females alone often in the lower axils. Fruiting perianths not compressed, enclosing the fruit, the orifice small, closed by small erect appressed valves.

25. **A. leptocarpa,** *F. Muell. in Trans. Phil. Inst. Vict.* ii. 74.—A perennial with a thick stock and herbaceous procumbent stems extending to 1 or 2 ft., the whole plant more or less hoary or white with a scaly tomentum. Leaves obovate or oblong, obtuse, entire when narrow, coarsely angular-toothed when broad, from under 1 in. to nearly 2 in. long. Flowers monœcious, all axillary, the males in little globular dense sessile heads of 1 to 1½ lines diameter, mostly 4-lobed, surrounded by several females or sometimes females only in the lower axils more or less stipitate. Fruiting perianth narrow tubular, cylindrical, 2 to 4 lines long, the lower portion (½ to ⅔) rather hard, enclosing the fruit which is more or less raised on the solid base of the tube, the upper portion above the fruit more herbaceous, green and elegantly veined, the orifice closed by 2 very short triangular entire valves not broader than the tube and without appendages. Radicle ascending or superior.

Queensland. Curriwillighie, *Dalton.*
N. S. Wales, *Leichhardt;* Castlereagh river, *C. Moore.*
S. Australia. Murray river, near Morunda, *F. Mueller.*

26. **A. limbata,** *Benth.* A procumbent or spreading perennial with the habit foliage and inflorescence of *A. leptocarpa,* but more scaly-to-mentose. Leaves obovate or oblong, entire or angular-toothed. Flowers monœcious, all axillary, the male clusters in the upper axils surrounded by females, the lower clusters all females. Fruiting perianth with a cylindrical tube enclosing the fruit as in *A. leptocarpa,* but sessile and usually rather larger and harder, and sometimes the solid base much elongated, extending the whole tube to ½ in. or more, but varying in this respect even in the same cluster, the orifice closed by 2 valves reduced to 3 lobes, of which the lateral ones are spreading incurved and hornlike about 1 line long, those of the 2 valves more or less united, the central lobe minute or almost obsolete, and alternating with the horns are 2 herbaceous spreading broad dorsal appendages, also about 1 line long, giving the apex of the perianth the appearance of a spreading 4-lobed limb, but with the disk closed. Radicle of the seed ascending or superior.

N. S. Wales. Darling river, *Victorian Expedition.* Included by F. Mueller in *A. leptocarpa,* but the difference in the fruiting perianth appears to be constant.

27. **A. halimoides,** *Lindl. in Mitch. Three Exped.* i. 285. A procumbent or diffuse perennial or undershrub, with the habit and inflorescence of *A. holocarpa,* but usually not so white and the leaves narrower mostly lanceolate or ovate-lanceolate and acute, but sometimes as tomentose and passing into the small rhomboidal form of that species. Fruiting perianth enlarged to 4 to 6 lines diameter, loosely fibrous and spongy with an inner and an outer membrane as in *A. holocarpa,* but broadly turbinate or almost hemispherical with a much depressed or flattened summit bordered by an annular horizontal wing or acute angle, the very small central orifice closed by small entire or 3-toothed erect valves as in the allied species. Fruit the same, except that the radicle

appears to be always inferior not superior.—*A. Lindleyi.* Moq. in DC.
Prod. xiii. ii. 100; *A. inflata,* F. Muell. in Trans. Phil. Inst. Vict. ii. 75.

Queensland. Burnett river, *F. Mueller;* Suttor and Bogan rivers', *Bowman;*
Curriwillighie, *Dalton.*

N. S. Wales. Darling desert, *Victorian Expedition, Mrs. Ford.*

Victoria. Wimmera, *Dallachy.*

S. Australia. Murray river and Cudnaka, *F. Mueller;* towards Cooper's Creek,
Howitt's Expedition.

Moquin changed Lindley's name on account of a previous *A. halimoides* of Tineo,
but that has never been otherwise published than as a name in a garden catalogue.

28. **A. holocarpa,** *F. Muell. Rep. Babb. Exped.* 19. A perennial with
a hard almost woody base and herbaceous diffuse or procumbent brancn-
ing stems, attaining from 6 in. to above 1 ft., softly mealy-tomentose.
Leaves on rather long petioles, obovate or rhomboidal, irregularly
toothed, from under ½ in. to above 1 in. long. Flowers monœcious, all
axillary, the males few in the upper axils surrounded by females, females
only and usually few together in most axils, very small and globular at
the time of flowering. Fruiting perianth obovoid-globular, scarcely
compressed, not flattened at the top, 4 to 6 lines diameter, of a loosely
fibrous and spongy consistence, with a thin membranous epidermis and
a thin inner membrane scarcely distinguishable from the pericarp and
sometimes (but not always) hardening over the seed as it ripens; the
summit of the perianth with a small central orifice closed by 2 erect
appressed, entire or 3-toothed valves, rarely above ½ line long. Seed
with the superior radicle of the majority of those species in which it is
enclosed in the perianth tube.

N. S. Wales. Murray and Darling desert, *Victorian Expedition, Mrs. Ford* and
others.

S. Australia. Eyre's Depot Creek, *Babbage's Expedition;* between Stokes
Range and Cooper's Creek, *Wheeler;* towards Spencer's Gulf, *Warburton.*

29. **A. spongiosa,** *F. Muell. in Trans. Vict. Inst.* ii. 74. A small
much-branched herb or undershrub, with numerous ascending or erect
stems, not above 6 in. high, more or less mealy-white as well as the
foliage or becoming glabrous when old. Leaves shortly petiolate, broadly
ovate obovate or orbicular, entire or sinuate-toothed, rather thick, ¼ to
½ in. long. Flowers monœcious and axillary as in *A. holocarpa,* but much
smaller and fewer together, the females mostly solitary or only 2 in each
axil. Fruiting perianth enlarged fibrous and spongy with a membranous
epiderm and the inner membrane inseparable from the pericarp as in
A. holocarpa, but much smaller, depressed globular, not exceeding 2 lines
diameter, the small orifice closed by 2 minute erect appressed triangular
valves. Seed of *A. holocarpa* with the radicle erect.—*A. semibaccata,* Moq.
in DC. Prod. xiii. ii. 97, not of R. Br.

N. Australia. Sturt's Creek, *F. Mueller.*

S. Australia. Lake Torrens, *F. Mueller* (I have not seen these specimens).

W. Australia, *Drummond, n.* 127.

SERIES 7 (or SECT. 2). THELEOPHYTON.—Monœcious prostrate crystalline herb. Flowers axillary. Perianth obovoid with short valves.
Seed compressed at right angles with, not parallel to, the valves.

30. **A. Billardieri,** *Hook. f. Fl. N. Zeal.* i. 215, *and Fl. Tasm.* i. 315, *t.* 95. A much-branched prostrate more or less succulent herb, spreading in masses of 1ft. diameter or more, the branches and foliage covered with watery shining papillæ like those of some *Mesembryanthema.* Leaves shortly petiolate, oblong obovate or ovate, obtuse, entire or slightly sinuate-toothed, ¼ to ½ in. long. Flowers monœcious, the males in small clusters of about 5 or 6 in the upper axils (without females?) the females in the lower axils solitary or 2 together and very minute. Fruiting perianths obovoid, membranous, scaly, slightly compressed at the base in a direction contrary to the valves, terete upwards, contracted at the orifice, the valves much shorter than the tube, appressed, entire or slightly toothed. Seed enclosed in the perianth-tube, slightly compressed. Embryo placed at right angles to the valves, with the radicle superior but not prominent.—*Obione Billardieri,* Moq. Chenop. Enum. 72; *Theleophyton Billardieri.* Moq. in DC. Prod. xiii. ii. 116; *Atriplex crystallina,* Hook. f. in Hook. Lond. Journ. vi. 279.

Victoria. Sands near high-water mark, Phillip Island and on the opposite coast, *F. Mueller;* E. Gipps' Land, *A. Taylor.*

Tasmania. Sands close to high-water mark near George Town, *Gunn;* South Port, *Oldfield.*

The exceptional direction of the embryo in this single species does not appear of itself sufficient to justify its separation from a genus which, with all the diversities of form assumed by the fruiting perianth, is, as a whole, a remarkably natural and well defined one.

TRIBE 2. CAMPHOROSMEÆ.—Branches continuous. Leaves narrow, entire, flat or terete, glabrous villous-tomentose or woolly. Testa membranous. Embryo curved round a mealy albumen.

5. ENCHYLÆNA, R. Br.

Flowers hermaphrodite. Perianth urceolate, at length depressed-globular, succulent or coriaceous, with 5 short broad lobes or teeth connivent and closing over the fruit, without any dorsal wings or appendages. Stamens 5 or fewer. Ovary depressed-globular. Styles 2 or 3, shortly connate at the base. Fruit depressed-globular, enclosed in the perianth, pericarp membranous. Seed more or less flattened, horizontal; testa membranous; embryo horseshoe-shaped or almost annular, enclosing a very scanty albumen.—Undershrubs or shrubs. Leaves linear-terete or linear-lanceolate, entire. Flowers solitary in the axils and sessile, without any or with one or two minute bracts.

The genus is limited to Australia. It only differs from *Kochia* in the fruiting perianth of a thicker consistence and often succulent, without any dorsal wings or appendages.

Leaves 1 to 2 lines long. Flowers numerous, mostly crowded in
terminal leafy spikes. Perianth not above 1 line diameter . . 1. *E. microphylla.*
Leaves mostly above ¼ in. long. Flowers distant.
Fruiting perianth globular, about ¾ line diameter, smooth, hairy
at the top. Plant very villous with soft fulvous hairs . . 2. *E. micrantha.*
Fruiting perianth depressed-globular, about 1½ lines diameter,
quite smooth, with very short teeth . . . . 3. *E. tomentosa.*

Fruiting perianth coriaceous, depressed-globular, about 2 lines
diameter, the lobes more or less gibbous outside . . . . 4. *E. villosa.*
Fruiting perianth broadly turbinate, very flat, with a nerve-like
edge, nearly 2 lines diameter, the tube 10-ribbed . . . . 5. *E. marginata.*

1. **E. microphylla,** *Moq. in DC. Prod.* xiii. ii. 128. A diffuse divaricate (or erect ?) shrub, with·numerous slender branches, slightly pubescent. Leaves linear-terete, fine or rather thick, 1 to 2 lines long. Flowers small, solitary in each axil as in the other species, but numerous and crowded into leafy spikes at or near the ends of the branches. Fruiting perianths ¾ to nearly 1 line diameter, shaped like those of *E. tomentosa,* but smaller, thinner (not succulent ?) and slightly angular. Styles usually 2.—*Suæda tamariscina,* Lindl. in Mitch. Trop. Austr. 239; Moq. in DC. Prod. xiii. ii. 461.

Queensland. Near Mount Kilsyth, *Mitchell;* Darling Downs, *Lau.*
N. S. Wales. Foot of Mount Flinders, *A. Cunningham.*

2. **E. ? micrantha,** *Benth.* A shrub, the branches in our specimens above a foot long, with numerous branchlets densely clothed as well as the foliage with soft fulvous silky or sometimes woolly hairs. Leaves rather crowded, linear, soft, 2 to 4 lines long. Flowers very small, solitary in the axils. Perianth, already much enlarged after flowering with the fruit nearly ripe, globular, glabrous or hairy especially the lobes, smooth and rather thick, scarcely ¾ lines diameter; lobes 5, short, obtuse, connivent over the fruit. Stamens 5, shortly exserted, with flattened filaments. Styles 3, connate at the base. Fruit depressed-globular, more or less hairy. Seed not seen quite ripe, but the embryo already large appears to be horizontal, annular, with the radicle prominent and somewhat ascending.

W. Australia, *Drummond, 4th coll. n.* 253, referred by F. Mueller, Fragm. vii. 12, with doubt to *Kochia villosa,* but I can see no trace of any wing or transverse prominence to the perianth.

3. **E. tomentosa,** *R. Br. Prod.* 408. A procumbent or divaricately branched undershrub, sometimes with ascending slightly branched stems under 6 in. long, sometimes much branched and attaining several feet, the branches hoary or silvery with a close or woolly tomentum, rarely glabrous or nearly so. Leaves linear-terete, entire, rarely above ½ in. long and sometimes under ¼ in. Flowers all axillary, solitary. and sessile and usually distant, with 1 or 2 minute bracts at the base. Perianth small at the time of flowering, 1½ lines diameter when in fruit and then depressed-globular, red and succulent when fresh, black when dry and perfectly smooth, the orifice closed by 5 short connivent teeth quite glabrous or minutely ciliate. Stamens very shortly exserted, the anthers very deciduous. Fruit enclosed in the perianth, the pericarp membranous and glabrous or scarcely hairy in the normal state. Styles usually 3 but sometimes 2.—Moq. in DC. Prod. xiii. ii. 128; Nees in Pl. Preiss. i. 635; *E. paradoxa,* R. Br. Prod. 408; Moq. l.c.; *E. pubescens,* Moq. l.c. (monstrous states, see below).

N. Australia. Sturt's Creek, *F. Mueller.*

Queensland. Burdekin river, *F. Mueller;* Bokhara Creek, *Leichhardt;* Rockhampton, *O'Shanesy;* Suttor river, *Bowman;* Armadilla, *Barton.*

N. S. Wales. Liverpool plains, *A Cunningham;* Castlereagh river, *C. Moore;* Murray desert and Goyinga mountains, *Victorian Expedition.*

Victoria. Murray river, *Herrgott.*

S. Australia. Islands off the S. Coast, *R. Brown;* from the Murray river to St. Vincent's Gulf, *F. Mueller;* Mount Searl, *Warburton;* Cooper's Creek, *Murray.*

W. Australia, *Drummond, n.* 717; Murchison river, *Oldfield;* Sharks Bay, *Gaudichaud;* Avon river, *Preiss, n.* 1935 (Moquin).

Var. *villosa.* Very densely fulvous-villous.—Cudnaka, *F. Mueller.*

Var.? *leptophylla.* Leaves very slender. Perianths very small.—Near Gainsford, Queensland, *Bowman.*—Perhaps a distinct species, but the specimens are very small.

Var. *glabra.* Stems and leaves quite glabrous.—Bay of Inlets, *Banks and Solander;* Brisbane river, *F. Mueller;* Darling river, *Victorian Expedition;* between Stokes Range and Cooper's Creek, *Wheeler.*

Besides the woolly globular galls to which this species is liable (like those of *Kochia villosa* and other Chenopodiaceæ), it is subject to a monstrosity, apparently caused also by an insect, by which the pericarp becomes densely enveloped in woolly intricate hairs proceeding from near the base and bursting through the apex of the perianth; whilst the ovary is abortive, and I have sometimes found its place occupied by a small grub. It is this monstrosity in the typical form that is described by Moquin as *E. pubescens,* and in the glabrous variety constitutes the *E. paradoxa,* Br.

4. **E. villosa,** *F. Muell. in Trans. Phil. Inst. Vict.* ii. 76. Stems branching at the base, procumbent or ascending, rarely exceeding 6 in., the whole plant or at least the inflorescence villous, or the lower part or nearly all glabrous. Leaves linear or linear-lanceolate, acute or obtuse, contracted at the base and sometimes petiolate, rather thick, rarely ½ in. and often not ¼ in. long. Flowers in the upper axils but not crowded. Fruiting perianth depressed as in *E. tomentosa,* but more angular, about 2 lines diameter, " coriaceous and not succulent," black when dry, shortly hirsute or rarely glabrous, the lobes connivent and closed over the fruit, larger and deeper than in *E. tomentosa;* 2 or 3 outer ones broad and thickened near the apex into 2 obtuse angles or lobes, or one of the outer ones irregular, 2 very rarely 3 inner ones flat and triangular with a thickened transverse line at the base outside. Styles 2 (or rarely 3 ?). Fruit and seed of *E. tomentosa.*

Queensland. Armadilla, *Barton.*

N. S. Wales. Peel river and near Cassilis, *Leichhardt;* Billabong, *Bissill.*

Victoria. Bacchus Marsh and Station Peak, *F. Mueller;* Little river, *Fullagar.*

S. Australia. Near Adelaide, *F. Mueller.*

This species connects in some measure *Enchylæna* with *Kochia,* for the transverse thickening of the perianth-lobes may be regarded as an incipient wing. The names both of *E. villosa* and *E. tomentosa* are unfortunately selected, as both are sometimes almost if not quite glabrous.

5. **E. marginata,** *Benth.* An undershrub branching at the base, with ascending stems not exceeding 6 in. in our specimens, very villous as well. as the young foliage with soft fulvous silky or woolly hairs. Leaves rather crowded, linear, obtuse, soft but flat, often above ½ in. long, very villous at first, becoming nearly glabrous with age. Flowers very small, solitary in the axils. Fruiting perianth sessile, with a flat circular base of nearly 1 line diameter, the tube shortly and broadly

turbinate, thick and apparently fleshy, obtusely 10-ribbed, the summit very flat with a nerve-like margin, 2 lines diameter, the lobes short and quite closed over the fruit. Styles apparently 2, but not seen perfect. Pericarp very flat. Embryo annular.

W. Australia. Swan river, *Drummond, 1st coll.* This is again in some measure intermediate between *Enchylæna* and the small flat-fruited species of *Kochia,* the nerve-like border (scarcely more than an angle), representing the narrow wing of *K. ciliata* and its allies.

6. KOCHIA, Schrad.

(Maireana, *Moq.* ; Sclerochlamys, *F. Muell.*)

Flowers hermaphrodite or polygamous. Perianth at first nearly glo-bular, at length depressed turbinate or pyramidal, not succulent, with 5 rarely 4 short broad lobes connivent and closing over the fruit, imbricate in the bud and 3 outer ones often rather larger than the 2 inner ones, bearing on their backs horizontal wings either distinct or united in a single annular wing surrounding the perianth. Stamens usually 5 or fewer by abortion. Styles 2 or 3, shortly connate at the base. Fruit depressed-globular, enclosed in the perianth; pericarp membranous. Seed more or less flattened, horizontal; testa membranous; embryo horseshoe-shaped or almost annular, enclosing a scanty albumen.— Undershrubs or shrubs, usually procumbent or spreading. Leaves linear or rarely oblong, usually small thick and often semiterete. Flowers solitary or very rarely 2 together in the axils, sessile, with very minute or without any bracts, the perianth very small at the time of flowering with the stamens and styles shortly exserted, but in most species there appear to be many female flowers without any perfect stamens. Fruit-ing perianth usually described as variously coloured red, from a pale pink to a rich crimson, but no colour remains in the dried specimens.

The genus is limited to the extratropical and subtropical regions of the Old World, the Australian species being apparently all endemic.

Perianth flat within the wings or nearly so.
  Leaves mostly ½ in. long, densely silky. Perianths enve-
    loped in long dense woolly hairs . . . . 7. *K. eriantha.*
  Leaves mostly ⅛ to ¼ in. long, linear or terete, tomentose
    or nearly glabrous (sometimes small and slender), .
    spreading. Perianth glabrous or tomentose . . . . 8. *K. villosa.*
  Leaves oblong or oblanceolate, flat, ¼ to ½ in. long. Pe-
    rianth of *K. villosa* . . . . 9. *K. planifolia.*
  Leaves oblong-clavate, almost terete, densely cottony, not
    exceeding ¼ in. Perianth of *K. villosa* . . . . . . 10. *K. sedifolia.*
  Leaves cottony, erect and appressed, rarely exceeding 1
    line. Perianth of *K. villosa* . . . . 11. *K. appressa.*
  Leaves minute, distant. Branches spinescent. Perianth
    of *K. villosa* . . . . 12. *K. aphylla.*
Fruiting perianth very flat at the top, surrounded by an annular
  more or less rigid horizontal border or thick wing, quite entire
  or regularly toothed.
  Annular border of the perianth entire, densely ciliate . . . . 13. *K. ciliata.*
  Annular border 5-angled, tube vertically 5-winged . . . . 14. *K. brachyptera.*
  Annular border with 10 to 12 radiating points, tube smooth . 15. *K. stelligera.*

1. **K. lobiflora,** *F. Muell. Herb.* A low much-branched undershrub
or shrub, our specimens not exceeding 1 ft., the branches and foliage
softly and densely silky-tomentose. Leaves sessile, linear, mostly
acute, ¼ to nearly ½ in. long, soft and thick but more or less flattened.
Flowers solitary in the upper axils. Fruiting perianth woolly-tomentose
all over, depressed and about 1½ lines diameter within the wings, the
lobes connate over the fruit nearly to the top and quite flat, with 5
separate dorsal wings broadly spathulate contracted into broad stipes
and horizontally spreading to a diameter of 5 or 6 lines, and alternating
with them there is in each sinus a narrow spathulate reflexed appendage,
shorter than the horizontal dorsal wings and concealed under them.
Styles usually 2.

N. S. Wales. High sandy banks of the Darling river, *Victorian Expedition.*

2. **K. lanosa,** *Lindl. in Mitch. Trop. Austr.* 88. An erect or spreading undershrub or low shrub, the branches and foliage silky-woolly as
in *K. lobiflora.* Leaves sessile, linear, mostly acute, thick and soft,
from under ¼ in. to nearly ½ in. long, flowers solitary in the axils.
Fruiting perianth more or less woolly all over, depressed and about 1
line diameter within the wings, the lobes obtuse and closed over the
fruit, the 5 dorsal wings thin and membranous, all distinct but not
stipitate, or more or less connate, spreading to from 4 to 6 lines diameter, with 5 linear acute appendages, alternating with the horizontal
wings, erect on their upper side and varying from ½ to 1½ lines in
length. Styles 3 or rarely 2.—Moq. in DC. Prod. xiii. ii. 461.

Queensland. Narran river, *Mitchell.*
N. S. Wales. Darling river, *Dallachy, Mrs. Ford.*
S. Australia. Murray desert, near Morunda, *F. Mueller;* towards Cooper's
Creek, *Neilson.*

Var. *minor.* Fruiting perianth smaller, the horizontal wings more connate, almost
as in *K. villosa,* but with the erect sinus-appendages of *K. lanosa.*

W. Australia. Murchison river, *Oldfield*

The erect appendages have been described as the lobes of the perianth, they will be found, however, like the reflexed ones of *K. lobiflora*, to alternate with the real lobes, which are flat, obtuse and closely connivent as in most species of the genus.

3. **K. triptera,** *Benth.* A low but stout diffuse or spreading shrub or undershrub, the foliage and often the branches also quite glabrous and somewhat glaucous. Leaves rather crowded, linear, semiterete, often acute, rather thick, ¼ to above ½ in. long. Flowers solitary in the axils. Fruiting perianth with a broadly turbinate tube, above 1 line long below the horizontal wings, with 3 or rarely 4 very prominent vertical wings, and more than half of it occupied by a thick solid base below the fruit, the upper part of the perianth within the wings flat and closing over the fruit as in most *Kochias*, the horizontal wings united in a single rigidly membranous ring expanding to 4 or 5 lines or even ½ in. diameter and quite concealing the vertical wings. Styles usually 3.

N. S. Wales. Darling river, *Victorian Expedition;* Mount Murchison, *Giles.*
W. Australia, *Drummond.*

Var. *erioclada.* Branches densely tomentose ; leaves glabrous as in the typical form, but more obtuse and terete.—Murray desert, *Herb. F. Mueller;* W. Australia, *Drummond, n.* 432.

4. **K. oppositifolia,** *F. Muell. in Trans. Vict. Inst.* 1855, 134, *and in Hook. Kew Journ.* viii. 205. A densely branched probably low shrub, more or less hoary or silky-white with a close tomentum. Leaves opposite or alternate, sessile, ovate or lanceolate, rather thick and prominently keeled, about 1 to 1½ lines long. Flowers solitary in the axils. Fruiting perianth much depressed, the tube very short and broad without longitudinal wings, the upper portion flat, with very short teeth closed over the fruit, bordered by three broad membranous veined wings, not connate though expanded into a circle of 3 to 4 lines diameter, with 2 inner smaller wings narrower and less spreading, corresponding to the inner perianth-lobes, and sometimes almost obsolete. Styles 2, connate at the base.

S. Australia. Seacoast opposite Lake Hamilton, *Wilhelmi;* Coorong desert and Spencer's Gulf, *F. Mueller;* Venus and Streaky Bay, *Babbage.*
W. Australia. King George's Sound, *R. Brown;* towards Cape Riche, *Harvey.* Drummond's specimens, 4th coll. n 251, without flowers, referred here with doubt by F. Mueller, appear to me to be more like *Didymanthus Roei,* in which the leaves are more constantly opposite and rather longer than in *Kochia oppositifolia.*

5. **K. brevifolia,** *R. Br. Prod.* 409. A much-branched rather slender shrub, the branches and foliage pubescent or tomentose with short woolly hairs or the leaves glabrous. Leaves alternate, sessile, linear or oblong, obtuse, thick and almost terete or somewhat flattened, about 1 line or rarely nearly 2 lines long. Flowers small and solitary in the axils. Fruiting perianth much depressed, glabrous or slightly pubescent, the tube hemispherical, the upper portion scarcely 1 line diameter within the wings, with 5 equal broad obtuse lobes horizontally closed over the fruit but rather thick and almost bullate, forming 5 distinct prominences, the perianth bordered by 5 horizontal mem-

branous veined wings, forming a complete circle of about 3 lines diameter but not united as in all the following species. Styles usually 2, rather short, united at the base.—Moq. in DC. Prod. xiii. ii. 131; *Salsola brachyphylla*, Spreng. Syst. i. 924; *Kochia thymifolia*, Lindl. in Mitch. Trop. Austr. 56, Moq. l.c. 461.

Queensland. Darling Downs, *Lau;* Armadilla, *Barton.*

N. S. Wales. Camden valley, Liverpool Plains, *A. Cunningham;* Morra Creek, Macquarrie river, *Mitchell;* Darling river, *Mrs. Ford.*

Victoria. Murray desert, *F. Mueller.*

S. Australia. Spencer's Gulf, *R. Brown;* Port Adelaide, *F. Mueller;* Burra-Burra, *Hinteracker.*

W. Australia. Murchison river, *Oldfield; Drummond, 6th coll. n.* 224.

6.? **K. pyramidata,** *Benth.* A divaricately branched shrub, with numerous rigid but scarcely spinescent branchlets, softly tomentose-pubescent or cottony as well as the foliage. Leaves alternate, very spreading, linear or terete, obtuse, 1 to 2 lines long, thick and soft, often clustered in the axils. Flowers small, solitary in the upper axils. Perianth about ¾ line long at the time of flowering, more ovoid and more deeply lobed than in other species, the stamens and 2 styles exserted. Fruiting perianth slightly pubescent and drying very black, the tube broadly turbinate the upper portion within the wing erect, pyramidal, at least 2 lines long and as much in diameter at the base, surrounded by an entire annular membranous wing from ½ to 1 line broad. Pericarp and seed as in the rest of the genus.

N. S. Wales. Lachlan river, *A. Cunningham;* sand hills near the Darling, occupying large tracts and giving a character to the country, *Beckler (Victorian Expedition);* Murray desert, *Herb. F. Mueller.*

F. Mueller thinks that this may be a state of *K. villosa* with a monstrously developed perianth, but besides some difference in habit and foliage (which approach those of *K. aphylla*) I find great uniformity in the enlarged perianths which are very numerous on the specimens, and in about half a dozen that I have examined I have always found perfectly normal pericarps, seeds and embryos.

7. **K. eriantha,** *F. Muell. Rep. Babb. Exped.* 20. Apparently a stout shrub, the branches woolly-tomentose. Leaves crowded, sessile, linear or lanceolate, obtuse or acute, thick and soft, densely clothed with silky fulvous hairs, mostly about ½ in. long. Flowers solitary in the axils but crowded along the branches, enveloped in long woolly hairs. Fruiting perianth of *K. villosa*, the horizontal wings connected in a ring spreading to about ½ in. diameter and woolly all over. Styles usually 2.

S. Australia. Elizabeth Creek, *Babbage's Expedition;* between Stokes Range and Cooper's Creek, *Wheeler.*

This has the foliage of *K. lanosa* and *K. loboptera*, but still more silky, with the perianth (except in the long woolly hairs) entirely of *K. villosa*, of which it might almost equally well be considered as a variety only.

8. **K. villosa,** *Lindl. in Mitch. Trop. Austr.* 91. An undershrub or shrub, erect spreading or decumbent, more or less silky-villous tomentose or woolly, or the foliage at length nearly glabrous. Leaves

alternate, linear, obtuse, thick and soft in the typical form, terete or flattened, from under ¼ in. to about ½ in. long. Flowers solitary in the axils. Fruiting perianth depressed, from quite glabrous except a slight pubescence on the edge of the lobes to tomentose all over including the wings, the tube short and broad without vertical wings, the summit flat within the wings, the lobes very short and closed over the fruit, the dorsal wings united in a single entire or rarely lobed horizontal ring, membranous and very finely veined, spreading to from ¼ to nearly ¾ in. diameter. Styles 2 or 3, usually long, united at the base.—Moq. in DC. Prod. xiii. ii. 461 ; *Maireana tomentosa,* Moq. in Ann. Sc. Nat. Ser. 2, xv. 97. t. 13 ; and in DC. Prod. xiii. ii. 130. *Kochia tomentosa,* F. Muell. Rep. Babb. Exped. 20 ; *K. pubescens,* Moq. in DC. Prod. xiii. ii. 131 as to the Australian but not as to the S. African plant.

N. Australia. Sturt's Creek, *F. Mueller.*
Queensland. Narran river, *Mitchell;* Suttor desert, *F. Mueller.*
N. S. Wales. Lachlan river and Liverpool plains, *A. Cunningham;* Bogan river, *Leichhardt;* from the Murray, Darling and Lachlan rivers to the Barrier Range, *Victorian and other Expeditions.*
Victoria. Murray river, *F. Mueller;* Little river, *Fullagar;* Wimmera, *Dallachy;* Skipton, *Whan.*
S. Australia. St. Vincent's Gulf, *R. Brown* (imperfect specimens referred with doubt to *K. brevifolia*) ; Flinders' Range and towards Spencer's Gulf, *F. Mueller.*
W. Australia, *Drummond, n.* 125, *4th coll. n.* 242 ; Sharks Bay, *Milne;* N. W. of the Great Bight, *Delisser.*

The species varies exceedingly in foliage and indumentum as well as in the size of the perianth-wing, which, moreover, although usually quite entire, is sometimes irregularly lobed. The following forms appear the most distinct.

Var. *humilis,* a low undershrub with ascending stems not exceeding 6 in.—Not unfrequent in the desert country of Victoria and adjoining portion of N. S. Wales.

Var. *microcarpa.* Branches very cottony. Leaves small (under ¼ in.), rather thick and nearly glabrous. Fruiting perianth 1½ lines diameter within the wing, 3 lines diameter including the wing.—Darling and Lachlan rivers, *Victorian and other Expeditions.*

Var.? *tenuifolia,* F. Muell. Nearly glabrous. Leaves fine, 2 to 4 lines long. Perianth rather small and late in developing the wing. Perhaps a distinct species.—Darling Downs, *Woolls;* Curriwillighie, *Dalton;* Armadilla, *Barton;* Cooper's Creek, *Murray;* also in *Leichhardt's* collection.

The preceding *K. eriantha,* and the following four species might almost equally well be considered as varieties of *K. villosa,* the lines of demarkation between them being often rather vague.

9. **K. planifolia,** *F. Muell. Fragm.* i. 213. An erect divaricately branched shrub of 2 to 3 ft. (*Oldfield*), the branches and young foliage covered with a soft and dense woolly tomentum which wears off from the older leaves. Leaves oblong or oblanceolate, obtuse, contracted into a distinct petiole, ¼ to ½ in. long rather thick but flat. Fruiting perianth precisely that of *K. villosa,* glabrous or tomentose, the wing generally entire, membranous and attaining 5 to 6 lines diameter.

W. Australia. Murchison river, *Oldfield* (*Herb. F. Mueller*). Perhaps a variety only of *K. villosa.*

10. **K. sedifolia,** *F. Muell. in Trans. Vict. Inst.* 1855, 134, *and in Hook. Kew Journ.* viii. 205. A stout very densely branched shrub at-

taining 2 to 3 ft., white or fulvous all over with a rather close dense cottony wool. Leaves oblong-clavate, obtuse, soft thick and often nearly terete, contracted at the base but sessile, mostly $1\frac{1}{2}$ to 3 lines long. Flowers rather crowded, often two in the same axil. Flowering perianth globular, densely tomentose, not 1 line diameter. Fruiting perianth of *K. villosa*, but the wing usually more regularly circular, glabrous or tomentose, expanding to 3 or 4 lines diameter. Styles usually 3.

N. S. Wales. Lachlan river and Mount Goningberi, *Victorian Expedition;* Darling river, *Mrs. Ford;* Mount Murchison, *Bonney.*
Victoria. Murray river, *Dallachy.*
S. Australia. Murray Scrub, *F. Mueller.*

11. **K. appressa,** *Benth.* A much-branched shrub, more or less clothed with a short but soft cottony wool. Leaves very small, linear or oblong, erect and appressed, rarely exceeding 1 line, thick and soft, imbricate on the young shoots. Flowers solitary in the axils. Fruiting perianth like that of some varieties of *K. villosa,* usually glabrous, the annular wing expanding to about 4 or 5 lines diameter, very thin, with the veins very fine and not very conspicuous.

Victoria ? Lake Tyrrell, Murray desert, *Herb. F. Mueller.*
S. Australia. Margaret Creek, *Babbage's Expedition.*
Like the two preceding species, this only differs from *K. villosa* in the foliage.

12. **K. aphylla,** *R. Br. Prod.* 409. A rigid divaricately branched scrubby shrub with rather slender spinescent branches, the whole plant white with a short soft woolly tomentum or becoming at length nearly glabrous. Leaves minute and deciduous, rarely above $\frac{1}{2}$ line long, although on some luxuriant barren branches they may exceed 1 line, the older branches usually glabrous and leafless. Fruiting perianth entirely that of *K. villosa,* of which F. Mueller considers this plant as a variety only. It appears to me however at least as distinct as either of the three preceding ones.—Moq. in DC. Prod. xiii. ii. 131.

Queensland. Armadilla, *Barton.*
N. S. Wales. Darling desert, *Victorian and other Expeditions.*
Victoria. N.W. part of the colony, *L. Morton ;* Murray river, *Herrgott.*
S. Australia. Spencer's Gulf, *R. Brown ;* Murray Scrub, *Behr. ;* Flinders' Range, *Howitt's Expedition ;* Gawler Ranges, *Sullivan.*

13. **K. ciliata,** *F. Muell. Rep. Babb. Exped.* 20. Apparently a decumbent undershrub, the ascending branches softly woolly-villous. Leaves alternate, linear or lanceolate, obtuse or acute, silky-villous on both sides, 2 to 3 lines long. Flowers solitary in the axils, but crowded into a terminal leafy raceme. Fruiting perianth very flat, clothed with long soft hairs, scarcely 2 lines diameter including the annular wing, which is thick and hard, quite entire and bordered by a dense fringe of long soft hairs. Styles 2.—*Sclerolæna uniflora,* Lindl. in Mitch. Trop. Austr. 72, not of R. Br.

N. S. Wales. Darling desert, *Dallachy ;* Macquarrie river, *Mitchell.*
S. Australia. Emu springs and Margaret Creek, *Babbage's Expedition.*

14. **K. brachyptera,** *F. Muell. 2nd Gen. Rep.* 15. A prostrate undershrub, spreading to 1 ft. or more, with shortly ascending branches, clothed as well as the foliage with long soft spreading hairs, not forming the cottony wool of most species. Leaves alternate, linear, sessile, $\frac{1}{4}$ to $\frac{1}{2}$ in. long, flat and thinner than in most species. Flowers small, solitary in the axils, the perianth ovoid with 5 short erect obtuse membranous lobes. Stamens 5. Styles 2, connate at the base. Fruiting perianth about $1\frac{1}{2}$ lines diameter including the wings, the tube below the horizontal wing hemispherical, shortly hollowed at the base, with 5 vertical wings adnate to the horizontal one and tapering to the base, the summit very flat, bordered by an exceedingly narrow rather thick horizontal 5-angled wing, the angles very acute being the small points of the vertical wings.—*Sclerochlamys brachyptera,* F. Muell. in Trans. Phil. Inst. Vict. ii. 76; *Echinopsilon brachypterus,* F. Muell. Fragm. vii. 13.

N. S. Wales. Lachlan, Murray and Darling rivers, *Victoria and other Expeditions.*
Victoria. Murray river, *Dallachy.*
S. Australia. Cudnaka and Lake Torrens, *F. Mueller ;* Cooper's Creek, *Howitt s Expedition.*

15. **K. stelligera,** *F. Muell. Fragm.* vii. 13. A diffuse or procumbent undershrub spreading to 1 ft. or more, with numerous rather slender ascending branches not above 6 in. high, clothed as well as the young foliage with a white cottony wool wearing off from the older leaves. Leaves alternate, sessile, narrow-linear, erect or spreading, very soft, 1 to 3 lines long. Flowers small, solitary in the axils, but often crowded in woolly leafy tufts at the ends of the branches, the perianth nearly globular, woolly outside, with short lobes, about $\frac{1}{2}$ line long. Stamens included. Styles 2, rather long, scarcely united at the base. Fruiting perianth nearly $1\frac{1}{2}$ lines diameter, of a thick hard consistence, the teeth hemispherical, quite smooth, without vertical wings, the basal hollow small, the summit very flat, bordered by a very narrow rather rigid horizontal wing, with 10 to 12 short rigid equally radiating teeth or points.—*Maireana stelligera,* F. Muell. Fragm. i. 139; *Echinopsilon stelligerus,* F. Muell. Fragm. vii. 13.

N. S. Wales. Sand hills and Clay flats, Darling desert, *Victorian Expedition, Dallachy and Goodwin.*

7. CHENOLEA, Thunb.

(Echinopsilon, *Moq.* Eriochiton, *F. Muell.*)

Flowers hermaphrodite or polygamous. Perianth depressed-globular, membranous, with 5 (or 4 ?) inflexed lobes closing over the fruit, and 5 or fewer dorsal divergent spines or soft horn-like appendages, often unequal, sometimes very small or quite obsolete. Stamens 5. Styles 2 or 3, connate at the base. Fruit enclosed in the perianth, more or less depressed. Pericarp membranous. Seed horizontal or oblique ; embryo annular, the radicle often ascending over the cotyledonar end.— Diffuse or spreading undershrubs or shrubs. Leaves alternate, narrow, usually soft and silky-villous or woolly, rarely glabrous.   Flowers sessile and solitary in the axils, enveloped in cottony wool or long hairs.

The genus is spread over the temperate regions of the Old World, the Australian species apparently all endemic.   The allied African and European species require, however, further examination and comparison.   The name of *Chenolea* was originally applied to a Cape species, in which the perianth spines are reduced to minute tubercles or are frequently quite inconspicuous and were omitted in Thunberg's description. Moquin, having ascertained their occasional presence, transferred Thunberg's species to his *Echinopsilon*, that is to the section of Brown's *Kochia* with spinescent perianth-appendages, and reserved Thunberg's name for a Canary Island plant unknown to Thunberg, in which these tubercles or spines have not yet been detected and perhaps never exist.   This development of the spines is, however, so vague in several species, that it seems hopeless to distinguish the two genera I have re united under Thunberg's name, which is the oldest   The spineless species differ from *Enchylæna* chiefly in the texture of the perianth, and in almost the whole genus the seed is not so perfectly flat as in *Enchylæna* and *Kochia*, the radicle being slightly ascending or erect, although not so much so as in *Sclerolæna*.

Fruiting perianth without any appendages, or the spines reduced
  to minute tubercles.   Flowers in terminal leafy spikes.
Leaves at length glabrous.   Fruiting perianths enveloped in
  long fulvous woolly hairs . . . . 1. *C. carnosa.*
Leaves hoary-tomentose.   Fruiting-perianths enveloped in
  short dense cottony wool . . . . 2. *C. Dallachyana.*
Fruiting perianth with 3 long radiating soft woolly horns obtuse
  and turned up at the end . . . . 3. *C. tricornis.*
Fruiting perianth with 5 radiating spines or awns.   Perianth
  without appendages above the spines.
Perianth with 5 awns or slender spines enveloped in fulvous
  silky hairs 4 or 5 lines long . . . . 4. *C. eurotioides.*
Perianth with 5 short radiating spines, enveloped in fulvous
  hairs not longer than the perianth . . . . 5. *C. Muelleri.*
Perianth with 5 membranous notched or bifid appendages and
  5 radiating spines lower down, enveloped in dense cottony
  wool . . . . 6. *C. sclerolænoides.*

1. **C. carnosa,** *Benth.*   A small undershrub, with a short branching hard base, covered with the remains of old leaves, and erect or ascending flowering stems of 3 or 4 in., slightly cottony or silky-villous or at length glabrous.   Leaves sessile, distant in the lower part of the flowering stems, linear, acute or obtuse, rather thick and fleshy, glabrous when full grown, mostly 3 to 4 lines long, the floral ones shorter, lanceolate or ovate-lanceolate, the upper ones not exceeding the flowers. Flowers sessile and solitary in each axil, but crowded into a dense terminal leafy spike of 1 in. or more, and densely enveloped in long intricate silky-woolly hairs, often very shining.   Fruiting perianth depressed, membranous, almost scarious, about 1 line diameter without the wool, nearly 3 lines with it, with 5 lobes closing over the fruit and surrounded by a slightly raised horizontal ridge, but without appendages of any sort. Stamens 5.   Styles 2 (or sometimes 3 ?) united to the middle.   Pericarp depressed, membranous, with a few long hairs.   Seed horizontal; embryo broadly annular, with the radicle shortly rising over the cotyledonar end.—*Echinopsilon? carnosus,* Moq. in DC. Prod. xiii. ii. 136; *Trichinium carnosum,* Moq. in Herb. Hook.

W. Australia.   *Drummond, 4th coll. n.* 246.

2. **C. Dallachyana,** *Benth.* A shrub or undershrub, the branches clothed with cottony wool. Leaves sessile, linear, obtuse, thick and soft, tomentose but with more appressed and less intricate hairs than the stems, 2 to 3 lines long, the floral ones rather broader and longer, but exceeding the flowers. Flowers solitary or 2 together in the axils, but crowded into terminal leafy spikes of about 1 in. Fruiting perianth (perhaps not quite ripe) nearly globular, membranous, densely woolly-tomentose, about 1½ lines diameter including the wool; lobes broad connivent over the fruit, without any dorsal appendage. Stamens 5. Styles 2, connate at the base. Pericarp depressed, glabrous. Seed horizontal; embryo annular with a shortly ascending radicle.

N. S. Wales or **Victoria.** Murray river, *Dallachy.*

3. **C. tricornis,** *Benth.* A diffuse or divaricately-branched shrub, densely clothed with a soft white cottony wool. Leaves sessile, linear, rather acute, very soft and densely ·silky-villous, mostly 3 to 5 lines long. Flowers solitary in the axils of the stem-leaves. Flowering perianth depressed globular, 5-lobed, with 3 obtuse horizontal protube-rances. Stamens (always ?) 3 only. Styles 3, shortly united at the base. Fruiting perianth depressed, of a thin texture, densely tomentose, slightly hollowed at the base, the lobes horizontally closing over the fruit, with 3 dorsal horizontally radiating soft horns, each fully 2 lines long, obtuse and turned up at the end, the base occupying the whole depth of the perianth-tube. Seed horizontal; embryo nearly annular, the radicle produced beyond the cotyledonar end, but not turned up; albumen very scanty.

N. S. Wales. Clay flats, Darling river, *Dallachy.* This plant at first sight re-sembles the three-flowered state of *Sclerolœna biflora,* but the three floral rays are the three appendages of one perianth, not three perianths united at the base.

4. **C. eurotioides,** *F. Muell.* An undershrub or perhaps a shrub, with rather slender branches, closely and softly tomentose. Leaves sessile, narrow-linear, acute, very soft and silky-villous, the floral ones similar. Flowers solitary in the upper axils, densely enveloped in long straight silky-fulvous hairs which attain 4 or 5 lines. Perianth already slightly enlarged, about 1 line long, membranous, with 4 obtuse lobes, and 3 or 4 long unequal slender dorsal awns. Ripe fruit not seen, but in the enlarged perianth the seed is oblique, and the embryo already formed with a slightly ascending radicle.—*Echinopsilon eurotioides,* F. Muell. Fragm. vii. 13.

W. Australia, *Drummond, 5th coll. suppl. n.* 83.

5. **C. Muelleri,** *Benth.* A shrub or undershrub, with the aspect of some forms of *Sclerolœna dicrantha,* the rather slender branches clothed with a soft tomentum, more silky on the foliage and passing into longer hairs about the inflorescence. Leaves sessile, linear, obtuse or acute, very soft, ¼ to ½ in. long, the floral ones rather shorter and broader. Flowers solitary in the axils, densely enveloped in soft silky hairs of 1 to 2 lines in length. Perianth small, membranous, with (5 ?) very

short obtuse lobes, and 4 or 5 dorsal awns, which in the fruiting perianth become divaricate spines, unequal in length, the longest but little more than 1 line. Stamens 5. Styles 2, united to near the middle. Seed horizontal or oblique, with an ascending radicle.

N. Australia. Sturt's Creek, *F. Mueller.*

6. **C. sclerolænoides,** *F. Muell.* A small undershrub, much branched at the base, usually under 6 in. high, densely clothed with a soft rather loose woolly tomentum. Leaves sessile, linear, obtuse, soft, 2 to 3 or rarely 4 lines long. Flowers solitary in the axils, densely involved in woolly hairs forming when in fruit a globular mass of 2 to 3 lines diameter. Perianth concealed in the wool, with 5 short flat lobes closing over the fruit, 5 dorsal erect membranous bifid appendages, and lower down 5 linear sometimes pungent and spine-like appendages radiating from near the base, all usually concealed under the wool or the points of the spines slightly protruding. Styles 2, united at the base. Seed horizontal; embryo flat, annular, the radicle either not at all or only very slightly rising above the cotyledonar end.—*Eriochiton scherolænoides,* F. Muell. Second Rep. 15; *Echinopsilon sclerolænoides,* F. Muell. in Trans. Phil. Inst. Vict. ii. 75.

N. S. Wales. Murray and Darling rivers, *Dallachy and Goodwin,*
Victoria. Murray river at the mouth of the Golgol, *F. Mueller.*
S. Australia. Cudnaka, Lake Torrens, *F. Mueller.*
W. Australia, *Drummond (Herb. F. Mueller).*

The *Sclerolæna uniflora,* Lindl. (not of Br.) referred here through some mistake by F. Mueller, Fragm. vii. 13, is the *Kochia ciliata.*

8. BABBAGIA, F. Muell.

Flowers hermaphrodite. Perianth urceolate, hard when in fruit, with 4 (or 5 ?) small membranous lobes, and 2 or 3 dorsal membranous stipitate wings more or less vertical. Stamens 4 (or 5 ?). Styles 2, connate at the base. Fruit enclosed in the perianth. Pericarp membranous. Seed horizontal; testa membranous; embryo nearly annular, enclosing a mealy albumen, the radicle slightly ascending above the cotyledonar end.—Diffuse glabrous undershrub or shrub. Leaves linear, alternate. Flowers solitary in the axils, sessile, without bracts.

The genus is limited to a single species, endemic in Australia.

1. **B. dipterocarpa,** *F. Muell. Rep. Babb. Exped.* 21. A small much-branched diffuse undershrub or spreading shrub, glabrous except sometimes a slight wool in the axils of the leaves, and more or less glaucous. Leaves linear or oblanceolate, thick or semiterete, under 3 lines long, often crowded on the young branches. Fruiting perianth rather more than 1 line long, the hollow base about ½ line long and as much diameter, closed under the fruit, the fruit-bearing part depressed-globular, about ¾ line diameter, the broad membranous semicircular wings expanding to about 3 lines diameter, each wing contracted into a thick base and placed obliquely or vertically with reference to the perianth,

apparently by the torsion of that base. Fruit much depressed. Seed very flat.

N. S. Wales. Mount Murchison (*Bonney?*); near Stokes Range, *Wheeler*.
S. Australia. Stuart's Creek and Elizabeth Creek, *Babbage ;* Cooper's Creek, *Howitt's Expedition.*

9. DIDYMANTHUS, Endl.

Flowers hermaphrodite. Perianth cylindrical, with 5 short broad thick lobes, closing over the fruit, and bearing on their backs distinct horizontal wings. Stamens 3 to 5. Styles 2, connate at the base. Fruit inclosed in the calyx. Pericarp membranous. Seed ovate, vertical ; testa membranous ; embryo horseshoe-shaped, almost annular, enclosing a small quantity of mealy albumen ; radicle erect.—Shrub or undershrub, more or less cottony. Leaves small, mostly opposite, entire. Flowers two together, sessile in the axils, connate by their base and horizontally divaricate.

The genus is limited to a single species endemic in W. Australia.

1. **D. Roei,** *Endl. Nov. Stirp. Dec.* 8, *and Iconogr. t.* 100. An erect branching shrub or undershrub, rarely above 1 ft. high, the branches and young leaves more or less hoary or white with close woolly hairs. Leaves mostly opposite or nearly so, sessile, lanceolate, acute, thick, about 2 lines long, silky-villous or nearly glabrous when full-grown. Fruiting perianths divaricate, forming as it were a single cylinder attached by the centre, each perianth varying from 1 to 2 lines in length, with 5 horizontal membranous wings spreading to a diameter of from $1\frac{1}{2}$ to 3 lines. Stamens in the flowers examined usually 3, rarely 4. Styles united to the middle.—Moq. in DC. Prod. xiii. ii. 124.

W. Australia, *Roe, Drummond, n.* 130, 148, 208. The fruiting perianths vary much in size, although the two of each pair are always similar. At the time of flowering they are very short and almost free from each other.

10. SCLEROLÆNA, R. Br.

(Kentropsis, *Moq.;* Dissocarpus, *F. Muell.*)

Flowers hermaphrodite. Perianth at first nearly globular, at length turbinate or depressed, somewhat compressed, not succulent and usually hard, with 5 short inflexed lobes, and 2 dorsal opposite divergent spines, either both equal or one smaller or scarcely developed. Stamens 5. Styles 2 or rarely 3, connate at the base. Fruit globular or depressed ; pericarp membranous. Seed usually globular or depressed at the base, with a more or less prominent ascending or erect rostellum. Testa membranous. Embryo almost annular, surrounding a mealy albumen, the radicle ascending above the cotyledonar end into the rostellum of the seed.—Undershrubs or shrubs, either prostrate decumbent or divaricately branched. Leaves alternate, narrow, usually soft and silky-villous or woolly. Flowers sessile in the axils, enveloped in cottony wool or soft hairs.

The genus is limited to Australia. It is nearly allied to *Anisacantha*, but the spines of the perianth (which I always find dorsal, not terminating the lobes as described by Moquin) are two only, and the seed is not so distinctly vertical.

Flowers solitary in the axils.
  Fruiting perianth 1 to 1½ lines long and usually as broad, tomen-
    tose or nearly glabrous.
    Leaves mostly linear-clavate, thick and obtuse. Spines of the
      perianth very short . . . . 1. *S. uniflora.*
    Leaves narrow-linear, rather acute. Spines 1 to 3 lines long,
      glabrous or nearly so . . . . 2. *S. diacantha.*
  Fruiting perianth 2 lines long, densely covered as well as the spines
    with long hairs . . . . 3. *S. lanicuspis.*
  Fruiting perianth 2 to 3 lines diameter, enveloped in a thick mass
    of white cottony wool. Spines ¼ to ½ in. long . . . . . . 4. *S. bicornis.*
Flowers 2 or 3 together united at the base and diverging horizon-
  tally . . . . 5. *S. biflora.*
Flowers several together united in a hard globular mass . . . . 6. *S. paradoxa.*

S. coriacea, Moq. in DC. Prod. xiii. ii. 123, from the Barren islands (off the S. coast?) *Herb. Mus. Par.,* is unknown to me, and having no spines to the perianth may not be a true congener. Branches densely woolly, hoary. Leaves imbricate, elliptical or oblong-elliptical, obtuse or rather acute, coriaceous, white and very tomentose, 3 to 4 lines long, 1 to 1½ lines broad. Flowers solitary, tomentose, ¼ line long. Fruiting perianth without spines. Styles much exserted, villous at the base with hispid hairs.

1. **S. uniflora,** *R. Br. Prod.* 410. A diffuse or decumbent much-branched undershrub or shrub, not exceeding 1 ft., covered with a dense hoary or fulvous tomentum. Leaves sessile, linear-clavate, very obtuse, thick and soft, mostly 2 to 4 lines long. Flowers small, solitary in the axils. Perianth oblique, very shortly and irregularly 4- or 5-toothed. Styles 2, connate to the middle. Fruiting perianth scarcely 1½ lines long including the broad hollow base or stipes, hard, tomentose or nearly glabrous, the mouth very oblique, bearing at the top of the tube 1 or 2 opposite very small dorsal spines, sometimes almost obsolete. Embryo horizontally annular at the base, with the radicle turning up over the cotyledonar end and erect.—Moq. in DC. Prod. xiii. ii. 123, partly.

S. Australia. Fowler's Bay, *R. Brown.*
W. Australia. Dirk Hartog's Island, *A. Cunningham;* Sharks Bay, *Milne.*

Moquin has correctly named Cunningham's West Australian specimens in Herb. Hooker, but in giving the station, Liverpool Plains, on Cunningham's authority, in the Prodromus from Herb. DC., he must have had in view some other plant, probably the following species, which may indeed prove to be only a variety of it.

2. **S. diacantha,** *Benth.* A diffuse or prostrate undershrub, densely clothed with a soft fulvous or white tomentum more silky on the foliage. Leaves sessile, linear, mostly acute, very soft, sometimes rather thick and ¼ to ½ in. long, sometimes longer and narrower, the floral ones not broader. Flowers solitary, the perianth broadly campanulate or almost urceolate, about ½ line long and ¾ line broad, the lobes very short membranous and slightly inflexed, the 2 opposite dorsal spines already long and nearly as deep as the perianth-tube. Anthers 5, half exserted.

Styles 2 (or 3 ?) very shortly connate at the base. Fruiting perianth hard, tomentose, depressed and slightly compressed at the top, about 1 line long, and the flat slightly hollow base 1 to 1½ lines diameter, closed at the orifice, the two opposite dorsal spines diverging or divaricate, nearly equal and varying in the typical form from 1 to nearly 2 lines in length. Seed globular or oblique, with an ascending or erect rostellum. Embryo horizontally annular with an erect radicle.—*Anisacantha diacantha*, Nees in Pl. Preiss. i. 635 ; *Kentropsis diacantha*, Moq. in DC. Prod. xiii. ii. 138, (both from Moquin's descr.) ; *Anisacantha kentropsidea*, F. Muell. in Trans. Vict. Inst. 1855, 133, and in Hook. Kew. Journ. viii. 204, reduced to *A. diacantha* in Fragm. vii. 14.

Queensland. Cape river, *Bowman;* Armadilla, *Barton;* Box Forest, *Leichhardt.*
N. S. Wales. Lachlan and Murray rivers, *Herb. F. Mueller.*
Victoria. Bacchus Marsh, *F. Mueller.*
S. Australia. Tumby and Holdfast Bays, *F. Mueller.*
W. Australia. Quangen plains, *Preiss, n.* 2379. I have not seen these specimens, but the descriptions given quite agree with the eastern plant. The Murchison river specimens referred here by F. Mueller, belong however to the nearly allied true *S. uniflora*, Br.

Var. *longispina.* ? Perianth rather longer and sometimes narrower, very tomentose or nearly glabrous. Spines slender, 2 to 4 lines long.—Darling desert, *Victorian Expedition ;* Murray river, *F. Mueller ;* Wimmera, *Dallachy ;* Gawler ranges, *Sullivan ;* towards Spencer's gulf, *Warburton.*

3. **S. lanicuspis,** *F. Muell.* A low undershrub or shrub, the stems rather stout and not exceeding 6 in. in our specimens, densely clothed with a loose cottony wool, more silky and appressed on the leaves. Leaves rather crowded, linear, thick and soft, often above ½ in. long. Fruiting perianth similar to that of *S. diacantha* but rather larger, 1½ to 2 lines long and very densely clothed with long silky or woolly hairs, the two dorsal opposite spines divergent, nearly equal, 1 to 2 lines long. —*Anisacantha lanicuspis,* F. Muell. Fragm. ii. 170 ; *Kentropsis eriacantha,* F. Muell. l.c. 140.

N. S. Wales. Darling river to the Barrier Range, *Victorian Expedition ;* Ballandool river, *Locker.*

4. **S. bicornis,** *Lindl. in Mitch. Three Exped.* ii. 47. A stout shrub, with divaricate rather thick branches, clothed with a short but soft cottony wool. Leaves very narrow-linear, acute, semiterete, often above ½ in. long, tomentose when young, becoming glabrous when full grown. Flowers solitary in the axils, the perianth 1½ lines long at the time of flowering. Fruiting perianth 2 to 3 lines diameter within the dense white cottony wool which covers it and almost doubles its size, very hard, nearly globular but slightly compressed, with 2 rigid divaricate dorsal spines, varying in length from ¼ to ½ in. or sometimes nearly ¾ in. Seed horizontal, with a long ascending rostellum.—Moq. in DC. Prod. xiii. ii. 123 ; *Kentropsis lanata,* Moq. Chenop. Enum. 83, and in DC. Prod. xiii. ii. 138 ; *Anisacantha bicornis,* F. Muell. Fragm. vii. 14.

N. Australia. Sturt's Creek and Plains of Promise, *F. Mueller.*
Queensland. Curriwillighie, *Dalton.*

N. S. Wales. Molle's Plains, *A. Cunningham;* between the Darling river and the Barrier Range, *Mitchell, Victorian Expedition;* Mount Murchison, *Bonney;* Ballandool river, *Locker.*
S. Australia. Cooper's Creek, *Murray.*

5. **S. biflora,** *R. Br. Prod.* 410. A procumbent or spreading branching shrub, clothed with a short close cottony wool, rather looser and more silky on the foliage of luxuriant shoots. Leaves sessile, narrow-linear, acute or obtuse, very soft, $\frac{1}{4}$ to $\frac{1}{2}$ in. long. Flowers 2 or rarely 3 together in the axils, connate at the base and divaricate at a very early stage. Perianth at the time of flowering about $\frac{3}{4}$ line long and broad, deeply divided into 5 lobes inflexed at the end, densely enveloped in cottony wool. Styles 2, connate at the base. Fruiting perianths horizontally diverging from and continuous with a common very broad hollow base or peduncle of about 1 line, the perianths themselves about $1\frac{1}{2}$ lines long, very hard and thick at the base, the orifice nearly closed by the inflexed thinner lobes without any or sometimes with 1 or 2 dorsal minute tubercles or short spines. Pericarp depressed. Seed horizontal or somewhat oblique with an ascending rostellum. Embryo annular with an ascending or erect radicle.—Moq. in DC. Prod. xiii. ii. 123 ; *Dissocarpus biflorus,* F. Muell. in Trans. Phil. Inst. Vict. ii. 75.

N. S. Wales. Sandhills, Darling river, *Victorian Expedition.*
Victoria. Murray river, *F. Mueller.*
S. Australia. Petrel Bay, *R. Brown.*

6. **S. paradoxa,** *R. Br. Prod.* 410. A decumbent much-branched undershrub, rarely exceeding 1 ft., densely clothed with a loose cottony wool. Leaves sessile, narrow-linear, obtuse, soft and woolly or rarely becoming glabrous with age, from under $\frac{1}{4}$ to nearly $\frac{1}{2}$ in. long. Flowers in dense axillary clusters, the perianths small, deeply lobed. Styles 2, connate to the middle. Fruiting perianths 10 to 20 together, very hard at the base, connate into a globular cottony or woolly mass of 5 or 6 lines diameter, each perianth with 1 or 2 small dorsal spines shortly protruding from or almost concealed in the wool. Seed nearly globular with a short ascending rostellum ; embryo forming a complete circle with the radicle turned upwards over the cotyledonar end.—Moq. in DC. Prod. xiii. ii. 123.

N. S. Wales. Lachlan river, *A. Cunningham* (not yet in flower, and therefore uncertain) ; *Herb. F. Mueller* (collector not named) ; Darling river, *Victorian Expedition ;* Mount Murchison, *Bonney.*
Victoria. Murray river, *Dallachy.*
S. Australia. St. Vincent's Gulf, *R. Brown ;* Murray Scrub, *F. Mueller.*

11. THRELKELDIA, R. Br.

(Osteocarpus, *F. Muell.*)

Flowers hermaphrodite. Perianth urceolate or cylindrical, hard when in fruit, with 4 or 5 short membranous lobes, without any dorsal appendages or in one species with 5 small erect spines. Stamens 5 or fewer. Styles 2 or 3, connate at the base. Fruit enclosed in the perianth, more or less depressed. Pericarp membranous. Seed horizontal or oblique. Testa membranous; embryo annular or nearly so, surrounding a mealy albumen; radicle ascending or level with the cotyledonar end or descending.—Diffuse procumbent or trailing undershrubs, quite glabrous or in one species scabrous. Leaves narrow, alternate. Flowers solitary in the axils, closely sessile but not obliquely adnate as in *Anisacantha*. Bracts none.

The genus is limited to Australia. It has the hard perianth of *Sclerolœna* and *Anisacantha*, but has either no spines or (in one species) very short erect ones, and differs moreover from the former in the want of any cottony wool, and from the latter in the seed. If, however, slight differences in the perianth and seed are taken into account, the four species here included might be regarded as forming as many genera.

Perianth without appendages or hollow base. Seed with an
    ascending rostellum.
    Fruiting perianth 1½ lines long, scarcely oblique at the top . . 1. *T. diffusa.*
    Fruiting perianth ¾ line long, very oblique and gibbous on one
      side at the top . . . . 2. *T. salsuginosa.*
Perianth with a large hollow base below the fruit.
    Perianth 1 to 1½ lines long, with 5 small dorsal erect spines.
      Seed horizontal . . . . 3. *T. brevicuspis.*
    Perianth 3 to 4 lines long, without spines. Seed very oblique,
      with a descending radicle . . . . 4. *T. haloragoides.*

1. **T. diffusa,** *R. Br. Prod.* 410. A prostrate diffuse or trailing undershrub, sometimes very small, sometimes extending to 1 or 2 ft., with shortly ascending branches, the whole plant glabrous and somewhat fleshy. Leaves rather crowded, linear, mucronate-acute or obtuse, thick and semiterete, contracted at the base, 2 to 4 lines or rarely ½ in. long. Flowers small, the perianth tubular, ¾ line long, obliquely contracted above the middle, with 4 or 5 broad membranous erect lobes shortly ciliate. Styles 2 or 3, connate to above the middle. Fruiting perianth ovoid, about 1½ lines long, hard and ribbed when dry, said to be fleshy when fresh, the orifice open or half-closed by the withered lobes, without appendages. Fruit enclosed in the perianth and not raised above the base, nearly globular. Embryo horseshoe-shaped or annular and horizontal at the base, with an ascending radicle.—Moq. in DC. Prod. xiii. ii. 127; Nees in Pl. Preiss. i. 635 ; Hook. f. Fl. Tasm. i. 315.

Victoria. Wilson's Promontory, *F. Mueller.*
Tasmania. Kent's Group, Bass's Straits, *R. Brown ;* seashore E. of George Town, *Gunn.*
S. Australia. Holdfast Bay, *F. Mueller.*
W. Australia. King George's Sound, *R. Brown ;* Bald Island, *Oldfield ;* N. W. of the head of the Great Bight, *Delisser ;* Swan river, *Preiss, n.* 1235 (*Moquin*).

Var. *latifolia.* Leaves flatter, broader and more petiolate, but small.—Lucky Bay, *R. Brown*; Dirk Hartog's Island, *Milne.*

2. **T. salsuginosa,** *F. Muell. Fragm.* vii. 12. A prostrate or diffuse undershrub, more slender than *T. diffusa,* and quite glabrous. Leaves narrow-linear, mucronate-acute or obtuse, contracted into a short petiole, under ¼ in. long. Flowers very small, the perianth not ½ line long, very oblique, with 4 short thin lobes. Styles 2, shortly united at the base. Fruiting perianth about ¾ line long, nearly globular but

very obl:que at the top with an obtuse hollow protuberance on one side, quite smooth even when dry and rather thick and hard. Seed globular at the base with a rostellum ascending into the protuberance of the perianth. Embryo annular at the base with an ascending radicle as in *T. diffusa.—Osteocarpum salsuginosum,* F. Muell. 2nd Gen. Rep. 15, and in Trans. Phil. Inst. Vict. ii. 77, also Pl. Vict. Lith. iii. t. 79, according to F. Muell. Fragm. vii. 12 (I have not seen the plate).

Victoria or **N. S. Wales.** Murray river, *F. Mueller, Dallachy.*

3. **T. brevicuspis,** *F. Muell.* A small glabrous undershrub with slender ascending branching stems not above 6 in. in our specimens. Leaves narrow-linear, semiterete, acute, under ½ in. long. Perianth very small and campanulate when in flower. Styles 2, connate to the middle. Fruiting perianth urceolate, about 1½ lines long, hard when dry, with 5 erect nearly equal dorsal spines shorter than the tube and connected by a narrow horizontal ring, constricted under the spines, the whole of the part below the constriction consisting of a hollow base, 10-ribbed outside, closed inside under the fruit, and again closed inside over the fruit, the short membranous lobes erect within the base of the spines. Fruit very flat within the base of the spines. Seed horizontal, with an annular embryo.—*Anisacantha brevicuspis,* F. Muell. Fragm. iv. 150; *Kentropsis brevicuspis,* F. Muell. l. c.

Queensland. Armadilla, *Barton;* Cape River, *Bowman.*

Notwithstanding the spines, this species appears to me to agree better with *Threlkeldia* than with any other genus. The habit is that of *T. diffusa* or of *Anisacantha echinopsila,* differing from the latter in the hollow base of the perianth, in the erect spines, and in the very flat horizontal seed.

4. **T. haloragoides,** *F. Muell. Herb.* Stems elongated, probably procumbent or ascending, much stouter than in the other species and at least 1 to 2 ft. long, more or less scabrous as well as the foliage with scattered asperities, otherwise glabrous. Leaves linear or linear-lanceolate, contracted at the base, flat, rather thick, from about ¼ in. to above ½ in. long. Flowers axillary or somewhat lateral, only seen when already somewhat advanced, narrow tubular and 2 lines long. Fruiting perianth 3 to 4 lines long, thick and very hard, constricted above a short depressed-globular broad hollow base, then cylindrical, with the orifice oblique and open bordered by the shrivelled lobes, closed inside at the constriction below the fruit, and again over the fruit below the top. Seed oblique or nearly vertical. Embryo folded or almost horseshoe-shaped with the radicle descending.

W. Australia, *Drummond, n. 55 and 5th coll. n.* 438.

Notwithstanding the reversed direction of the embryo this species appears to me to be much more closely related to *Threlkeldia* than to any other genus.

12. ANISACANTHA, R. Br.

Flowers hermaphrodite. Perianth urceolate or ovoid, hard when in fruit, obliquely attached at the base, with 4 or 5 short membranous lobes and 3 to 5 dorsal divergent unequal spines, 1 usually much smaller than the others or reduced to a tubercle. Stamens 5 or fewer. Styles 2 or 3, connate at the base in a column usually persistent and hardened. Fruit enclosed in the perianth, usually ovoid. Pericarp membranous. Seed vertical, somewhat compressed; testa membranous; embryo annular or nearly so, surrounding a mealy albumen, the radicle erect.—Intricately branched shrubs or diffuse undershrubs, glabrous or very rarely villous, especially the young shoots. Leaves linear, alternate, sessile. Flowers solitary in the axils. Fruiting perianths closely sessile and often almost adnate at the base to the stem and to the subtending leaf. Bracts none.

The genus is limited to Australia.

Leaves linear or linear-lanceolate, flat, contracted at the base.
  Perianth-spines 5 or 4, one often very small . . . . 1. *A. muricata.*
Leaves small, linear, rather flat, with a broad persistent hardened
  base. Perianth-spines 3 . . . . 2. *A. Drummondii.*
Leaves linear-terete or semiterete
  Perianth-spines 3 rarely 4, one sometimes very small.
    Shrubby. Perianth above 1 line long.
      Smallest spine above 1 line long . . . . 3. *A. divaricata.*
      Smallest spine reduced to a tubercle or rarely nearly 1 line
        long . . . . 4. *A. bicuspis.*
    Small diffuse undershrub. Perianth not exceeding 1 line,
      the spines short and slender . . . . 5. *A. glabra.*
  Perianth-spines 5, short. Small diffuse undershrub. Perianth
    under 1 line long . . . . 6. *A. echinopsila.*

1. **A. muricata,** *Moq. Chenop. Enum.* 84, *and in DC. Prod.* xiii. ii. 122. A broad bushy or spreading shrub of 2 or 3 ft., with numerous intricate flexuose branches, the typical form quite glabrous and somewhat glaucous, or the young shoots slightly villous. Leaves linear, flat but rather thick, mucronate-acute, contracted at the base, from scarcely above ¼ in. to nearly 1 in. long. Fruiting perianth adnate by a broad oblique base, the hard tube rarely above 1 line long, the membranous lobes short, the dorsal spines 4 or 5, very unequal and spreading, the longest 3 to 6 lines long, the smallest very short, and often the 2 smallest united at the base.—*A. quinquecuspis,* F. Muell. in Trans. Vict. Inst. 1855, 134, and in Hook. Kew Journ. viii. 204.

Queensland. Armadilla, *Barton.*
N. S. Wales. Lachlan river, Liverpool and Molle's plains, *A. Cunningham;* Colroy Creek, *Leichhardt;* Liverpool plains, *C. Moore;* Darling and Murray desert, where the old plants detached by the winds and rolling over the desert plains have received the name of " Roley-poley," *F. Mueller* and others.

Var. *villosa.* The whole plant, at least in young specimens, softly villous.—*A. gracilicuspis,* F. Muell. Fragm. ii. 170.—Mackenzie Downs, *F. Mueller.*

2. **A. Drummondii,** *Benth.* A small much branched shrub, glabrous except a few long spreading hairs on the leaves. Leaves linear, flat but thick, rarely almost terete, not 3 lines long in the only specimen seen, with a broad hard base which persists after the leaf has fallen.

Fruiting perianth glabrous, about ½ line long with 3 slender spreading spines, of which 2 are from 2 to 3 lines long, the third much smaller.

W. Australia, *Drummond* (*Herb. F. Mueller*).

A. divaricata, Moq. in DC. Prod. xiii. ii. 122, described by Moquin from a specimen of Drummond's in De Candolle's herbarium received from Kew Gardens, but certainly not Brown's plant of that name, is most probably the above species. There is now no *Anisacantha* from Drummond in the Kew herbarium, but several Chenopodiaceæ which had been lent from it to Dr. Bunge were unfortunately lost by a shipwreck in the Baltic.

3. **A. divaricata,** *R. Br. Prod.* 410. A diffuse or divaricately branched shrub, glabrous and somewhat glaucous like *A. muricata,* but usually more compact and more densely beset with the prickles of the perianths. Leaves linear-terete, mucronate-acute, often above ½ in. long. Fruiting perianth closely sessile with a broad oblique base, 1 to 1½ lines long, with 3 or 4 very unequal divaricate spines, rather finer than in *A. muricata,* the longest often ⅓ in. long, but sometimes none above ¼ in., the smallest only 1 to 2 lines, the fourth when present very slender and small; lobes of the perianth usually erect connivent, minutely ciliate.—*A. erinacea,* Moq. in DC. Prod. xiii. ii. 122 ; *A. tricuspis,* F. Muell. in Trans. Vict. Inst. 1855, 133, and in Hook. Kew Journ. viii. 204.

Queensland. Suttor river, *F. Mueller, Bowman ;* Bokhara Creek, *Leichhardt ;* Darling Downs, *Lau.*

N. S. Wales. Lachlan river and Molle's plains, *A. Cunningham.*

Victoria. Murray river near the Golgol, *F. Mueller.*

S. Australia. Head of Spencer's Gulf, *R. Brown ;* Murray river near Morunda, *F. Mueller.*

4. **A. bicuspis,** *F. Muell. in Trans. Vict. Inst.* 1855, 133, *and in Hook. Kew Journ.* viii. 204. A rigid stout but compactly branched shrub or undershrub, the specimens seen not above 6 in. high and quite glabrous. Leaves linear, semiterete, acute, from ¼ in. to above ½ in. long. Flowers closely sessile and semi-adnate as in the preceding species but longer. Fruiting perianth 2 to 3 lines long, the tube ovoid, very hard, with 3 diverging spines, of which 2 (either equal or unequal) varying from ¼ to 1 in. long, the third small, sometimes reduced to a tubercle, sometimes 1 line long ; perianth-lobes membranous, obtuse, erect or inflexed. Styles 2, united to above the middle into a hard column. Pericarp usually hardened. Seed obliquely erect with a superior radicle.

S. Australia. Salt plains, Cudnaka, *F. Mueller* (with the perianth about 2 lines long and the longest spines not above ½ in.) ; between Stokes Range and Cooper's Creek, *Wheeler* (with the perianth fully 3 lines long and the longest species ¾ to 1 in.). Both are single specimens (in Herb. F. Mueller), and may prove to be varieties only of *A. divaricata.*

5. **A. glabra,** *F. Muell. Herb.* A small undershrub, closely re-sembling *A. echinopsila,* with the same foliage and inflorescence, but the fruiting perianth, although scarcely longer, has a much broader, less oblique base, somewhat compressed at the top, with 2 opposite diverging

very unequal spines, 1½ to 2 lines long, and a third very small one, the perianth thus approaching in form that of *Sclerolæna,* but the seed is vertical or slightly oblique as in *A. echinopsila.—Kentropsis glabra,* F. Muell. Fragm. i. 139.

N. Australia. Upper Victoria river and Sturt's Creek, *F. Mueller.*

6. **A. echinopsila,** *F. Muell. Fragm.* vii. 14. A much-branched diffuse or prostrate undershrub, spreading to above 1 ft. diameter, the branches ascending to near 6 in., or sometimes the whole plant not exceeding 2 or 3 in., glabrous and somewhat glaucous, or very rarely the young shoots slightly pubescent. Leaves narrow-linear, semiterete, mucronate-acute or obtuse, mostly about ¼ in., rarely ½ in. long. Flowers very small, closely sessile, with an oblique base. Styles 2, united to the middle into a column hardened at the base. Fruiting perianth hard, scarcely 1 line long, the tube often produced below its insertion into a small protuberance or short spur ; lobes 5, short, membranous, with 5 dorsal radiating unequal spines slightly connected in a ring round the summit of the tube, the longest rarely above 1 line long. Seed vertical or slightly oblique, with a superior radicle.—*Echinopsilon anisacanthoides,* F. Muell. in Trans. Phil. Inst. Vict. ii. 76.

Queensland. Desert of the Suttor, *F. Mueller ;* Crocodile Creek, *Bowman ;* Rockhampton, *O'Shanesy.*

N. S. Wales. Darling desert, *Dallachy ;* Ballandool river, *Locker.*

Tribe 3. Salicornieæ.—Branches articulate, fleshy. Leaves none. Flowers more or less immersed. Testa various. Embryo curved or folded, with little or no albumen.

13. SALICORNIA, Linn.

(Halocnemum, *Bieb. ;* Arthrocnemum, *Moq.*)

Flowers hermaphrodite or polygamous. Perianth thin and membranous or at length thickened and fungous, with 2 to 5 teeth or lobes. Stamens 1 or 2. Styles 2 (rarely 3) united in a column or cone at the base. Fruit enclosed in the unchanged or slightly enlarged perianth. Seed ovoid or nearly globular, often compressed, oblique or vertical. Testa crustaceous or thin. Embryo folded or semicircular, either without albumen or with a small quantity, either lateral or within the curve of the embryo.—Succulent herbs with a hard base or shrubs. Branches articulate, leafless, each article usually concave at the upper end and often dilated into a circular border or into 2 opposite protuberances or lobes (rudiments of opposite leaves) and receiving the next article in the concavity, the articles becoming at length united into a continuous woody stem in the shrubby species ; the flowering articles shorter, usually more dilated, forming more or less compact terminal spikes. Flowers usually 3 together, rarely 5 or 7, on each side of each article and more or less immersed in its base, without bracts or bracteoles.

The genus, which with Hooker and others I take in the Linnean sense, including the whole tribe of Salicorrnieæ as defined by Moquin in DC. Prod., has a wide range over the seacoasts and saline marshes of most parts of the globe, more especially in the Old World. Of the seven Australian species, one extends to New Zealand, another is possibly the same as an Asiatic one, the remaining five appear to be endemic. The species, however, require much further investigation from living plants before the value of the differences in the flowers, which are considerable, can be properly appreciated. A large proportion of the dried specimens before me are not in a state to be satisfactorily examined.

SECT. 1. **Halocnemum.**—*Perianths not dilated at the top, usually narrow. Spikes usually short. Flowers in threes, all or the central one hermaphrodite.*

Spikes ovoid or shortly oblong, with few articles, the margins forming opposite thick triangular-conical lobes.
Stout shrub. Spikes 5 to 6 lines thick . . . . 1. *S. robusta.*
Intricately branched shrub. Spikes not 2 lines thick . . . . . 2. *S. arbuscula.*
Spikes oblong-cylindrical, the articles numerous, very short and closely imbricate, the margins dilated into semicircular opposite scarious scales . . . . 3. *S. cinerea.*

SECT. 2 **Arthrocnemum.**—*Perianths at length dilated into a flat transverse or oblique top. Spikes cylindrical, often elongated.*

Flowers in threes, all or the central one hermaphrodite.
Spikes continuous, the margins of the articles broad obtuse and not prominent. 4. *S. leiostachya.*
Spikes interrupted by the very prominent 2-lobed margins of the articles . . . . 5. *S. bidens.*
Flowers in threes, unisexual (diœcious?). Articles of the spikes with prominent 2-lobed margins . . . . 6. *S. tenuis.*
Flowers in fives or sevens, mostly hermaphrodite. Articles of the spikes with slightly prominent annular margins . . . . . . 7. *S. australis.*

Besides the above, I am unable to identify the *Halocnemum australasicum*, Moq. Chenop. Enum. 110, and in DC. Prod. xiii. ii. 149, from King George's Sound, *Herb. Mus. Par*, which I have not seen. It is certainly not *S. indica*, Br. It is described as having herbaceous ascending stems, with the articles of the branches 2-lobed, those of the spikes broad concave and very obtuse (as in *S. leiostachya ?*), but the spikes are said to be very large, conico-cylindrical and 6 to 15 lines long, which does not agree with any of our species. The perianth is described as that of the section *Halocnemum.*

SECT. 1. HALOCNEMUM.—Perianths not dilated at the top, usually hyaline and narrow. Spikes usually short.

1. **S. robusta,** *F. Muell. Fragm.* vi. 251, *and Pl. Vict. t.* 83, *ined.* (*F. Mueller*). Shrubby and much stouter than the other species, the articles ½ to ¾ in. long, with two opposite prominent angles ending in opposite thick triangular lobes or rudimentary leaves projecting to nearly 2 lines. Spikes ovoid or nearly globular, ½ in. diameter, with very thick prominent points of the floral scales or lobes of the floral articles. Flowers in threes, partially immersed and shorter than the subtending scales. Perianths adnate to the article above them at the base, with narrow hyaline lobes, the two lateral perianths 2-lobed enclosing each 1 stamen only, the central one (always?) 3-lobed, enclosing the pistil and 1 stamen. Seed nearly globular; embryo vertical, enclosing a mealy albumen, the radicle ascending beyond the cotyledonar end.— *Arthrocnemum triandrum*, F. Muell. Fragm. i. 139.

N. S. Wales. Salt desert near the junction of the Darling and Murray, *F. Mueller.* **Victoria.** Lake Victoria, Murray river, *F. Mueller.*

I have not yet received the plate above referred to.

2. **S. arbuscula,** *R. Br. Prod.* 411. A bushy erect shrub, from under 6 in. to 2 or rarely 3 or 4 ft. high, with numerous rather slender short but intricate branches. Articles often ¼ in. long in the principal branches, 1 to 2 lines in the smaller ones, dilated at the top but without prominent lobes. Spikes terminal, thick, ovoid or oblong, rarely above ¼ in. long but varying much in diameter, consisting of 2 to 6 articles deeply excavated and cup-shaped at the top with more or less prominent lobes or scales. Flowers in threes, all monandrous and hermaphrodite, at first shorter than the scale but at length somewhat exserted, all cohering with each other at the base and immersed in and adnate to the article above them, with short free thin lobes, opening in fruit so that after flowering each ovary appears to be in a separate excavation of the rhachis, with a membranous fringed border. In fruit the lateral ovaries are often abortive, and the central pericarp grows out into a prominent beak (the thickened base of the styles), projecting horizontally considerably beyond the subtending scale.—Hook. f. Fl. Tasm. i. 316; *Arthrocnemum arbuscula,* Moq. Chenop. Enum. 113, and in DC. Prod. xiii. ii. 152; *Arthrocnemum halocnemoides,* Nees in Pl. Preiss. i. 632; Moq. in DC. l.c.

N. Australia ? N.W. coast, *Bynoe in herb. Hook.* (possibly some error).

Victoria. Seacoast near Melbourne, *Adamson;* Port Phillip and Point Lonsdale, *F. Mueller;* Wimmera, *Dallachy;* Murray river, *Herrgott;* N.W. part of the colony, *L. Morton.*

Tasmania. Port Dalrymple, *R. Brown;* salt marshes near Hobarton and Clarence plains, *J. D. Hooker.*

W. Australia. Sand flats N. of Stirling range, *F. Mueller;* Vasse river, *Oldfield;* Swan river, *Drummond, 1st coll.;* Sharks Bay, *Milne* (the two latter sets of specimens very small in all their parts).

3. **S. cinerea,** *F. Muell. Fragm.* vi. 251. Apparently annual, branching at the base, with several single or slightly branched ascending stems of 6 to 8 in., the whole plant of an ashy grey colour when dry. Articles of the branches ¼ to ½ in. long, slightly dilated at the top. Spikes terminal, oblong-cylindrical, rounded at both ends, very compact, ½ to ¾ in. long, 2 to 2½ lines diameter, the articles very numerous, closely imbricate, the margins dilated into opposite semicircular scarious scales, without projecting points but forming projecting acute lines. Flowers in threes, all apparently hermaphrodite and monandrous. Perianths immersed in and adnate to the rhachis at the base, thin and free at the top, very shortly toothed. Seed ovate, erect, but the radicle not always superior.—*Halocnemum cinereum,* F. Muell. Fragm. i. 139.

N. Australia. Point Pearce and Sturt's Creek, *F. Mueller.*
Queensland. Cape York, *M'Gillivray.*

SECT. 2. ARTHROCNEMUM.—Perianths at length dilated into a flat transverse or oblique top. Spikes cylindrical, often elongated, compact or loose.

4. **S. leiostachya,** *Benth.* A spreading much-branched shrub of 2 or 3 ft., the articles of the branches cylindrical, ¼ to ½ in. long, slightly

thickened but not lobed at the top. Spikes numerous, nearly sessile and opposite at the nodes or terminal, cylindrical, compact, $\frac{1}{2}$ to nearly 1 in. long; articles numerous, at first rather distinct with their obtuse margins slightly prominent, but at length very closely packed into an apparently continuous spike of 2 lines diameter, without prominent scales, the separation of the articles only marked by slightly depressed transverse lines. Flowers in threes, wholly immersed and closely packed side by side (not in a triangle as in *S. herbacea*), all hermaphrodite and monandrous, but often only the central one perfecting its seed. Perianths thickened upwards, with a narrow triangular obliquely truncate top. Seed apparently compressed and vertical but not seen very perfect.

N. Australia. Sandflats about Providence Hill and between M'Adam Range and Point Pearce, *F. Mueller;* Kyejeron Creek, Central Australia, *M'Douall Stuart's Expedition.*

W. Australia, *Drummond, (Herb. F. Mueller.).*

5. **S. bidens,** *Benth.* Stems procumbent, hard and apparently woody, with ascending or erect branches, from under 6 in. to above 1 ft. high. Articles mostly under $\frac{1}{2}$ in. long, dilated at the top into 2 opposite lobes with prominent keels to a breadth of 2 lines or rather more; in the older stems the articles are somewhat thickened with the lobes scarcely or not at all prominent. Spikes terminal, at length exceeding 1 in., the articles at first but little more than 1 line long, but growing out to $1\frac{1}{2}$ lines when in fruit, the lobes or scales at their ends very prominent. Flowers in threes, at first almost in a line, more in a triangle when in fruit, all usually hermaphrodite and monandrous. Perianths partially immersed, at first shorter than the subtending scale, when in fruit longer, quite distinct and free from the rhachis, thickened upwards and obliquely truncate and flat at the top. Seed vertical and compressed.—*Arthrocnemum bidens,* Nees in Pl. Preiss. i. 632; Moq. in DC. Prod. xiii. ii. 151.

W. Australia. Salt lagoons N. of Stirling Range, *F. Mueller;* Swan river, *Preiss,* n. 1261; Murchison river, *Oldfield;* Sharks Bay, *Milne.*

This species seems to be closely allied to the E. Indian *S. brachiata,* Roxb., and may possibly prove not to be really distinct from it, but the precise structure of the flowers and seeds requires further investigation in both.

6. **S. tenuis,** *Benth.* A divaricately-branched or diffuse shrub, more slender than the other Australian species. Articles of the branches $\frac{1}{4}$ to $\frac{1}{2}$ in. long, the upper end dilated into a membranous sometimes scarious margin. Flowers apparently diœcious, both sexes in threes. Male spikes short, but only commencing to flower in the specimens seen, probably at length elongated, the articles larger and broader than in the females, with scarious margins. Perianths all three distinct and slender, with 1 stamen in each, and no trace of pistil. Female spikes slender, 1 to 2 in. long, the articles $\frac{1}{2}$ to above 1 line long. Perianths free, at first almost in a line, at length in a triangle, clavate, turbinate, when in fruit white, much dilated at the top which is obliquely truncate

and flat, each of the three containing a perfect vertical seed, and I have found no trace of any stamen, but the specimens are all past flower.

N. S. Wales. Darling desert, *Victorian Expedition* (♂), and probably the same region, *Mitchell* (♀).

S. Australia. In the interior, *Howitt's Expedition* (♂) and *M'Douall Stuart's Expedition* (♀).

The specimens are very few and I do not feel certain that the male and the fruiting ones are correctly matched. They all differ from the other Australian species in their slender articles with the more or less membranous margins to the upper end.

7. **S. australis,** *Soland. in Forst. Prod.* 88 (*name only*). Stems procumbent, hard, sometimes woody at the base, with ascending or erect branches rarely above 6 in. high, the articles varying from a little more than ¼ to above ½ in. high, not much or not at all dilated at the end and either quite terete or obscurely 2-lobed. Spikes cylindrical, 1 to 2 in. long, usually thicker than the stems, the articles about 1 line long, dilated at the top into slightly prominent rings but not lobed. Flowers 5 or 7 together side by side, all hermaphrodite or the ovaries of the lateral ones abortive, mostly if not all diandrous. Perianths free and distinct, shortly immersed at the base, the central ones rather longer than the lateral, but all shortly prominent beyond the subtending ring, clavate, obliquely truncate and flat at the top and almost closed by the minute connivent teeth.—*S. indica*, R. Br. Prod. 411; Hook. f. Fl. Tasm. i. 317, not of Willd.

Queensland? Fitzroy river, *Thozet* (not in flower).
N. S. Wales? Hastings river, *Beckler ;* Clarence river, *Wilcox* (not in flower).
Victoria. Common in wet marshy places on the seashore, Portland, *Robertson;* Port Phillip and Station Peak, *F. Mueller* ; salt marshes, Streatham, *Lau.*
Tasmania. Abundant in stony places near the sea and in all salt marshes, *J. D. Hooker.*
S. Australia. Fowler's and Petrel Bays, *R. Brown.*
W. Australia. King George's Sound, *R. Brown.*

The species is common in New Zealand, whence proceeded Forster's original specimen in Herb. Mus. Brit. The true *S. indica* is insufficiently described and pictured by Willdenow (Neu. Schr. Gesellsch. Natur. Fr. Berlin, ii. (not v.) 111, t. 4); but it is probably the same as a species not uncommon on the shores of India and S. Africa, better figured by Wight, Ic. 737, and described by Moquin under the name of *Arthrocnemum indicum*, which has always the flowers in threes, whilst in *S. australis*, besides other differences, they are always in fives or sevens. This species appears to have been unknown to Moquin. His *Halocnemum australasicum*, to which he refers Brown's plant (which he had not seen), must be a very different one. In his diagnosis he omits the character of the 5 to 7 flowers expressly given by Brown.

The northern stations given above must remain doubtful until flowering specimens shall have been observed.

TRIBE 4. SALSOLEÆ.—Branches continuous (not articulate). Leaves narrow, flat, or terete, entire. Testa of the seed various. Embryo spirally coiled, without albumen.

14. SUÆDA, Forsk.
(Chenopodina, *Moq.*)

Flowers mostly hermaphrodite. Perianth depressed-globular, herbaceous or slightly fleshy, with 5 broad lobes connivent over the fruit, without appendages or with a slight horizontal protuberance or thick scale on the back at the base of each lobe. Stamens 5. Styles 2 or 3, rarely more; free or shortly united at the base. Fruit enclosed in the perianth; pericarp membranous, very thin but separable from the seed. Seed flat, horizontal or vertical; testa crustaceous with a thin inner membrane. Embryo flat, spirally twisted, without any or with scarcely any albumen.—Glabrous herbs or undershrubs. Leaves alternate, sessile, linear, thick or terete. Flowers small, sessile, solitary or clustered in the axils. Bracts and bracteoles very small and scarious.

The genus, consisting of a small number of species, is widely diffused over the sea-coasts and saline districts of both the New and the Old World, the only Australian species being the most common one over nearly the whole area of the genus.

1. **S. maritima,** *Dumort. Fl. Belg.* 22 (*Moq.*). A much-branched herb, erect and attaining 1 ft. or more, or low and spreading, quite glabrous, somewhat succulent, with a hard almost woody base, but usually annual or biennial. Leaves linear-terete or semi-cylindrical, usually acute, ½ to above 1 in. long. Flowers very small, clustered or rarely solitary in the axils, the fruiting perianth usually about 1 line diameter, with a very flat horizontal seed, but occasionally I have found one perianth in the cluster narrower with a vertical seed. Under each flower there are usually 1 bract and 2 bracteoles, all small transparent scales, but sometimes one of the three is wanting. Seed shining.—Hook. f. Fl. Tasm. i. 316; *Chenopodina maritima,* Moq. in DC. Prod. xiii. ii. 161; *Chenopodium australe,* R. Br. Prod. 407; *Suæda australis,* Moq. in Ann. Sc. Nat. ser. 1, xxiii. 318; *Chenopodina australis,* Moq. in DC. Prod. xiii. ii. 163.

Queensland. Cleveland Bay, *Bowman;* Fitzroy river, *Thozet.*
N. S. Wales. Hunter's river, *Leichhardt;* Ash island, *Woolls.*
Victoria. Glenelg river, Portland Bay, *Robertson*
Tasmania. Abundant on mud and shingle beaches close to high-water mark, *J. D. Hooker.*
W. Australia. Lucky Bay and Goose Island Bay, *R. Brown;* Cape Naturaliste, *Lay and Collie;* Murchison river, *Oldfield.*

The species is common on the seacoasts of most temperate and subtropical regions both in the New and the Old World. The Australian plant is usually distinguished from the common northern one by its suffrutescent habit, but it is doubtful whether its duration exceeds the second year, which the European plant is said frequently to attain. In both, the base of the stem becomes very hard and more or less woody. I can, no more than J. D. Hooker, detect any other difference between the two.

15. SALSOLA, Linn.

Flowers hermaphrodite. Perianth of 5 rarely 4 distinct segments when in fruit, bearing each on their backs a horizontal wing or protuberance, their points closed over the fruit. Stamens 5 or rarely fewer. Styles 2, rarely 3, united at the base or above the middle. Fruit enclosed in the perianth. Pericarp membranous. Seed depressed or nearly globular, testa membranous; embryo coiled in a conical or doubly convex spire, without albumen.—Herbs or undershrubs usually hard or fleshy. Leaves narrow-linear or terete, entire. Flowers axillary, sessile, solitary within each floral leaf (or subtending bract), with 2 opposite bracteoles.

The genus is widely spread over the temperate regions of the globe in more or less saline situations. The only Australian species is the most common one over nearly the whole area of the genus.

1. **S. Kali,** *Linn.; Moq. in DC. Prod.* xiii. ii. 187. A hard procumbent or divaricately-branched herb, glabrous or slightly pubescent, usually under 1 ft. but sometimes extending to 2 ft. Leaves alternate or rarely here and there opposite, sessile, hard and rigid in the typical form, the lower ones terete or dilated at the base, from ½ in. to above 2 in. long, the upper ones shorter, thicker, and often more flattened above, but sometimes all terete, the lower floral ones similar to the stem-leaves, the upper ones gradually smaller and sometimes, especially on side branches, reduced to thick triangular or lanceolate bracts not exceeding the calyx, all as well as the bracteoles ending in rigid pungent points. Flowers sessile and solitary in the axil of each bract, but often, owing to the reduction of the flowering branch, clustered in the axils of the primary floral leaves. Bracteoles similar to the floral leaf or subtending bract, but usually smaller. Segments of the fruiting perianth forming at the base a hard or thin campanulate or turbinate tube rarely much above 1 line long, surrounded at the top by the 5 horizontal wings which are either all equal or 2 narrower than the others, each one sometimes 2 lines long and broad, thin and scarious, sometimes very small and thick or in some flowers scarcely perceptible, the summit of each perianth-segment within the wing acute scarious and closing over the fruit. Pericarp with the upper portion flat circumsciss and deciduous. Embryo spiral, the two cotyledons in separate coils one over the other, with the radicle coiled horizontally round the lowest coil or between the two.—*S. australis,* R. Br. Prod. 411; Moq. in DC. Prod. xiii. ii. 188; *S. macrophylla,* R. Br. l.c.; Moq. l.c. 187; Nees in Pl. Preiss. i. 637.

Queensland. Bay of Inlets, *Banks and Solander;* Maria island, *Dallachy;* in the interior, *Mitchell;* Cape and Suttor rivers, *Bowman;* Armadilla, *Barton;* Curriewillighie, *Dalton.*

N. S. Wales. Botany Bay, *Banks and Solander;* Clay flats and saline places from the Murray and Darling to the Barrier Range, *Victorian and other Expeditions.*

Victoria. Lake Hindmarsh, *F. Mueller.*

S. Australia. Petrel Bay, *R. Brown;* St. Vincent's and Spencer's Gulfs, *F. Mueller;* between Stokes Range and Cooper's Creek, *Wheeler.*

W. Australia. Drummond, *n.* 244, 245; Swan river, *Preiss, n.* 2396; Murchison river, *Oldfield.*

The species is widely distributed over the temperate regions of the New as well as the Old World in more or less saline districts, extending not unfrequently to within the tropics. I can discover nothing to separate the Australian specimens from the European form even as a variety. Moquin cites both as growing together in Timor.

Var. *leptophylla.* Leaves slender, almost filiform, but pungent when full grown. —Queensland and N. S. Wales.

Var. *strobilifera.* Flowers densely clustered in globular heads with the points of the subtending bracts protruding like the scales of a pine-cone.—Darling desert and Mount Murchison.

Var. *brachypteris.* Wings or appendages of the perianth reduced to prominent transverse ribs, in all or nearly all the flowers.—*S. brachypteris,* Moq in DC. Prod. xiii. ii. 189.—Rockingham Bay, *Dallachy;* Curtis island, *Thozet.* The size of the perianth-wings is as variable in European as in Australian specimens, and in some flowers of most specimens and in nearly all of other specimens they remain, in both countries, very short or undeveloped as in *S. brachypteris;* in this state *S. Kali* can always be readily distinguished from *S. Soda* by the pungent leaves.

ORDER XCVII. AMARANTACEÆ.

Perianth-segments 5, free or shortly united at the base, rigid and scarious or coloured at least on the margin and tips, imbricate in the bud. Stamens 5 or fewer, opposite the perianth-segments, free or united at the base, with or without intervening scales or teeth (stami-nodia of some authors); anthers 1- or 2-celled, the cells parallel. Ovary 1-celled, with 1 or several ovules attached to a filiform funicle erect from the base of the cavity. Style simple and entire, with a capitate stigma or more or less divided into 2 or 3 branches or separate styles, stigmatic at the end or along the inner edge. Fruit a membranous in-discent utricle or rarely a circumsciss capsule or a succulent berry, en-closed in or resting on the persistent perianth. Seed usually vertical, orbicular or ovate and compressed, testa crustaceous. Embryo horse-shoe-shaped or annular, enclosing a mealy albumen.—Herbs or under-shrubs rarely shrubs or woody climbers. Leaves alternate or opposite, entire, without stipules. Flowers rarely solitary in the axils, more fre-quently in axillary or terminal simple or paniculate spikes or rarely cymes or clusters, each flower sessile or rarely pedicellate, within 2 sca-rious bracteoles, and subtended by a scarious bract or rarely by a floral leaf.

A considerable Order, spread over the temperate and warmer regions of both the New and the Old World, disappearing in high latitudes and in alpine districts. Of the nine Australian genera, three are widely dispersed over the warmer regions of the globe, each including two or three common tropical weeds, and one of them represented by several species in more temperate districts, a fourth extends over tropical Asia and more sparingly into Africa, another belongs to the tropical American Flora, the remaining four are endemic.

TRIBE 1. **Celosieæ.**—*Anthers 2-celled.* *Ovary with several ovules. Leaves alter-nate.*

Woody climbers. Fruit succulent . . . . 1. DEERINGIA.

TRIBE 2. **Achyrantheæ.**—*Anthers 2-celled. Ovary with a single ovule.*

Leaves alternate.

Stamens shortly united at the base. Flowers axillary, solitary.
  Maritime plants with linear succulent leaves . . . . 2. HEMICHROA.
Stamens free. Flowers in axillary or terminal and paniculate
  cymes or clusters. Leaves flat . . . . 3. AMARANTUS.
Stamens shortly united at the base. Flowers in dense single
  terminal spikes, often shortened into heads.
    Perianth-segments with the laminæ more or less plumose with
      dorsal articulate hairs, the tips alone glabrous . . . . . . 4. TRICHINIUM.

Perianth-segments with the whole lamina scarious, coloured
and glabrous . . . . 5. PTILOTUS.
Leaves opposite. Stamens united in a cup at the base, with
truncate teeth or lobes between the filaments.
Flowers 5-merous, in terminal elongated spikes. Bracts and
perianth-segments acute . . . . 6. ACHYRANTHES.
Flowers 4-merous, in axillary or terminal spikes shortened into
heads. Bracts and perianth-segments spinescent . . . . 7. NYSSANTHES.

TRIBE 3. **Gomphreneæ.**—*Anthers 1-celled. Ovary with a single ovule. Leaves
opposite.*

Stigma capitate. Spikes often shortened into heads, axillary (rarely
also terminal) . . . . 8. ALTERNANTHERA.
Stigma 2-lobed. Spikes often shortened into heads, terminal or
rarely axillary . . . . 9. GOMPHRENA.

Besides the above, the following species described by F. Mueller cannot be referred
to its proper genus until it shall again have been observed. It is evidently not a *Psilotrichum* from its inflorescence. The other species formerly referred by F. Mueller to
Psilotrichum is now correctly placed in *Trichinium*, from which the following- plant
differs essentially in its opposite leaves. The teeth of the staminal cup or tube in pairs
between each two filaments might, with the opposite leaves, refer it to *Gomphrena*, but
from its having been placed in *Psilotrichum* we are led to suppose that the anthers were
bilocular.

Psilotrichum capitatum, F. Muell. Fragm. i. 238. A glabrous erect or ascending
undershrub of 1 ft. or less. Leaves opposite, ovate or ovate-lanceolate, acute, contracted
into a petiole, rather thick, ½ to 1 in. long. Flower-heads (spikes?) terminal, sessile,
nearly globular. Bracteoles finely pointed, somewhat shorter than the perianth, the
bract slightly exceeding the bracteoles. Perianth-segments 3 to 4 lines long, glabrous,
the inner ones bearded inside at the base. Filaments capillary, separated by small
divaricate teeth in pairs.—Stirling's Creek, *F. Mueller.*—Described by F. Mueller from
notes taken at the time of gathering it, the specimens having been lost in the passage
home.

TRIBE 1. CELOSIEÆ.—Anthers 2-celled. Ovary with several ovules.
Leaves alternate.

1. DEERINGIA, R. Br.

Flowers hermaphrodite or diœcious. Perianth of 5 equal glabrous
segments spreading under the fruit. Stamens 5, shortly united in a
ring at the base, without intervening staminodia. Anthers 2-celled.
Ovary with several ovules. Styles 3, rarely 4, few and stigmatic from
the base. Fruit enlarged, succulent, indehiscent. Seeds several.—Tall
woody climbers, glabrous or rusty-pubescent. Leaves alternate. Flowers
numerous, loosely spicate, the spikes in axillary or terminal panicles.
Bracts and bracteoles small, scarious or petal-like.

The genus is widely spread over tropical Asia, more sparingly extending into Africa.
Of the two Australian species one is the same as the commonest Asiatic one, the other
is endemic.

Quite glabrous. Flowers hermaphrodite . . . . 1. *D. celosioides.*
Branches and young leaves pubescent. Flowers diœcious (the ovaries
of the males without ovules) . . . . 2. *D. altissima.*

1. **D. celosioides,** *R. Br. Prod.* 413. A woody glabrous climber
scrambling over bushes to the height of 10 or 12 ft. (according to Aus-

tralian collectors). Leaves petiolate, ovate or ovate-lanceolate, acuminate, entire, mostly 2 to 3 in. long. Flowers hermaphrodite, solitary or somewhat clustered, in slender interrupted spikes varying from 2 or 3 to 8 or 10 in. long, either in the upper axils or in a loose terminal panicle, each flower nearly sessile in the axil of a small linear-lanceolate acute bract, shorter than the perianth except sometimes in the lower part of the spike, and accompanied by two smaller bracteoles. Perianthsegments ovate, scarcely 1 line long, of a greenish white when fresh, black with a whitish margin when dry. Stamens united at the base in a prominent ring or very short cap. Berry red, nearly globular, 3-furrowed, usually about 2 lines diameter or rather larger.—Bot. Mag. t. 2717; Endl. Iconogr. t. 62; Wight Ic. Pl. t. 728; *D. baccata,* Moq. in DC. Prod. xiii. ii. 236.

Queensland. Shoalwater and Keppel Bays and Broad Sound, *R. Brown;* Endeavour river, *A. Cunningham;* Barnard Isles, *M'Gillivray;* Port Denison, *Fitzalan;* Rockingham Bay, *Dallachy;* Rockhampton, *Dallachy* and others; Darling Downs, *Lau;* Brisbane river, Moreton Bay, *F. Mueller* and others.

N. S. Wales. Port Jackson, *R. Brown, Woolls;* northward to Macleay, Hastings, and Clarence rivers, *Beckler;* New England, *C. Stuart;* southward to Kiama, *Harvey.*

The species is common in E. India and the Archipelago, extending to New Caledonia.

2. **D. altissima,** *F. Muell. Fragm.* ii. 92, vi. 251. A woody climber resembling *D. celosioides* but larger, ascending to the tops of the tallest trees, the young parts clothed with a soft rusty crisped pubescence which disappears from the upper surface of the older leaves, persisting underneath or at least leaving some traces along the midrib. Leaves ovate or oblong, rarely ovate-lanceolate, very shortly acuminate, thicker than in *D. celosioides,* and attaining 3 to 6 in. Flowers diœcious, more numerous than in *D. celosioides,* in denser spikes and larger panicles, the males much whiter and more petaloid than in that species, the stamens united in a scarcely prominent ring, the ovary ovoid and conical externally perfect as well as the styles but without any ovules; the female flowers smaller, not so white, in more compact and less branched panicles, the ovary more globular with about 10 to 15 ovules, the stamens more or less imperfect or altogether abortive. Berry globular, red, scarcely 1½ lines diameter.—*Lagrezia altissima,* Moq. in DC. Prod. xiii. ii. 253; *Lestibudesia arborescens,* R. Br. Prod. 414; *Celosia arborescens,* Spreng. Syst. i. 815; Moq. in DC. Prod. xiii. ii. 243.

Queensland. Endeavour river, *Banks and Solander;* Port Denison, *Fitzalan;* Rockingham and Edgecombe Bays, *Dallachy;* Brisbane river, Moreton Bay, *A. Cunningham* and others; Port Mackay, *Nernst.*

N. S. Wales. Hastings river, *Beckler;* Richmond river, *Henderson;* Tweed and Clarence rivers, *C. Moore.*

The male specimens being the handsomest are the most frequently collected and the only ones seen by R. Brown and by Moquin. The ovary in them is so large that it has not the appearance of being abortive, but I have opened a large number without ever finding any ovules. From its membranous appearance Brown thought the fruit might be capsular, and Moquin that it might be monospermous, which accounts for their not having associated the plant with *Deeringia.*

TRIBE 2. ACHYRANTHEÆ.—Anthers 2-celled. Ovary with a single ovule. Leaves alternate or opposite.

2. HEMICHROA, R. Br.

Flowers hermaphrodite. Perianth-segments 5, free, erect, glabrous, rigid, white at least inside. Stamens 5 or fewer, the filaments united in a short cup at the base, without intervening teeth or scales. Ovary uniovulate. Style with 2 very short or rather long stigmatic branches. Fruit an indehiscent utricle. Seed vertical.—Maritime prostrate herbs or low spreading shrubs. Leaves alternate, linear, fleshy. Flowers sessile and solitary in the axils, between 2 rigid scarious bracteoles.

The genus is limited to Australia, but is scarcely distinct from the European and Asiatic genus *Polycnemum*, differing indeed only in the thick succulent leaves and longer style.

Prostrate herb. Stamens 5. Style bifid . . . . 1. *H. pentandra.*
Small divaricate shrub. Stamens 2, unilateral. Style with a very
   short bifid stigma . . . . 2. *H. diandra.*

1. **H. pentandra,** *R. Br. Prod.* 409. A glabrous perennial, with prostrate stems and numerous shortly ascending branches, forming dense patches of ½ to 1 ft. diameter. Leaves sessile, linear, mucronate, thick or semiterete and succulent, ¼ to ¾ in. long, the upper floral ones somewhat dilated and concave at the base, but otherwise similar to the others. Bracteoles ovate, acute, scarious with a prominent midrib, from ½ to ¾ the length of the perianth. Perianth about 1½ lines long, the segments acute, more or less scarious and coloured (white ?) inside, the outer ones almost ovate and 3-nerved, the inner ones narrower, more lanceolate, 1- or 2-nerved. Stamens 5, much shorter than the perianth, united at the base in a short annular cup. Style in the flowers examined divided to the middle or lower down, but described by others as very shortly lobed. Fruit ovoid, shorter than the perianth, not oblique; pericarp rather thickly membranous. Seed with a crustaceous shining testa. Embryo semiannular according to Moquin, forming almost a complete ring in the seeds examined.—Moq. in DC. Prod. xiii. ii. 334; Hook. f. Fl. Tasm. i. 311.

Victoria. In rather muddy places on the seashore from Gipps Land to the western frontier, *F. Mueller.*

Tasmania. On the shore, Port Dalrymple, *R. Brown;* Great Swan Port, *Backhouse;* George Town, *Gunn.*

S. Australia. Port Adelaide, *F. Mueller.*

W. Australia. King George's Sound, *R. Brown.*

2. **H. diandra,** *R. Br. Prod.* 409. A small glabrous much-branched spreading plant, evidently shrubby and apparently less prostrate than *H. pentandra.* Leaves linear and succulent as in that species but not so thick and the floral ones more dilated at the base. Flowers rather larger and more scarious. Perianth about 2 lines long, and the bracteoles not much shorter, segments rather obtuse, 1-nerved, the inner ones not

much narrower than the outer. Stamens 2, not much shorter than the perianth, the filaments much dilated below the middle, and united at the base on the gibbous side of the ovary, the staminal cup interrupted on the other side, without any rudimentary stamens. Ovary very gibbous on the side next the stamens. Style undivided, with a small 2-lobed stigma. Fruit not seen ripe.—Moq. in DC. Prod. xiii. ii. 334.

N. Australia? A fragment from Nichol Bay, N.W. coast, in herb. F. Muell., appears to belong rather to this species than to *H. pentandra*, but is insufficient for determination.

S. Australia. Fowler's Bay, *R. Brown;* hills near Lake Hamilton, *Wilhelmi;* head of Spencer's Gulf, *F. Mueller, Warburton.*

W. Australia. Salt marshes, Sharks Bay, *Milne.*

3. AMARANTUS, Linn.

(Sarratia, *Moq.;* Amblogyne *and* Euxolus, *Rafin.*)

Flowers usually monœcious. Perianth-segments 3 to 5, erect with scarious margins or (especially when in fruit) more or less dilated at the end into spreading scarious laminæ. Stamens 3 to 5, free, without intervening staminodia. Anthers 2-celled. Ovary uniovulate. Styles 2 or 3, free and stigmatic from the base. Fruit a membranous utricle, either circumsciss when ripe or indehiscent with the pericarp loose or adnate to the seed. Seed solitary.—Herbs mostly annual, glabrous or nearly so, green or red. Leaves alternate. Flowers small, in dense cymes or clusters, the clusters all axillary or collected in terminal spikes which are either simple or branching into dense panicles. Bracts and bracteoles small, green or scarious. The female flowers are usually much more numerous than the males, which are in the same clusters, chiefly in the upper parts of the inflorescence, with the same number or with fewer perianth-segments than the females.

The genus has a very extensive range over the New and the Old World, some of the larger species much cultivated for the seed and several others very common weeds of cultivation. Of the ten species here enumerated one is evidently an escape from cultivation, two others are common weeds of cultivation, one chiefly in the Old World, the other in the warmer regions of both the New and the Old World. The remaining seven are endemic, but two of them bear a remarkable resemblance to the West Indian and Central American group of *Amblogyne.*

The characters derived from the dehiscent or indehiscent pericarp or from the more or less spreading laminæ of the fruiting perianth, are so little in accord with habit or with any other character, that I have thought it better to retain the collective genus as a natural and very fairly defined one, than to adopt the purely artificial disseverances proposed by Moquin and others.

SECT. 1. **Euamarantus.**—*Pericarp circumsciss.*

Perianth-segments mostly 5, erect. Tall erect plant with an
  ample panicle, the points of the bracts very prominent .   .   .   1. *A. paniculatus.*
Perianth-segments mostly 3, erect or slightly dilated at the end.
  Clusters all axillary or the upper ones in a short dense terminal
  spike   .   .   .   .   2. *A. Blitum.*
Perianth-segments mostly 4 or 5, erect or slightly dilated at the
  end. Clusters axillary and in a long loose terminal shortly-
  branched spike .   .   .   .   3. *A. leptostachyus.*
Perianth-segments mostly 5, with dilated scarious spreading
  laminæ (when in fruit). Terminal spikes usually paniculate .   4. *A. pallidiflorus.*

SECT. 2. **Euxolus.**—*Pericarp membranous, indehiscent or bursting irregularly.*

Pericarp separate from the seed, shorter than or not much longer
   than the perianth.
   Segments of the fruiting perianth 5, with dilated and scarious
      laminæ.   Cymes axillary.   Pericarp longitudinally ribbed . 5. *A. Mitchellii.*
   Segments of the fruiting perianth erect or slightly dilated and
      spreading. Clusters axillary and in a terminal spike.   ·
         Fruiting perianths mostly 5-merous . . . . 6. *A. interruptus.*
         Fruiting perianths mostly 3-merous . . . . 7. *A. viridis.*
Pericarp separate from the seed, oblong, at least twice as long as
   the perianth.   Clusters all axillary . . . . 8. *A. macrocarpus.*
Pericarp small, globular, very thin and not readily separable from
   the seed.   Leaves narrow.   Clusters all axillary.   Perianth-
   segments mostly 4.
      Perianth-segments twice as long as the fruit . . . . . . 9. *A. tenuis.*
      Perianth-segments about as long as the fruit . . . . . . 10. *A. enervis.*

SECT. 1. EUAMARANTUS.—Pericarp circumsciss when ripe, the lower
half with the perianth usually (but not always) remaining attached after
the seed has fallen.   Male perianths usually of as many segments as the
females.

1. **A. paniculatus,** *Linn.; Moq. in DC. Prod.* xiii. ii. 257.   An erect
stout annual sometimes attaining 5 or 6 ft., the foliage and inflorescence
often assuming a reddish hue and sometimes the panicle a rich crimson.
Leaves on long petioles, ovate or ovate-lanceolate, 2 to 4 in. long.   Flowers
in dense cylindrical spikes, the lower ones axillary, the upper ones
forming a dense terminal panicle, 6 in. to 1 ft. long, the central spike
thicker and longer than the lateral ones, all appearing more or less
echinate by the fine points of the bracts and bracteoles which usually
exceed the perianth.   Perianth-segments mostly 5, erect, tapering into
a fine point, 1 to 1¼ line long.   Pericarp membranous, rugose, circumsciss.   Styles 2 or 3.—*A. frumentaceus,* Roxb. Fl. Ind. iii. 699; Wight,
Ic. t. 720.

N. S. Wales. Port Jackson, *R. Brown;* Clarence river, *Beckler.*
Evidently an escape from cultivation as suggested by R. Brown, and therefore
omitted from his Prodromus.   Extensively cultivated in India, and probably a native
of Asia.

There is also a single specimen from Darling Downs, *Lau,* in Herb. F. Mueller, of
what appears to be *A. caudatus,* Linn., another cultivated species, nearly allied to the
last, but usually with a longer and narrower panicle, and the very short points of the
bracts not exceeding the perianth, whence the aspect of the plant is very different.

2. **A. Blitum,** *Linn.; Moq. in DC. Prod.* xiii. ii. 263.   An erect or
procumbent branching annual, attaining from 1 to 2 ft.   Leaves on
long petioles, ovate or rhomboidal, obtuse or obtusely acuminate, mostly
1 to 2 in. long in good soils, under 1 in. in drier places.   Flowers chiefly
in dense sessile axillary clusters, but also forming sometimes a short
dense terminal spike.   Bracts and bracteoles not exceeding the perianth.
Perianth-segments almost always 3, mucronate-acute, lanceolate, scarious
with a green midrib, about ¾ line long, erect or when in fruit slightly

dilated and spreading at the end. Pericarp membranous, slightly rugose, about as long as the perianth, circumsciss. Styles 2 or 3.

N. S. Wales, *Leichhardt.* Perhaps introduced.

3. **A. leptostachyus,** *Benth.* An erect annual, from under 1 ft. to about 1½ ft. high. Leaves on rather long petioles, ovate, obtuse, under 1 in. long, with the primary veins often remarkably prominent. Clusters of flowers rather loose, very numerous, the lower ones axillary, the upper ones forming a long terminal interrupted spike leafy at the base only like that of *A. interruptus.* Perianth-segments erect or scarcely spreading and persistent with the lower portion of the circumsciss pericarp as in *A. Blitum,* but usually 4 or 5 instead of 3 only.

N. Australia. Port Darwin, *Schultz.*
Queensland. Islands off Cape Flattery, *M'Gillivray.*

4. **A. pallidiflorus,** *F. Muell. Fragm.* i. 140. An erect or decumbent annual of 1 to 2 ft. Leaves on long petioles, ovate, obtuse, 1 to 3 in. long, rather thin and of a pale green. Lower cymes or clusters of flowers axillary, the upper ones in dense spikes forming a terminal panicle of ½ ft. or more, the central spike very long, the lateral ones short. Bracts and bracteoles scarious, lanceolate or ovate-lanceolate, very acute, about as long as the claws of the perianth-segments or rarely as long as the whole perianth. Perianth-segments 5, lanceolate at the time of flowering, those of the females, when in fruit with broad erect claws of about ½ line and expanded into broadly ovate mucronate-acute spreading and scarious laminæ, rather longer than the claws. Pericarp very rugose, membranous, circumsciss, with a thick apex and 3 fine styles.

N. Australia. Nichol Bay, *Walcott;* Victoria river, *F. Mueller.*
W. Australia. Port Walcott, *Harper.*

This species has precisely the aspect of and is closely allied to *A. scariosus,* Benth., from Central America, and with that species would be referred to *Sarratia* as defined by Moquin, or to *Amblogyne* as defined by A. Gray, Proc. Amer. Acad. Sc. v. 168. The American plant is indeed only to be distinguished from the Australian by the longer points to the bracts and by the retuse or emarginate laminæ of the fruiting perianth.

SECT 2. EUXOLUS.—Pericarp membranous, indehiscent or bursting irregularly, loose and separate from or close and adhering to the seed. Male flowers usually but not always trimerous or tetramerous. Fruiting perianth of 3, 4 or 5 segments usually falling off with the fruit.

5. **A. Mitchellii,** *Benth.* Apparently erect, rather stout and rigid, branching but not tall. Leaves on long petioles, ovate-lanceolate or oblong, obtuse, narrowed at the base, 1 to 2 in. long. Flowers all axillary and numerous, in sessile or shortly pedunculate cymes often ½ in. broad, rarely reduced to close clusters. Bracts scarious, nearly as long as the perianth, with a prominent midrib ending in a sharp point, the bracteoles similar but rather smaller. Segments of the fruiting perianth 5, with a rigid erect stipes of about ½ line, and a broad scarious spreading lamina at least as long, the midrib produced into a rigid point. Perianth globular, membranous, with 12 to 15 prominent undulate

longitudinal ribs, indehiscent or bursting irregularly, with a thick summit projecting beyond the perianth, and 3 short subulate stigmatic styles.—*A. undulatus,* Lindl. in Mitch. Trop. Austr. 102, not of R. Br.

Queensland. Narran river, *Mitchell;* Flinders river, *Sutherland;* Charlesville, *Giles;* Armadilla, *Barton.* Used as a vegetable, *Sutherland, Giles.*

N. S. Wales. Between the Darling and Cooper's Creek, *Neilson;* Ballandool river, *Locker.*

With the radiating fruiting perianth of *A. pallidiflorus,* this has the habit of *A. Blitum* or almost of *A. crassipes (Scleropus,* Schrad.), with the pericarp indehiscent as in *Euxolus,* but differing in its prominent ribs from all *Amaranti* known to me.

6. **A. interruptus,** *R. Br. Prod.* 414. Erect or decumbent, from 6 in. to nearly 2 ft. high. Leaves petiolate, ovate or almost rhomboidal, obtuse, ½ in. to near 2 in. long. Cymes or clusters dense or at length rather loose, the lower ones axillary, the upper ones forming a long loose spike leafy at the base, either simple or rarely with a few short branches. Bracts and bracteoles shorter than the perianth or about as long, scarcely pointed. Segments of the fruiting perianth 5, narrow, erect, slightly spathulate, shortly but finely pointed, about ¾ line long, white and scarious on the margins, dark in the centre, those of the male flowers usually 3 only and not dilated upwards. Pericarp membranous, rugose, not ribbed, indehiscent or bursting irregularly, loose over the seed with a short thick summit about as long as the perianth. Styles 2 or rarely 3.—*A. undulatus, A. rhombeus,* and *A. lineatus,* R. Br. l.c.; *Euxolus undulatus, E. rhombeus* and *E. interruptus,* Moq. in DC. Prod. xiii. ii. 272 and 275 ; *E. lineatus,* Moq. l.c. 276 as to the Australian, but not the Sandwich Island plant.

N. Australia. Arnhem N. Bay and neighbouring parts of the N. coast, *R. Brown;* Sandy islands, Victoria river, *F. Mueller ;* N. coast, *Landsborough.*

Queensland. Rockhampton, *O'Shanesy;* Brisbane river, *Leichhardt;* Port Mackay, *Nernst.*

This species has the aspect nearly of *A. Blitum,* with the fruit of *A. viridis,* and is readily distinguished from both by the segments of the fruiting perianth almost constantly 5, not 3 ; they fall off with the fruit as in most species of the section *Euxolus.* Brown's four species appear to me to be scarcely even varieties of a single one. The specimens of *A. undulatus* are young, with small broad leaves on long petioles slightly crisped on the margin, the terminal spike still dense and commencing flowering. Those of *A. interruptus* are older, the spike long and loose, and most of the fruits already fallen off; the leaves are narrower than in *A. undulatus. A. rhombeus* is, as it were, intermediate between the two. The specimens of *A. lineatus* appear to me to be old ones of *A. interruptus* which have been eaten down, or have otherwise lost the upper part of their main stem, which has shot up branches giving it a different aspect. All are from the same localities, probably sandy or arid. Nernst's specimens from Port Mackay are very luxuriant, with broad leaves twice the size of those of most others, but not otherwise different.

7. **A. viridis,** *Linn.* An erect or decumbent annual of 1 to 2 ft. Leaves petiolate, ovate or ovate-lanceolate, obtuse, rather thin but the pinnate veins usually prominent underneath, 1 to 2 in. long. Flowers small, green with an obtuse appearance, the lower ones in small axillary sessile cymes or close clusters, the upper ones in rather loose or interrupted spikes, forming a short terminal panicle, the central spike 1 to

3 in. long, the lateral ones few and short. Bracts and bracteoles narrow, not exceeding the perianth. Perianth-segments 3, narrow, erect, scarcely ¾ line long, falling off with the fruit. Pericarp rugose, indehiscent, free from the seed, about as long as the perianth. Styles usually 3.—*Euxolus viridis*, Moq. in DC. Prod. xiii. ii. 273.

Queensland. Brisbane river, Moreton Bay, *F. Mueller;* Rockhampton, *Dallachy* and others; Nerkool Creek, *Bowman.*

N. S. Wales. Glendon, Cassilis, *Leichhardt.*

W. Australia, *Drummond, n.* 105.

Common in waste and cultivated places in the warmer regions of Europe, Asia, and Africa, and now frequent in several parts of America. Possibly introduced only into Australia.

8. **A. macrocarpus,** *Benth.* A small diffuse or decumbent plant, none of our specimens above 6 in. and some, although the whole plant, much smaller. Leaves on long petioles, obovate or oblong, very obtuse or emarginate, about ½ in. long or rather more. Flowers in dense axillary sessile nearly globular clusters, chiefly females in our specimens, the males not seen perfect. Bracts and bracteoles shorter than the perianth. Fruiting perianth of 3 rarely 5 narrow-linear erect very pointed segments sometimes slightly dilated below the point. Fruit oblong or almost bottle-shaped, 1¼ to 1½ lines long, much exceeding the perianth, the pericarp inflated, membranous and reticulate-rugose, drying black in the common form. Seed erect, obovoid, much smaller than the cavity of the fruit.

Queensland. Armadilla, *Barton;* Dawson river, *Leichhardt.*

N. S. Wales. Junction of the Murray and Darling, *F. Mueller;* Darling river, *Woolls.*

Var. *pallida.* Fruit of a pale green when dry. Perianth smaller and often reduced to 1 or 2 segments.—Curriwillighie, *Dalton.*

9. **A. tenuis,** *Benth.* Stems in our specimens single, erect, slender, 6 to 9 in. high. Leaves narrow-lanceolate, acute, the radical ones contracted into a long petiole, the lamina usually 1 to 1½ in. long, the upper ones more sessile. Flowers small, in dense axillary clusters, mostly females in our specimens, the males not seen perfect. Bracts and bracteoles very small or quite obsolete. Fruiting perianth of 4, rarely 3 or 5, oblong or oblong-lanceolate segments about 1 line long. Fruit scarcely half so long, slightly tubercular, rugose, indehiscent, and the very thin pericarp scarcely separable from the seed. Styles 2 or 3.

N. S. Wales. Lower Darling river, *Herb. F. Mueller.*

10. **A. enervis,** *F. Muell.* A small annual, branching at the base, with decumbent or erect stems not exceeding 6 in. Leaves linear-lanceolate, obtuse or acute, contracted at the base but scarcely petiolate, rather thick, nerveless or the midrib scarcely prominent, ¼ to ½ in. long. Flowers small, all in axillary clusters, but crowded in the upper part of the stem with small floral leaves forming a terminal leafy spike, chiefly females. Bracts and bracteoles rather smaller than the perianth. Male flowers few, with 3 or 4 perianth-segments and stamens. Fruiting

perianth about ½ line long, the segments usually 4, rather broad but unequal, erect, acute. Fruit about as long as the perianth, tubercular-rugose, the pericarp very thin and scarcely separable from the seed. Styles 2 (or 3 ?).—*Euxolus enervis*, F. Muell. Fragm. i. 140.

N. S. Wales. Murray and Darling rivers, *Dallachy.*

4. TRICHINIUM, R. Br.

(Goniotriche, *Turcz.*; Hemisteirus *and* Arthrotrichum, *F. Muell.*)

Flowers hermaphrodite. Perianth with a short turbinate hard tube, reduced sometimes to a slight expansion of the peduncle; segments 5, all equal or the three inner ones rather smaller, linear, rigid, usually 3-ribbed at the base, scarious at the tips or also along the margins of the upper portion, covered outside either entirely or rarely along the centre only with straight more or less distinctly articulate (several-celled) hairs giving them a plumose appearance, the short tips alone glabrous. Stamens normally 5, but usually 1, 2, or 3 of them small and without anthers, or entirely abortive, and all the filaments unequal, or more rarely all equal and antheriferous, united at the base in a mem-branous cup adnate to the perianth-tube or shortly free from it, without or rarely with intervening scale-like teeth or lobes, which, when present, are very thin and transparent; anthers 2-celled. Ovary uniovulate. Style simple, rigid, with a small capitate stigma. Fruit an indehiscent utricle, usually obovoid or contracted into a stipes at the base and oblique at the top, with the persistent style more or less excentrical. Seed vertical.—Herbs undershrubs or rarely shrubs, glabrous or hairy with crisped articulate woolly or stellate hairs. Leaves alternate, narrow or rarely obovate. Flowers in dense globular ovoid or cylindri-cal spikes, very rarely elongated and interrupted. Bracts and bracteoles scarious and shining, nerveless or with a more or less prominent mid-rib produced into a fine or short point. Perianths usually pink or straw-colour. Stamens and ovary often enveloped in dense wool or long hairs proceeding either from the lower part or claws of the inner perianth-segments or from the outside of the staminal cups.

The genus is limited to Australia, for the opposite-leaved *T. Zeyheri* from S. Africa, admitted by Moquin, must be referred to *Sericocoma*, the presence or absence of scales between the stamens being by no means of absolute generic importance.

This and the following genus have been united by Poiret and F. Mueller under the name of *Ptilotus*, by Sprengel under that of *Trichinium*, and they might perhaps be really better considered as sections of one genus than as two distinct ones, were it not for the useless confusion which would result in the nomenclature of species. At any rate, if the union be adopted it would be better to follow Sprengel in preferring the name of *Trichinium* for the united genus, as being that which belongs to four-fifths of the species, and entails therefore the least change, besides that it is the most familiar of the two from the number of species that have been cultivated or figured. Neither name has the right of priority, both being of the same date, and both are equally apposite for the groups they designate, and equally inappropriate for the combined species, for *Trichinium* means "a clothing of hairs," *Ptilotus* "having featherless wings."

In the subdivision of the genus I have been unable to establish any natural well-characterized sections. Even the presence of the teeth or lobes of the staminal cup, considered by Moquin as at least a generic if not a tribal character, separates species which in other respects are almost identical. The groups here adopted as the best which have hitherto been suggested are founded chiefly upon the nature and position of the different kinds of hairs. With the exception of the short stellate tomentum of the foliage in the first series, the hairs are all so-called articulate, that is, consisting of several cells, sometimes very conspicuously so, with more or less prominent denticulations at the joints or almost plumose, sometimes very fine with the articulations visible only under a very high power. These hairs are sometimes (always so on the backs of the laminæ of the perianth) straight, at first appressed afterwards spreading, sometimes, on the backs of the lower part of the perianth-segments or inside the inner ones, long and intricate forming masses of white wool, sometimes, especially on the branches, short and crisped.

SERIES 1. **Astrotricha.**—*Foliage hoary or white with a stellate tomentum (glabrous or with crisped or woolly or silky hairs in all the other series).*
Spikes dense, globular ovoid or shortly cylindrical, not exceeding 1 in.
Spikes ½ to 1 in. diameter. Laminæ of perianth-segments linear.
Leaves mostly broad, rather thick and densely tomentose. Spikes globular or at length ovoid. Bracts glabrous or nearly so . . . . 1. *T. obovatum.*
Leaves mostly narrow, thick and densely tomentose. Spikes ovoid, at length cylindrical. Bracts woolly . . . . . 2. *T. incanum.*
Leaves mostly narrow, rather thin, less densely tomentose. Spikes ovoid, at length cylindrical. Bracts glabrous or slightly woolly . . . . 3. *T. parviflorum.*
Spikes not above ¼ in. diameter. Laminæ of perianth-segments almost ovate . . . . 4. *T. astrolasium.*
Spikes cylindrical, above 2 in. long and 1½ in. diameter. Leaves orbicular, very densely woolly . . . . 5. *T. rotundifolium.*
Spikes elongated with distant flowers. Leaves oblong or lanceolate, the stellate hairs short and scattered . . . . . . 6. *T. dissitiflorum.*
(The foliage is also very densely silky-woolly in 46, *T. helichrysoides,* but the hairs not stellate.)

SERIES 2. **Straminea.**—*Spikes cylindrical or elongated or rarely globular, 1 to 2 in. diameter. Flowers more or less yellow or greenish, not red. Inner segments without internal dense wool, but the stamens usually surrounded by a few long hairs.*
Spikes elongated with distant flowers. Leaves filiform . . . 7. *T. distans.*
Spikes dense, at length long and cylindrical.
Leaves linear. Bracts wholly transparent. Bracteoles broad without prominent midribs. Perianth under ¾ in. . . . 8. *T. alopecuroideum.*
Leaves obovate or oblong. Bracts opaque in the centre. Bracteoles oblong or lanceolate with prominent keels.
Perianth above ¾ in. long. . . . . 9. *T. nobile.*
Perianth not exceeding ½ in.. . . . . 10. *T. polystachyum.*
Spikes ovoid or shortly cylindrical, 2 in. diameter. Bracts transparent. Leaves linear . . . . 11. *T. macrocephalum.*
Spikes globular or rarely ovoid. Perianth-segments rather broad, the dorsal hairs very short . . . . 12. *T. corymbosum.*

SERIES 3. **Rhodostachya.**—*Spikes globular ovoid or rarely cylindrical, 1 to 2 in. diameter, terminating simple or rarely branched stems. Perianth straight, pink or red (white in* T. esquamatum), *the inner segments woolly inside towards the base.*
Spikes 1½ to 2 in. diameter.
Perianth-segments very rigid with short narrow tips. Stems erect. Spikes about 1½ in. diameter.
Spikes at length elongated and cylindrical . . . . . . 13. *T. exaltatum.*

Spikes globular or shorter than broad . . . . 14. *T. semilanatum.*
Perianth-segments with conspicuous coloured obtuse glabrous
    tips. Spikes about 2 in. diameter.
Stems short, decumbent. Radical leaves spathulate, the
    others linear . . . . 15. *T. Manglesii.*
Stems short erect. Leaves spathulate, all crowded at the
    base of the stem . . . . 16. *T. Beckerianum.*
Spikes about 1 in. diameter. Stems erect, simple, with small
    narrow leaves.
Radical leaves oblong-spathulate. Bracts ovate-lanceolate,
    brown. Perianth pale pink . . . . 17. *T. gomphrenoides.*
Leaves all small. Bracts broad, transparent, pale-coloured.
    (Perianth white?) . . . . 18. *T. esquamatum.*
SERIES 4. **Incurva.**—*Spikes globular, ¾ to 1½ in. diameter, terminating simple
stems. Perianths curved upwards (straight or curved downwards in all other series),
the inner segments woolly inside at the base. Leaves linear.*
Spikes sessile within the last leaves . . . . 19. *T. declinatum.*
Upper leaves distant, usually reduced to scarious scales . . . 20. *T. erubescens.*
SERIES 5. **Polycephala.**—*Stems mostly branched or rarely some of them long de-
cumbent and simple, glabrous or with crisped woolly hairs. Spikes mostly globular,
¾ to 1 in. diameter.*
Inner perianth-segments very woolly inside towards the base
    (less so in *T. helipteroides*). Bracts rather loose.
Shrubby with divaricate branches. Leaves linear, rigid.
    Spikes about 1 in. diameter . . . . 21. *T. divaricatum.*
Herbaceous with decumbent, ascending or erect stems.
    Spikes about ¾ in. diameter.
Leaves narrow. Stems more or less silky or woolly.
    Bracts and bracteoles very thin, nearly as long as the
        perianth. . . . . 22. *T. helipteroides.*
    Bracts and bracteoles rather rigid, not half as long as the
        perianth . . . . 23. *T. Stirlingii.*
Leaves broad. Stems glabrous or nearly so except the
    young shoots.
    Stamens 2, long, dilated and connate at the base on one
        side of the ovary . . . . 24. *T. laxum.*
    Stamens 3 or 4 perfect, the filaments all dilated at the
        base forming a complete ring or cup . . . . . . 25. *T. axillare.*
Inner perianth-segments nearly glabrous inside, the wool pro-
    ceeding chiefly from the staminal cup. Bracteoles closely
    embracing the perianth.
Undershrub with divaricate branches. Leaves few, narrow
    and small. Panicle divaricate . . . . 26. *T. striatum.*
Herb with large obovate or oblong leaves chiefly radical.
    Stem simple at the base with a compact panicle . . . . 27. *T. auriculifolium.*
Inner perianth-segments nearly glabrous inside. Staminal cup
    surrounded by long straight hairs.
Leaves obovate oblong or lanceolate, usually glabrous. Pe-
    rianth straight. Stamens 3 or 4 perfect, connate at the
    base, in a complete cup or ring . . . . 28. *T. sericostachyum.*
Leaves oblong or lanceolate, loosely villous underneath. Pe-
    rianth recurved. Stamens 2, the filaments dilated and
    connate on one side of the ovary . . . . 29. *T. roseum.*
Leaves linear-filiform. Perianth straight. Stamens 3 or 4
    perfect, the filaments forming at the base a complete
    cup or ring.
    Perennial with a thick rootstock. Spikes about 1 in. dia-
        meter . . . . 30. *T. fusiforme.*
    Annual. Spikes about ¾ in. diameter . . . . 31. *T. gracile.*

SERIES 6. **Squamigera.**—*Spikes globular or cylindrical, $\frac{1}{2}$ to 1 in. diameter. Staminal cup with transparent scale-like teeth or lobes between the filaments (wanting in all the other series). Leaves narrow.*

Perennial with simple stems of 1 ft. or more. Spikes globular, about $\frac{3}{4}$ to 1 in. diameter . . . . 32. *T. Drummondii.*
Annual with slender branching stems of 1 ft. or more. Spikes cylindrical, about $\frac{1}{4}$ in. diameter . . . . 33. *T. calostachyum.*
Perennial with a branching stock and slender stems of about 6 in. Spikes globular, under $\frac{1}{2}$ in. diameter . . . . 34. *T. Fraseri.*

SERIES 7. **Spathulata.**—*Perennials with short decumbent stems leafy to the spike. Spikes sessile, within the last leaves, globular ovoid or cylindrical, $\frac{3}{4}$ in. diameter or more. Leaves mostly spathulate.*

Spikes ovoid, at length cylindrical, the hairs round the base of the perianth shorter than the segments.
Bracteoles acute, half concealed by the very plumose perianths.
Perianth-tube $\frac{1}{2}$ to $\frac{3}{4}$ line long . . . . 35. *T. spathulatum.*
Bracteoles broad, obtuse, conspicuous. Perianth-segments free to the base . . . . 36. *T. pyramidatum.*
Spikes globular, at length ovoid, the bracteoles and the hairs surrounding the perianths nearly as long as the segments . . 37. *T. holosericeum.*

SERIES 8. **Parviflora.**—*Spikes globular, ovoid or cylindrical, $\frac{1}{4}$ to $\frac{1}{2}$ in. diameter.*

Stems erect, branching, glabrous or slightly hairy. Annual.
Leaves linear. Spikes conical or cylindrical, 2 or 3 together, sessile on a terminal peduncle . . . . . . 38. *T. Cunninghamii.*
Leaves linear. Spikes globular or ovoid, solitary on a terminal peduncle. Perianth hairs short . . . . 39. *T. leucocoma.*
Prostrate woolly-hairy annual. Leaves lanceolate. Spikes solitary, ovate-conical. Perianth very woolly . . . . . 40. *T. villosum.*
Perennials. Branches woolly or villous, at least when young.
Spikes numerous, sessile or shortly pedunculate.
Perianth surrounded by long wool concealing the bracts and segments. Spikes cylindrical . . . . 41. *T. brachyanthum.*
Bracts and bracteoles nearly as long as the perianth and very conspicuous.
Branches and foliage villous. Spikes narrow-cylindrical . 42. *T. arthrolasium.*
Young shoots woolly. Leaves glabrous, broad. Spikes ovoid. Perianth-segments glabrous inside . . . . 43. *T. œrvoides.*
Branches closely woolly. Leaves broad, obtuse. Spikes ovoid. Inner perianth-segments woolly inside . . . . 44. *T. Roei.*
Glabrous undershrub, with a densely tufted leaf-stock. Leaves small, nearly terete. Spike very short. Bracts conspicuous 45. *T. cæspitulosum.*

SERIES 9. **Helichrysoidea.**—*Low densely tufted thick perennial, closely covered with thick silky-woolly leaves. Spikes nearly globular, sessile, $\frac{3}{4}$ in. diameter.*

Single species . . . . 46. *T. helichrysoides.*

Species insufficiently known.

Stem slender, branching. Leaves ovate, about 1 line long.
(Spikes globular?) . . . . 47. *T. parvifolium.*

SERIES. 1 ASTROTRICHA.—Foliage hoary or white with a stellate tomentum, sometimes dense soft and woolly, sometimes short and scattered. Erect branching perennials undershrubs or shrubs.

1. **T. obovatum,** *Gaudich. in Freyc. Voy. Bot.* 445, t. 49. An erect undershrub, from under 1 ft. to 3 or even 4 ft. high, paniculately

branched sometimes from the base sometimes at the top only, clothed with a soft dense stellate tomentum, intermixed occasionally with longer denticulate hairs but with fewer of the latter than in *T. incanum.* Leaves obovate or oblong, very obtuse or rarely mucronate-acute, contracted into a petiole rather long in the lower leaves, short in the upper ones, the larger ones attaining 2 in. but mostly under 1 in. long. Spikes nearly globular or scarcely ovoid, ½ to ¾ in. diameter, sessile or shortly pedunculate, in terminal corymbose panicles, which are sometimes compact and leafless, sometimes more spreading and leafy at the base. Bracts and bracteoles usually rather brown, obtuse or shortly mucronate-acute, glabrous or nearly so, under 2 lines long. Perianth 3 to 4 lines long in the typical form, the tube about ¼ line, segments rigid, plumose with long white hairs, the glabrous tips short obtuse and denticulate in the outer segments, the inner ones shorter and more acute, glabrous inside. Stamens unequal, 3 or 4 perfect, filaments dilated downwards, scarcely united above the perianth-tube, surrounded by a ring of long articulate hairs. Ovary usually hairy on the top. Style excentrical.—Moq. in DC. Prod. xiii. ii. 286; *Ptilotus obovatus,* F. Muell. Fragm. vi. 228; *T. incanum,* Moq. l.c. 285 not of R. Br.; *T. sessilifolium,* Lindl. in Mitch. Three Exped. ii. 13; Moq. l.c. 284; *T. lanatum,* Lindl. in Mitch. Three Exped. ii. 123; Moq. l.c. 285; *Ptilotus Lindleyi,* F. Muell. Fragm. vi. 233 (from the synonym); *T. atriplicifolium,* A. Cunn. in Moq. l.c. 286, F. Muell. Pl. Vict. t. 78 (the plate not yet received); *T. variabile,* F. Muell. in Linnæa xxv. 436; *Gomotriche tomentosa,* Turcz. in Bull. Soc. Imp. Nat. Mosq. 1849, 37 (corrected to *Goniotriche,* l.c. 1852, ii. 181).

N. Australia. N. W. coast, *Bynoe;* in the interior, *M'Douall Stuart's Expedition.* **Queensland.** Narran river, *Mitchell.* **N. S. Wales.** Bogan and Murray rivers, *Mitchell;* Mount Caley, Mount Flinders, &c., *A. Cunningham;* Murray and Darling rivers to the Barrier Range, *Victorian and other Expeditions.*

Victoria. Murray river, *F. Mueller.* **S. Australia.** Flinders' Range, *F. Mueller, Howitt's Expedition;* Spencer's Gulf and Mount Searle, *Warburton;* Lake Gillies, *Burkitt*

W. Australia, *Drummond, n.* 74, 233; throughout the interior, *Harper;* Murchison river, *Oldfield;* Sharks Bay, *A. Cunningham, M. Brown.*

Var. *grandiflorum.* Perianth ½ in. long, the hairs surrounding the stamens more woolly and more attached to the inner perianth-segments.—Harrington plains, *Fraser, A. Cunningham;* Murray Desert, *F. Mueller, Victorian Expedition,* &c.; Cudnaka, *F. Mueller;* Gawler Ranges, *Sullivan.* Both varieties included by Moquin under *T. atriplicifolium.*

2. **T. incanum,** *R. Br. Prod.* 415, *not of Moq.* Stems erect or ascending, divaricately branched, hard and almost woody at the base, 6 in. to above 1 ft. high, the whole plant densely and softly tomentose or woolly with stellate and plumose hairs. Leaves from broadly elliptical to narrow oblong, obtuse or mucronate, contracted into a short petiole, thick and soft, the larger ones on the main stem sometimes above 2 in., those of the side branches under 1 in. long. Spikes ovoid or at length cylindrical, ½ to 1 in. long and scarcely ½ in. diameter, sessile or nearly so, forming an irregular leafy panicle, with some lateral spikes much lower

down.   Bracts and bracteoles very thin and transparent, 1 to 1½ lines long, the bracts very woolly outside.   Perianth scarcely above 3 lines long, the tube about ¼ line long, hirsute with short hairs, the segments rigid, plumose with long white dorsal hairs, the glabrous tips obtuse and slightly denticulate in the outer ones, shorter and acute in the inner ones, all glabrous inside except that the long hairs which, as in *T. obovatum*, surround the stamens are rather more on the base of the inner segments than in that species.   Stamens 3 or 4 perfect, filaments unequal, dilated at the base but scarcely united above the perianth-tube. Ovary glabrous ; style excentrical.—*Ptilotus incanus*, Poir. Dict. Suppl. iv. 620 ; F. Muell. Fragm. vi. 228 ; *T. gnaphalodes*, A. Cunn. ; Moq. in DC. Prod. xiii. ii. 285.

N. Australia.   N. W. coast ? *Baudin's Expedition ;* Dampier's Archipelago, *A. Cunningham ;* Nichol Bay, *F. Gregory's Expedition ;* Sturt's Creek, *F. Mueller.*

It is evident from Moquin's character as well as from Cunningham's specimens named by him, that his *T. incanum* is a form or state of *T. obovatum.* It is true that by a note of exclamation he indicates having seen an authentic specimen of Brown's, but that must be a mistake ; Brown never gathered either species himself, but described *T. incanum* from specimens of Baudin's, corresponding with those described by Moquin under the name of *T. gnaphalodes.*

3. **T. parviflorum,** *Lindl. in Mitch. Three Exped.* ii. 13.   A perennial or undershrub with a thick woody stock and erect branching stems, hoary as well as the foliage with a stellate tomentum, closely allied to *T. obovatum* and *T. incanum.*   Lower leaves sometimes obovate, but mostly lanceolate or almost linear, obtuse, thinner than in those two species, the stellate tomentum disappearing with age on the upper surface, and not very dense on the lower.   Spikes at first short, but lengthening out to about 1 in., and about ½ in. diameter, sessile or shortly pedunculate in a loose divaricate panicle.   Bracts and flowers of *T. incanum*, but much less woolly.—Moq. in DC. Prod. xiii. ii. 284 ; *T. virgatum*, A. Cunn. ; Moq. l.c. 286.

Queensland.   Flinders' river, *Bowman ;* Curriwillighie, *Dalton ;* Armadilla, *Barton.*

N. S. Wales.   Inundated plains, Lachlan river, *Mitchell, A. Cunningham.*

4. **T. astrolasium,** *F. Muell.*   A perennial or undershrub with a thick rhizome and several erect stems of 1 ft. or more, paniculately branched in the upper part and stellately tomentose as well as the foliage. Leaves obovate or oblong, smaller than in *T. obovatum.*   Spikes sessile or very shortly pedunculate, globular or at length ovoid, 3 to 4 lines diameter, forming a terminal corymbose leafy panicle.   Bracts and bracteoles ovate, acute, scarious, about 1 line long, the bracts loosely tomentose outside.   Perianth about 1¼ lines long, the segments free from the base, the lower portion rigid with the dorsal hairs very dense white and straight, the scarious lamina broad, almost ovate, more sparingly hairy on the back.   Staminal cup truncate, surrounded by long intricate woolly hairs ; filaments scarcely dilated at the base.—*Ptilotus astrolasius*, F. Muell. Fragm. vi. 233.

N. Australia.   Sturt's Creek, *F. Mueller ;* N. W. coast, *Hughan.*

5. **T. rotundifolium,** *F. Muell. Fragm.* iii. 122.   An erect shrub of 2 or 3 ft., the branches and foliage covered with a stellate tomentum very soft and dense, almost woolly.   Leaves on short petioles, nearly orbicular, very obtuse, soft and thick, about 1 to 1½ in. diameter.   Spikes at first conical, becoming cylindrical, 2 or 3 in. long, and at least 1½ in. diameter.   Bracts broad, acute and mucronate, scarious with dark tips, woolly outside, shorter than the bracteoles.   Bracteoles at least 4 lines long, very broad and thin, shortly mucronate, glabrous or with very few woolly hairs at the base.   Perianth nearly ¾ in. long, the dorsal hairs long fine and almost silky, the tube about ⅓ line long, the segments narrow, scarious, obtuse, the tips not at all or very shortly glabrous outside; the three inner ones rather shorter and very densely woolly inside near the base.   Stamens all antheriferous and equal or nearly so.—*Ptilotus rotundifolius,* F. Muell. Fragm. vi. 230.

N. Australia.   Near Hammersley range, N. W. coast, *F. Gregory's Expedition.*

6. **T. dissitiflorum,** *F. Muell. Fragm.* iv. 89.   Erect and branching, hard and almost woody at the base, the branches and foliage more or less hoary with a minute stellate tomentum, the older foliage black when dry.   Leaves oval oblong or shortly lanceolate, obtuse or acute, rather thick, ½ to ¾ in. long or the upper ones smaller, contracted into a short petiole.   Spikes interrupted, close above the last leaves, 1 to 3 in. long, the flowers more or less distant.   Bracts ovate or lanceolate, acute, about 2 lines long, the upper ones brown and scarious, the lower ones thicker and tomentose; bracteoles rather broader and more scarious.   Perianth 6 to 7 lines long, the tube nearly 1 line long and hirsute with short white hairs, the segments narrow, rigid, plumose outside, the glabrous tips not 1 line long, the three inner ones rather smaller and woolly inside towards the base.   Staminal cup very woolly-hairy outside, with a few hairs also on the filaments.   Ovary glabrous; style excentrical.

N. Australia.   Gulf of Carpentaria, *F. Mueller.*

SERIES 2.   STRAMINEA.—Spikes cylindrical elongated or rarely globular, 1 to 2 in. diameter.   Flowers more or less yellow (usually a pale greenish yellow or straw colour), not red.   Inner segments without the dense internal wool of the *Rhodostachya,* but the stamens usually surrounded by a few long hairs.

7. **T. distans,** *R. Br. Prod.* 415.   A perennial with a hard stock at length woody, and erect virgate slender simple or branched stems of 1 to 2 ft., glabrous as well as the foliage.   Leaves narrow-linear, almost filiform, the lower ones sometimes 2 in. long, the others much smaller and distant.   Spikes terminal, slender and interrupted, 3 or 4 in. long, the flowers all distant, or in luxuriant specimens twice as long with the upper flowers more crowded.   Bracts and bracteoles narrow-ovate or oblong, scarious and shining, 2 to 3 lines long, all similar or the bracteoles smaller narrower and more acute.   Perianth about ¼ in. long, the

tube about ½ line, the segments narrow, rigid, plumose on the back with small narrow glabrous tips, all glabrous inside. Staminal cup shortly free, with copious articulate hairs outside more or less continued on the filaments. Ovary hirsute on the top.—Moq. in DC. Prod. xiii. ii. 297; *Ptilotus distans*, Poir. Dict. Suppl. iv. 620; F. Muell. Fragm. vi. 228.

N. Australia. Arnhem S. Bay, *R. Brown;* Victoria river, Macadam and Sea Ranges, *F. Mueller;* S. Goulburn Island, *A. Cunningham.*
Queensland, *A. Cunningham;* Rockingham Bay, *Dallachy;* Cape river, *Bowman.*

8. T. alopecuroideum, *Lindl. in Mitch. Three Exped.* ii. 13, *but not of Bot. Reg.*—A perennial with ascending or erect slightly-branched stems of 1 to 3 ft., the young shoots and foliage often sprinkled with short crisped hairs, becoming at length glabrous. Leaves linear or lanceolate, acute, the lower ones often several in. long and contracted into a long petiole, the upper ones few small and more sessile. Spikes on long terminal peduncles, becoming very soon cylindrical, attaining sometimes 6 in. or more and from a little more than 1 in. to above 1½ in. diameter. Bracts and bracteoles broadly ovate or orbicular, obtuse or with a small point, wholly scarious and shining, with the central nerve scarcely conspicuous, 1½ to 3 lines long. Perianth pale yellow or straw colour, the tube exceedingly short, the segments ½ to ¾ in. long, narrow, obtuse, but appearing acute from the involution of the margins at the tip, the dorsal hairs not so copious as in some species and all glabrous inside. Staminal cup shortly prominent, surrounded by long straight hairs sometimes very few sometimes copious; stamens very unequal, one or two of the filaments usually short and without anthers. Ovary glabrous.—Moq. in DC. Prod. xiii. ii. 296; *Ptilotus alopecuroideus*, F. Muell. Fragm. vi. 227; *T. giganteum*, A. Cunn.; Moq. l.c. 296; *T. pallidum*, Moq. l.c. 295 (very tall and stout specimens); *T. Preissii*, Nees in Pl. Preiss. i. 629; Moq. l.c. 295; *T. candicans*, Nees in Pl. Preiss. i. 629; Moq. l.c. 296 (with rather broader leaves and the stems somewhat procumbent at the base).

N. Australia. Water island, Montague Sound, *A. Cunningham;* Usborne's Harbour, *Beagle Voyage;* Glenelg district, *Martin* (with a very long drawn out spike).
Queensland. Armadilla, *Barton;* Curriwillighie, *Dalton.*
N. S. Wales. Lachlan river, *Mitchell;* Lower Darling river, *Mrs. Ford;* between the Darling and the Barrier Range, *Victorian Expedition;* New England, *C. Stuart.*
Victoria. Murray river, *F. Mueller, Dallachy.*
S. Australia. Murray river to St. Vincent's Gulf, *F. Mueller and others;* Lake Gairdner, *Babbage;* Cooper's Creek, *Howitt's Expedition, Neilson.*
W. Australia. Swan river, *Fraser, Drummond, n.* 434, *Preiss, n.* 1370, 1371; Murchison river, *Oldfield, Drummond,* 6th *coll. n.* 221.

T. conicum, Lindl. in Mitch. Trop. Austr. 363; Moq. in DC. Prod. xiii. ii. 462, not of Spreng. is the *T. alopecuroideum* in a young state just coming into flower.

9. T. nobile, *Lindl. in Mitch. Three Exped.* ii. 22. A stout erect, glabrous perennial, the stems simple or slightly-branched upwards, 1 to 3 ft. high. Leaves from broadly obovate to oblong, rarely lanceolate, the lower ones chiefly radical on long petioles, obtuse or mucronate, the upper ones narrower and more sessile. Spikes terminal, oblong, attaining 3 to 6 in. in length and nearly 2 in. diameter, the rhachis very densely hairy. Bracts 3 to 5 lines long, ovate or oblong, the midrib prominent and projecting into a point, scarious but with a dark centre especially towards the tip; bracteoles similar but rather smaller and often with a few dorsal hairs. Perianth usually about 1 in. long, of a greenish yellow, the tube 1 to $1\frac{1}{2}$ lines long and densely hirsute with short hairs, the segments narrow, copiously plumose with dorsal hairs, the glabrous tips short, obtuse in the outer segments more acute in the inner, all without any wool inside but a few long straight hairs round the base of the stamens, which are not united above the perianth-tube, the filaments very unequal, and 1 or 2 without anthers. Ovary glabrous.— Moq. in DC. Prod. xiii. ii. 286; *T. densum,* A. Cunn.; Moq. l.c. 289.

N. S. Wales. Lachlan river, *Mitchell;* Strangford Plains, *A. Cunningham;* from the Lachlan, Murray, and Darling rivers to the Barrier Range, *Victorian and other Expeditions.*

Victoria. Murray river, *F. Mueller.*

S. Australia. Murray desert, St. Vincent's Gulf, Flinders' Range, Cudnaka, *F. Mueller ;* near Adelaide, *Whittaker.*

F. Mueller includes also under *Ptilotus nobilis* (Fragm. vi. 227) the *T. exaltatum* and *T. semilanatum,* which have similar bracts but usually smaller flowers, red not yellow, and with copious wool inside the lower part of the inner segments.

10 ? **T. polystachyum,** *Gaudich. in Freyc. Voy. Bot.* 445. Stem herbaceous, erect, paniculately branched. Leaves lanceolate, obtuse or shortly mucronate, contracted at the base, green and pubescent, $1\frac{1}{2}$ to 2 in. long. Spikes several, distant, ovate-oblong and $\frac{3}{4}$ to 1 in. long (*Moquin*), cylindrical (*Gaudichaud*), about $\frac{3}{4}$ in. diameter. Bracts $\frac{1}{3}$ the length of the perianth, ovate-lanceolate, acuminate; bracteoles rather longer, elliptical, obtuse, all 1-nerved, glabrous, pale brown. Perianth 5 lines long, yellowish (*Moquin*), the segments linear-spathulate with glabrous tips and short rigid white dorsal hairs, the outer ones 2-toothed at the end, the inner ones somewhat acute. Filaments filiform.—Moq. in DC. Prod. xiii. ii. 283.

W. Australia. Sharks Bay, *Gaudichaud.*

I have no specimens answering to the above character. *Ptilotus polystachyus,* F. Muell. Fragm. vi 230, to which he refers Gaudichaud's plant, includes *T. Stirlingii,* *T. roseum,* and *T. laxum,* all of which have globular spikes and pink flowers. Gaudichaud's character comes nearest to that of *T. nobile,* but with much smaller flowers. Neither he nor Moquin describe the wool or hairs, if any, surrounding the stamens.

11. **T. macrocephalum,** *R. Br. Prod.* 415, *not of others.* Stems from a hard perennial base erect or ascending, simple, stout and rigid, 1 to 2 ft. high, usually glabrous as well as the foliage. Leaves few at the base of the stem, linear or narrow-lanceolate, acute or rarely obtuse, contracted into a long petiole, 2 to 4 in. long, the upper ones smaller few and distant. Spikes solitary, at first ovoid, at length cylindrical, attaining 4 or 5 in. in length and at least 2 in. diameter. Bracts scarious and very shining, obtuse or mucronate, without prominent midribs or dark colour, about $\frac{1}{2}$ in. long and the bracteoles nearly as

large.   Perianth yellow, $\frac{3}{4}$ to above 1 in. long, the tube very short, the segments narrow, rigid, densely plumose outside, with short glabrous tips, all nearly equal without any internal wool, although a few of the marginal hairs at the base of the inner ones may be turned inside round the stamens.   Filaments very unequal, filiform, scarcely dilated at the base, very shortly united above the perianth-tube, and surrounded by a few long hairs, the shorter filaments usually without any anthers.   Ovary glabrous, but a few hairs often on the style.—*Ptilotus macrocephalus*, Poir. Dict. Suppl. iv. 620 ; *T. angustifolium*, and *T. pachocephalum*, Moq. in DC. Prod. xiii. ii. 293, 294 ; *Ptilotus pachocephalus*, F. Muell. Fragm. vi. 228 ; *T. fusiforme*, Lindl. in Mitch. Trop. Austr. 383, and A. Cunn. Herb., not of R. Br.

Queensland. In the interior, *Mitchell ;* Newcastle Range, Burnett and Dawson rivers, *F. Mueller ;* Bowen river, *Bowman ;* Kennedy district, *Daintree.*
N. S. Wales. Liverpool Plains, *A. Cunningham, Leichhardt* (with rather smaller flowers) ; New England, *C. Stuart ;* Darling river, *Neilson.*
Victoria. "Received by Sir J. Banks, probably from Port Phillip," *R. Brown ;* Port Phillip, *Gunn ;* Glenelg river, *Robertson, Allitt ;* Bacchus Marsh, Wimmera, Station peak, *F. Mueller ;* Skipton, *Whan.*

12. **T. corymbosum,** *Gaudich. in Freyc. Voy. Bot.* 444, *not of Spreng.* A glabrous perennial (or sometimes annual ?) with rigid ascending or erect simple or branched stems of 1 to 2 or even 3 ft.   Leaves linear or linear-lanceolate, mucronate-acute, sessile or contracted into a petiole, the larger ones 1 to 2 in. long, but mostly small and distant.   Spikes globular or ovoid or rarely at length cylindrical, about 1 in. diameter, on rather long peduncles, forming a loose irregular panicle.   Bracts and bracteoles broad, obtuse, thinly scarious, rather brown but without prominent midribs, not half as long as the perianth.   Perianth about $\frac{1}{4}$ in. long, the segments free almost from the base, all nearly equal, and glabrous inside, with broad scarious white margins, the green centre alone hirsute outside with articulate hairs much shorter than in any allied species. Stamens very unequal, the filaments shortly dilated at the base and very shortly united, 1 or 2 of the shorter ones without anthers, surrounded by a few woolly hairs proceeding chiefly from their base. Ovary glabrous.   Style excentrical.—Moq. in DC. Prod. xiii. ii. 291 ; *T. Gaudichaudii*, Steud. Nom. Bot. ed. 2 ; *Hemisteirus psilotrichoides*, F. Muell. in Linnæa, xxv. 435 ; *Ptilotus hemisteirus*, F. Muell. Fragm. iv. 90, vi. 231.

N. Australia. Hammersley range, Nichol Bay, N.W. coast, *F. Gregory's Expedition.*
S. Australia, Lake Gillies, *Burkitt.*
W. Australia. Sharks Bay, *M. Brown ;* Murchison river, *Oldfield ;* Swan river, *Fraser, Drummond, n.* 432 ; Gordon and Blackwood rivers, *Oldfield.*

Var. *parviflora.* Perianth scarcely above 4 lines long.—Cudnaka, *F. Mueller.*
When the spike elongates, the species bears much resemblance to *T. alopecuroideum*, but the leaves are much narrower, and the shortness of the perianth-hairs gives the spike an almost glabrous aspect.

SERIES 3. RHODOSTACHYA.—Spikes globular ovoid or rarely cylindrical, 1 to 2 in. diameter, terminating simple or rarely branched stems.

Perianth pink or red (white in *T. esquamatum?*), the inner segments woolly inside towards the base or below the middle.

13. **T. exaltatum,** *Benth.* A stout perennial, with a thick stock and erect stems, attaining 2 or 3 ft., usually branching in the upper portion, glabrous or hirsute with spreading hairs. Radical and lower leaves oblong-lanceolate, attaining 3 to 5 in., rather thick, contracted into a long petiole, the upper ones small, sessile, broad or narrow, often undulate or with crisped margins. Spikes erect, on long peduncles, at first ovoid-conical, at length oblong-cylindrical, about 2½ in. diameter. Bracts and bracteoles rarely half as long as the perianth, ovate-lanceolate, mucronate, scarious with a brown midrib and sometimes broadly brown towards the end, the bracteoles usually rather shorter than the bracts. Perianth rarely above ¾ in. long and sometimes rather shorter, the tube above 1 line long, the segments narrow, rigid and almost acute, plumose outside with long articulate hairs, the short glabrous tips of a dull red colour, the inner ones with dense wool inside below the middle. Stamens unequal, the filaments dilated but scarcely united at the base, or 2 of the shorter ones without anthers. Ovary usually but not constantly glabrous in the Western specimens, hairy in the Eastern ones, contracted into a rather long stipes.—*Ptilotus exaltatus*, Nees in Pl. Preiss. i. 630 (from the character given); Moq. in DC. Prod. xiii. ii. 281; *T. macrocephalum*, Moq. l.c. 290, not of R. Br.; *T. alopecuroideum*, Lindl. Bot. Reg. 1839, t. 28, but not the plant originally described in Mitch. Three Exped.; *Ptilotus nobilis*, F. Muell. Fragm. vi. 227, partly.

N. Australia. Careening Bay, N.W. coast, *A. Cunningham;* Depuech island, *Bynoe;* Victoria river, *F. Mueller;* Nichol Bay, *F. Gregory's Expedition.*

Queensland. Suttor river, *Sutherland;* Cape river, *Bowman;* Armadilla, *Barton;* Curriwillighie, *Dalton.*

N. S. Wales. Bengalla, *Leichhardt;* between Darling river and Cooper's Creek, *Neilson.*

Victoria. Avoca and Murray rivers, *F. Mueller;* Wimmera, *Dallachy* (all with very hirsute branches).

S. Australia. Lake Gairdner, *Babbage;* Gawler Ranges, *Sullivan.*

W. Australia, *Drummond, n.* 44, 437; Murchison river, *Oldfield;* Salt river and Cape Knob, *Maxwell.*

The western specimens are mostly tall, stout, and glabrous, or nearly so, the spikes often elongated, the wool inside the perianth very copious, and the ovary almost always glabrous. The Queensland and N. S. Wales specimens are often rather hairy, the ovary almost always so, and the spikes usually shorter, showing an approach to the following species (or variety?).

14. **T. semilanatum,** *Lindl. in Mitch. Trop. Austr.* 45. A perennial with a tufted stock and erect simple or slightly branched stems of 6 in. to 1 ft., more or less pubescent as well as the foliage with short crisped hairs or quite glabrous. Leaves linear or linear-lanceolate, acute, the radical ones not persistent at the time of flowering, the lower ones petiolate, 1 to 2 in. long, the upper ones more sessile and smaller, the margins flat or undulate-crisped. Spikes at first depressed or hemispherical, at length globular, about 1½ in. diameter. Bracts about 2 lines long, ovate-lanceolate, mucronate, more or less brown in the centre and tips;

bracteoles broader, more obtuse and not brown. Perianth about $\frac{3}{4}$ in. long, the tube about $\frac{3}{4}$ line, the segments narrow, rigid, plumose on the back, with short glabrous pink tips; the inner ones with long wool inside below the middle not very copious. Filaments dilated at the base, shortly connate, 1 or 2 short and without anthers. Ovary hairy on the top.—Moq. in DC. Prod. xiii. ii. 462; *T. pulchellum,* A. Cunn. and *T. setigerum,* A. Cunn.; Moq. l.c. 290; *Ptilotus nobilis,* F. Muell. Fragm. vi. 227 partly.

Queensland, *Mitchell;* head of the Gilbert river, *F. Mueller;* Wide Bay, *Bidwill;* Rockhampton, *O'Shanesy;* Midge Creek, *Bowman;* Warwick, *Beckler;* Darling Downs, *Lau.*

N. S. Wales. Bogan river, *Mitchell.*

The species is very near *T. exaltatum,* with which F. Mueller unites it, and from some specimens of which it is difficult to distinguish it. The spikes appear to be always short, the bracts smaller and more scarious and the foliage different.

15. **T. Manglesii,** *Lindl. Bot. Reg.* 1839 *under n.* 28. A perennial with a short hard tufted stock and decumbent ascending or rarely erect stems of $\frac{1}{2}$ to 1 ft., simple or rarely with 1 or 2 branches, the whole plant except the inflorescence glabrous. Radical leaves on long petioles, ovate obovate oblong narrow-spathulate or linear, obtuse or acute, 1 to 3 in. long, the stem-leaves few narrow and very acute. Spikes globular or ovoid, above 2 in. diameter, conspicuous for the coloured tips of the perianths protruding from the long white hairs. Bracts and bracteoles broadly lanceolate, acutely acuminate, the outer ones at least more or less brown in the centre, from $\frac{1}{2}$ to $\frac{3}{4}$ the length of the perianth. Perianth $\frac{3}{4}$ to 1 in. long, the tube narrow, about 1 line long, hirsute outside with short hairs; segments with glabrous obtuse pink or whitish tips of 2 to 4 lines, the remainder plumose outside with long hairs; the inner ones rather smaller and narrower, with long woolly hairs inside below the middle chiefly from the margins. Filaments dilated at the base but scarcely united above the perianth-tube, 1, 2 or 3 of them short without anthers. Ovary glabrous, contracted into a long stipes; style very excentrical, quite glabrous.—Field and Gardn. Sert. Pl. t. 52; Moq. in DC. Prod. xiii. ii. 289; *Ptilotus Manglesii,* F. Muell. Fragm. vi. 230; *T. spectabile,* Field and Gardn. l.c. t. 53; Moq. l.c. 289; *T. macrocephalum,* Nees in Pl. Pr. i. 627, not of R. Br.

N. Australia. Glenelg and Roebuck Bays, N.W. coast, *Martin.*

W. Australia. Swan river, *Drummond, 1st coll. n.* 435, 436, *Preiss, n.* 1358, 1359, and many others; northward to Murchison river and southward to Kalgan river, *Oldfield.*

The broad and narrow-leaved specimens, distinguished as *T. Manglesii* and *T. spectabile,* are so much intermixed and connected by intermediates that they cannot be reckoned as marked varieties. The cultivated specimen figured Bot. Mag. t. 5448 has the spike much more elongated than I have seen it in any of the numerous wild ones I have had before me.

16. **T. Beckerianum,** *F. Muell. in Linnæa,* xxv. 436. A perennial with a short branching stock and erect simple stems not exceeding 6 in. in our specimens, glabrous as well as the foliage. Leaves crowded

at the base of the stem, oblong-lanceolate, acute, ¾ to nearly 2 in. long, the lower ones contracted into a petiole, the greater part of the stem bearing only a few scarious scales or very small leaves. Spikes globular or at length ovoid, about 1½ in. diameter, much resembling those of *T. Manglesii* but smaller and the pink glabrous tips not so long. Bracts and bracteoles ovate, very scarious, slightly mucronate, with the midrib prominent, 3 to 4 lines long, the outer ones not so brown in the centre as in *T. Manglesii.* Perianth about ¾ in. long, the tube scarcely ½ line, the segments nearly equal, the dorsal hairs not so long as in *T. Manglesii*, and the glabrous tips scarcely 1 line, the inner segments with long woolly hairs outside near the base. Filaments shortly dilated at the base but scarcely united above the perianth-tube, 1 or two short and without anthers. Ovary glabrous; style with a few long spreading hairs.— *Ptilotus Beckeri,* F. Muell. Fragm. vi. 233.

S. Australia. Scrub near Spencer's Gulf, *Wilhelmi, Warburton.* Very near *T. Manglesii*, but besides the difference in foliage the flowers are smaller with less prominent tips and the styles hairy.

17. **T. gomphrenoides,** *Moq. in DC. Prod.* xiii. ii. 287. A perennial with a tufted stock and erect simple stems of 1 to 1½ ft., glabrous or sprinkled upwards with a few short hairs. Radical leaves oblong-spathulate, obtuse or mucronate, 1 to 1½ in. long, contracted into a long petiole; stem-leaves few small and distant, sessile, linear or lanceolate, very acute. Spikes solitary, at first globular, more ovoid when fully out, about 1 in. diameter or rather more. Bracts brown with scarious margins, lanceolate or almost ovate, very acute, several empty ones often crowded at the base of the spike; bracteoles as long and equally acute, but broader and the midrib alone brown. Perianth about ½ in. long, the turbinate tube about ⅓ line, the segments plumose with fine hairs, the outer ones rather broad, scarcely ribbed, with obtuse glabrous tips, the inner ones smaller, with acute tips and woolly hairs inside below the middle. Staminal cup very short, free from the perianth-tube; filaments short, unequal, the larger ones much dilated at the base, one very short without any anther. Ovary slightly hirsute. Style excentrical.

S. Australia. S. coast, *Strutt (Herb. Hook.).*

18. **T. esquamatum,** *Benth.* A glabrous perennial, with the branching stock, the erect simple rigid stems of 1 to 2 ft., and the linear mucronate acute small distant leaves of *T. Drummondii*, but without the lobes of the staminal cup of that species. Radical leaves not persistent at the time of flowering as in *T. gomphrenoides.* Spikes globular or ovoid, rather above 1 in. diameter. Bracts and bracteoles ovate, mucronate, thin and shining, with slightly prominent midribs. Perianth-tube exceedingly short, segments 5 to 6 lines long, rigid, 3-nerved, with narrow scarious margins, densely plumose outside with very fine long hairs, the glabrous scarious tips broader and more obtuse in the outer than in the inner ones, glabrous inside except the long woolly hairs

surrounding the stamens some of which proceed from the base of the
inner segments. Staminal cup very short; filaments rather short, not
very unequal, scarcely dilated at the base, but without intervening
teeth or lobes. Ovary woolly-hirsute.

W. Australia, *Drummond,* probably Swan river.

Series 4. Incurva.—Spikes globular, ¾ to 1½ in. diameter, terminating simple stems. Perianths curved upwards, the inner segments
woolly inside at the base. No teeth or lobes to the staminal cup between the filaments. Leaves linear.

19. **T. declinatum,** *Moq. in DC. Prod.* xiii. ii. 293. Stems from a
tufted stock prostrate or shortly ascending, in some specimens only 2
or 3 in. long, in others attaining 9 or 10 in., glabrous as well as the
foliage or sprinkled with a few woolly hairs, and generally a few woolly
tufts on the stock. Leaves linear or narrow-lanceolate, often rather
crowded, from under ½ in. to above 1 in. long, those close under the
spikes often the longest. Spikes nearly globular, 1 to 1½ in. diameter,
closely sessile within the last leaves. Flowers not numerous, more or
less incurved as in *T. erubescens,* but larger. Bracts and bracteoles
thin, broad, mucronate-acute, 3 to 4 lines long. Perianth ¾ to 1 in.
long, the segments free almost from the base, narrow, rigid, plumose
outside with long fine hairs, the glabrous tips short and acute; the
inner segments rather smaller, densely woolly inside at the base.
Staminal cup very short, glabrous, truncate; filaments scarcely or not
at all dilated at the base. Ovary woolly or nearly glabrous.—*Ptilotus
declinatus,* Nees in Pl. Preiss. i. 631; *T. eriocephalum,* Moq. in DC. Prod.
xiii. ii. 293.

W. Australia, *Drummond, n.* 429, *Preiss, n.* 1362; Murchison river, *Oldfield.*

20. **T. erubescens,** *Moq. in DC. Prod.* xiii. ii. 293. Stems several
from a thick rhizome or densely tufted stock, simple, erect or ascending,
6 in. to above 1 ft. high, glabrous as well as the foliage or sprinkled
with a few woolly hairs and interspersed on the stock with tufts of hairs
usually straight. Leaves linear, acute, the radical ones often 2 or 3 in.
long on long petioles, the stem ones much smaller and sessile or nearly
so, the uppermost distant from the spike and sometimes passing into
scarious bracts. Spikes solitary, nearly globular or at length scarcely
ovoid, ¾ to 1 in. diameter. Bracts and bracteoles broad, thin, with
short points, closely embracing the flowers, 3 to 4 lines long, the midribs scarcely prominent. Perianth more or less curved upwards, especially when in bud, ½ to ¾ in. long, the tube about ½ line long or
sometimes scarcely more than a slightly expanded disk, the segments
narrow, plumose outside, the short glabrous tips obtuse in the outer
ones, the inner segments with more acute tips and densely woolly inside below the middle. Filaments dilated and very shortly united at
the base. Ovary hairy or glabrous.—Dietr. Fl. Univ. Ser. 2. t. 14;

Ptilotus erubescens, Schlecht. in Linnæa, **xx.** 575; F. Muell. Fragm. vi.
229; *T. linifolium,* A. Cunn.; Moq. in DC. Prod. xiii. ii. 292.

N. S. Wales. Barren rocky country W. of Wellington valley, *A. Cunningham.*
Victoria. Grampians? *Mitchell;* Avoca river, *F. Mueller;* Skipton, *Whan;*
Glenelg river, *Robertson.*
S. Australia. Gawlertown, *Behr.;* Lofty Range, Salt Creek, *F. Mueller;* Port
Adelaide, *Blandowski.*

SERIES 5. POLYCEPHALA.—Stems mostly branched or rarely some
of them long decumbent and simple, glabrous or with crisped woolly
hairs. Spikes mostly globular, ¾ to 1 in. diameter. Perianths straight
or recurved. Filaments without intervening teeth or lobes.

21. **T. divaricatum,** *Gaudich. in Freyc. Voy. Bot.* 445. A glabrous
shrub of 1 to 2 ft., with rigid striate spreading branches. Leaves rather
distant, sessile or nearly so, linear or linear-lanceolate, obtuse or acute,
rather thick, ½ to 1 in. long. Spikes globular or ovoid, nearly 1½ in.
diameter, more or less pedunculate, forming compact leafy panicles.
Bracts and bracteoles 3 to 3½ lines long, thinly scarious and shining,
without prominent midribs, closely enveloping the perianths. Perianth
5 to 7 lines long, the tube fully ½ line long, shortly hairy outside, the
segments plumose outside with long fine hairs, the outer ones with
very short obtuse glabrous tips, the inner ones much shorter, with
narrow tips scarcely glabrous and with dense long woolly hairs inside
below the middle. Filaments unequal, not long, dilated at the base
and very shortly united above the perianth-tube. Ovary stipitate,
glabrous. Style excentrical.—Moq. in DC. Prod. xiii. ii. 291; *Ptilotus
divaricatus,* F. Muell. Fragm. vi. 229.

W. Australia. Champion Bay and Murchison river, *Oldfield;* Sharks Bay (*Gaudichaud*), *M. Brown.*

22. **T. helipteroides,** *F. Muell. Fragm.* iii. 122. Apparently annual,
with several erect or decumbent simple or branched stems of 6 in. to
1 ft., clothed as well as the foliage with silky-woolly hairs. Leaves
linear or linear-lanceolate, the lower ones petiolate 1 to 2 in. long, the
upper ones nearly sessile and smaller, not numerous. Spikes at first
globular, at length ovoid, ¾ to 1 in. long. Bracts and bracteoles ovate,
thin and transparent, the bracts acute and as long as the perianth or
nearly so, the bracteoles much shorter and more obtuse. Perianth 3½
to 4 lines long, the united disk very short, but the claws forming an
erect tube of about 1 line, the segments narrow and rather rigid, the
dorsal hairs very fine and not dense, the inner ones rather shorter and
more acute than the outer. Filaments much dilated at the base, but
scarcely united above the disk, surrounded by woolly hairs not very
copious proceeding from the base of the inner segments, unequal, 1 or
2 of the shortest without anthers.—*Ptilotis helipteroides,* F. Muell.
Fragm. vi. 231; *T. brachytrichum,* F. Muell. Fragm. iii. 161.

N. Australia. Nichol Bay, N. W. Coast, *Gregory's Expedition;* sandy plains of
the interior, *M'Douall Stuart's Expedition.*

23. **T. Stirlingii,** *Lindl. Bot. Reg.* 1839, *under n.* 28. A perennial with long procumbent or ascending simple or branched stems more or less clothed as well as the foliage with white crisped woolly hairs, sometimes dense especially on the lower part of the stems, sometimes small and rare in the upper part and leaves or accompanied by a short glandular pubescence on the peduncles. Leaves lanceolate oblong or almost linear, the lower ones obtuse and contracted into a short petiole, the upper ones small, more acute and sessile, broad or narrow. Spikes globular, solitary at the ends of the stems or loosely paniculate, ¾ to 1 in. diameter. Bracts and bracteoles scarious but rather rigid, mucronate-acute, with the midrib more or less prominent, the bracteoles about 2 lines long, the bracts usually shorter. Perianth about 5 to 6 lines long, the tube about ½ line long and shortly hispid, the segments plumose with fine hairs, long in the lower half, shorter and not so dense higher up, the outer segments with broad dentate glabrous pink tips, the inner ones with narrower tips and long woolly hairs inside near the base chiefly from the margins. Filaments more or less dilated and shortly united at the base, either all anther-bearing, or 1, 2 or 3 of them short and without anthers, or sometimes even the longer ones scarcely dilated and the anthers imperfect. Ovary stipitate, glabrous; style slightly excentrical.—Moq. in DC. Prod. xiii. ii. 297; *T. carneum,* Moq. l.c. 291.

W. Australia. Swan river, *Fraser;* Murchison river and Champion Bay, *Old-field;* Sharks Bay, *M. Brown.*

24. **T. laxum,** *Benth.* Perennial (?) with procumbent or ascending loosely branched stems, glabrous or sprinkled with a few short crisped hairs. Leaves broadly ovate or obovate, obtuse or mucronate, the margins slightly crisped, green and not thick, contracted into a short petiole, sometimes above 1 in. long, the upper ones smaller. Spikes globular ovoid or at length shortly cylindrical, rather under 1 in. diameter, all pedunculate in a loose leafy panicle. Bracts and bracteoles ovate or oblong, obtuse or scarcely mucronate, the midrib usually prominent, the outer ones more or less brown or red at least in the centre, 2 to nearly 3 in. long, the bract usually smaller than the bracteoles. Perianth 5 to nearly 6 lines long, the tube cylindrical, about ¾ line long, resembling a thick pedicel but hollow, enclosing the stipes of the ovary; segments plumose outside with long fine hairs, the outer ones with broad coloured denticulate glabrous tips, the inner ones rather shorter, with narrower tips, and woolly hairs inside near the base chiefly from the margins. Staminal cup very short and oblique, the two upper filaments long, much dilated at the base with oblong anthers, the others short and without anthers. Ovary glabrous; style excentrical.

W. Australia. Between Cape Le Grand and Cape Paisley, *Maxwell.*

25. **T. axillare,** *F. Muell. Herb.* A perennial with prostrate or ascending branching stems of about 1 ft., the young shoots with long white woolly hairs, otherwise glabrous. Leaves ovate or elliptical, very acute,

contracted into a rather long petiole, $\frac{1}{2}$ to 1 in. long. Spikes nearly globular, about 1 in. diameter, on short axillary peduncles or flowering branchlets, usually with a few small leaves close under the spike. Bracts and bracteoles very acute and mucronate, about 3 lines long, very thin but the outer ones rather brown with prominent midribs. Perianth about 6 lines long, the short tube shortly hirsute, the segments narrow, plumose outside with long fine hairs, with short glabrous pink truncate or denticulate tips, the inner segments smaller with narrower tips and with a few woolly hairs inside at the base. Filaments 5, unequal and some without anthers, but all dilated towards the base and united in a short cup. Ovary glabrous; style quite lateral.

N. Australia. Nichol Bay, N. W. coast, *F. Gregory's Expedition.*

26. **T. striatum,** *Moq. in Herb. Hook.* A glabrous undershrub, with rather slender but rigidly divaricate striate branches. Leaves few and distant, sessile or nearly so, linear, $\frac{1}{2}$ to $\frac{3}{4}$ in. long or the upper ones smaller. Spikes at first hemispherical but at length somewhat elongated, all pedunculate forming a loose irregular leafy panicle. Bracts and bracteoles obtuse, rather broad, closely enveloping the perianth, scarious and shining without prominent midribs, about 2 lines long. Perianth about 5 lines long, very deciduous leaving the bracts persistent, the segments free almost from the base, narrow and rigid, plumose almost to the tips with long fine soft hairs, the outer ones often rather longer than the inner with more prominent tips, all glabrous inside. Staminal cup very short, surrounded by very dense long and intricate woolly hairs proceeding mostly from the cup itself; filaments unequal, the longer ones dilated downwards. Ovary glabrous; style excentrical.

W. Australia, *Drummond, n.* 430; Port Gregory, *Oldfield ;* Dirk Hartog's island, *Milne* (the last two very imperfect specimens and somewhat doubtful).

Drummond's specimens were afterwards referred by Moquin to *T. divaricatum,* to which they bear some resemblance, but from which they differ in a much looser divaricate panicle, the flowers smaller, the perianth segments free to the base, and the wool surrounding the stamens proceeding from the staminal cup and not from the inner segments.

27. **T. auriculifolium,** *A. Cunn.; Moq. in DC. Prod.* xiii. ii. 287. A perennial, probably with a tufted stock. Radical leaves obovate or obovate-oblong, very obtuse, glabrous, rather thick, with undulate margins, $1\frac{1}{2}$ to 2 in. long, on rather short petioles. Flowering stems erect, 6 in. to 1 ft. high in our specimens, nearly glabrous below, clothed upwards with soft hairs, and bearing a few small distant leaves. Spikes nearly globular, 4 or 5 together nearly sessile and crowded in a terminal dense ovate panicle. Bracts broadly ovate, acute, scarcely 2 lines long ; bracteoles twice as long, obtuse, very broad and enveloping the flowers, all very thin and transparent. Perianth 5 to nearly 6 lines long, the turbinate tube about $\frac{3}{4}$ line ; segments narrow, nearly equal, densely plumose with rather long hairs, with short glabrous tips rounded or truncate in the outer ones, narrower in the inner ones, all glabrous inside. Staminal cup shortly free, surrounded by long woolly hairs reach-

ing to the top of the stamens and some of them proceeding from the
filaments themselves; anthers often all 5 perfect. Ovary on a long
stipes, woolly at the top.

N. Australia. Dampier's Archipelago, N.W. coast, *A. Cunningham.*

28. **T. sericostachyum,** *Nees in Pl. Preiss.* i. 627. Stems from
a thick rhizome procumbent ascending (or erect ?), branching, 1 to 1½
ft. long, green and glabrous as well as the foliage or sprinkled with a
few woolly hairs. Leaves oblong or lanceolate, acute or obtuse, the
larger ones 1 to 2 in. long, much smaller on the branches, all contracted
into a petiole. Spikes at first globular, at length more ovoid, about ¾
in. diameter, sessile within the last leaves of the numerous branches.
Bracts lanceolate, about 2 lines long, very thin but with a prominent
green centre produced into a fine point; bracteoles about as long but
broader with the midrib only slightly prominent. Perianth-segments
free almost from the base, narrow, rigid, 4 to 4½ lines long, densely
plumose outside with fine white hairs, the narrow glabrous tips obtuse
in the outer segments, the inner segments shorter with very short acute
tips and very few hairs inside below the middle. Staminal cup short,
free from the perianth, surrounded by articulate straight hairs; filaments
slightly dilated at the base, very unequal, 1, 2 or 3 without anthers.
Ovary glabrous.—Moq. in DC. Prod. xiii. ii. 284; *Ptilotus sericostachyus,*
F. Muell. Fragm. vi. 230; *T. floribundum,* Moq. l.c. 283; *Ptilotus
floribundus,* F. Muell. Fragm. vi. 233.

W. Australia. Swan river, *Collie, Drummond, n.* 149; *Preiss, n.* 1372, *Oldfield.*

29. **T. roseum,** *Moq. in DC. Prod.* xiii. ii. 284. A perennial with
decumbent or ascending loosely branched stems of above 1 ft., more or
less hirsute with crisped but rather spreading hairs, not so white as in
T. Stirlingii. Leaves ovate oblong or elliptical, acute or obtuse, contracted at the base and the lower ones petiolate, green and usually
loosely villous underneath; the larger ones 1 in. long, the upper ones
small and distant. Spikes globular or at length ovoid, 1 in. diameter or
rather more. Bracts and bracteoles thin, mucronate-acute, with prominent midribs, about 3 lines long or the outer ones smaller. Perianth
5 to 6 lines long, recurved, plumose outside with fine but not very long
hairs, the outer ones with glabrous tips slightly dentate, the inner ones
shorter with small acute tips, all glabrous inside. Staminal cup very
oblique, surrounded by hairs on the upper side, the two upper filaments
long shortly dilated at the base, with perfect anthers, the others very
small without anthers or quite obsolete. Ovary glabrous; style very
excentrical.

W. Australia. Swan river, *Drummond, 1st coll. n.* 433; Murray river? *Oldfield;* Vasse river, *Mrs. Molloy* (a more glabrous form, with rather smaller flowers).

30. **T. fusiforme,** *R. Br. Prod.* 415. A perennial with a fusiform
rhizome and slender erect branching stems of 1 to 2 ft. Leaves narrow-
linear or almost filiform, the lower ones often 2 in. long, the upper ones
few small and distant. Spikes ovoid, about 1 in. diameter, on long

slender branches or peduncles. Bracteoles broad, very obtuse, closely enveloping the perianth, very thin and shining, about 2 lines long, the bracts shorter and more acute. Perianth 5 to 6 lines long, the segments free almost or quite to the base, narrow, rigid, 3-nerved, plumose outside with rather rigid long hairs, with short glabrous tips rather longer and more obtuse in the outer than the inner segments, all glabrous inside. Staminal cup very short, densely covered outside with long straight hairs; filaments unequal, all antheriferous or one without an anther. Ovary densely hairy.—Moq. in DC. Prod. xiii. ii. 294; *Ptilotus fusiformis*, Poir. Dict. Suppl. iv. 619.

N. Australia. Islands of the Gulf of Carpentaria, *R. Brown;* Dampier's Archipelago, N.W. coast, *A. Cunningham;* Victoria river, *F. Mueller.*

31. **T. gracile,** *R. Br. Prod.* 415. Very near *T. fusiforme,* and the structure of the flowers the same, but an annual with still more slender branching stems, the leaves filiform, the spikes globular not above ¾ in. diameter, and the perianth only about 4 lines long with much shorter glabrous tips.—Moq. in DC. Prod. xiii. ii. 294; *Ptilotus gracilis,* Poir. Dict. Suppl. iv. 620.

N. Australia. Islands of the Gulf of Carpentaria, *R. Brown;* N.W. coast, *Bynoe.*

Series 6. Squamigera.—Spikes globular or cylindrical, ½ to 1 in. diameter. Staminal cup with transparent scale-like teeth or lobes between the filaments.

The presence of the lobes of the staminal cup, or *staminodia* of Moquin, would technically remove the three following species from the genus, but the character is so purely artificial, as not even to constitute a good section, these species being perhaps each of them more nearly allied to corresponding species in other groups than to each other.

32. **T. Drummondii,** *Moq. in DC. Prod.* xiii. ii. 292. A glabrous perennial, the stock branching at the base into several erect simple rigid but not stout stems of 1 to 2 ft., the radical leaves not persisting as in *T. gomphrenoides.* Stem-leaves linear, mucronate-acute, ½ to 1 in. long, sessile or nearly so, or the lower ones longer and petiolate, the uppermost small and distant. Spikes globular or at length ovoid, ¾ in. diameter or rather more. Bracts and bracteoles broad, obtuse or shortly mucronate, thin and shining, 2 to 3 lines long. Perianth 4 to 5 lines long, the tube turbinate about ⅓ line, the segments rigid, scarcely ribbed, densely plumose outside with very fine hairs, with scarious margins and glabrous tips, the outer ones obtuse, the inner ones rather shorter and woolly inside below the middle. Filaments not very unequal, flat and tapering to the top, very shortly united above the perianth-tube, with oblong fringed exceedingly thin and transparent scales between them, surrounded by loose woolly hairs besides those proceeding from the inner segments. Ovary glabrous.—*Ptilotus Drummondii,* F. Muell. Fragm. vi. 229; *T. fusiforme,* Nees in Pl. Preiss. i. 626, not of R. Br.

W. Australia. Swan river, *Collie, Drummond,* 1st *coll. n.* 431, *Preiss, n.* 1374; Champion Bay, *Oldfield;* Walker's Brook, *Maxwell.*

The general resemblance of this plant to *T. esquamatum* is so close that it is not readily distinguished without examining the flowers. The spikes are, however, usually but not quite constantly, considerably smaller.

33. **T. calostachyum,** *F. Muell.* An erect slender slightly branched annual of 1 to 2 ft. Leaves few, very narrow-linear almost filiform, the lower ones 1 to 2 in. long, the upper ones small and distant. Spikes on slender peduncles, at first conical, at length oblong-cylindrical, 1 to 1½ in. long. Bracts and bracteoles ovate, mucronate, very thin and shining, 1½ to 2 lines long. Perianth pink, 2½ to 3 lines long, the segments free from the base, scarious, densely plumose outside with fine hairs short in the lower part, longer in the upper half, with shortly glabrous obtuse tips, the 3 inner ones rather shorter, with woolly nairs inside towards the base but on the segments and not on the staminal cup. Filaments slightly unequal, all bearing anthers, united in a short cup with linear or lanceolate exceedingly thin scales between them fringed or glandular on the margin. Ovary glabrous.—*Arthrotrichum calostachyum,* F. Muell. in Trans. Bot. Soc. Edinb. vii. 500 ; *Ptilotus calostachyus,* F. Muell. Fragm. vi. 231.

N. Australia. Islands of the Gulf of Carpentaria, *R. Brown;* Upper Victoria river, Hooker's and Sturt's Creeks, *F. Mueller;* Nichol Bay, *Walcott;* Roebuck Bay, *Martin.*

The habit approaches that of some of the annual *Ptiloti,* the scales between the stamens are somewhat variable but present in all the flowers examined, usually about ¼ line long, the other characters are entirely those of *Trichinium.*

34. **T. Fraseri,** *A. Cunn.*; *Moq. in DC. Prod.* xiii. ii. 295. Stems from a woody but rather slender branching base, erect, slender, about 6 in. high, glabrous as well as the foliage. Leaves small, very narrow-linear, almost terete. Spikes small, probably globular when perfect. Bracts and bracteoles thin, acute, 2 to 2½ lines long. Perianth scarcely exceeding the bracts, the segments free from the base, plumose outside, the tips shortly glabrous ; the inner segments rather smaller with a few of the marginal hairs below the middle turned inside. Staminal cup glabrous ; filaments rather short, filiform, all nearly equal and bearing anthers, with oblong transparent fringed scales between them.

N. S. Wales, *Fraser,* the precise station not given.

SERIES 7. SPATHULATA.—Perennials with short decumbent stems leafy to the spike. Spikes globular ovoid or cylindrical, sessile within the last leaves. Perianths straight. Filaments without intervening teeth or scales. Leaves mostly spathulate.

35. **T. spathulatum,** *R. Br. Prod.* 415. A perennial with a thick woody rhizome and spreading prostrate stems of 3 to 6 in. without the spike, glabrous as well as the foliage or nearly so. Radical leaves ovate or spathulate, obtuse, ½ to 1 in. long and more or less decurrent on the long petiole ; stem-leaves smaller narrow more acute and scarcely petiolate, those immediately under the spike again rather larger. Spikes ovate, at length cylindrical, sessile within the last leaves, 2 to 4 in. long

and $\frac{3}{4}$ to 1 in. diameter, of a yellowish hue.   Bracts and bracteoles thin
and shining, mostly acute, about 3 lines long.   Perianth 4 or 5 lines
long, the tube narrow, about $\frac{3}{4}$ line, the segments densely plumose outside, with long hairs denticulate at the points the outer ones with very
short more or less truncate and denticulate glabrous tips, the inner ones
rather shorter, tapering into entire or scarcely toothed tips.   Filaments
dilated and very shortly united at the base, surrounded by articulate
hairs not very copious, all bearing anthers or one of them short and
without any anther.   Ovary glabrous; style excentrical.—Moq. in DC.
Prod. xiii. ii. 287.; Hook. f. Fl. Tasm. i. 310, t. 94 ; *Ptilotus spathulatus*,
Poir. Dict. Suppl. iv. 620 ; *Trichinium mucronatum*, Nees in Pl. Preiss.
i. 628 ; Moq. l.c. 288.

Victoria.   Murray and Avoca rivers, *Dallachy, F. Mueller;* Melbourne, *Harvey*
and others ; Skipton, *Whan ;* Little river, *Fullagar.*
Tasmania.   Derwent river, *R. Brown;* abundant on dry plains near Ross, *J. D.
Hooker.*
S. Australia.   Enfield, Barossa and Lofty Ranges, *F. Mueller* (the specimens
from the latter locality luxuriant with branching stems of nearly 1 ft.); Venus Bay,
Warburton ; Gawler Ranges, *Sullivan.*
W. Australia, *Drummond, n.* 428 ; Mount Brown, York district, *Preiss, n.* 1373 ;
Vasse river, *Mrs. Molloy.*

36. **T. pyramidatum,** *Moq. in DC. Prod.* xiii. ii. 288.   A small
plant probably perennial, with a tufted stock and ascending or erect
stems not above 1 in. without the spike in our specimens, glabrous as
well as the foliage or woolly-hairy under the spike.   Leaves small,
oblong-spathulate, the radical ones petiolate, the others sessile.   Spikes
terminal, solitary or 2 or 3 together, conical and about 1 in. long in our
specimens but probably at length longer and cylindrical.   Bracts and
bracteoles broad, very thin, about 2 lines long, the bracteoles very obtuse, the bracts more acute with an opaque midrib.   Perianth 4 lines
long or rather more, the segments free from the base, densely plumose
outside with soft white hairs, the inner ones smaller and more acute
than the outer.   Staminal cup very short, surrounded by straight articulate hairs proceeding from the base of the inner segments; filaments
unequal, 1, 2 or 3 of them short and without anthers or quite deficient.
Ovary glabrous.—*Ptilotus pyramidatus*, F. Muell. Fragm. vi. 230.

W. Australia, *Drummond, n.* 99, 221. Perhaps a depauperated state of a species
usually larger.

37. **T. holosericeum,** *Moq. in DC. Prod.* xiii. ii. 287.   A perennial
with the thick rhizome tufted stock and short procumbent stems of
T. spathulatum, glabrous as well as the foliage or the young shoots
sprinkled with a few silky hairs.   Radical leaves obovate or oblong-
spathulate, $\frac{1}{2}$ to 1 in. long, on long petioles, the stem-leaves few and
small.   Spikes sessile within the last leaves, globular or at length
scarcely ovoid, about $\frac{3}{4}$ in. diameter, whiter and more shining than in
T. spathulatum, the broad shining bracts being larger in proportion to
the perianth and more conspicuous.   Perianth about 5 lines long, the
tube about $\frac{1}{2}$ line long, the segments narrow, densely plumose outside

with silky hairs less prominently articulate than in *T. spathulatum* and not denticulate, all very long and the lower ones as long as the whole perianth; the outer segments with very short rounded glabrous tips, the inner ones more acute, with long wool inside below the middle. Filaments slightly dilated and very shortly united at the base, all nearly equal or 1 or 2 short and without anthers. Ovary glabrous.—*Ptilotus holosericeus*, F. Muell. Fragm. vi. 229.

W. Australia, *Drummond, n.* 75 *and* 232.

SERIES 8. PARVIFLORA.—Spikes globular ovoid or cylindrical, ¼ to ½ in. diameter, filaments without intervening teeth or scales.

38. **T. Cunninghamii,** *Benth.* An erect rather flaccid slightly branched annual, attaining 1 ft. or more, glabrous or sprinkled as well as the foliage with a few soft hairs. Leaves linear or linear-lanceolate, acute or obtuse, contracted into a petiole, the lower ones often 2 in. long, the upper ones distant. Spikes solitary or in a cluster of 2 or 3, at first conical, at length oblong (or cylindrical?) above ½ in. long and about ¼ in. diameter, but the old ones of our specimens imperfect. Bracts and bracteoles very thin and transparent, the bracteoles broad and about ½ line long, the bracts narrower more acute and rather longer. Perianth 1½ lines long, the segments free almost from the base, transparent, narrow, densely clothed outside with white hairs, woolly on the lower half, straight on the upper half, all glabrous inside but the marginal wool towards the base of the inner ones slightly turned inwards. Staminal cup short; filaments slender, rather unequal but all bearing anthers, without intervening teeth or scales. Ovary glabrous. —*Ptilotus lanatus*, A. Cunn.; Moq. in DC. Prod. xiii. ii. 281.

N. Australia. Point Cunningham, Cygnet Bay, N. W. Coast, *A. Cunningham.* I have not adopted Cunningham's specific name because in transferring the plant to *Trichinium* it is wholly inappropriate, and, moreover, might create confusion owing to there having been a *T. lanatum*, Lindl. (now reduced to *T. obovatum*), besides a *T. semilanatum*, Lindl., still retained.

39. **T. leucocoma,** *Moq. in DC. Prod.* xiii. ii. 292. Probably herbaceous, erect or ascending, rather slender, slightly branched, glabrous or nearly so, our specimens not exceeding 6 in., with linear or linear-lanceolate flaccid leaves not above 1 in. long. Spikes at first globular at length oblong, ½ to ¾ in. long and nearly ½ in. diameter. Bracteoles broad obtuse or with small points, thin and shining, 1 to 1¼ lines long; bracts usually not so broad and more pointed, and often shorter. Perianth about 2¼ lines long, the tube rather thick ½ line long, the segments rigid with scarious margins, the dorsal hairs not so dense nor so long as in most species, the tips shortly glabrous truncate and denticulate in the outer segments, more acute but not shorter in the inner ones, all glabrous inside. Staminal cup shortly free near the base of the perianth-tube, surrounded by woolly hairs; filaments scarcely dilated. Ovary glabrous.

S. Australia. Great marsh of the interior, *Strutt* (*Herb. Hook.*)

40. ? **T. villosum,** *Nees in Pl. Preiss.* i. 628. An annual with prostrate simple slender villous stems. Leaves petiolate, the radical ones spathulate, the stem ones narrow-lanceolate, acute, undulate, clothed with rather long woolly hairs, about 1 in. long. Spikes solitary, ovate-conical, obtuse, 5 lines long. Bracts ovate, finely pointed, 1 nerved, whitish transparent and shining, loosely villous, as long as the perianth. Perianth greenish white, about 3 lines long, the segments mucronate-acute, very densely woolly.—Moq. in DC. Prod. xiii. ii. 285.

W. Australia. Swan river, *Preiss. n.* 1365 (*Nees*). I have seen no specimen corresponding with the above character. The plant is said to be allied to *T. incanum,* but does not appear to have any stellate tomentum, and the habit must be widely different. The internal structure of the flower and the precise position of the wool are not described.

41. **T. brachyanthum,** *F. Muell. in Herb. Hook.* A perennial with a hard almost woody rhizome and several erect or ascending simple or branched stems under 1 ft., clothed with a short soft woolly tomentum, the foliage glabrous or nearly so. Leaves from ovate-oblong to almost linear, very obtuse, contracted into a petiole, rather thick and soft, rarely above 1 in. long. Spikes terminal, solitary and shortly pedunculate, or 2 together and almost sessile, at first globular, at length oblong-cylindrical, ½ to ¾ in. long and 4 to 5 lines diameter. Bracts and bracteoles broad, thin, shining, obtuse or with minute points, scarcely above 1 line long. Perianth 1½ to 2 lines long, the tube reduced to a very small open disk, but very densely clothed outside with long woolly hairs enveloping the whole fruiting perianth; segments plumose to the top with long straight hairs, the inner ones woolly inside towards the base. Staminal cup nearly ½ line long, glabrous, truncate; filaments unequal, 1 or 2 sometimes without anthers. Ovary glabrous.

N. Australia, *F. Mueller.* Included by F. Mueller, Fragm. vi. 233, as a lanuginous variety in *T. arthrolasium,* but appears to me to be distinct in habit and inflorescence as well as in the wool.

42. **T. arthrolasium,** *F. Muell.* A perennial or undershrub with a thick rhizome and erect much branched stems under 1 ft. high, hard and almost woody at the base, clothed as well as the foliage with articulate crisped hairs, usually dense and fulvous on the branches. Leaves lanceolate or oblong-linear, obtuse or acute, contracted into a short petiole, rather thick and soft, ½ to 1 in. long. Spikes at first conical but very soon elongated, ½ to ¾ in. long, rather numerous, shortly pedunculate, forming a corymbose leafy panicle. Bracts and bracteoles thin and shining, mucronate-acute, glabrous or the outer ones slightly hairy, 1 to 1½ lines long. Perianth about 1½ lines long, the very short tube densely surrounded by straight hairs longer than the whole perianth; segments narrow-lanceolate, nearly equal, acute, scarious, plumose outside with straight hairs very much shorter than the basal ones, all glabrous inside, but a few of the marginal hairs of the inner ones turned inside. Staminal cup free, truncate, not sur-

rounded by hairs. Ovary glabrous.—*Ptilotus arthrolasius,* F. Muell. Fragm. vi. 232.

N. Australia. Sturt's Creek, *F. Mueller.*

43. **T. ærvoides,** *F. Muell. Fragm.* iii. 123. Probably perennial and procumbent. Stems branching, the young shoots and peduncles clothed with white woolly hairs. Leaves ovate or ovate-lanceolate, acute, contracted into a rather long petiole, $\frac{1}{2}$ to 1 in. long. Spikes numerous, solitary or 2 or 3 together on short axillary or terminal peduncles, ovoid or conical (or at length cylindrical?), 3 to 4 lines diameter. Bracts ovate, acute, with brown tips, the bracteoles more transparent broader and more obtuse, all about as long as the perianth and usually bearing dorsal hairs at the base. Perianth scarcely 2 lines long, the segments free from the base, rigid, acute, plumose outside with short glabrous tips, glabrous inside. Filaments unequal, united at the base in a short glabrous truncate cup, 1, 2 or 3 of them without anthers. Ovary densely villous on the top.—*Ptilotus ærvoides,* F. Muell. Fragm. vi. 231.

N. Australia. Nichol Bay, N.W. Coast, *F. Gregory's Expedition.*

44. **T. Roei,** *F. Muell. Herb.* Probably perennial and procumbent, closely allied to *T. ærvoides,* the branches white with a close cottony wool. Leaves petiolate, obovate or orbicular, obtuse or the upper ones mucronate, $\frac{1}{2}$ to 1 in. long. Bracts and bracteoles more rigid and villous than in *T. ærvoides,* nearly as long as the perianth. Perianth nearly 3 lines long, the segments rigid and erect in the lower half, the upper half lanceolate spreading, plumose outside with glabrous tips, the inner ones densely woolly inside below the middle. Filaments unequal, 1 or 2 without anthers. Ovary glabrous.

W. Australia. Lake Barlee, *Forrest's Expedition.*

45. **T. cæspitulosum,** *F. Muell.* A perennial with a densely tufted stock, the crowded short branches covered with the imbricate persistent remains of old leaves, the flowering branches slender, erect, simple, 3 to 5 in. high. Leaves crowded on the short cæspitose barren branches, distant on the flowering ones, linear-terete, mucronate-acute, $1\frac{1}{2}$ to 3 lines long, with a small thick callous and persistent base. Spikes terminal, shortly conical, 4 to 6 lines long. Bracts and bracteoles nearly 2 lines long, very broad, closely enveloping the perianths, thin and shining, the midrib produced into a small point, glabrous except a few hairs at the base. Perianth-tube short, the segments shortly exceeding the bracts, very obtuse, densely plumose outside with straight hairs, the glabrous ends nearly $\frac{1}{3}$ of the whole segment. Perfect stamens 2 only, with very short filaments. Style short.— *Ptilotus cæspitulosus,* F. Muell. Fragm. vi. 232.

W. Australia, *Drummond, n.* 189. The specimens are not in a good state, and I have been unable to ascertain the precise form of the staminal cup.

SERIES 9. HELICHRYSOIDEA.—Characters the same as of the species.

46. **T. helichrysoides,** *F. Muell.* A low perennial with a hard thick densely-branched stock covered by the withered remains of old leaves, the leafy flowering stems not above a few in. high, densely clothed with a woolly tomentum more silky shining and silvery on the fresh foliage. Leaves crowded up to the spike, sessile, elliptical, oblong or almost ovate, obtuse with a short fine point, thick and silky tomentose on both sides, about ½ in. long. Spikes sessile within the last leaves, depressed-globular, ¾ in. diameter or rather more. Bracts and bracteoles broadly ovate, about 2 lines long, scarious and shining but usually with a tuft of woolly hairs on the back. Perianth about 4 lines long, the tube turbinate, about ½ line long, the segments densely plumose outside with fine hairs rather long and silky on the lower portion, shorter on the upper portion, the glabrous tips very short; all glabrous inside. Filaments slightly dilated towards the base and very shortly united, surrounded by very few hairs only, but the ovary very densely hirsute.— *Psilotrichum helichrysoides,* F. Muell. Fragm. i. 237 ; *Ptilotus helichrysoides,* F. Muell. Fragm. vi. 231.

W. **Australia,** *Drummond ;* hills near Baker's Well, Port Gregory, *Oldfield.*— The species has no immediate affinity with any other one known.

Species insufficiently known.

47. **T. parvifolium,** *F. Muell. Rep. Babb. Exped.* 19. Of this the only specimen is a slender branching fragment, not 6 in. long, pubescent with a few short woolly hairs. Leaves few, minute, ovate or lanceolate, acute, none of them above 1 line long. No perfect spikes on the specimen, but only a very few flowers apparently like those of *T. semilanatum* but smaller.—*Ptilotus parvifolius,* F. Muell. Fragm. vi. 229.

S. **Australia.** Stuart's Creek, *Babbage's Expedition.*

5. PTILOTUS, R. Br.

Flowers hermaphrodite. Perianth-segments 5, linear, free or united in a very short tube at the base, rigid, the lower portion usually 3-ribbed and glabrous, or covered outside with articulate hairs or intricate wool, the upper moiety a glabrous coloured lamina, all glabrous inside, or the inner ones with woolly hairs below the lamina. Stamens 5, one or two of them sometimes small without anthers, all united in a short cup or ring at the base, without intervening teeth or lobes ; anthers 2-celled. Ovary uniovulate ; style central or slightly excentrical. Fruit an indehiscent utricle. Seed vertical.—Herbs mostly (or always ?) annual and glabrous except the inflorescence. Flowers in globular conical or cylindrical spikes, with a woolly rhachis. Bracts and bracteoles scarious.

Like *Trichinium,* the genus is probably limited to Australia. *P. corymbosus* is indeed said to be found also in the island of Flores in the Moluccas, but from Blume's short character it is doubtful whether it be the same as the Australian plant of that name, or even a congener. *P. amabilis*, Span., from Timor, has never been described ; *P. ovatus*, Moq., from E. India, with opposite leaves, is a *Psilotrichum*, *P. Sandwicensis*, A. Gray, from the Sandwich Islands, is an *Achyranthes*.

The genus only differs from some of the smaller flowered *Trichinia*, in the absence of the dorsal hairs which, in the latter genus, give the laminæ of the perianth-segments a plumose appearance.

Perianths glabrous outside except a few hairs round the base.
Leaves linear.
Spikes globular or scarcely ovate.
Filaments dilated under the anthers . . . . 1. *P. conicus.*
Filaments filiform except at the base.
Perianth not exceeding 2 lines. Bracts mostly acute
and appressed . . . . 2. *P. corymbosus.*
Perianth 3 to 4 lines long. Bracts broad, mostly obtuse
and loose . . . . 3. *P. grandiflorus.*
Spikes at first conical, at length cylindrical . . . . . 4. *P. spicatus.*
Perianths enveloped in dense white cottony wool proceeding
from the lower half. Leaves oblong-lanceolate or obovate.
Spikes cylindrical. Leaves oblong.
Spikes sessile . . . . 5. *P. Murrayi*
Spikes pedunculate . . . . 6. *P. gomphrenoides.*
Spikes globular. Leaves obovate . . . . 7. *P. latifolius.*
Perianths enveloped in long dense articulate hairs proceeding
from the lower half. Leaves narrow.
Spikes globular or ovoid. Leaves lanceolate or oblong.
Stout plant. Spikes ½ in. diameter . . . . 8. *P. macrotrichus.*
Small slender plant. Spikes ¼ in. diameter . . . . . 9. *P. villosiflorus.*
Spikes cylindrical. Leaves linear . . . . 10. *P. humilis.*

1. **P. conicus,** *R. Br. Prod.* 415. An erect glabrous annual, closely resembling *P. corymbosus*, but usually more rigid, 1 to 2 feet high, with elongated branches. Leaves very narrow-linear as in that species. Spikes few on long peduncles, larger than in *P. corymbosus*, globular and 5 lines diameter, or at length ovoid and 7 or 8 lines long. Bracts narrow, acute or aristate, shorter than the perianth. Perianth-segments all scarious, or the inner ones more rigid and slightly ribbed at the base, 2½ to nearly 3 lines long. Filaments much dilated towards the base, and again shortly dilated and obcordate under the anthers.—Moq. in DC. Prod. xiii. ii. 282 ; *Trichinium conicum*, Spreng. Syst. i. 816.

N. Australia. Islands of the Gulf of Carpentaria, *R. Brown ;* Goulburn islands, *A. Cunningham ;* Port Essington, *Armstrong.*

2. **P. corymbosus,** *R. Br. Prod.* 415. An erect slender glabrous annual of about 1 ft. or rather more, loosely and corymbosely branched at the top or nearly from the base. Leaves very narrow-linear, acute, the lower ones often nearly 2 in. long, those of the branches very small. Spikes small, at first hemispherical, at length globular or almost ovoid, glabrous outside. Bracts and bracteoles ovate, scarious, minutely mucronate, much shorter than the perianth. Perianth-segments all equal, about 2 to 2½ lines long, acute, the two outer ones scarious almost from the base, glabrous inside as well as out, the three inner ones more rigid, ribbed, and woolly inside in the lower half or claw.

Filaments filiform to the top, slightly dilated at the base and united in a very short cup.—Moq. in DC. Prod. xiii. ii. 282 ; *Trichinium corymbosum,* Spreng. Syst. i. 816 not of Gaudich.

N. Australia. Islands of the Gulf of Carpentaria, *R. Brown;* N.W. Coast, *Bynoe ;* Victoria river and Sturt's Creek, *F. Mueller.*

Var. *acutiflorus.* Perianth-segments more acute; bracts and bracteoles almost aristate.—Arnhem's Land, *M'Kinlay.*

3. **P. grandiflorus,** *F. Muell. Fragm.* i. 237. An erect slender rather weak branching annual of 6 in. to 1 ft., glabrous or nearly so. Leaves linear or linear-lanceolate, acute, the larger ones 1 to 1½ in. long and contracted into a petiole, the upper ones small and distant. Spikes pedunculate, at first nearly globular, at length somewhat ovoid, about ¾ in. diameter when fully developed. Bracts and bracteoles thin and transparent, mucronate, 2 to 2½ lines long, the bract rather broadly ovate, the bracteoles narrower. Perianth 4 to 5 lines long, surrounded by straight articulate hairs a few of which are on the perianth itself at the base, otherwise quite glabrous, the segments free from the base, obtuse, scarious, pink, the inner ones scarcely shorter, without any wool inside. Staminal cup short, surrounded by articulate straight hairs, the filaments rather short, all nearly equal, with perfect anthers. Ovary glabrous.

W. Australia. Champion Bay and Murchison river, *Oldfield.*

Var. *lepidus.* Spikes and flowers smaller, the perianth scarcely above 3 lines long. I can detect no other difference.—*P. lepidus,* F. Muell. Fragm. iv. 89.—Sharks Bay, *M. Brown.*

4. **P. spicatus,** *F. Muell. Herb.* An erect glabrous annual (or with a perennial rhizome ?) of 1 to 2 ft., with long branches bearing usually each a single spike as in *P. conicus.* Leaves narrow linear or rarely linear-lanceolate acute, the larger ones 2 in. long. Spikes at first shortly conical, at length cylindrical and above 1 in. long. Bracts and bracteoles narrow, acute or aristate, shorter than the perianth. Perianth about 2½ lines long, surrounded at the base by a dense ring of rigid hairs some of which are also on the lower portion or claws of the segments ; segments free from the base, the upper half scarious coloured (pink or red), obtuse and quite glabrous, the 3 inner ones woolly inside below the middle. Staminal cup very short, the filaments not very unequal, filiform, scarcely dilated at the base.

N. Australia. Victoria river, *F. Mueller.* Included in *P. corymbosus* by F. Mueller, Fragm. iii. 125.

Var. *leianthus.* Claws or lower portion of the perianth-segments without any or scarcely any dorsal hairs.

N. Australia. Gulf of Carpentaria, *Leichhardt ;* Attack Creek, *M'Douall Stuart's Expedition.*

Queensland. Flinders' river, *Bowman, Sutherland.*

5. **P. Murrayi,** *F. Muell. Fragm.* iii. 145. A small apparently prostrate branching annual (or perennial ?) our specimens not exceeding 2 or 3 in. but not the entire plant. Leaves oblong, obtuse, under ½ in.

long, contracted into a petiole, glabrous as well as the branches. Spikes axillary and terminal, sessile, at first globose, at length oblong or cylindrical and about ½ in. long and 2½ to 3 lines diameter, the pink tips of the perianths just appearing above the white wool. Bracts and bracteoles ovate, obtuse, scarious, glabrous, scarcely above ½ line long. Perianth about 1¼ lines long, with a very short turbinate base, the segments thinly scarious with a red centre, glabrous in the upper half, the lower half covered outside with a long dense intricate white wool. Filaments slender, nearly as long as the perianth, united at the base in a truncate ring, slightly prominent from the perianth-tube ; anthers all 5 perfect (or one sometimes abortive ?). Fruit glabrous ; style rather excentrical.

S. Australia. Flooded tracts of Wills' Creek, *Howitt's Expedition.*

6. **P. gomphrenoides,** *F. Muell. Fragm.* vi. 233 (*name only*). Apparently erect, slender, glabrous, slightly branched, our specimens 6 in. long, but not the whole plant. Leaves oblong-lanceolate, ½ to 1 in. long, contracted into a long petiole. Spikes at length cylindrical and ½ in. long, the pink tips of the perianths just appearing above the white wool that envelops their base and the internal structure of the flowers entirely as in *P. Murrayi*, except that the spikes are borne on slender peduncles and the flowers perhaps a trifle larger.

N. Australia. Hammersley Range, N.W. coast, *F. Gregory's Expedition.* The accompanying label however in Herb. F. Muell. has evidently been by some accident ·misplaced, for it indicates a " tree or shrub 18 in. to 2 ft. high."

7. **P. latifolius,** *R. Br. App. Sturt's Exped.* 25. Stems erect, much branched, herbaceous (annual ? or from a thick rhizome ?) " attaining 2 ft." the branches and young foliage covered with an intricate white cottony wool, the older leaves becoming glabrous. Leaves obovate, very obtuse, rather thick, with slightly crisped margins, contracted into a rather long petiole, the largest leaves in the specimens scarcely 1 in. long, the upper ones much smaller. Spikes nearly globular, fully ½ in. diameter, numerous, sessile or shortly pedunculate, terminal or in the upper axils, usually with one or two small herbaceous leaves close under them. Bracts and bracteoles thinly scarious, white and shining, very broadly ovate or almost orbicular, obtuse, loose or spreading, 2½ to 3 lines long. Perianth not 2 lines long, the base a very short open disk, the segments with a narrow base densely clothed outside with long woolly hairs, glabrous inside, the lamina or upper half rather broader obtuse (pink ?) and glabrous. Stamens all perfect and nearly equal in the flowers examined. Ovary glabrous.—F. Muell. Fragm. vi. 232.

S. Australia. Sand ridges, Wills' Creek, *Howitt's Expedition.* I have not seen Sturt's specimens described by R. Brown.

8. **P. macrotrichus,** *F. Muell. Fragm.* iv. 90, vi. 232. Erect and branching with the habit of *P. latifolius,* 1 ft. high or more, the branches and foliage sprinkled with a few woolly hairs and sometimes almost glabrous. Leaves on long petioles, lanceolate or oblong, obtuse or acute,

rather thick, the lower ones above 1 in. long, the upper ones smaller. Spikes nearly globular, fully ½ in. diameter, numerous, sessile or shortly pedunculate, terminal or in the upper axils, with 1 or 2 small herbaceous leaves close under them. Bracts and bracteoles thinly scarious, white and shining, very broadly ovate, obtuse, loose or spreading, about 2 lines long. Perianth not 2 lines long, the segments free almost to the base, the lower half opaque and densely covered outside with long white silky-woolly hairs exceeding the perianth, the upper half or lamina obtuse scarious coloured and glabrous, the three inner ones with a small quantity of wool inside below the middle. Stamens short, scarcely united at the base. Ovary glabrous.

W. Australia, *Drummond, 6th coll. n.* 222 ; Sharks Bay, *M. Brown.*

9. **P. villosiflorus,** *F. Muell. Fragm.* iii. 125. A small slender annual with erect or decumbent branches not above 3 in. high in the single specimen known, the branches loosely woolly. Leaves small, linear or lanceolate, obtuse, contracted into a long petiole, glabrous or slightly woolly. Spikes globular, numerous and sessile or nearly so as in *P. macrotrichus,* but only ¼ in. diameter. Bracts and bracteoles as long as the perianth, and perianth-segments with long silky-woolly hairs on the lower half with the upper half or lamina glabrous, all entirely as in *P. macrotrichus* but scarcely above half the size.

N. Australia. Nichol Bay, *F. Gregory's Expedition.* The single specimen may be only a young starved and small-flowered state of *P. macrotrichus.*

10. **P. humilis,** *F. Muell. Fragm.* vi. 229. A small glabrous annual, branching at the base, with several prostrate slender stems, usually simple and 2 to 4 in. long, but sometimes ascending slightly branched and attaining at least 6 in. Leaves linear or linear-spathulate, obtuse or the upper ones mucronate-acute, mostly about ½ in. long besides the petiole, which is long in some specimens, very short in others. Spikes at first ovate, at length oblong, attaining fully 1 in. in length and about ½ in. diameter, the central one usually sessile at the base of the stems, the others terminal and sessile within the last leaves of the stems or their branches. Bracts very thin and transparent, broadly ovate or orbicular, nearly as long as the perianth. Perianth 2½ to nearly 3 lines long, the very short entire base or disk surrounded by a dense tuft of long articulate hairs sometimes as long as the perianth ; segments 3-nerved in the lower half, the upper half or lamina scarious and rather broad towards the top, all glabrous or with very few hairs on the lower half chiefly marginal. Staminal cup very short, surrounded by woolly hairs proceeding from the base of the inner segments. Ovary glabrous. —*Trichinium humile,* Nees in Pl. Preiss. i. 628 ; Moq. in DC. Prod. xiii. ii. 288 ; *T. nanum,* F. Muell. Fragm. iii. 161.

W. Australia, *Drummond, n.* 421 (or 427 ?) ; York district, *Preiss, n.* 1363 ; Kalgan river, *Oldfield ;* N. of Stirling Range, *F. Mueller ;* W. or N.W. coast, *Bynoe.*

Var. *parviflora.* Small and slender but branching, with several spikes much narrower than in the typical form, the perianth-segments not 2 lines long and broader in proportion than in the type, with the opaque pink centre extending higher up, and the surrounding hairs shorter.—W. Australia, *Burgess.*

6. ACHYRANTHES, Linn.

Flowers hermaphrodite. Perianth-segments 5, slightly unequal, hardened after flowering and erect, enclosing the fruit, usually glabrous. Stamens 5, united in a cup at the base, with as many small scales or staminodia between them. Anthers 2-celled. Ovary uniovulate. Style simple, with a capitate stigma. Fruit a membranous indehiscent utricle enclosed in the perianth. Seed solitary.—Herbs. Leaves opposite. Flowers green and rigid or rarely scarious, reflexed, in terminal heads or long spikes. Bracteoles subulate, rigid and often spinescent.

The genus is widely distributed over the tropical and subtropical regions of the Old World. The only Australian species is a common weed over the whole range of the genus.

1. **A. aspera,** *Linn.; Moq. in DC. Prod.* xiii. ii. 314. An erect or spreading annual or biennial, with a hard almost woody base and branching stems of 2 or 3 ft., more or less hoary as well as the foliage with a soft pubescence. Leaves, shortly petiolate ovate, ovate-oblong or almost oblong, obtuse or shortly acuminate and acute, usually 1 to 2 in. but sometimes 3 in. long. Flowers of a shining green, in long slender but rigid terminal spikes. Perianth 1½ to nearly 2 lines long, closely reflexed after flowering. Bracteoles rigidly subulate or spinescent (but not spreading as in *Nyssanthes*), usually nearly as long as the perianth, but variable in length, dilated and scarious at the base.—Wight, Ic. Pl. t. 1777; *A. australis,* R. Br. Prod. 417; Moq. in DC. Prod. xiii. ii. 313.

N. Australia. Islands of the Gulf of Carpentaria, *R. Brown;* Goulburn islands, *A. Cunningham;* Victoria river, *F. Mueller;* Escape Cliffs, *Hulls;* Nichol Bay, N.W. coast, *Ridley's Expedition;* Port Darwin, *Schultz.*

Queensland. Broad Sound, *R. Brown;* Albany island, *F. Mueller;* Cape York, *Daemel;* Rockingham Bay, *Dallachy;* Rockhampton, a common weed, *O'Shanesy;* Nerkool Creek, *Bowman;* Moreton Bay, *Leichhardt, F. Mueller.*

N. S. Wales. Lord Howe's island, *M'Gillivray, Milne.*

A. canescens, R. Br. Prod. 417, Moq. in DC. Prod. xiii. ii. 312, is a more pubescent or hoary-villous variety, with thicker leaves and rather larger perianths, passing very gradually into the commoner forms; the most marked specimens are Brown's from the Carpentaria islands and F. Mueller's from Victoria river.

A. argentea, Lam.; Moq. l.c. 315, is another variety or form only to be distinguished from the common one by the more acuminate leaves, but is not generally so common or so well marked in Australia as in Africa and in S. Europe. Some specimens, however, such as those of Schultz's from Port Darwin, are quite characteristic.

7. NYSSANTHES, R. Br.

Flowers hermaphrodite. Perianth-segments 4, of which 2 upper ones smaller, all hardened after flowering and erect, enclosing the fruit, all or the 2 outer ones more or less spinescent. Stamens 2 or 4, united in a cup at the base, with as many short scales or staminodia between them. Anthers 2-celled. Ovary uniovulate. Style with a capitate stigma. Fruit a membranous indehiscent utricle, enclosed in the perianth. Seed solitary.—Herbs. Leaves opposite. Flowers green and rigid, very spreading or reflexed after flowering, in sessile head-like spikes or clusters, the bracts and bracteoles spinescent and very spreading.

The genus is limited to Australia, differing slightly from *Achyranthes* in its inflorescence, spreading bracts and constantly 4-merous flowers.

Stamens 4 . . . . 1. *N. erecta.*
Stamens 2 . . . . 2. *N. diffusa.*

1. **N. erecta,** *R. Br. Prod.* 418.   Erect and probably 2 ft. high or more, the upper flowering portion with spreading opposite dichotomous branches, the whole plant more or less pubescent with soft appressed hairs.   Leaves elliptical-oblong or almost lanceolate, mostly acute, contracted into a short petiole, those on the main stem and the lower floral ones 1 to 3 in. long, the upper floral ones very small.   Flowers in dense clusters or short spikes sessile in the upper axils and forks of the panicle, each flower sessile within a spinescent bract shortly dilated and scarious at the base, and 2 similar but smaller bracteoles very divaricate on the fruiting perianth.   Perianth with 2 outer segments about 1½ lines long but unequal, hairy, lanceolate, tapering into a rigid spinescent point, with 2 or 3 more or less conspicuous nerves on each side of the prominent midrib, the 2 inner ones smaller glabrous and scarcely nerved.   Stamens 4, the filaments short, the intervening scales or staminodia broad, truncate, about half as long as the filaments.   Fruit nearly globular, about ¾ line long, membranous except the depressed summit which is harder.—Moq. in DC. Prod. xiii. ii. 309.

Queensland, *Bowman;* Dawson and Brisbane rivers, *F. Mueller.*
N. S. Wales. Nepean river, *R. Brown;* New England, *C. Stuart.*

I have seen but few specimens.   The perianths are usually but not always more thickened at the base and reflexed than in *N. diffusa.* The spinescent bracts and summits of the perianth-segments are very variable in length and relative proportions, but they are usually, especially the bracteoles, shorter than in *N. diffusa.* The difference in the stamens appears to be constant.

2. **N. diffusa,** *R. Br. Prod.* 418.   An annual or biennial closely allied to *N. erecta,* but usually more branched from the base, more slender, and the parts smaller.   Stems attaining 1 to 3 ft., the greater portion consisting of a broad leafy panicle, the branches dichotomous or the lower ones trichotomous.   Leaves ovate or oblong, obtuse or acute, rarely much above 1 in. and mostly small.   Spikes or clusters of flowers very short, in the axils and in the forks of the panicle, the flowers usually smaller than in *N. erecta,* but the segments as well as the bracts and bracteoles even more variable in size and relative proportions than in that species; sometimes all three bracts are subulate almost from the base and 3 to 4 lines long, more frequently the bracteoles are much shorter, but all three very divaricate; one of the outer perianth-segments usually long, the other much shorter, both spinescent, the inner ones lanceolate and tapering into a shorter spine.   Stamens always 2 only, with short broad truncate scales or staminodia between them.—

Moq. in DC. Prod. xiii. ii. 308; *N. media*, R. Br. Prod. 418; Moq. l.c. 309.

Queensland. Shoalwater Bay, *R. Brown;* Brisbane river, *F. Mueller;* Rockhampton, *Sutherland, O'Shanesy;* Nerkool Creek, *Bowman.*

N. S. Wales. Port Jackson to the Blue Mountains, *R. Brown, Leichhardt* and others; Clarence river, *Beckler.*

N. media, Br., appears to me to be rather a luxuriant state than a variety of *N. diffusa.*

8. ALTERNANTHERA, R. Br.

(Telanthera, *Moq.*)

Flowers hermaphrodite. Perianth divided to the base into 5 segments, all equal or the outer ones larger, ovate or lanceolate, scarious and coloured (usually white) glabrous or with long woolly hairs at the base. Stamens united at the base into a short exceedingly thin cup; filaments short with or without intervening teeth or lobes, unequal, 2 or 3 of them often without anthers and reduced to short teeth. Anthers small. Ovary uniovulate; style short sometimes scarcely any, with a capitate stigma. Fruit an indehiscent utricle usually compressed. Seed vertical. —Annual or perennial herbs, mostly prostrate, glabrous or softly hairy. Leaves opposite. Flowers small, in axillary sessile or pedunculate spikes (very rarely also terminal), usually short or oblong. Bracts and bracteoles scarious.

The genus is widely spread over the warmer regions of both the New and the Old World, including three of our common tropical weeds. Of the Australian species one is a common one in tropical Asia and Africa, another is closely allied to but perhaps not quite identical with a still more generally diffused species, the remaining six appear to be endemic, but the circumscription of the species as well as of the genus itself requires much further investigation. Moquin attributes to the whole genus (including *Telanthera*) 5 stamens with intervening staminodia (teeth or lobes of the staminal cup). In those species which he refers to *Alternanthera* proper, I can see no trace of these staminodia unless we consider as such the two or three of the five filaments which are often reduced to small teeth. The extreme tenuity of the staminal cup renders it exceedingly difficult to ascertain its form, unless examined in the bud before the enlargement of the ovary.

Staminal cup without teeth between the filaments.
  Perianth perfectly glabrous.
    Plant glabrous or slightly pubescent at the nodes or in two decurrent lines. Leaves narrow.
      Perianth-segments and bracteoles (above 1¼ lines long) with fine points. Spikes at length several together in dense globular clusters . . . . 1. *A. nodiflora.*
      Perianth-segments and bracteoles (1 line long or under) broad with short points. Spikes small, at length cylindrical and scarcely clustered . . . . 2. *A. denticulata.*
    Plant more or less hairy or rarely glabrous. Leaves mostly broad. Rhachis of the spike woolly . . . . 3. *A. nana.*
  Perianth-segments enveloped in long woolly hairs proceeding from the base (often concealed in the young spike by the bracteoles).
    Bracteoles shorter than the perianth. Perianth-segments all equal (about 1 line long) and glabrous inside . . . . . . 4. *A. angustifolia.*

Bracteoles as long as the perianth. Perianth-segments 1½ to 2
    lines long, the inner ones smaller and woolly inside at the
    base.
Spikes all axillary, ovoid. Bracteoles and perianth-seg-
    ments very acute. Anthers 5 . . . . 5. *A. decipiens.*
Spikes terminal and axillary, globular. Bracteoles and peri-
    anth-segments scarcely mucronate. Anthers usually 3 . . 6. *A. polycephala.*
Staminal cup with prominent teeth or lobes between the filaments.
    Perianth-segments hairy on the back to above the middle. Leaves
    linear.
Spikes axillary, shortly pedunculate, ovoid. Lobes of the staminal
    cup much shorter than the filaments . . . . 7. *A. leptophylla.*
Spikes on long peduncles, globular. Lobes of the staminal cup
    rather longer than the filaments . . . . 8. *A. longipes.*

1. **A. nodiflora,** *R. Br. Prod.* 417. Stems prostrate, decumbent or
ascending, 6 in. to 1 or even 2 ft. long, glabrous or slightly pubescent
in decurrent lines. Leaves linear or lanceolate, shortly contracted at
the base, 1 to 2 in. long. Spikes globular, about 4 lines diameter when
fully out, but often clustered many together into dense globular masses
sometimes above 1 in. diameter, interspersed with a few small floral
leaves. Bracts, bracteoles and perianth-segments narrow, acuminate
with fine points, usually about 1½ lines long or rather more. Stamens
very short (about ¼ line), the filaments unequal, dilated at the base and
united into a minute open cup, 2 or 3 bearing anthers, the others reduced to small teeth. Style distinct though very short. Utricle not
half as long as the perianth, much compressed, broad, deeply notched
with obtuse thickened margins.—Moq. in DC. Prod. xiii. ii. 356.

N. Australia. Sturt's Creek, *F. Mueller ;* Victoria river, *Flood ;* in the interior,
M'Douall Stuart's Expedition; Albert river, *Henne.*
    Queensland. Broad Sound, *R. Brown;* Armadilla, *Barton;* in the interior,
Mitchell.
    N. S. Wales. Gwydir river, *Leichhardt;* New England, *C. Stuart;* Ballandool
river, *Locker;* Murray and Darling desert, *F. Mueller, Victorian Expedition.*
    Victoria. Murray river, *F. Mueller ;* Skipton, *Whan.*
    S. Australia. S. of Wills' Creek, *Howitt's Expedit.on.*
    W. Australia. *Drummond, n.* 220.
    The species appears to be widely spread over E. India and Africa, but is not always
easy to distinguish from *A. denticulata, A. sessilis,* and perhaps some others. The
Australian specimens when first in flower are very much like those of *A. denticulata.*
When fully developed the perianths and bracts are much longer and more acuminate,
the fruit shorter in proportion and broader than in *A. denticulata,* and the notch,
although variable, usually much deeper.

2. **A. denticulata,** *R. Br. Prod.* 417. Stems prostrate, creeping
and rooting at the lower nodes, often extending to 2 or 3 ft. and shortly
ascending, glabrous or minutely pubescent in decurrent lines. Leaves
linear or linear-lanceolate, obtuse or mucronate, shortly contracted at
the base, mostly ¾ to 1½ in. long. Spikes globular or at length ovoid
or oblong, about ¼ in. diameter, closely sessile in the axils and sometimes 2 or 3 together but not so densely clustered as in *A. nodiflora,*
usually quite glabrous. Perianth-segments under 1 line long, very
acute though shorter pointed than in *A. nodiflora.* Stamens and style

entirely of that species. Utricle shorter than the perianth ; compressed and broadly obcordate or truncate, but longer in proportion and less notched than in *A. nodiflora,* shorter in proportion than in *A. sessilis.*— Moq. in DC. Prod. xiii. ii. 356 ; *Illecebrum denticulatum,* Spreng. Syst. i. 820 ; *A. sessilis,* Br. var. Hook. f. Fl. Tasm. i. 310.

Queensland. Keppel Bay, *R. Brown;* Port Denison, *Fitzalan;* Rockhampton, *O'Shanesy;* Gilbert river, *Daintree;* Armadilla, *Barton;* Darling Downs, *Lau.*
N. S. Wales. Hunter's river, *A. Cunningham;* Clarence river, *Beckler.*
Victoria. Emu Creek, *Whan.*
Tasmania. Port Dalrymple, *R. Brown;* near Launceston, *Gunn.*

Sometimes very difficult to distinguish from *A. nodiflora,* especially when first coming into flower, and on the other hand very near some varieties of the widely-distributed *A. sessilis,* Br., with which J. D. Hooker, perhaps not incorrectly, unites it. In general it differs slightly in the narrower leaves, glabrous spikes, more acute flowers and shorter utricles.

Var ? *micrantha.* Smaller and more slender and sometimes slightly pubescent. Leaves linear, ½ to 1 in. long. Spikes 1½ lines diameter, with a few hairs on the rhachis. Perianth-segments ½ to ¾ line long, scarcely mucronate, the flowers very deciduous, leaving the bracts persistent.

N. Australia. Arnhem's Land, *F. Mueller.*

3. **A. nana,** *R. Br. Prod.* 417. Stems prostrate or ascending, loosely pubescent as well as the foliage, and often woolly at the nodes. Leaves oblong lanceolate or almost linear, obtuse or acute, tapering at the base and shortly petiolate, ¾ to 1½ in. long or in the broader leaved specimens under ¾ in. Spikes sessile in the axils, about ¼ in. diameter, at first depressed-globular but at length ovoid or shortly cylindrical, 4 to 5 lines long and very shining, the rhachis woolly. Perianth-segments oblong or lanceolate with a small point, thickened and hardened at the base when in fruit, 1¼ to above 1½ lines long, glabrous as well as the short bracts and bracteoles. Stamens very short, the filaments filiform or slightly dilated at the base, usually 3 bearing anthers and 2 reduced to small teeth. Utricle scarcely half s̀o long as the perianth.—Moq. in DC. Prod. xiii. ii. 360 ; *Illecebrum nanum,* Spreng. Syst. i. 819.

N. Australia. Nicholson river and Sturt's Creek, *F. Mueller.*
Queensland. Broad Sound, *R. Brown;* Brisbane river, *F. Mueller.*
N. S. Wales. "Near Mr. Scott's and everywhere in the dry bed of the river," *Leichhardt;* New England, *C. Stuart;* Ballandool river, *Lockhardt.*

Var. *major.* Larger and more hairy, leaves longer, perianth-segments acute.—Rockhampton, *O'Shanesy.*

The species is very variable, especially as to the size of the flowers, and some specimens come very near some forms of *A. sessilis,* but always with the utricle much shorter in proportion to the perianth. Brown's specimens as well as some of F. Mueller's and of Leichhardt's have the flowers much smaller than in the others.

4. **A. angustifolia,** *R. Br. Prod.* 417, *but not of Moq.* Prostrate, glabrous or slightly hoary-pubescent. Leaves linear in the typical form, above 1 in. long and much like those of *A. denticulata* or even narrower. Spikes sessile, solitary or rarely clustered, seldom above ¼ in. diameter and mostly smaller and very short, the rhachis more or less woolly. Bracts very acute, glabrous, shorter than the perianth.

Perianth-segments lanceolate, acute, 1 line long or a little more, with long woolly hairs outside; the inner ones narrower than the outer. Filaments short, only 3 bearing anthers, all dilated at the base. Stigma capitate, sessile in the flowers examined.—*Illecebrum angustifolium,* Spreng. Syst. i. 818.

N. Australia. Islands of the Gulf of Carpentaria, *R. Brown;* Sturt's Creek, *F. Mueller.*

Var. *lanata.* More woolly. Leaves narrow-oblong, under ½ in. long. Spikes more woolly.—Arnhem's Land, *F. Mueller.*

5. **A. decipiens,** *Benth.* Apparently a small annual, the specimens under 6 in., branching and softly hirsute. Leaves ovate or ovate-lanceolate, rather acute, contracted into a rather long petiole. Spikes numerous, axillary, closely sessile, ovoid and about ¼ in. long in the specimens seen but still very young. Bracteoles 2 lines long, narrow and tapering into a fine point. Perianth of *A. polycephala,* but the segments narrower and tapering into fine points. Stamens all 5 perfect in the specimens examined. Ovary of *A. polycephala.*

Queensland? Subtropical Australia, *Mitchell (Herb. Hook.).*

This may possibly prove to be a variety of *A. polycephala* but the long points to the bracteoles and perianth-segments give it the aspect of the *A. achyrantha,* Br., from which however it differs in the long hairs enveloping the perianth and other characters.

6. **A. polycephala,** *Benth.* A very much branched annual of 6 in. to 1 ft., the stems and foliage pubescent, or woolly when young. Leaves lanceolate, contracted into a petiole, rarely above ½ in. long. Spikes globular or ovoid, 4 to 5 lines diameter, sessile, axillary and terminal, so numerous as at length almost to conceal the leaves, the shining bracts at first, and later the wool very conspicuous; the rhachis woolly. Bracteoles glabrous, ovate, obtuse or slightly mucronate, as long as the perianth. Perianth enveloped in long silky-woolly hairs proceeding from its base, 1½ to 2 lines long, thin and transparent with the centre of the lower half of the segments opaque, the 3 inner segments smaller with long woolly hairs inside at the base. Filaments slender, united at the base into a very short cup partially adnate to the perianth, only 3 of them antheriferous as far as I could ascertain. Style about half as long as the ovary, with a capitate stigma. Utricle shorter than the perianth, compressed but rounded at the top and not notched.—*A. angustifolia,* Moq. in DC. Prod. xiii. ii. 354, not of R. Br.

N. Australia. Greville island, Regent's harbour, N.W. Coast, *Bynoe.*—The flowers are far advanced in the specimens, and I had great difficulty in ascertaining the structure of the staminal apparatus owing to its extreme tenuity and to the copious wool in which it is enveloped, but I believe the above account of it to be correct.

7. **A. leptophylla,** *Benth.* Stems slightly bifariously pubescent. Leaves narrow-linear, glabrous. Spikes ovoid (probably at length cylindrical) very shortly pedunculate in the axils, about ¼ in. diameter, the rhachis slightly woolly. Bracteoles obtuse, glabrous, nearly as long as the perianth, the bracts shorter. Perianth 1 to 1¼ lines long, the segments with long woolly hairs on the back, and short glabrous obtuse

white tips, all glabrous inside, the inner ones rather smaller. Filaments filiform, united in a short cup, all antheriferous, and alternating with short broad scale-like truncate or jagged teeth or lobes. Style longer than the ovary, with a capitate stigma. Utricle compressed but not notched.

N. Australia. Sturt's Creek, *F. Mueller;* a single small specimen in Herb. F. Mueller.

8. ? **A. longipes,** *Benth.* An annual with slender erect slightly branched glabrous stems of about 1 ft. Leaves sessile, linear, acute, contracted at the base, slightly hairy, 1½ to 2 in. long, the younger ones hoary-villous. Spikes globular, about 4 lines diameter, solitary on filiform rigid peduncles of 1 to 2 in. Bracteoles ovate, acuminate, glabrous, hyaline, nearly as long as the perianth, bracts rather shorter. Perianth 1½ lines long, enveloped at the base in long flexuose hairs, the segments linear-lanceolate, acute, green in the centre, white on the margin. Filaments linear-subulate, alternating with subulate ligulate staminodia, rather longer than them and entire or jagged.—*Telanthera longipes,* Moq. in DC. Prod. xiii. ii. 370.

N. Australia. N.W. Coast, *Bynoe.*—I have not seen the specimens, nor does it appear in what herbarium Moquin examined them. I have taken the above from his character and description.

9. GOMPHRENA, Linn.

(Philoxerus, *R. Br.*)

Flowers hermaphrodite. Perianth divided to the base into 5 segments, all equal or the outer ones rather larger, linear, scarious and coloured with the centre green at least at the base, woolly outside below the middle or glabrous. Stamens united at the base in a long or short tube, the free part of the filaments short, with or without intervening teeth or scale-like lobes; anthers 1-celled. Ovary uniovulate; style short or filiform, with 2 short, often minute, linear stigmatic lobes. Fruit an indehiscent utricle. Seed vertical.—Annual or perennial herbs, glabrous or with more or less of soft woolly hairs. Leaves opposite. Flowers in terminal or rarely axillary spikes usually dense, either shortened into globular or hemispherical heads or more or less lengthened ovoid-oblong or cylindrical. Bracts and bracteoles scarious, glabrous, the bracteoles more or less complicate and keeled. Ovary glabrous.

The genus comprises a considerable number of species, the extra-Australian ones all from the warmer regions of America, one of which, of early cultivation in gardens, is now a common weed in E. India. The Australian species appear to be all endemic.

The circumscription of the genus is in a very unsatisfactory state. If Moquin's technical characters were strictly followed, the Australian species would be distributed among at least four of his genera. It appears to me, however, that the presence or absence of the teeth or lobes of the staminal tube between the filaments is of no more value here than in the case of *Tricninium,* and I have followed Brown in distinguishing *Gomphrena* from *Alternanthera* chiefly by the 2-lobed stigma. Moreover, the shortness of the staminal tube, by which Brown separated *Philoxerus,* can scarcely hold if

G. lanata, Br. (*G. Brownii*, Moq.) is to be retained in *Gomphrena*, and is at the best rather a sectional than a generic character.

Staminal tube longer than the ovary. Spikes globular or ovoid, usually large. Filaments flat. Leaves linear.

Staminal tube with filiform teeth between the filaments. Spikes globular. Perianth-segments acute. Annual . . . . . 1. *G. canescens.*

Staminal tube without teeth or lobes between the filaments.
    Annual of 1 to 2 ft. Spikes at length ovoid or oblong. Perianth-segments obtuse . . . . 2. *G. flaccida.*
    Annual under 1 ft. Spikes hemispherical or globular. Perianth-segments acute . . . . 3. *G. affinis.*
    Perennial under 1 ft. Spikes hemispherical or globular. Perianth-segments rather obtuse . . . . 4. *G. humilis.*

Staminal tube shorter than the ovary. Spikes rarely above ¼ in. diameter.

Spikes hemispherical or globular. Perianth very woolly outside below the middle.
    Staminal tube with teeth or lobes between the filaments.
        Leaves linear. Bracts much shorter than the bracteoles . 5. *G. Brownii.*
        Leaves lanceolate. Bracts nearly as long as the bracteoles 6. *G. brachystylis.*
    Staminal tube without teeth or lobes between the filaments.
        Leaves broadly lanceolate. Filaments with a minute tooth at the top on each side . . . . 7. *G. leptoclada.*
        Leaves narrow-lanceolate. Filaments filiform at the top without lateral teeth.
            Spikes ½ in. diameter or more. Perianth woolly to above the middle. Bracteoles shorter than the perianth . . 8. *G. Maitlandi.*
            Spikes 4 to 5 lines diameter. Perianth woolly only below the middle. Bracteoles longer than the perianth 9. *G. pusilla.*
        Leaves filiform. Spikes and perianth of *G. pusilla* . . 10. *G. tenella.*

Spikes ovoid or oblong-cylindrical. Perianth woolly below the middle.
    Spikes 5 lines diameter. Leaves all linear. Filaments broad and 2- or 3-toothed at the end . . . . 11. *G. conica.*
    Spikes 3 or 4 lines diameter, densely crowded with broadly-lanceolate floral leaves. Filaments tapering at the end . 12. *G. conferta.*

Spikes at length narrow-cylindrical. Perianth small, quite glabrous.
    Leaves hairy, lanceolate or oblong . . . . 13. *G. diffusa.*
    Leaves glabrous or nearly so, narrow-linear . . . . . . 14. *G. parviflora.*

1. **G. canescens**, *R. Br. Prod.* 416. An erect more or less branching annual, usually stout and hard, 1 to nearly 2 ft. high, more or less hoary with long soft hairs, the older parts rarely glabrous. Leaves linear or linear-lanceolate, acute, the larger ones 2 to 3 in. long, the margins usually recurved. Spikes globular, sessile between the last leaves, about 1 in. diameter, the rhachis thick, ovoid or globular, woolly. Bracts and bracteoles thinly scarious, lanceolate, acute, about 3 lines long. Perianth-segments about 5 lines long, narrow, acute, 1-nerved, slightly woolly on the back below the middle, glabrous inside. Filaments united in a tube variable in length but always much longer than the ovary, the free portion short, flattened, with intervening filiform teeth or lobes, sometimes as long as the anthers, but often shorter. Style filiform.—Moq. in DC. Prod. xiii. ii. 398; *Philoxerus canescens*, Poir. Dict. Suppl. iv. 393.

N. Australia. Mainland of the Gulf of Carpentaria, *R. Brown, Henne;* Depuech island, N.W. coast, *Bynoe;* Nichol Bay, *Gregory's and Ridley's Expeditions;* Victoria river and Sturt's Creek, *F. Mueller;* Goulburn islands, *A. Cunningham;* Port Darwin and several other points along the coast, *Schultz* and others; Attack Creek in the interior, *M'Douall Stuart's Expedition.*

2. **G. flaccida,** *R. Br. Prod.* 416. An erect annual of 1 to 1½ ft., simple or branched, usually more slender than *G. canescens,* but sometimes as stout, the young parts woolly, becoming at length nearly glabrous. Leaves linear or linear-lanceolate, 1 to 2 in. or rarely longer, those under the spike much smaller. Spikes at first globular but soon becoming ovoid, solitary or 2 or 3 in a close cluster at the ends of the branches, about ¾ in. diameter and sometimes at length nearly 1 in. long, the rhachis woolly. Bracts 2½ to 3 lines long, the bracteoles much complicate and keeled. Perianth about 4 lines long, much flattened when old, the segments rather obtuse, slightly woolly outside near the base. Staminal tube varying in length as in *G. canescens,* and always longer than the ovary, the filaments shortly free and flattened but without the intervening teeth of that species.—Moq. in DC. Prod. xiii. ii. 398; *Philoxerus flaccidus,* Poir. Dict. Suppl. iv. 392; *G. firma,* F. Muell. Fragm. iii. 123.

N. Australia. Arnhem N. Bay, *R. Brown;* Regent's river and Cambridge Gulf, N.W. coast, *A. Cunningham;* Usborne harbour, *Voyage of the Beag'e;* Victoria river, *Bynoe, F. Mueller;* Glenelg river, *Martin;* Port Essington, *Armstrong;* Port Darwin, *Schultz.*

Queensland. Cape York, *Daemel.*

The filaments are often broad and sometimes irregularly jagged or toothed towards the end, but different in different flowers of the same specimen, and not regularly 3-toothed as described by Moquin in the genus generally, and in this and other species specially.

3. **G. affinis,** *F. Muell. Herb.* An erect rigid much-branched annual, under 1 ft. high, hoary with silky-woolly hairs. Leaves linear or linear-lanceolate, with recurved margins, ¾ to 1½ in. long. Spikes sessile within the last pair of leaves, hemispherical or at length globular, ¾ in. diameter or rather more. Bracts and bracteoles very acute or almost aristate, nearly as long as the perianth. Perianth-segments acute, 4 to 5 lines long, very woolly to above the middle. Staminal tube longer than the ovary, the filaments broad, obtuse or truncate, with the anther on a minute central tooth, without any teeth between the filaments.

N. Australia. Upper Victoria river, *F. Mueller.*

4. **G. humilis,** *R. Br. Prod.* 416. A perennial, with a woody stock often bearing tufts of wool, the stems erect, branching, 6 to 9 in. high, glabrous as well as the foliage or with a few silky-woolly hairs, especially about the nodes. Leaves narrow-linear, mostly 1 to 1½ in. long, with smaller ones often clustered in the axils. Spikes depressed, almost hemispherical, ½ to 1 in. diameter, mostly pedunculate above the last leaves, or rarely with a pair of small leaves close under them, the rhachis woolly. Bracts and bracteoles glabrous, very thin and transparent, finely 1-nerved, the bracteoles usually about 2 lines long, broad and rather obtuse, the bracts smaller narrower and more acute, but both

variable. Perianth-segments 3 to 4 lines long, pink, rather broad, obtuse or scarcely acute, woolly outside to above the middle. Staminal tube longer than the ovary, usually bearing a few woolly hairs outside, the filaments lanceolate, acuminate, broad or narrow, without intervening teeth.—Moq. in DC. Prod. xiii. ii. 418; *Philoxerus humilis,* Poir. Dict. Suppl. iv. 392.

N. Australia. Albert river, *Henne.*
Queensland. Broad Sound, *R. Brown;* Port Denison, *Fitzalan;* Suttor and Bowen rivers, *Bowman.*

5. **G. Brownii,** *Moq. in DC. Prod.* xiii. ii. 397. An erect branching annual of 6 to 9 in., hoary with rather long silky or woolly hairs. Leaves linear, mostly ¾ to 1 in. long, acute and soft. Spikes globular or depressed, 4 to 5 lines diameter, sessile or very shortly pedunculate within the last leaves, or pedunculate in the upper axils without floral leaves. Bracteoles transparent, rather broad, acute, as long as the perianth or rather longer, the bracts much shorter. Perianth-segments scarcely 1½ lines long, acute, with a narrow opaque centre, densely woolly outside with long hairs. Staminal tube not longer than the ovary though not much shorter, the filaments short, with oblong or lanceolate entire or denticulate teeth or lobes between them as long as the anthers.—*G. lanata,* R. Br. Prod. 416, not of Poir; *Philoxerus lanatus,* Poir. Dict. Suppl. iv. 392; *Alternanthera Baueri,* Moq. in DC. Prod. xiii. ii. 354.

N. Australia. Islands of the Gulf of Carpentaria, *R. Brown.*
Queensland. Suttor river, *Bowman.*

6. **G. brachystylis,** *F. Muell. Fragm.* iii. 124. Stems branching, above 1 ft. long, clothed as well as the foliage with soft woolly hairs. Leaves sessile, lanceolate or linear-lanceolate, acute, soft, the margins undulate or flat and not recurved, ½ to 1 in. long. Spikes globular, about ½ in. diameter, sessile between the last leaves or here and there on long peduncles without floral leaves. Bracteoles mucronate-acute, rather longer than the perianth; bracts but little shorter. Perianth-segments 2 to 2½ lines long, densely covered outside with long woolly hairs, the glabrous tips very pale pink. Staminal tube shorter than the ovary, the filaments flat but narrow, acuminate, with long anthers; the intervening teeth or lobes as long as the filaments but broader and denticulate at the end. Style rather short.

N. Australia. Hooker's Creek, *F. Mueller.*

7. **G. leptoclada,** *Benth.* A slender much-branched annual, of 6 in. to 1 ft., the young plants densely clothed with white woolly hairs, the older stems glabrous and red. Leaves sessile, broadly or narrow lanceolate, acute, green above with rather long straight hairs, white underneath with woolly hairs, the lower ones ½ to 1 in. long, the upper ones smaller. Spikes globular or depressed, 4 to 5 lines diameter, sessile between the last pair of leaves or here and there on long peduncles without floral leaves. Bracteoles white, very acute, about as long

as the perianth, the bracts rather shorter. Perianth-segments scarcely 2 lines long, with long woolly hairs outside at the base, the upper half glabrous and bright pink. Staminal tube shorter than the ovary, the filaments rather broad, minutely 3-toothed at the apex, the anther borne on the rather larger central tooth; no teeth or lobes to the tube between the filaments. Style short.

N. Australia. Glenelg district, N.W. coast, *Martin.*

8. **G. Maitlandi,** *F. Muell. Fragm.* iii. 124. *t.* 23. An erect branching annual of about 6 in., more or less woolly, the older leaves nearly glabrous. Leaves lanceolate or narrow-oblong, obtuse or acute, contracted at the base and sometimes shortly petiolate, ½ to 1½ in. long. Spikes depressed-globular, ¼ to nearly ¾ in. diameter, sessile or shortly pedunculate above the last leaves. Bracts and bracteoles rather broad, very acute, nearly as long as the perianth. Perianth-segments about 2 lines long, acute, rather unequal, densely clothed outside to above the middle with long woolly white or ferruginous hairs. Staminal tube shorter than the ovary, without teeth or lobes between the filaments.— *Iresine Cunninghamii,* Moq. in DC. Prod. xiii. ii. 342.

N. Australia. Dampier's Archipelago, N.W. coast, *A. Cunningham;* Pyramid hill, Nichol Bay, *F. Gregory's Expedition.*

9. **G. pusilla,** *Benth.* A slender branching annual, under 6 in. high, with the loose wool, linear-lanceolate leaves and globular sessile spikes 4 to 5 lines in diameter of *G. Brownii.* Bracteoles acute, longer than the perianth, bracts rather shorter. Perianth-segments scarcely above 1½ lines long, woolly outside below the middle, the upper half scarious and white, the outer ones very obtuse, the inner ones narrower. Staminal tube very short and truncate, the filaments slightly dilated, not toothed at the end, and without intervening teeth or lobes.

N. Australia. Foul Point, N.W. coast, *Voyage of the Beagle.*

10. **G. tenella,** *Benth.* A very slender branching annual of ½ to 1 ft., glabrous or slightly woolly under the spikes. Leaves filiform, acute, 1 to 2 in. long. Spikes globular, 4 to 5 lines diameter, on slender peduncles. Bracteoles acute, about as long as the perianth, bracts shorter. Perianth-segments 1½ lines long, with long woolly hairs outside near the base, the upper half glabrous white and scarious, the green centres reaching to about ⅔ of the segment, the inner segments smaller and narrower. Staminal tube or cup very short; truncate, the filaments rather short, scarcely dilated, without intervening teeth or lobes. Style very short.—*Iresine tenella,* Moq. in DC. Prod. xiii. ii. 343.

N. Australia. Cygnet Bay, *A. Cunningham;* Foul Point, N.W. coast, *Voyage of the Beagle.*

11. **G. conica,** *Spreng. Syst.* i. 824. An erect branching annual of 1 to 1½ ft. with the aspect of *G. flaccida,* slightly hoary woolly or glabrous. Leaves linear, with recurved margins, 1 to 2 in. long. Spikes at first ovoid, at length cylindrical, about 5 lines diameter and attain-

ing nearly 1 in. in length, pedunculate and solitary, or (in R. Brown's specimens) frequently in clusters of 2 or 3 and more conical. Bracts and bracteoles acute, scarcely half as long as the perianth. Perianth-segments 2½ to 3 lines long, covered outside with long dense wool, with white obtuse glabrous tips. Staminal tube or cup shorter than the ovary; filaments broad, 2-toothed with the anther sessile between the teeth, without intervening lobes or teeth to the cup. Style short.—*Philoxerus conicus,* R. Br. Prod. 416; *Iresine conica,* Moq. in DC. Prod. xiii. ii. 342; *Gomphrena breviflora,* F. Muell. Fragm. iii. 125.

N. Australia. Islands of the Gulf of Carpentaria, *R. Brown;* Upper Victoria river and Sturt's Creek, *F. Mueller;* Lara station, *Kennedy.*

12. **G. conferta,** *Benth.* Erect hard stout and probably tall, but apparently annual, the specimens very imperfect, the branches bearing a few white woolly hairs and linear or linear-lanceolate leaves of 1 to 2 in. Spikes ovoid or cylindrical, 3 to 4 lines diameter and some of them above ½ in. long, sessile and crowded on very short axillary branchlets and surrounded by broadly lanceolate herbaceous softly villous floral leaves about as long as the spikes. Bracteoles very broad and obtuse, rather longer than the perianth; bracts shorter and more acute. Perianth-segments scarcely 1½ lines long, woolly outside to above the middle, the glabrous white tips very obtuse in the outer ones, less so and narrower in the inner. Staminal tube short; filaments dilated at the base, acuminate, without intervening teeth or lobes. Style short.—*Iresine macrocephala,* Moq. in DC. Prod. xiii. ii. 342.

N. Australia? Victoria river? *Bynoe.*
Queensland. Cape Flinders, *A. Cunningham.*

13. **G. diffusa,** *Spreng. Syst.* i. 824. Stems from a perennial often woody stock procumbent, branching, slender, 1 to 2 ft. long, the branches and foliage softly pubescent. Leaves lanceolate, acute, contracted into a very short petiole, ½ to 1 in. long. Spikes axillary, shortly pedunculate, at first ovoid or oblong but lengthening to ½ or ¾ in. and not ¼ in. diameter, the rhachis pubescent. Bracts and bracteoles obtuse, as long as the perianth, the bracteoles broad and readily splitting. Perianth-segments about 1 line long, quite glabrous, obtuse, scarious with a narrow opaque centre not reaching either to the summit or to the base. Staminal tube rather shorter than the ovary; filaments short, much dilated, tapering to a point, without intervening teeth or lobes.—*Philoxerus diffusus,* R. Br. Prod. 416; *Iresine Brownii,* Moq. in DC. Prod. xiii. ii. 341.

N. Australia. Islands of the Gulf of Carpentaria, *R. Brown.*

This and the following species have the habit almost as much of *Alternanthera* as of *Gomphrena,* but the style is decidedly lobed.

14. **G. parviflora,** *Benth.* Stems long and slender, probably diffuse, turning red, the whole plant quite glabrous or with a few long hairs on the young shoots. Leaves sessile, linear or linear-lanceolate, acute, with recurved margins, ½ to 1 in. long. Spikes pedunculate,

axillary and terminal, quite glabrous, about 2 lines diameter, at first short and conical, at length cylindrical and ½ in. long. Bracts short broad and persistent, bracteoles erect, more than half as long as the perianth and deciduous with it. Perianth-segments quite glabrous, scarcely 1 line long, obtuse, white, shortly green at the base. Staminal tube shorter than the ovary, truncate; filaments but slightly dilated, without intervening teeth or lobes. Style very short.

N. Australia. Regent river, N.W. coast, *A. Cunningham ;* Port Darwin, *Schulz.*

Order XCVIII. PARONYCHIACEÆ.

Perianth-segments or lobes 5, rarely 4 or 3, usually scarious at the margins, imbricate in the bud. Stamens as many as perianth-segments and opposite to them or fewer by abortion, with or without intervening teeth lobes or staminodia, usually filiform, rarely petal-like or perfect stamina. Ovary 1-celled with 1 ovule attached to a funicle erect from the base of the cavity. Style more or less divided into 2 or 3 branches or separate styles, stigmatic at the end or along the inner edge. Fruit a membranous indehiscent utricle enclosed in or resting on the persistent perianth. Seed usually vertical, orbicular or ovate and compressed ; testa crustaceous or membranous. Embryo curved or annular, enclosing a mealy albumen.—Herbs usually small and much branched. Leaves opposite or rarely alternate, entire, accompanied by small scarious stipules or connected by a raised line or narrow membrane. Flowers small, in axillary or terminal cymes, sometimes reduced to dense clusters or rarely solitary. Bracts small, usually scarious ; bracteoles only under solitary flowers or the terminal flower of the cymes.

A small Order, widely dispersed over the temperate and cooler regions of the globe, more rare within the tropics and there chiefly in mountainous or sandy districts. The only Australian genus is common to both the northern and southern extratropical regions of the Old World. The Order is undistinguishable from Amarantaceæ by any positive character, differing chiefly in the stipules or connecting lines of the leaves like those of Caryophylleæ. The teeth or lobes intervening between the stamens (staminodia), which in Amarantaceæ are exceedingly thin and transparent, are in Paronychiaceæ very various, sometimes thin and transparent, sometimes filiform or developed into stamens (in the European *Scleranthi*) or into petals (in *Corrigiola*) ; in both Orders they are frequently deficient.

1. SCLERANTHUS, Linn.

(Mniarum, *Forst.*)

Perianth-tube as long as the lobes, somewhat enlarged and hardened after flowering. Stamens 5 or fewer, opposite the lobes, connected by a membrane or raised line at the mouth of the perianth-tube, with or without intervening teeth or filaments, or in species not Australian perfect stamens. Styles 2, slender. Fruit a membranous utricle enclosed in the perianth-tube.—Small densely branched herbs. Leaves opposite, narrow, connected by a raised line or sheathing membrane, without stipules. Flowers solitary or several together, sessile or nearly so within a pair of scarious bracts at the end of axillary peduncles.

The genus is represented by a few species in Europe, temperate and subtropical Asia and northern Africa. Of the four Australian species one is also in New Zealand, the three others are endemic. The specimens of this genus from the Melbourne herbarium have been accidentally omitted from those sent over to me for examination, but have been fully described by F. Mueller in the Plants of the Colony of Victoria, i. 214 to 216.

Leaves narrow-linear, not pungent. Flowers green, not above 1 line
  long.
Flowers solitary. Stamen 1 . . . . 1. *S. mniaroides.*
Flowers 2 together. Stamen 1 . . . . 2. *S. biflorus.*
Flowers several together. Stamens 2 . . . . 3. *S. diander.*
Leaves rather broader, rigid and pungent-pointed. Flowers scarious,
  2 lines long, 2 or several together. Stamens 5, with intervening
  filaments . . . . 4. *S. pungens.*

1. **S. mniaroides,** *F. Muell. Pl. Vict.* i. 215, t. 12. A perennial with densely branched decumbent stems of 3 to 4 in., resembling *S. biflorus* both in habit and foliage. Leaves linear, scarcely acute, entire, 2 to 4 lines long. Peduncles at first very short, longer than the leaves when in fruit as in *S. biflorus,* but never bearing more than a single flower, which is sessile between 2 minute broad concave bracts. Perianth scarcely above ½ line long when in flower, nearly 1 line when in fruit, the tube then ovoid and thickened, crowned by 4 or 5 short herbaceous lobes not enlarged. Stamen 1, inserted opposite one of the lobes in a notch in an annular membrane at the mouth of the tube. Fruit of *S. biflorus.*—*Mniarum singuliflorum,* F. Muell. in Trans. Phil. Soc. Vict. i. 13 and in Hook. Kew. Journ. viii. 69.

Victoria. Glacier ranges of the Australian Alps, not descending below 5000 ft., *F. Mueller.*

2. **S. biflorus,** *Hook. f. Fl. N. Zeal.* i. 74 *and Fl. Tasm.* i. 42. A low tufted perennial, with short decumbent stems forming dense masses of a few inches diameter or rarely looser and lengthening to 6 in., glabrous or with a minute pubescence on the branches. Leaves crowded, narrow-linear, acute, entire or minutely scabrous, serrulate, mostly 3 to 4 lines long. Peduncles axillary, at first very short and sometimes remaining so, but more frequently lengthening before or after flowering and exceeding the leaves when in fruit, each bearing 2 small flowers sessile within a pair of ovate acute concave bracts, and one of the flowers (the latest in expanding) with a pair of smaller bracteoles. Perianth when in flower ½ to ¾ line long, with 4 or 5 herbaceous lobes and a single stamen opposite one of the lobes inserted in a scarcely prominent annular membrane at the mouth of the tube. Fruiting perianth nearly 1 line long, the hardened tube ovoid, almost closed by the staminal membrane, crowned by the unenlarged persistent lobes and enclosing the membranous utricle.—F. Muell. Pl. Vict. i. 214, *Ditoca muscosa,* Banks in Gærtn. Fr. ii. 196, t. 126; *Mniarum biflorum,* Forst. Gen. 2, t. 1; R. Br. Prod. 412; DC. Prod. iii. 378; *M. pedunculatum,* Labill. Pl. Nov. Holl. i. 8, t. 2.

N. S. Wales. Blue Mountains, *Miss Atkinson*; Ben Lomond, Mount Mitchell and sources of Clarence river, *Beckler*, (*F. Mueller*), I have seen none of these specimens.

Victoria. Abundant on the alpine and subalpine plateaux and valleys throughout the chain of the snowy alps, descending to 3000 or rarely to 2000 ft., *F. Mueller.*

Tasmania. Port Dalrymple, *R. Brown ;* common on dry grassy pastures on stones, &c., *J. D. Hooker.*

The species is also in New Zealand. *Mniarum fasciculatum,* R. Br. Prod. 412, Hook. Ic. Pl. t. 283, DC. Prod. iii. 378, or *Scleranthus fasciculatus,* Hook. f. Fl. Tasm. i. 42, appears to be rather a state of the plant with the peduncles remaining shorter than the leaves than a distinct variety.

3. S. diander, *R. Br. Prod.* 412. A perennial with a densely tufted almost woody stock and much branched erect or decumbent stems of 2 to 4 in. Leaves crowded, linear, very acute but not pungent, 2 to 3 lines long. Flowers several together, sessile in little heads shortly pedunculate in the upper axils or at the end of the branches. Bracts very acute, the outer ones leafy often longer than the flowers, the inner ones smaller, one to each flower or 2 to the terminal one of each branch of the head or cyme. Perianth about 1 line long, with 5 obtuse scarious lobes as long as the tube, slightly enlarged in fruit. Stamens 2, the filaments slightly dilated, the connecting membrane irregular or obsolete.—DC. Prod. iii. 378 ; Hook. f. Fl. Tasm. i. 42 ; F. Muell. Pl. Vict. i. 215.

N. S. Wales. Argyle County, *Backhouse.*

Victoria. Subalpine meadows through most ramifications of the Australian alps, *F. Mueller.*

Tasmania. Port Dalrymple, *R. Brown ;* dry banks, pastures, &c , Launceston, *Gunn.*

F. Mueller mentions a desert variety flowering the first year, along the Murray and Wimmera rivers and Lake Alexandrina, which I have not seen.

4. S. pungens, *R. Br. Prod.* 412. A densely tufted or much-branched perennial, larger and more rigid than any other species, attaining 6 in. Leaves very rigid and pungent-pointed, linear or linear-lanceolate, $\frac{1}{4}$ to $\frac{1}{2}$ in. long. Peduncles in the upper axils or terminal, very short, bearing a head of several sessile flowers. Bracts lanceolate, pungent-pointed, shorter than the flowers. Perianth about 2 lines long, the lobes lanceolate, very acute, longer than the tube, scarious at least at the margin. Stamens 5, the filaments filiform, opposite the perianth-lobes and alternating with as many filiform or flattened staminodia as long as the stamens. Ovary tapering into a rather long beak. —DC. Prod. iii. 378 ; F. Muell. Pl. Vict. i. 216.

Victoria. Barren ridges and wastes of the north-western district, *F. Mueller.*

S. Australia. Memory Cove, *R. Brown ;* Murray Scrub, Flinders' Range and Spencer's Gulf, *F. Mueller.*

W. Australia ? King George's Sound, *Herb. Hooker.,* but possibly some mistake in the station.

I have not seen any of F. Mueller's specimens.

Order XCIX. **POLYGONACEÆ.**

Perianth-segments 6, 5 or fewer, free or shortly united at the base, imbricate in the bud and more or less in 2 rows, regular and equal or the inner ones enlarged. Stamens variable in number, usually 9, 8, 6 or fewer, alternate with the perianth-segments or having no definite relation to them; anthers with 2 parallel cells. Ovary free, with a single erect orthotropous ovule. Styles or style-branches 2, 3 or rarely more, the stigmas terminal, capitate or dilated, entire or fringed. Fruit a small seed-like nut, usually with as many angles as styles, enclosed in or scarcely protruding from the persistent perianth. Seed erect with a membranous testa. Embryo in a mealy albumen, straight and central or more frequently curved and lateral; radicle superior.—Herbs shrubs woody climbers or rarely, in species not Australian, trees. Leaves alternate. Stipules usually thin and scarious, brown or silvery, forming a sheath or ring round the stem. Flowers small, herbaceous or coloured, clustered in the axils of the leaves, or within small sheathing bracts or even without bracts along the rhachis of simple or paniculate spikes or racemes, without bracteoles on the pedicels.

A considerable Order dispersed over every part of the globe. Of the four Australian genera, two have a very wide distribution especially in temperate regions of both the New and the Old World, another extends through New Zealand to Antarctic and Andine America, the fourth belongs to the Mediterranean and South African regions of the Old World, and may possibly be of recent introduction into Australia.

Perianth-segments 6, the 3 inner ones closing over the fruit.
    Stamens 6. Styles 3. Stigmas fringed.
    Fruiting perianth hard, with a 3-angled tube, the outer segments
      spreading and spinescent, the inner short and erect . . . . 1. Emex.
    Fruiting perianth with the outer segments small and recurved,
      the inner enlarged and closed over the fruit . . . . . . 2. Rumex.
Perianth-segments 5, nearly equal. Stamens 8 or fewer.
    Flowers mostly hermaphrodite. Styles or style-branches 2 or 3;
      stigmas entire . . . . 3. Polygonum.
    Flowers more or less dioecious. Styles or style-branches 3, stig-
      mas more or less fringed, or rarely entire . . . . 4. Muhlenbeckia.

1. **EMEX,** Necker.

Flowers monoecious. Male fl.: Perianth-segments 5, equal, spreading. Stamens 4 to 6. Pistil rudimentary. Female fl.: Perianth with a triangular turbinate tube and 6 lobes, enlarged and hardened in fruit, the 3 outer lobes corresponding to the angles of the tube rigid spreading and spinescent, the three inner ovate erect and connivent over the fruit. Ovary small, 3-angled, styles 3, short, with large fringed stigmas.—Herbs. Leaves ovate. Stipules brown and scarious, sheathing but soon loose and torn or jagged. Flowers in whorl-like clusters, the females axillary, the males distant on axillary peduncles, the lower clusters including sometimes a few females.

Besides the Australian species, which is identical with a South African one and perhaps introduced from thence, there is one other closely allied to it from the Mediterranean region of the northern hemisphere.

1. **E. australis,** *Steinh. in Ann. Sc. Nat. Ser.* 2, ix. 195, *t.* 7. Stems diffuse, rather thick, 1 to 1½ ft. long, glabrous as well as the whole plant. Leaves on long petioles, broadly ovate, very obtuse, truncate or broadly cordate at the base, 1 to 2 in. long. Fruiting perianth very hard, triquetrous, 4 to 5 lines long, with 3 rather long rigid thick spinescent and divaricate lobes, the three inner ones short, broad, erect, obtuse or mucronate.—Miq. in Pl. Preiss. i. 625 ; Meissn. in Linnæa xxvi. 363 ; *E. centropodium,* Meissn. in Linnæa xiv. 490, in Pl. Preiss. ii. 273, and in DC. Prod. xiv. 40.

S. Australia. Near Adelaide and Holdfast Bay, *F. Mueller.*
W. Australia, *Drummond, 2nd. coll. n.* 290 ; *Preiss, n.* 1895.

A common maritime plant in S. Africa, differing slightly from the Mediterranean species (*E. spinosa,* Campd.) in the larger fruiting perianth less rugose, the spinescent segments longer, and the inner erect ones broader and more rounded.

2. RUMEX, Linn.

Flowers hermaphrodite or unisexual. Perianth of 6 segments, the three inner ones enlarged after flowering and closing over the fruit, the three outer much smaller narrow and recurved. Stamens 6. Styles 3, shortly filiform, with large fringed stigmas. Nut triangular, enclosed in the persistent perianth. Embryo lateral, the radicle superior.—Herbs or rarely, in species not Australian, shrubs, usually glabrous. Flowers small, herbaceous or the males petal-like, all often turning red, usually on recurved pedicels, in whorl-like clusters, either axillary or in terminal racemes or panicles. Stipules sheathing, scarious, usually brown, at first entire but almost always very soon torn or jagged.

A considerable genus, widely distributed over most parts of the globe, but more especially in temperate regions, and a few species are amongst the roadside weeds which the most readily establish themselves in new countries. Of the 8 species here enumerated, three are certainly and a fourth possibly introduced from Europe or South Africa, the remaining four appear to be endemic.

Flowers all or mostly hermaphrodite.
  Inner perianth-segments with entire margins, one or all bearing a tubercle on the midrib.
    Inner perianth-segments broad, mostly cordate. Panicle dense with short crowded branches . . . . 1. *R. crispus.*
    Inner perianth-segments narrow. Panicle with elongated branches, the clusters of flowers distant . . . . . . 2. *R. conglomeratus.*
  Inner perianth-segments bordered by long teeth.
    Flower-clusters distant along the branches, without floral leaves except to the lower ones.
      Teeth of the inner perianth-segments five, with revolute points . . . . 3. *R. Brownii.*
      Teeth of the inner-perianth-segments rigid and straight or rarely curved at the point . . . . 4. *R. pulcher.*
    Floral leaves all longer than the flowers.
      Panicle very much divaricately branched. Clusters fewflowered. Fruiting perianth rather large, without tubercles . . . . 5. *R. dumosus.*

Branches erect. Flowers numerous and very small in
dense clusters. Fruiting perianth small, with a tubercle on each segment . . . . 6. *R. halophilus.*
Flowers monœcious. Stems mostly simple, the upper clusters
males, often without floral leaves, the lower clusters females,
mostly axillary . . . . 7. *R. bidens.*
Flowers diœcious. Small plant, with narrow leaves mostly
sagittate. Stipules silvery-shining . . . . 8. *R. acetosella.*

Besides the above, I find reference to a *Rumex crystallinus*, as described by Lange
in the Index of seeds of the Garden of Copenhagen for the year 1861, from plants
raised from seeds received from F. Mueller. I have been unable to find a copy of the
seed-catalogue for that year, and am quite unacquainted with the plant in question,
which, if really Australian, is probably identical with some one of those above enumerated.

1. **R. crispus,** *Linn.; Meissn. in DC. Prod.* xiv. 44. A perennial
with a thick rhizome, and erect furrowed stems 2 to 3 ft. high, the
branches few and short. Radical leaves narrow, usually 6 to 8 in. long,
much undulate and crisped at the edges, the upper ones smaller, passing
gradually into bracts. Flower-clusters numerous, and when in fruit
much crowded into a long narrow and dense terminal panicle. Inner
segments of the fruiting perianth broadly ovate, entire, 2 to 3 lines
long, with a coloured tubercle on the midrib.

N. S. Wales. Hunter's river, *Oldfield.*
Victoria. On roadsides about Melbourne, *Adamson, F. Mueller.*
S. Australia. Barker town, *F. Mueller.*

Common in Europe and temperate Asia, and now naturalized in many other parts of
the globe. The Australian specimens, like others grown in warm and dry climates,
have tubercles on all three inner perianth-segments. More northern specimens have
them frequently only on one segment of each perianth.

2. **R. conglomeratus,** *Murr.; Meissn. in DC. Prod.* xiv. 49. An
erect perennial of 2 or 3 ft. more branched than *R. crispus.* Radical
leaves on long petioles, mostly acute, rounded or even cordate at the
base, sometimes 8 to 10 in. long, the upper ones smaller. Panicle with
long rather spreading branches, the clusters of flowers distinct or distant even when in fruit. Fruiting perianth smaller than in *R. crispus,*
the inner segments narrow-ovate, entire, with an oblong tubercle on
the midrib of each.—*R. acutus,* Sm. and some others.

Queensland. About Brisbane, *C. Stuart.*
N. S. Wales. Port Jackson, *Woolls.*
Victoria. About Melbourne, *F. Mueller.*

Like *R. crispus,* this species is indigenous in Europe and temperate Asia, and is now
become naturalized in many other parts of the globe.

3. **R. Brownii,** *Campd. Monogr. Rum.* 81. A perennial with a thick
rhizome and erect simple or slightly branched stems of 1 to 2 ft. Radical and lower leaves on long petioles, often cordate or hastate, oblong
and obtuse, the stem-leaves mostly lanceolate and acute, the floral ones
reduced to small bracts or quite deficient. Clusters remote, many-
flowered, forming long simple or slightly-branched racemes, the fruiting pedicels slender or thick, 1 to 3 lines long. Inner segments of the
fruiting perianth broadly triangular, 1½ to 2 lines long, bordered on
each side by 4 to 6 bristles, much hooked or almost involute at the end,

the midrib prominent but without any distinct tubercle.—Meissn. in
DC. Prod. xiv. 61; Hook. f. Fl. Tasm. i. 305; *R. fimbriatus*, R. Br.
Prod. 421 not of Poir.

Queensland. Rockhampton, *O'Shanesy.*
N. S. Wales. Port Jackson, *R. Brown, Woolls;* Newcastle everywhere where
sheep have been, *Leichhardt:* Macleay river, *Beckler.*
Victoria. Plenty river, *F. Mueller;* Wendu Vale, *Robertson.*
Tasmania. Abundant in marshy places, Circular Head, &c., *J. D. Hooker.*
S. Australia. Adelaide and Torrens river, *F. Mueller;* Kangaroo island, *Hinzenrode.*

4. **R. pulcher,** *Linn.; Meissn. in DC. Prod.* xiv. 58. Stems erect,
short below the inflorescence, the flowering branches very spreading,
forming a broad panicle of above 1 ft. Radical and lower leaves on
long petioles, oblong-lanceolate or almost linear, 2 to above 6 in. long,
slightly crisped on the margins, often contracted in the middle and
obtuse or cordate at the base, the lower floral leaves linear and sometimes above 1 in. long, all the others reduced to small bracts or obsolete. Clusters distant, the flowers more numerous than in *R. dumosus,*
less so than in *R. Brownii.* Pedicels shorter or longer than the perianth.
Inner segments of the fruiting perianth lanceolate or triangular, 2 to 3
lines long, with a rigid point and 2 or 3 rigid subulate teeth on each
side, straight or rarely slightly hooked at the extremity, the midrib
bearing a prominent callous tubercle.—*R. oxysepalus,* Meissn. in Pl.
Preiss. i. 625; *R. Drummondii,* Meissn. in Pl. Preiss. ii. 272 and in
DC. Prod. xiv. 61.

Victoria? Near Melbourne, *Adamson, Robertson* (specimens in flower only).
W. Australia, *Drummond, n.* 27, 108, 207, 703; *Preiss, n.* 1357; Kalgan river,
Oldfield; N. of Stirling range, *F. Mueller.*
The species is common in a great part of Europe, Western Asia, and Northern Africa,
extending to South Africa, and possibly introduced from thence into West Australia. The
Melbourne specimens, as well those above quoted as those of F. Mueller, correspond-
ing to others previously sent to London and described by Meissner as *R. Muelleri* (in
DC. Prod. xiv. 61), are not far enough advanced to determine accurately, and may belong to a luxuriant form either of *R. pulcher* or of *R. dumosus.* The lower leaves are
sometimes 10 in. long.

5. **R. dumosus,** *A. Cunn.; Meissn. in DC. Prod.* xiv. 62. "A divaricate bushy plant" spreading to 2 or 3 ft., the flowering branches
when in an advanced state numerous intricate flexuose and dichotomous.
Lower leaves long, but not seen perfect, the upper ones small, lanceolate, acuminate, but all even the uppermost floral ones longer than
the flowers. Clusters distant, composed of very few, often only 2 or
3 flowers, the pedicels 1 to 2 lines long, sometimes hardened after the
fall of the fruit. Inner segments of the fruiting perianth about 2 lines
long, rigid acute and prominently reticulate, without tubercles, bordered
by rigid subulate almost spinescent teeth not hooked, the small outer
segments sometimes muricate on the keel.—*R. Brownii,* Schlecht. in
Linnæa. xx. 576 (from the descr.), not of Campd.

N. S. Wales. Liverpool plains, and plains subject to inundation on the Cujeegong
and Goulburn rivers, *A. Cunningham.*
S. Australia. Torrens river, *F. Mueller.*

6. **R. halophilus,** *F. Muell. Fragm.* iv. 48. An erect branching herb of about 1 ft., some specimens appearing annual. Leaves linear or lanceolate, acute or obtuse, the lower ones on long petioles sometimes cordate at the base and 2 to 3 or even 4 in. long, the upper ones small, but all or nearly all longer than the flowers. Flowers very small and very numerous, in dense axillary clusters crowded on the greater part of the plant. Inner segments of the fruiting perianth sometimes under 1 line long, with long fine points and marginal teeth, sometimes rather longer and broader with shorter and more rigid points and teeth, always with a very prominent ovoid tubercle on the midrib.

N. Australia. Gulf of Carpentaria, *F. Mueller.*
Queensland. Flinders and Burnett rivers, *F. Mueller;* Darling Downs, *Lau.*
N. S. Wales. Darling river, *Goodwin and Dallachy,* Mrs. *Ford;* Murray river, *F. Mueller.*

This plant has entirely the aspect of the European *R. maritimus,* Linn., and may be a variety only. The teeth of the fruiting perianth-segments although fine and long are however always much less so than in the northern plant.

7. **R. bidens,** *R. Br. Prod.* 421. A perennial with a thick stem, rooting at the joints and creeping in the mud, throwing up numerous erect thick flowering branches of 1 to 2 ft. Leaves lanceolate, the lower ones often 8 to 10 in. long, rather broad and obtuse, the upper ones narrower and more acute, passing into the small almost linear floral leaves. Flowers densely clustered, unisexual, but the two sexes mixed in the same clusters, the lower ones chiefly females the upper ones chiefly or entirely males. Perianth-segments at the time of flowering almost petaloid and not very unequal. Inner segments of the fruiting perianth variable in size, usually 2 to 3 lines long, very broad, with 1 or 2 rigid teeth on each side near the base, the midrib more or less thickened into a tubercle.—Meissn. in DC. Prod. xiv. 73; Hook. f. Fl. Tasm. i. 305.

Victoria. On the Yarra about Melbourne, *Harvey, F. Mueller.*
Tasmania. Port Dalrymple and Lagoon Beach, *R. Brown;* common in marshes in the northern part of the island, *J. D. Hooker.*
S. Australia. Murray river, *Behr.;* near Morunda, *F. Mueller;* with a small starved, small-flowered variety.

*8. **R. acetosella,** *Linn.; Meissn. in DC. Prod.* xiv. 63. A slender plant with a creeping rhizome and erect stems from a few in. to nearly 1 ft. high, often turning red. Leaves narrow-lanceolate or linear, some or all hastate or sagittate at the base; stipules usually silvery and very thin. Flowers small, diœcious, in slender terminal panicles, the clusters numerous, few-flowered, without floral leaves. Perianth-segments broad, entire, not very unequal, the inner ones in the fruiting perianth closed over the seed but scarcely enlarged.

A plant probably of European and Asiatic origin, now common in most temperate and subtropical regions of the globe, and evidently introduced only into Australia, where it has appeared in **N. S. Wales, Victoria, S. Australia,** and **W. Australia.** It is in R. Brown's collection from the neighbourhood of Port Jackson, but omitted from the Prodromus as an introduced plant.

3. POLYGONUM, Linn.

Flowers usually hermaphrodite. Perianth of 5, rarely fewer, segments, all equal or the 2 or 3 outer ones enlarged. Stamens 5 to 8, varying often in the same species. Styles or style-branches 2 or 3, with entire terminal stigmas. Nut flattened or triangular, enclosed in or surrounded by the persistent perianth. Embryo lateral, more or less curved, the radicle superior.—Herbs varying much in habit. Stipules in the majority of species thinly scarious, closely sheathing, the lower portion adnate to the petiole, the greater part connate within it; in the section *Avicularia* looser jagged and silvery. Flowers small, pale green or red with white edges, either clustered or rarely solitary in the axils, or in little clusters within a sheathing bract arranged in simple or paniculate spikes.

A large genus, with almost the cosmopolitan range of the Order. Of the thirteen Australian species, four are common European species extending more or less generally over the temperate and subtropical regions of the globe, five are distributed over the warmer portions of Asia, one or two of them being also found in South America, another appears to be also S. American although not yet identified with any Asiatic species; one more extends at least to New Caledonia, and one to New Zealand; the only remaining one may be endemic, although not yet perhaps sufficiently known.

SECT. 1. **Avicularia.**—*Flowers in axillary clusters. Stems prostrate.*

Stems elongated and wiry. Perianth about 1½ lines long. Nut
minutely granular-striate . . . . 1. *P. aviculare.*
Stems short and compact. Perianth under 1 line long. Nut
smooth and shining . . . . 2. *P. plebeium.*

SECT. 2. **Echinocaulon.**—*Flowers in very short spikes on axillary dichotomous peduncles. Stems usually scrambling, with short reversed bristles on the angles.*

Single Australian species . . . . 3. *P. strigosum.*

SECT. 3. **Persicaria.**—*Flowers spicate, the spikes in terminal panicles, or rarely solitary and terminal or axillary. Embryo curved towards the edge of the cotyledons. Nut flat or convex on both sides or triquetrous.*

Spikes solitary, mostly axillary. Stem prostrate . . . . . 4. *P. prostratum.*
Spikes terminal, long, slender and interrupted, usually single.
Stem slender, erect. Perianth densely glandular-dotted . . 5. *P. hydropiper.*
Spikes usually continuous and 2 or more in a terminal panicle.
Perianth without any or with few glandular dots. Stem
erect or ascending.
Stipules bordered by bristles usually long, at least the upper
ones.
Glabrous. Spikes few and slender . . . . 6. *P. minus.*
Strigose-hirsute. Spikes slender, pedunculate. Leaves
nearly sessile . . . . 7. *P. subsessile.*
Strigose-hirsute. Spikes compact, on short peduncles.
Bristles of the stipules very long . . . . . 8. *P. barbatum.*
Stipules truncate, without any or with a few fine short cilia.
Strigose-hirsute. Spikes compact . . . . 9. *P. articulatum.*
Glabrous or nearly so . . . . 10. *P. lapathifolium.*
Cottony-white . . . . 11. *P. lanigerum.*

SECT. 4. **Amblygonon.**—*Flowers spicate, the spikes in terminal panicles. Embryo curved towards the back of the cotyledons. Nut usually flat on both sides, with obtuse margins. Stems tall and erect.*

Stipules, at least the lower ones, dilated and green at the top   . 12.   *P. orientale.*
Stipules all closely sheathing and scarious to the top . . . . 13.   *P. attenuatum.*

Muhlenbeckia Cunninghamii has almost the styles and stigmas of *Polygonum,* but the flowers are diœcious.

SECT. 1. AVICULARIA, Meissn.—Flowers in axillary clusters. Stems (in the Australian species) prostrate.

1. **P. aviculare,** *Linn. Meissn. in DC. Prod.* xiv. 97. A glabrous annual, much branched at the base, sometimes erect or ascending when young, but the stems soon prostrate, wiry, extending to 1 or 2 ft. or even more. Leaves shortly petiolate, elliptical oblong lanceolate or rarely linear, the larger ones above 1 in. long but mostly smaller. Stipules broad, more or less brown at the base, the remainder silvery and split into finely pointed lobes. Flowers axillary, solitary or in clusters of 2 to 5, very shortly pedicellate. Fruiting perianth above 1 line long, the segments green in the centre, white on the margins. Styles 3. Nut triangular, black but opaque (not shining) owing to a minute granulation visible under a strong lens.

Victoria. Near Melbourne, *Adamson, Hildebrand;* Skipton, *Whan.*
S. Australia. Near Adelaide, *F. Mueller.*
W. Australia, *Drummond, n. 231.*

A common weed, apparently of European or Asiatic origin, but now found over a great part of the globe, especially in temperate regions, and probably introduced only in Australia.

2. **P. plebeium,** *R. Br. Prod.* 420. A much branched prostrate annual, much more compact than *P. aviculare,* and rarely above 1 ft. long, glabrous or the branches slightly hoary. Leaves linear, narrow-oblong or slightly spathulate, rarely above ½ in. long. Stipules as in *P. aviculare* short silvery and ragged at the edges. Flowers very small, in clusters of 2 to 5 in the axils of most of the leaves. Fruiting perianth under 1 line long, the segments green, with a narrow white edge. Nuts triangular, very smooth and shining.—Meissn. in DC. Prod. xiv. 94 ; Benth. Fl. Hong̃k. 287, with the synonyms there adduced.

Queensland. Broad Sound, *R. Brown;* Brisbane river, Moreton Bay, *A. Cunningham, F. Mueller* and others ; Gilbert river, *F. Mueller;* Rockhampton, *O'Shanesy;* Wide Bay, *Bidwill;* Kennedy district, *Daintree.*
N. S. Wales. Glendon, *Leichhardt;* Murray and Darling rivers, *Dallachy, Mrs. Ford* and others ; Stokes Range to Cooper's Creek, *Wheeler.*
Victoria. Murray river, *F. Mueller;* Wimmera, *Dallachy.*
S. Australia. Bugle Range, *F. Mueller;* Cooper's Creek, *Howitt's Expedition.*

Exceedingly common all over tropical Asia, extending into Africa. Very near *P. aviculare,* but amidst all its variations it seems to me constantly to differ from that species in its compact habit and small flowers, and in the nuts always smooth and shining.

SECT. 2. ECHINOCAULON, Meissn.—Flowers in very short spikes on axillary dichotomous peduncles. Stems usually scrambling, with short reversed bristles on the angles.

3. **P. strigosum,** *R. Br. Prod.* 420. Stems weak, erect or straggling, 2 to 3 ft. long, with reflexed bristles on the angles of the branches and on the petioles and midribs of the leaves, and a short glandular pubescence on the peduncles, otherwise the whole plant glabrous. Leaves petiolate, lanceolate, acuminate, hastate or broadly sagittate at the base, the larger ones 2 to 4 in. long. Stipules sheathing, shortly ciliate-toothed or entire. Peduncles axillary, loosely dichotomous, usually longer than the leaves, the branches terminating in short rather dense but few-flowered spikes. Bracts denticulate and ciliate, nearly 2 lines long. Perianth-segments about 1½ lines long, slightly enlarged when in fruit, completely enclosing the smooth nut. Style-branches and angles of the nut more frequently 3 than 2.— Meissn. in DC. Prod. xiv. 134; Hook. f. Fl. Tasm. i. 307; *P. pedunculare,* Wall.; Meissn. l.c. 133; *P. muricatum,* Meissn. l.c.

Queensland. Brisbane river, Moreton Bay, *A.Cunningham, F. Mueller;* Rockingham Bay, *Dallachy.*

N. S. Wales. Port Jackson, *R. Brown, Woolls;* Newcastle, *Leichhardt;* New England, *C. Stuart;* Hastings and Clarence rivers, *Beckler* and others.

Victoria. Ovens and Plenty rivers, *F. Mueller.*

Tasmania. Port Dalrymple, *R. Brown;* northern parts of the island, *J. D. Hooker;* Launceston, *Gunn;* South Esk river, *C. Stuart.*

The species extends over the Archipelago and Eastern India to S. China, varying with the spikes dense or slender and interrupted, and with the pistils 2-merous or 3-merous on the same plant.

SECT. 3. PERSICARIA.—Flowers spicate, the spikes in terminal panicles, or rarely solitary and terminal or axillary. Embryo curved towards the edge of the cotyledons (accumbent) or sometimes oblique. Nut flat or convex on both sides or triquetrous.

4. **P. prostratum,** *R. Br. Prod.* 419. A prostrate branching perennial, often woody at the base and rooting at the lower nodes, extending to 1 or 2 ft., sprinkled with slender appressed hairs or nearly glabrous. Leaves lanceolate, contracted into a short petiole 1 to 1½ or rarely 2 in. long. Stipules sheathing, ciliate with a few long hairs on the back and margins. Spikes shortly pedunculate, axillary or rarely terminal, about ½ in. or rarely ¾ in. long, rather dense. Perianth scarcely 1 line long when in flower, slightly enlarged afterwards. Stamens 5 or 6. Style-branches 2, not enclosed in the perianth, lenticular, both sides very convex.—Meissn. in DC. Prod. xiv. 116; Hook. f. Fl. Tasm. i. 307.

Queensland. Rockhampton, *O'Shanesy.*

N. S. Wales. Nepean river, *R. Brown, Woolls;* New England, *C. Stuart;* Argyle county, *A. Cunningham.*

Victoria. Glenelg river, *Robertson;* Yarra, Sandy Creek, Ballarat, *F. Mueller;* Murray river, *Herrgott;* Emu Creek, *Whan.*

Tasmania. Port Dalrymple and Derwent river, *R. Brown;* not uncommon in various parts of the island, *J. D. Hooker.*

S. Australia. Near Morunda, *F. Mueller.*

W. Australia, *Drummond, 5th coll. n.* 230.

The species is also in New Zealand.

5. **P. hydropiper,** *Linn.; Meissn. in DC. Prod.* xiv. 109. A slender erect or decumbent glabrous annual, 1 to 2 ft. high. Leaves lanceolate, acuminate. Stipules sheathing, rather short, ciliate at the top. Spikes slender and interrupted, often several in. long, solitary or nearly so at the ends of the stem or branches and often nodding. Perianth and often the bracts also dotted with prominent glands. Style-branches usually 2. Nuts flat.—*P. gracile,* R. Br. Prod. 419 ; Meissn. in DC. Prod. xiv. 109.

N. S. Wales. Port Jackson, *R. Brown, Woolls;* Hunter's river, *Leichhardt ;* Clarence river, *Beckler.*

Victoria. Yarra-Yarra, *Robertson, F. Mueller* ; Goulburn river, *F. Mueller.*

The species is widely diffused over Europe, temperate and subtropical Asia and North America I can perceive no difference between the Australian and many of the northern specimens.

6. **P. minus,** *Huds. ; Meissn. in DC. Prod.* xiv. 111. Rather slender, erect or decumbent, smaller and less branched than *P. lapathifolium,* rarely exceeding 2 ft., quite glabrous in the typical form. Leaves shortly petiolate, lanceolate, acuminate. Stipules closely sheathing, the margins ciliate. Spikes 1 to 1½ in. long, few in a terminal panicle, more slender than in *P. lapathifolium,* much less so than in *P. hydropiper.* Bracts small, naked or shortly ciliate. Perianth small, not glandular. Stamens 5 or 6. Style-branches 2 or 3, varying often in the same spike, more frequently 3 than 2 in the Australian specimens.—Hook. f. Fl. Tasm. i. 306; *P. decipiens,* R. Br. Prod. 420 ; Meissn. in DC. Prod. xiv. 104.

Queensland. Brisbane, Burnett, and Burdekin rivers, *F. Mueller ;* Moreton island, *M'Gillivray ;* Rockhampton, *O'Shanesy ;* Rockingham Bay, *Dallachy ;* in the interior, *Mitchell.*

N. S. Wales. Port Jackson, *R. Brown, Woolls;* Argyle county, *Backhouse ;* New England, *C. Stuart.*

Victoria. Murray and Yarra rivers, *F. Mueller, Robertson ;* Portland, *Allitt ;* Emu Creek, *Whan.*

Tasmania. Common in the northern parts of the island, *J. D. Hooker.*

S. Australia. Murray and Torrens rivers and Holdfast Bay, *F. Mueller.*

W. Australia, *Drummond, n.* 20; Moore river, *Oldfield.*

The species is very common in the temperate, subtropical, and even tropical regions of the Old World, and varies much. In Australia some specimens agree with the commonest European forms, in others, chiefly from Queensland, the cilia of the stipules are longer, there are a few small strigose hairs on the under side of the leaves, and the pistil is almost always 3-merous. These constitute probably (with corresponding European forms) the *P. serrulatum,* Lag. cited from Australia and New Zealand by Meissn. in D.C. Prod. xiv. 110.

7. **P. subsessile,** *R. Br. Prod.* 419. A perennial with an almost woody rhizome and erect loosely branched stems attaining 2 or 3 ft., the whole plant more or less strigose with appressed hairs, short on the upper side of the leaves, longer underneath. Leaves lanceolate, acuminate, mostly 3 or 4 in. long, contracted into a very short petiole. Stipules sheathing, much longer than the petiole, bordered by long cilia. Spikes rather slender, 1 to 2 in. long, usually 2 to 4 on long peduncles in a loose terminal panicle. Perianth-segments petal-like,

not glandular. Stamens 5. Style-branches 2, rather long. Nut lenticular, the sides very convex.—Meissn. in DC. Prod. xiv. 113; Hook. f. Fl. Tasm. i. 306.

Queensland. Brisbane river, Moreton Bay, *F. Mueller*, *C. Stuart;* Rockhampton, *O'Shanesy;* Rockingham Bay, *Dallachy.*

N. S. Wales. Port Jackson, *R. Brown, Woolls;* New England, *C. Stuart.*

Victoria. Goulburn, Yarra-Yarra, and Tara rivers, *F. Mueller.*

Tasmania. Port Dalrymple, *R. Brown;* northern parts of the island, *J. D. Hooker.*

The species appears to be also in New Caledonia, if I am right in identifying with it Viellard's n. 1075.

8. **P. barbatum,** *Linn. ; Meissn. in DC. Prod.* xiv. 104. Stems ascending or erect, glabrous or nearly so, 2 to 3 ft. high. Leaves lanceolate, contracted into a short petiole although not so short as in *P. subsessile,* 3 to 6 in. long, sprinkled with appressed hairs never wanting on the midrib underneath. Stipules sheathing hairy, outside, bordered by very long cilia. Spikes in the common form compact, 1 to 1½ in. long, on short peduncles in a narrow terminal panicle, in some Asiatic varieties drawn out and slender. Bracts always ciliate on the margin. Stamens 5 or 6. Style-branches (always?) 3. Nut triangular.

Queensland. Roper and Burnett rivers, *F. Mueller ;* Port Curtis, *M'Gillivray.*

The species is common in tropical Asia and Africa.

9. **P. articulatum,** *R. Br. Prod.* 420. Erect and rather stout, the stems glabrous below, hirsute in the upper part with appressed hairs. Leaves lanceolate, tapering into long points and contracted into rather long petioles, 3 to 6 in. long, hirsute on both sides with appressed hairs short and strigose on the upper surface, longer and more silky underneath, rigid on the margins. Stipules sheathing, long and truncate, without any or only with few very short marginal cilia. Spikes few, rather dense, 1 to 2 in. long. Bracts shortly ciliate. Styles 2. Nut compressed, the sides rather convex. Curvature of the embryo in the two seeds examined rather oblique with relation to the cotyledons, but more accumbent than incumbent.—Meissn. in DC. Prod. xiv. 117 ; *P. australe,* Spreng. Syst. ii. 258.

Queensland. Broad Sound and Shoalwater Bay, *R. Brown.*

I have not seen this in any other collection, and have not been able to match it with any Asiatic species. The habit is that of *P. barbatum* and *P. glabrum,* differing from the former in the want of the long cilia or bristles to the stipules, from the latter in the indumentum, which is that of *P. subsessile,* from which it is removed by the long petioles and the want of cilia to the stipules.

10. **P. lapathifolium,** *Linn. ; Meissn. in DC. Prod.* xiv. 119. Tall erect and glabrous, except sometimes very short strigose hairs on the margins and midrib of the leaves, and in some varieties a very slight hoariness on their under surface. Leaves lanceolate or ovate-lanceolate, acuminate, 3 to 6 in. long or even larger, contracted into a petiole usually exceeding the stipules, the glandular dots of the under surface more conspicuous than in many species. Stipules sheathing, without marginal bristles or rarely with a few very small cilia. Spikes rather

slender, from under 1 to above 2 in. long, in a terminal branching more or less leafy panicle. Bracts small, truncate or shortly acuminate. Perianth, at least the outer segments, more or less glandular and the prominent glands extending sometimes to the bracts and peduncles. Stamens 5 or 6. Style short with 2 long branches. Nut very flat, the sides concave or rarely slightly convex. Radicle curved towards the edge of the cotyledons (accumbent).—*P. nodosum,* Pers.; Meissn. in DC. Prod. xiv. 118; *P. glandulosum,* R. Br. Prod. 419; Meissn. l.c. 116; *P. adenophorum,* Spreng. Syst. ii. 258; *P. elatius,* R. Br. l.c. 419; Meissn. l.c. 121.

Queensland. Brisbane river, *C. Stuart;* Nerkool Creek, *Bowman;* Armadilla, *Barton.*

N. S. Wales. Port Jackson to the Blue Mountains, *R. Brown* and others; New England, *C. Stuart;* Argyle County, *A. Cunningham;* Darling river, *Mrs. Ford.*

Victoria. From the Glenelg to Gipps' Land, *Robertson, F. Mueller,* and others.

Tasmania. Port Dalrymple, *R. Brown.*

Common in the temperate and subtropical regions of the New and the Old World, especially in the northern hemisphere. The *P. nodosum,* to which Meissner refers the Australian specimens, is distinguished by him from the typical *P. lapathifolium* chiefly by the comparative absence of glands or glandular-pubescence on the peduncles; but in this respect the Australian specimens vary as much as the European ones. Brown's *P. elatius* included the broad-leaved, *P. glandulosum* the narrow-leaved specimens.

11. **P. lanigerum,** *R. Br. Prod.* 419. Stems erect, slightly branched, attaining 2 or 3 ft., the whole plant white with a close woolly or arachnoid tomentum, or the upper surface of the leaves and the lower part of the stem at length glabrous. Leaves shortly petiolate, lanceolate, acuminate, 3 to 4 in. long or even larger. Stipules sheathing, usually long, bordered by few fine cilia. Spikes narrow but rather dense, 1 to 1½ in. long, several in a terminal panicle. Bracts short, denticulate-ciliate. Perianth often glandular. Style short, with 2 long branches. Nut flat.—Meissn. in DC. Prod. xiv. 117.

Queensland. Burdekin river, *F. Mueller.*

N. S. Wales. Hawkesbury, *R. Brown;* also in *Leichhardt's* collection.

Victoria. Wheat fields, Snowy river, *F. Mueller.*

The species is common in E. India, and extends to S. Africa. Although nearly allied to *P. lapathifolium,* of which F. Mueller considers it as a variety, it appears constantly to differ from the var. *incanum* of that species, in the abundance of the white indumentum, and is said to be perennial not annual.

SECT. 4. AMBLYGONON, Meissn.—Flowers spicate, the spikes in terminal panicles. Embryo curved towards the back of the cotyledons (incumbent). Stems tall and erect. Nuts usually flat on both sides with obtuse margins.

12. **P. orientale,** *Linn.; Meissn. in DC. Prod.* xiv. 123. An erect softly pubescent or hirsute annual of 2 to 5 ft. Leaves on rather long petioles, ovate or the upper ones lanceolate, acute or acuminate, 3 to 6 in. long. Stipules very hairy, closely sheathing and scarious at the base, but all except the uppermost expanded at the top into a green spreading limb. Spikes (especially in the hirsute form) rather slender

continuous or slightly interrupted, several in a loose terminal panicle. Bracts hairy and ciliate. Flowers rather large. Stamens usually 7. Style with 2 slender branches. Nut flat or the sides convex, the margin obtuse. Radicle curved towards the back of the cotyledons.— R. Br. Prod. 420.

Queensland. Rockingham Bay, *Dallachy;* Rockhampton, common on the edges of lagoons, &c., *Dallachy, O'Shanesy, Bowman;* Archer's station, *Leichhardt;* Moreton Bay, *C. Stuart.*

N. S. Wales. Port Jackson, *R. Brown;* Blue Mountains, *Woolls.*

Most of the Australian specimens belong to the var. *pilosum,* Meissn. (*P. pilosum,* Roxb.), which extends over E. India to S. China and which I am also unable to distinguish from the S. American *P. hispidum.* H. B. K., or at least from the Brasilian specimens quoted and figured by Meissner in Mart. Fl. Bras. Polyg. 13. t. 2, in which I find the embryo of *Amblygonon* and not of *Persicaria.* Of the more glabrous broad-leaved variety, with shorter, more nodding spikes, considered as the typical Asiatic plant, I have seen no Australian specimens except Woolls's, which may possibly have been introduced.

13. **P. attenuatum,** *R. Br. Prod.* 420. A tall species, very nearly allied to *P. orientale,* the stems and foliage more or less sprinkled with rather fine appressed hairs, hoary when young, or sometimes nearly glabrous. Leaves lanceolate, tapering into a very long point and contracted into a rather long petiole, mostly above 6 in. long. Stipules entirely sheathing, truncate and shortly ciliate, without the green limb of *P. orientale.* Spikes continuous, 1½ to 3 in. long, few on rather long peduncles in a terminal panicle. Bracts rather thick, truncate, shortly ciliate or entire. Perianth, stamens, style, nut and embryo entirely of *P. orientale.*—Meissn. in DC. Prod. xiv. 117.

N. Australia. Islands of the Gulf of Carpentaria, *R. Brown;* Upper Victoria river, Sturt's Creek, Wentworth, Flinders and Macarthur rivers, *F. Mueller.*

N. S. Wales? Near the Darling river, *Victorian Expedition* (a very imperfect specimen).

S. Australia. Cooper's Creek, *Howitt's Expedition;* Cooper's and Thomson's rivers, *A. C. Gregory.*

I have not identified this with any Asiatic species, but I am unable to distinguish it from the Brasilian *P. spectabile,* Mart.; Meissn. in DC. Prod. xiv. 119, and in Mart. Fl. Bras. Polygon. 13. t. 3, in the seeds of which I find the radicle incumbent as in *Amblygonum.* Meissner has not figured the embryo either in this or in *P. hispidum.*

4. MUHLENBECKIA, Meissn.

Flowers mostly diœcious. Perianth of 5 segments, all equal or the outer ones slightly enlarged. Stamens 8 or rarely fewer, filaments short, the anthers oblong in the males, small and imperfect or abortive in the females. Ovary in the females 3-angled, with a very short trifid style and 3 more or less fringed stigmas, small and rudimentary in the males. Nut triangular or nearly globular, enclosed in the persistent membranous or succulent perianth.—Undershrubs shrubs or woody twiners, rarely herbaceous from the base; all at least of the Australian species glabrous. Stipules brown and scarious, loosely sheathing, very soon torn or ragged. Flowers small, green or whitish, in whorl-like clusters, axillary or within small sheathing bracts, in axillary or terminal simple or paniculate interrupted spikes. Radicle usually curved towards the back of the cotyledons (incumbent), rarely here and there towards their edge (accumbent).

The genus extends to New Zealand, extratropical S. America, and along the Andes to Mexico. Of the seven Australian species, one is identical with a New Zealand one, another is closely allied as well to a New Zealand as to a S. American species, the others appear to be endemic.

Leaves more or less cordate hastate or sagittate, usually broad and above 1 in. long. Stem prostrate or climbing.
Fruit globular, obscurely angled, not rugose, with the perianth succulent. Spikes mostly axillary.
    Leaves rather thick, obtuse or shortly acuminate . . . . 1. *M. adpressa.*
    Leaves thin, with a long point. Spikes very slender.
    Flowers small . . . . 2. *M. gracillima.*
Fruit ovoid, more or less 3-angled, very rugose. Leaves rather thick. Spikes mostly in a large terminal panicle . 3. *M. rhyticarya.*
Leaves ovate or rounded, under ½ in. long. Stems short, prostrate, and densely matted. Fruit prominently 3-angled . . 4. *M. axillaris.*
Leaves narrow, tapering at the base. Erect or diffuse undershrubs or shrubs.
    Stems stout, decumbent, not much branched. Leaves broadly lanceolate. Spikes axillary. Fruit globular . . . . . 5. *M. polybotrya.*
    Shrub. Leaves on luxuriant branches lanceolate or rarely hastate, on the others small and linear. Clusters axillary, with few small flowers. Fruit globular, 3-angled . . . . 6. *M. polygonoides.*
    Shrub. Leaves on luxuriant branches linear, on the others very small or none. Clusters mostly in terminal spikes. Fruit prominently 3-angled . . . . 7. *M. Cunninghamii.*

1. **M. adpressa,** *Meissn. Gen. Pl. Comm.* 227, *in Pl. Preiss.* ii. 272, *and in DC. Prod.* xiv. 145. Stems woody at the base, prostrate and spreading or climbing. Leaves petiolate, from orbicular-cordate to broadly lanceolate, hastate, obtuse or scarcely acuminate, rarely under 1 in. long except on the smaller branchlets, and in luxuriant specimens 3 in. or more, the margins often undulate-crisped. Spikes interrupted, 1 to 2 or even 3 in. long, solitary or clustered in the axils, scarcely paniculate at the end of the branches, the flower-clusters distant, the lowest usually at the base of the rhachis and sometimes the whole spike reduced to a sessile cluster. Bracts 1 to 2 lines long. Flowers few in each cluster, the males more numerous than the females. Fruiting perianth globular, more or less succulent. Nut completely enclosed in the perianth, nearly globular, the three angles very obtuse or obscure, not rugose. Seed with 3 longitudinal furrows.—Hook f. Fl. Tasm. i. 308; *Polygonum adpressum,* Labill. Pl. Nov. Holl. i. 99, t. 127; R. Br. Prod. 420 (partly); *Sarcogonum adpressum,* Don. in Sweet Hort. Brit. ed. 3, 577 (*Meissner*).

The three following forms of this plant appear to be nearly constant enough to be reckoned as distinct species.

a. rotundifolia. Stems trailing on the ground or climbing on bushes, rarely above 6 ft. high. Leaves broad and short. Perianth not very succulent.

N. S. Wales? Medway river near Berrima, *Miss Atkinson.* (Uncertain, the specimens in leaf only).

Victoria. Wendu vale, *Robertson;* mouth of the Glenelg, *Allitt;* Wimmera, *Dallachy;* Yowaka river, *F. Mueller.*

Tasmania. Port Dalrymple, *R. Brown;* common along the seacoast, *J. D. Hooker.*

S. Australia. Seacoast, Spencer's and St. Vincent's gulfs, *F. Mueller* and others; Kangaroo islands, *Waterhouse ;* Gawler ranges, *Sullivan.*

W. Australia. *Labillardiere ;* Swan river, *Drummond, n.* 204, *Preiss, n.* 1205, *Oldfield;* Kalgan and Gordon rivers, *Oldfield;* Stirling and Plantagenet ranges to Cape Arid, *Maxwell.*

b. hastifolia, Meissn. Leaves broadly lanceolate-hastate, twice or three times as long as broad, mostly acuminate. Stems climbing over the tops of tall trees. Flowers and fruiting perianths large, the latter very succulent.—*M. Gunnii,* Hook. f. in Hook. Lond. Journ. vi. 278; *Polygonum adpressum,* Hook. Bot. Mag. t. 3145.

Tasmania. In humid forests on rich soils, *Gunn.*

S. Australia. Near Port Adelaide, *F. Mueller.*

c. flexuosa. Leaves scarcely 1 in. long, broad, obtuse or shortly acuminate. Flowers smaller than in the type and chiefly in axillary clusters. Fruit not seen.—*M. flexuosa,* Meissn. in Pl. Preiss. i. 624, and in DC. Prod. xiv. 148.

W. Australia. Middle island, *A. Cunningham;* Goderich district, *Preiss, n.* 1355; Gordon and Kalgan rivers, *Oldfield;* Albany, *F. Mueller.*

Hooker (Handb. Fl. N. Zeal. 236) and some others have considered this and the *M. australis,* Forst., a native of New Zealand and Norfolk Island, as one species. In the latter, however, the inflorescence is more branched, the nut more shining, very prominently 3-angled, and usually exceeding the scarcely succulent perianth, and it seems to me to be at least as distinct as the *M. gracillima.*

2. **M. gracillima,** *Meissn. in DC. Prod.* xiv. 145. A tall herbaceous twiner, much more slender than *M. adpressa* and *M. rhyticarya.* Leaves on long petioles, ovate-cordate or broadly sagittate, acuminate, membranous, the margins usually crisped, 1 to 2 or rarely 3 in. long. Spikes very slender, almost filiform, simple or paniculately branched, the males much longer and more branched than the females, the flower-clusters small and distant. Perianth not above 1 line diameter when fully spread. Stigmas large and copiously fringed. Fruiting perianth globular and succulent, enclosing an opaque obscurely 3-angled nut as in *M. adpressa,* but considerably smaller.—*Polygonum adpressum,* R. Br. Prod. 420 (partly).

Queensland. Moreton Bay, *Leichhardt ;* Rockhampton, *O'Shanesy ;* Dawson and Burnett rivers, *F. Mueller.*

N. S. Wales. Port Jackson to the Blue Mountains, *R. Brown* and others ; New England, *C. Stuart;* Glendon, *Leichhardt;* Macleay, Clarence and Hastings rivers, *Beckler ;* southward to Illawarra, *Herb. Hooker and F. Mueller.*

3. **M. rhyticarya,** *F. Muell. Fragm.* v. 92. Stems climbing, woody at the base. Leaves mostly broadly ovate-cordate, obtuse or shortly acuminate, 1 to 2 in. long, often rather thick and glaucous, resembling those of the broad-leaved form of *M. adpressa.* Spikes long and interrupted in the upper axils, and forming a large loose terminal panicle. Flowers rather larger than in *M. adpressa.* Fruiting perianth almost' membranous, enclosing an ovoid-triquetrous nut often 2 lines long and deeply rugose. Seed scarcely furrowed.

Queensland. Cape York, *Daemel;* Port Denison, *Fitzalan ;* Edgecombe and Rockingham Bays, *Dallachy.*

N. S. Wales. New England, *C. Stuart;* Shoalhaven Gullies, *C. Moore.*

R. Brown's specimens of *Polygonum adpressum,* from Keppel and Shoalwater Bays, appear to belong to *M. rhyticarya,* but they are not in fruit.

4. **M. axillaris,** *Hook. f. in Hook. Lond. Journ.* vi. 278; *Fl. Tasm.* i. 308. A small diffuse or prostrate shrub, forming matted patches of a few in. diameter, or sending out weak flexuose stems of 6 in. to nearly 1 ft. Leaves on slender petioles, ovate oblong or almost orbicular, under ½ in. long. Flowers small, solitary or 2 or 3 together in the upper axils or at the ends of the branches, on very short pedicels. Stigmas at length denticulate but without the long fringes of most species. Nuts as long as the scarcely succulent fruiting perianth, smooth and prominently 3-angled. Seed not seen perfect.—Meissn. in DC. Prod. xiv. 147; *M. parvifolia*, Meissn. in Linnæa, xxvi. 362.

N. S. Wales, *Anderson, Vicary;* naked rocky margins of Fish river, *A. Cunningham;* Lord Howe's Island, *C. Moore.*
Victoria. Gravelly banks of the Mitta-Mitta, Maroka valley and Snowy river, *F. Mueller.*
Tasmania. Moist places in various parts of the colony, ascending to 4000 ft., *J. D. Hooker.*

This species is also in New Zealand.

5. **M. polybotrya,** *Meissn. in Pl. Preiss.* i. 623, *and in DC. Prod.* xiv. 145. Stems woody at the base, decumbent or ascending, 1 to 2 ft. long, rather stout and not much branched. Leaves oblong-lanceolate, acute, narrowed into a rather long petiole, the larger ones 2 to 3 in. long. Spikes interrupted, the males usually numerous and longer than the leaves, the females short. Male perianths about 1 line long, on rather long pedicels, females larger on shorter pedicels. Stigmas more or less fringed. Fruiting perianth scarcely thickened. Nut nearly globular, obscurely 3-angled. Seed with 6 prominent longitudinal obtuse angles separated by 3 deep and narrow, and 3 alternate shallow furrows. Embryos observed incumbently and accumbently incurved in different seeds from the same specimen.—F. Muell. Fragm. iv. 130, also named *Polygonum polybotryum.*

W. Australia. Swan river, *Drummond, Oldfield, Preiss, n.* 1353, and others.

6. **M. polygonoides,** *F. Muell. Fragm.* v. 73. A diffuse or erect undershrub or shrub, attaining several ft., with numerous branches elongated and leafy when young or luxuriant, clustered wiry and appearing sometimes almost leafless on old stems. Leaves linear or lanceolate, contracted at the base, rarely broad or occasionally hastately 3-lobed on luxuriant shoots, small and narrow on most flowering branches. Flowers small, very few together and the females often almost solitary, all axillary. Perianth under 1 line long when in flower, 1 to 1½ lines when in fruit but scarcely succulent. Style shortly 3-lobed, the stigmas at first somewhat dilated and entire, larger and lobed or fringed when fully out. Nut nearly globular, obtusely 3-angled, somewhat shining but often slightly rugose. Seed not furrowed.— *Polygonum diclinum,* F. Muell. in Hook. Kew Journ. viii. 203, and in Trans. Phil. Soc. Vict. i. 23 : Meissn. in DC. Prod. xiv. 85 ; *Muhlenbeckia stenophylla,* F. Muell. Fragm. i. 138; *Polygonum angustissimum,* F. Muell. Fragm. v. 92.

T 2

N. S. Wales. Murray river near the mouths of the Murrumbidgee and Darling rivers, *F. Mueller.*
Victoria. Rocky mountains on the Macalister and Snowy rivers, *F. Mueller;* Wimmera, *Dallachy.*

7. **M. Cunninghamii,** *F. Muell. Fragm.* v. 91. A shrub, sometimes low and bushy, sometimes attaining 6 to 8 ft. or even taller, with numerous terete sulcate or angular branches, the young and luxuriant ones often with a few linear or linear-lanceolate leaves of 1 to 3 in. contracted into a rather long petiole, but most of the branches rigid or rushlike often clustered and either quite leafless or with a few linear leaves. Stipules deciduous. Flower-clusters solitary or in short spikes at the leafless nodes or in the axils of the small leaves forming long terminal interrupted spikes. Perianth rather above 1 line long. Style 3-branched, with broad peltate terminal stigmas more like those of a *Polygonum* than of a *Muhlenbeckia,* but the flowers quite diœcious. Nut prominently 3-angled, smooth and shining, enclosed in the ovoid slightly thickened perianth which is enlarged to from 1½ to 2 lines in length. Seed scarcely furrowed.—*Polygonum Cunninghamii,* Meissn. in Linnæa, xxvi. 364, and in DC. Prod. xiv. 85; *Muhlenbeckia florulenta,* Meissn. in Linnæa, xxvi. 362; *Polygonum junceum,* A. Cunn.; Lindl. in Mitch. Trop. Austr. 85.

N. Australia. Victoria river, *F. Mueller.*
Queensland. Rockhampton, *O'Shanesy;* Suttor river, *Bowman;* Bokhara Creek, *Leichhardt;* Curriwillighie, *Dalton;* Darling Downs, *Lau.*
N. S. Wales. Liverpool Plains, *Leichhardt, C. Moore;* Lachlan river, *A. Cunningham;* Macquarrie river, *Mitchell;* Murray and Darling rivers to the Barrier Range, *Victorian and other Expeditions.*
Victoria. Wendu Vale, *Robertson;* Melbourne, *Adamson;* Bacchus Marsh and Station Peak, *F. Mueller;* Creswick, *Whan.*
S. Australia. Murray river to St. Vincent's Gulf, *F. Mueller;* Salt Creek, *Behr;* Dombey Bay, *Wilhelmi;* towards Spencer's Gulf, *Warburton.*
W. Australia. Between Moore and Murchison rivers, *Drummond, 6th coll. n.* 218.

ORDER C. **NYCTAGINEÆ.**

Perianth simple, inferior, the lower portion persistent and enclosing the ovary and fruit, the upper portion variously shaped, with 5 rarely 4 angles folds teeth or lobes, deciduous or withering. Stamens either of the same number as the folds or teeth of the perianth or fewer or more, often inconstant in the same species, and never more than 20, inserted on (or united at the base with) a narrow or cupshaped disk more or less adnate to the stipes of the ovary within the perianth (or rarely free from the base?); filaments slender, usually exserted; anthers 2-celled, the cells attached back to back and opening longitudinally round the outer margin. Ovary shortly stipitate within the base of the perianth, 1-celled, with 1 erect anatropous ovule. Style terminal, simple, undivided, with a single stigma. Fruit 1-seeded, enclosed in the persistent tough or hardened base of the perianth which falls off with it having the appearance of a pericarp, the real pericarp thin and membranous more or less adherent to and often inseparable from the equally thin testa of the seed. Embryo curved transversely folded or longitudinally convolute around or within a mealy albumen, radicle inferior.—Herbs shrubs or trees, the nodes often tumid and articulate. Leaves usually opposite, often unequal in each pair, rarely alternate, usually entire or scarcely sinuate. Flowers solitary clustered or umbellate, the bracts in many genera not Australian forming an involucre round them sometimes large and coloured, but often small, and in the Australian genera the bracts all very small and deciduous.

A small Order of which the genera are all American, and exclusively so with the exception of the two found in Australia, which are both of them widely dispersed over the tropical and subtropical regions of both the New and the Old World.

Herbs with small flowers. Upper portion of the perianth deciduous.
Stamens 1 to 4. Embryo folded . . . . 1. Boerhaavia.
Trees, shrubs, or woody climbers. Limb of the perianth persistent
on the fruit. Stamens usually 6 to 8. Embryo straight, cotyledons
convolute . . . . 2. Pisonia.

1. BOERHAAVIA, Linn.

Flowers hermaphrodite. Upper portion of the perianth campanulate or funnel-shaped, truncate and plicate or very shortly 5-lobed, falling off after flowering, leaving the lower portion persistent and enclosing the ovary and fruit. Stamens 1 to 4, usually exserted. Stigma dilated peltate or oblique. Fruit completely enclosed in the somewhat hardened 5-ribbed base of the perianth. Embryo transversely folded at the base of the cotyledons.—Herbs usually perennial. Leaves opposite.—Flowers small, in little clusters or umbels on simple or branched axillary peduncles. Bracts and bracteoles very small.

The genus is widely diffused over the tropical and subtropical regions of the New and the Old World. The Australian species are both Asiatic, one a common weed in Africa and Asia, the other perhaps confined to Asia; both, however, require further comparison with some of the American species.

Leaves orbicular or ovate to narrow-lanceolate. Deciduous portion of
the perianth campanulate, not exceeding 1 line . . . . 1. B. diffusa.
Leaves cordate-ovate to lanceolate. Deciduous portion of the perianth
3 to 4 lines long, funnel-shaped, with a narrow tube . . . . . 2. B. repanda.

1. **B. diffusa,** Linn. Chois. in DC. Prod. xiii. ii. 452. A perennial with procumbent diffuse or ascending stems extending sometimes to several ft., the whole plant glabrous, or the lower portion and foliage pubescent or hirsute, with a few longer articulate hairs and sometimes glandular-viscid. Leaves of each pair unequal or nearly equal, petiolate, the lower ones often broadly ovate or orbicular ½ to 1 in. diameter, the upper ones usually smaller or narrower, sometimes nearly all narrow-lanceolate, from very obtuse to acute, entire sinuate or crisped on the margins especially when hirsute, the smaller ones sometimes clustered in the axils, the floral ones· passing into minute bracts. Flowers very small, pale pink, sessile or on very short pedicels somewhat lengthened

under the fruits, in little heads or umbels on filiform peduncles more
or less branched into compound umbels or irregular panicles axillary
or terminal. Bracts under the pedicels small and lanceolate, with a
pair of minute bracteoles under the perianth. Perianth constricted
above the ovary, the lower portion about ⅓ line long, rather thick, 5-
ribbed, studded with stipitate glands, the upper deciduous portion campanulate, petal-like, scarcely lobed, from ¾ to rather above 1 line long.
Stamens usually 3, sometimes 1, 2 or 4, inserted in a narrow cup-shaped
toothed disk adnate to the stipes of the ovary, which as well as the
whole ovary is entirely enclosed in the base of the perianth; filaments
as long as or rather longer than the perianth. Fruiting base of the
perianth oblong-turbinate, glandular-muricate when dry, becoming enveloped in mucilage in hot water, from 1 to nearly 2 lines long.—*B.
mutabilis*, Br. Prod. 422, Chois. in DC. Prod. xiii. ii. 455, (there placed
in a wrong division); Nees in Pl. Preiss. i. 622; F. Muell. Pl. Vict.
ii. t. 67, (the insertion of the stamens and style incorrect); *B. pubescens*,
R. Br. Prod. 422; *B. procumbens*, Roxb.; Wight, Ic. t. 874.

N. Australia. Victoria river, *F. Mueller;* Islands of the Gulf of Carpentaria, *R.
Brown;* (glabrous, glandular-pubescent and very villous-tomentose varieties), *Henne;*
Adams Bay, *Hulls;* Port Essington, *Armstrong;* Port Darwin, *Schultz.*
Queensland. Keppel Bay, *R. Brown;* Port Molle and Isles off Cape Flattery,
M'Gillivray; Howick's group, *F. Mueller;* Rockingham Bay, *Dallachy;* common
about Rockhampton, *Dallachy* and others; Nerkool and Amity Creeks, &c., *Bowman;*
Warwick, *Beckler;* on the Maranoa, *Mitchell* (some of the specimens very hirsute, with
the leaves white underneath); Armadillo, *Barton;* Curriwillighie, *Dalton.*
N. S. Wales. Mudgee, *Woolls;* from the Darling river to the Barrier Range,
Victorian and other Expeditions; New England, *C. Stuart.*
Victoria. Murray river, Mount Hope, *F. Mueller.*
S. Australia. Spencer's Gulf, *R. Brown;* Flinders Range, Taberton, *F. Mueller;*
Cooper's Creek, *Howitt's Expedition.*
W. Australia. Swan river, *Drummond, Preiss, n.* 2389; Murchison river, *Old-
field;* Port Walcott, *Harper.*

The glabrous and hirsute states are generally found in the same localities, the former,
however, by far the most frequent. The species is a common weed in the warmer
regions of Asia and Africa.

2. **B. repanda,** *Willd.*; *Chois. in DC. Prod.* xiii. ii. 455. A stouter
and coarser plant than *B. diffusa,* usually glabrous and spreading.
Leaves petiolate, cordate-ovate or lanceolate, usually acuminate, the
margins undulate, rather thick, often above 1 in. long, but sometimes small. Flowers pedicellate in umbels on axillary or terminal
peduncles and very frequently with a whorl of 6 to 8 pedicellate flowers
at some distance from the terminal umbel. Pedicels 2 to 6 lines long,
or more usually unequal in the same umbel, not so slender as in *B.
diffusa.* Lower portion of the perianth under 1 line long, glandular
muricate as in *B. diffusa,* upper deciduous portion funnel-shaped, 3 to 4
lines long, the slender part of the tube as long as the campanulate summit. Stamens usually 2 or 3, sometimes 4, exserted. Fruit considerably longer than in *B. diffusa.*—Wight, Ic. t. 1766.

Queensland. Lady Elliott's Island, *Capt. Robertson.*

W. Australia. Sharks Bay *Milne, Maitland Brown;* between Moore and Murchison rivers, *Drummond, 6th coll. n.* 217.

The species has a considerable range in East India, the Indian Archipelago, &c., chiefly in maritime districts.

2. PISONIA, Linn.

Flowers hermaphrodite or unisexual. Perianth in the hermaphrodite flowers contracted above the ovary, in the males campanulate, in the females ovoid or cylindrical, the margin in all 5-angled or 5-toothed. Stamens usually 6 to 8, longer than the perianth, none in the female flowers. Ovary entirely wanting or imperfect in the male flowers. Stigma in the females dilated, oblique. Fruiting perianth oblong or elongated, 5-ribbed, smooth or glandular-muricate, crowned by the small withered limb of the perianth. Seed oblong with a deep longitudinal furrow. Albumen scanty. Embryo straight, the cotyledons convolute over the intruded testa.—Trees shrubs or woody climbers. Leaves opposite or scattered. Flowers in dense or loose cymes, often arranged in terminal panicles. Bracts and bracteoles very small or none.

The species are numerous in tropical and subtropical America, but there are also a few in southern Asia and in the island region from the S African coast to the Pacific. Of the three Australian species one is widely spread over the tropical regions both of the New and the Old World, another extends to Norfolk Island and New Zealand, and possibly to some of the islands of the Eastern Archipelago and South Pacific, the third may be endemic, but closely resembles if it be not identical with a Pacific island species, the synonymy, however, notwithstanding the researches of Seemann, remains exceedingly confused, and the specimens now in our herbaria are wholly insufficient to clear it up.

Tall woody climber with axillary spines. Flowers unisexual. Fruiting perianth muricate . . . . 1. *P. aculeata.*
Unarmed trees.
 Flowers unisexual. Perianth shortly villous, muricate when in
  fruit . . . . 2. *P. inermis.*
 Flowers (all ?) hermaphrodite. Perianth glabrous, elongated and
  smooth or minutely papillose-scabrous when in fruit . . . . 3. *P. Brunoniana.*

1. **P. aculeata,** *Linn.; Chois. in DC. Prod.* xiii. ii. 440. A tall woody climber, forming impenetrable masses on the borders of forests, reduced to a low straggling bush in open places, glabrous or rarely pubescent, often armed with stout recurved axillary prickles (abortive peduncles). Leaves opposite or here and there alternate, petiolate, ovate, often broad, or rarely oblong or lanceolate, obtuse, entire, rarely exceeding 3 in. and often all under 2 in. long. Flowers dioecious, in small dense cymes or globular clusters, of which several are usually collected into small panicles in the upper axils, the common peduncle rarely exceeding the leaves and often very short. Male perianths shortly pedicellate, campanulate, shortly and broadly 5-toothed, the bud clavate and prominently 5-angled at the top, opening to about 2 lines diameter. Female perianths nearly sessile, ovoid, about 1 line long, obscurely 5-toothed, enclosing the ovary, the style shortly protruding, with a deeply lobed or fringed stigma. Fruits in loose cymes forming often large

panicles, the pedicels lengthening to above ½ in., the enlarged perianth oblong or linear-clavate, ¼ to ½ in. long, 5-ribbed, glandular-muricate.— Wight, Ic. t. 1763, 1764 ; *P. villosa,* Poir. ; Chois. in DC. Prod. xiii. ii. 440 ; *P. limonella,* Blume ; Chois. l.c. 446.

Queensland. Cape York, *Daemel ;* Burdekin river, *Fitzalan ;* Edgecombe and Rockingham Bays, *Dallachy ;* Broad Sound and Nerkool Creek, *Bowman ;* Rockhampton, *O'Shanesy.*

N. S. Wales. Clarence river, *Beckler ;* Tweed river, *C. Moore.*

The species is very widely distributed over the tropical regions of the New and the Old World, especially in maritime districts.

2. **P. inermis,** *Forst. Prod.* 75. A tall tree, glabrous except the inflorescence. Leaves petiolate, ovate or oblong, acuminate or almost obtuse, rounded or narrowed at the base, often 6 to 8 in. long. Flowers diœcious, small, collected in small cymes forming a terminal corymbose rather compact panicle, usually shortly pedunculate and much shorter than the leaves. Perianths both male and female narrow, ovoid-oblong, scarcely 2 lines long at the time of flowering the males rather longer and more dilated at the orifice than the females, all pubescent or villous. Stamens shortly exserted. Style scarcely protruding from the females. Fruiting perianths in a loose panicle on more or less elongated pedicels, the perianth about ½ in. long, very prominently muricate along the angles.—*P. grandis,* R. Br. Prod. 422; Chois. in DC. Prod. xiii. ii. 441 ; Endl. Iconogr. t. 30.

N. Australia. Islands of the Gulf of Carpentaria, *R. Brown.*
Queensland. Northumberland islands, *R. Brown ;* Rockingham Bay, *Dallachy.*

The species is also in the Pacific islands, for, as far as can be judged from Forster's specimens and from the description quoted by Guillemin as mentioned below, the Taitian and the Australian plant appear to be identical. I have therefore followed Seemann in restoring to it Forster's name, the previously published *P. inermis,* Jacq. being referred by Choisy to *P. nigricans,* Sw. by Seemann to *P. mitis,* Linn.

3. **P. Brunoniana,** *Endl. Prod. Fl. Norf.* 43, *but not of Chois.* A tree attaining sometimes a great height, quite glabrous or the inflorescence minutely pubescent, but never so much so as in *P. inermis.* Leaves mostly opposite, the upper ones sometimes irregularly alternate or approximate so as to appear verticillate, petiolate, from broadly ovate to obovate elliptical or almost oblong, obtuse, contracted or rarely rounded at the base, often 8 in. long or even more, but sometimes not half that size, somewhat coriaceous when old. Flowers mostly hermaphrodite, although the ovary in some individuals and the stamens in others may be imperfect or less perfect, all shortly pedicellate in small cymes collected into terminal leafless panicles. Perianth when fully out usually about 3 lines long, funnel-shaped, the tube distinctly contracted above the ovary when that is perfect, scarcely so when it is more or less abortive, expanded and obscurely lobed at the top. Stamens usually shortly exserted. Fruiting perianth narrow and above 1 in. long, more or less 5-angled, glabrous and smooth or the angles slightly papillose or tubercular, contracted upwards into a narrow neck crowned by the withered

border of the perianth.—Hook. f. Handb. N. Zeal. Fl. 229 ; *P. Sinclairii,*
Hook. f. Fl. N. Zeal. i. 209, t. 50 ; *P. Mooriana,* F. Muell. Fragm. i. 20.

Queensland. Rockingham Bay, *Dallachy.*

N. S. Wales. Richmond river, *Richards, Henderson;* Illawarra, *A. Cunning-*
ham, C. Moore; also Sydney woods, Paris Exhibition, 1855, *Macarthur, n.* 197.

The species is also in Norfolk island and New Zealand, all other stations doubtful,
for the identification of the Australasian and Polynesian specimens of *Pisonia* is often
impossible from their incompleteness. The foliage affords no marks, and the flowers
and fruits are often mismatched. Choisy in DC. Prod. xiii. ii. 441, has described as
P. Brunoniana the Pacific island *P. procera,* Bert., figured in Deless. Ic. Sel. iii. t. 87,
and (as *P. Forsteriana*) in Pl. Meyen. t. 51, a species .not unlike the *P. Brunoniana*
when seen in fruit, but with very different flowers, these being diœcious in all the specimens I have seen, the male perianth campanulate and prominently 5-lobed, the female
perianth much smaller and ovoid-oblong. This has been correctly identified by Seemann with the *Ceodes umbellifera,* Forst., and if Horsfield's specimens are rightly determined, with *P. excelsa,* Blume, a name which under all the circumstances ought, I
think, to have the preference over *P. umbellifera,* Seem. ; nor can I agree with Seemann
in referring to it the Australian plant. With regard to the Norfolk Island and New
Zealand species which Seemann separates from the Australian one and refers to *P.*
inermis, Forst., he has been evidently misled by Guillemin, who Zeph. Tait. in Ann. Sc.
Nat. Sér. 2, vii. 191, having before him only the *P. procera,* Bert. (*Ceodes,* Forst.) refers to it Forster's manuscript description of *P. inermis,* Forst.

ORDER CI. **MYRISTICEÆ.**

Flowers diœcious, regular. Perianth deciduous, 3-lobed or rarely
2- or 4-lobed, the lobes valvate in the bud. Male fl. Stamens united
in a central column ; anthers 3, 6, or more adnate to the column at the
apex, or in a ring immediately below the column, each with 2 parallel
cells opening longitudinally. Female fl. Ovary free within the perianth,
with a single erect anatropous ovule ; stigma sessile or nearly so, capitate or depressed. Fruit succulent, opening tardily in 2 valves. Seed
erect, sessile, more or less covered with a lobed or jagged often scarlet
arillus (or arillodium) proceeding from the base of the seed. Albumen
remarkably ruminate. Embryo very small, at the base of the seed, with
divaricate cotyledons.—Trees, often aromatic. Leaves alternate, entire,
usually dotted, penniveined, without stipules. Flowers small, in axillary
or supra-axillary racemes or panicles, more numerous in the males than
in the females. Bracts minute or none.

The Order is limited to the single genus *Myristica.*

1. **MYRISTICA,** Linn.

Characters those of the Order.

The genus is entirely tropical, most abundant in the Eastern Archipelago, with a few
species from Continental India or from the Mascarene and South Pacific islands, and
several from tropical America. The only Australian species may be endemic, but is
closely allied to an Indian one.

1. **M. insipida,** *R. Br. Prod.* 400. A fine tree of 60 to 70 ft. or more
(*Dallachy*), glabrous but the young branches and petioles often ferru-

ginous. Leaves oval-elliptical or oblong, shortly acuminate, rounded
or cuneate at the base, all under 4 in. long in some specimens, all above
6 in. in others, and often very variable in size and relative breadth on
the same specimens, pale and shining above with the veins impressed,
more or less glaucous underneath, with from 7 to 20 prominent primary
veins on each side of the midrib. Male flowers few or rather numerous,
in sessile axillary clusters. Pedicels shorter than the perianth, with a
small broad ciliolate bract close under the flowers. Perianth cylindrical,
2 to 2½ lines long, with 3 lobes scarcely above ½ line long. Staminal
column included, not dilated at the top; anthers 6, linear, adnate in a
ring below the top of the column and occupying ⅔ of its length. Female
flowers not seen. Fruits solitary or 2 together on very short thick
axillary pedicels, ovoid or ovoid-oblong, about 1 in. long, rusty-tomentose or nearly glabrous. Seeds normal; embryo with very small thick
divaricate cotyledons quite entire.—A. DC. Prod. xiv. 206; *A. cimici-
fera*, R. Br. Prod. 400; A. DC. l.c. 191.

N. Australia. Islands of the Gulf of Carpentaria, *R. Brown;* Brunswick Bay,
A. Cunningham; Melville island, *Fraser;* Port Essington, *Armstrong.*

Queensland. Endeavour river, *Banks and Solander;* Albany island, *W. Hill;*
Rockingham Bay, *Dallachy;* near Rockhampton, *Thozet.*

The species is very near to *M. Zeylanica*, A. DC., united by Thwaites with *M. laurifolia*, Hook. f. and Thoms. from Ceylon, and is indeed scarcely to be distinguished
from it, as far as our specimens show, but by the narrower perianth and apparently
fewer anthers.

ORDER CII. MONIMIACEÆ.

Perianth regular, usually at first globular or nearly so and more or
less adnate to or continuous with the expanded receptacle or staminal
disk, the limb of 4 or more connivent lobes or segments in 2 or rarely
more rows but all of similar texture, calycine or scarcely petaloid,
deciduous or persistent. Stamens either definite and opposite the perianth-segments or more frequently indefinite and irregularly arranged
in several rows; filaments very short; anthers adnate, usually extrorse,
the cells opening in separate valves or in longitudinal slits, either distinct or confluent at the apex. Gynœcium of several carpels, free and
distinct, rarely reduced to a single one, each with a single ascending
or pendulous anatropous ovule. Style terminal, usually oblique excentrical or almost lateral, filiform or very short or almost none, with
a small or pulvinate terminal stigma. Fruit of several (or rarely only
one) 1-seeded drupes or nuts, resting on the expanded receptacle or persistent portion of the perianth-tube or enclosed in the enlarged perianth.
Seed with a membranous testa and fleshy albumen. Embryo usually
very small, with divaricate or appressed cotyledons and a short or long
radicle next the hilum.—Trees shrubs or woody climbers, usually
glandular-dotted and aromatic. Leaves opposite, entire or toothed,
without stipules. Flowers solitary or in trichotomous cymes or definite
raceme-like or thyrsoid panicles, axillary or rarely terminal. Bracts
usually very small.

The Order is chiefly South American, tropical with a few extratropical species, and is also represented in the Mascarene islands and more sparingly in tropical Asia, New Zealand, and the islands of the South Pacific. Of the eight Australian genera one is specially American, another is Asiatic, a third is represented in New Zealand and extratropical South America, a fourth in New Zealand and the South Pacific, and the remaining four are endemic.

Anther-cells distinct, opening each in a separate valve from the base
  upwards.
  Ovule ascending. Peduncles short with 1 or 3 rather large
    flowers. Style bearded and persisting.
    Flowers hermaphrodite. Anther-connective with a long subu-
      late appendage . . . . 1. DORYPHORA.
    Flowers unisexual. Anther-connective not produced . . . 2. ATHEROSPERMA.
  Ovule pendulous. Flowers hermaphrodite, small, in thyrsoid
    panicles. Anther-connective not produced . . . . 3. DAPHNANDRA.
Anther-cells opening longitudinally and confluent in a single reni-
  form or horse-shoe cell. Flowers unisexual, small. Ovule
  pendulous. Perianth 4-lobed.
  Stamens indefinite, few or numerous, irregularly lining the peri-
    anth-tube. Flowers usually diœcious . . . . 4. MOLLINEDIA.
  Stamens 4 opposite the perianth-lobes, with frequently 1 to 3
    smaller ones within. Flowers usually monœcious . . . . . 5. KIBARA.
Anthers, at least the outer ones, with 2 distinct cells opening longi-
  tudinally. Flowers unisexual or polygamous.
  Carpels numerous. Anthers almost sessile.
    Fruiting-perianth circumsciss or expanding under the carpels
      and scarcely enlarged . . . . 6. HEDYCARYA.
    Fruiting-perianth enlarged, irregularly globular, completely en-
      closing the carpels. Climber . . . . 7. PALMERIA.
  Carpels solitary. Filaments filiform, nearly as long as the anthers.
    Climber . . . . 8. PIPTOCALYX.

1. DORYPHORA, Endl.

(Learosa, *Reichb.*)

Flowers hermaphrodite. Perianth-tube campanulate, segments 6 in 2 rows. Stamens usually 6 perfect, opposite the perianth-segments round the orifice of the tube, with 6 to 12 staminodia within them; filaments short, with a wing-like appendage on each side, anthers extrorse, with 2 distinct cells opening from the base upwards in convex valves, the connective produced into a long linear-subulate appendage. Carpels several at the base of the tube with one ascending ovule in each. Style long, slightly lateral. Fruiting carpels included in the persistent perianth-tube, the segments deciduous, each carpel growing out laterally so that the long plumose style appears almost basal. Seed not seen perfect.—Tree. Leaves toothed. Flowers 3 together on short axillary peduncles. The whole plant highly aromatic.

The genus is limited to a single species endemic in Australia.

1. **D. sassafras,** *Endl. Iconogr.* t. 10. A tree of considerable size but of irregular growth, glabrous except the inflorescence or the young shoots hoary-tomentose. Leaves petiolate, ovate elliptical or oblong-lanceolate, acuminate, coarsely toothed, narrowed at the base, 2 to 4 in.

long, nearly smooth on the upper side, prominently penniveined and reticulate underneath. Peduncles 2 to 3 lines long, with a pair of very deciduous bracts of 3 or 4 lines close under the flowers. Perianth-tube about 1 line long when in flower, enlarged and irregularly split when in fruit, segments about 4 lines long, lanceolate, very acute. Anther-appendages nearly as long as the perianth-segments. Carpels slightly hairy, the styles lengthening after fecundation into long plumose awns. —Tul. Monogr. Monim, 424; A. DC. Prod. xvi. ii. 676; Baill. Hist. Pl. i. 318, fig. 357 to 359.

N. S. Wales. Port Jackson to the Blue Mountains, *Woolls* and others ; Clarence river, *Beckler ;* Manning river, *C. Moore ;* southward to the Illawarra, *A. Cunningham, Ralston.*

Although the embryo has been described by Endlicher in his Genera Plantarum, it is doubtful whether he had seen it, for throughout that work the tribual characters are repeated under each genus without having always verified them in each case, and the seed is not figured in his above-quoted Icones. In the 2nd Suppl. to the Genera, p. 35, he proposes to substitute the name of *Learosa* (Reichb. Nomencl. 2612, a work I can find no record of) for *Doryphora*, the latter being pre-engaged by Zoologists, a plea now considered insufficient for changing a botanical name.

2. ATHEROSPERMA, Labill.

Flowers diœcious. Perianth-tube campanulate, lobes 8 or rarely 10 in 2 rows. Stamens usually from as many to twice as many as perianth-lobes, without staminodia in the males; filaments flat, with a wing-like appendage on each side; anthers short, extrorse, with 2 distinct cells opening from the base upwards in convex valves, the connective truncate. Carpels in the females numerous in several rows, the outer ones imperfect, the inner ones with one erect ovule in each and taper-ing into long styles. Fruiting perianth-tube persistent, slightly en-larged, enclosing numerous narrow dry carpels, the long persistent terminal plumose styles exserted, the lobes deciduous. Pericarp and testa of the seeds thin; albumen fleshy. Embryo very small, with short erect cotyledons.—Trees. Leaves toothed or entire. Flowers axillary, in the Australian species solitary, in others in cymes of 3 to 7. The whole plant highly aromatic.

Besides the Australian species which is endemic, the genus comprises one from New Zealand, and (if *Laurelia* be regarded as a true congener) one from extratropical S. America.

1. **A. moschata,** *Labill. Pl. Nov. Holl.* ii. 74, t. 224. A tree attain-ing a large size, the young branches tomentose. Leaves ovate elliptical oblong or lanceolate, acute, coarsely and irregularly toothed or entire, contracted into a rather short petiole, 1½ to 3 in. long, coriaceous, glabrous above, glabrous glaucous or white-tomentose underneath, the primary somewhat branched veins alone conspicuous. Flowers solitary on axillary peduncles of ¼ to ½ in. long, at length recurved. Bracts 2, close under the flower, ovate, acute, 4 or 5 lines long, silky inside and out, very deciduous. Male perianth-tube ovate-campanulate, about 3

lines long, silky outside, glabrous inside, the lobes 4 to 5 lines long, the outer ones broader than the inner. Stamens about 12, inserted round the top of the tube in about 3 rows, without staminodia or imperfect carpels. Female perianth with rather smaller more silky lobes than the males, the tube broader and villous inside. Carpels very numerous, villous, lining the whole disk and tube in numerous rows, the two or three outer rows abortive, considered by some as staminodia, but with the shape and hairs of the carpels and tapering into a minute glabrous rudimentary style, the numerous inner carpels with elongated styles. Fruiting perianth with a persistent broadly campanulate tube 4 or 5 lines diameter, the lobes deciduous, but the outer abortive carpels somewhat enlarged and assuming the appearance of 2 or 3 rows of teeth to the perianth.—A. DC. Prod. xvi. ii. 676; Baill. Hist. Pl. i. 320, f. 360 to 364; Hook. f. Fl. Tasm. i. 12; F. Muell. Pl. Vict. i. 24.

Victoria. Rare in dense swampy forest gullies towards Cape Otway, more frequent at the sources of the Yarra in the Dandenong ranges, and in the southern part of Gipps' Land, *F. Mueller.*

Tasmania. Common in Beech forests throughout the island, *J. D. Hooker.*

3. DAPHNANDRA, Benth.

Flowers hermaphrodite. Perianth-tube short, segments about 15 in about three rows. Stamens 4 or 5, opposite the inner segments; filaments flat, with a wing-like appendage on each side; anthers short, extrorse, with 2 distinct cells opening from the base upwards in convex valves, the connective truncate; staminodia (or abortive carpels?) 5 to 12 between the stamens and carpels. Carpels several in 2 or 3 rows with one pendulous ovule in each, and tapering into the style. Fruit unknown.—Tree. Leaves serrate. Flowers small, in axillary thyrsoid panicles.

The genus is limited to a single species, endemic in Australia. With the stamens of *Atherosperma* it has the inflorescence and pendulous ovules of *Mollinedia,* and the plant cannot well be associated with either, although in the absence of fruit, the generic character is as yet incomplete.

1. **D. micrantha,** *Benth.* A handsome tree of moderate size, quite glabrous or the young inflorescence minutely hoary. Leaves petiolate, oblong-lanceolate or elliptical, acuminate, more or less serrate, contracted at the base, 3 to 4 in. long, green on both sides, the primary veins oblique and anastomosing. Panicles shorter than the leaves, the flowers not numerous. Bracts scarcely 1 line long, very deciduous. Perianth-tube short and broad, outer segments broad, about 1 line diameter, inner ones narrower and more petal-like. Stamens not exceeding the perianth. Carpels of the ovary glabrous or slightly hairy, sessile in the hairy receptacle.—*Atherosperma micranthum,* Tul. Monogr. Monim. in Archiv. Mus. Par. viii. 421, t. 34; Alph. DC. Prod. xvi. ii. 676.

Queensland. Moreton Bay, *Fraser, W. Hill.*

N. S. Wales. Clarence, Richmond, and Lansdowne rivers, *C. Moore.*

4. MOLLINEDIA, Ruiz and Pav.

Flowers unisexual, usually diœcious. Perianth ovoid globular or turbinate, nearly closed by 4 short connivent lobes or rarely, in species not Australian, more deeply divided and spreading. Stamens in the males indefinite, few or many, irregularly lining the inside of the perianth-tube; anthers sessile or nearly so, the cells confluent at the apex and opening longitudinally round the apex and to the base on each side, without staminodia or abortive carpels. Female perianth circumsciss after flowering; staminodia none. Carpels usually numerous, covering the receptacle in many rows, with one pendulous ovule in each; style very short and obtuse or the stigma sessile. Fruiting carpels several, ovoid, drupaceous, sessile or stipitate on the expanded receptacle. Seed with a fleshy albumen; embryo small, with small erect cotyledons, the radicle superior.—Trees or shrubs. Leaves entire or toothed. Flowers small, in axillary lateral or almost terminal cymes or thyrsoid panicles. Bracts very small.

The genus is rather numerous in tropical America, but unknown elsewhere besides the Australian species, which are endemic.

Leaves coriaceous, much reticulate underneath, the petioles short.
　Young parts often pubescent. Ovary and young fruits villous　.　1. *M.Huegeliana*.
Leaves membranous or chartaceous, obtuse, the veins scarcely pro-
　minent. Petioles rather long. Flowers and fruit glabrous　.　.　2. *M. Wardellii*.

Species insufficiently known.

Leaves nearly of *M. Wardellii*. Stamens unknown. Carpels of the
　fruit obtusely and obliquely acuminate .　.　.　.　.　.　.　.　.　3. *M. loxocarya*.
Leaves nearly of *M. Wardellii*, but acutely acuminate. Stamens
　crowded on the disk. Fruit unknown .　.　.　.　.　.　.　.　.　.　4. *M. acuminata*.

1. M. Huegeliana, *Tul. in Ann. Sc. Nat. Sér.* 4. iii. 45, *and in Archiv. Mus. Par.* viii. 399. A small tree, the young shoots inflorescence and underside of the leaves usually pubescent. Leaves on short petioles, ovate-elliptical to oblong-lanceolate, obtuse or shortly acuminate, entire or bordered by short rigid teeth, rounded or rarely acute at the base, mostly 3 to 4 in. long, but varying to short and broad or long and narrow, more coriaceous and shining above than other species, strongly reticulate. Flowers apparently diœcious, in little cymes or thyrsoid panicles very short and sessile or nearly so in the axils, pedicels rather long in the males with a very short common peduncle, the pedicels shorter in the females with a longer peduncle or rhachis, almost forming a few-flowered raceme of about an inch. Male perianth glabrous, nearly globular, about 1 line diameter, the small orifice almost closed by 4 minute broad lobes, 2 inside the 2 others and 1 or 2 outer lobes (or adnate bracts ?) opposite the inner ones. Stamens 8 to 14, irregularly lining the perianth as in American *Mollinediæ*. Female perianth larger, very villous inside, circumsciss after flowering. Carpels numerous and villous. Drupes sessile on the base of the perianth or disk, expanded to a diameter of 3 or 4 lines, ovoid-globose, ½ in. long, pubescent or at length glabrous.— *Wilkiea Huegeliana*, A. DC. Prod. xvi. ii. 669.

Queensland. Brisbane river, Moreton Bay, *A. Cunningham, Fraser, F. Mueller.*
N. S. Wales. Hastings river, *Beckler;* Richmond river, *Henderson;* Cook's river, *U. S. Exploring Expedition;* Kiama, *Harvey.*

The female and fruiting specimens are, when glabrous, very difficult to distinguish from those of *Kibara macrophylla.* The stamens are very different. It is this species that F. Mueller alludes to under *Wilkiea calyptrocalyx,* Fragm. v. 3, as having as many as 13 stamina.

2. **M. Wardellii,** *F. Muell. Fragm.* v. 155. A glabrous shrub. Leaves ovate elliptical or obovate-oblong, obtuse, obscurely crenulate, contracted into a rather long petiole, mostly about 2 in. long, but varying from 1½ to 3 in., of a much thinner consistence than in *M. Huegeliana,* and the primary veins scarcely conspicuous. Panicles thyrsoid, loose, often as long as the leaves. Pedicels 2 to 3 lines long. Male perianth obovoid, globular, 1½ lines diameter, with 4 short broad connivent lobes. Stamens 15 to 20, irregularly lining the perianth as in the American *Mollinediæ.* Female flower not seen. Fruiting perianth circumsciss, the carpels or drupes ovoid-globular, obtuse, glabrous, 4 to 5 lines long, very shortly stipitate on the flat expanded perianth-base or disk.—*Wardellia paniculata,* F. Muell. l.c.

Queensland. Rockingham Bay, *Dallachy.*

3. **M. ? loxocarya,** *Benth.* A glabrous shrub, apparently very nearly allied to *M. Wardellii,* the branches often compressed under the leaves as in *Kibara macrophylla.* Leaves elliptical-oblong, contracted into a rather long petiole as in *M. Wardellii* and of the same consistence, obtuse or shortly acuminate, quite entire, 3 to 4 in. long. Flowers not seen. Fruiting panicles 1½ to 2 in. long, resembling those of *M. Wardellii,* but with a pair of small leaves under the first pair of peduncles and the carpels or drupes closely sessile, ovoid and shortly obliquely and obtusely acuminate.

Queensland. Rockingham Bay, *Dallachy.*

4. **M. ? acuminata,** *F. Muell. Fragm.* v. 155. A small tree, quite glabrous or the inflorescence slightly pubescent. Leaves ovate-lanceolate or elliptical, acutely acuminate, contracted into a rather long petiole, entire or obscurely and irregularly denticulate, 2 to 3 in. long, rather thinner even than in the last two species but not seen in an old state. Panicles thyrsoid, shorter than the leaves, axillary or terminal. Flowers small and drying black as in the Australian *Mollinediæ.* Male perianth glabrous, depressed-globular, about 1½ lines diameter, with 2 short broad bracteoles sometimes adnate at the base; lobes 8 in 2 rows, the 4 outer ones orbicular and much imbricate, the 4 inner narrower and almost valvate. Anthers rather numerous, nearly sessile, occupying the whole disk or with a few abortive hairy carpels in the centre; the outer ones with 2 parallel dorsal cells opening longitudinally as in *Hedycarya,* the inner ones with the cells confluent at the apex as in *Mollinedia.* Female flowers and fruits unknown.

Queensland. Rockingham Bay, *Dallachy.*

This plant, with the habit and general aspect nearly of *M. Wardellii*, is evidently not a congener, the perianth and andrœcium being much nearer to, although certainly not identical with, those of *Hedycarya.* In the absence, however, of female flowers and fruits its real genus cannot be determined, and I have therefore refrained from giving it a new name to be rejected hereafter as another superfluous synonym.

5. KIBARA, Endl.

(Brongniartia, *Blume ;* Sciadicarpus, *Hassk. ;* Sarcostigma, *Griff. ;* Wilkiea, *F. Muell*)

Flowers unisexual, usually monœcious. Perianth ovoid globular or hemispherical, nearly closed by 4 short connivent lobes or teeth, usually (but not in all species) inflexed, or thickened inside in the females. Male fl. Stamens 4 opposite the perianth-lobes, the filaments flattened and more or less adnate to the tube, and above their union either distinct or united in a cup or ring, always shorter than the perianth, and usually 1, 2, or 3 shorter stamens within the four ; anther-cells confluent in a single terminal cell opening longitudinally. No rudimentary carpels. Female perianth circumsciss. Staminodia none (unless the thickened interior of the perianth-lobes be regarded as such). Carpels several, often numerous, with 1 pendulous ovule in each. Style short and obtuse. Fruiting carpels several, ovoid, drupaceous, sessile or stipitate on the expanded receptacle. Seed with a fleshy albumen ; embryo small, with small erect cotyledons, the radicle superior.—Trees or shrubs. Leaves entire or acutely toothed. Flowers small, in axillary or lateral cymes or panicles. Bracts very small.

Besides the Australian species, which are endemic, the genus comprises one or perhaps 3 or 4 from the Malayan Peninsula and the Indian Archipelago. Tulasne indicated its close affinity to *Mollinedia.* Baillon, in uniting it with that genus, adds also *Matthœa*, Blume, which I have not seen, but which, if Blume's figure is correct, appears nearer to *Hedycarya* in the insertion and form though not in the number of stamens. The fruits of *Mollinedia, Kibara,* and *Hedycarya* supply no generic distinctive characters.

Flowers and the whole plant glabrous.
  Leaves narrow, rigidly coriaceous. Petioles short. Inflorescence
    very short. Pedicels 2 to 3 lines long . . . . 1. *K. macrophylla.*
  Leaves broad, thinly coriaceous. Inflorescence loose. Pedicels
    1 in. long or more . . . . 2. *K. longipes.*
Inflorescence and often the foliage pubescent.
  Panicles loose. Carpels stipitate . . . . 3. *K. laxiflora.*
  Panicles short. Carpels sessile . . . . 4. *K. pubescens.*

1. **K. macrophylla,** *Benth.* A tree of considerable size, perfectly glabrous in all its parts. Leaves on very short petioles, oblong or oblong-lanceolate, shortly acuminate, bordered by short pungent-pointed teeth or almost or quite entire, cuneate rounded or cordate at the base, rigidly coriaceous, shining and reticulate. Inflorescence simple or branched, axillary, very short, rarely above ¾ in. long. Pedicels slender, 2 to 3 lines long, with a small bract close under the flower. Perianth globular, quite glabrous, about 1 line diameter, the females with a minute orifice very minutely 4-toothed with thick reflexed fleshy

glands (points of the lobes? or staminodia?) inside, the males rather less closed and not thickened. Stamens in the males 4 opposite the lobes, shortly free, with very short broad filaments, and 1 or 2 smaller ones inside. Carpels in the females 12 to 15, perfectly glabrous, with broad thick flat sessile stigmas. Drupes closely sessile, ovoid, smooth and glabrous, about ½ in. long.—*Hedycarya macrophylla*, A. Cunn. in Ann. Nat. Hist. Sér. 1, i. 215; *Mollinedia macrophylla*, Tul. in Ann. Sc. Nat. Sér. 4, iii. 45, and in Archiv. Mus. Par. viii. 401; *Wilkiea calyptrocalyx*, F. Muell. in Trans. Phil. Inst. Vict. ii. 64 and Fragm. v. 3 (partly); *W. macrophylla*, A. DC. Prod. xvi. ii. 669.

Queensland. Endeavour river, *A. Cunningham;* Brisbane river, Moreton Bay, *A. Cunningham, F. Mueller,* and others; Rockhampton, *Dallachy, Thozet, O'Shanesy.*
N. S. Wales. New England, *C. Stuart.*

The fruiting specimens sometimes resemble some nearly glabrous forms of *Mollinedia Huegeliana,* with which F. Mueller unites this and the three following species as varieties of one *Wilkiea calyptrocalyx.*

2. **K. longipes,** *Benth.* A tree of considerable size, perfectly glabrous in all its parts. Leaves elliptical oblong or ovate, shortly acuminate, entire or with a few irregular teeth, acute at the base, 4 to 8 in. long, smooth and somewhat shining but much thinner than in *K. macrophylla.* Peduncles slender, axillary or supra-axillary, divaricately branched, forming loose panicles often half as long as the leaves and very broad. Pedicels often 1 in. long, slightly thickened upwards. Bracteoles 1 or 2, minute, close under the perianth. Flowers monœcious, the males and females in the same panicle, and the latter much the more numerous in our specimens. Perianths nearly globular, 1 to 1½ lines diameter, the males with 4 connivent lobes, the 2 outer broader than the 2 inner, and not thickened inside; females with a small orifice closed by 4 small lobes, 2 outer and 2 inner, each with a thick reflexed gland-like scale inside. Stamens 4, more or less connate, with 1 or 2 small ones inside. Carpels in the females 11 to 13, oblong, conical, pubescent, with small glabrous stigmatic tips. Perianth circumsciss after flowering. Drupes ovoid, about ½ in. long, glabrous or smooth, very shortly stipitate on the expanded disk or base of the perianth.

Queensland, *Dallachy.* Very closely allied to, but perhaps really distinct from the Malayan *K. coriacea.*

3. **K. laxiflora,** *Benth.* A small tree, the young branches and foliage more or less pubescent. Leaves elliptical or oblong, acuminate or obtuse, rounded or acute at the base, entire or slightly toothed, 2 to 4 in. long. Panicles axillary or from leafless nodes, trichotomous and loose, but usually shorter than the leaves, the males more branched and with more flowers than the females, but sometimes both sexes in the same panicle. Pedicels 2 to 4 lines long, bearing often a small bract at some distance from the flower. Male perianth depressed-globular, 1 to 1¼ lines diameter, the orifice almost closed by 4 small lobes. Stamens 4, united in a broad fleshy cup or tube, pubescent inside, without any additional inner stamens in the flowers examined.

Female perianth similar to the male, but twice the size, with a more minute orifice, and circumsciss after flowering, neither male nor female with any inflexed point or scale within the lobes. Carpels above 30, hispid, with short glabrous stigmatic styles, connivent and almost coherent at the tips. Drupes nearly globular or shortly ovoid, nearly ½ in. long, distinctly stipitate on the expanded disk or base of the perianth.

Queensland. Rockingham Bay, *Dallachy.*

4. **K. pubescens,** *Benth.* A tree, the branches inflorescence and young foliage more or less pubescent, the older leaves often glabrous or nearly so. Leaves from broadly ovate-elliptical to oblong-lanceolate, obtuse or shortly and obtusely acuminate, shortly and rigidly toothed or nearly entire, rounded at the base, becoming at length coriaceous with the veins prominent underneath, 2 to 4 in. long, on petioles of ¼ to ½ in. Flowers in small cymes or clusters sometimes lengthened into short thyrsoid panicles, which are sessile or pedunculate, but usually scarcely ¼ so long as the leaves. Pedicels 2 to 3 lines long. Bracts adnate to the perianth. Perianth about 1 line diameter, nearly globular, the orifice closed with 4 small lobes, thin in the males, thickened inside in the females. Stamens 4 opposite to the perianth lobes, the filaments shortly free and distinct, with 1 or 2 smaller stamens within them. Female perianth circumsciss. Carpels numerous, villous, with thick glabrous nearly sessile stigmas. Drupes closely sessile on the dilated disk or base of the perianth, ovoid, glabrous or nearly so, 3 to 4 lines long.

Queensland. Rockingham Bay, *Dallachy* (with the flowers chiefly males).
N. S. Wales. Richmond river, *C. Moore ;* Hastings and Clarence rivers, *Beckler* (with female flowers and fruits).

The Queensland and N. S. Wales specimens have every appearance of belonging to one species, but require further investigation. The specimens I have seen are rather numerous, but probably each of the four gatherings from a single tree.

6. HEDYCARYA, Forst.

Flowers dioecious. Perianth hemispherical or flat, with about 8 (7 to 10) small inflexed lobes. Stamens in the males numerous, covering the whole disk or leaving a small villous centre, without rudimentary carpels ; anthers sessile or nearly so, the cells usually lateral, distinct, opening longitudinally. Carpels in the females numerous, sessile, occupying the whole disk without staminodia, with one pendulous ovule in each, and terminating in a short thick obtuse style. Fruiting carpels drupaceous, small numerous crowded and almost coherent in the Australian species, larger fewer and distinct (as in *Mollinedia* and *Kibara*) in other species. Seed with a fleshy albumen. Embryo small, with erect cotyledons.—Trees or shrubs. Leaves toothed or entire. Flowers in short axillary cymes or raceme-like panicles. Bracts small or none.

Besides the Australian species which is endemic, there is one from New Zealand and another from the islands of the S. Pacific. The above-described *Mollinedia ? acuminata* may possibly prove to be a second Australian species of *Hedycarya.*

1. **H. angustifolia,** *A. Cunn. in Ann. Nat. Hist. Sér.* 1, i. 215. A tall shrub or small tree, the young shoots and inflorescence slightly hoary-pubescent, the adult parts glabrous. Leaves on rather long petioles, from ovate-elliptical to oblong-lanceolate, shortly acuminate, acute or rarely rounded at the base, irregularly serrate-crenate or almost entire, mostly 3 to 4 in. long, rigidly membranous, penniveined and broadly reticulate. Flowers in short axillary raceme-like cymes, the pedicels very variable in length. Bracts usually very small or none, rarely larger and leaf-like. Perianth flatly hemispherical, $2\frac{1}{2}$ to 3 lines diameter, with 8 or 10 small inflexed lobes. Stamens very numerous, covering the whole disk or concealing a very small central space. Female perianth at first smaller than the male, with smaller inflexed lobes, which become reflexed as the disk and carpels enlarge. Carpels very numerous, surrounded by a few abortive ones (or staminodia ?). Drupes 10 to 20, nearly globular and succulent, each 1 to $1\frac{1}{2}$ lines diameter, all closely packed and almost connate in a globular fruit of 3 or 4 lines diameter. Endocarps crustaceous, minutely rugose.—*H. Cunninghamii,* Tul. in Arch. Mus. Par. viii. 408 ; *H. dentata,* var. *australasica,* Sond. in Linnæa xxviii. 228 ; *H. australasica,* A. DC. Prod. xvi. ii. 673 ; *H. pseudomorus,* F. Muell. in Trans. Phil. Inst. Vict. ii. 63, Pl. Vict. i. 23, t. suppl. 2.

N. S. Wales. Hastings river, *Fraser ;* Hastings and Clarence rivers, *Beckler ;* Hastings and Macleay rivers, *C. Moore ;* New England, *C. Stuart ;* Blue Mountains, *R. Cunningham, Miss Atkinson ;* southward to Illawarra, *Backhouse ;* Twofold Bay, *F. Mueller.*

Victoria. In moist forest gullies of the fern-tree country from Cape Otway to Mount Disappointment, and through the Western Port district and Gipps' Land to Wilson's Promontory, *F. Mueller.*

The leaves are so variable in breadth on the same specimen, that it is impossible on that ground to separate two distinct varieties.

7. PALMERIA, F. Muell.

Flowers diœcious. Male perianth hemispherical, with 4 or 5 connivent lobes. Stamens numerous, without staminodia or rudimentary carpels ; anthers sessile or nearly so ; the cells distinct, opening longitudinally. Female perianth nearly globular, with a minute orifice, staminodia none. Carpels numerous, with 1 pendulous ovule in each ; styles filiform slightly protruding through the orifice of the perianth. Fruiting perianth enlarged, irregularly globular or pear-shaped, completely enclosing the drupes. Seed pendulous, but not seen ripe.— Woody climbers (or trees ?). Leaves entire. Flowers small, in axillary raceme-like cymes or panicles. Bracts none.

The genus is endemic in Australia.

Stamens about 20, surrounding a small hairy disk . . . . 1. *P. scandens.*
Stamens about 60, occupying the whole disk . . . . 2. *P. racemosa.*

1. **P. scandens,** *F. Muell. Fragm.* iv. 152, v. 2. A tall woody climber, the branches minutely tomentose or woolly pubescent. Leaves

u 2

shortly petiolate, ovate or broadly elliptical, acuminate, 3 to 5 in. long, hoary-tomentose on both sides or at length glabrous above. Male inflorescences somewhat branched and half as long as the leaves, female shorter and more simple. Male perianth about 2 lines diameter, flat on the top, the lobes connivent and produced into long points inflexed over the stamens into the centre of the disk before the flower expands. Stamens in the flowers examined 16 to 20, surrounding in several rows a small hairy disk; anthers oblong, slightly hairy, not longer than the perianth. Female perianth about 1½ lines diameter, densely villous inside. Carpels 10 or more, glabrous, tapering into the style. Fruiting perianth irregularly shaped, more or less fleshy, about ½ in. diameter. Drupes glabrous, with a thin exocarp and hard bony endocarp. Seed with a membranous testa, pendulous from near the apex of the cavity, with a rather long hilum, the inside dried up and the embryo undeveloped in all the drupes opened.—A. DC. Prod. xvi. ii. 657.

Queensland. Rockingham Bay, *Dallachy.*

2. ? **P. racemosa,** *A. DC. Prod.* xvi. ii. 657. A tree (*Tulasne*) with the foliage inflorescence and indumentum of *P. scandens,* or the leaves rather more contracted at the base. Male flowers also the same, except that the stamens are 60 or more, and occupy the whole disk of the flower. Female flowers and fruits unknown.—*Hedycarya racemosa,* Tul. in Ann. Sc. Nat. Sér. 4, iii. 45, and in Archiv. Mus. Par. viii. 409, t. 34.

Queensland ? or **N. S. Wales ?** "New Holland," *Baume, Huegel;* probably from Moreton Bay, or from the northern districts of N. S. Wales. I have not seen the specimens, but if the description is correct, notwithstanding the general resemblance to *P. scandens* shown by the figure, it must be a distinct species if a real congener.

Specimens of a tall climber from Paramatta and the Blue Mountains, *Woolls* (in Herb. F. Muell.), in leaf only, may belong to *P. scandens* or to some allied species. Other similar specimens from M'Leod's Creek, *C. Stuart,* said to be from a shrub of 12 to 15 ft., have some female flowers in very young bud, which in that state resemble those of *P. scandens,* but are not sufficiently advanced for accurate identification.

8. PIPTOCALYX, Oliv.

Flowers polygamous. Perianth-tube or disk very short, segments about 6, nearly equal, in 2 rows, very deciduous. Stamens indefinite, filaments filiform; anthers oblong-linear, adnate, the cells distinct and parallel, opening longitudinally; no rudimentary carpels in the male flowers. Hermaphrodite flowers with stamens as in the males and a single carpel, with a single pendulous anatropous ovule and a sessile broad stigma. Fruit unknown.—Woody climber. Leaves opposite, entire. Flowers in simple racemes, the males opposite with the terminal flower hermaphrodite.

The genus consists of the single Australian species. The habit is that of *Palmeria,* but the flowers are very different, and the fruit being unknown the affinity is doubtful. I am unable, however, to trace any closer connection with any other Order than that which it evidently bears to *Monimiaceæ.*

1. **P. Moorei,** *Oliv. in Herb. Kew.* A woody climber of 30 to 40 ft. (*C. Moore*), the branches inflorescence and principal veins on the

underside of the leaves ferruginous with rather long soft hairs. Leaves petiolate, ovate-lanceolate, with a long narrow point, entire or obscurely crenate, rounded at the base, 3 to 4 in. long, rather thin, shining above, pale opaque and minutely glandular-dotted underneath, the veins conspicuous on both sides. Racemes shorter than the leaves, the pedicels very short. Bracts small and very deciduous. Perianth-segments very deciduous, the outer ones ovate-orbicular, 1½ lines long, hirsute outside, the inner ones elliptical or obovate-oblong, thin, all more or less dotted with immersed glands. Stamens about 15, about 1 line long, the filaments rather shorter than the anthers; anthers tipped by a short appendage to the connective. Ovary glabrous, oblong, about 1 line long.

N. S. Wales. Hastings river, *C. Moore.*

Order CIII. **LAURINEÆ.**

Perianth regular, the tube very short or none at the time of flowering, sometimes enlarged over or under the fruit, or rarely adnate to the ovary and fruit; segments 6 or rarely fewer, all equal or 3 outer ones smaller, imbricate in the bud. Stamens normally twice as many as perianth-segments, 6 opposite them with introrse anthers, 6 within and opposite to the outer ones with the anthers extrorse in some genera, introrse in others, but in many genera 3 or sometimes all of the inner stamens, and in others 3 or all of the outer ones reduced to short staminodia or wanting, and in some flowers the stamens abnormally and irregularly increased or diminished in number; there are also frequently a pair of sessile or stipitate rather large glands, one on each side of the filament either of the inner or rarely of the outer of the two stamens opposed to the inner perianth-segments; anthers adnate, with 2 collateral cells or 2 superposed pairs of cells, each cell opening in a valve from the base upwards or in *Hernandia* from the inner to the outer side. Ovary free, or in *Hernandia* adnate, consisting of a single carpel with one ovule suspended from the apex of the cavity from a funicle adnate to its side, or rarely with a second small abortive ovule. Style simple, often very short, with a capitate disk-shaped or obliquely dilated stigma, entire or shortly and irregularly lobed. Fruit a berry or drupe, rarely dry or nearly so, the perianth entirely deciduous, or the tube enlarged and disk-shaped or cup-shaped under the fruit or more or less succulent and closing over it or adnate to it. Seed pendulous, with a membranous or (in *Hernandia*) a hard testa, without albumen. Embryo with thick fleshy cotyledons filling the seed, enclosing the plumula and short superior radicle.—Trees or shrubs with alternate or rarely irregularly opposite leaves, more or less glandular-dotted and aromatic (except in *Hernandia*) usually entire and evergreen, or, in *Cassytha*, leafless parasitical twiners. Stipules none. Flowers usually small, in cymes reduced sometimes to clusters umbels or heads, the cymes solitary or arranged in racemes, clusters, or panicles; or, in *Cassytha*, the flowers singly arranged in spikes or racemes.

A considerable Order, abundant in tropical Asia and America, less so in Africa, with a very few species penetrating into more temperate regions both in the northern and southern hemisphere. Of the eight Australian genera seven are Asiatic, three of them exclusively so (besides the Australian species), the four others with a few American representatives and two of them also African; the sixth genus has besides only two New Zealand species, but it is nearly allied to an Asiatic genus, and as far as known, still closer to a small West Indian one. The large, more especially the American genera, are none of them represented in Australia.

SUBORDER I. **Laureæ.**—*Trees or shrubs with perfect leaves. Anther-valves opening upwards.*

Three stamens, belonging to the inner series, with extrorse anthers.
Anthers 2-celled.
   Stamens of the outer series (usually 6) perfect.
     Ovary more or less immersed in the perianth-tube, which
      completely encloses the fruit . . . .   1. CRYPTOCARYA.
     Ovary not immersed. Perianth completely deciduous   . .   2. NESODAPHNE.
   Stamens of the outer series reduced to a thick prominent ring
    or entirely deficient. Perianth deciduous or scarcely en-
    larged under the fruit . . . .   3. ENDIANDRA.
Anthers 4-celled. Stamens of the outer series (usually 6) perfect.
   Perianth scarcely enlarged under the fruit . . . .   4. CINNAMOMUM.
All the stamens with introrse anthers. Flowers diœcious, in heads
   or umbels, with an involucre of about 4 bracts.
   Perfect stamens 9 or more. Perianth usually 6-merous   . . .   5. TETRANTHERA.
   Perfect stamens 6 or fewer. Perianth usually 4-merous   . . .   6. LITSÆA.

SUBORDER II. **Cassytheæ.**—*Leafless parasitical twiners. Anther-valves opening upwards.*

Single genus . . . .   7. CASSYTHA.

SUBORDER III. **Hernandieæ.**—*Trees with perfect leaves. Anther-valves opening laterally.*

Single genus . . . .   8. HERNANDIA.

SUBORDER 1. LAUREÆ.—Trees or shrubs with perfect leaves. Flowers in cymes umbels or clusters, which are clustered or arranged in racemes or panicles, rarely solitary. Anther-valves opening upwards. Seeds with distinct thick and fleshy cotyledons.

1. CRYPTOCARYA, R. Br.

(Caryodaphne, *Blume.*)

Flowers hermaphrodite. Perianth-segments 6, equal or nearly so. Stamens of the outer series 6, all perfect with introrse anthers, of the inner series 3 with extrorse anthers, alternating with 3 short staminodia; glands 6 at the base of the inner perfect stamens or almost as near to the outer ones opposed to them. Anthers all 2-celled. Ovary immersed in the perianth-tube which after flowering closes over the ovary, and finally becomes more or less fleshy or succulent, completely enclosing and usually consolidated with the fruit, the limb of the perianth deciduous leaving a small scar at the apex or rarely persistent.—Trees or tall shrubs. Flowers small, in cymes arranged in axillary racemes or panicles, the upper ones often forming an apparently terminal panicle with the subtending leaves very small or deficient. Fruiting perianths globular ovoid or oblong, having the appearance of inferior fruits.

The genus is chiefly Asiatic, with a few species from S. Africa and S. America. The Australian species appear to be all endemic.

Nees and Meissner distinguish *Caryodaphne* from *Cryptocarya* by the triplinerved leaves and the adherence of the fruiting perianth-tube to the pericarp, but I can find no difference in the latter respect between the fruits of the typical *C. glaucescens* and those of *C. triplinervis*, and the triplinerved venation, though well marked in two species and in some leaves of *C. triplinervis*, passes gradually into the penniveined arrangement in other leaves of the latter species, and in no case draws any natural line of demarcation between the two.

F. Mueller, Fragm. v. 170, observes that the generic name of *Cryptocarya* must give way to the older name of *Peumus* established by Molina in his Natural History of Chili ; but if he had turned to that work, he would have at once seen why the so-called genera there proposed are in most cases inadmissible. Molina gives no generic characters, and in the present instance, under the name of *Peumus* he includes three or four species belonging to at least three genera and two natural orders.

Inflorescence pubescent or villous.
  Leaves penniveined, with the reticulations very conspicuous
      underneath, scarcely so above.
    Panicles very villous, compact. Leaves 6 to 10 in. long.
      Fruit nearly globular . . . . 1. *C. Murrayi.*
    Panicles tomentose-pubescent, loose and many-flowered.
      Leaves 4 to 8 in. long. Fruit ovoid . . . . 2. *C. Mackinnoniana.*
    Panicles tomentose-pubescent, loose, few-flowered. Leaves
      2 to 5 in. long, softly pubescent underneath till old, the
      reticulations less raised than in the two preceding
      species. Fruit oblong . . . . 3. *C. patentinervis.*
  Leaves penniveined, the reticulations faint or equally prominent on both sides. Fruit globular.
    Leaves thickly coriaceous, smooth and shining above, the
      primary veins very prominent underneath, the reticulations inconspicuous . . . . 4. *C. obovata.*
    Leaves more or less coriaceous, the reticulations fine, conspicuous or obsolete on both sides . . . . 5. *C. glaucescens.*
  Leaves more or less triplinerved or quintuplinerved.
    Leaves usually pubescent underneath, triplinerved, passing
      into penniveined. Fruit ovoid . . . . 6. *C. triplinervis.*
    Leaves glaucous or white underneath, prominently triplinerved. Fruit globular . . . . 7. *C. cinnamomifolia.*
Inflorescence glabrous. Panicles small and loose.
  Leaves penniveined, the veins scarcely prominent . . . . 8. *C. Meissneri.*
  Leaves prominently triplinerved . . . . 9. *C. australis.*

1. **C. Murrayi,** *F. Muell. Fragm.* v. 170. A large tree, the young branches stout and softly tomentose-villous, more or less ferruginous. Leaves shortly petiolate, oblong oval or elliptical, obtuse or shortly acuminate, 6 to 10 in. long, glabrous above when full grown or the midrib minutely pubescent, very prominently penniveined and reticulate underneath, the veins rusty-pubescent, somewhat glaucous between them. Panicles short compact and villous. Flowers sessile in the cymes. Perianth hirsute, the segments rather longer than the tube (about 1 line), almost acute. Glands scarcely connected with the inner stamens, stipitate ; staminodia thick, sessile, rather acute. Fruiting perianth ovoid or nearly globular, shining, about ½ in. diameter.

Queensland. Rockingham Bay, *Dallachy.*

2. **C. Mackinnoniana,** *F. Muell. Fragm.* v. 169. A noble tree 100 ft. high (*Dallachy*) or a tall shrub of 12 ft. (*W. Hill.*), the young branches petioles and inflorescence minutely rusty-pubescent. Leaves oblong or almost ovate, obtuse or shortly acuminate, rounded or cuneate at the base, 4 to 8 in. long, rather thick, at first minutely pubescent, at length glabrous and almost shining above, very prominently penniveined and reticulate underneath, the veins rusty-pubescent, often glaucous between them. Panicles loosely corymbose or thyrsoid. Pedicels very short. Perianth scarcely above 1 line long, the segments about as long as the tube. Glands large, free from the inner stamens, nearly sessile. Fruiting perianth ovoid or oblong, about ¾ in. long.

Queensland. Cape York, *W. Hill;* Rockingham Bay, *Dallachy.*

3. **C. patentinervis,** *F. Muell. in DC. Prod.* xv. i. 508 *and Fragm.* v. 166. A small tree, the branches and inflorescence ferruginous or hoary with a close tomentum. Leaves ovate to oblong-lanceolate, acuminate, 2 to 4 in. long, glabrous above, softly or minutely pubescent and more or less glaucous underneath, the primary veins prominent, the reticulate veinlets also somewhat conspicuous. Cymes sometimes solitary in the axils, more frequently several in short terminal thyrsoid panicles, the flowers not numerous, hoary-pubescent. Perianth 1½ lines long, the tube turbinate, rather shorter than the lobes. Glands of the inner stamens shortly stipitate. Staminodia sessile, thick, acuminate. Fruiting perianth ellipsoid-oblong, nearly ½ in. long, the pericarp rather more distinct from it than in most species.—*C. rigida,* Meissn. in DC. Prod. xv. i. 508.

N. S. Wales. Bellinger and Tweed rivers, *C. Moore;* Hastings and Clarence rivers, *Beckler.*

4. **C. obovata,** *R. Br. Prod.* 402. A fine bushy-headed tree (*Dallachy*), the young shoots and inflorescence minutely tomentose and more or less ferruginous. Leaves oblong to obovate, very obtuse and 2 to 4 in. long in the typical form, larger broader and sometimes shortly and obtusely acuminate in some northern specimens, rather thick, the margins often recurved, glabrous with the veins scarcely conspicuous above, often glaucous or even very minutely pubescent when young underneath, with the primary pinnate veins very prominent, the reticulations scarcely conspicuous. Panicles loosely thyrsoid, numerous and many flowered, the upper ones forming a terminal panicle. Flowers rather larger than in *C. glaucescens.* Perianth-segments as long as the oblong tube. Glands stipitate, appearing to belong as much to the outer as to the inner staminal series. Staminodia sessile, acuminate. Fruiting perianth globular, about ½ in. diameter.—Meissn. in DC. Prod. xv. i. 73, 507; *C. hypospodia,* F. Muell. Fragm. v. 170.

Queensland. Rockingham Bay, *Dallachy* (with large leaves); Brisbane river, Moreton Bay, *A. Cunningham, F. Mueller, W. Hill.*
N. S. Wales. Hunter's river, *R. Brown;* Clarence and Richmond rivers, *C. Moore, Beckler, Wilcox;* Glendon, *Leichhardt.*

5. C. glaucescens, *R. Br. Prod.* 402. A tree of 40 ft. and upwards, the young branches and petioles minutely pubescent when young but soon becoming glabrous, the inflorescence usually hoary-pubescent. Leaves ovate elliptical or oblong, obtuse or shortly acuminate, shortly contracted at the base, flat, not usually so rigid as in *C. obovata,* penniveined and reticulate but the veins rather fine and the reticulations little conspicuous or in some varieties conspicuous or even prominent on both sides, both surfaces green or somewhat glaucous, rarely above 4 in. long in flowering specimens, but larger on luxuriant shoots. Flowers numerous, shortly pedicellate, in thyrsoid panicles shorter or longer than the leaves, the upper ones often forming a large terminal panicle. Perianth 1 to 1¼ lines long, hoary-pubescent, the tube oblong when fully out, turbinate when young, the segments about as long as the tube or rather longer. Glands stipitate, appearing to belong as much to the outer as to the inner filaments. Staminodia acuminate. Fruiting perianth depressed-globular, ½ to nearly ¾ in. diameter.— Meissn. in DC. Prod. xv. i. 72.

Queensland. Rockingham Bay, *Dallachy;* Rockhampton, *Thozet, Dallachy;* Brisbane river, Moreton Bay, *A. Cunningham, F. Mueller,* and others.

N. S. Wales. Hawkesbury, *R. Brown;* Hastings river, *Beckler;* New England, *C. Stuart;* Port Jackson, *Woolls;* Illawarra, *A. Cunningham* and others; Sydney woods, Paris Exhibition, 1855, *Macarthur, n.* 6 *and* 30; Northern woods, n. 4, and Southern woods, n. 115, London Exhibition, 1862, *C. Moore.*

Var. *Cunninghamii.* Scarcely distinguishable from the eastern form, except by rather larger flowers and the perianth-tube rather shorter.—*C. Cunninghamii,* Meissn in· DC. Prod. xv. i. 73.

N. Australia. Hunter's river, Brunswick Bay, N.W. coast, *A. Cunningham.*

C. hypoglauca, Meissn. l.c. from N. W. Australia, which I have not seen, but is supposed to be from A. Cunningham's collection, is probably the same species.

Var. *reticulata,* Meissn. Veins of the leaves more conspicuous on both sides but fine.—*C. microneura,* Meissn. l.c.—Seaview Range, *Fraser;* Hastings river, *C. Moore;* Illawarra, *A. Cunningham;* Sydney woods, Paris Exhibition, 1855, *C. Moore,* n. 5, *Macarthur, n.* 198, 224, 234.

Var. *coriacea.* Leaves thick, rigid, and more prominently reticulate, often very glaucous underneath.—Rockingham Bay, *Dallachy.*

Var. *nitida.* Leaves coriaceous and shining with fine veins, green on both sides or scarcely glaucous.—*C. Moretoniana,* Meissn. l.c. 74.—Moreton Bay, *A. Cunningham;* Archer's Creek, *Leichhardt.*

C. Bidwillii, Meissn. l.c. 74, of which the specimens from Wide Bay, *Bidwill,* are in young fruit only, and not satisfactory, may be the same variety.

All the above varieties run much one into the other, and in view of the large number of specimens I have had before me, it seems impossible to consider any of them as species permanently distinct.

6. C. triplinervis, *R. Br. Prod.* 402. A tall tree. Leaves ovate-elliptical or oblong-lanceolate, acuminate, glabrous above, more or less pubescent underneath, rarely above 4 in. long, triplinerved or more or less irregularly penniveined with 2 to 5 primary veins on each side of the midrib, sometimes very prominent underneath sometimes fine, the reticulations not very conspicuous. Panicles dense short and thyrsoid in the axils, or the upper ones much branched forming a terminal panicle with numerous flowers, hoary-pubescent with appressed hairs or more frequently hirsute with spreading hairs. Flowers nearly sessile. Perianth-tube cylindrical, about 1 line long, the segments narrow and nearly as long. Stamens nearly as long as the perianth, the glands stipitate, as near to the outer as to the inner stamens; staminodia rather narrow, acuminate. Fruiting perianth ovoid, about ½ in. long.—*Caryodaphne Brownniana*, Nees, Syst. Laurin. 230; Meissn. in DC. Prod. xv. i. 78.

Queensland. E. coast, *R. Brown;* Edgecombe and Rockingham Bays, *Dallachy;* Port Denison, *Fitzalan;* Rockhampton, *Dallachy, O'Shanesy,* and others; Cape river, Crocodile and Nerkool Creeks, *Bowman;* Archer's Creek, *Leichhardt;* Brisbane river, Moreton Bay, *Leichhardt, F. Mueller, W. Hill,* and others.

N. S. Wales. Richmond and Clarence rivers, *C. Moore, Beckler.*

There are two or three marked forms of foliage, but with occasional intermediates and sometimes the leaves of the principal branches different as to venation from those of the younger branches of the same specimen. In the more typical forms the leaves are rather thick, prominently triplinerved or quintuplinerved and softly pubescent underneath, in other forms the primary veins are more numerous, and in some of the northern specimens the leaves are often almost glabrous and almost as penniveined as in *C. glaucescens.* The fruit appears to be always differently shaped from that of *C. glaucescens,* although the perianth-tube is equally consolidated with the pericarp in both species.

The specimen of Milne's from Lord Howe's island, referred by Meissner to *Caryodaphne densiflora,* Blume, is in old leaf only and cannot be determined with any certainty. It appears to me to be much more like some varieties of *Cryptocarya triplinervis* than any specimen of the Javanese *C. densiflora.*

7. **C. cinnamomifolia,** *Benth.* A tree of 40 ft. (*Dallachy*), the young shoots and inflorescence minutely ferruginous-tomentose. Leaves ovate or broadly elliptical, acuminate, cuneate at the base, rigidly coriaceous, prominently triplinerved, glabrous above, glaucous or white underneath, with the reticulations conspicuous, 3 to 6 in. long. Panicles axillary or in terminal panicles, short and dense in the specimens seen but as yet only in young bud and the form of the perianth not ascertained. Stamens of *Cryptocarya,* but as yet very young. Fruit depressed globular, above ½ in. diameter, very similar to that of *C. glaucescens.*

Queensland. Rockingham Bay, *Dallachy.*

8. **C. Meissneri,** *F. Muell. Fragm.* v. 170. A small or large tree, quite glabrous in every part. Leaves elliptical or oblong-lanceolate, obtusely acuminate, contracted into a short petiole, mostly about 2 in. and rarely above 3 in. long, rather thick, penniveined but the veins irregular and even the primary ones not very prominent. Panicles short, axillary and terminal, the flowers not numerous and quite glabrous outside. Perianth-tube narrow, at first turbinate, ovoid and contracted at the top when fully out and ¾ line long, the lobes nearly as long, minutely hairy inside. Stamens much shorter than the perianth, the glands sessile but as near to the outer as to the inner stamens. Ovary immersed in the perianth-tube. Fruit not seen.— *C. hypoglauca,* var. *attenuata,* Meissn. in DC. Prod. xv. i. 508 (F. Mueller).

N. S. Wales. Hastings, Macleay and Bellinger rivers, *C. Moore, Beckler.*

9. **C. australis,** *Benth.* A large shrub or small tree, quite glabrous in all its parts. Leaves ovate elliptical or almost oblong, acuminate, contracted into a short petiole, coriaceous or thin, pale underneath but not at all white, prominently triplinerved, 2 to 4 in. long. Panicles very loose, few-flowered, always much shorter than the leaves and quite glabrous. Pedicels rather long. Perianth-tube turbinate, nearly 1 line long, the lobes at least as long, ovate and more spreading than in other species. Stamens short, especially the 3 inner ones. Ovary immersed in the tube. Fruiting perianth obovoid, pear-shaped, nearly ¾ in. long, usually crowned by the remains of the perianth-limb.— *Laurus Bowiei,* Hook. Journ. Bot. iv. 419. t. 23; *Oreodaphne Bowiei,* Walp. Ann. i. 576; *Laurus australis* A. Cunn.; Hook. Bot. Mag. under n. 3931; *Caryodaphne australis* A. Braun; Meissn. in DC. Prod. xv. i. 77.

Queensland. Brisbane river, Moreton Bay, *A. Cunningham* and many others; Rockingham Bay, *Dallachy.*

N. S. Wales. Clarence river, *Beckler;* Richmond river, *Henderson.*

The specific name *Bowiei* was originally given to this plant under a mistake as to the origin of the seeds from which it was raised at Kew, and was therefore afterwards suppressed by Hooker himself, and Cunningham's name adopted.

2. NESODAPHNE, Hook. f.

Flowers hermaphrodite. Perianth-segments 6, equal or nearly so. Stamens of the outer series 6, all perfect with introse anthers, of the inner series 3 with extrorse anthers, alternating with 3 short staminodia; glands 6, at the base of the inner perfect stamens. Ovary not immersed, the whole perianth deciduous. Berry free, resting on the apex of the slightly thickened peduncle.—Trees or tall shrubs. Leaves alternate. Flowers small in broad cymes arranged in panicles either terminal or in the upper axils. Bracts usually oblong.

Besides the Australian species which is endemic, there are two from New Zealand. The genus is moreover closely allied to the East Indian *Beilsckmiedias,* scarcely differing except in the absence of the peculiar substance intruded between the cotyledons in the latter genus, which has been called a false dissepiment in the fruit, but appears rather to be an intrusion of the testa. The West Indian *Hufelandia* is perhaps still closer to or even inseparable from *Nesodaphne.*

1. **N. obtusifolia,** *Benth.* A large and handsome tree, the young shoots and inflorescence sprinkled with minute appressed hairs, otherwise glabrous. Leaves elliptical oblong or oval-oblong, obtuse or obtusely acuminate, narrowed at the base into a short petiole, rather rigid, green on both sides, penniveined and loosely reticulate, 2 to 4 in. long. Cymes large and rather broad, the upper ones in a broad pyramidal or almost corymbose terminal panicle. Pedicels thick and ½ line long above the small bracteoles. Perianth-tube very short and broad, the segments 1 line long. Stamens opposed to the outer segments broad, those opposed to the inner ones narrow; glands stipitate or almost sessile, attached to the inner stamens; staminodia sessile,

thick, acute. Berry elliptical-oblong, the whole perianth deciduous.
—*Cryptocarya obtusifolia;* F. Muell.; Meissn. in DC. Prod. xv. i. 508.

Queensland. Rockingham Bay, *Dallachy;* Fitzroy river, *Bowman;* Rockhampton, *Thozet;* Archer's Creek, *Leichhardt.*
N. S. Wales. Clarence river, *Wilcox.*

3. ENDIANDRA, R. Br.

(Dictyodaphne, *Blume.*)

Flowers hermaphrodite. Perianth-segments 6, equal or nearly so. Stamens of the outer series reduced to a thick prominent ring below the perianth-segments or entirely wanting, of the inner series 3, with extrose or semi-extrose anthers, alternating sometimes with 3 small staminodia which are often deficient; glands either 6 at the base of the perfect stamens but free from them or none. Ovary not immersed. Berry free, resting on the wholly or partially persistent but not at all or scarcely enlarged perianth.—Trees. Leaves alternate. Flowers in axillary panicles, sometimes almost reduced to single cymes. Fruits oblong or globular.

The genus extends over the Indian Archipelago to the eastern provinces of India. The Australian species appear to be all endemic. The genus has been incorrectly placed in the tribe of Cryptocaryeæ as characterised by Nees and by Meissner, for neither the ovary nor the fruit are enclosed in the perianth-tube.

Perianth-tube small and turbinate, the limb broad and very open or when closed much broader than the tube.
  Outer stamens replaced by a thick fleshy ring round the base of the perianth-segments.
    Panicles thyrsoid. Flowers 2 to 2½ lines diameter. Bracts very deciduous. Leaves glabrous . . . . 1. *E. glauca.*
    Panicles narrow, raceme-like. Flowers 3 to 4 lines diameter. Bracts persistent. Leaves rusty-pubescent or villous underneath . . . . 2. *E. hypotephra.*
  Outer stamens entirely deficient. Small staminodia usually present between the inner ones.
    Leaves white underneath, the primary veins prominent . . 3. *E. discolor.*
    Leaves green and reticulate on both sides . . . . 4. *E. Sieberi.*
Perianth-tube thick and fleshy, broader than the connivent segments. Leaves green on both sides. Flowers small.
  Leaves glabrous. Staminal glands present.
    Leaves mostly oblong, much reticulate. Perianth not 1 line diameter . . . . 5. *E. virens.*
    Leaves mostly elliptical-ovate, less reticulate, the primary veins more prominent. Perianth 1½ lines diameter . . . 6. *E. Muelleri.*
  Leaves pubescent or villous underneath. No staminal glands . 7. *E. pubens.*

1. **E. glauca,** *R. Br. Prod.* 402. A small tree, the young shoots and inflorescence minutely ferruginous-tomentose. Leaves elliptical-oblong, acuminate, contracted at the base, mostly 3 to 5 in. long, glabrous and green above, glaucous or white underneath, otherwise glabrous or the very prominent primary veins minutely ferruginous-tomentose. Panicles thyrsoid, rather loose, shorter than the leaves, the bracts small

and very deciduous. Pedicels short. Perianth-tube turbinate, much narrower than the limb, which spreads to 2 or 2½ lines diameter. Staminodia of the outer row forming a thick fleshy crenulate ring bordering the orifice of the tube and enclosing the ovary, the 3 inner stamens protruding shortly from the ring.—Meissn. in DC. Prod. xv. i. 509.

Queensland. Endeavour river, *Banks and Solander;* Cape Grafton, *A. Cunningham;* Rockingham Bay, *Dallachy.*

2. **E. hypotephra,** *F. Muell. Fragm.* v. 166, *partly.* A moderate sized tree, the young branches petioles and inflorescence softly ferruginous-tomentose or villous. Leaves ovate ovate-elliptical or oval-oblong, shortly acuminate or rarely obtuse, 3 to 5 in. long, coriaceous, glabrous above with impressed veins, glaucous underneath, with the primary raised veins ferruginous-tomentose or villous. Flowers much larger than in any other species, few together in dense cymes on very short peduncles forming interrupted raceme-like panicles, usually shorter but sometimes as long as the leaves. Bracts small, but larger and more persistent than in the other species. Pedicels very short. Perianth-tube small, turbinate, the limb spreading to 3 or 4 lines diameter. Staminodia of the outer row forming a thick fleshy crenulate ring bordering the orifice of the tube and enclosing the ovary, the 3 inner stamens shortly protruding from the ring. Fruit oblong, ¾ in. long, resting on the persistent but not enlarged perianth.

Queensland. Rockingham Bay, *Dallachy.* F. Mueller includes both *E. glauca* and the present species under the name of *E. hypotephra,* but his description applies more especially to the present one.

3. **E. discolor,** *Benth.* A shrub or tall tree, the young branches and inflorescence minutely hoary-tomentose. Leaves ovate or elliptical, obtuse or shortly acuminate, 2 to 3 in. long, green and glabrous above, glaucous or white underneath, with few prominent primary veins, the smaller reticulations conspicuous on both sides in the full-grown leaf. Panicles thyrsoid, much shorter than the leaves, the pedicels very short. Perianth-tube small and turbinate, the limb very open, about 1¼ lines diameter, with ovate segments. Stamens of the outer row entirely deficient, of the inner series 3, with sessile glands at their base and alternating with 3 small staminodia. Fruit ovoid-oblong, ¾ to 1 in. long, resting on the unenlarged remains of the perianth.

Queensland. Albany island and Cape York, *W. Hill;* Rockingham Bay, *Dallachy.*

N. S. Wales. Macleay, Richmond and Hastings rivers, *C. Moore.*

4. **E. Sieberi,** *Nees. Syst. Laurin.* 194. A shrub or tree, glabrous except a minute pubescence scarcely perceptible on the inflorescence. Leaves ovate-lanceolate or oblong, obtuse or obtusely acuminate, 2 to 3 in. long, contracted into a short petiole, rather thin, green on both sides, with the smaller reticulations as prominent as the rather numerous and fine primary veins. Panicles thyrsoid, shorter than the leaves.

Perianth-tube very small, the limb very open. Stamens of the outer series entirely deficient, 3 of the inner series perfect with a gland on each side at the base, the intervening staminodia very small or obsolete. Young fruit globular, resting on the apex of the peduncle, the whole perianth deciduous. Ripe fruit not seen.—*Meissn. in DC. Prod.* xv. i. 79.

N. S. Wales. Port Jackson to the Blue Mountains, *Sieber, n.* 275, and many others; Sandy soil near the seacoast, *Leichhardt;* Richmond river, *C. Moore.*

5. **E. virens,** *F. Muell.; Meissn. in DC. Prod.* xv. i. 509. A tall shrub or a tree attaining a considerable height, glabrous in all its parts. Leaves oblong, usually narrow, rarely broader and elliptical, obtuse, contracted at the base, 2 to 3 in. long in some specimens, twice that size in others, not thick, green and reticulate on both sides, somewhat shining above, a few of the primary veins more prominent than the others. Panicles slender, glabrous, much shorter than the leaves. Pedicels rather long. Flowers small. Perianth-tube very thick, broadly turbinate, nearly 1 line diameter, forming a thick projecting ring round the base of the lobes which are shorter than the tube, broad and connivent, the 3 inner ones rather smaller than and quite enclosed in the outer ones. Stamens of the outer row entirely deficient, of the inner row 3, without glands at their base, but alternating with sessile staminodia. Fruit globular, ½ to ¾ in. diameter, resting on thickened pedicels of ½ in. or more.

N. S. Wales. Clarence and Richmond rivers, *Beckler, C. Moore.*

In some specimens the perianth-tube enlarges, apparently from the effect of some insect, into a hard globular verrucose gall of 1 in. diameter or more.

6. **E. Muelleri,** *Meissn. in DC. Prod.* xv. i. 509. A moderate-sized tree, glabrous except a minute ferruginous tomentum on the inflorescence and sometimes on the young shoots. Leaves ovate elliptical or broadly oblong, acuminate, cuneate at the base, green on both sides, the primary veins more prominent and the reticulations less so than in *E. Sieberi*, mostly 3 to 5 in. long. Panicles axillary, loose, much shorter than the leaves. Pedicels longer than the flower. Perianth-tube thick and fleshy, broadly turbinate, 1½ lines diameter, the lobes small and connivent. Stamens 3 with broad flattened glands; staminodia of the outer series deficient, of the inner series small or obsolete. Fruit not seen.

N. S. Wales. Hastings and Macleay rivers, *Beckler.* A specimen from Bellinger river, *C. Moore,* may belong to it also, but is only in bud and doubtful. The species is near *E. virens,* but with broader, less reticulate leaves, and the flowers twice as large. The glands in the flower of this and some other species described sometimes as staminodia appear to me to be precisely the same as the glands of the inner stamens of so many other Laurineæ.

7. **E. pubens,** *Meissn. in DC. Prod.* xv. i. 509. A large tree, the branches and petioles more or less velvety-tomentose and ferruginous. Leaves oval to elliptical-oblong, obtusely acuminate or almost obtuse, narrowed at the base, 4 to 8 in. long, glabrous above, prominently

veined and pubescent or villous underneath. Panicles axillary, broadly thyrsoid, usually about 1 in. long, sometimes more branched and half as long as the leaves, sometimes very short, more or less ferruginous-hirsute. Bracts narrow, the lower ones often 1 line long, those in the cymes smaller. Pedicels scarcely so long as the flower. Perianth nearly 1½ lines long, the tube thick, turbinate, broader than the limb, the lobes small and erect. Stamens 3, usually exserted, the filaments rather narrow, without glands, but alternating with small staminodia, the outer series quite deficient. Fruit globular, ½ to ¾ in. diameter.— *Cryptocarya Muelleri,* Meissn. l.c. 73.

Queensland. Brisbane river, Moreton Bay, *F. Mueller.*
N. S. Wales. Clarence river, *Beckler;* Richmond river, *Henderson;* Bellinger river, *C. Moore.*

Var. *glabriflora.* Perianth rather larger, glabrous. Bracts ovate, concave.—Richmond river, *Henderson.*

4. CINNAMOMUM, Burm.

Flowers usually more or less unisexual. Perianth-segments 6, equal or nearly so. Stamens of the outer series 6, all perfect with introrse anthers, of the inner series 3 perfect, with extrorse anthers, alternating with 3 short staminodia; anthers 4-celled, or the inner ones rarely 2-celled; glands 6, at the base of the inner perfect stamens. Ovary not immersed. Berry seated on the somewhat enlarged truncate or 6-lobed perianth-tube, the segments wholly or partially deciduous.—Trees or shrubs. Leaves opposite or often more or less alternate on the same tree, 3-nerved or rarely 5-nerved. Flowers in axillary panicles, more or less unisexual, the females usually rather larger and fewer in the panicle with the stamens slightly imperfect, the males smaller and more numerous with a sterile ovary. The numbers of parts of the perianth and of the stamens liable to occasional variation, especially in the females.

The genus extends over tropical and subtropical Asia as far as Japan, the only Australian species being the same as an E. Indian one.

1. **C. Tamala,** *Th. Nees.; Nees. Syst. Laurin.* 56. A large tree with a smooth almost white bark (*Dallachy*). Leaves opposite or here and there alternate, oblong-lanceolate or ovate-lanceolate, obtusely acuminate, acute at the base, 3 to 6 in. long, triplinerved, silvery underneath when young with small closely appressed hairs, the older ones glabrous or nearly so and showing underneath the fine reticulate veinlets. Peduncles in the uppermost axils, bearing in the Australian specimens (probably from a fertile tree) a loose panicle rather shorter or rather longer than the leaves more or less silvery-pubescent. Pedicels about as long as the perianth. Perianth-tube about 1 line, the segments or lobes 2 lines long. Stamens shorter than the perianth, some or all apparently perfect but unequally so in the flowers examined. Stigma broadly peltate.—Miq. Rev. Cinnam. in Ann. Mus. Lugd. Bat. i. 268; Meissn. in DC. Prod. xv. i. 17; *Laurus Tamala,* Hamilt. in Trans. Linn.

Soc. xiii. 555; *L. Cassia,* Roxb. Fl. Ind. ii. 297, not of Willd; *Cinnamomum Laubatii,* F. Muell. Fragm. v. 165.

Queensland. Sea-view Range, Rockingham Bay, *Dallachy.*

These specimens appear to me to agree perfectly with E. Indian ones of the fertile tree, accurately described by Roxburgh, who gives it as a native of various mountains of the Indian Continent. The figure of Th. Nees, in the Supplement to his Officinal Plants, represents the male or semimale form, with smaller, more numerous flowers in larger panicles. The Australian specimens have only very few of the flowers fully out, and no fruit. In Indian ones the fruiting perianth has 6 short truncate lobes, the upper portion of the lobes being alone deciduous. There are several other Indian *Cinnamoma* described as species which, as observed by both Miquel and Meissner, are very difficult to distinguish in all their various forms from *C. Tamala,* and may be hereafter united with it; but even then, as far as I have been able to ascertain, Hamilton's specific name of *Tamala* will have to be retained as the oldest.

5. **TETRANTHERA,** Jacq.

(Cylicodaphne, *Nees.*)

Flowers diœcious. Perianth-segments 6, equal or nearly so, or rarely unequal or fewer by abortion. Stamens of the outer series usually 6, perfect, of the inner series either 6 perfect, or 3 perfect alternating with 3 staminodia or (in species not Australian) the whole number more or less increased; anthers all introrse, 4-celled; glands usually 6, one on each side of 3 inner stamens; the stamens reduced to staminodia in the female flowers, but the glands usually present. Ovary imperfect or wholly abortive in the males, free in the females; stigma usually dilated and toothed or lobed. Berry resting on the more or less enlarged flat concave or cup-shaped perianth-tube, the segments deciduous.—Trees. Leaves alternate or rarely irregularly opposite, penniveined. Flowers in cymes reduced to small umbels heads or clusters within an involucre of 4 concave deciduous bracts, arranged in short racemes or clusters in the axils or at the leafless nodes.

A large genus, chiefly tropical Asiatic, extending in few species to Japan and Western America. Of the four Australian species, one has a wide range over tropical Asia, the others, as far as I have been able to ascertain, are endemic.

Sect. 1. **Tomex.**—*Perianth-tube slightly enlarged under the fruit, flat or slightly concave.*

Leaves hoary-pubescent or glabrous, usually large, the reticulations not prominent. Flowers rather large . . . . 1. *T. laurifolia.*

Sect. 2. **Cylicodaphne.**—*Perianth-tube more enlarged under the fruit, more concave or cupshaped.*

Leaves large, broad, very obtuse, glabrous, not reticulate . . . 2. *T. Bindoniana.*
Leaves broad or narrow, obtuse, ferruginous underneath, not reticulate . . . . 3. *T. ferruginea.*
Leaves rather thin, glabrous, conspicuously reticulate on both sides . . . . 4. *T. reticulata.*

Sect. 1. Tomex.—Perianth-tube slightly enlarged under the fruit, flat or slightly concave.

1. **T. laurifolia,** *Jacq., Meissn. in DC. Prod.* xv. i. 178. A small tree, the branches foliage and inflorescence more or less hoary-pubescent or the older leaves glabrous. Leaves petiolate, ovate obovate-elliptical or oblong, obtuse, shortly acuminate or rather acute, 4 to 8 in. long on the flowering branches, often larger on barren ones, green on both sides or glaucous underneath, the primary veins prominent on the underside. Peduncles 3 to 6 lines long, clustered or racemose, the common rhachis sometimes very short and usually shorter than the peduncles. Bracts of the involucres about 2 lines diameter, enclosing 5 or more sessile or very shortly pedicellate villous flowers. Perianth-segments very irregular, 1 to 6 or entirely abortive, and always very deciduous. Stamens in the Australian specimens usually 10 or 11, in some Asiatic ones more numerous ; filaments filiform, hairy, 3 or 4 of the inner ones with a pair of glands at the base. Fruit globular, 3 or 4 lines diameter, resting on the more or less thickened pedicel with the persistent perianth-tube somewhat enlarged but flat or slightly concave. —*T. apetala*, Roxb. Pl. Corom. ii. 26, t. 147 ; R. Br. Prod. 403.

N. Australia. Islands of the Gulf of Carpentaria, *R. Brown, Henne;* Port Darwin, *Schulz.*

Queensland. Port Denison, *Fitzalan, Dallachy;* Elliott river, *Bowman.*

The Australian specimens belong chiefly to Meissner's var *citrifolia*, with acuminate or almost acute leaves, but Brown's include also some with very obtuse leaves.

SECT. 2. CYLICODAPHNE.—Perianth-tube more enlarged under the fruit than in *Tomex*, more concave or cup-shaped.

2. **T. Bindoniana,** *F. Muell. Fragm.* v. 167. A small tree, the young branches and petioles minutely ferruginous-tomentose or at length glabrous. Leaves on rather long petioles, broadly ovate obovate or orbicular, obtuse, 5 to 8 in. long, firmly coriaceous, glabrous above, glaucous or somewhat ferruginous underneath, with the primary veins much raised, the smaller transverse ones not numerous. Male flowers not seen. Female peduncles 2 to 3 lines long, in almost sessile clusters, ferruginous-pubescent as well as the bracts. Flowers 5, sessile. Perianth nearly 2 lines long, villous ; segments 6, narrow, as long as the tube. Staminodia 6, outer ones without glands, 3 inner each with a pair of glands. Ovary pubescent. Style thick, villous, with a flat spreading somewhat lobed stigma. Fruit (not seen ripe) protruding from the enlarged persistent cup-shaped truncate perianth-tube.— *Cylicodaphne Bindoniana*, F. Muell. l.c.

Queensland. Summit of M'Alister hill, Rockingham Bay, *Dallachy.*

3. **T. ferruginea,** *R. Br. Prod.* 403. A tree of 30 ft. or more, the branches and petioles ferruginous-pubescent or villous. Leaves from broadly ovate to elliptical-oblong, acuminate or rarely obtuse, rounded or cuneate at the base, 3 to 5 in. long, rather firm, glabrous and shining above, ferruginous-pubescent underneath, with raised primary veins, and transverse veinlets. Peduncles clustered in the axils or at the old nodes, 3 to 6 lines long in the males, shorter in the females.

Bracts orbicular, enclosing 5 or 6 flowers on very short thick pedicels. Perianth-segments 6, lanceolate ciliate and very deciduous in the females, broader more obtuse and sometimes fewer in the males. Perfect stamens 2, twice as long as the perianth, the filaments hairy; staminodia in the females 12, short. Fruit (not seen quite ripe) ovoid, resting in the enlarged cup-shaped truncate perianth-tube which attains 3 to 4 lines diameter,—Meissn. in DC. Prod. xv. i. 192; *Cylicodaphne Leefeana,* F. Muell. Fragm. v. 169.

Queensland. Cape Grafton and Endeavour river, *Banks and Solander, A. Cunningham;* Rockingham Bay, *Dallachy;* between Cleveland and Rockingham Bays, *W. Hill.*

Var. *lanceolata,* Meissn. Leaves oblong or oblong-lanceolate. Male flowers as in the typical form, females unknown.—*T. nesogena,* F. Muell. Fragm. v. 169.—Family island, Rockingham Bay, *Dallachy;* Brisbane river, *C. Moore* (Sydney woods, Paris Exhibition, 1855, n. 15).

It seems very doubtful whether *Hexanthus* of Loureiro, from Cochinchina, usually referred to *T. ferruginea,* is really the same species.

4. **T. reticulata,** *Meissn. in DC. Prod.* xv. i. 192. A tree of considerable size, glabrous except the flowers, or the young shoots minutely silky-pubescent. Leaves obovate-oblong or oblong-elliptical, obtuse or scarcely acuminate, narrowed into the petiole, mostly 3 to 4 in. long, not thick, green on both sides, the primary veins not very prominent, the reticulations much more conspicuous on the upper than on the lower surface. Peduncles 3 to 5 lines long in the males, much shorter in the females, usually in short irregular racemes or clusters, on a common rhachis at first very short but sometimes lengthening to ½ in., glabrous as well as the bracts. Perianth-segments narrow, acute, silky-villous. Stamens in the males 6 outer ones rather longer than the perianth and without glands, 3 inner rather shorter, each with a pair of glands. Staminodia in the females shorter than the perianth. Ovary glabrous; stigma broad and lobed. Fruit ovoid, resting in the enlarged truncate cup-shaped perianth-tube.—*Cylicodaphne Fawcettiana* or *Tetranthera Fawcettiana,* F. Muell. Fragm. v. 168.

Queensland. Rockingham Bay, *Dallachy;* Sydney woods (probably from Brisbane river), Paris Exhibition, 1855, *Macarthur, n.* 24, 192.

6. LITSÆA, Juss.

Flowers diœcious. Perianth-segments usually 4, equal or nearly so. Stamens of the outer series usually 4, perfect, of the inner series 2 perfect, without staminodia; glands 4, one on each side of the 2 inner stamens; anthers all 4-celled introrse; stamens in the females reduced to staminodia. Ovary imperfect or abortive in the males, free in the females; stigma disk-shaped. Berry resting on the flat somewhat or scarcely dilated perianth-tube, the segments persistent or deciduous.— Trees. Leaves alternate, frequently crowded and almost whorled at the ends of the shoots, usually penniveined but with few primary veins and the lower pair more prominent so as often to appear triplinerved.

Flowers in sessile or nearly sessile clusters surrounded by several very deciduous imbricate bracts.

The genus extends over tropical Asia. Of the two Australian species one is a common Indian one, the other appears to be endemic.

Glabrous or the inflorescence slightly pubescent. Fruiting perianth
entire (the segments deciduous) . . . . . . 1. *L. zeylanica.*
More or less ferruginous-villous, at least the young shoots and inflorescence. Fruiting perianth 4-toothed (the segments wholly or partially persistent) . . . . 2. *L. dealbata.*

1. **L. zeylanica,** *Nees, frr. Cinnam. Disput. in Amœn. Bot. Bonn.* i. 58, t. 5. A large tree, the branches and inflorescence quite glabrous or scarcely hoary with a very minute tomentum. Leaves ovate-elliptical or elliptical-oblong, acuminate, contracted at the base, 3 to 5 in. long, glabrous and green above, white or glaucous underneath, penniveined but with few primary veins, the lowest pair more prominent than the others. Flowers in sessile clusters in the axils or at the old nodes, on pedicels of 1 to 2 lines usually glabrous as well as the perianths. Perianth-segments ovate-oblong, obtuse. Filaments exserted, with a few hairs about the base; glands of the two inner ones stipitate. Berry globular or slightly ovoid, larger than in *L. dealbata,* resting on the per-sistent perianth-tube expanded into an entire or slightly angular flat disk of 2½ to 3 lines diameter, the segments entirely deciduous.—Meissn. in DC. Prod. xv. i. 226 ; Wight, Ic. t. 132 and 1844.

Queensland. Lizard island, *Banks and Solander* (the specimens imperfect and therefore omitted by R. Brown) ; Port Denison, *Fitzalan;* Rockingham Bay, *Dallachy;* between Cleveland and Rockingham Bays, *W. Hill;* Rockhampton, *Thozet.*

The species has a wide range over tropical Asia, including *L. foliosa,* Nees, Meissn. in DC. Prod. xv. i. 222, *L. consimilis,* Nees, Meissn. l.c. 223, *L. pulchella,* Meissn. l.c. 224, and probably some others of the Prodromus. The details of the flower in Nees's plate are very indifferent, and rather coarse in Wight's figures.

2. **L. dealbata,** *Nees. Syst. Laurin.* 630. A moderate-sized tree, the young shoots softly ferruginous-villous. Leaves petiolate, ovate-ellip-tical or almost oblong, acuminate, contracted at the base, 3 to 6 in. long, glabrous above with the primary veins slightly prominent, glau-cous underneath, the primary veins more prominent and glabrous or villous, few in number and the lowest pair usually longer and thicker. Flowers in sessile clusters, axillary or at the old nodes, the pedicels thick, 1 to 2 lines long. Perianth-segments 4, lanceolate, 1 to 1½ lines long, villous outside and the margins fringed with long ferruginous hairs. Filaments filiform, longer than the perianth, bearded with a few hairs below the middle ; the staminodia in the females small and irre-gular. Ovary slightly hairy; stigma broad and oblique. Fruit globular, 3 to 4 lines diameter, resting on the persistent perianth-tube enlarged into a small flat disk, 4-toothed with the persistent remains of the segments. —Meissn. in DC. Prod. xv. i. 224 ; *Tetranthera dealbata,* R. Br. Prod. 403.

Queensland. Rockingham Bay, *Dallachy.*
N. S. Wales. Hawkesbury and Hunter's river, *Beckler;* Hastings river, *Beckler;* Richmond river, *Henderson, Fawcett;* Clarence river, *C. Moore;* Illawarra, *A. Cunning-ham;* Sydney woods, Paris Exhibition, 1855, *M'Arthur, n.* 101.

Var. *rufa.* The rufous hairs more abundant and persistent on the branches and underside of the leaves. Brisbane river, Moreton Bay, *Fraser, F. Mueller, W. Hill, Leichhardt;* Blue Mountains, *Miss Atkinson.*

SUBORDER 2. CASSYTHEÆ.—Leafless parasitical twiners.   Flowers in spikes or racemes.   Anther-valves opening upwards.

The suborder is limited to a single genus.

7. CASSYTHA, Linn.

Flowers hermaphrodite (or semi-diœcious ?).   Perianth-segments 6, the 3 outer ones short broad and resembling the bracts, the three inner ones (when full grown) at least twice as long, almost valvate in the bud.   Stamens of the outer row 6, all perfect with introrse anthers, or rarely 3 of them (opposite the inner segments) reduced to linear or spathulate staminodia ; of the inner series 3 perfect with extrorse anthers (opposite the outer perianth-segments), alternating with 3 staminodia ; anthers all 2-celled ; glands 6, one on each side at the base of the inner perfect stamens.   Ovary free, scarcely immersed in the perianth-tube till after flowering ; stigma small, obtuse or capitate.   Fruit drupaceous, completely enclosed in the enlarged persistent and succulent perianth-tube, usually crowned by the small persistent segments.   Seed with a membranous testa.   Embryo with thick fleshy cotyledons, distinct at an early stage, but completely consolidated when ripe, assuming the appearance of a fleshy albumen at the base of which the plumula simulates an embryo.—Leafless parasites with filiform or wiry twining stems attaching themselves to living plants (usually shrubs or trees) by means of small *haustoria* (suckers).   Leaves replaced by minute scales.   Flowers small, sometimes very minute, in pedunculate or rarely sessile spikes, which are either reduced to close heads or elongated and interrupted, or in racemes ; each flower sessile or pedicellate within a minute scale-like bract, with 2 similar bracteoles close under the perianth.

The genus is chiefly Australian and more or less maritime, and the species enumerated are all endemic, with the exception of one which extends also over the tropical regions of Africa, Asia, and America.   There are besides one or two species from S. Africa, and one from Borneo which may be distinct, but require further investigation.

The anomalous habit of *Cassytha,* so exactly that of *Cuscuta,* has induced Lindley and others to propose it as a distinct natural Order, but the structure of the flower is so precisely that of *Cryptocarya,* that it has been again reunited with Laurineæ as a sub-order.   The fruit is also the same with the exception of the hard endocarp, and the cotyledons are so completely consolidated in the ripe seed, that Gærtner described them as a fleshy albumen, mistaking the plumula, which is at least as much developed as in other Laurineæ, for the embryo.   R. Brown pointed out this error, and Griffith and others figured the real embryo with a distinct line of separation between the two cotyledons.   In the dried fruits I had at my disposal I could not detect any such demarcation, and I should have followed Gærtner in describing the seed as albuminous had it not been for Brown's very decided contradiction, more especially as Colonel Champion in some sketches made from the living plant in Hong Kong seemed to confirm Gærtner's view.   On writing, however, to Dr. Thwaites in Ceylon, he has kindly examined fresh seeds, and fully corroborates Brown's and Griffith's statements, explaining the discrepancies by the circumstance that it is only at an early stage that the cotyledons are clearly distinct, the line of demarcation becoming obliterated long before maturity.

In several species the specimens show two forms of flower, always on different individuals, in the one the perianth-tube is exceedingly short, in the other it is globular and as long as the segments. In both, the stamens as well as the ovary appear to be perfect, usually more advanced in the latter than in the former; but I have not as yet found any intermediate state, a circumstance which suggests a certain degree of unisexuality. Nees has generally and Meissner occasionally considered the two forms as distinct species, the one with a rotate, the other with an urceolate perianth, in which view, however, I cannot concur.

Flowers sessile in a spike or head.
  Flowers capitate, very small (the spikes contracted into ovoid
     globose or few-flowered heads).
    Spikes sessile. Flowers very few . . . . 1. *C. nodiflora.*
    Spikes pedunculate. Flowers few, glabrous . . . . 2. *C. glabella.*
    Spikes pedunculate. Flowers rather numerous, densely pu-
     bescent . . . . 3. *C. flava.*
    (The spikes of 9, *C. micrantha*, 4, *C. pubescens*, and 8, *C. me-*
    lantha, are sometimes capitate when in bud.)
  Flowers spicate, the spikes when fully out oblong or elongated,
    the lower or all the flowers more or less distant.
    Flowers pubescent or villous. Ovary hirsute.
     Spikes short, almost capitate when young. Fruiting perianth
      globular . . . . 4. *C. pubescens.*
     Spikes elongated. Flowers all distant. Fruiting perianth
      obovoid or pear-shaped . . . . 5. *C. phæolasia.*
    Flowers glabrous or nearly so. Ovary glabrous.
     Flowers 1 to 1½ lines long. Spikes usually elongated, the
      flowers distant.
      Fruiting perianth with 6 raised ribs . . . . 6. *C. panicula a.*
      Fruiting perianth smooth, the ribs scarcely visible . . . 7. *C. filiformis.*
     Flowers 2 lines long. Spikes very short . . . . . . 8. *C. melantha.*
     Flowers ½ line long. Spikes short but slender . . . . 9. *C. micrantha.*
Flowers (when fully out) pedicellate in a raceme, sometimes almost
  shortened into an umbel.
  Stamens opposite the inner perianth-segments reduced to linear
    staminodia . . . . 10. *C. racemosa.*
  Stamens all perfect . . . . 11. *C. pomiformis.*

I have been unable to recognise *C. coronata*, Nees in Pl. Preiss. i. 620, from W. Australia, Preiss, n. 1627 (Meissn. in DC. Prod. xv. i. 254), described as diœcious, densely pubescent, with short recurved peduncles, short dense spikes, the stamens of the female flowers all reduced to broad petal-like staminodia.

1. **C. nodiflora,** *Meissn. in DC. Prod.* xv. i. 252. Quite glabrous Stems slender. Spikes closely sessile at the nodes, reduced to heads of 2 to 6 rather small sessile flowers quite glabrous. Bracts broad, obtuse, the margins often slightly fringed. Perianth nearly 1 line long. Fruit ovoid, nearly 2 lines long, but not seen perfect.

W. Australia, *Drummond, n.* 149 (in young fruit), *5th coll. n. 226 and 228* (in flower).

2. **C. glabella,** *R. Br. Prod.* 404. Glabrous. Stems filiform. Spikes pedunculate, reduced to heads of 3 to 6 or rarely more very small flowers. Perianth ¾ line long, quite glabrous as well as the bracts; in some specimens the tube undeveloped but the ovaries perfect, in others the flowers rather longer and the tube enclosing the ovary but the sta-

mens apparently perfect. Fruit ovoid, about 2 lines long.—Meissn. in
DC. Prod. xv. i. 254; Hook. f. Fl. Tasm. i. 318; F. Muell. Pl. Vict.
ii. t. 68; *C. dispar*, Schlecht in Linnæa, xx. 578, Meissn. l.c. 253; *C.
microcephala*, Meissn. l.c. 253; *C. casuarinæ*, Nees in Pl. Preiss. i. 619;
Meissn. l.c. 253.

N. Australia. Islands of the gulf of Carpentaria, *R. Brown;* Sturt's Creek,
F. Mueller ; Port Darwin, *Schulz.*
N. S. Wales. Port Jackson, *R. Brown, J. D. Hooker, Clowes, Woolls ;* Castle-
main, *Leichhardt;* Twofold Bay, *F. Mueller.*
Victoria. Portland and Glenelg river, *Robertson ;* near Melbourne, *Adamson;*
Dandenong Grampian and Victoria Ranges, *F. Mueller.*
Tasmania. Port Dalrymple, *R. Brown ;* abundant on the north coast, densely
matted over bushes, etc., *J. D. Hooker.*
S. Australia. Kangaroo Island, *Seeley.*
W. Australia. King George's Sound and adjoining districts, *R. Brown* and many
others, *Drummond* (3rd coll.?) *suppl. n.* 64, 68, *Preiss, n.* 1624; Murchison river,
Oldfield.

3. **C. flava,** *Nees in Pl. Preiss.* i. 620. Stems slender, frequently
filiform, more or less pubescent with whitish hairs, which are spreading
or rarely appressed under the spikes. Spikes or heads globular or ovoid,
consisting of very small closely sessile flowers usually more numerous
than in *C. glabella* and the peduncles longer. Bracts and outer segments of the perianth ciliate and hirsute ; inner segments under ¾ line
long, pubescent outside with yellowish hairs. Ovary hirsute. Fruit
not seen.—Meissn. in DC. Prod. xv. i. 253.

W. Australia, *Drummond,* (3rd coll.?) *suppl. n.* 62 *and* 67 ; Swan river, *Preiss,*
n. 1622 ; near Cape Riche, *Harvey.*

4. **C. pubescens,** *R. Br. Prod.* 404. Stems more robust than in
C. glabella, less so than in *C. melantha,* but variable, the young branches
and inflorescence shortly pubescent or villous. Spikes short, forming
when in bud close heads of few flowers, but lengthening out sometimes
to ¾ in., with the lower flowers rather distant, the peduncle always short.
Flowers usually about 1¼ lines long, pubescent outside, the inner segments appearing narrower in the fruiting specimens than in those in
which the tube remains undeveloped. Ovary villous. Fruiting perianth
ovoid, pubescent, about 2 lines long.—Schlecht. Linnæa, xx. 577;
Meissn. in DC. Prod. xv. i. 255; Hook. f. Fl. Tasm. i. 318; *C. piligera,*
Schlecht. Linnæa, xxi. 446; *C. tasmanica,* Meissn. l.c. 252; *C. rugulosa,*
Meissn. l. c. 255.

Queensland. Hervey Bay and Sandy Cape, *R. Brown.*
N. S. Wales. Port Jackson, very common, *R. Brown, J. D. Hooker, Fraser,
Woolls,* and others; Hastings river, *Beckler ;* New England, *C. Stuart.*
Victoria. Glenelg river, *Robertson;* Port Phillip, *Gunn;* Melbourne, *Adamson;*
Wilson's Promontory, *F. Mueller.*
Tasmania, *R. Brown.* Abundant in many parts of the island, *J. D. Hooker.*
S. Australia. From the Murray to St. Vincent's Gulf, *F. Mueller ;* Port Lincoln,
Wilhelmi.
W. Australia, *Drummond, n.* 147 ; Murchison river, *Oldfield* (the latter speci-
mens and others from *Drummond, n.* 10, very bad and therefore doubtful).

5. **C. phæolasia,** *F. Muell. Fragm.* v. 167. Very near *C. pubescens*

and *C. filiformis*, differing from the former in inflorescence, from the latter in indumentum and from both perhaps in the form of the fruit. Stems glabrous or nearly so. Spikes pubescent, interrupted, 1 to 2 in. long, the flowers all distant. Perianth in the specimens with unenlarged tube not 1 line long, hirsute, in those with enlarging ovary, the tube at first globular densely ferruginous-hirsute, the segments shorter than the tube, pubescent or nearly glabrous. Fruit obovoid or pear-shaped, nearly 2 lines long, but not seen quite ripe.

N. S. Wales. Twofold Bay, *F. Mueller.*
Victoria. Yarra river, Portland, and near Brighton, *F. Mueller.*

6. **C. paniculata,** *R. Br. Prod.* 404. Quite glabrous or with a scarcely perceptible pubescence on the inflorescence. Spikes pedunculate, elongated and sometimes branched, the flowers usually smaller and more distant than those of *C. filiformis*, but sometimes difficult to distinguish from them. Perianth quite glabrous or rarely with a few hairs. Fruiting perianth globular, with 6 raised longitudinal ribs.— Meissn. in DC. Prod. xv. i. 256; *C. glabella,* Sieb. Pl. Exs. not of R. Br.

Queensland. Moreton island, *F. Mueller.*
N. S. Wales. Port Jackson to the Blue Mountains, *R. Brown, Sieber, n.* 218, and others; Port Macquarrie, *Backhouse;* Clarence river, *Beckler.*

Var. *remotiflora.* Inflorescence very slightly pubescent.—*C. remotiflora,* F. Muell.; Meissn. in DC. Prod. xv. i. 256.—To this belong the Moreton island specimens.

Specimens from New Zealand may possibly belong to this species, but they are not in fruit.

7. **C. filiformis,** *Linn.; Meissn. in DC. Prod.* xv. i. 255. Glabrous or the young shoots and inflorescence slightly pubescent. Spikes interrupted, ½ to nearly 2 in. long, the flowers all distant and sessile or nearly so. Perianth 1 to 1½ lines long, glabrous except short cilia on the margins of the outer segments, the inner ones broad, the 3 outer stamens opposite the outer segments (as in most species) much broader than those opposite the inner ones, but all perfect. Ovary glabrous. Fruiting perianth globular, 2½ to 3 lines diameter, smooth, without prominent ribs.—*C. guineensis,* Schum.; Meissn. in DC. Prod. xv. i. 255; *C. americana,* Nees; Meissn. l.c.

N. Australia? Some specimens from Victoria river, *F. Mueller,* in an imperfect state may probably belong to this species.
Queensland. Albany island, Howick's group, and between Dawson and Burnett rivers, *F. Mueller;* Rockhampton, *O'Shanesy, Bowman.*
The species is widely spread over tropical Asia, Africa, and America, chiefly in maritime districts, for I am unable to detect any difference between our numerous African and American specimens and the Asiatic ones. The New Zealand specimens referred by Meissner to *C. paniculata* may possibly belong also to *C. filiformis;* they are in flower only, and without the fruit the two species cannot be distinguished with certainty. Both have occasionally branched spikes, although this be more frequent in *C. paniculata* than in *C. filiformis.*

8. **C. melantha,** *R. Br. Prod.* 404. Stems glabrous, stouter and flowers larger than in any other species. Spikes very short and shortly pedunculate, sometimes almost reduced to heads especially when in bud

and few-flowered, the fruiting rhachis varying from 3 to 6 lines long, the flowers sessile. Perianth about 2 lines long, pubescent with short appressed hairs or nearly glabrous. Fruiting perianth ovoid-globular, 3 to 4 lines long, the whole plant especially the flowers usually drying very black.—Meissn. in DC. Prod. xv. i. 254; Hook. f. Fl. Tasm. i. 317; *C. robusta*, Meissn. l. c. 255.

N. S. Wales. Port Jackson, *R. Brown*.

Victoria. Near Melbourne, *Harvey;* Corner Inlet, Dandenong and Buffalo Ranges, *F. Mueller*.

Tasmania. Abundant near Launceston, chiefly on Acacias, *J. D. Hooker*.

S. Australia. Lake Victoria, Mount Baker, Flinders Ranges, *F. Mueller;* Gawler Ranges, *Sullivan* (the latter specimens bad and doubtful).

W. Australia. S.W. Bay, on Acacias near the sea, *Oldfield*.

9. **C. micrantha,** *Meissn. in DC. Prod.* xv. i. 256. As *C. melantha* is the stoutest largest-flowered species so this is the most slender and smallest-flowered one, quite glabrous, with filiform stems. Peduncles 3 to 6 lines long, with few flowers either close together at the end or the lower ones distant. Perianth scarcely ½ line long. Fruit not seen.

W. Australia. *Harvey, Drummond* (3rd coll. ?) *suppl. n.* 61 *and n.* 152.

10. **C. racemosa,** *Nees in Pl. Preiss.* i. 621. Glabrous in the typical form, with slender stems. Spikes or racemes pedunculate with few flowers, sometimes crowded at the end of the peduncle, more frequently distant, pedicellate or rarely nearly sessile, small and glabrous. Perianth under 1 line long. Three stamens of the outer row, those opposite the inner perianth-segments, reduced to linear staminodia, nearly as long as the perfect stamens and sometimes slightly dilated above the middle, but always without anther-cells, the other stamens as in the rest of the genus. Fruiting perianth globular ovoid or obovoid, obscurely 6-ribbed, about 2 lines diameter.—Meissn. in DC. Prod. xv. i. 257; *C. Muelleri*, Meissn. l. c. 257; *C. digitata*, Nees in Pl. Preiss. i. 620; Meissn. l. c. 257.

Queensland. Moreton island, *F. Mueller*. Although from a station so distant from that of the rest of the species, and therefore distinguished under the name of *C. Muelleri*, I am unable to discover any character to separate it even as a variety.

W. Australia. King George's Sound and adjoining districts, *Harvey, Preiss, n.* 1623, *Drummond, n.* 133, 226, 229, *and suppl. n.* 68, *F. Mueller*.

Var. *pilosa*. Stems more or less pubescent, with loose spreading hairs. Racemes short, the pedicels rather long.—*C. subcapitata*, Meissn. in DC. Prod. xv. i. 253, partly. —W. Australia, *Drummond, n.* 203; King George's Sound and Gordon river, *Oldfield*.

When the pedicels are very short, *C. racemosa* might be mistaken for *C. glabella*, but it is always readily distinguished in all its forms by the staminodia instead of stamens opposite the inner perianth-segments, which I have not observed in any other species.

C. umbellata, Meissn. in DC. Prod. xv. i. 257, from towards Cape Riche, *Harvey*, seems to me to be rather a half-monstrous state of *C. racemosa*, var. *pilosa*, than a distinct species. The pedicels are most of them much elongated and thickened, sometimes 4 or 5 lines long, and irregularly clustered at the apex of a very short or of a long peduncle, but here and there in the specimens are a few racemes almost or quite normal.

11. **C. pomiformis,** *Nees in Pl. Preiss.* i. 620.　Rather slender and the typical form glabrous.　Spikes or racemes short and rigid, at first dense forming a small head, at length interrupted with the flowers rather distant and borne on short pedicels, the rhachis and pedicels much thickened after flowering.　Perianth about 1 line long.　Stamens opposite the inner segments narrow but perfect, those opposite the outer segments broad and almost petal-like.　Fruiting perianth small, ovoid, not ribbed.—Meissn. in DC. Prod. xv. i. 253 ; *C. ceratopoda*, Meissn. l.c. 257.

N. Australia, *Drummond, n* 151, and perhaps *n.* 150 (the latter in very young bud and doubtful) ; Swan river, *Preiss, n.* 1625 ; King George's Sound, *Harvey, F. Mueller.*

This typical form is near *C. glabella,* but as the flowering advances, the longer spike and thickened pedicels will readily distinguish it.

Var. *pubiflora.*　Flowers pubescent, with the short yellowish hairs of *C. flava,* but pedicellate and more or less racemose.—*C. subcapitata,* Meissn. in DC. Prod. xv. i. 253, partly ; *C. multiflora,* Nees, in Pl. Preiss. i. 621 ; Meissn. l. c. 253.—W. Australia, *Drummond, suppl. n.* 63 ; King George's Sound, *Preiss,* n. 2629.

Suborder 3.　Hernandieæ.—Trees or shrubs with perfect leaves. Flowers monœcious, the females with an involucel which enlarges and encloses the fruit.　Anther-valves opening laterally.　Seeds without albumen.

Formerly associated with *Inocarpus,* on grounds which now appear quite unintelligible, in a distinct order, *Hernandia* has been left isolated, and I had thought, from the characters given, that it had been related to *Euphorbiaceæ.*　The examination of the flowers, however, at once shows the correctness of the more recent views placing it in close connection with Laurineæ, with which Order Meissner was unwilling actually to unite it on account of the dehiscence of the anthers.　The difference is, however, in this respect but very small.　The shape of the stamens, their basal glands, the innate anther-cells closed by deciduous valves, are precisely the same, the only distinction being that in Laurineæ generally the valves are detached from the base upwards, and in *Hernandia* from the inner to the outer side.　Another difference consists in the inferior ovary, the perianth-tube being from the first adnate, whilst in *Laurineæ* it only becomes so as the fruit enlarges (e.g. in *Cryptocarya* and in *Cassytha*) or remains quite free.　On the whole, therefore, it appears to me that *Hernandia* and the closely allied if not congener *Hernandiopsis* are best placed as a Suborder of Laurineæ, separated by characters of scarcely more importance than those which distinguish *Cassytha.*

8. HERNANDIA, Linn.

Flowers monœcious.　Perianth-segments in two rows, valvate in each row in the bud, 3 or 4 in each row in the males and 4 or 5 in the females. Male fl.　Stamens as many as the outer perianth-segments and opposite to them, with a gland on each side at the base (or in species not Australian on one side only or none) ; anthers 2-celled, introrse, the valves separating laterally from the inner to the outer edge.　Female fl. inserted in a cup-shaped or lobed involucel.　Glands or staminodia as many as outer perianth-segments and opposite to them.　Ovary inferior, fleshy ; style short, thick, with a dilated irregularly toothed or lobed stigma.　Fruit somewhat fleshy or coriaceous, indehiscent, enclosed in

the enlarged fleshy or thickly membranous involucel.   Seed globular ;
testa thick and hard, without albumen.   Embryo with thick fleshy
deeply-lobed cotyledons.—Trees.   Leaves alternate, peltate or palmately
nerved.   Flowers in loose panicles on lateral peduncles at the ends of
the branches, each branch of the panicle terminating in an involucre of
4 or 5 verticillate bracts enclosing 3 flowers, the central one female,
sessile within the cup-shaped involucel, the 2 lateral ones males and
pedicellate.

The genus contains but few species, chiefly maritime, extending over the tropical
regions of the New as well as the Old World.   Of the two Australian species one has
a considerable range in the Old World, the other appears to be endemic.

Leaves peltate.   Involucel of the female flowers and fruits entire,
  truncate.   Male flowers 3-merous, females 4-merous .   .   .   .   .   .   1. *H. peltata.*
Leaves not peltate.   Involucel of the female flowers and fruit deeply
  2-valved.   Male flowers 4-merous, females 5 merous   .   .   .   .   .   2. *H. bivalvis.*

1. **H. peltata,** *Meissn. in D C. Prod.* xv. i. 263.   A large tree, with
a spreading head, glabrous or the inflorescence very slightly hoary-
tomentose.   Leaves on long petioles, broadly ovate, acuminate, peltately
attached near the base, 5- to 9-nerved and remotely penniveined, the
larger ones nearly 1 ft. long, the upper ones much smaller.   Panicles
shorter than the leaves, the flowers almost clustered on the branches,
one terminal female between two males within a whorl of 4 bracts, and
sometimes one or two males lower down with a small bract under each
pedicel.   Male perianth slightly pubescent, the segments 3 in each row,
almost petal-like, veined, about 2 lines long.   Stamens 3, shorter than
the segments with short filaments.   Female flowers with a cup-shaped
entire truncate involucel a little below the ovary and 1½ lines long at
the time of flowering, but soon enlarged and growing over the ovary
or perianth-tube.   Perianth-tube from the first completely adnate to
the fleshy ovary, segments 4 in each row, the outer ones ovate 2 lines
long, the inner ones narrow.   Glands 4, large and nearly globular.   Style
villous, thickened upwards, with a dilated oblique irregularly lobed
glabrous stigma, the whole style deciduous with the perianth-lobes.
Fruit completely enclosed in the involucel which has become inflated
globular, smooth and fleshy, above 1½ in. diameter, with a circular entire orifice of about ½ in. diameter.   Fruit about 1 in. diameter, more or
less distinctly marked with 8 broad raised longitudinal ribs, with a
raised terminal umbo.   Seed very hard, about ¾ in. diameter.   Embryo
divided into 4 or 5 thick fleshy ruminate lobes.—Seem. Fl. Vit. 205,
t. 32.

Queensland. Frankland islands, *M'Gillivray;* Dunk island, *Dallachy.*

The species extends over the seacoasts of the South Pacific and Eastern Archipelago,
westward to the Mascarene islands and northward to the Philippine islands and Loo
Choo.

2. **H. bivalvis,** *Benth.*   Leaves on long petioles, ovate or ovate-
lanceolate, acuminate, rounded or slightly cordate at the base but not
peltate, 3- or rarely 5-nerved, 4 to 6 in. long.   Inflorescence of *H.*

peltata, the involucre of 4 or 5 bracts, the central female flowers sessile, the two lateral male ones on short pedicels articulate below the middle. Involucel of 2 distinct broad concave bracts enclosing the perianth-tube or ovary. Perianth-segments usually 8 in the males, 10 in the females, about 3 lines long, in 2 rows (one of the inner rows deficient in one flower examined). Stamens 4, the filaments slender, with 2 glands. Style slender, glabrous, with a broad crenate stigma. Involucel enclosing the fruit nearly 2 in. long, very broad, cordate at the base, much inflated, of an almost membranous texture when dry and reticulate, but drying black, divided nearly to the base into 2 valves. Fruit about 10-ribbed, with a very small terminal umbo. Seed as in *H. peltata.*

Queensland. Brisbane river, *Fraser;* Wide Bay, *Bidwill;* Moreton Bay, *Herb. F. Mueller.*

ORDER CIV. **PROTEACEÆ.**

Flowers hermaphrodite or rarely partially unisexual. Perianth regular or irregular, deciduous, consisting of 4 segments valvately united in the bud, the claws forming a tube cylindrical or dilated towards the base, the laminæ short, forming a globular ovoid or rarely elongated limb; the segments at length separating either from the base upwards or revolute from the laminæ downwards, leaving a portion of the tube entire or open on one side, the laminæ sometimes cohering long after the segments have separated lower down. Stamens 4, opposite the perianth-segments and usually inserted on them, either with the filaments wholly adnate leaving the anthers sessile at the base of the laminæ, or the filaments shortly free below the laminæ, or very rarely the stamens entirely free from the perianth; anthers various, all perfect or rarely partially abortive, most frequently with 2 parallel cells adnate to a connectivum continuous with the filament. Hypogynous or perigynous glands or scales in many genera 4, alternating with the stamens, but in some genera variously united or reduced in number or wholly deficient. Ovary 1-celled, sessile or stipitate, more or less excentrical, with a single terminal undivided style, variously shaped at the end, with a small terminal oblique or lateral stigma. Ovules either solitary, or 2 collaterally attached or slightly superposed, or several imbricate in 2 contiguous rows, either pendulous and orthotropous, or more frequently laterally attached and more or less amphitropous, rarely erect and anatropous, the micropyle always inferior and frequently prominent from the incomplete development of the primine. Fruit either an indehiscent nut or drupe, or a more or less dehiscent coriaceous or woody follicle, very rarely a completely 2-valved capsule; either 1-celled and 1-seeded, or when 2 seeds are ripened in a drupe sometimes really 2-celled from the growth of the endocarp between as well as round the seeds, or when 2 or more seeds ripen in a follicle, apparently 2- or more-celled by the consolidation of the external coating of the 2 adjoining seeds into a membranous or woody plate detaching itself from the remainder of the seed. Seeds without albumen, the testa usually thin, rarely coriaceous or hard ; embryo straight, with fleshy cotyledons and a short inferior radicle.—Shrubs or trees, rarely undershrubs or even perennial herbs. Leaves alternate or scattered, in a very few genera strictly opposite or verticillate, but often crowded under the inflorescence so as to appear verticillate, usually coriaceous, often vertical with stomata on both sides, or in the same genera horizontal or narrow and terete, entire toothed or variously divided, without stipules. Flowers axillary or terminal, solitary or in racemes or spikes, often condensed into umbels heads or cones, each flower or pair of flowers subtended by a bract, very deciduous in some genera and perhaps sometimes really deficient, the pedicels always without bracteoles.

Proteaceæ, with their chief seat in Australia and South Africa, extend on the one hand to New Caledonia, the Indian Archipelago, and tropical Asia, chiefly eastern, to Japan, and on the other to South America. The seven tribes of the Order are all in Australia. Of the first four, constituting the Nucamentaceæ, the two principal ones, Proteeæ and Personieæ, are also in South Africa, but represented by different genera, the nine Australian ones being, as well as the four constituting the small tribes Conospermeæ and Franklandieæ, all endemic with the exception of a single New Zealand species of *Persoonia*, and a New Caledonian *Cenarrhenes*. None of the Nucamentaceæ are either in America or Asia, for the South American *Andripetalum* and *Guevina*, referred by Meissner to Persoonieæ, belong with *Helicia* and *Macadamia* to the Grevilleæ. Of the Australian genera of this tribe of Grevilleæ, *Helicia* is chiefly Asiatic, *Adenostephanus* is tropical American, with one New Caledonian species, and the large genus *Grevillea* has also a few New Caledonian species, the remaining eight genera are endemic. Of the Australian Embothrieæ, *Lomatia* extends to the Andes of South America, where it is accompanied by two nearly allied genera, and *Stenocarpus* to New Caledonia ; the two remaining genera, as well as the two which constitute the tribe Banksieæ, are endemic in Australia.

The clavate fusiform or disk-shaped end of the style in Proteaceæ is usually described as the stigma, and where it is more or less constricted it is said to be articulate, but I have never found any real articulation, and although the thickened style-end may be an essential aid in the collection or dissemination of the pollen, its surface is not stigmatic, the real stigma being usually very small, either on the point terminating the style-end, or in the centre of the disk, or quite lateral. The diversified mode in which in different genera the conformation of this style-end and its relation to the anthers promotes the dissemination of pollen whilst it impedes self-fertilization, upon which I have drawn up a few notes for the Linnean Society founded on the examination of dried specimens, would be an interesting study for local botanists who have the means of examining and watching the plants living in their native stations.

In the distribution of the numerous species of this most natural Order into tribes, genera, and sections, I have only had to follow, with slight modifications, the admirable arrangement proposed by Brown and further developed by Endlicher and Meissner; but in the great subdivision into Nucamentaceæ and Folliculares, these terms must not be taken strictly in their literal sense, for indehiscent drupes occur in both divisions. Taking however the fruit generally, in conjunction with the arrangement of the ovules and the inflorescence, neither of them again strictly constant, we have very fairly definite characters for two large groups which are both natural and to a certain degree geographical. For although both are abundant in Australia, the Nucamentaceæ alone are in Africa, and the Folliculares alone in Asia and America.

SUBORDER 1. **Nucamentaceæ.**—*Fruit an indehiscent nut or drupe. Flowers usually solitary within each bract.*

TRIBE 1. **Proteeæ.**—*Anthers all perfect or very rarely the upper one abortive, with 2 parallel cells adnate to the connectivum, inserted at the base of the short spreading laminæ of the perianth. Ovule 1 or rarely 2. Stigma terminal. Fruit a dry nut.*

Flowers in dense cone-like spikes or heads with imbricate scale-
like bracts, with few or many outer empty bracts forming an
involucre. Anthers free.
Cone-scales firmly adhering to the rhachis and opening for the
emission of the more or less flattened nuts . . . .   1. Petrophila.
Cone-scales either very deciduous or remaining closely imbricate after flowering till they fall off with the nuts which are
not flattened . . . .   2. Isopogon.
Flowers solitary within an involucre of 4 to 8 bracts . . . . .   3. Adenanthos.
Flowers in small heads with very small bracts. Anthers cohering
round the style and the adjoining cells of two different anthers
applied face to face in the bud forming a single cell . . . .   4. Stirlingia.

Tribe 2. **Conospermeæ.**—*Anthers : one with 2 perfect cells, two with 1 perfect
and 1 abortive cell, the fourth abortive, the perfect cells broad, concave, erect, without
any connective, the adjoining ones of distinct anthers applied face to face in the bud
forming a single cell, all on very short thick filaments at the base of the laminæ or
at the summit of the tube of the perianth. Ovule 1. Fruit a dry nut.*

Upper anthers abortive, replaced by a short membrane connecting
the filament with the disk-shaped stigma. Nut ovoid or oblong.
Leaves mostly divided . . . .   5. Synaphea.
Lower anther abortive. Stigma raised above the stamens on the
beak-like end of the style. Nut turbinate, flat and comose on
the top. Leaves entire . . . .   6. Conospermum.

Tribe 3. **Franklandieæ.**—*Anthers all perfect with parallel adnate cells enclosed in and adnate to the slender perianth-tube. Ovule 1. Fruit a dry nut with a
pappus-like cone.*

Single genus . . . .   7. Franklandia.

Tribe 4. **Persoonieæ.**—*Anthers all perfect, with parallel cells adnate to the connective, the stamens inserted at or below the middle of the perianth-segments. Ovules
2 or sometimes 1. Fruit a drupe or rarely a dry nut or membranous.*

Leaves divided or lobed. Flowers in interrupted spikes or racemes.
Ovules 2. Fruit dry, indehiscent.
Filaments inserted on the perianth, converging and united in a
ring round the style. Fruit a nut . . . .   8. Symphyonema.
Stamens free at the base of the perianth-segments. Fruit mem-
branous, flattened . . . .   9. Bellendena.
Leaves entire. Flowers in interrupted axillary spikes. Ovule
1. Fruit a 3-winged nut . . . . 10. Agastachys.
Leaves toothed. Flowers in interrupted axillary spikes. Ovule 1.
Fruit a drupe . . . . 11. Cenarrhenes.
Leaves entire. Flowers axillary or rarely forming a terminal or
infra-terminal raceme by the abortion of the floral leaves.
Ovules 2 or 1. Fruit a drupe . . . . 12. Persoonia.

Suborder 2. **Folliculares.**—*Fruit dehiscent, follicular or 2-valved, rarely (in the
first 2 genera), drupaceous and indehiscent. Flowers usually in pairs, with a single
bract to each pair, rarely (in Carnarvonia, Lambertia and Stenocarpus), the inflorescence anomalous.*

Tribe 5. **Grevilleeæ.**—*Ovules 2 or 4, collateral. Seeds without any intervening
substances or separated by a thin lamina or mealy substance. Flowers in racemes or
clusters, with deciduous or abortive bracts, or with an involucre of imbricate bracts.*

Ovules 2. Perianth regular or nearly so, small (under ½ in. except
in one species), the anthers on short filaments attached below
the laminæ Style cylindrical or clavate at the end.
Flowers pedicellate, in pairs, in racemes. Fruit with a thick
woody indehiscent pericarp or putamen.
Leaves alternate. Ovules ascending . . . . 13. Helicia.

Leaves verticillate. Ovules descending . . . . 14. MACADAMIA.
Flowers sessile, in pairs, in cylindrical spikes. Fruit thick
  and woody, tardily dehiscent. Leaves opposite . . . . 15 XYLOMELUM.
Flowers solitary or clustered on irregularly branched peduncles. Fruit a follicle. Leaves alternate, compound . 16. CARNARVONIA.
Flowers sessile or nearly so, in pairs, in cylindrical or oblong
  spikes. Fruit a follicle. Leaves alternate . . . . . 17. ORITES.
Ovules 2. Perianth long and narrow. Anthers linear. Flowers
  solitary, or 7 together in an involucre of persistent imbricate
  bracts. Leaves verticillate . . . . 18. LAMBERTIA.
Ovules 2. Perianth revolute in the bud or rarely straight and
  regular. Anthers short and sessile within the concave
  laminæ. Leaves alternate.
Ovules orthotropous, pendulous. (Fruit a drupe?) . . . . 19. ADENOSTEPHANUS.
Ovules amphitropous, laterally attached. Fruit a follicle.
  Seeds without wings or the wings short at both ends or
    annular. Inflorescence terminal, rarely also axillary . . 20. GREVILLEA.
  Seeds winged, chiefly or entirely at the upper end. Inflo-
    rescence axillary . . . . 21. HAKEA.
Ovules 4, collateral. Perianth revolute in the bud or straight
  and regular. Anthers short and sessile within the concave
  laminæ. Fruit a follicle. Leaves alternate.
Perianth revolute in the bud. Hypogynous gland unilateral
  or semiannular. Follicle short and broad . . . . . . 22. BUCKINGHAMIA.
Perianth straight. Hypogynous glands 4. Follicle oblong,
  recurved . . . . 23. DARLINGIA.

TRIBE 6. **Embothrieæ.**—*Ovules several, imbricate in 2 rows. Seeds usually sepa-*
rated by thin laminæ or a mealy substance.

Flowers in short compact racemes, surrounded by an involucre
  of imbricate coloured bracts . . . . 24. TELOPEA.
Flowers in loose racemes. Bracts small or deciduous.
  Hypogynous glands 3. Ovules imbricate upwards. Seeds
    winged at the upper end . . . . 25. LOMATIA.
  Hypogynous glands 4. Ovules imbricate downwards. Seeds
    with narrow wings all round . . . . 26. CARDWELLIA.
Flowers in umbels without bracts . . . . 27. STENOCARPUS.

TRIBE 7. **Banksieæ.**—*Ovules 2, collateral. Seeds separated either by a hard*
usually woody substance or by a membrane rarely wanting. Flowers in dense cones or
heads.

Flowers in ovoid or cylindrical cones, without any involucre . . 28. BANKSIA.
Flowers in heads surrounded by an involucre of imbricate bracts
  and floral leaves . . . . 29. DRYANDRA.

SUBORDER 1. NUCAMENTACEÆ.—Fruit an indehiscent nut or drupe, either 1-seeded or if 2-seeded the seeds separated by a complete dissepiment continuous with the endocarp. Flowers usually solitary within each bract, in cones or spikes or solitary, very rarely racemose, the spikes often shortened into heads.

TRIBE 1. PROTEEÆ.—Anthers all perfect, or very rarely the upper one abortive, with 2 parallel cells adnate to the connectivum, inserted at the base of the short spreading laminæ of the perianth. Ovule 1 or in a very few species a second one more or less developed. Stigma at the point of the straight style end. Fruit a dry nut.

1. PETROPHILA, R. Br.

Flowers hermaphrodite. Perianth regular, the tube slender, separating into 4 segments from the base or (in two sections) remaining united, the limb of 4 linear laminæ. Anthers all perfect and free, sessile at the base of the laminæ, usually linear, the connective produced into a small appendage. No hypogynous scales. Ovary sessile, with a single or very rarely 2 collateral ovules, pendulous from near the apex of the cavity, and orthotropous or slightly amphitropous. Style filiform, either dilated and truncate towards the end under a slender or continuous and fusiform *brush*, always glabrous below the brush, the brush usually shortly hispid or papillose, at least before the expansion of the flower, with a glabrous tip and terminal stigma. Fruit a small dry and indehiscent nut, usually compressed, sometimes winged, with a coma of long hairs on the margins or from the base only or also on one very rarely on both faces.—Shrubs with rigid entire or divided leaves, terete or if flat usually narrow. Flowers usually white or yellow, in dense spikes or *cones*, each flower sessile within a bract or scale; the cones globular ovoid oblong or rarely cylindrical, terminal or rarely axillary, the receptacle or rhachis woolly and usually cylindrical, the scales broad and hardened after flowering, persistent, at least at the base, and imbricate but not so closely so as in *Isopogon*, opening for the emission of the fruit, the thinner points of the scales often falling off after flowering. At the base of the cone are also several imbricate empty bracts forming an involucre sometimes larger than the scales and concealing them, usually smaller, persistent or deciduous. In several species new leaves and shoots form in the axils of the innermost of these empty bracts, which ultimately fall away, leaving the old cones sessile in the forks of the branches without empty outer bracts. Nuts usually shorter than the scales, the points rarely but the coma frequently protruding.

The genus is limited to extratropical Australia and is chiefly Western. Like the closely allied *Isopogon*, it differs chiefly from the South African genera of the same tribe in the absence of hypogynous scales. The part of the style which is here termed the brush, is usually considered as an upper article of the stigma, but I have never observed any real articulation separating it from the rest of the style, and it does not appear to be ever stigmatic except at the point.

Sect. 1. **Arthrostigma.**—*Leaves undivided. Cones terminal, usually large. Style thickened and truncate below the narrow villous or hirsute brush.*

Leaves terete (slightly grooved in *P. acicularis*).
　Scales of the cone not striate. Leaves usually rather thick.
　　Style-brush very densely and closely tomentose-villous, much longer than broadly turbinate style-end below it. Outer bracts free and narrow . . . . 1. *P. teretifolia.*
　　Style-brush densely hirsute with spreading hairs, rather longer than the narrow-turbinate style-end below it. Outer bracts rigid, linear-lanceolate . . . . 2. *P. longifolia.*
　　Style-brush loosely hirsute with spreading hairs, much longer than the broadly-turbinate style-end below it. Outer bracts very numerous, linear-subulate . . . . 3. *P. media.*

Scales of the cone more or less ribbed or striate. Leaves
    long and slender. Style and bracts of *P. media* . . .   4. *P. acicularis.*
Leaves flattened, rigidly linear. Style and bracts of *P. media*   5. *P. linearis.*

SECT. 2. **Xerostole.**—*Leaves flat, ternately divided or rarely entire. Cones
axillary, ovoid. Perianth-tube slender, usually falling off entire. Style thickened and
usually truncate under the narrow nearly glabrous brush.*

Nuts with broad wing-like margins, shortly comose at the base
    only.
  Leaves undivided or 3-fid, 2 to 4 in. long. . . . .   6. *P. heterophylla.*
  Leaves all divided, the segments mostly lobed or again
    divided.
    Leaf-segments broad and short (rarely narrow-linear in the
      lower leaves). Cone-scales villous. Perianth 8 to 10
      lines long. Style-end long and narrow below the brush   7. *P. biloba.*
    Leaf-segments linear. Cone-scales villous. Perianth 4 to
      5 lines long. Style-end shortly turbinate below the brush   8. *P. propinqua.*
    Leaf-segments linear. Cone-scales glabrous. Perianth 4
      to 5 lines long. Style-end only slightly thickened below
      the brush . . . .   9. *P. squamata.*
Nuts comose on the faces as well as on the margins, tapering
    upwards (uncertain in *P. colorata*). Style-end shortly
    turbinate below the brush.
  Outer cone-scales larger than the empty bracts, coloured and
    glabrous . . . .   10. *P. colorata.*
  Outer cone-scales small and villous, concealed under the
    large coloured and glabrous empty bracts . . . . .   11. *P. striata.*

SECT. 3. **Serrurioides.**—*Leaves divided, the segments terete or flat. Cones axillary, ovoid. Perianth-tube slender, usually falling off entire. Style continuous, fusiform.*

Leaf-segments terete, rigid, pungent, divaricate, the whole
    leaf usually 2 to 3 in. broad . . . .   12. *P. divaricata.*
Leaf-segments terete but grooved above, the whole leaf not
    exceeding 1 in.
  Leaf-segments erect or divaricate, fine but often pungent .   13. *P. Serruriæ.*
  Leaf-segments very close, compact and erect, not pungent .   14. *P. inconspicua.*
Leaf-segments flat, few, rather broad, the whole leaf usually
    2 to 4 in. long . . . .   15. *P. trifida.*

SECT. 4. **Symphyolepis.**—*Leaves flat, lobed or divided. Cones axillary or rarely
also terminal. Perianth-segments usually falling off separately. Style continuous,
fusiform.*

Leaves sessile, oblong-lanceolate, pinnatifid. Perianth-limb
    glabrous . . . .   16. *P. carduacea.*
Leaves petiolate, divided. Perianth glabrous. Cones often
    long.
  Nut broad, obtuse, the inner face glabrous . . . . . .   17. *P. Shuttleworthiana.*
  Nut tapering at the end, comose on both faces . . . . .   18. *P. macrostachya.*
Leaves variously divided. Perianth villous . . . . .   19. *P. diversifolia.*

SECT. 5. **Petrophyle.**—*Leaves divided or rarely simple, the segments terete (or flat
but narrow in the first two species). Cones terminal (or in the Eastern species also
axillary). Perianth-segments falling off separately. Style continuous, fusiform.*

Leaves flat, divided. Cones ovoid or oblong. (See above, 17.
    P. Shuttleworthiana and 18. *P. macrostachya*.)
Leaves flat, with 3 or 5 segments, or the lower ones entire.
    Cones broadly globular.

Leaves glabrous; segments long and divaricate. Perianth
　glabrous. Nut with comose margins . . . . 20. *P. biternata.*
Leaves plumose-hirsute ; segments small, on a long petiole.
　Perianth villous. Nut comose at the base only . . . . 21. *P. plumosa.*
Leaves or leaf-segments terete.
　Leaves crowded, ¼ to ¾ in. long.
　　Leaves undivided. Perianth glabrous. Nuts with comose
　　　margins . . . . 22. *P. ericifolia.*
　　Leaves pinnate. Perianth villous. Nuts comose at the
　　　base only . . . . 23. *P. chrysantha.*
　Leaves more than 1½ in. long
　　Eastern species. Cones usually 2 or 3 at the end of the
　　　branches, often axillary when old. Leaves not pungent.
　　　Cones pedunculate. Perianth glabrous . . . . . . 24. *P. pedunculata.*
　　　Cones sessile. Perianth silky-villous.
　　　　Foliage glabrous. Cones oblong . . . . 25. *P. pulchella.*
　　　　Young shoots silky or hoary. Cones ovoid . . . . 26. *P. sessilis.*
　　Western or Southern species. Cones solitary, terminal or
　　　in the forks.
　　　Perianth glabrous. Cones ovoid or globular.
　　　　Leaf-segments numerous, fastigiate, not pungent . . 27. *P. fastigiata.*
　　　　Leaf-segments divaricate, pungent-pointed, the lower
　　　　　leaves sometimes entire . . . . 28. *P. seminuda.*
　　　Perianth villous. Leaf-segments divaricate and pungent-
　　　　pointed.
　　　　Outer bracts large and imbricate, concealing the cone-
　　　　　scales. Cone nearly flat-topped, 1½ in. diameter . 29. *P. circinata.*
　　　　Cones large, ovoid-globular, glutinous.
　　　　　Perianth viscid. Fruiting cones globular . . . . 30. *P. Drummondii.*
　　　　　Perianth not viscid. Fruiting cones ovoid or oblong 31. *P. crispata.*
　　　　Cones ovoid, oblong or cylindrical, not viscid.
　　　　　Cones ovoid, chiefly terminal. Branches glabrous . 32. *P. rigida.*
　　　　　Cones ovoid-oblong or cylindrical, chiefly in the
　　　　　　forks.
　　　　　　Leaf-segments numerous and short. Branches
　　　　　　　glabrous . . . . 33. *P. multisecta.*
　　　　　　Leaf-segments few and long. Branches tomentose 34. *P. conifera.*

SECT. 6. **Hebegyne.**—*Leaves terete, simple or 2- or 3-lobed at the end. Cones
terminal or in the forks. Perianth-segments falling off separately. Style pubescent,
thickened towards the end, but scarcely fusiform.*

Single species . . . . 35. *P. semifurcata.*

P. Roei, Endl. Gen. Suppl. iv. 75, has never been described, and is most probably
the same as one of the species here enumerated.

SECT. 1. ARTHROSTIGMA, Endl.—Leaves undivided. Cones terminal, usually large. Style thickened and truncate below the narrow
villous or hirsute brush. Perianth-segments usually separating from
the base.

1. **P. teretifolia,** *R. Br. in Trans. Linn. Soc.* x. 68, *Prod.* 364. A
shrub, either erect and tall or sometimes diffuse, glabrous except the
cones. Leaves terete, not grooved, somewhat thickened upwards or
the lower ones slender, 1½ to 2 in. long in some specimens, 4 to 8 in.
long in others. Cones terminal, sessile or very shortly pedunculate
above the last leaves, nearly globular or at length almost ovoid, ½ to ¾

in. diameter without the perianths. Outer bracts not numerous, not exceeding the scales, a few of the outermost lanceolate-subulate and rigid, the others broader and passing into the cone-scales, which are broad, scarcely acuminate, glabrous except at the base or shortly ciliate, 2 lines diameter when in flower, 3 to 4 lines when in fruit. Perianth about 8 lines long, villous with hairs at first silky at length spreading, the segments usually falling off separately. Style-end glabrous, broadly turbinate and truncate below the narrow-cylindrical, very densely and closely tomentose-villous brush, the glabrous tip short. Nut broad and much flattened, the long marginal coma protruding beyond the cone-scales, the inner face hairy, the outer one glabrous.—Meissn. in Pl. Preiss. i. 492 and in DC. Prod. xiv. 268; *P. crassifolia,* R. Br. Prot. Nov. 5; Meissn. ll. cc.

W. Australia. Lucky Bay, *R. Brown, Baxter ;* towards Cape Riche, *Preiss, n.* 630, *Drummond,* 4*th coll. n.* 259 *and* 260; Stirling and Russell Ranges, Cape Arid and Israelite Bay, *Maxwell.*

The differences observed in Baxter's specimens, on which *P. crassifolia* was founded, appear to me to be owing to the specimens being in a more advanced fruiting state.

2. **P. longifolia,** *R. Br. Prot. Nov.* 5. A shrub with the leaves terete and undivided as in *P. teretifolia,* but usually 6 to 10 in. long. Cones terminal and sessile within the last long leaves, broadly ovoid-conical, ¾ to 1 in. diameter without the perianths. Outer bracts rigid, linear-lanceolate, 4 to 5 lines long. Cone-scales broad, not striate, glabrous except at the base, more or less acuminate and not ciliate. Perianth rather slender, 8 or 9 lines long, hirsute, the segments falling off separately. Style-end below the brush glabrous, oblong-turbinate, nearly as long as the brush which is densely hirsute but with hairs more distinct and spreading than in *P. teretifolia.* Nut as broad as long, with comose margins, the outer face glabrous, the inner one convex and covered with long hairs.—Meissn. in Pl. Preiss. i. 493 and in DC. Prod. xiv. 269.

W. Australia. Dry stony and gravelly places, King George's Sound and adjoining districts, *Baxter, Drummond,* 2*nd coll. n.* 241, *Preiss, n.* 623, 625, *F. Mueller,* and others ; eastward to the Mount Barren ranges, *Maxwell.*

Var. *tenuifolia.* Leaves longer and more slender, the cones and flowers smaller, the cone-scales broader and less acuminate.—Kalgan river, *Oldfield.*

3. **P. media,** *R. Br. Prot. Nov.* 5. A shrub with the habit of *P. teretifolia.* Leaves similarly terete and rather thick, varying very much in length even on different branches of the same specimen, mostly 2 to 3 in. long but in some specimens all under 2 in., in others 4 to 10 in. long, obtuse or when long with recurved points. Cones terminal, sessile, ovoid, 6 to 7 lines diameter without the perianths when full-grown. Outer bracts linear-subulate, much more numerous and longer than in *P. teretifolia,* and often whitish in the dried state. Cone-scales ovate-lanceolate, not at all or very obscurely ribbed, with acuminate often reflexed or at length deciduous summits, not ciliate. Perianth 7 or 8 lines long, villous, the segments falling off separately. Style-end below

the brush glabrous, broadly turbinate, much shorter than the narrow brush which is hirsute, with the hairs short and spreading as in *P. longifolia*, but not nearly so dense, and the glabrous tip longer.— Meissn. in Pl. Preiss. i. 492, ii. 245 and in DC. Prod. xiv. 268; *P. brevifolia*, Lindl. Swan Riv. App. 35; Meissn. in Pl. Preiss. i. 491 and in DC. Prod. xiv. 268.

W. Australia. King George's Sound, Mount Gardner, *Baxter;* N. of Stirling Range, *F. Mueller*, and from thence to Swan river, *Drummond, n.* 14, 39, *2nd coll. n.* 293, *3rd coll. n.* 240 (*or* 241 ?); *Preiss, n.* 628, 629; Champion Bay and Port Gregory, *Oldfield*. In all the above specimens the leaves are mostly under 4 in. long and some-times under 2 in.

Var. *juncifolia*. Leaves rather more slender, 6 to 10 in. long and often with a hooked or recurved point.—*P. juncifolia*, Lindl. Swan Riv. App. 35; Meissn. in Pl. Preiss. i. 493; and in DC. Prod. xiv. 269.—Swan river, *Drummond, 1st coll. n.* 553 ; *Preiss, n.* 621, 622, 624.

The species is readily distinguished from the two preceding ones by the outer bracts. The style-end below the brush is short as in *P. teretifolia*, but the brush itself is thinly hirsute, not densely and closely villous or *tow-like* (*stuposus*) as in that species.

4. **P. acicularis,** *R. Br. in Trans. Linn. Soc.* x. 69, *Prod.* 364. A low erect nearly simple or tufted shrub of about 2 ft., glabrous except the cones. Leaves undivided, usually slender, terete and more or less distinctly grooved on the upper side, obtuse or with a short straight or curved point, 2 to 6 in. long. Cones terminal and sessile, nearly glo-bular, about ¾ in. diameter, the outer bracts rigidly linear, slightly broad at the base, often numerous. Cone-scales from broadly ovate to ovate-lanceolate, more or less acuminate, the summits sometimes de-ciduous, the hardened portion striate with several nerves especially prominent after flowering. Perianth of *P. media*. Style-end shortly turbinate below the brush, scarcely half as long as the narrow brush which is not very densely hirsute with short spreading hairs. Nut broad and flat, the margins comose with long hairs, the outer face glabrous, the inner villous.—Meissn. in Pl. Pr. i. 494 and in DC. Prod. xiv. 268; Bot. Mag. t. 3469; *P. filifolia*, R. Br. ll. cc., Meissn. ll. cc.

W. Australia. King George's Sound, common, *R. Brown, A. Cunningham, Baxter, Drummond, n.* 181 (*or* 184?), *3rd coll. n.* 242, *Preiss, n.* 626, *Oldfield, F. Mueller*.

This species has the bracts and the style nearly of *P. media*, but the longitudinal ribs of the cone-scales are always more prominent and the leaves more slender.

5. **P. linearis,** *R. Br. Prot. Nov.* 6. An erect shrub of about 2 ft., glabrous except the cones or hoary-glaucous. Leaves linear, flat but thick, 1 to 2 or rarely 3 lines broad above the middle, contracted into a terete base or petiole, mostly incurved towards the end or falcate, obtuse or with a straight or curved point, 1½ to 4 in. long. Cones globular or ovoid, sessile at the ends of the branches, ½ to ¾ in. diameter with or without the perianths. Outer bracts as in *P. media*, linear-subulate and often numerous. Cone-scales broad, not ciliate, smooth or when old slightly striate. Perianth at least 1 in. long, very densely villous with silky-white or ferruginous hairs, the segments falling off separately. Style-

end narrow-turbinate below the brush, rather shorter than the brush, which is not very densely hirsute with short spreading hairs. Nut broad and flat, the margins comose with long hairs, the outer face glabrous, the inner villous.—Meissn. in Pl. Preiss. i. 494 and in DC. Prod. xiv. 267.

W. Australia. Swan river, *Fraser, Drummond, 1st coll. n. 558, Preiss, n.* 636; Vasse and Swan rivers, *Oldfield.*

Var. *anceps.* Leaves straighter with thinner margins; flowers rather smaller. –*P. anceps,* R. Br. Prot. Nov. 5; Meissn. in DC. Prod. xiv. 267.—W. Australia, *Drummond, 4th coll. n.* 261, *5th coll. n.* 394; King George's Sound, *Baxter;* foot of Stirling Range, *F. Mueller.*

SECT. 2. XEROSTOLE.—Leaves flat, ternately divided or rarely entire. Cones axillary, ovoid. Perianth-tube slender, usually falling off entire or shortly splitting into four at the base. Style thickened and usually truncate at the end under the narrow nearly glabrous brush.

The inflorescence and perianth are nearly those of the section *Serrurioides,* but the style is that of *Arthrostigma,* except that the brush is only very minutely papillose-pubescent or quite glabrous, although it appears to collect the pollen as in other sections.

6. **P. heterophylla,** *Lindl. Swan Riv. App.* 35. A rigid shrub, glabrous except the cones. Leaves linear or linear-lanceolate, more or less flattened, entire and acute or dilated towards the end and then often truncate or notched with a small point in the notch, or deeply divided into 2 or 3 lobes, the whole leaf 2 to 4 in. long, rigid and veined, much narrowed to the base. Cones all axillary, sessile, ovoid-oblong, ¼ to ½ in. long without the perianths, the old ones often ¾ in. long. Outer bracts numerous and imbricate, the outermost small, the inner ones passing into the cone-scales of which the outer ones are broadly ovate and ciliate only, the inner ones smaller and more villous. Perianth slender, 9 to 10 lines long, hairy, the tube falling off entire. Ovary glabrous except the hairs at the base. Style-end clavate truncate and 4-angled below the narrow glabrous brush. Nut flat, dilated into wing-like margins, often notched at the top, glabrous except a coma of short hairs at the base.—Meissn. in Pl. Preiss. i. 501, ii. 246, and in DC. Prod. xiv. 274.

W. Australia. Swan river to King George's Sound, *Drummond, 1st. coll. n.* 571, *3rd coll. n.* 244; Stirling Range, *F. Mueller.*

7. **P. biloba,** *R. Br. Prot. Nov.* 7. A shrub, with the branches not very stout, the young shoots tomentose-pubescent and villous with spreading hairs, the older foliage and branches glabrous. Leaves, of the flowering branches at least, usually small, very rigid and flat, petiolate, 3-lobed or pinnately 4-lobed, the rhachis terminating in a point or small lobe, the lower lobes sometimes again lobed and the lobes all very obliquely ovate rhomboid or oblong, pungent pointed, under ½ in. long, but the lower leaves in some specimens and nearly all in others crowded and more divided into very narrow linear lobes. Cones small, ovoid, sessile in the axils, often numerous and crowded along the branches,

scarcely above ¼ in. long when in fruit. Outer bracts small and not numerous. Cone-scales silky-villous, with small glabrous tips. Perianth very villous with spreading hairs, rather slender, 8 to 10 lines long, the tube usually falling off whole. Style-end below the brush long and narrow, rather clavate than turbinate, the brush shorter, filiform, minutely papillose or glabrous. Nut flat with wing-like margins, obovate-orbicular, glabrous except a short coma at the base.—Meissn. in Pl. Preiss. i. 500, and in DC. Prod. xiv. 273.

W. Australia. Swan river, *Fraser, Drummond, 1st coll. n. 566, Preiss, n.* 656 ; Mount Toodyay, *Oldfield.*

8. **P. propinqua,** *R. Br. Prot. Nov.* 7. A shrub of 3 or 4 ft., glabrous except the cones, or the young shoots minutely hoary, the branches rather slender. Leaves with long petioles, twice trifid or pinnate with the lower pinnæ again divided, the segments flat, linear or linear-lanceolate, mostly pungent-pointed, about ½ in. long or rather longer when narrow. Cones small, ovoid or at length globular, sessile in the axils, not ½ in. long without the perianths. Outer bracts nearly glabrous, small, acute, rigid. Cone-scales villous or with very small glabrous tips. Perianth 4 to 5 lines long, very villous with spreading hairs, the tube falling off entire. Style-end broadly turbinate 4-angled and truncate under the narrow terete almost glabrous brush. Nut flat with broad wing-like margins, broadly obovate, 2 lines long, glabrous except a few hairs forming a short coma at the base.—Meissn. in Pl. Preiss. i. 501, and in DC. Prod. xiv. 273.

W. Australia. Swan river, *Fraser, Drummond, 1st coll n.* 567.

Var. *sericiflora.* A stouter shrub. Leaves more divided, rigid, pungent-pointed, the segments ¼ to 1 in. long. Perianth rather smaller and more silky-villous.—East Shoal Cape and Cape Arid, *Maxwell.*

9. **P. squamata,** *R. Br. in Trans. Linn. Soc.* x. 70, *Prod.* 365. A shrub of 2 or 3 ft., glabrous except the cones, or the young branches slightly tomentose. Leaves on rather long petioles, once or twice ternately divided, with flat linear rigid pungent-pointed segments, sometimes very short and rather broad, sometimes very narrow and ¼ to ½ in. long. Cones small, ovoid, sessile in the axils, almost globular when in fruit, not above ½ in. long. Outer bracts small. Cone-scales acute, rigid but smooth and sometimes almost scarious, glabrous or slightly ciliate, pubescent at the base only. Perianth slender, silky-villous with short hairs, under ½ in. long, the tube falling off entire. Style-end somewhat thickened and glabrous under the rather long filiform nearly glabrous brush. Nut flat, broad, with wing-like margins, slightly pubescent, with a very short coma at the base.—Meissn. in DC. Prod. xiv. 272 ; *P. Cunninghamii,* Meissn. in Pl. Preiss. i. 499, ii. 245 ; *P. gracilis,* A. Cunn. Herb.

W. Australia. King George's Sound and adjoining districts, *Menzies, Fraser, Preiss, n.* 651, 652, *F. Mueller ;* Vasse and Gordon rivers, *Oldfield ;* Clay flats, Willy-ungup, *Maxwell.*

Meissner's varieties *major* and *gracilis* appear to me to be old and young plants or branches of the same plant rather than distinct varieties.

10. **P. colorata,** *Meissn. in Pl. Preiss.* ii. 246, *and in DC. Prod.* xiv. 273. A glabrous shrub with the habit and foliage of *P. squamata.* Cones also as in that species ovoid and sessile in the axils ; outer bracts ovate-lanceolate, acuminate, pale-coloured, much smaller than the scales; cone-scales much longer than in *P. squamata,* coloured and glabrous or ciliate towards the base, the outer ones ovate, the inner lanceolate, the larger ones fully 3 lines long. Perianth slender, silky-villous, the tube falling off entire. Style-end rather broadly turbinate angular and truncate under the filiform brush as in *P. propinqua.* Nut not seen ripe, when young it appears to be comose on the margins and inner face.

W. Australia, *Drummond,* 2nd *coll. n.* 296.

11. **P. striata,** *R. Br. Prot. Nov.* 6. A rigid shrub of 1 to 2 ft., the young shoots pubescent and sprinkled with long fine hairs, otherwise glabrous except the cones. Leaves petiolate, once or twice pinnate, the segments linear or cuneate, narrow or broad, entire or lobed, divaricate rigid and pungent-pointed, the whole leaf with the petiole 1½ to 2½ in. long, and nearly as broad. Cones sessile in the upper axils, ovoid and under ½ in. long without the perianths, more globular and ½ in. diameter when in fruit. Outer bracts numerous, imbricate, glabrous, almost membranous, gradually enlarged, the inner ones 3 or 4 lines long concealing the scales. Cone-scales villous, the outer ones shortly ovate, the inner longer and lanceolate, Perianth ¾ in. long, silky-villous, the tube long and slender, falling off entire, the lobes tipped with horn-like glabrous appendages nearly 1 line long. Style-end turbinate clavate angled and truncate under the slender nearly glabrous brush. Nuts lanceolate, tapering into a long beak, comose all over near the base, often 4 or 5 lines long including the beak.—Meissn. in Pl. Preiss. i. 502, ii. 246, and in DC. Prod. xiv. 275.

W. Australia. Swan river, *Fraser, Drummond,* 1st *coll. n.* 565, *Preiss, n.* 639, 640, *Clarke.*

SECT. 3. SERRURIOIDES.—Leaves divided, the segments terete or flat. Cones axillary, ovoid. Perianth-tube slender, usually falling off entire. Style-end continuous, fusiform, usually shortly hirsute at the angles with reflexed hairs.

The plants of this section show the nearest approach to *Isopogon,* the perianth and style are nearly the same as in some species of that genus, but the cone-scales and nuts are those of *Petrophila.*

12. **P. divaricata,** *R. Br. Prot. Nov.* 7. A shrub attaining 3 to 6 ft., the branches and young leaves often bearing long fine spreading hairs, otherwise glabrous except the cones. Leaves twice pinnate, the pinnæ and segments very divaricate, terete, pungent-pointed, rigid, but not thick. Cones sessile in the upper axils or within the last leaves, ovoid or oblong, usually about ¾ in. long and ½ in. diameter without the perianths. Outer bracts deciduous, rather broad, acutely acuminate, glabrous outside, silky inside, the young shoots often protruding from within them whilst the cone is still in flower. Cone-scales broad, villous

with rather long acute silky-hairy points. Perianth rather slender, $\frac{1}{2}$ to $\frac{3}{4}$ in. long, conspicuously silky-villous with short yellow hairs, the tube usually falling off entire. Style-end continuous, fusiform with reflexed hairs on eight prominent longitudinal lines. Fruiting cones with the scales closely imbricate, becoming more glabrous, but retaining short points. Nuts with wing-like margins, truncate on the top, ciliate with short hairs especially at the base, otherwise glabrous.—Meissn. in Pl. Preiss. i. 498, and in DC. Prod. xiv. 272 ; *P. intricata,* Lindl. Swan. Riv. App. 35.

W. Australia. Swan river, *Drummond, 1st coll. n.* 568 ; King George's Sound and adjoining districts, *Baxter, Preiss, n.* 646, *Oldfield, Maxwell, F. Mueller.*

13. **P. Serruriæ,** *R. Br. Prot. Nov.* 6. An erect shrub of 3 or 4 ft., not much branched, the young branches and leaves silky-pubescent or sprinkled with long fine spreading hairs, the older foliage more glabrous. Leaves small, crowded, twice or thrice pinnate with the lower segments close to the base, the segments numerous, slender, terete, grooved above, erect or divaricate, softly acute or pungent-pointed, the whole leaf rarely above 1 in. long and broad. Cones ovoid or at length globular, sessile or shortly pedunculate in the upper axils, often crowded in terminal clusters, scarcely above $\frac{1}{2}$ in. diameter when in fruit. Outer bracts few and small, glabrous outside. Cone-scales villous at the base, with glabrous deciduous tips. Perianth slender, silky or ferruginous-villous, about 5 or 6 lines long, the tube usually falling off entire, the laminæ short, tipped with small glabrous points or stipitate glands sometimes very prominent, sometimes concealed within the hairs. Style-end continuous, fusiform, more or less hirsute with reflexed hairs in longitudinal lines. Nut rather narrow, tapering into the style, glabrous on the back, the inner face and sides comose with long hairs chiefly from the base.—Meissn. in Pl. Preiss. i. 497 and in DC. Prod. xiv. 271 ; *P. glanduligera,* Lindl. Swan. Riv. App. 35 ; Meissn. in Pl. Preiss. i. 498, and in DC. Prod. xiv. 271 ; *P. axillaris,* Meissn. in Hook. Kew Journ. vii. 68, and in DC. Prod. xiv. 275.

W. Australia. King George's Sound and adjoining districts, *Baxter, Fraser, F. Mueller, Oldfield, Maxwell ;* and thence to Vasse and Swan rivers, *Oldfield, Drummond, 1st coll. n.* 569, *Preiss, n.* 641 ; between Moore and Murchison rivers, *Drummond, 6th coll. n.* 166.

Some of the northern specimens (*P. glanduligera*) have the leaves more silky-hairy and less pungent, but others are quite like the southern ones.

14. **P. inconspicua,** *Meissn. in Hook. Kew Journ.* vii. *68, and in DC. Prod.* xiv. 272. A shrub with the aspect almost of an *Adenanthos* or of *Isopogon adenanthoides,* the young shoots tomentose-pubescent and sprinkled with long fine spreading hairs. Leaves short, crowded, pinnate with compact narrow terete segments, grooved on the upper side, minutely pointed but not pungent, the whole leaf about $\frac{1}{2}$ in. long. Cones in our specimens crowded in leafy tufts at the ends of the branches, but all in an advanced state, the outer bracts apparently fallen off. Cone-scales linear or lanceolate, thin and flat, villous outside. Perianth very

slender, hirsute, nearly 1 in. long, the tube falling off entire except the lower glabrous portion which is more persistent as in *Isopogon*. Style-end continuous, fusiform, slightly thickened at the base, and hirsute with a few reflexed hairs disappearing after the flowering is over. Nut, according to Meissner, nearly flat, oval, with a short obtuse terminal wing, glabrous with ciliolate margins. I have only found young fruits which appeared to me to be comose all over, as in *Isopogon*.

W. Australia. Between Moore and Murchison rivers, *Drummond, 6th coll. n.* 172. This species has so much of the character of *Isopogon*, that I should at once have transferred it to that genus were it not for the uncertainty which prevails about the shape and indumentum of the nut, besides that the style is much more that of the section *Serrurioides* of *Petrophila* than of *Isopogon adenanthoides*, which is the nearest to the present species in *Isopogon*.

15. **P. trifida,** *R. Br. in Trans. Linn. Soc. x. 70, Prod.* 365. A rigid shrub, the young shoots bearing a few fine spreading hairs, otherwise glabrous except the cones. Leaves on long flattened petioles, cuneately 3-fid or deeply pinnatifid, with few usually broad rigid pungent-pointed segments, the lower ones sometimes 1 in. long when narrow, the whole leaf with the petiole 2 to 4 in. long. Cones small, ovoid or nearly globular, sessile in the axils or terminating very short axillary branches, not ½ in. long without the perianths, or, when in fruit, ½ to ¾ in. long. Outer bracts glabrous small and narrow. Cone-scales broad, glabrous in the lower part, densely villous round the obtuse end. Perianth silky-villous, about ½ in. long, the tube slender, usually falling off entire. Style-end continuous, fusiform, hirsute with a few reflexed hairs in longitudinal rows. Nut flat, ovate, 2 to 3 lines long, with broad wing-like margins, glabrous except a short coma at the base.—Meissn. in Pl. Preiss. i. 501, and in DC. Prod. xiv. 273, but not the plate quoted of Lodd. Bot. Cab.

W. Australia. Lucky Bay, *R. Brown;* towards Cape Riche? *Drummond, 1st coll. n.* 576.

Sect. 4. SYMPHYOLEPIS, Endl.—Leaves flat, lobed or divided, the segments broad or also narrow. Cones axillary or rarely also terminal. Perianth-segments usually falling off separately. Style-end continuous, fusiform.

This section has the essential characters of *Petrophile*, differing generally but not absolutely in foliage and inflorescence and in the axis of the cones and the base of the cone-scales usually more hardened when in fruit.

16. **P. carduacea,** *Meissn. in DC. Prod.* xiv. 274. A shrub with the young branches tomentose-pubescent or villous with spreading hairs, the adult foliage glabrous. Leaves sessile, oblong-lanceolate, pinnatifid or deeply toothed and undulate, the lobes or teeth broadly triangular, pungent-pointed, the lowest pair of lobes rather smaller, more deeply separated and occasionally toothed, having the appearance of stipules. Cones axillary, pedunculate, at first small and globular, ovoid or ovoid-oblong when in fruit, above 1 in. long and ¾ in. diameter. Bracts small along the peduncles and a few close under the cone, all as

well as the cone-scales glabrous.　Perianth scarcely above 4 lines long, the tube silky-villous and readily separating into segments, the limb broader and glabrous.　Ovary hairy.　Style-end continuous, fusiform, with a few reflexed hairs.　Scales of the fruiting cone very broad.　Nut very flat, with wing-like margins, 3 lines long and broad, pubescent with short hairs, with a short coma at the base.

W. Australia.　*Drummond, 4th coll. n.* 262 ; Stirling Range, *F. Mueller.*

17. **P. Shuttleworthiana,** *Meissn. in Pl. Preiss.* ii. 246, *and in DC. Prod.* xiv. 275.　A rigid shrub, glabrous except the cones.　Leaves on long petioles, flattened upwards, cuneate and deeply 3-fid, the segments broad or narrow, often above 1 in. long, entire or 2-3-fid, the lobes all very rigid and pungent-pointed.　Cones sessile or shortly pedunculate, terminal or in the upper axils, oblong or cylindrical, $\frac{3}{4}$ to 1 in. long, $\frac{1}{4}$ in. diameter without the perianths, or when in fruit and perfect twice as long and thick, but often partially abortive and remaining short.　Outer bracts very deciduous.　Cone-scales at first small, densely villous outside, glabrous inside with small lanceolate glabrous deciduous tips, very broad and glabrous in the old cones.　Perianth glabrous, 4 or 5 lines long, the segments falling off separately.　Style-end continuous, fusiform, shortly and sparingly hirsute.　Nut broad, not winged but the margins acute, the margins and outer face comose, the inner glabrous.

W. Australia.　*Drummond, 2nd coll. n.* 298 ; Murchison river, *Oldfield.*

18. **P. macrostachya,** *R. Br. Prot. Nov.* 7.　A shrub of 2 or 3 ft., the young branches tomentose or villous, the adult foliage glabrous. Leaves on rather long petioles, once twice or three times deeply trifid, the lobes or segments broad or narrow, veined, rigid, pungent-pointed, the whole leaf 1½ to 3 in. long.　Cones sessile in the axils and sometimes also terminal, cylindrical, about 1 in. long when in flower, 1½ to 2 in. long, and nearly ½ in. diameter when in fruit.　Outer bracts not numerous, glabrous, shorter than the scales.　Cone-scales villous at the base, with glabrous acuminate deciduous ends.　Perianth glabrous, about ½ in. long, the segments falling off separately, each tipped with a small point.　Style-end continuous, fusiform, shortly and sparingly hirsute.　Nut broad or narrow, not winged, acuminate, hairy all over but the marginal coma longer than the hairs of the faces.—Meissn. in Pl. Preiss. i. 502, and in DC. Prod. xiv. 275.

W. Australia.　Swan river, *Fraser, Drummond, 1st coll. n.* 575 ; *Preiss, n.* 638 ; W. coast, *Baudin's Expedition.*

19. **P. diversifolia,** *R. Br. in Trans. Linn. Soc.* x. 70, *Prod.* 365.　A shrub attaining 6 or 8 ft., the young branches often pubescent or villous, the adult foliage glabrous and sometimes shining and veined like the fronds of some ferns.　Leaves petiolate, pinnate, the pinnæ sometimes few broadly cuneate toothed or pinnatifid, giving the whole leaf a broadly triangular form of 1 to 2 in., sometimes numerous and lanceolate, the lower ones pinnatifid the upper ones gradually smaller and more entire, the whole leaf ovate-lanceolate in form and 2 or 3 in. long,

the segments and teeth mucronate-acute and sometimes pungent-pointed, but less so than in *P. striata* and *P. macrostachya.* Cones axillary, pedunculate, ovoid, ½ in. long when in flower, 1 in. when in fruit. Outer bracts glabrous. Cone-scales villous, with small glabrous often deciduous tips, and the whole scale becoming nearly glabrous when in fruit. Perianth densely villous, about ½ in. long, the segments falling off separately, tipped with prominent slender glabrous points. Style-end continuous, fusiform, nearly glabrous. Nut flat, broader than long, expanded at the top into 3 wing-like lobes, the lateral ones diverging, the central one bearing the style, glabrous except a short coma at the base.—Meissn. in Pl. Preiss. i. 499, ii. 246, and in DC. Prod. xiv. 274.

W. Australia. King George's Sound and adjoining districts, *R. Brown, A. Cunningham, Preiss, n.* 637, *F. Mueller,* and others ; Vasse river, *Oldfield;* Darling Range, *Drummond,* 2nd *coll. n.* 297, (3rd *coll. ?*) *n.* 267.

SECT. 5. PETROPHILE, Endl.—Leaves divided or rarely simple, the segments terete or, in *P. biternata* and *P. plumosa,* flat but narrow. Cones terminal, or in *P. pedunculata* and *P. pulchella,* also axillary. Perianth-segments falling off separately. Style-end continuous, fusiform.

20. **P. biternata,** *Meissn. in Hook. Kew Journ.* vii. 69, *and in DC. Prod.* xiv. 275. A stout rigid shrub, glabrous except the cones or the branches minutely hoary-tomentose. Leaves pinnate with 3 or 5 segments or the lower ones again 2- or 3-lobed, all flat but narrow, thick and rigid, pungent-pointed, ½ to 1½ in. long. Cones ovoid-globular, above 1 in. diameter, terminal and almost sessile above the last leaves. Outer bracts broad, short, hard, glabrous and shining. Cone-scales broad, the outer ones 3 or 4 lines long, acuminate, rigid, woolly at the base only, the inner ones smaller, very woolly, with small glabrous tips. Perianth scarcely above ½ in. long, slender, glabrous or slightly viscid. Style-end continuous, fusiform, shortly papillose-hirsute. Nut broadly obovate, the margins comose, both faces glabrous.

W. Australia. Between Moore and Murchison rivers, *Drummond,* 6th *coll. n.* 168.

21. **P. plumosa,** *Meissn. in Hook. Kew Journ.* vii. 69, *and in DC. Prod.* xiv. 273. An erect shrub of 1 or 2 ft., the branches virgate, hoary-tomentose and hirsute as well as the leaves with long fine spreading hairs, the older foliage becoming nearly glabrous. Leaves linear-spathulate, dilated at the end and entire or shortly divided into 2 or 3 rigid pungent-pointed flat lobes, the whole leaf ¾ to 1½ in. long. Cones terminal, sessile, depressed-globular, ¾ in. diameter without the perianths. Outer cone-scales ovate-oblong, villous at the base, the deciduous upper portion glabrous with ciliate margins, the inner ones narrow and villous. Perianth apparently nearly 1 in. long but not seen very perfect, very villous with long fulvous hairs, the segments falling off separately. Style end continuous, narrow fusiform, nearly

glabrous. Nuts flat, with broad wing-like margins, obovate, truncate, 3 lines long, minutely pubescent and very shortly comose at the base.

W. Australia. Moore river, *Drummond, 6th coll. n.* 164.

22. **P. ericifolia,** *R. Br. Prot. Nov.* 5. An erect shrub of 2 or 3 ft., with virgate branches more or less tomentose or woolly, but usually almost concealed under the foliage which is usually glandular or scabrous-pubescent, with a few longer glandular hairs, or sometimes almost or quite glabrous and smooth. Leaves terete, not thick, ¼ to ½ in. long, erect and crowded along the branches, mostly terminated by a small point or oblique gland or quite obtuse. Cones terminal, sessile, at first broadly ovoid, at length globular, ½ to ¾ in. diameter. Outer bracts ovate-lanceolate, glutinous, imbricate and persistent. Cone-scales with the upper portion lanceolate acute very glutinous and imbricate on the young cones, but deciduous, leaving the enlarged base broadly ovate tomentose and villous in the fruiting cone. Perianth slender, 7 or 8 lines long, glutinous and hirsute with spreading hairs, the segments often separating but not so readily as in some species. Style-end continuous, narrow-fusiform, almost glabrous. Nut rather broad, the margins comose, both faces glabrous.—Meissn. in Pl. Preiss. i. 494, and in DC. Prod. xiv. 267.

W. Australia. King George's Sound and adjoining districts, *R. Brown, Baxter, Drummond, Preiss, n.* 650, and many others.

Var. *scabriuscula.* Leaves rather longer and more scabrous, cones larger; flowers more villous.—*P. scabriuscula,* Meissn. in Pl. Preiss. i. 495, and in DC. Prod. xiv. 268.—Swan river? *Drummond, 1st coll. n.* 557.

Var. *glabriflora.* Perianth glabrous.—Stirling Range, *F. Mueller.*

P. phylicoides, R. Br. Prot. Nov. 6; Meissn. in DC. Prod. xiv. 268, described from some specimens of Baxter's from Lucky Bay, without perianths, appears to me to be the same as the more glabrous forms of *P. ericifolia,* a species which as a whole is very distinct from any other. The small crowded leaves give it some outward resemblance to *P. inconspicua* or to *P. chrysantha,* but these leaves are all simple, never divided as in the latter two.

23. **P. chrysantha,** *Meissn. in Hook. Kew Journ.* vii. 68, *and in DC. Prod.* xiv. 271. A shrub of 2 or 3 ft., with erect branches, the young shoots tomentose-pubescent and sprinkled with long fine hairs, becoming at length nearly glabrous. Leaves short and crowded along the branches, simply pinnate, with terete pungent-pointed segments grooved above, the lowest pair proceeding from near the base of the petiole, the whole leaf not exceeding ¾ in. Cones terminal, sessile, ovoid, 3 to 4 lines diameter without the perianths. Outer bracts broad, obtuse or with minute points, imbricate, glabrous except the ciliate margins; outer cone-scales similar, the inner ones gradually narrower, more concave, hirsute outside with long hairs, glabrous inside. Perianths about 5 lines long, very densely hirsute with yellow or fulvous hairs, the segments falling off separately. Style-end continuous, fusiform, at first bearing a few reflexed hairs, but at length nearly glabrous. Nuts expanded in the upper part into 2 flat truncate wings, 2 lines long and

1½ lines broad, densely comose at the base, the remainder hirsute with shorter hairs.

W. Australia. Between Moore and Murchison rivers, *Drummond, 6th coll. n.* 165; near Dandaroga, *Oldfield.*

24. **P. pedunculata,** *R. Br. in Trans. Linn. Soc.* x. 70, *Prod.* 364. A tall glabrous shrub. Leaves pinnate with much divided 2- 3-chotomous pinnæ, the ultimate segments numerous, rather fine, rigid but not pungent, terete and grooved above. Cones axillary, ovoid or oblong, ¾ to 1 in. long, on peduncles of ¼ to ½ in. with small empty bracts at the base of the peduncle. Cone-scales glabrous, broad, hard, with a short persistent erect point. Perianth glabrous, about 5 lines long, the segments falling off separately. Style-end continuous, fusiform, angular, minutely pubescent or glabrous. Nut broad, the margins comose, both faces glabrous.—Meissn. in DC. Prod. xiv. 269; Guillem. Ic. Pl. Austral. t. 18.

N. S. Wales. Blue Mountains, *R. Brown, Sieber, n.* 20, *A. and R. Cunningham* and others.

25. **P. pulchella,** *R. Br. in Trans. Linn. Soc.* x. 69, *Prod.* 364. A shrub of 6 to 8 ft., glabrous as well as the foliage or minutely pubescent when young. Leaves twice or thrice pinnate, the segments numerous, not spreading, terete, grooved above, rather slender and not pungent, the whole leaf 1½ to 2¼ in. long, the petiole as long as the divided part. Cones terminal, sessile, solitary or with one or two axillary ones close below it, oblong or cylindrical, 1 to 1½ in. long, and ½ to ¾ in. diameter without the perianths. Outer bracts few and small. Cone-scales broad, the outer ones very shortly acuminate and pubescent, the inner more silky at the base with lanceolate deciduous points, all at length broad hard and glabrous. Perianth silky-pubescent, 6 to 7 lines long, the segments falling off separately. Style-end continuous, narrow-fusiform, sparingly and shortly hirsute. Nut broad, copiously comose on the margins, more sparingly hirsute on the inner face, glabrous on the back.—Meissn. in DC. Prod. xiv. 270; *Protea pulchella,* Schrad. Sert. Hannov. 15, t. 7; Cav. Ic. t. 550; Bot. Mag. t. 796; *Protea fucifolia,* Salisb. Prod. 48; *Petrophila fucifolia,* Knight. Prot. 92; *Protea dichotoma.* Cav. Ic. vi. 34, t. 551.

N. S. Wales. Port Jackson to the Blue Mountains, *R. Brown, Sieber, n.* 19, *Fl. Mixt. n.* 479, and many others.

26. **P. sessilis,** *Sieb. in Roem. and Schult. Syst. Veg.* iii. *Mant.* 262. A shrub attaining 8 to 12 ft., closely allied to *P. pulchella,* and as suggested by R. Brown, perhaps a variety with a more rigid foliage, the segments divaricate and the young shoots hoary-tomentose or almost silky. Cones rather broader and shorter. Perianths and style and nuts the same.—R. Br. Prot. Nov. 6; Meissn. in DC. Prod. xiv. 270; *P. canescens,* A. Cunn. in R. Br. Prot. Nov. 6; Meissn. l.c. 270.

Queensland. Moreton Bay, *A. Cunningham, Fraser.*
N. S. Wales. Blue Mountains, *Caley, Sieber, n.* 21; New England, *C. Stuart, C. Moore;* Sydney woods, Paris Exhibition, 1855, *Macarthur, n.* 214.

27. **P. fastigiata,** *R. Br. in Trans. Linn. Soc.* x. 70, *Prod.* 364. A shrub of about 3 ft., glabrous except the cones or the branches slightly tomentose. Leaves twice or thrice ternately divided, the petiole as long as the divided portion, the segments terete, slender but obtuse or nearly so, all erect and attaining about the same height. Cones terminal, ovoid, sessile, ¾ to 1 in. long when in fruit. Outer bracts numerous, imbricate, silky inside, broadly triangular and acute, at length deciduous. Cone-scales woolly at the base, broad, with small deciduous glabrous tips. Perianth glabrous, about 5 lines long, the segments falling off separately. Style-end continuous, fusiform,very shortly hirsute. Nuts narrow, acute, comose on the margins and more sparingly so on the back, glabrous on the front or inner face.—Meissn. in DC. Prod. xiv. 270 ; F. Muell. Fragm. vi. 245.

W. Australia. Lucky Bay, *R. Brown, Baxter;* near Eyre's Range, *Maxwell.*

28. **P. seminuda,** *Lindl. Swan Riv. App.* 34. A bushy shrub of 2 or 3 ft., quite glabrous except the cones. Leaves once or twice 3-fid, with terete segments of ½ to 1 in., not very thick but rigid and pungent-pointed. Cones terminal, sessile, ovoid, attaining ½ in. diameter. Outer bracts numerous, ovate-lanceolate or lanceolate, acute, at length deciduous. Cone-scales with a broad woolly-tomentose base, and small deciduous lanceolate glabrous tips. Perianth glabrous, about ½ in. long, not very slender, the segments falling off separately. Style-end continuous, fusiform, very shortly and sparingly hirsute. Nut ovate, with comose margins, the back hairy, the inner face glabrous.—Meissn. in Pl. Preiss. i. 495 and in DC. Prod. xiv. 269.

W. Australia. Swan river, *Drummond,* 1st coll. n. 561, 562, 572, *Preiss, n.* 634, 635 ; Stirling Range, *F. Mueller;* Cape Le Grand, *Maxwell.*

Var *indivisa.* Leaves 1 to 2 in. long, mostly undivided or here and there a few bifid or trifid at the end.—W. Australia, *Drummond.*

29. **P. circinata,** *Kipp. ; Meissn. in Hook. Kew Journ.* vii. 67, *and in DC. Prod.* xiv. 272. A stout bushy shrub, glabrous except the cones or the branches slightly pubescent. Leaves crowded, twice pinnate, with numerous divaricate intricate terete and pungent-pointed segments, the whole leaf 3 to 5 in. long. Cones terminal, large, depressed-globular. Outer bracts numerous, broad, shortly acuminate, coriaceous, glabrous or minutely hoary outside, silky-villous inside, mostly above ½ in. long, imbricate, forming an involucre of 1 to 1½ in. diameter, concealing the scales. Cone-scales lanceolate or linear, apparently deciduous. Perianths exceedingly numerous, villous with minute glabrous tips, projecting about ¼ in. beyond the outer bracts, forming an almost flat top of 1½ in. diameter, the segments readily separating. Style-end continuous, narrow-fusiform, very sparingly hirsute or nearly glabrous. Nut ovate, acute, shortly comose on the back as well as the margins.

S. Australia ? N. of Adelaide, *Whittaker (Herb. Hook.),* but possibly some mistake.

W. Australia, *Drummond,* 5th coll. Suppl. n. 3.

The cone-scales appear to be deciduous and the involucre large and persistent as in *Isopogon latifolius*, and others of the section *Hypsanthus* of that genus; the perianth, style and nut (the latter not seen perfect) are rather those of *Petrophila*.

30. **P. Drummondii,** *Meissn. in Pl. Preiss.* i. 496, *and in DC. Prod.* xiv. 270. An erect shrub, glabrous except the cones or the upper branches and leaves softly pubescent. Leaves twice or thrice ternately divided, or pinnate with the lower pinnæ twice or thrice divided, the segments terete and pungent-pointed, $\frac{1}{2}$ to $\frac{3}{4}$ in. long. Cones terminal, sessile, ovoid or nearly globular, $\frac{3}{4}$ to 1 in. diameter. Outer bracts rather numerous, ovate-lanceolate, glutinous but otherwise glabrous. Cone-scales with a broad woolly-villous base and ovate-lanceolate or lanceolate deciduous ends, the outer scales in the old cones becoming glabrous. Perianth (about $\frac{3}{4}$ in. long?) hirsute with spreading viscid hairs, the segments falling off separately. Style-end continuous, rather long, fusiform, hirsute. Nut broad, not acuminate and at length notched at the end, comose on the margin and outer face, the inner face glabrous.

W. Australia, *Drummond, 1st coll. n.* 570.

P. triternata, Kipp.; Meissn. in Hook. Kew Journ. vii. 67, and in DC. Prod. xiv. 270, from Drummond's 2nd coll. n. 2, appears to me to be the same plant with cones in a more advanced state, with the leaves of young shoots growing out from within the outer bracts. In neither are the perianths in a very good state.

31. **P. crispata,** *R. Br. Prot. Nov.* 6. A rigid shrub, very closely allied to *P. Drummondii* and to *P. rigida,* and perhaps a variety of the latter. Leaves twice or thrice divided, with terete pungent-pointed segments, shorter and more rigid than in *P. Drummondii,* longer than in *P. rigida.* Cones terminal, sessile, ovoid or at length oblong and sometimes above 1 in. long when in fruit, and not much above 1 in. diameter. Outer bracts glutinous and deciduous. Cone-scales with a broad tomentose-villous base and short glabrous persistent or deciduous ends. Perianth not seen very perfect, the segments falling off separately, densely villous outside with short hairs, not viscid or at least not so much so as in *P. Drummondii.* Nut broad, truncate, the margins comose, both faces glabrous or slightly pubescent.—Meissn. in Pl. Preiss. i. 496, ii. 245 and in DC. Prod. xiv. 271.

W. Australia. King George's Sound, *Baxter, Preiss, n.* 647, *Drummond, 3rd coll. n.* 248.

32. **P. rigida,** *R. Br. in Trans. Linn. Soc.* x. 69, *Prod.* 364. A rigid divaricate shrub of 2 or 3 ft., quite glabrous except the cones. Leaves very rigid, trichotomously bipinnate, the ultimate segments in the typical form very short but in other specimens $\frac{1}{2}$ to 1 in. long, all terete divaricate and pungent-pointed. Cones terminal or in the forks of the branches, sessile, nearly globular, about $\frac{1}{2}$ in. diameter without the flowers, or rather larger when in fruit. Outer bracts lanceolate, acuminate, glabrous outside, silky inside, deciduous. Cone-scales broad, scarcely acuminate, silky-villous outside at the base, otherwise glabrous. Perianth nearly $\frac{3}{4}$ in. long, very villous almost plumose, the

segments falling off separately. Style-end continuous, fusiform, with few short hairs, glabrous and 4-angled at the base. Nut not seen ripe, the young ones densely comose.—Meissn. in Pl. Preiss. i. 497, and in DC. Prod. xiv. 271.

W. Australia. King George's Sound and adjoining districts, *R. Brown, A. Cunningham, Drummond, 1st coll., Preiss, n.* 645, 648, 649, and others.

33. **P. multisecta,** *F. Muell. Fragm.* vi. 242. A densely branched bushy shrub, glabrous except the cones and somewhat glaucous, closely allied to and perhaps a variety of *P. rigida.* Leaves trichotomously divided or pinnate with dichotomous pinnæ, the segments terete, rigid, pungent-pointed, not so short as in some varieties of *P. rigida,* but quite like those of other forms of that species. Cones ovoid or oblong, ¾ to 1 in. long and about ½ in. diameter without the perianths, sessile in the forks of the branches. Outer bracts not numerous, lanceolate, acuminate, glabrous. Cone-scales broad, silky-tomentose outside, with prominent persistent acuminate and glabrous ends. Perianth silky-villous, about ½ in. long, the segments falling off separately. Style-end continuous, fusiform, shortly pubescent. Nut broad, densely comose on the margins, more sparingly hirsute on both faces.

S. Australia. Kangaroo Island, *Waterhouse.*

34. **P. conifera,** *Meissn. in Hook. Kew Journ.* vii. 67, *and in DC. Prod.* xiv. 271. A much-branched bushy shrub of 1 to 3 or 4 ft. the younger branches tomentose-pubescent, almost woolly, the older ones and foliage glabrous. Leaves trifid or the middle segment again divided, the segments divaricate, terete, rigid, pungent-pointed, ½ to ¾ in. long. Cones mostly in the forks of the branches, sessile, oblong-cylindrical, attaining about 1 in. in length and 4 to 6 lines diameter without the perianths. Outer bracts very deciduous. Cone-scales softly tomentose, at first acuminate, but the points deciduous and in the old cones the scales very broad, about 3 lines diameter, hard, tomentose, few for the length of the cone although closely imbricate. Perianth hirsute, 6 to 7 lines long, the segments falling off separately. Style-end continuous, narrow-fusiform, hirsute with few short hairs. Nut ovate, almost acuminate, comose all over, but more densely so on the margins and back than on the inner face.

W. Australia. Murchison river, *Oldfield, Drummond, 6th coll. n.* 167. The species is very nearly allied to *P. rigida,* but with tomentose branches and longer cones.

SECT. 6. HEBEGYNE.—Habit and characters of *Petrophile* except that the style has not a distinct fusiform end.

35. **P. semifurcata,** *F. Muell. Herb.* A bushy shrub with the habit and tomentose branches of *P. conifera.* Leaves glabrous, rigid, terete, 3 to 5 in. long, entire or divided at the end into 2 or 3 short segments, obtuse or shortly pointed but not pungent. Cones terminal or in the forks of the branches, shortly pedunculate, oblong-cylindrical, ¾ to

1½ in. long, and about ½ in. diameter. Outer bracts very deciduous. Cone-scales broad, rigid, tomentose, rounded or obtuse with very small points, not numerous but closely imbricate as in *P. conifera.* Perianth silky-villous, nearly ¾ in. long, the segments falling off separately, not seen however very perfect. Style pubescent from the base, tomentose-villous and slightly thickened upwards but not distinctly fusiform at the end.

W. Australia. Murchison river, *Oldfield.*

2. ISOPOGON, R. Br.

Flowers hermaphrodite. Perianth regular, the tube slender, the upper portion falling off entire with the 4 linear or oblong segments of the limb, leaving a persistent base which finally splits or is cast off as the fruit ripens. Anthers all perfect and free, sessile within the segments of the limb, the connective tipped with a small appendage. No hypogynous scales. Ovary sessile, with a single orthotropous or slightly amphitropous ovule, pendulous from near the apex of the cavity. Style filiform, usually more or less dilated or clavate towards the end, and separated from the narrow often bulbous-based brush by a short neck or constriction, the clavate portion usually papillose-pubescent, rarely the style-end continuous and slender, the stigma terminal. Fruit a small dry and indehiscent nut, usually ovoid-conical, scarcely compressed and not winged, hirsute all over, the lower hairs or nearly all forming a long coma.—Shrubs with the habit of *Petrophila.* Leaves rigid, entire or divided, terete or flat and sometimes broad. Flowers yellow, pink or lilac, in dense spikes or *cones,* each flower sessile within a bract or *scale,* the cones hemispherical globular or ovoid, terminal or rarely axillary, the receptacle or rhachis woolly, flat convex conical or cylindrical, the scales tomentose or villous outside, glabrous inside, imbricate, deciduous after flowering or if long persistent and retaining the seed, readily detached and always falling off with the seed, or in a few species leaving a very short persistent base. At the base of the cone are also, as in *Petrophila,* several imbricate empty bracts, forming an involucre, larger or smaller than the cone-scales and usually more persistent, the cones are also almost always closely surrounded by floral leaves. Nuts shorter than the cone-scales, and very little varied in the whole genus.

The genus is limited to extratropical Australia, and is chiefly Western. Although the majority of the species differ from *Petrophila* in the mode of breaking up the perianth, in the form of the style-end, and in the shape and indumentum of the nut, all these characters have exceptions, and perhaps the most constant one is that of the cone-scales, which in *Petrophila* remain firmly attached to the receptacle, opening spontaneously or by force for the emission of the nuts, whilst in *Isopogon* they separate from the receptacle either with the nuts or previously.

SECT. 1. **Hypsanthus.**—*Cone-scales acuminate, not very closely imbricate, the inner on's narrow. Receptacle flat, convex or rarely oblong. Leaves flat, except in* I. adenanthoides.

Leaves flat, all quite entire.
Leaves mostly 2 to 6 in. long. Cones ½ to 1½ in. diameter.
Perianth glabrous or with tufts of hairs only at the ends of
the segments.
Cones large, solitary, with numerous outer bracts. Pe-
rianth 1 in. long or more.
Involucre 1½ in. diameter, exceeding the cone-scales.
Perianth 1½ in. long. . . . . 1. *I. latifolius.*
Involucre 1 in. diameter, not exceeding the cone-scales.
Perianth 1 in. long . . . . 2. *I. cuneatus.*
Cones under 1 in. diameter, often clustered. Perianth
about ½ in. long. Leaves narrow.
Outer bracts longer than the cone-scales . . . . . 3. *I. linearis.*
Outer bracts shorter than the cone-scales . . . . . 4. *I. polycephalus.*
Perianth-segments hirsute from the middle. Cones under
1 in. diameter, often clustered. Involucral bracts lanceo-
late, acuminate . . . . 5. *I. attenuatus.*
Perianth segments densely plumose-villous.
Stems erect, leafy, villous. Terminal cones large, solitary
or clustered . . . . 6. *I. sphærocephalus.*
Stems dwarf or scarcely any. Cones clustered at the
base of the elongated petioles . . . . 7. *I. uncinatus.*
Leaves mostly under 1 in. long or the lower ones 2 in. Cones
small, ovoid. Perianth segments plumose.
Cones clustered at the ends of the branches . . . . . 8. *I. buxifolius.*
Cones axillary along the branches . . . . 9. *I. axillaris.*
Leaves cuneate, mostly 3-toothed . . . . 10. *I. tridens.*
Leaves undulate, broadly cuneate and dentate or broadly twice
or thrice 3-lobed . . . . 11. *I. Baxteri.*
Leaves linear or linear-cuneate, once or twice ternately divided 12. *I. roseus.*
Leaves slender, terete, trifid, crowded, under 1 in. long . . . 13. *I. adenanthoides.*
(See also *Petrophila circinata*, which has the cones nearly of
Hypsanthus with the flowers of *Petrophila*, and *P. incon-
spicua*, which much resembles *I. adenanthoides*.)

SECT. 2. **Eustrobilus.**—*Cone-scales all with broad dilated or truncate ends, closely
imbricate after flowering. Receptacle convex, conical, or cylindrical.*
Perianth silky pubescent or villous. Leaves flat.
Leaves cuneate or spathulate, mostly 3-toothed or shortly
3-lobed . . . . 14. *I. trilobus.*
Leaves linear or linear-lanceolate, mostly divided into 3 seg-
ments of ½ to 1 in. . . . . 15. *I. tripartitus.*
Leaves linear or oblanceolate, 4 to 8 in. long, entire or with 1
or 2 long linear lobes . . . . 16. *I. longifolius.*
Perianth silky-pubescent or villous. Leaves terete.
Outer bracts few besides the floral leaves, tomentose. Cone-
scales plumose-villous.
Leaves all undivided. Cones large, terminal, depressed-
globular . . . . 17. *I. Drummondii.*
Leaves undivided or 2- or 3-lobed. Cones terminal, ovoid-
globular. . . . . 18. *I. heterophyllus.*
(See also 29, *I. scabriusculus.*)
Leave 2- 3-chotomous, very long. Stems dwarf or scarcely
any. Cones large, ovoid-oblong, sessile amongst the leaves 19. *I. villosus.*
Outer bracts imbricate, broad, glabrous.
Leaves rigid, entire, or once, twice, or thrice divided. Cones
mostly nodding. Western species . . . . 20. *I. teretifolius.*
Leaves slender, twice or thrice divided. Cones erect.
Eastern species . . . . 21. *I. anethifolius.*

Perianth glabrous or with a tuft of hairs at the end of the seg-
    ments. Leaves much divided or rarely entire, flat and
    veined, or if nearly terete g noved or channelled above.
Eastern species. Leaves much divided, with flat pungent-
    pointed segments. Flowers yellow.
  Outer bracts shorter than the cone-scales.
    Petioles 2 to 3 in. long. Perianth about 4 lines . . . 22. *I. petiolaris.*
    Petioles 1 to 1½ in. long. Perianth 5 or 6 lines . . . 23. *I. anemonifolius.*
  Outer bracts numerous, longer than the cone-scales. Leaf-
    segments very numerous, divaricate, and pungent-
    pointed . . . . 24. *I. ceratophyllus.*
Western species. Leaf-segments or leaves narrow and con-
    cave or nearly terete but grooved. Flowers red or lilac
    (colour unknown in *I scabriusculus*).
  Leaves crowded, short, with narrow-linear segments. Cones
    small, often crowded at the ends of the branches . . . 25. *I. asper.*
  Leaves mostly once or twice divided into linear-cuneate
    segments. Cones terminal, rather large . . . . . . 26. *I. crithmifolius.*
  Leaves much divided, with short nearly terete pungent-
    pointed segments. Cones terminal . . . . 27. *I. formosus.*
  Leaves once or twice divided, with slender nearly terete
    segments. Cones terminal. Perianth 1 in. long . . . 28. *I. divergens.*
  Leaves terete or linear and thick, 2 to 4 in. long, entire or
    3 lobed. Cones terminal, small . . . . 29. *I. scabriusculus.*

I. pedunculatus, R. Br. Prot. Nov. 7, Meissn. in DC. Prod. xiv. 277, was founded
on two specimens of Fraser's from Swan river, with the cones in too imperfect a state
to establish their generic affinity with certainty. If the plant is a true *Isopogon,* it is
most probably a variety of *I. divergens,* but the cone not being so closely surrounded
by floral leaves as in most species of *Isopogon,* it is more probably a *Petrophila,* and in
that case referable to *P. seminuda.*

SECT. 1. HYPSANTHUS, Endl.—Cone-scales acuminate, not very
closely imbricate, the inner ones narrow, often plumose-villous and
very deciduous. Receptacle flat convex or rarely oblong. Leaves
flat, often entire, sometimes broad, divided into few flat segments in a
few species, with terete segments in *I. adenanthoides.*

The two divisions proposed by Brown and established as sections by Endlicher, appear to me much more definite than the three founded chiefly on the foliage by Meissner.

1. **I. latifolius,** *R. Br.* Prot. Nov. 8. A tall stout species attaining
10 ft., and from a distance assuming the aspect of a *Rhododendron* (*F.
Mueller*), the branches pubescent towards the end, the foliage glabrous.
Leaves obovate to elliptical-oblong, obtuse with a small callous point,
entire, narrowed into a very short petiole or nearly sessile, 3 to 4 in.
long, obscurely veined. Cones large, terminal, depressed-globular.
Outer bracts ovate to ovate-lanceolate, the inner ones ½ in. long, numerous and imbricate in several rows in an involucre of 1½ in. diameter
concealing the scales, which are woolly outside, glabrous inside, the outer
ones lanceolate, the inner linear. Perianth-tube filiform, glabrous, 1½ in.
long; laminæ linear, about ¼ in. long, tipped with a small tuft of silky hairs.
Ovary crowned by a tuft of short hairs. Style-end oblong-clavate
pubescent and suddenly contracted under the fusiform brush, which
bears reflexed hairs in 8 longitudinal rows. Receptacle conical or
almost cylindrical, nearly 1 in. long after the fall of the fruit.—Meissn.

in DC. Prod. xiv. 282, as to Brown's typical plant only; *I. protea,* Meissn. l.c. 283; F. Muell. Fragm. vi. 237.

W. Australia. King George's Sound or the immediate neighbourhood, *Baxter, Drummond, 5th coll. n.* 398; Summit of Mongyrup, Stirling Range, *F. Mueller.* Baxter appears to have gathered only a single specimen in fruit, preserved in Brown's supplemental herbarium, which was probably not shown to Meissner when he went through the Proteaceæ of Brown's own collecting.

2. **I. cuneatus,** *R. Br. in Trans. Linn. Soc.* x. 73, *Prod.* 366. A stout shrub, attaining 7 or 8 ft. but flowering sometimes when quite low, glabrous except the cones, or the young shoots silky-villous. Leaves from obovate-oblong to lanceolate or oblanceolate, obtuse with a small callous point, contracted into a short broad petiole or almost sessile and often dilated and half stem-clasping at the base, rather thick, obscurely veined, 3 to 4 in. long, and varying in breadth in the same specimen from ½ to 1½ in. Cones terminal, depressed-globular, 1 to 1¼ in. diameter without the perianths. Outer bracts broad, glabrous or nearly so, obtuse, shorter than the scales. Cone-scales numerous, the outer ones ovate, the inner ones lanceolate or linear, all very villous outside. Perianth pale purple, about 1 in. long, glabrous or with small tufts at the tips of the laminæ. Style-end clavate and glabrous except the obtuse villous extremity below the narrow reflexed-hairy brush. Receptacle hemispherical or shortly and obtusely conical.—Meissn. in DC. Prod. xiv. 283; *I. Loudoni,* Baxt. in R. Br. Prot. Nov. 8; Bot. Mag. t. 3421; Meissn. l.c.; F. Muell. Fragm. vi. 238; *I. latifolius* var. *Preissii* and var. *lanceolatus,* Meissn. in Pl. Preiss. i. 508, and in DC. Prod. xiv. 282, 283.

W. Australia. King George's Sound and adjoining districts, *Menzies, Baxter, Drummond, 5th coll. n.* 397, *Preiss, n.* 664, and many others.

The pubescence of the young shoots is very variable, and neither that nor the breadth of the leaves afford characters for separating distinct varieties.

3. **I. linearis,** *Meissn. in Hook. Kew Journ.* vii. 69, *and in DC. Prod.* xiv. 282. An erect shrub of 1 to 2 ft., the branches and young leaves softly pubescent, the older foliage glabrous. Leaves linear, with a callous point, contracted into a short petiole, mostly 1½ to 2½ in. long, thick, with more or less distinct nerve-like margins and a very few oblique veins. Cones nearly globular, ¾ to 1 in. diameter, terminal and solitary or in a cluster of 2 or 3. Outer bracts not numerous but rather large and imbricate, ovate-lanceolate, silky-pubescent or at length nearly glabrous, the inner ones ½ in. long. Cone-scales shorter, the outer ones broad the inner ones narrow-lanceolate, all very woolly-villous outside. Perianth quite glabrous, rather above ½ in. long. Style-end slightly clavate and minutely pubescent, separated by a narrower neck from the pubescent bulbous base of the otherwise glabrous narrow brush. Receptacle ovoid-conical.—F. Muell. Fragm. vi. 236.

W. Australia. Gardiner's Range north of Dundiragan, towards Moore river, *Drummond, 6th coll. n* 169.

4. **I. polycephalus,** *R. Br. in Trans. Linn. Soc.* x. 73, *Prod.* 366. An erect or spreading shrub of 1 to 3 ft., the young shoots tomentose or villous the adult foliage glabrous. Leaves linear-oblong or oblanceo-

late, obtuse with a small callous point, contracted into a short petiole, 2 to near 4 in. long, or in some specimens nearly twice as long, thick and veinless or obscurely veined. Cones sessile and usually clustered 2 or 3 together at the ends of the branches, about ½ in. diameter or the terminal one larger. Outer bracts few and tomentose. Cone-scales densely villous outside, a few of the outer ones broad, all the others lanceolate or linear and mostly terminating in subulate densely plumose points. Perianth nearly ½ in. long, quite glabrous. Style slightly thickened and shortly hairy towards the end, or quite glabrous, without any constriction or distinct brush.—Meissn. in DC. Prod. xiv. 281; F. Muell. Fragm. vi. 236.

W. Australia. King George's Sound, *R. Brown, Baxter;* Stirling Range, *F. Mueller;* Gales Brook and E. Shoal Cape, *Maxwell.*

5. **I. attenuatus,** *R. Br. in Trans. Linn. Soc.* x. 73, *Prod.* 366. A shrub of 2 or 3 ft., glabrous except the cones, or the young shoots slightly pubescent. Leaves oblong-spathulate to almost linear, with a small straight or hooked point, much narrowed into the petiole, mostly 4 to 6 in. long, thick and almost veinless. Cones terminal or in the upper axils, sessile, depressed-globular, ¾ to 1 in. diameter without the perianths. Outer bracts not numerous, lanceolate, rigid, as long as or longer than the scales, the outer ones passing into small floral leaves. Cone-scales lanceolate or the inner ones linear, the outer ones villous at the base and ciliate on the margins, the inner more villous all over the back. Perianth "pale yellow" not ½ in. long, the laminæ villous outside, the tube glabrous or nearly so. Style slightly thickened towards the end as in *I. polycephalus,* without any distinct constriction or brush.—Bot. Mag. t. 4372; Meissn. in DC. Prod. xiv. 281; F. Muell. Fragm. vi. 237.

W. Australia. King George's Sound and adjoining districts, *R. Brown, Baxter, Drummond,* 2nd coll. n. 294, *Preiss,* n. 663, and others.

Var. *latebracteata.* Outer bracts broadly lanceolate and thinner than in the typical form.—Swan river, *Fraser;* Gordon river, *Oldfield.*

6. **I. sphærocephalus,** *Lindl. Swan Riv. App.* 34. An erect shrub of several ft., the branches and younger leaves pubescent and clothed or sprinkled with long spreading hairs. Leaves linear or almost lanceolate, obtuse with a short callous point, slightly contracted towards the base but sessile, 2 to 4 in. long, the margins often recurved and the midrib prominent underneath. Cones solitary and terminal or 2 or 3 crowded at the ends of the branches, globular, ½ to ¾ in. diameter without the perianth. Outer empty bracts not numerous, imbricate, lanceolate, villous, not exceeding the scales. Outer cone-scales ovate, inner ones narrow, all villous outside, with small recurved points. Perianth above ½ in. long, the tube glabrous, the laminæ densely hirsute with yellow hairs. Style-end turbinate, densely pubescent and separated by a short constriction from the somewhat bulbous base of the linear almost glabrous brush. Receptacle ovoid-oblong.—Meissn. in Pl. Preiss. i. 508, and in DC. Prod. xiv. 281; Bot. Mag. t. 4332.

W. Australia. Swan river, *Drummond, 1st coll. n.* 559, *Preiss, n.* 688; Swan and Vasse rivers, *Oldfield.*

7. **I. uncinatus,** *R. Br. Prot. Nov.* 8. Stems very short or scarcely any, bearing a cluster of 2 to 4 sessile cones in a tuft of long leaves, thus assuming the aspect of *Conospermum petiolare.* Leaves linear or lanceolate, terminating in a hooked callous point, involute when young, contracted into a long petiole, the longer ones attaining 8 to 12 in., the broader ones much shorter, finely hairy when young, at length glabrous. Outer bracts few, nearly glabrous. Cone-scales villous, lanceolate, the outer ones rather broad, the inner very narrow. Perianth about ½ in. long, the laminæ and upper part of the tube densely hirsute with yellowish hairs, the lower portion alone glabrous. Style narrow-fusiform towards the end and slightly pubescent in the lower portion of the thickened part, but not divided by any distinct constriction.—Meissn. in Pl. Preiss. i. 509, ii. 247, and in DC. Prod. xiv. 281.

W. Australia. King George's Sound or neighbouring districts, *Baxter, Drummond, 3rd coll. n.* 243, *Preiss, n.* 758.

8. **I. buxifolius,** *R. Br. in Trans. Linn. Soc.* x. 74, *Prod.* 367. A bushy and leafy shrub of 3 or 4 ft., the branches and young shoots pubescent, the adult foliage glabrous. Leaves very variable, usually small and crowded, rarely 1 in. long, flat or concave. Cones sessile and solitary or clustered at the ends of the branches in a tuft of floral leaves, ovoid, 4 or 5 lines long without the perianths. Outer bracts few, lanceolate, ciliate. Outer cone-scales like the outer bracts but villous also on the back, inner ones linear. Perianth about ½ in. long, the segments separating far below the laminæ, leaving the entire tube short, glabrous except terminal tufts of hairs which sometimes extend half way down the laminæ. Style slightly thickened into a narrow fusiform brush marked with longitudinal lines of reflexed hairs, but without any dilatation or constriction below the brush.—Meissn. in DC. Prod. xiv. 282; *I. spathulatus,* R. Br. Prot. Nov. 8; Meissn. in Pl. Preiss. i. 509, ii. 247, and in DC. Prod. xiv. 282.

W. Australia. King George's Sound and adjoining districts, *R. Brown,* and many others.

The forms assumed by the leaves in different specimens are so different that the following varieties might be easily taken for distinct species.

a. spathulatus. Leaves obovate-spathulate, contracted into a distinct petiole, mostly about ½ in. long —*Drummond, 3rd coll n.* 249.

b. obovatus, Br. Leaves obovate or oblong, more or less contracted at the base but not petiolate, ½ to 1 in. long.—*Drummond, 5th coll. n.* 396.

c. typicus. Leaves broadly sessile, ovate, with short recurved points, 3 to 4 lines long.—Only seen in Herb. R. Brown.

d. linearis, Br. Leaves narrow-oblong or linear, sessile, ½ to ¾ in. long.—Bot. Mag. t. 3450.—Apparently the most common variety, occurring in the collections of *Baxter, Drummond, 5th coll. n.* 395, *Oldfield, Maxwell,* and *F. Mueller.*

9. **I. axillaris,** *R. Br. in Trans. Linn. Soc.* x. 74, *Prod.* 367. A shrub with erect virgate branches, glabrous except the cones. Leaves linear or oblanceolate, obtuse with a small callous point, contracted into a

short petiole or the smaller ones sessile, the lower ones often 3 or 4 in. long, the floral ones sometimes all under 1 in., all thick and veinless. Cones small, ovoid, sessile and axillary, rarely ½ in. long without the perianths. Outer bracts ovate, obtuse, glabrous or with shortly ciliate margins, concave and imbricate, concealing the scales. Cone-scales silky-villous outside, a very few of the outer ones nearly ovate, the inner ones linear. Flowers often not above 10 or 12 in the cone. Perianth-tube filiform, glabrous, at least 1 in. long, lobes narrow, 4 or 5 lines long, plumose-villous outside above the middle. Style-end elongated clavate pubescent, separated by a constriction from the somewhat bulbous pubescent base of the brush, which is otherwise only minutely pubescent along the angles.—Meissn. in Pl. Preiss. i. 510, and in DC. Prod. xiv. 282; Guillem. Ic. Pl. Austral. t. 19; Hook. Ic. Pl. t. 438.

W. Australia. King George's Sound and adjoining district, *R. Brown, Baxter, Drummond, Preiss, n.* 653, and many others; Vasse river, *Oldfield.*

10. **I. tridens,** *F. Muell. Fragm.* vi. 239. A shrub with the habit and nearly the foliage of the shorter-lobed forms of *I. trilobus*, but a very different cone. Young shoots slightly pubescent, adult foliage and branches glabrous. Leaves narrow-cuneate, mostly 3-toothed, contracted into a rather long petiole, thick and obscurely veined, the whole leaf 1½ to 3 in. long. Cones terminal, sessile, depressed-globular, about ¾ in. diameter without the perianths. Outer bracts broad, tomentose outside, numerous and closely imbricate, forming an involucre of ¾ in. diameter. Cone-scales acuminate, the outer ones ovate-lanceolate, the inner ones narrow, all very densely villous on the back with long hairs, fulvous in the lower concealed portion, white on the exposed tips. Perianths not seen. Receptacle convex. Fruit of *Isopogon.*— *I. trilobus* var. *tridens*, Meissn. in Hook. Kew Journ. vii. 70 and in DC. Prod. xiv. 280.

W. Australia. Sandy plains near Diamond Spring, Moore river, *Drummond, 6th coll. n.* 170.

11. **I. Baxteri,** *R. Br. Prot. Nov.* 9. An erect shrub of several ft., the young shoots softly villous, the adult foliage glabrous. Leaves from broadly cuneate undulate and toothed only at the end, to twice or thrice 3-lobed, the lobes or teeth all broad undulate and pungent-pointed, the whole leaf 1 to 2 in. long and often as broad at the end as long, contracted at the base or almost petiolate. Cones depressed-globular, terminal, often clustered amongst numerous floral leaves, the innermost of which have hard dilated bases and small laminæ, passing into the few outer bracts. Cone-scales linear or linear-lanceolate, very villous with long silvery or fulvous hairs. Perianth pink, very villous, ¾ in. long. Style with a long clavate pubescent end, under a short almost glabrous brush.—Meissn. in Pl. Preiss. ii. 247, and in DC. Prod. xiv. 280; Bot. Mag. t. 3539; F. Muell. Fragm. vi. 240.

W. Australia. King George's Sound and neighbouring districts, *Baxter, Drummond, 3rd coll. n.* 245; Stirling Range, *F. Mueller.*

12. **I. roseus,** *Lindl. Bot. Reg.* 1842, *Misc.* 39. A bushy shrub of 1 to 3 or 4 ft., the young shoots tomentose-pubescent or sometimes densely villous, the adult leaves usually but not always glabrous. Leaves once or twice ternately divided or shortly pinnate, the segments linear or cuneate, entire or 3-lobed, rigid, flat, concave or channelled, acute but scarcely pungent, the whole leaf in some specimens scarcely 1 in., in others 2 to 3 in. long, including the petiole, which is often as long as the divided part. Cones terminal, globular, solitary and ¾ to 1 in. diameter, or clustered and scarcely above ½ in. Outer bracts numerous, ovate-lanceolate, acuminate, imbricate, the inner ones almost concealing the scales. Outer cone-scales lanceolate, the inner ones linear, densely woolly outside but tapering into long glabrous or slightly hairy points. Perianth pink, 1 in. long, glabrous, tipped with small tufts of hairs. Style-end linear-clavate, papillose-pubescent, separated by a short neck from the slightly bulbous base of the short nearly glabrous brush. Receptacle convex.—Meis n. in. DC. Prod. xiv. 279; F. Muell. Fragm. vi. 240; *I. scaber,* Meissn. in Pl. Preiss. i. 506, Bot. Mag. t. 4037, not of Lindl.; *Petrophila dubia,* R. Br. Prot. Nov. 7; Meissn. in DC. Prod. xiv. 276.

W. Australia. Swan river, *Fraser, Drummond,* 1st coll. *n.* 564, *Preiss, n.* 682, 686; Dundaragan and Toodyay, *Oldfield.*

13. **I. adenanthoides,** *Meissn. in Hook. Kew Journ.* vii. 69 *and in DC. Prod.* xiv. 278. A shrub with the aspect of an *Adenanthos* near *A. sericea* or of *Petrophila inconspicua,* the branches virgate, hirsute as well as the foliage with long fine spreading hairs. Leaves crowded, trifid, linear-terete, slender, acutely mucronate, ½ to ¾ in. long. Cones terminal, densely surrounded by the floral leaves, depressed-globular, 4 to 5 lines diameter without the perianths. Outer bracts ovate, acute, softly villous outside, passing into the cone-scales of which the inner ones are narrow from slightly spathulate to linear-acuminate. Perianth glabrous, about 1 in. long. Style-end long-clavate, densely papillose-pubescent, with a slight constriction under the pubescent bulbous base of the narrow tapering almost glabrous brush. Receptacle convex.— F. Muell. Fragm. vi. 241.

W. Australia. Hills west of Moore river, *Drummond,* 6th coll. *n.* 171.

SECT. 2. EUSTROBILUS.—Cone-scales all with broad dilated or truncate ends, closely imbricated after flowering in an areolated globular or ovoid mass, often long persistent, but breaking up when the fruits fall. Receptacle convex conical or cylindrical.

14. **I. trilobus,** *R. Br. in Trans. Linn. Soc.* x. 72, *Prod.* 366. A rigid shrub of 1 to 2 ft., the branches and young shoots hoary-tomentose, the adult foliage glabrous or glaucous. Leaves on long petioles, cuneate and broadly 3- or 5-toothed at the end, or more or less deeply 3-lobed with broad and short lobes, all thick and obscurely veined, the whole leaf including the petiole 2 to 3 in. long. Cones terminal, sessile, ovoid-globular, very closely imbricate tomentose and ¾ to 1 in. diameter

after flowering. Outer bracts not numerous, broad, acute or acuminate, shorter than the scales. Cone-scales acute when very young, after flowering broadly cuneate, truncate, thick and hard, 3 lines broad at the top, convex and densely woolly outside. Perianth shortly silky-pubescent, about 4 lines long. Style-end continuous and slightly fusiform, minutely and sparingly hairy on the angles, but scarcely forming a distinct brush. Receptacle oblong-conical.—Meissn. in DC. Prod. xiv. 280; *I. trilobus* var. *eloba*, F. Muell. Fragm. vi. 239; *Petrophila trifida*, Lodd. Bot. Cab. t. 1883, not of R. Br.

W. Australia. Lucky Bay, *R. Brown, Baxter;* Cape Riche, Thomas Brook, and E. Mount Barren, *Maxwell.*

15. **I. tripartitus,** *R. Br. Prot. Nov.* 8. A shrub of 2 or 3 ft., glabrous except the cones, closely allied to *I. trilobus* and *I. longifolius,* with the same inflorescence, cones and flowers, and intermediate between the two in foliage, the leaves being nearly all deeply 3-lobed, with narrow lobes from ½ to 1 in. long.—Meissn. in Pl. Preiss. ii. 247, and in DC. Prod. xiv. 280; *I. trilobus* Meissn. in Pl. Preiss. i. 507; F. Muell. Fragm. vi. 239.

W. Australia. King George's Sound, *Baxter, Drummond, 3rd coll. n.* 246; north of Stirling Range, *F. Mueller.*

This species, united by F. Mueller with *I. trilobus,* seems to pass rather more gradually into *I. longifolius,* and the three might well be considered as varieties of one species, although in the majority of specimens they appear very distinct.

16. **I. longifolius,** *R. Br. in Trans. Linn. Soc.* x. 73, *Prod.* 366. A shrub of 2 to 8 ft., glabrous except the cones or the young shoots minutely hairy. Leaves long, linear or oblanceolate, obtuse with a small callous or acute point, narrowed into a long petiole, thick, longitudinally veined, entire or deeply 2- or 3-lobed, mostly 4 to 6 in. long and sometimes twice as long including the petioles, the lower ones often short. Cones terminal, sessile, ovoid or at length globular, ¾ to 1 in. diameter after flowering. Outer bracts not numerous, acuminate, shorter than the scales. Cone-scales when very young acuminate with narrow points, but after flowering broad and truncate with short points, thick and hard, very numerous and closely imbricate, densely tomentose on the convex back. Perianth yellow, silky-villous, about 5 lines long. Style-end continuous and narrow-fusiform, the thickened part 4-angled and glabrous at the base, the upper portion or brush minutely pubescent in longitudinal lines or glabrous. Receptacle oblong, often 1 in. long.—Meissn. in Pl. Preiss. i. 507, and in DC. Prod. xiv. 281; Bot. Reg. t. 900; F. Muell. Fragm. vi. 237.

W. Australia. King George's Sound and adjoining districts, *R. Brown, Baxter*, *Drummond, n.* 26, *Preiss, n.* 665, and many others ; eastward to Salt river, *Maxwell.* Those specimens in which most of the leaves are 3-lobed only differ from *I. tripartitus* in their greater length. The inflorescence flowers and fruit are the same in *I. trilobus,* *I. tripartitus,* and *I. longifolius.*

17. **I. Drummondii,** *Benth.* A shrub with the branches and young shoots tomentose, the adult foliage glabrous. Leaves undivided, terete

with a callous point, rather thick, resembling those of *Petrophila teretifolia*, attenuate at the base, 1½ to 3 in. long. Cones terminal, at first depressed at length globular, ¾ to 1 in. long, surrounded by numerous floral leaves. Outer bracts lanceolate, not numerous, not exceeding the scales and shorter than them in the fruiting cone. Cone-scales narrow, especially the inner ones and shortly acuminate, but more or less cuneate, densely villous outside, and after flowering their convex summits closely imbricate in a globular mass as in others of this section. Perianth scarcely 4 lines long, the tube usually pubescent, the laminæ glabrous except a small tuft of hairs at the end. Style-end slightly clavate, minutely papillose-pubescent, separated by a very slight constriction from the pubescent slightly bulbous base of the otherwise glabrous but furrowed brush. Receptacle oblong.—*I. petrophiloides*, Meissn. in Pl. Preiss. i. 503, and in DC. Prod. xiv. 276, partly, but not of Br.

W. Australia. Swan river, *Drummond, 1st coll., Preiss.* The foliage of this species is nearly that of the undivided states of *I. teretifolia* and *I. scabriuscula*, with the former of which (the *I. petraphiloides*, Br.) it may have been confounded by Meissner, as he quotes Baxter's specimens as well as Drummond's and Preiss's.

18. **I. heterophyllus,** *Meissn. in Pl. Preiss.* i. 504, *and in DC. Prod.* xiv. 278. Glabrous when in fruit except a slight pubescence below the cone. Leaves terete, usually thickened upwards and incurved, acute, entire, bifid or trifid, rarely with one or two of the segments again divided, 2 to 3 in. long including the petiole. Cones terminal, sessile, ovoid-globular and ¾ in. diameter when in fruit. Outer bracts broad, not numerous, villous outside. Cone-scales broadly cuneate, somewhat hardened and truncate when in fruit, very densely villous outside. Perianth not seen. Receptacle oblong. Coma of the nuts very long.

W. Australia. *Drummond, n.* 731, *Preiss, n.* 672. (I have only seen Drummond's specimens.)

19. **I. villosus,** *Meissn. in DC. Prod.* xiv. 277. Stems very short, thick and woody, rarely 6 in. high, densely tomentose-villous. Leaves terete, rigid, repeatedly forked, 8 to 10 in. long including the long petioles, softly tomentose or at length almost glabrous, the segments divaricate and almost pungent-pointed. Cones ovoid, closely sessile within the leaves in a cluster of 3 or 4, each cone about 1 in. long without the perianths. Outer bracts few and short. Cone-scales cuneate, densely woolly outside, with long lanceolate-subulate plumose deciduous points. Perianth 8 to 9 lines long, very densely hirsute with spreading hairs. Style-end continuous, very shortly thickened and minutely pubescent under the nearly glabrous long and slightly thickened brush. Receptacle oblong, sometimes nearly 1 in. long.—F. Muell. Fragm. vi. 241.

W. Australia, *Drummond, 5th coll. n.* 399.

20. **I. teretifolius,** *R. Br. in Trans. Linn. Soc.* x. 71, *Prod.* 365. A shrub of 2 to 4 ft., the young shoots silky-pubescent, the adult foliage glabrous. Leaves terete, rigid, in a few specimens all or nearly all

simple and 2 or 3 in. long, but usually once twice or even three times
bifid or trifid, the segments usually divaricate, with callous or scarcely
acute points. Cones terminal, sessile or nearly so but almost always
more or less oblique or cernuous, rarely quite erect, depressed-globular,
¾ to 1 in. diameter after flowering. Outer bracts broad, obtuse or with
small recurved points, closely imbricate but rather shorter than the
scales. Cone-scales obovate-cuneate, the convex closely imbricate ends
densely tomentose, the remainder densely villous outside with fulvous
hairs. Perianths very numerous, about ½ in. long, more or less villous,
the laminæ tipped with longer hairs. Style-end shortly clavate and
pubescent, constricted into a short neck below the bulbous base of the
glabrous linear or slightly tapering brush. Receptacle conical.—Meissn.
in Pl. Preiss. i. 504, and in DC. Prod. xiv. 277 ; F. Muell. Fragm. vi.
241 ; *I. petrophiloides*, R. Br. Prot. Nov. 7 (specimens with all or nearly
all the leaves entire) ; *I. cornigerus*, Lindl. S. R. App. 34 (specimens
with the leaves but little divided).

W. Australia. King George's Sound and adjoining districts, frequent, *R. Brown,*
Baxter, and many others ; eastward to the Mounts Barren, *Maxwell;* northward to
Quangen plains, *Preiss,* and towards Swan river, *Drummond, 1st coll. (Preiss, n.* 662,
668, 669, 675, and perhaps 681).

Amidst all the variations of foliage from simple to much divided, which being sometimes met with on one and the same bush, cannot serve to characterize definite
varieties, this species may usually be at once recognised by the cernuous heads which I
have not observed in any other *Isopogon.* There are however a few specimens in
which this character is not very decided or in which the heads are quite erect, possibly
from having become straightened in drying. In Brown's original specimens (from
Baxter) of *I. petrophiloides* the heads are very cernuous, in those of his own collecting
of *I teretifolius* they are erect, in all others that I have seen with divided leaves, excepting one or two of Preiss's, they are decidedly cernuous.

21. **I. anethifolius,** *Knight. Prot.* 94. An erect shrub of 3 or 4 ft.,
glabrous except the cones. Leaves once or twice pinnate, with rather
slender terete usually erect and crowded segments, acute but not pungent, often above 1 in. long. Cones terminal, sessile or shortly
pedunculate within the floral leaves, ovoid, globular, ½ to ¾ in. diameter or
even more when in fruit. Outer bracts numerous but small, mostly
glabrous, more or less acuminate, the inner ones broad and with shorter
points, all shorter than the scales. Cone-scales very numerous, their
broad truncate tomentose ends closely imbricate in the fruiting cone,
the concealed portion densely villous on the back with fulvous hairs.
Perianth yellow, about ½ in. long, sparingly and shortly silky-hairy,
with a tuft of longer hairs towards the end of the laminæ. Style-end
clavate, minutely and densely pubescent, separated by a short but
rather deep constriction from the bulbous base of the brush. Receptacle cylindrical.—R. Br. in Trans. Linn. Soc. x. 71, Prod. 365; Meissn.
in DC. Prod. xiv. 277 ; *Protea anethifolia,* Salisb. Prod. 48 ; *Protea*
acufera, Cav. Ic. vi. 33, t. 549 ; *Protea divaricata,* Andr. Bot. Rep.
t. 465.

N. S. Wales. Port Jackson to the Blue Mountains, *R. Brown, Sieber, n.* 17,
and many others.

22. I. petiolaris, *A. Cunn. in R. Br. Prot. Nov.* 8. A low bushy or procumbent shrub, nearly allied to *I. anemonifolius*, the young shoots tomentose-pubescent, the adult foliage becoming glabrous. Leaves flat and ternately or pinnately divided as in *I. anemonifolius*, but more rigid and striate, the petioles usually 2 to 3 in. long, the segments divaricate, often pungent-pointed. Cones globular, ½ to ¾ in. diameter without the perianths, or from ¾ to 1 in. when in fruit. Outer bracts few, rather broad, acuminate. Cone-scales broadly cuneate, very woolly outside but with longer points than in *I. anemonifolius*. Perianth scarcely 4 lines long, glabrous except the small terminal tufts, the tube short. Styles of *I. anemonifolius*.—Meissn. in DC. Prod. xiv. 279.

Queensland. North of Macintyre's Brook, *A. Cunningham.*
N. S. Wales. Paramatta, *Woolls ;* New England, *(.' Stuart;* Reedy Creek, *C. Moore;* between the Bogan and Buree rivers, *A. Cunningham.*

23. I. anemonifolius, *Knight. Prot.* 93. A shrub of 4 to 6 ft., glabrous except the cones or the branches and young shoots pubescent. Leaves on rather long petioles, once or twice trifid or pinnately divided, with linear or linear-cuneate entire or 2- or 3-lobed segments, usually diverging or falcate, mostly pungent-pointed, rather rigid and obscurely veined, the whole leaf 2 to 4 in. long and nearly as broad. Cones sessile, solitary or in clusters of 2 or 3 at the ends of the branches, nearly globular, ½ to ¾ in. diameter. Outer bracts numerous but mostly small and narrow. Cone-scales very numerous, woolly outside, the expanded truncate imbricate ends becoming glabrous with very minute points. Perianth yellow, 5 to 6 lines, glabrous except the terminal tufts of short hairs. Style-end clavate, minutely papillose-pubescent, separated by a short constriction from the bulbous base of the nearly glabrous brush. Receptacle oblong or cylindrical.—R. Br. in Trans. Linn. Soc. x. 72, Prod. 366 ; Meissn. in DC. Prod. xiv. 279 ; F. Muell. Fragm. vi. 238 ; Lodd. Bot. Cab. t. 1337 ; *Protea anemoni-folia*, Salisb. Prod. 48 ; Bot. Mag. t. 697 ; Andr. Bot. Rep. t. 332 ; *P. tridactylites*, Cav. Ic. vi. 33, t. 548.

N. S. Wales. Port Jackson to the Blue Mountains, *R. Brown, Caley* (with pubescent leaves), *Sieber, n.* 18, and *Fl. Mixt., n.* 480, and many others.

Var. *tenuifolius*, F. Muell. Leaf-segments narrow-linear, short, channelled above like those of *I. formosus.*—Twofold Bay, *F. Mueller*, the specimens in fruit only.

Var.? *pubiflorus.* Leaf-segments numerous, erect, long. Perianth slightly hirsute. —Sydney? *Bynoe.*

24. I. ceratophyllus, *R. Br. in Trans. Linn. Soc.* x. 72, *Prod.* 366. A low glabrous shrub, usually forming dense very prickly tufts under 1 in. high, but sometimes attaining 1 to 2 ft. Leaves crowded, on rather long petioles, flattened but undulate, ternately or pinnately divided into linear rigid intricately divaricate pungent-pointed segments, quite smooth or obscurely striate. Cones surrounded by numerous leaves, globular, about ½ in. diameter or nearly ¾ in. when in fruit. Outer bracts ovate, glabrous, rather thin, imbricate and almost concealing the scales at the time of flowering. Cone-scales broad, villous

outside, the outer ones with short broad glabrous ends, all closely imbricate after flowering. Perianth rather above ½ in. long, the tube very
slender and the laminæ small, glabrous or with minute terminal tufts
of hairs. Style-end clavate, minutely papillose-pubescent, contracted
into a short neck below the bulbous base of the nearly glabrous brush.
Receptacle ovoid-conical, rather short.—Meissn. in DC. Prod. xiv. 279;
Hook. f. Fl. Tasm. i. 319; F. Muell. Fragm. vi. 238.

Victoria. Port Phillip, *R. Brown ;* from the Glenelg river, *Robertson,* to Gipps'
Land, *F. Mueller ;* Wimmera, *Dallachy.*

Tasmania. Flinders' Island, *Gunn;* isles of Bass's Straits, *Bynoe.*

S. Australia. Mount Barker, *Whittaker ;* St. Vincent's Gulf, *Blandowski ;* Lofty
Range, Guichen and Encounter Bays, *F. Mueller.*

W. Australia? King George's Sound, *M'Lean in Herb. Hooker,* but perhaps
some mistake.

25. **I. asper,** *R. Br. Prot. Nov.* 8. A shrub, sometimes low, with
erect nearly simple branches of 1 to 2 ft., (*Preiss* and others), sometimes
more branched and attaining several ft. (*Oldfield*), the branches pubescent, the foliage slightly scabrous. Leaves crowded, pinnate with the
lower segments forked or 3-lobed, all the segments rigid, linear, flat or
channelled, mostly acute, the whole leaf rarely above 1 in. long. Cones
depressed-globular, ½ in. diameter without the perianths, terminal or
on short axillary branches, forming dense leafy clusters at or near the
ends of the branches. Floral leaves numerous, the inner ones with dilated petioles and smaller segments, passing into the ovate acuminate
outer bracts, and the inner ones of these passing into the obovate
spathulate cone-scales, which are villous outside with spreading ovate
coloured glabrous points, the inner scales gradually narrower, all exceedingly numerous and closely imbricate after flowering. Perianth
" red," glabrous, about ½ in. long or rather more. Style-end long and
clavate, densely papillose-pubescent, separated by a short constriction
from the slightly bulbous pubescent base of the brush which is minutely
hirsute in longitudinal lines. Receptacle nearly globular.—Meissn. in
Pl. Preiss. i. 505, and in DC. Prod. xiv. 278; *I. scaber,* Lindl. Swan
Riv. App. 34, not of Bot. Mag.

W. Australia. Swan river, *Drummond, 1st coll. n.* 574; Colonial Church Grant,
Preiss, n. 689; Hampden, *Clarke ;* Gordon and Canning rivers, *Oldfield.*

26. **I. crithmifolius,** *F. Muell. Fragm.* vi. 239. Very closely allied to *I. formosus,* and perhaps one of its numerous varieties, but the
leaves are, as in *I. roseus,* flattened though concave, once or twice ternately divided into linear or linear-cuneate entire or 2- or 3-lobed segments, sometimes very short but more frequently the petiole and the
divided portion each from ½ to 1 in. long. Cones and flowers entirely
of *I. formosus.*

W. Australia. Swan river, *Drummond, 1st coll. n.* 563 (with narrow leaf-segments), *J. S. Roe* (with short broad leaf-segments). Drummond's specimens are referred by Meissner to *I. roseus,* he having inadvertently, as pointed out by F. Mueller,
overlooked the sectional difference in the structure of the cones.

27. **I. formosus,** *R. Br. in Trans. Linn. Soc.* x. 72, *Prod.* 366. A shrub low and bushy, or erect less branched and attaining 4 to 6 ft., the young shoots sometimes densely villous with soft spreading hairs, the adult foliage usually glabrous, the branches more or less tomentose. Leaves rather crowded, once, twice or three times ternately divided into narrow segments, terete or grooved, sometimes short divaricate rigid and pungent-pointed, sometimes longer more erect and acute only, the whole leaf rarely above 2. in. long. Cones terminal or rarely in the upper axils, sessile, globular or at length ovid, ½ to ¾ in. diameter without the perianths, usually very villous. Outer bracts lanceolate or ovate-lanceolate, not exceeding the scales. Cone-scales cuneate, very villous outside, scarcely mucronate, closely imbricate after flowering. Perianth red, glabrous, but with small terminal tufts of hairs, about ¾ in. long. Style-end narrow-clavate, contracted into a short neck below the pubescent bulbous base of the brush. Receptacle oblong, ¾ to nearly 1 in. long.—Bot. Reg. t. 1288; Meissn. in Pl. Preiss. i. 506, ii. 247, and in DC. Prod. xiv. 278; F. Muell. Fragm. vi. 240.

W. Australia. King George's Sound and adjoining districts, very frequent, *R. Brown, A. Cunningham, Drummond,* n. 182, 185, *coll.* 2, n. 295, *coll.* 3, n. 247; *Preiss,* n. 683, 687, and many others, extending to Vasse river, *Oldfield,* and eastward to Cape Arid, *Maxwell,* the latter with rather smaller cones in the upper axils.

28. **I. divergens,** *R. Br. Prot. Nov.* 7. A glabrous shrub, either spreading and 1 to 1½ ft. high, or more bushy and attaining 3 or 4 ft. Leaves once or twice pinnately divided into rather slender though rigid terete segments, obtuse or mucronate, erect or spreading, the whole leaf rarely under 3 in. and often above 4 in. long. Cones terminal, ovoid, conspicuous for their long purple flowers, but the cones themselves never much above ½ in. diameter and ¾ in. long. Outer bracts few and short. Cone-scales broadly cuneate, villous outside, the broad ends becoming glabrous in the old cones and closely imbricate. Perianth fully 1 in. long when well developed, shorter in a few specimens, glabrous except small terminal tufts. Style-end clavate, minutely but densely pubescent, separated by a slight constriction from the broadly bulbous base of the brush which is prominently ribbed and nearly glabrous. Receptacle oblong-cylindrical.—Meissn. in Pl. Preiss. i. 505, and in DC. Prod. xiv. 277; F. Muell. Fragm. vi. 241.

W. Australia. Swan river, *Fraser, Drummond,* 1st *coll.* n. 560, 573, *Preiss,* n. 667; Culjong, Murchison river, *Oldfield.*

29. **I. scabriusculus,** *Meissn. in DC. Prod.* xiv. 276. A much-branched rigid shrub, glabrous except the cones or the branches minutely pubescent. Leaves linear, terete or somewhat flattened and grooved or concave, mucronate, thick, undivided or very rarely shortly forked, 3 to 6 in. long. Cones globular or at length shortly ovoid, about ½ in. diameter, terminal or rarely also in the upper axils. Outer bracts broad, closely imbricate, tomentose outside, persistent and often hardening after flowering, passing into the scales which are narrower, very densely villous outside and mostly with minute glabrous tips.

Perianth rather above ½ in. long, glabrous or minutely pubescent besides the small tufts of hairs at the tips of the laminæ. Style-end slightly clavate, minutely papillose-pubescent, separated by a slight constriction from the pubescent slightly bulbous base of the nearly glabrous brush. Receptacle ovoid-conical.—F. Muell. Fragm. vi. 240.

W. Australia. *Drummond, 4th coll. n.* 263. This species is in many respects allied to *I. Drummondii,* but the leaves are less terete, the perianth longer and more glabrous, although the cones themselves are smaller.

3. ADENANTHOS, Labill.

Flowers hermaphrodite. Perianth regular or nearly so, the tube slender, usually splitting more or less on the lower side; laminæ equal, the perianth usually falling off entire, leaving a very short persistent annular base. Anthers all perfect and free or the lower one linear and sterile, sessile within the segments of the limb, the connective tipped with a small appendage. Hypogynous scales or glands 4, often shortly adnate at the base to the persistent perianth-ring but protruding beyond it. Ovary sessile, with a single laterally attached amphitropous ovule. Style elongated, usually arched and protruding above the middle from the slit of the perianth-tube before the end is set free by the opening of the limb, finally erect and longer than the perianth, usually attenuate below the end, which is more or less thickened or dilated elliptical or linear, with a stigmatic slit descending from the apex to the middle or nearly to the base of the lower side. Fruit a small oblong or rarely ovoid obtuse indehiscent nut (or drupe?) with a single erect seed.— Shrubs sometimes almost growing into small trees, sometimes low and prostrate, often silky-villous. Leaves entire or divided, often rather small and crowded, flat or terete, rarely rigid and pungent-pointed. Flowers red or greenish, terminal or axillary, each flower sessile within a short involucre of 4 to 8, usually 6, imbricate bracts, the inner ones the longest, the involucres solitary or in clusters of 3 or 4, shortly pedunculate or nearly sessile. Perianth usually pubescent or villous outside, the laminæ bearded inside behind the anthers or in a few species almost beardless. Torus with a tuft of hairs round the ovary within the glands.

The genus is limited to Western extratropical Australia, and is not closely allied to any other one hitherto known, although with the inflorescence of the uniflorous species of *Lambertia.*

SECT. 1. **Eurylæma.** — *Perianth-tube obliquely dilated and recurved above the middle. Lower anther linear and sterile. Style-end ovate or elliptical. Leaves flat, entire. Flowers axillary.*

Leaves elliptical, oblong, or lanceolate, ¾ to 2 in. long .   .   .   .   1. *A. barbigera.*
Leaves obovate, ½ to ¾ in. long   .   .   .   .   2. *A. obovata.*

SECT. 2. **Stenolæma.**—*Perianth-tube nearly straight, not enlarged above the middle. Anthers all four perfect. Style-end slightly thickened.*

Flowers axillary. Young shoots hoary-tomentose.
Leaves flat, cuneate, toothed at the broad end .   .   .   .   .   3. *A. cuneata.*

Leaves divided into narrow linear, obtuse, flat, or concave segments . . . . 4. *A. Cunninghamii.*
Leaves terete, rigid, pungent-pointed, entire, bifid or trifid   . 5. *A. pungens.*
Flowers terminal. Leaves entire.
  Leaves sessile, obovate or broadly elliptical, ½ to ¾ in. long   . 6. *A. venosa.*
  Leaves petiolate, oblong-linear or spathulate, under ½ in. long   7. *A. Dobsoni.*
  Leaves narrow-linear, ½ to 1½ in. long. . . . . 8. *A. linearis.*
Flowers terminal. Leaves divided into narrow terete, not pungent segments, usually crowded, at least round the flowers.
  Laminæ of the perianth densely bearded inside behind the
    anthers. Shrubs usually tall and erect.
    Leaves not very dissimilar. Perianth fully 1 in. long.
      Perianth silky-villous outside . . . . 9. *A. sericea.*
      Perianth sparingly glandular-pubescent outside   . . . 10. *A. Meissneri.*
    Floral leaves usually twice as long as those on the branches,
      all filiform. Perianth ¾ in. long, the tube nearly glabrous, the laminæ hairy . . . . 11. *A. filifolia.*
  Laminæ of the perianth glabrous inside, or with few hairs behind the anthers. Shrubs usually procumbent.
    Perianth 1 in long, sparingly glandular-pubescent . . . 10. *A. Meissneri.*
    Perianth ¾ in. long, pubescent or villous. Stem-leaves
      short and appressed; floral ones twice as long . . . . 12. *A. terminalis.*
    Perianth ¾ in. long, villous, the laminæ yellow-plumose.
      Leaves very silky . . . . 13. *A. flavidiflora.*
    Perianth ½ in. long, villous with short hairs. Leaves very
      fine, the floral ones much longer than the others . . . 14. *A. apiculata.*

SECT. 1. EURYLÆMA.—Perianth-tube very obliquely dilated and recurved above the middle. Lower anther (on the back of the style) linear and sterile. Style-end ovate or elliptical, compressed, the stigmatic slit descending along the centre of the upper face. Leaves flat, entire. Flowers axillary.

1. **A. barbigera,** *Lindl. Swan Riv. App.* 36. Stems erect, nearly simple and 1 to 2 ft. high, or with several erect virgate branches and attaining 3 or 4 ft., tomentose-pubescent and hirsute with long fine hairs, the adult foliage often glabrous. Leaves from elliptical-oblong and under 1 in. to lanceolate and 2 in. or linear-lanceolate and nearly 3 in. long, obtuse or with a callous point, contracted into a very short petiole, prominently veined, the primary veins few and almost longitudinal. Peduncles solitary in the axils, 1 to 3 lines long. Bracts lanceolate, acute, villous, the inner ones often ¼ in. long. Perianth villous with fine hairs, rather above 1 in. long, the tube dilated and recurved above the middle, the short laminæ long-cohering, the 3 upper segments ultimately separating to about ⅓ of the perianth, each with a perfect anther in the lamina, the lower segment with a sterile anther and separating lower down. Style glabrous or sparingly bearded with fine hairs, the dilated end elliptical, compressed but thick, the stigmatic slit descending to about half way down the inner face and bordered by slightly raised margins.—Meissn. in Pl. Preiss. i. 510 and in DC. Prod. xiv. 311.

W. Australia. Swan river, *Drummond, 1st coll. n.* 591, *Preiss, n.* 792 ; *Harvey ;* Gordon and Harvey rivers, *Oldfield.*

2. **A. obovata,** *Labill. Pl. Nov. Holl.* i. 29, t. 37. A shrub of 3 or 4 ft. with erect virgate branches, glabrous or minutely hoary-pubescent when young. Leaves rather crowded, entire, obovate, obtuse or with a callous point, contracted at the base but usually sessile, $\frac{1}{2}$ to $\frac{3}{4}$ in. long, obscurely 3-nerved, the nerves converging at the apex and usually visible only on the under side. Peduncles axillary, solitary, 1 to 2 lines long. Inner bracts 2 to 3 lines long and almost acute, outer ones short and obtuse. Perianth about 1 in. long, silky-pubescent or villous, the tube dilated above the middle, then recurved and constricted at the base of the laminæ. Lower anther linear and sterile. Style bearded with few hairs, the dilated end broadly elliptical, compressed but thick, the stigmatic slit descending about half way down the upper face and bordered by raised margins. Fruit oblong, obtuse, about 3 lines long, glabrous or nearly so.—R. Br. in Trans. Linn. Soc. x. 151; Prod. 367; Meissn. in Pl. Preiss. i. 511, and in DC. Prod. xiv. 311.

W. Australia. King George's Sound and neighbouring districts, *Labillardière, R. Brown, A. Cunningham,* and many others; Blackwood river, *Oldfield;* Swan river? *Drummond, 1st coll. n.* 592; near Guildford, *Preiss, n.* 790.

SECT. 2. STENOLÆMA.—Perianth-tube nearly straight, not enlarged above the middle. Anthers all four perfect. Style-end slightly thickened, not compressed, the stigmatic slit or line descending down the upper side.

Meissner describes one anther as abortive in *A. cuneata* and in *A. Meissneri,* which must have been accidental in the flowers examined. I have found all four perfect in all the buds I opened in both species as in all others of this section.

3. **A. cuneata,** *Labill. Pl. Nov. Holl.* i. 28, t. 36. A shrub of 3 to 6 ft., the branches and foliage silky-tomentose. Leaves cuneate, the broad end truncate, with 3 to 7 obtuse crenatures, contracted at the base into a short petiole, the whole leaf $\frac{3}{4}$ to 1 in. long, rather thick, veinless or obscurely 3- or 5-nerved. Peduncles solitary in the axils, slender, often longer than the petioles. Bracts acute, the inner ones enlarged to 3 lines long under the fruit. Perianth about 1 in. long, silky-pubescent, the tube slender and straight or slightly enlarged below the middle after flowering, the laminæ bearded inside behind the anthers which are all perfect. Style-end scarcely thickened. Fruit oblong, about $\frac{1}{4}$ in. long.—R. Br. in Trans. Linn. Soc. x. 152; Prod. 367; Meissn. in Pl. Preiss. i. 511, ii. 247 and in DC. Prod. xiv. 312; *A. flabellifolia,* Knight, Prot. 96; *A. crenata,* Willd. in Spreng. Syst. i. 472.

W. Australia. King George's Sound and adjoining districts, *Labillardière, R. Brown, Drummond, 3rd coll. n.* 245, *Preiss, n.* 793, and others; eastward to Phillip's river and Eyre's Relief, *Maxwell.*

4. **A. Cunninghamii,** *Meissn. in Pl. Preiss.* i. 513, *and in DC. Prod.* xiv. 313. A tall erect shrub, the branches and foliage tomentose and often sprinkled with a few fine spreading hairs, the older leaves less tomentose but hoary. Leaves crowded, once or twice trifid or pinnate

with few rather long linear segments, narrow but flat or concave and mostly obtuse, the whole leaf 1 to 2 in. long including the short petiole. Involucres solitary in the axils, on peduncles of 2 to 3 lines. Bracts acute, silky-hairy, the inner ones 2 lines long. Perianth about 1 in. long, silky-villous, the tube straight. Anthers all perfect. Style-end slender.

W. Australia. King George's Sound, *Fraser ;* in the interior, *Preiss, n.* 2621 (*Meissner*). I have not seen Preiss's specimen ; all others, which I have seen in different collections, appear to have originated in a shrub raised in the Sydney Botanic Garden from Fraser's seeds.

5. **A. pungens,** *Meissn. in Pl. Preiss.* i. 515, ii. 248, *and in DC. Prod.* xiv. 313. A rigid bushy or spreading shrub, the young shoots hoary-tomentose, the adult foliage glabrous. Leaves terete, slightly grooved above, entire or divaricately bifid or trifid above the middle, rather thick, rigid and pungent-pointed, ¾ to 1½ in. long. Involucres solitary in the upper axils, sometimes rather crowded towards the ends of the branches, on peduncles of 1 to 2 lines. Bracts softly silky-villous, the inner ones about 1½ lines long. Perianth rather under 1 in. long, silky-villous. Anthers all perfect. Style slightly hairy, the end scarcely thickened. Fruit oblong, 2 to 3 lines long.—*A. armata,* Meissn. in DC. Prod. xiv. 313.

W. Australia, *Drummond,* 3rd coll. n. 256 (with the leaves mostly, but not all, undivided), 5th coll. n. 400 (with the leaves mostly, but not all, 3-fid) ; in the interior, *Preiss, n.* 671 (*Meissner*). Meissner distinguishes *A. armata* from the divided-leaved specimens of *A. pungens,* by the leaves divided to below the middle and by the laminæ of the perianth-segments longer in proportion, neither of which characters holds good in our specimens, all from Drummond ; I have not seen Preiss's.

6. **A. venosa,** *Meissn. in DC. Prod.* xiv. 311. A bushy shrub of 3 or 4 ft., the branches and young shoots softly and often densely villous, the older foliage nearly glabrous. Leaves crowded under the flowers, more distant along the branches, sessile, entire, obovate or broadly elliptical, shortly acuminate or mucronate-acute, narrowed at the base, often ciliate on the margins, ½ to ¾ in. long, often coriaceous, more or less prominently marked with almost longitudinal veins, the floral ones often rather larger than those below them. Involucres in terminal clusters or umbels of 3 to 6, rarely solitary, on short peduncles ; bracts 4 to 6. Perianth about 1 in. long, slender, glandular-hirsute. Anthers all perfect. Style-end scarcely thickened.

W. Australia, *Drummond,* 4th coll. n. 264 ; in the interior from Cape Le Grand and summit of W. Mount Barren, *Maxwell.*

7. **A. Dobsoni,** *F. Muell. Fragm.* vi. 204. A prostrate much-branched shrub, spreading to 1 or 2 ft. diameter, the young shoots silky-hairy, the older foliage glabrous. Leaves entire, oblong-linear or spathulate, mostly under ½ in. long, very obtuse, narrowed into a distinct petiole. Involucres terminal, solitary, on very short peduncles ; bracts obtuse, the inner ones about 1½ lines long. Perianth slender,

8 to 9 lines long, silky-pubescent. Anthers all perfect. Style-end scarcely thickened.

W. Australia. Point Malcolm, *Maxwell.*

8. **A. linearis,** *Meissn. in DC. Prod.* xiv. 311. Apparently procumbent, with slender branching stems of above 1 ft., the young shoots silky-pubescent and hirsute with long fine hairs, the older foliage glabrous. Leaves entire, narrow-linear, obtuse, attenuate at the base, rather thick but flat, ½ to 1½ in. long. Involucres solitary or 2 or 3 together at the ends of the branches, on peduncles of about 1 line ; inner bracts nearly 2 lines long. Perianth 6 to 7 lines long, softly hairy. Anthers all perfect. Style sparingly bearded, the end narrow-oblong.

W. Australia, *Drummond, 4th coll. n.* 265.

9. **A. sericea,** *Labill. Pl. Nov. Holl.* i. 29. *t.* 38. A tall shrub or small tree of 10 to 20 ft., the branches and foliage softly silky-pubescent or villous with soft appressed or longer or spreading hairs. Leaves crowded, very shortly petiolate, twice ternately or pinnately divided into linear-terete almost filiform segments, often ending in small glabrous gland-like tips and sometimes the lower segments short, as if mutilated, with dilated almost peltate gland-like tips, the whole leaf 1 to 1½ in. long, the floral ones often rather longer than the others. Involucres terminal, solitary or rarely 2 or 3 together, almost concealed by the foliage, on peduncles of about 1 line, the bracts silky-pubescent, the inner ones 2 lines long. Perianth above 1 in. long, silky-villous, slender, the laminæ densely bearded inside behind the anthers which are all perfect. Style glabrous, the end slightly thickened.—*R. Br. in Trans. Linn. Soc.* x. 152, *Prod.* 367 ; Meissn. in Pl. Preiss. i. 513, ii. 248, and in DC. *Prod.* xiv. 312 ; *A. apiculata,* Meissn. in Pl. Preiss. i. 514, and in DC. l.c. 313, not of R. Br.

W. Australia. King George's Sound and adjoining districts, *Labillardière, R. Brown,* and many others, and thence towards Swan river, *Drummond, 1st coll. 3rd coll. n.* 255, *Preiss, n.* 787, 788, and others, and eastward to Cape Arid, *Maxwell.*

The specimens of Drummond's and Preiss's referred by Meissner to *A. apiculata* appear to me to be undistinguishable from the common *A. sericea,* except perhaps in the rather more rigid foliage with more spreading hairs, but even this distinction is very inconstant. I have not seen in any of them the truly lateral gland at the ends of the leaf-segments as in the true *A. apiculata,* Br. (*A. procumbens,* Meissn.).

Var. ? *brevifolia.* Leaves rather shorter but silky-villous and the perianth-laminæ densely bearded inside as in the typical *A. sericea.*—*A. barbata,* F. Muell. Herb.

S. Australia. Kangaroo Island, *F. Mueller, Waterhouse.*

The four following species may perhaps hereafter prove to be varieties only of *A. sericea.*

10. **A. Meissneri,** *Lehm. Pl. Preiss.* i. 512, ii. 248. A procumbent or irregularly spreading shrub of 3 or 4 ft., the branches pubescent or villous, the foliage hirsute pubescent or almost glabrous. Leaves mostly twice trifid but varying either more divided or less so, with terete rather rigid segments, more spreading than in *A. sericea* and

mostly short, the whole leaf often scarcely above ½ in. long, those clustered round the flowers however usually twice as long and often plumose at the base. Involucres terminal, usually 3 or 4 together. Perianth ¾ to 1 in. long, glandular-pubescent and not silky, the laminæ with few hairs inside behind the anthers which are all perfect.—Meissn. in DC. Prod. xiv. 312.

W. Australia, *Drummond, 2nd coll. n.* 301, *Preiss, n.* 791; Point d'Entrecasteaux, *Walcott;* Cape Leschenault and near Bunbury, *Oldfield.*

Var. *velutina.* Softly and densely villous, leaves rather longer and the laminæ of the perianth more bearded inside, showing an approach to *A. sericea,* but the perianth glandular-pubescent only outside as in the typical *A. Meissneri.*—*A. velutina,* Meissn. in DC. Prod. xiv. 312.—W. Australia, *Drummond, 4th coll. n.* 266.

11. **A. filifolia,** *Benth.* A shrub of 5 or 6 ft. with pubescent branches. Leaves glabrous or nearly so, twice or even thrice pinnately divided into filiform segments, those of the stem-leaves short, those of the floral leaves much longer and slightly plumose at the base, the whole leaf on the branches not above ½ in. long, round the flowers ¾ to 1 in., all glandular at the point. Involucres terminal, solitary or 2 or 3 together. Perianth fully ¾ in. long, the tube somewhat angular, glabrous or slightly hairy in the upper part, the laminæ darker coloured, hairy outside, bearded inside behind the anthers which are all perfect.

W. Australia. Stirling Range, *F. Mueller;* Kojonerup hills, *Maxwell.*

Var. *sericifolia.* Leaves silky-pubescent.—W. Australia, *Drummond, n.* 69.

12. **A. terminalis,** *R. Br. in Trans. Linn. Soc.* x. 152, *Prod.* 367. A procumbent shrub, extending to 3 or 4 ft., tomentose-pubescent and more or less sprinkled with fine spreading hairs. Leaves divided into 3 to 7 linear-terete segments, those along the branches usually appressed and 3 to 5 lines long, those around the flowers crowded and twice as long. Involucres terminal, solitary or 2 or 3 together, the inner floral leaves less divided with a dilated ciliate petiole, or even reduced to a simple filiform leaf. Bracts plumose at the base. Perianth about ¾ in. long, hirsute outside. Anthers all perfect, with very few hairs on the perianth-laminæ behind them. Style-end slender.—Meissn. in DC. Prod. xiv. 313; Endl. Iconogr. t. 110.

Victoria. Wimmera, *Dallachy;* N.W. districts, *L. Morton.*
S. Australia. Port Lincoln, *R. Brown;* Onkaparinga and Encounter Bay, *F. Mueller;* Penola, *Woods;* Kangaroo Island, *F. Mueller.*

Preiss's West Australian specimens here included by Meissner are probably referrible to some of the varieties of *A. sericea.* The only ones I have seen are in leaf only.

13. **A. flavidiflora,** *F. Muell. Fragm.* i. 157. A procumbent much branched shrub, the branches and foliage silvery-tomentose and more or less hirsute with spreading hairs. Leaves divided into 3 to 7 linear-terete obtuse segments without terminal glands, those along the branches ¼ to nearly ½ in. long, those crowded round the flowers nearly ¾ in. Involucres terminal, solitary or clustered, on very short pedicels, the bracts silky-hairy. Perianth ¾ in. long, villous outside, the laminæ

A A 2

densely plumose outside with yellow hairs, glabrous inside or nearly so. Anthers all perfect. Style-end narrow.

W. Australia. North of Stirling Range, *F. Mueller;* W. Mount Barren, *Maxwell.*

14. **A. apiculata,** *R. Br. Prot. Nov.* 9, *not of Meissn.* A procumbent shrub spreading to 2 or 3 ft., the branches slender, silky-villous when young. Leaves divided into 3 to 5 filiform segments usually with a depressed lateral gland at the end, those of the branches often short and nearly glabrous, the floral leaves crowded, often 1 in. long, and ciliate with a few long fine glabrous hairs. Involucres 2 or more together in terminal clusters and nearly sessile, the bracts nearly glabrous. Perianth not above $\frac{1}{2}$ in. long, villous with short spreading hairs, the laminæ glabrous inside or with very few hairs behind the anthers which are all perfect. Style-end oblong.—*A. procumbens,* Meissn. in Pl. Preiss. i. 512, ii. 248, and in DC. Prod. xiv. 312; *A. Drummondii,* Meissn. in Pl. Preiss. i. 514, and in DC. Prod. xiv. 313.

W. Australia. King George's Sound or to the eastward, *Baxter, Preiss, n.* 589; towards Cape Riche, *Harvey;* between Swan river and King George's Sound, *Drummond, 1st coll. n.* 593, *3rd coll. n.* 253.

Independently of the fine nearly glabrous foliage and lateral glands (which are not quite constant), this species is readily distinguished from the four preceding ones by the short flowers.

4. STIRLINGIA, Endl.

(Simsia, *R. Br. not of Pers.*)

Flowers hermaphrodite or male by abortion. Perianth regular, the tube cylindrical, at length separating into distinct segments, recurved above the middle. Anthers all perfect, erect on short thick filaments below the base of the laminæ, cohering round the style when the flower first opens, at length recurved with the perianth-segments, the cells of each anther separated by a broad connective, and the two adjoining cells of two adjoining anthers applied face to face in the bud so as to form a single cell. No hypogynous scales. Ovary sessile, with a single anatropous ovule erect from the base; style filiform with a terminal obtuse or dilated and peltate stigma. Fruit a small dry indehiscent nut, usually broadly obovoid or obconical with a convex or nearly flat top, hirsute all over, the upper hairs usually longer forming a coma.— Undershrubs or shrubs usually glabrous, branching and leafy at the base. Leaves dichotomous· or rarely trifid only. Peduncles terminal, leafless, long and simple or more or less branched and paniculate. Flowers small, in globular spikes or heads terminating the branches of the panicle, each flower sessile within a small bract, the rhachis or receptacle cylindrical ovoid or short, usually villous.

The genus is limited to extratropical W. Australia. By the curious conformation of the anthers it connects the *Proteeæ* with the *Conospermeæ.*

Leaf-segments terete, filiform or rigid.

　Bracts narrow, from half as long to nearly as long as the perianth-tube. Peduncles single-headed or rarely divided into 2 or 3 single-headed branches.

　　Peduncles solitary or few, 1 to 1$\frac{1}{2}$ ft. long . . . . 1. *S. simplex.*

Peduncles usually several, 2 to 4 in. long . . . . 2. *S. abrotanoides.*
Bracts broad, ciliate, imbricate in the young spike, as long as
   the perianth-tube. Panicle loose . . . . 3. *S. teretifolia.*
Bracts minute, broad. Panicle loose or many-headed . . . . 4. *S. tenuifolia.*
Leaf-segments flat, linear to oblong-lanceolate. Panicles much-
branched, the ultimate peduncles short . . . . 5. *S. latifolia.*

1. **S. simplex,** *Lindl. Swan Riv. App.* 30. Leafy stems short
Leaves several times di- or tri-chotomous, with slender filiform seg
ments, exceedingly fine and erect when young, but at length more rigid
and spreading, the whole leaf 1 to 2 in. diameter on a petiole of 2 to 6
in. Peduncles simple or with one or two branches near the base, 1 to
1½ ft. long, bearing a single spike of numerous flowers condensed into
a globular head of ½ to ¾ in. diameter. Bracts very small, lanceolate,
acuminate, the inner ones almost subulate. Perianth about 4 lines long,
the laminæ but little more than 1 line. Stigma not so broad as in some
species.—Meissn. in Pl. Preiss. i. 516, and in DC. Prod. xiv. 326; *S.
capillifolia,* Meissn. in Hook. Kew Journ. vii. 70, and in DC. l.c. (some
specimens of the latter with the long peduncle rather more branched).

W. Australia. Swan river, *Drummond, 1st coll. n.* 586, *Preiss, n.* 772 ; between
Moore and Murchison rivers, *Drummond, 6th coll. n.* 173.

2. **S. abrotanoides,** *Meissn. in Pl. Preiss.* i. 517, *and in DC. Prod.*
xiv. 326. Stems rather slender, leafy to the inflorescence, simple or
branched, about 1 ft. high. Leaves smaller and less divided than in
the other species, on short petioles, the segments terete, slender, erect,
the whole leaf rarely exceeding 1 in. Peduncles terminal and in the
upper axils, single-headed but often numerous, 2 to 4 in. long. Flowers
rather numerous in the spike or head. Bracts lanceolate, acuminate, at
least half as long as the perianth-tube.

W. Australia. Swan river, *Drummond, 1st coll. n.* 587, *Preiss, n.* 2622 ; Cabin-
yong, *Oldfield* (in a very imperfect state). The above character is taken from Drummond's specimens quoted by Meissner, in which the bracts are certainly narrow.
Meissner describes them as ovate and minute, probably from Preiss's specimen which I
have not seen. There may be therefore some doubt as to the identity of the two, at
least as varieties.

3. **S. teretifolia,** *Meissn. in Pl. Preiss.* i. 515, *and in DC. Prod.* xiv.
325. Stems erect or ascending, branching and leafy at the base. Leaves
dichotomous, with rather rigid erect terete segments, the divided part
of the leaf 1 to 2 in. long, on a petiole about as long. Spikes or heads
globular, not ½ in. diameter, in a loose but rather rigid panicle, the ultimate peduncles often several in. long and always longer than the spike.
Flowers rather numerous. Rhachis ovoid, villous. Bracts ovate, rigid,
often ciliolate, as long as the perianth-tube and a few of the outer ones
empty or with sterile flowers, forming an involucre under the expanded
spike. Perianth about 2 lines long, the tube scarcely longer than the
laminæ. Stigma slightly peltate. Summit of the nut convex, with
silvery shining hairs.—*S. affinis,* Meissn. in Pl. Preiss. i. 516, and in
DC. l.c.

W. Australia. King George's Sound or to the eastward, *Baxter, Drummond,*
4th coll. n. 267, *Preiss ,n.* 770.

4. **S. tenuifolia,** *Endl. Gen. Pl. Suppl.* iv. 81. Leaves on rather long petioles, crowded at the base of the plant, or on a more or less elongated leafy stem, repeatedly dichotomous, the segments terete, divaricate, very fine in the typical form, sometimes all under ¼ in. long, more frequently about ½ in. or longer. Panicle leafless, sometimes few-headed and scarcely exceeding the leaves, more frequently rather loose and 6 in. to 1 ft. long. Spikes or heads on slender peduncles, rather small. Flowers pale yellow, 8 to 20 in the spike. Bracts very small, ovate, acute. Perianth 2 to 2½ lines long, constricted under the limb. Stigma capitate or slightly peltate. Nuts densely comose.—Meissn. in DC. Prod. xiv. 326; *Simsia tenuifolia,* R. Br. in Trans. Linn. Soc. x. 152, Prod. 368; *Stirlingia anethifolia,* Endl. Iconogr. t. 23, Meissn. in Pl. Preiss. i. 516, and in DC. Prod. xiv. 326.

W. Australia. King George's Sound and adjoining districts, *R. Brown, Harvey, Drummond, 4th coll. n.* 268, *Preiss, n.* 771, *Oldfield, F. Mueller;* Mount Melville, *F. Mueller* (small specimens not above 6 in. high and flowering the first year so as to appear annual).

Var. *anethifolia.* Leaves more rigid, panicle of fewer spikes on shorter peduncles, but the bracts not perceptibly different.—*Simsia anethifolia,* R. Br., in Trans. Linn. Soc. x. 153, Prod. 368; *Stirlingia intricata,* Meissn. in DC. Prod. xiv. 325.—Towards Cape Riche, *Baxter, Harvey, Drummond, 4th coll. n.* 269; Lucky Bay, *R. Brown;* thence to Cape Arid, *Maxwell.*

5. **S. latifolia,** *Steud. Nom. Bot. ed.* 2. An undershrub, the leafy stems rarely 1 ft. high, simple or branching, the leafless peduncle including the panicle 1 to 1½ ft. long. Leaves once or twice bifid or trifid, with flat rigid vertical segments, broadly linear or narrow-lanceolate and 2 to 4 in. long in the typical specimens, the whole leaf then 6 in. to 1 ft. long, narrow-linear and 1 to 2 in. long in some Swan River specimens, cuneate-oblong 2 to 4 in. long and ½ to 1 in. broad in others, all with a small callous point but rounded at the end when broad. Panicle oblong, usually much branched, with minute bracts under the branches. Spikes or heads globular, very numerous, on peduncles of 1 to 3 lines. Bracts very short, broad, truncate. Perianth varying in different specimens from scarcely 2 lines to fully 3 lines long, " of a greenish yellow, reddish at the base." Stigma broadly peltate, undulate. Nut broadly turbinate, densely comose.—Meissn. in Pl. Preiss. i. 517, and in DC. Prod. xiv. 326; *Simsia latifolia,* R. Br. Prot. Nov. 9; *Stirlingia paniculata,* Lindl. Swan Riv. App. 30; Meissn. ll. cc.

W. Australia. King George's Sound, *Baxter, Preiss, n.* 769, *F. Mueller,* and others, and thence to Swan river, *Drummond, 1st coll., Preiss, n.* 767, *Oldfield;* Murchison river, *Oldfield.*

The species is very variable as to ramification, the size and breadth of the leaf-segments and the size of the flowers; the extreme forms I have seen are represented by Preiss's, n. 767, from Swan river, with rather small linear-lobed leaves and small flowers, and by Drummond's from the same locality, with large broad-lobed leaves and large flowers; the typical King George's Sound specimens are intermediate between the two, perhaps nearer to the latter than to the former, and there are many intermediates. In several flowers I observed the ovary abortive, with a short style and no stigmatic dilatation.

Tribe 2. CONOSPERMEÆ.—Anthers: one with 2 perfect cells, two with 1 perfect and 1 abortive cell, the fourth abortive, the perfect cells broad, concave, erect, without any connective, the adjoining ones of distinct anthers applied face to face, so as to form in the bud one cell; all on very short thick filaments at the base of the laminæ or summit of the perianth-tube. Ovule 1. Fruit a dry nut.

5. SYNAPHEA, R. Br.

Flowers hermaphrodite. Perianth oblique or incurved, the tube short, the segments separating, the upper one with an erect ovate or oblong lamina, the three others usually shorter and more spreading. Stamens inserted at the base of the laminæ, the filaments short and thick. Anthers of the lowest stamen with two distinct cells, of the lateral stamens with one cell each, the cells concave, each one of the lateral anthers when in bud facing the adjoining one of the lower anthers and forming but one cell with it, but separating as the flower opens; the upper anther abortive and replaced by a small membrane connecting the filament with the posterior margin of the stigma. Ovary 1-celled, crowned by a tuft of gland-like hairs, with one laterally attached ovule. Style filiform, dilated at the end into an oblique disk, stigmatic on its upper surface which is turned towards the upper perianth-lobe and retained in that position by the membrane connecting it with the filament, the anterior margin of the disk often lobed or 2-horned. Fruit a small indehiscent nut.—Shrubs or undershrubs. Leaves all, or in one species only the lower ones, on long petioles with a sheathing scale-like dilatation at the base, the lamina entire or divided, with few primary veins, pitted all over by minute reticulations. Flowers small, yellow, in spikes often at first dense at length elongated, each one sessile within a small concave bract, the common peduncle simple or branched, often very long, inserted in the axil of a rather large sheathing scale, being the base of an abortive leaf.

The genus is limited to extratropical West Australia, very distinct as a whole from all others, but difficult as to the discrimination of species. With the exception of *S. polymorpha* and *S. pinnata*, the foliage is almost as variable in a single individual as in the whole group of species, and the habit, inflorescence, perianths, and stamens are nearly uniform; there remains therefore, besides minor differences in indumentum and the size of the flowers, very little of specific distinction except the modifications of the stigma or stigmatic end of the style, and even these are sometimes not very well defined.

Spikes simple, not exceeding the shortly petiolate floral leaves . . 1. *S. polymorpha.*
Leaves all on long petioles. Flowering branches long, leafless,
    and usually branched.
    Stigma 2-horned.
        Base of the petioles hirsute. Spike pubescent. Leaves mostly
          entire or shortly lobed . . . . 2. *S. dilatata.*
        Whole plant glabrous or the base of the petioles slightly silky.
          Leaves except the lowest deeply lobed or divided . . . . 3. *S. favosa.*
    Stigma produced into a single oblong incurved entire or 2-lobed
        appendage. Leaves with long diverging lobes . . . . 4. *S. Preissii.*

Stigma produced into a short broad notched or 2-lobed appendage. Leaf-lobes short, divaricate, pungent-pointed. Flowers small . . . . 5. *S. acutiloba.*
Stigma with 2 broad lateral lobes, sometimes shortly confluent. Flowers small. Leaf-lobes long.
   Glabrous or nearly so . . . . 6. *S. petiolaris.*
   Base of the petiole hirsute and spike pubescent as in *S. dilatata* . . . . 7. *S. decorticans.*
Stigma broad without lobes or appendages. Leaf-segments long, distinct, almost petiolulate . . . . 8. *S. pinnata.*

1. S. polymorpha, *R. Br. in Trans. Linn. Soc.* x. 156, *Prod.* 370. Stems leafy, 1 to 2 ft. high, rigid, usually more or less silky especially about the base of the petioles, the adult foliage glabrous. Lower leaves on long petioles, entire or cuneately 3-lobed as in several of the following species, but the upper ones numerous, shortly petiolate, once or twice deeply divided into 2- or 3-lobed or toothed segments, the whole leaf spreading to 2 or 3 in. diameter, the lobes mostly pungent-pointed, broad or narrow, the small reticulations less prominent than in most species. Spikes simple, pubescent, rarely exceeding the leaves. Perianth 2½ to 3 lines long. Stigma produced into an oblong or linear entire or emarginate incurved appendage as in *S. Preissii.* Nut obovoid-oblong, shortly stipitate.—Meissn. in Pl. Preiss. i. 529, and in DC. Prod. xiv. 315; *S. brachystachya,* Lindl. Swan Riv. App. 32; Meissn. in Pl. Preiss. i. 530, and in DC. l.c. 316.

W. Australia. King George's Sound and adjoining districts, *R. Brown* and many others, and from thence to Swan river, *Drummond, 1st coll. n.* 590, *Preiss, n.* 774, 775, and others, and to Murchison river, *Oldfield;* eastward to Cape Arid, *Maxwell.*

2. S. dilatata, *R. Br. in Trans. Linn. Soc.* x. 156, *Prod.* 370, *and App. Flind. Voy.* ii. 606, *t.* 7. Stems very short or decumbent and lengthening out to 1 or even 1½ ft., more or less clothed as well as the petioles, at least when young, with long spreading hairs. Leaves all on long petioles, from cuneate-oblong and entire to broadly cuneate and once or twice 3-lobed or rarely irregularly pinnatifid, the lamina 2 to 4 in. long, usually 1- or 3-nerved when entire, the small reticulations conspicuous. Spikes simple or branched, sometimes only 2 or 3 in., sometimes above 1 ft. long including the peduncle, always more or less silky-villous. Flowers at first dense, but remote when the rhachis elongates. Bracts broad, 1 to 1½ lines long. Perianth pubescent, 3 to 4 lines long. Ovary crowned by a tuft of thick transparent hairs. Stigma anteriorly produced into 2 rather long erect horn-like appendages. Nut small, oblong.—Meissn. in Pl. Preiss. i. 527, ii. 251, and in DC. Prod. xiv. 314; Endl. Iconogr. t. 32; *Conospermum reticulatum.* Sm. in Rees' Cycl. ix.; *Synaphea Drummondii,* Meissn. in DC. Prod. xiv. 315.

W. Australia. King George's Sound and adjoining districts, *Menzies, Baxter, Fraser, Oldfield, Drummond, n.* 21, *2nd coll. n.* 303, *3rd coll. n.* 259, *Preiss, n.* 773, 776.

3. S. favosa, *R. Br. in Trans. Linn. Soc.* x. 156, *Prod.* 369. Stems short or decumbent, the whole plant glabrous or with a short silky pubescence at the base of the petioles and rarely a few short hairs on the spike. Leaves on long petioles, a few of the outer ones entire but mostly divided nearly to the base into 3 entire or 2- or 3-lobed segments, the whole leaf 3 to 10 in. long, including the petiole. Flowering stems leafless, slightly branched, longer than the leaves, the flowers rather numerous, and at length distant. Bracts small. Perianth 2 to 2½ lines long. Stigma 2-horned but the horns not so long as in *S. dilatata.* Nut ovoid, contracted into a stipes nearly as long as itself.— Meissn. in Pl. Preiss. ii. 251, and in DC. Prod. xiv. 314.

W. Australia. King George's Sound, *R. Brown, Baxter, Drummond, 3rd coll. n.* 258; heaths north of Albany, *F. Mueller.* Drummond's 2nd coll. n. 302, referred by Meissner to *S. petiolaris,* and Preiss, n. 780, referred to *S. decorticans,* have certainly, in the specimens examined, the 2-horned stigma of *S. favosa.*

Var. *divaricata.* Leaves shorter, twice or even three times divided into divaricate lobes. Flowering stems shorter and the flowers rather smaller than in the type, but in the specimens the inflorescence is not yet fully developed. The stigma is 2-horned as in the type.—Eyre's Relief, *Maxwell,* and specimens from King George's Sound, *Fraser,* are apparently the same, but not in flower.

4. **S. Preissii,** *Meissn. in Pl. Preiss.* i. 529, ii. 251, *and in DC. Prod.* xiv. 315. Stems short or decumbent, quite glabrous or the dilated base of the petioles very shortly silky-pubescent. Leaves all on long petioles, the lower ones sometimes entire but mostly with long divaricate lobes, the whole leaf sometimes 1 ft. long and the lobes 2 or 3 in., obtuse or acute. Flowering stems long and leafless, slightly branched, glabrous. Perianths usually about 2⅓ lines long, the segments rather narrow. Stigma produced anteriorly into an oblong truncate or emarginate appendage, at least as long as broad and incurved. Nut ovoid, about 2 lines long.

W. Australia. King George's Sound, *Preiss, n.* 779, *Drummond, 3rd coll. n.* 257, *Harvey, Oldfield, Maxwell;* Blackwood and Gordon rivers, *Oldfield.*

5. **S. acutiloba,** *Meissn. in Pl. Preiss.* i. 528, *and in DC. Prod.* xiv. 315. Stems short or decumbent, quite glabrous. Leaves all on long petioles, mostly once twice or thrice ternately divided into short divaricate undulating mostly pungent-pointed lobes, the whole lamina 2 to 3 in. long and broad or sometimes broader than long. Flowers small as in *S. petiolaris,* but not so much incurved. Stigma produced anteriorly into a short broad shortly 2-lobed appendage.

W. Australia. Swan river, *Drummond, 1st coll. n.* 589, *Preiss, n.* 777, 782. Perhaps a variety of *S. petiolaris.*

6. **S. petiolaris,** *R. Br. in Trans. Linn. Soc.* x. 156, *Prod.* 370. Stems short or decumbent, glabrous or slightly silky about the petioles and sometimes a few short hairs on the spikes. Leaves all on long petioles, mostly once or twice or even three times divided into spreading lobes, long and narrow when few, shorter when more divided, obtuse or with short points, the whole leaf including the petiole from a few in. to above 1 ft. long, the lowest leaves as in the allied species usually entire. Flowering stems long and leafless, usually branched, the flowers

small and distant. Perianth more incurved than in other species, not exceeding 2 lines. Stigma anteriorly produced on each side into a broad semicircular auricle or short broad lobe. Nut ovoid, about 2 lines long.—Meissn. in Pl. Preiss. i. 528, and in DC. Prod. xiv. 315.

W. Australia. King George's Sound and adjoining districts, *R. Brown, Baxter, A. Cunningham, Preiss, n.* 781, *Drummond, Oldfield, F. Mueller.*

Var. *gracillima.* Leaf-segments long and narrow. Flowers very small and more curved in slender spikes —*S. gracillima,* Lindl. Swan Riv. App. 32 ; Meissn. in DC. Prod. xiv. 315.—Swan river, *Drummond,* 1st coll. *n.* 588, and a still more slender elongated form, Murchison river, *Oldfield.*

7. **S. decorticans,** *Lindl. Swan Riv. App.* 32. Stems short or decumbent, hirsute as well as the petioles with spreading hairs as in *S. dilatata,* or rarely nearly glabrous. Leaves also as in that species cuneate, undulate, once or twice 3-lobed at the end, 3 to 4 in. long including the petioles. Flowering branches long and slender, perianths scarcely 2 lines long and stigma with short lateral rounded lobes as in *S. petiolaris,* without the horns of *S. dilatata.*—Meissn. in DC. Prod. xiv. 314, partly.

W. Australia. Swan river, *Drummond,* 1st coll.

8. **S. pinnata,** *Lindl. Swan Riv. App.* 32. Leafy stems in our specimens exceedingly short or scarcely any, the whole plant quite glabrous and somewhat glaucous or the spike slightly pubescent. Leaves radical, on long petioles, divided at the end into 3 digitate segments, or rarely pinnate with 5 segments, the lowest pair distant, the segments all contracted at the base, quite distinct, lanceolate, acute, 1½ to 3 in. long, entire or divided into 3 more or less decurrent or confluent segments, the first leaves sometimes undivided. Flowering stems leafless, slender, often above 1 ft. long, with a few long branches. Flowers not numerous, towards the end of the branches, a few of the lower ones distant. Bracts 1 to 2 lines long, broad, acute. Perianth nearly 3 lines long, the claws very oblique and at least as long as the laminæ, and the upper lamina not so broad as in the other species. Stigma broad, concave, without lobes or appendages.—Meissn. in Pl. Preiss. i. 530, and in DC. Prod. xiv. 316.

W. Australia. Swan river, *Drummond,* 1st coll., *Preiss, n.* 783 (*Meissner*). I have only seen Drummond's specimens.

6. CONOSPERMUM, Sm.

Flowers hermaphrodite. Perianth-tube straight, entire ; limb of 4 nearly equal spreading lobes or 2-lipped, the upper lip very broad, concave, shortly acuminate or with recurved margins, the lower with 3 narrow lobes. Stamens inserted in the gibbous apex of the tube or concave base of the limb ; filaments short, thick ; anther of the uppermost stamen with 2 perfect cells, of the lateral stamens with 1 perfect and 1 abortive cells, of the lowest stamen with 2 abortive cells, the perfect cells stipitate erect concave, each one of the lateral anthers when in bud facing the adjoining one of the upper anther and forming with it but one cell, but separating as the flower opens, the abortive cells usually subulate. Ovary obconical, crowned by a tuft of long hairs, 1-celled with 1 pendulous orthotropous ovule. Style filiform at the base, more or less thickened and curved on a level with the anthers and terminating in an oblong or narrow beak with a lateral stigma close to the end elastically turned down towards the lower lobe of the perianth as the limb expands. Fruit a small indehiscent turbinate or obconical nut, the apex broad flat or concave, covered with a coma of usually long hairs, the sides villous with shorter hairs.—Shrubs or undershrubs. Leaves quite entire. Flowers blue lilac pink or white (not yellow), in short dense spikes, which are either sessile in dense compound heads, or solitary on axillary peduncles or variously paniculate on axillary or terminal peduncles, each flower sessile within a broad sheathing persistent bract, the rhachis of the spike often somewhat lengthened and thickened as the flowering advances.

The genus is limited to Australia, and the greater number of species to extratropical W. Australia. Among the Eastern species, the most common one extends to within the tropics. The anthers, style, ovary and fruit are remarkably uniform in the whole genus, and are therefore not mentioned in the following descriptions, although they have been examined in every species of which the specimens were sufficient.

Sect. 1. **Isomerum.**—*Perianth-lobes as long as or longer than the tube, nearly equal and spreading, the cavity in which the anthers are placed forming the summit of the tube and rather more gibbous on the upper side.*

Spikes in a dense compound head, sessile at the base of very
  long leaves terminating a dwarf stem. Perianth villous.
  Leaves linear. Perianth-lobes about as long as the tube . . 1. *C. capitatum.*
  Leaves linear-lanceolate. Perianth-lobes much longer than the
    tube . . . . 2. *C. petiolare.*
Spikes in leafless panicles. Leaves only at the base of the stem.
  Perianth glabrous.
  Leaves terete, rush-like. Spikes in a compact corymbose
    panicle. Perianth ¾ in. long . . . . 3. *C. teretifolium.*
  Leaves flat, linear or lanceolate. Spikes or heads in an intri-
    cately branched divaricate flexuose panicle. Perianth ¼ in.
    long . . . . 4. *C. flexuosum.*

Sect. 2. **Euconospermum.**—*Perianth-limb 2-lipped, as long as or shorter than the tube, the upper lip very broad, concave over the anthers, the lower with 3 narrow lobes.*

Flowers glabrous or pubescent, not woolly.
  Stems leafy to the inflorescence. Spikes not corymbose.
    Western species.
    Peduncles all axillary, short and single-spiked.
      Leaves 1 to 2 in. long, terete, rigid and pungent-pointed.
        Perianth white, lobes as long as the tube . . . . . 5. *C. acerosum.*
      Leaves under ¾ in. long, linear-terete, not pungent. Pe-
        rianth blue, lobes short . . . . 6. *C. amœnum.*
    Peduncles terminal, or if in the upper axils leafy at the
      base, single-spiked and short.
      Leaves flat, oblong . . . . 7. *C. nervosum.*
      Leaves linear-terete, grooved above . . . . 8. *C. diffusum.*

Peduncles terminal and axillary, usually leafy at the base,
  slender, simple or branched.   Bracts large and coloured,
  concealing the very small perianth . . . .   9. *C. glumaceum.*
Stems leafy at the base only, with long terminal simple or
  paniculate leafless peduncles.   Western species (except
  C. longifolium).
Spikes several, sessile along the simple peduncle.
  Leaves terete and rush-like . . . . 10. *C. ephedroides.*
Spikes numerous and small, in a large leafless panicle.
  Leaves almost filiform . . . . 11. *C. polycephalum.*
Spikes not numerous, in a loose panicle. Stems decumbent.
  Leaves oblong or oblanceolate . . . . 12. *C. cœruleum.*
  Leaves narrow-lanceolate or linear . . . . 13. *C. debile.*
Spikes single at the end of a long leafless peduncle.
  Leaves oblanceolate, hirsute, with long spreading hairs . 14. *C. scaposum.*
  Leaves narrow-linear or subulate, glabrous . . . . . 15. *C. Huegelii.*
  Leaves crowded, filiform, hirsute with long spreading hairs 16. *C. densiflorum.*
Spikes several in a compact corymbose panicle at the end
  of the long leafless peduncle.
  Leaves crowded, filiform, hirsute with long spreading hairs 16. *C. densiflorum.*
  Leaves cuneate lanceolate or obovate-oblong, glabrous,
    under 2 in. long . . . . 17. *C. Brownii.*
  Leaves lanceolate oblong-lanceolate or linear, 3 to 6 in.
    long . . . . 18. *C. longifolium.*
Stems leafy to the inflorescence.   Peduncles several, terminal
  or in the upper axils, each with several spikes, forming a
  corymbose panicle.   Eastern species.
Perianth-limb about as long as the tube.
  Leaves very narrow, 3 to 6 in. long. Inflorescence loose 19. *C. tenuifolium.*
  Leaves rigidly linear, crowded, erect, 2 to 3 in. long. In-
    florescence compact . . . . 20. *C. Mitchellii.*
Perianth-limb not above half as long as the tube.
  Leaves rigidly linear, crowded, erect, 2 to 3 in. long . . 21. *C. sphacelatum.*
  Leaves crowded, under 1 in. long (except in one var. of
    C. taxifolium).
    Leaves linear or linear-lanceolate, very spreading . . 22. *C. patens.*
    Leaves linear linear-oblong or lanceolate, erect or
      slightly spreading . . . . 23. *C. taxifolium.*
    Leaves very narrow-linear . . . . 24. *C. ericifolium.*
    Leaves elliptical or oblong-cuneate . . . . 25. *C. ellipticum.*
Flowers very densely woolly-villous except the minute upper lip.
Leaves terete or semiterete.
  Spikes simple in the upper axils. Leaves subulate, crowded. .
    Leaves 1½ to 3 in. long . . . . ; 26. *C. distichum.*
    Leaves under ¾ in. long, very spreading and incurved . 27. *C. floribundum.*
  Spikes racemose or paniculate on a terminal peduncle.
    Leaves slender, crowded, spreading, incurved, ½ to ¾ in.
      long . . . . 28. *C. incurvum*
    Leaves slender, crowded, 1½ to 3 in. long . . . . . 29. *C. brachyphyllum.*
    Leaves rigid, terete or semiterete and channelled, 3 to 6
      in. long or more . . . . 30. *C. stœchadis.*
  Leaves lanceolate, 3-nerved. Spikes paniculate . . . . 31. *C. triplinervium.*
Flowering spikes very densely villous with long spreading silky
  hairs.   Lobes of the perianth as long as the tube.
Leaves at the base of the stem petiolate, obovate, 3-nerved.
  Stem-leaves short, ovate, stem-clasping. Spikes in the
  upper axils flexuose . . . . 32. *C. bracteosum.*
Leaves at the base of the stem very long, with a prominent
  midrib.   Scapes leafless, with a large dense corymbose
  panicle . . . . 33. *C. crassinervium.*

SECT. 1. **Isomerum,** R. Br.—Perianth-lobes as long as or longer than the tube, all nearly equal, linear and spreading, the cavity in which the anthers are placed forming rather the summit of the tube than the base of the lobes, and rather more gibbous on the upper or posterior side.

R. Brown restricted the section *Isomerum* to the *C. flexuosum,* and united the three other species under *Chilurus,* characterized by the longer and more slender perianth-lobes. It appears to me, however, that *C. teretifolium* is much more removed in habit inflorescence and .perianth from *C. capitatum* and *C. petiolare,* than from *C. flexuosum,* and that the four species make one well-marked section which if broken up at all, must be divided into three.

1. **C. capitatum,** *R. Br. in Trans. Linn. Soc.* x. 155, *Prod.* 369. A dwarf shrub or undershrub, resembling at first sight *Isopogon attenuatus.* Stems very short and woody. Leaves crowded, linear, flexuose but rigid, 6 in. to 1 ft. long, with nerve-like margins, contracted into a slender petiole. Flowers sessile amongst the leaves, in dense terminal compound heads of ½ to 1 in. diameter, with numerous imbricate broadly lanceolate acute bracts, black when dry, the common rhachis thick and conical, the partial ones silky-pubescent, lengthening out to from ¼ to ½ in. Perianth slightly pubescent, about 1 in. long, the tube contracted above the middle, nearly equally dilated at the top round the anthers or rather more gibbous on the upper side, slightly contracted over the anthers by the thickened base of the laminæ, which are all equal, linear-subulate and as long as the tube. Style much thickened on a level with the anthers.—Meissn. in Pl. Preiss. i. 526, ii. 251, and in DC. Prod. xiv. 324.

W. Australia. King George's Sound and adjoining districts, *R. Brown, Drummond, 3rd coll. n.* 251, *Preiss, n.* 759, 760, and others.

2. **C. petiolare,** *R. Br. Prot. Nov.* 11. A dwarf shrub or under-shrub with the habit and inflorescence of *C. capitatum.* Stems woody, sometimes very short, sometimes proliferous and 6 to 8 in. high. Leaves linear-lanceolate or oblong-lanceolate, often hooked at the end, contracted into a long petiole, coriaceous, with more or less prominent nerve-like margins, 6 in. to 1 ft. long or a few of the outer ones short and broad. Flower-heads compound, terminal and sessile amongst the leaves, larger than in *C. capitatum,* but with similar imbricate bracts. Perianth villous, the tube ¼ to ½ in. long, very gibbous at the top over the anthers especially on the upper side, the laminæ all equal, almost filiform, about 1 in. long.—Meissn. in Pl. Preiss. i. 525, ii. 250, and in DC. Prod. xiv. 524.

W. Australia. King George's Sound, *Baxter, Drummond, 3rd coll. n.* 250, *Preiss, n.* 757, *F. Mueller.*

3. **C. teretifolium,** *R. Br. in Trans. Linn. Soc.* x. 155, *Prod.* 369. A glabrous erect undershrub, attaining 2 ft. or rather more. Leaves in the lower part of the stem terete, rigid, rush-like, often 6 in. to 1 ft. long. Upper part of the plant leafless forming a terminal corymbose panicle, with numerous flowers in short spikes at the ends of the

branches, the leaves all reduced to small scales. Bracts broadly sheathing, truncate, about 1½ lines long. Perianth glabrous, the tube about
4 lines long, slightly gibbous at the top on the upper side; laminæ
narrow-linear, 6 to 7 lines long, all equal and slightly thickened inside
along the centre.—Meissn. in Pl. Preiss. i. 525, and in DC. Prod. xiv.
324; Endl. Iconogr. t. 46.

W. Australia. King George's Sound, *R. Brown, A. Cunningham, Drummond,
2nd coll. n.* 311, *Preiss, n.* 785, and many others; E. Mount Barren, *Maxwell.*

4. **C. flexuosum,** *R. Br. Prot. Nov.* 11. An undershrub attaining
3 or 4 ft. (*Oldfield*), the greater part occupied by a broad leafless
panicle, with numerous intricately divaricate very flexuose prominently
angled branches. Leaves radical or at the base of the stem, long-
lanceolate, obtuse or with a callous point, narrowed into a long petiole,
rather rigid, with prominent margins, 6 in. to nearly 1 ft. long including
the petiole. Flowers small, whitish, quite glabrous, in little spikes or
heads of 2 to 6 at the ends of the branchlets. Bracts sheathing, obtuse, nearly as long as the perianth-tube. Perianth-tube about 1 line
long, gibbous over the anthers on the upper side; laminæ all equal,
spreading, narrow-oblong, 1½ to 2 lines long, thickened inside along
the centre.—Meissn. in Pl. Preiss. i. 526, ii. 251, and in DC. Prod.
xiv. 324.

W. Australia. King George's Sound and adjoining districts, *Baxter, Drummond, 2nd coll. n.* 309, 310, *5th coll. n.* 402, *Preiss, n.* 753, *Oldfield, F. Mueller ;* Cape
Naturaliste and Vasse river, *Oldfield.*

Sect. 2. Euconospermum.—Perianth-limb 2-lipped, as long as or
shorter than the tube, the upper lip very broad, concave over the
anthers, the end and margins more or less flat and erect or recurved,
lower lip more or less deeply divided into 3 narrow lobes, often
thickened along the centre.

5. **C. acerosum,** *Lindl. Swan Riv. App.* 30. An erect rigid glabrous
shrub, attaining 3 or 4 ft. Leaves terete, rigid, acute and often
pungent-pointed, mostly 1 to 2 in. long. Flowers in axillary pedunculate spikes much shorter than the leaves, or the upper spikes crowded,
longer, and on longer peduncles so as almost to conceal the shorter
leaves. Bracts broad, sheathing, half as long as the perianth-tube.
Perianth glabrous, about 4 lines long, the lobes about as long as the
tube, the upper one broad and gibbous at the base over the anthers, the
lower ones shortly united in a lower lip.—Meissn. in Pl. Preiss. i. 522,
and in DC. Prod. xiv. 318.

W. Australia. Swan river, *Drummond, 1st coll., Preiss, n.* 786 ; between Moore
and Murchison rivers, *Drummond, 6th coll. n.* 174; Murray and Murchison rivers,
Oldfield.

6. **C. amœnum,** *Meissn. in Pl. Preiss.* i. 522, *and in DC. Prod.* xiv.
318. An erect or spreading shrub of 1 or 2 ft., the branches and inflorescence usually hoary-pubescent, the foliage glabrous. Leaves

numerous, linear-terete, mostly acute but not pungent, $\frac{1}{4}$ to $\frac{1}{2}$ in. or rarely $\frac{3}{4}$ in. long. Flowers in axillary spikes, usually few in the spike but the spikes crowded in the upper part of the branches and often exceeding the leaves, the rhachis and bracts minutely or densely pubescent. Bracts broad, sheathing, coloured, more than half as long as the perianth-tube. Perianth 3 to $3\frac{1}{2}$ lines long, retaining the blue colour when dry, nearly glabrous or hoary-tomentose but never woolly as in *C. distichum* and its allies, the concave upper lip as broad as the three lobes of the lower lip.—*C. cærulescens*, F. Muell. Fragm. i. 157.

W. Australia. King George's Sound, *Milne;* Kalgan river and Cooginup, *Oldfield;* Swan river, *Drummond,* 1*st coll. n.* 583, *Preiss, n.* 745 ; Salt river and Cape Knob, *Maxwell.*

7. **C. nervosum,** *Meissn. in Hook. Kew Journ.* vii. 71, *and in DC. Prod.* xiv. 321. Stems leafy, simple at the base, paniculately branched in the upper part and minutely hoary-tomentose. Leaves oblong, obtuse or with a small recurved point, the lower ones several in. long and contracted into a rather long petiole, the others nearly sessile and mostly under 1 in. long, all rigid, veined and with an intramarginal nerve conspicuous on the under side. Spikes short, nearly globular, shortly pedunculate in the upper axils and shorter than or scarcely exceeding the leaves. Bracts broad, acuminate, shorter than the perianth-tube, shortly ciliate and sparingly pubescent as well as the rhachis. Perianth about $3\frac{1}{2}$ lines long, the tube slightly pubescent, the limb as long as the tube, the upper segment or lip concave with recurved margins, the lower about as long and shortly 3-lobed.

W. Australia. Between Moore and Murchison rivers, *Drummond,* 6*th coll. n.* 175. The two varieties mentioned in the Prodromus may both be found on one specimen.

8. **C. diffusum,** *Benth.* A much-branched spreading or diffuse shrub, glabrous or the branches minutely hoary-tomentose. Leaves linear, terete, grooved above, obtuse or with a small recurved point, mostly about 1 in. long. Spikes nearly globular, shortly pedunculate in the upper axils or terminating short axillary branches and shorter than the leaves. Bracts glabrous or minutely ciliate, very broadly sheathing, shortly acuminate. Perianth blue, about 3 lines long, glabrous, the limb as long as the tube, the upper segment or lip concave with recurved margins and the lower lip very shortly 3-lobed, as in *C. nervosum.*

W. Australia, *Drummond.*

9. **C. glumaceum,** *Lindl. Swan Riv. App.* 30. A shrub or undershrub of 3 or 4 ft., quite glabrous. Leaves crowded, linear or linear-lanceolate, acute or with a callous point, with nerve-like margins, $\frac{3}{4}$ to $1\frac{1}{2}$ in. long. Peduncles very numerous terminating short axillary branchlets, slender, simple or branched, 4 in. to above 1 ft. long, forming a large leafy panicle. Spikes terminating the peduncles or branches, remarkable for the thin coloured broadly lanceolate acute bracts, 3 to 5

lines long, and concealing the small flowers. Rhachis slightly hirsute. Perianth glabrous, about 1½ lines long, on a very short pedicel adnate to the base of the bract, the tube obliquely obovate, the upper lip very broad and concave, much shorter than the tube, the lower lip as long as the tube, deeply and narrowly 3-lobed. Coma of the nut short.— Meissn. in Pl. Preiss. ii. 249, and in DC. Prod. xiv. 323; *C. lupulinum*, Endl. Gen. Pl. Suppl. iv. 80; Meissn. in Pl. Preiss. ii. 249.

W. Australia. Swan river, *Drummond, 1st coll. n. 585, Preiss, n. 855.*

10. **C. ephedroides,** *Kipp.; Meissn. in Hook. Kew Journ.* vii. 70, *and in DC. Prod.* xiv. 323. An undershrub with erect rushlike stems of 1 to 2 ft., slightly branched and minutely hoary-silky. Leaves in the lower part only, terete, rush-like, rather thick, 2 to 6 in. long, the upper ones all reduced to small scales. Flowers small, in short spikes sessile and distant along the upper part of the stems. Bracts broadly ovate, acuminate, hirsute at the base, as long as the perianth-tube. Perianth-tube hirsute, cylindrical, a little above 1 line long, the limb glabrous, 2 lines long, the upper lip very broad, concave, obtuse, the lower of 3 narrow convex lobes.

W. Australia. Between Swan river and King George's Sound, *Gilbert, Drummond, n. 25.*

11. **C. polycephalum,** *Meissn. in Pl. Preiss.* ii. 249, *and in DC. Prod.* xiv. 323. An undershrub or shrub of 2 to 3 ft., glabrous except the spikes. Leaves in the lower part of the stem or branches terete, almost filiform, 3 to 6 in. long, or here and there still longer. Spikes numerous, almost globular, in a long leafless much-branched panicle often exceeding 1 ft. Bracts broad, truncate with a small point, shorter than the perianth-tube, more or less pubescent or hirsute in the typical form as well as the rhachis of the spike. Perianth blue, about 3 lines long or rather more, the tube minutely and sparingly pubescent, the limb glabrous, the upper lip broad and concave, about as long as the tube, the lower lip with narrow lobes scarcely exceeding the upper lip.

W. Australia. *Drummond, 2nd coll. n.* 305. Some specimens in young bud from Darling range and Canning river, *Oldfield,* may also possibly belong to the same species.

Var. *leianthum.* Spikes quite glabrous.—Stokes Inlet and Esperance Bay, *Maxwell.*

12. **C. cæruleum,** *R. Br. in Trans. Linn. Soc.* x. 154, *Prod.* 369. An undershrub with a thick woody base and decumbent or ascending flowering stems of 1 to 1½ ft. Leaves at the base of the stems oblong or oblong-lanceolate, 2 to 6 in. long and contracted into a petiole at least as long in the typical form, slightly veined, with an intramarginal or almost marginal nerve conspicuous underneath; there are also sometimes a few smaller narrower leaves below the middle of the stem, the greater part of which is a long narrow leafless panicle with few branches, each bearing a short ovoid or oblong spike of deep blue flowers, the rhachis and bracts white with a silky wool. Bracts broad, with a glabrous point as long as or rather longer than the perianth-tube. Peri-

anth 3 to 4 lines long, the tube slightly hirsute, the lips nearly glabrous, longer than the tube. Apex of the nut very broad and concave.— Meissn. in Pl. Preiss. i. 520, and in DC. Prod. xiv. 322.

W. Australia. King George's Sound, *R. Brown, A. Cunningham, Preiss, n.* 734, *Drummond, Oldfield, F. Mueller.*

Var. *marginatum.* Leaves much smaller and more numerous, the lower ones 2 to 3 in. long including the long petiole. Spikes few, much less woolly or nearly glabrous.— *C. marginatum,* Meissn. in Pl. Preiss. ii. 248, and in DC. Prod. xiv. 323.—W. Australia, *Drummond, 2nd coll. n.* 306 ; Vasse river, *Oldfield.*

Var. *spathulatum.* Leaves still more numerous and smaller, oblong-spathulate, mostly under 1 in. including the short petiole. Spikes woolly as in the typical form.— Between King George's Sound and Swan river, *Harvey.*

13. **C. debile,** *Kipp. ; Meissn. in Hook. Kew Journ.* vii. 70, *and in DC. Prod.* xiv. 322. Stems slender, decumbent or procumbent, 1 ft. long or more, glabrous as well as the foliage. Lower leaves on long petioles, linear or linear-lanceolate, those along the stems not numerous, narrow-linear and sessile, 1 to 2 in. long. Panicle terminal, loose, but slightly branched, with short spikes and flowers similar to those of *C. cæruleum* or rather smaller.

W. Australia. *Gilbert, n.* 164, *Drummond.* Possibly an extreme form of *C. cæruleum.*

14. **C. scaposum,** *Benth.* Apparently herbaceous, the petioles and lower part of the stems hirsute with long fine spreading hairs, the older leaves nearly glabrous. Leaves radical or at the base of the stems, ½ to 1½ in. long and contracted into a petiole about as long, lanceolate, with a callous point and thickened nerve-like margins. Scapes or flowering stems simple or slightly branched, ½ to 1½ ft. high, with a single small nearly globular hirsute spike terminating each branch. Bracts broad, acuminate, ciliate, longer than the perianth-tube. Perianth hirsute with rather long hairs, about 2¼ lines long, the limb rather longer than the tube, the upper lip broad and concave, the lower with three narrow lobes.

W. Australia. Between Swan river and King George's Sound, *Drummond.*

15. **C. Huegelii,** *R. Br. in Endl. Nov. Stirp. Dec.* 58. An under-shrub with the leafy part of the stem very short, glabrous except the spike. Leaves crowded, narrow-linear, from almost subulate and 1 to 2 in. long to 6 or 8 in. long and 1 line broad. Peduncles erect, simple, leafless, often above 1 ft. long, bearing a single terminal ovoid or oblong spike of blue flowers. Bracts ovate, acuminate, villous at the base as well as the rhachis. Perianth glabrous, about 3½ lines long, the limb 2-lipped, shorter than the tube.—Meissn. in Pl. Preiss. i. 521, and in DC. Prod. xiv. 323.

W. Australia. Swan river, *Huegel, Drummond, 1st coll. n.* 584, *Preiss, n.* 735.

16. **C. densiflorum,** *Lindl. Swan Riv. App.* 32. An undershrub, woody branched and leafy at the base, the stems and foliage hirsute with long fine spreading hairs. Leaves densely crowded in the lower

part of the stem, filiform, 1 to 2 in. long. Peduncles leafless, erect, above 1 ft. long, simple with a single terminal spike or bearing a compact terminal corymb of 3 or 4 spikes, all short dense globular or ovoid and hirsute. Bracts acuminate, hirsute with long hairs. Perianth about 5- lines long, the tube shortly and sparingly hirsute, the limb glabrous, shorter than the tube, the very broad concave upper lip shorter than the narrow lobes of the lower lip.—Meissn. in. Pl. Preiss. i. 521, and in DC. Prod. xiv. 324.

W. Australia. Swan river, *Drummond*, 1*st coll. n.* 582, *Preiss, n.* 2301, *b.* (I have only seen Drummond's specimens.)

17. **C. Brownii,** *Meissn. in Pl. Preiss.* ii. 248, *and in DC. Prod.* xiv. 324. Flowering stems apparently simple, leafy in the lower part, glabrous and glaucous as well as the foliage. Leaves lanceolate cuneate or obovate-oblong, almost acute, 1½ to 2 in. long, contracted into a short petiole dilated at the base, rigid, 3-nerved. Peduncle terminal, 6 in. to 1 ft. long, leafless and simple except at the top, where it bears a short compact corymbose panicle of numerous small spikes quite glabrous. Bracts short, broad, obtuse, of a deep blue, the upper ones imbricate. Perianth glabrous, the tube fully 4 lines long, the upper lip broad, concave, about 1 line long, the lower one rather longer and 3-lobed.

W. Australia, *Drummond,* 2*nd coll. n.* 304.

18. **C. longifolium,** *Sm. Exot. Bot.* ii. 45, *t.* 82. A shrub or undershrub, glabrous except the inflorescence or the branches tomentose. Leaves in the typical form lanceolate or oblong-lanceolate, acute, 3 to 6 in. long and narrowed into a long petiole, veined and with nerve-like margins. Peduncles terminal or terminating short branchlets in the upper axils, often 1 ft. long, branched towards the end into a compact corymbose panicle. Spikes at first short and capitate but lengthening to 1 in. or more, the rhachis silky-tomentose. Bracts short, acuminate. Perianth usually pubescent, about 4 lines long, the tube at least twice as long as the limb, the upper lip short broad and concave, the lower somewhat longer with rather broad lobes.—R. Br. in Trans. Linn. Soc. x. 154, Prod. 369; Meissn. in DC. Prod. xiv. 321; *C. Smithii,* Pers. Syn. i. 116.

N. S. Wales. Port Jackson, *R. Brown, Sieber, n.* 41, and others.

Var. *angustifolium,* R. Br. Prot. Nov. 10. Leaves all narrow-linear, the peduncles not usually so long as in the typical form, but the two forms, though at first sight very distinct, are connected by numerous intermediates —*C. tenuifolium,* Sieb. Pl. Exs. not of R. Br.; *C. commutatum,* Roem. and Schult. Syst. iii. Mant. 275.—Port Jackson, *R. Brown, Sieber, n.* 40, and others.

C. acinacifolium, Grah. in Edinb. Philos. Journ. 1826, 171, Meissn. in DC. Prod. xiv. 320, raised from Fraser's seeds, would appear from the detailed description given, to be the same narrow-leaved variety of *C. longifolium.*

19. **C. tenuifolium,** *R. Br. in Trans. Linn. Soc.* x. 154, *Prod.* 369. Stems from a woody base procumbent ascending or erect, often above 1 ft. long, usually glabrous. Leaves numerous, very narrow linear or

almost terete, grooved above, mostly with an incurved point, 3 to 6 in.
long or sometimes much longer.　Peduncles terminal and in the upper
axils, slender, almost filiform, mostly about 6 in. long, bearing each
about 2 to 6 shortly pedunculate spikes of small flowers, forming a terminal corymb.　Bracts broad, shortly acuminate, nearly as long as the
perianth-tube.　Perianth "lilac," pubescent, about 2 lines long, the
limb as long as the tube or rather longer, the lips nearly equal, obtuse,
the upper one concave, the lower one shortly 3-lobed.—Meissn. in DC.
Prod. xiv. 321 ; *C. repens*, Sieb. in Roem. and Schult. Syst. iii. Mant.
276.

N. S. Wales. Port Jackson to the Blue Mountains, *R. Brown, Sieber, n.* 45
A. Cunningham, and others ; Illawarra, *A. Cunningham, Shepherd.*

Meissner describes the perianth-lobes as twice as short as the tube ; this can only
apply to the lobes of the lower lip, the lips themselves are usually rather longer than
the tube.

20. **C. Mitchellii,** *Meissn. in DC. Prod.* xiv. 320.　An erect shrub,
with the crowded erect linear rigid leaves general habit and compact
terminal corymbs of *C. sphacelatum,* of which F. Mueller considers it as a
variety, but the perianths are more densely and softly pubescent, only
3 lines long and the lips as long as the tube, differences which are quite
constant in all the specimens I have seen.— *C. Dallachyi,* F. Muell. Ann.
Rep. 1858 (name only).

Victoria. Grampians, *Mitchell, F. Mueller ;* Wimmera and Lutitt Bay, *Dallachy ;*
Glenelg river, *Robertson, Allitt.*

21. **C. sphacelatum,** *Hook. in Mitch. Trop. Austr.* 342.　An erect
shrub, the branches and young leaves silky or hoary-tomentose,
the older foliage glabrous, the inflorescence pubescent.　Leaves
crowded, erect, linear, rigid, with a small callous point, obscurely 1-
nerved, mostly 2 to 3 in. long.　Peduncles in the upper axils longer
than the leaves, bearing each several spikes, and forming a compact
broad terminal corymb.　Bracts broad, shortly acuminate.　Perianth
shortly pubescent, about 5 lines long, the limb about half as long as
the tube, the lips nearly equal, the upper one broad, concave, shortly
acuminate, the lower one divided to below the middle into 3 narrow
lobes.—Meissn. in DC. Prod. xiv. 320.

Queensland. Near Mount Pluto, *Mitchell.*

22. **C. patens,** *Schlecht. in Linnæa,* xx. 587.　An erect shrub, minutely
hoary-tomentose or the foliage at length glabrous.　Leaves numerous,
spreading, linear or linear-lanceolate, acute, contracted below the middle,
mostly $\frac{1}{2}$ to $\frac{3}{4}$ in. long.　Peduncles several in the upper axils, 3 to 5
in. long, bearing each a small corymb of pedunculate spikes.　Bracts
broad, acuminate, rarely as long as the perianth-tube.　Perianth hoary-
pubescent, about $2\frac{1}{2}$ lines long, the limb about half as long as the tube,
the upper lip very broad, the lower rather longer, divided to the middle
into 3 narrow lobes.—Meissn. in DC. Prod. xiv. 320 ; F. Muell. Pl Vict.
ii. t. 70.

N. S. Wales? Twofold Bay, *F. Mueller* (specimens almost passing into *C. taxifolium*, from which *C. patens* differs chiefly in its loose habit and spreading leaves).
Victoria. Grampians, *F. Mueller;* Wimmera, *Dallachy;* N. W. districts, *L. Morton;* Glenelg river, *Robertson.*
S. Australia. Bethanie, St. Vincent's Gulf, *Behr; F. Mueller*, and others; Kangaroo Island, *Waterhouse.*

23. **C. taxifolium,** *Sm. in Rees' Cycl.* ix. An erect shrub of several ft., with virgate branches, minutely hoary-tomentose or glabrous, the inflorescence usually pubescent. Leaves crowded, linear or lanceolate, acute, rigid, erect or slightly spreading, contracted at the base, mostly $\frac{1}{2}$ to $\frac{3}{4}$ in. long, but in a few specimens nearly 1 in. and the lower ones even still longer. Peduncles in the upper axils usually rather numerous, 1 to 3 in. long, rarely longer, each bearing several pedunculate spikes, the whole forming a more or less corymbose panicle. Bracts broad, acuminate, shorter than the perianth-tube. Perianth pubescent, $2\frac{1}{2}$ to 3 lines long, the limb much shorter than the tube, the upper lip broad and concave, the lower rather longer, divided to the middle into narrow lobes.—R. Br. in Trans. Linn. Soc. x. 154, Prod. 368; Meissn. in DC. Prod. xiv. 319; Hook. f. Fl. Tasm. i. 319; Bot. Mag. t. 2724; *C. falcifolium*, Knight, Prot. 95 (*R. Br.*); *C. affine*, Roem. and Schult. Syst. iii. Mant. 274; *C. spicatum*, R. Br. Prot. Nov. 10; Meissn. in DC. l.c.; *C. propinquum*, R. Br. l.c.; Meissn. l.c.; *C. lavandulifolium*, A. Cunn.; Meissn. in Pl. Preiss. i. 519, and in DC. l.c.

Queensland. Moreton island, *M'Gillivray, F. Mueller;* Estuary of the Burdekin, *Herb. F. Mueller.*
N. S. Wales. Port Jackson to the Blue Mountains, *R. Brown, Sieber, n. 42 and Fl. Mixt. n.* 471; New England, *C. Stuart;* southward to Illawarra, *A. Cunningham.*
Tasmania. Spring Bay, East coast, *Backhouse, Gunn.*

Var. *lanceolata.* Leaves mostly under $\frac{1}{4}$ in. long.—*C. lanceolatum*, R. Br. Prot. Nov. 10; Meissn. in DC. Prod. xiv. 320.—Hunter's river, *R. Brown, Backhouse, Beckler;* Hastings river, *Beckler;* Richmond river, *Henderson.*

Var. *linifolium.* Leaves more spreading and inflorescence looser, forming almost a passage into *C. patens.*—*C. linifolium*, A. Cunn.; Meissn. in Pl. Preiss. i. 518, and in DC. Prod. xiv. 320.—Peel's Island and Red Cliff Point, Moreton Bay, *A. Cunningham.*

Var. ? *leianthum.* Leaves narrow. Bracts and perianths perfectly glabrous and rather smaller than in the typical form.—Tasmania, *Story.*

24. **C. ericifolium,** *Sm. in Rees' Cycl.* ix. An erect shrub of several ft., minutely hoary-tomentose or nearly glabrous, closely allied to *C. taxifolium*, with similar virgate branches, crowded erect short leaves, corymbose inflorescence and the same flowers, and only differing in its much narrower leaves, mostly $\frac{1}{4}$ to $\frac{1}{2}$ in. long, rarely $\frac{3}{4}$ in. or rather more, and about $\frac{1}{2}$ line broad or sometimes quite filiform.—R. Br. in Trans. Linn. Soc. x. 154, Prod. 368; Rudge in Trans. Linn. Soc. x. 292, t. 17; Meissn. in DC. Prod. xiv. 319; Bot. Mag. t. 2850; Endl. Iconogr. t. 31; *C. erectum*, Grah. Edinb. Phil. Journ. 1828, 171 (*Meissn.*).

N. S. Wales. Port Jackson, *R. Brown, Sieber, n.* 43, and many others.

25. **C. ellipticum,** *Sm. in Rees' Cycl.* ix. A shrub with the virgate branches and erect leaves of *C. taxifolium*, but the branches softly villous

and the leaves much broader, varying however from broadly lanceolate to oval-elliptical or oblong-cuneate, obtuse or acute, from under ½ in. to nearly ¾ in. long. Inflorescence corymbose as in *C. taxifolium*, but the peduncles shorter, more villous and the spikes fewer. Perianth villous, 2½ to 3 lines long, the tube but little longer than the lips.—R. Br. in Trans. Linn. Soc. x. 153, Prod. 368; Meissn. in DC. Prod. xiv. 322; *C. rigidum*, Knight, Prot. 95.

N. S. Wales. Port Jackson, *R. Brown*, and others.

Var. *imbricatum*. Leaves more closely imbricate and shorter, mostly about ¼ in. long. —*C. imbricatum*, Sieb. in Spreng. Syst. Cur. Post. 46; R. Br. Prot. Nov. 9; Meissn. in DC. Prod. xiv. 322.—Port Jackson or Blue Mountains, *Sieber*, *n.* 44; Illawarra, *A. Cunningham*.

26. **C. distichum,** *R. Br. in Trans. Linn. Soc.* x. 155, *Prod.* 369, *not of Meissn.* A tall erect bushy shrub, glabrous except the inflorescence or the young shoots minutely tomentose. Leaves rather crowded, linear-terete, slender, sometimes filiform, 1½ to 3 in. long, the floral ones shorter. Spikes shortly pedunculate in the upper axils, 1 to 2 in. long, the rhachis tomentose, the flowers at length distant, very densely silky-woolly. Bracts very small, ovate, the margins woolly-ciliate, the surface glabrous. Perianth about 4 lines long, the very short broad concave upper lip nearly glabrous, but only very shortly protruding from the dense wool which covers the remainder of the perianth including the 3-lobed lower lip.—*C. procerum*, F. Muell. Fragm. i. 157.

W. Australia. King George's Sound or more probably to the eastward? *Baxter ;* Swan river? *Drummond, 1st. coll. n.* 585; Cape Arid, *Maxwell.*

27. **C. floribundum,** *Benth.* A shrub of 2 or 3 ft., closely allied to *C. distichum*, but bearing the same relation to it that *C. incurvum* does to *C. brachyphyllum*. It is usually more bushy and the foliage often assumes a somewhat silvery aspect. Leaves crowded, very narrow linear, almost terete, very spreading and incurved, of a nearly uniform length, rather under ½ in. in some specimens and always under ¾ in. Spikes in the upper axils 1 to 2 in. long, simple as in *C. distichum*, but owing to the number of flowering branches forming a broad compact corymbose panicle. Bracts glabrous, dark-coloured and very conspicuous on the very young spikes, but the larger lower ones very deciduous, and the upper ones which alone remain when the inflorescence is fully developed are all very small. Flowers usually but not always smaller than in *C. distichum*, similarly clothed with a dense silky wool. Perianth as in that species 4 lines long with a very small nearly glabrous upper lip.—*C. distichum*, Meissn. in Pl. Preiss. i. 522, and in DC. Prod. xiv. 318, not of R. Br.

W. Australia. Swan river, *Drummond, 1st coll. n.* 580, *Preiss, n.* 740 ; Stirling Range, *F. Mueller.*

28. **C. incurvum,** *Lindl. Swan Riv. App.* 30. An erect branching shrub, the stems minutely pubescent, the foliage glabrous. Leaves crowded, very narrow linear, almost terete, spreading and incurved, ½ to 1 in. long. Peduncles terminal, 6 to 10 in. long including the in-

florescence, bearing at the base a few small closely appressed erect and subulate leaves or bracts, the remainder a long narrow dense raceme-like panicle. Spikes numerous along the rhachis, nearly sessile, $\frac{1}{2}$ to 1 in. long, very densely silky-woolly. Bracts small and deciduous. Perianth about 3 lines long. The tube slender, the lips very short, the upper one pubescent only and very shortly protruding from the dense silky wool which covers the rest of the perianth.—Meissn'. in Pl. Preiss. i. 523, ii. 250, and in DC. Prod. xiv. 318.

W. Australia. Swan river, *Drummond, 1st coll. n.* 579 ; Perongerup, *Mrs. Knight* (a very imperfect and therefore doubtful specimen).

29. **C. brachyphyllum,** *Lindl. Swan Riv. App.* 31. Very near *C. incurvum* and probably only a long-leaved variety, the young shoots sometimes softly hirsute, the adult foliage glabrous. Leaves more crowded than in *C. incurvum*, filiform, 1 to 3 in. long, the raceme-like panicles sometimes flowering from the base, sometimes supported on a long peduncle. Perianths densely woolly like those of *C. incurvum* but rather longer, mostly about 4 lines long and the small glabrous upper lip rather more conspicuous.—Meissn. in Pl. Preiss. i. 524, and in DC. Prod. xiv. 318 ; *C. filifolium,* Meissn. in Pl. Preiss. i. 523, and in DC. l.c.

W. Australia. Swan river, *Drummond, 1st coll. n.* 578, *Preiss, n.* 2624. The specific name is unfortunately chosen, as the leaves are longer than those of its nearest allied species, although much shorter than in *C. stœchadis*. Meissner's name is better, but of more recent date.

Var. *larifolium*. Leaves more crowded at the base of the stem, the panicle with its long peduncle often above 1 ft. long, and the spikes more developed. Perianths at least 5 lines long —Swan river, *Drummond*. This is the form which Meissner considers as the typical *C brachyphyllum*.

Var.? *rigidum*. Leaves very narrow-linear, but rigid, crowded, erect and 1½ to 3 in. long. Panicle very long and somewhat branched, the spikes short and dense along the branches, as on the rhachis of the typical form.—W. Australia, *Drummond, n.* 35.

30. **C. stœchadis,** *Endl. in Ann. Wien. Mus.* ii. 208, *and Nov. Stirp. Dec.* 60. An erect rigid shrub of 3 or 4 ft., the young shoots silky-tomentose, the adult foliage glabrous. Leaves terete, rigid, 3 to 6 in. long or in a few specimens still longer, rather slender and scarcely channelled above in the typical form. Peduncles in the upper axils usually branched, 6 to 8 in. long, densely velvety-villous, the spikes few long and interrupted. Bracts short, broad, acuminate, tomentose. Perianth 3 to 4 lines long, densely woolly-hirsute, except the very small upper lip, which is pubescent only or almost glabrous.—Meissn. in Pl. Preiss. i. 524 ; *C. sclerophyllum,* Lindl. Swan Riv. App. 30 ; Meissn. in DC. Prod. xiv. 317.

W. Australia. Swan river, *Drummond, 1st. coll. n.* 581 ; *Preiss, n.* 736, 741, 744.

Var. *canalicu'ata*. Leaves longer, rather broader (but still very narrow-linear), more evidently channelled above or concave. Panicle on a longer peduncle, and the flowers rather larger.— *C. canaliculatum,* Meissn. in Pl. Preiss. ii. 250, and in DC. Prod. xiv. 317.—W. Australia, *Drummond, 2nd coll. n.* 307. Some specimens of Preiss's n. 742, appear also to belong rather to this variety than to the typical form.

31. **C. triplinervium,** *R. Br. Prot. Nov.* 11.　A shrub of 2 to 3 ft., the branches erect, glabrous or minutely silky when young.　Leaves in the typical form lanceolate, rather broad, acute or with a callous point, 3-nerved, contracted into a short or rather long petiole, glabrous or silvery-silky, 1½ to 3 in. long, but varying from that to almost linear and 3 or 4 in. long.　Peduncles terminal or in the upper axils, from under 6 in. to nearly 1 ft. long, more or less tomentose, simple or branched, bearing several interrupted spikes of 1 to 3 in.　Bracts small, acuminate.　Perianth 2 to 3 lines long, densely woolly except the very small broad upper lip, which is pubescent only or nearly glabrous.— Meissn. in Pl. Preiss. i. 519, and in DC. Prod. xiv. 316 ; *C. laniflorum,* Endl. in Ann. Wien. Mus. ii. 208, and Nov. Stirp. Dec. 59 ; *C. undulatum,* Lindl. Swan Riv. App. 31 ; Meissn. in Pl. Preiss. i. 520, and in DC. Prod. xiv. 317.

W. Australia.　King George's Sound, *Baxter,* and thence to Swan river, *Drummond, 1st coll. n.* 577 ; *Preiss, n.* 738, 739 ; Kalgan river, *Oldfield;* Salt and Fitzgerald rivers, *Maxwell.*　The undulation of the leaves in the specimens distinguished under the name of *C. undulatum,* appears to me to be accidental only, and I can discover no other character.

Var. *minus,* Meissn.　Leaves 1 to 2 in. long, very shortly petiolate and silvery-silky. —W. Australia, *Drummond, 5th coll. n.* 401.

32. **C. bracteosum,** *Meissn. in Pl. Preiss.* i. 518, ii. 248, *and in DC. Prod.* xiv. 317.　Stems hard, simple or slightly branched, 1 to 1½ ft. high, more or less silky-villous, the young leaves also silky but becoming glabrous when old.　Radical leaves and a few at the base of the stem petiolate, obovate orbicular or spathulate, very obtuse, 3-nerved. ½ to 1 in. long, contracted into a petiole at least as long ; stem-leaves bract-like, sessile, stem-clasping and closely appressed, ovate, shortly acuminate or obtuse, about ½ in. long.　Spikes from the upper axils 2 to 3 in. long, densely silky-villous, the rhachis very flexuose, the flowers distant and very spreading.　Bracts ovate, acute, shorter than the perianth, silky and ciliate.　Perianth recurved, about 3 lines long, the lips more than half as long as the tube, both of them as well as the tube very densely clothed with long spreading silky hairs.

W. Australia.　*Drummond, 3rd coll. n.* 252 ; east from Salt river, *Maxwell ;* also *Preiss, n.* 746 (*Meissn.*), whose specimen I have not seen.

33. **C. crassinervium,** *Meissn. in DC. Prod.* xiv. 317.　Stems forming a short woody base or stock, covered with the imbricate almost distichous remains of old leaves.　Leaves radical or at the ends of the short branches of the stock, linear or linear-lanceolate, 6 in. to above 1 ft. long, acute, silky-pubescent or villous, the margins thick and nerve-like, the midrib very prominent underneath, with a few transverse raised veins when the leaf is broad enough, or the midrib and margins occupying the whole under surface when narrow, contracted into a long petiole dilated and imbricate at the base as in *Synaphea.*　Scapes 1 to 2 ft. high, leafless except small ovate acute spreading scales under the branches, bearing at the end a compact corymbose panicle about 6 in.

diameter, very densely villous with spreading silky hairs. Spikes short and dense terminating the very numerous branches. Bracts under the flowers obovate or cuneate, acute, often 2 lines long besides a long plumose point, the whole bract densely silky-villous outside, glabrous inside. Perianths almost concealed by the bracts, villous with long silky hairs only on the lobes, the lips as long as the tube, the upper one broad and concave, the lower one divided to the base into 3 narrow lobes. Coma of the nut very short.

W. Australia, *Drummond, 4th coll. n.* 270; near the Murra-murra, *Oldfield.*

TRIBE 3. FRANKLANDIEÆ.—Anthers all perfect with adnate parallel cells, enclosed in and adnate to the slender perianth-tube. Ovule 1. Fruit a dry nut with a pappus-like coma.

7. FRANKLANDIA, R. Br.

Flowers hermaphrodite. Perianth regular, the tube long and slender, the lobes spreading. Anthers all perfect, linear, included in and adnate to the perianth tube. Perigynous scales inserted in the perianth-tube below the middle at first united in a ring round the style, at length free from each other and erect. Ovary sessile, crowned by a ring of long hairs or by 3 plumose awns; style filiform with a terminal dilated stigma; ovule solitary, pendulous, orthotropous. Fruit a narrow nut crowned by a pappus-like coma of long hairs or of 3 plumose awns. Embryo with the cotyledons much shorter than the radicle.—Shrubs. Leaves alternate, dichotomously divided into terete segments. Flowers long, " yellow," in racemes either terminal or in the upper axils, solitary within small bracts.

The genus is limited to Western extratropical Australia.

Nut tapering into a short neck crowned by a concave disk bordered
   by a ring of long hairs. Perianth-tube 1 to 1½ in. long . . . . 1. *F. fucifolia.*
Nut tapering into a long neck crowned by 3 long plumose awns.
   Perianth-tube 2 in. long . . . . 2. *F. triaristata.*

1. **F. fucifolia,** *R. Br. in Trans. Linn. Soc.* x. 157, *Prod.* 370, *and App. Flind. Voy.* ii. 604, t. 6. An erect glabrous often glaucous shrub of 2 to 5 ft., the foliage and flowers and sometimes the whole plant sprinkled with glandular tubercles. Leaves petiolate, repeatedly forked, with erect terete rather thick segments of ½ to 1 in., the whole leaf 2 to 6 in. long. Racemes terminal or in the upper axils, 3 to 6 in. long, the flowers distant, shortly pedicellate. Bracts ovate, about 1 line long. Perianth-tube slender, slightly contracted above the middle, 1 to 1½ in. long, the lobes linear-lanceolate spreading, about ¾ in. long. After flowering the segments (including the upper part of the claws or tube) fall off to the base of the anthers, and separate without falling to the insertion of the scales at about ⅓ of the original tube. Anthers adnate to the top of the cells, the connective shortly produced and free above them. Ovary crowned by a ring of long hairs reaching to the top of the scales, with short hairs within them. Style bearded to the level of the top of the

coma, densely villous immediately above it, the remainder glabrous with a dilated stigma on a level with the free tips of the anthers. Nut fusiform, glabrous, contracted into a short neck crowned by a dilated concave disk, sometimes 2 lines diameter, bearing on its margin the long coma of simple hairs resembling the pappus of Compositæ.—Meissn. in Pl. Preiss. i. 530, and in DC. Prod. xiv. 327 ; Endl. Iconogr. t. 52.

W. Australia. King George's Sound and adjoining districts, *R. Brown, Baxter, Drummond, 4th coll. n.* 271, *Preiss, n.* 755, and others ; Tone river, *Oldfield ;* eastward beyond Eyre's Range, *Maxwell.*

2. **F. triaristata,** *Benth.* An erect shrub with the habit and nearly the foliage and inflorescence of *F. fucifolia,* the leaves rather less divided and the ultimate segments shorter. Flowers much larger, the perianth-tube nearly 2 in. long, tapering into a long pedicel, the laminæ lanceolate with a fine point, about 1 in. long. Stamens and perigynous scales the same as in *F. fucifolia,* but the coma of the ovary already consisting of 3 slender awns densely plumose with long hairs. Nut on a densely villous stipes of about ¼ in., the nut itself narrow-oblong, nearly ½ in. long and quite glabrous, tapering into a spirally plumose slender neck attaining 2 to 2½ in., and then branching into 3 plumose awns, also 2 to 2½ in. long when fully developed.

W. Australia, *Drummond ;* Tone and Capel rivers, *Oldfield.*

TRIBE 4. PERSOONIEÆ.—Anthers all perfect, with parallel cells adnate to the connective, the stamens inserted at or below the middle of the perianth-segments. Ovules 2, or sometimes 1. Fruit a drupe or rarely a dry nut or membranous.

8. SYMPHYONEMA, R. Br.

Flowers hermaphrodite. Perianth regular, cylindrical in the bud, the segments free or nearly so. Filaments inserted near the base of the segments, free but incurved and united at the end in a ring round the style, the anthers erect and free, the connective very shortly produced beyond the cells. No hypogynous glands. Ovary shortly stipitate ; style filiform, with a capitate or slightly dilated terminal stigma ; ovules 2, pendulous, orthotropous. Fruit an oblong nut, ripening usually a single seed.—Perennials or undershrubs. Leaves scattered or the lower ones opposite, trichotomously divided into narrow segments. Flowers small, yellow, in rather slender spikes, each one sessile within a small bract.

The genus is limited to Eastern extratropical Australia.

Leaf-segments flat, linear or linear-lanceolate . . . . 1. *S. montanum.*
Leaf-segments very narrow, semi-terete . . . . 2. *S. paludosum.*

1. **S. montanum,** *R. Br. in Trans. Linn. Soc.* x. 158, *Prod.* 371. A perennial or undershrub, with erect or shortly decumbent stems of 1 to 1½ ft., glabrous or the inflorescence very slightly glandular-pubescent.

Leaves shortly petiolate, twice or three times trifid, with short flat linear or linear-lanceolate mucronate-acute segments, the whole leaf 1 to 1½ in. long. Spikes terminal and in the upper axils forming a terminal panicle of 1 to 2 in., the flowers not very close and at length distant. Bracts very small, broad, acuminate. Perianth nearly 2 lines long. Nut oblong, a little more than 1 line long, obtuse, quite glabrous.—Meissn. in DC. Prod. xiv. 328; Reichb. Iconogr. Exot. t. 107 ; Endl. Iconogr. t. 12.

N. S. Wales. Grose river, *R. Brown;* Blue Mountains, *Sieber, n.* 63, *A. Cunningham, Fraser, Woolls,* and others ; Castlereagh, *C. Moore.*

2. **S. paludosum,** *R. Br. in Trans. Linn. Soc.* x. 158, *Prod.* 371. A glabrous perennial or undershrub, closely resembling *S. montanum,* and perhaps a variety only. It is more diffuse, the leaves rather less divided and the segments very narrow, either semiterete and grooved above or concave, rarely almost flat or the lower leaves even quite flat. Flowers rather smaller and more slender than in *C. montanum.* Fruit the same as in that species.—Meissn. in DC. Prod. xiv. 327; *S. abrotanoides,* Sieb. in Spreng. Syst. Cur. Post. 46, and in Roem. and Schult. Syst. iii. Mant. 274.

N. S. Wales. Port Jackson, *R. Brown, Sieber, n.* 61, 62 ; Argyle County, *Fraser;* Illawarra, *Shepherd.*

9. BELLENDENA, R. Br.

Flowers hermaphrodite. Perianth regular, the segments free, spreading. Stamens inserted at the base of the perianth-segments, but free from them ; filaments erect, anthers all perfect, the connective not produced beyond the cells. No hypogynous scales. Ovary shortly stipitate, tapering into a short thick style with a terminal stigma; ovules 2, pendulous, orthotropous. Fruit membranous, compressed, indehiscent, bordered by a very narrow wing, the style reflexed upon one margin.— Shrub. Leaves scattered, toothed at the end, or entire. Flowers small, in a terminal pedunculate dense raceme, without bracts, the pedicels singly scattered, not in pairs.

The genus is limited to a single exclusively Tasmanian species.

1. **B. montana,** *R. Br. in Trans. Linn. Soc.* x. 166, *Prod.* 374. A low glabrous shrub, sometimes under 6 in. high and bushy or tufted, sometimes decumbent and extending to 1½ or two ft. Leaves usually cuneate, broad or narrow, with 3 obtuse crenatures or short rounded terminal lobes, sometimes again broadly crenate, the whole leaf ¾ to above 1 in. long, tapering into a short petiole, flat but rather thick and sometimes glaucous; in some specimens the leaves are much narrower and almost entire, and in one variety mostly oblong-linear and quite entire. Peduncles terminal, much longer than the leaves, bearing a short dense raceme of small white flowers on pedicels of 2 to 3 lines, the rhachis and sometimes the pedicels minutely hoary-pubescent.

Perianth about 1¼ lines long, the stamens nearly as long. Ovary glabrous. Fruit obovate, 4 to 5 lines long, rounded at the end, but the style quite lateral, reflexed, and almost indented into the upper margin. —Meissn. in DC. Prod. xiv. 348; Hook. f. Fl. Tasm. i. 322; Guillem. Ic. Pl. Austral. t. 7.

Tasmania. Mount Wellington, *R. Brown;* abundant on Mounts Wellington, Ben Lomond, Surrey hills, &c. at an elevation of 3000 to 5000 ft. *J. D. Hooker,* and others, the specimens with entire narrow leaves from Ben Lomond, *Milligan, Gunn.*

10. AGASTACHYS, R. Br.

Flowers hermaphrodite. Perianth regular, cylindrical in the bud, the segments free, recurved. Anthers all perfect, on short filaments inserted below the middle of the perianth-segments, the connective shortly produced beyond the cells. No hypogynous glands. Ovary sessile, 3-angled; style rather short, with a thick oblong unilateral stigma; ovule solitary, laterally attached at or near the top. Fruit apparently dry and indehiscent, bordered by 2 broad lateral wings and one narrow dorsal one.—Shrub. Leaves crowded, entire. Flowers white, in axillary elongated spikes, each one sessile within a persistent bract.

The genus is limited to a single species, endemic in Tasmania, and quite exceptional in the Order in the form of the ovary style and fruit.

1. **A. odorata,** *R. Br. in Trans. Linn. Soc.* x. 158, *Prod.* 371. A stout bushy shrub attaining from 5 to 9 ft., quite glabrous. Leaves crowded, linear-oblong, obtuse, contracted into a very short petiole, rather thick, smooth and shining, veinless or the midrib scarcely conspicuous, 1½ to 3 in. long. Spikes numerous, solitary in the upper axils, flowering from the base, 3 to 5 in. long, the upper ones crowded into an erect terminal panicle. Flowers sweet-scented. Bracts erect, lanceolate, from half as long to as long as the perianth. Perianth 3 to 3½ lines long, the segments linear. Style reaching to the base of the anthers, the lateral stigma as long as the rest of the style below it. Fruit not seen quite ripe, but when far advanced and perhaps fully formed it is as long as the subtending persistent bract, with 2 longitudinal rather broad wings almost embracing the rhachis, and one dorsal narrow wing.—Meissn. in DC. Prod. xiv. 328; Hook. f. Fl. Tasm. i. 320.

Tasmania. Adventure Bay, *R. Brown;* S. and W. coasts, Recherche Bay to Port Macquarrie, *Gunn, Milligan,* and others.

11. CENARRHENES, Labill.

Flowers hermaphrodite. Perianth regular, ovoid, acuminate in the bud, the segments free, spreading. Stamens inserted at the base of the segments; filaments short, recurved; anthers incurved, broad, the connective produced into a fine point. Hypogynous scales obovate.

Ovary sessile; style short, filiform, with a small terminal stigma; ovule solitary, pendulous. Fruit a drupe, with a succulent exocarp and a hard endocarp.—Shrub or tree. Leaves alternate, toothed. Flowers in spikes, axillary or terminal, each one sessile within a small bract.

The genus is limited to a single species, endemic in Tasmania, it is, however, closely allied to *Persoonia*, differing chiefly in inflorescence and in the toothed leaves.

1. **C. nitida,** *Labill. Pl. Nov. Holl.* i. 36, *t.* 50. A tall shrub or small tree, attaining rarely 20 to 30 ft. (*C. Stuart*), quite glabrous, of a bright green, fœtid when bruised, turning black in drying. Leaves obovate-oblong or oblong-lanceolate, obtuse, coarsely toothed, contracted into a short petiole, the midrib prominent, otherwise veinless smooth and shining, 3 to 6 in. long. Spikes in the upper axils or several at the ends of the branches, much shorter than the leaves, the rhachis often flexuose but rigid angular and quite glabrous, the flowers rather distant. Bracts small, ovate-triangular, concave. Perianth about 2 lines long, the segments lanceolate, acuminate. Stamens much shorter than the perianth. Ovary short, thick, with a broad pendulous ovule. Drupe very succulent, globular, about ½ in. diameter.—R. Br. in Trans. Linn. Soc. x. 159, Prod. 371; Meissn. in DC. Prod. xiv. 328; Hook. f. Fl. Tasm. i. 320.

Tasmania. Shaded woods, Recherche Bay, Macquarrie harbour and Mountains of the interior, *A. Cunningham, Gunn, Milligan,* and others, but not gathered by R. Brown.

12. PERSOONIA, Sm.

(Linkia, *Cav.*)

Flowers hermaphrodite. Perianth regular, cylindrical in the bud or constricted above the base, the segments free or nearly so, recurved in the upper portion, the laminæ scarcely broader than the claws. Anthers all perfect (except in one species) on short filaments inserted at or below the middle of the perianth-segments, the cells adnate to the connective. Hypogynous scales or glands usually small. Ovary stipitate, (the stipes in a few species very thick and short), with a terminal style either short and inflexed or elongated and filiform, the stigma terminal; ovules 2 or rarely 1, orthotropous, pendulous with short funicles and not strictly collateral, one ovule with a longer funicle or attached lower down than the other. Fruit a drupe, with a succulent exocarp and a hard very hard endocarp, either 1-celled and 1-seeded, or obliquely 2-celled with a single seed in each cell.—Shrubs or small trees. Leaves entire, alternate or rarely here and there almost whorled. Flowers yellow or white, solitary in the axils or owing to the abortion or reduction of the floral leaves forming short racemes at first terminal or axillary, or at length at the base of a leafy branch, rarely in slender terminal 1-sided racemes.

With the exception of a single New Zealand species the genus is limited to Australia.

SECT. 1. **Pycnostyles.**—*Style short, often as thick as the ovary, incurved or hooked at the end, burying the stigma in a cavity of the upper perianth-segment below the anther. Species all Western except P. falcata.*

Leaves terete.
  Leaves rigid, grooved underneath. Perianth glabrous, 5 lines
    long, the upper segment saccate, the upper anther abortive   1. *P. hakeæformis.*
  Leaves rather rigid, not at all or irregularly grooved. Pe-
    rianth pubescent, the upper segment concave but not saccate   2. *P. teretifolia.*
  Leaves slender, more or less distinctly grooved underneath.
    Perianth pubescent, 6 lines long, the upper segment saccate.
    Anthers all perfect . . . . 3. *P. saccata.*
Leaves flat.
  Leaves very narrow-linear, 3 to 6 in long, rigid and doubly
    grooved underneath . . . . 4. *P. Saundersiana.*
  Leaves narrow-cuneate, 1-nerved or longitudinally veined.
    Perianth upper segment saccate . . . . 5. *P. comata.*
  Leaves linear-cuneate, 1-nerved. Perianth upper segment
    concave but not saccate . . . . 6. *P. brachystylis.*
  Leaves long, falcate, narrow or broad, 1-nerved. Tropical
    species . . . . 7. *P. falcata.*

Sect. 2. **Acranthera.**—*Style elongated beyond the anthers, with a terminal
stigma. Connective of the anthers produced into an appendage beyond the cells.
Species all Western.*
Perianth villous, usually ferruginous. Ovary villous (always?)
  1-ovulate.
  Leaves mostly oblong-lanceolate or spathulate.
    Leaves mostly 3-nerved on both sides, not twisted. Flowers
      clustered. Anther-appendages short . . . . 8. *P. trinervis.*
    Leaves 1-nerved above, 3-nerved underneath, twisted.
      Flowers solitary. Anther-appendages long . . . . . 9. *P. tortifolia.*
  Leaves narrow-linear, almost terete.
    Young shoots slightly hoary. Leaves rigid, striate, 1½ to
      3 in. long . . . . 10. *P. angustiflora.*
    Young shoots hirsute with spreading hairs. Leaves
      crowded, 1 to 1½ in. long, channelled above . . . . . 11. *P. rudis.*
Perianth glabrous or pubescent. Ovary glabrous, 2-ovulate (ex-
  cept in *P. striata* and *P. quinquenervis,* and perhaps in *P.
  acicularis*).
  Leaves long, linear-terete . . . . 12. *P. microcarpa.*
  Leaves linear-subulate, pungent-pointed.
    Leaves mostly ¾ to 1 in. long. Anther-appendages long
      and narrow . . . . 13. *P. sulcata.*
    Leaves rarely above ½ in. Anther-appendages very short
      and thick . . . . 14. *P. acicularis.*
  Leaves narrow-linear, not pungent, with revolute margins.
    Perianth pubescent. Style much bent at the base. Leaves
      mostly above ¼ in. . . . . 15. *P. scabrella.*
    Perianth glabrous. Style nearly straight. Leaves rarely
      above ¼ in. . . . . 16. *P. dillwynioides.*
  Leaves linear or linear-lanceolate, prominently 5-nerved or
    rarely 3-nerved. Ovary 1-ovulate.
    Leaves narrow-linear. Anther-appendages rather long   . 17. *P. striata.*
    Leaves broadly linear-spathulate or oblong-lanceolate.
      Anther-appendages rather short . . . . 18. *P. quinquenervis.*
  Leaves linear-lanceolate or oblong-spathulate, 1-nerved.
    Leaves thick, 1 to 2 in. long. Perianth ferruginous-villous.
      Ovary nearly sessile . . . . 19. *P. rufiflora.*
    Leaves crowded, scabrous, under 1 in. Perianth glabrous
      or scarcely pubescent . . . . 20. *P. scabra.*
    Leaves 6 to 8 in. long. Flowers small, glabrous, in slender
      1-sided racemes . . . . 21. *P. graminea.*

SECT. 3. **Amblyanthera.**—*Style elongated beyond the anther-cells, with a terminal stigma. Connective of the anthers not produced beyond the cells.*

* *Western species. Ovary glabrous, the stipes articulate above the base. Leaves flat.*

Leaves linear or linear-lanceolate, falcate, 5 to 8 in. long . . . 22. *P. longifolia.*
Leaves oblong-lanceolate, straight, 3 to 6 in. long . . . . . 23. *P. articulata.*
Leaves broadly ovate or elliptical, 1½ to 3 in. long . . . . . 24. *P. elliptica.*

** *Eastern species. Stipes of the ovary inarticulate or articulate at the very base.*

Ovary villous (rarely almost glabrous in *P. media*).
  Leaves glabrous, flat, ovate, obovate, elliptical or broadly lanceolate.
    Perianth ferruginous-hirsute . . . . 25. *P. ferruginea.*
    Perianth pubescent with appressed hairs.
      Leaves mostly lanceolate. Perianth-segments tipped with
        dorsal points . . . . 26. *P. media.*
      Leaves mostly elliptical. Perianth-segments without
        points . . . . 27. *P. cornifolia.*
      Leaves mostly obovate. Perianth-segments tipped with
        dorsal points . . . . 28. *P. marginata.*
  Leaves pubescent or silky-villous, flat or with recurved margins, from lanceolate to obovate.
    Leaves mostly obovate or oblong-spathulate . . . . . 29. *P. sericea.*
    Leaves mostly narrow . . . . 30. *P. Mitchellii.*
  Leaves scabrous or hispid, with revolute margins.
    Leaves narrow-linear, spreading, incurved, ¾ to 1½ in. long 31. *P. fastigiata.*
    Leaves oblong, rarely exceeding ½ in. . . . . 32. *P. hirsuta.*
  Leaves smooth, linear-subulate, with recurved margins, about
    ¼ in. long . . . . 33. *P. chamæpitys.*
Ovary glabrous. Flowers erect. Leaves flat, veined, mostly
  about 1¼ in., elliptical, falcate, lanceolate or linear, usually
  glabrous.
  Perianth 8 or 9 lines long (6 lines or under in all the following species) . . . . 34. *P. arborea.*
  Leaves mostly falcate, 4 to 8 in. long. Pedicels 2 to 4 lines
    long . . . . 35. *P. salicina.*
  Stems prostrate or trailing. Leaves usually short and broad.
    Pedicels short . . . . 36. *P. prostrata.*
  Stems erect. Leaves rarely above 4 in. when narrow, always
    shorter when broad.
    Leaves mostly lanceolate or elliptical. Pedicels very short.
      Flowers solitary or rarely 2 together . . . . . . 37. *P. lanceolata.*
      Flowers in axillary short racemes or clusters of 6 to 10 . 38. *P. confertiflora.*
        (See also 26. *P. media*).
    Leaves linear-lanceolate, acute, rather long. Pedicels 1 to
      3 lines long . . . . 39. *P. lucida.*
    Leaves linear, often very narrow. Flowers axillary. Ovary
      2-ovulate . . . . 40. *P. linearis.*
    Leaves filiform. Flowers in dense racemes with short floral
      leaves. Ovary 1-ovulate . . . . 41. *P. pinifolia.*
Ovary glabrous. Flowers erect on very short pedicels. Leaves
  with recurved margins or sometimes flat, usually obtuse,
  veinless, glabrous or silky underneath, not exceeding 2 in.
  Perianth about 5 lines long.
    Leaves narrow-linear. Perianth-segments tipped with subulate points . . . . 42. *P. Caleyi.*
    Leaves oblong-linear. Perianth-segments without points . 43. *P. ledifolia.*
    Leaves from obovate-oblong to oblong-lanceolate, very obtuse 44. *P. revoluta.*
  Perianth about 9 lines long. Leaves flat, thick, very obtuse . 45. *P. Gunnii.*

Ovary glabrous. Flowers erect, almost sessile. Leaves with
much recurved or revolute margins, narrow, acute, nerve-
less, silky underneath.
Leaves spreading, smooth above, lanceolate or linear-lanceo-
late, 1½ to 2½ in. long . . . . 46. *P. mollis.*
Leaves incurved, scabrous above, ¾ to 1½ in. long.
Leaves spathulate or linear-spathulate. Perianth villous . 47. *P. rigida.*
Leaves narrow-linear. Perianth pubescent . . . . . 48. *P. curvifolia.*
Ovary glabrous (rarely with a few hairs in *P. oblongata*).
Flowers spreading or nodding. Leaves flat or with recurved
margins, the midrib conspicuous, under 1½ in. long.
Leaves ovate-lanceolate (1 to 1½ in.). Pedicels about ½ in.
long. Perianth glabrous, without points . . . . . 49. *P. oblongata.*
Leaves broad or lanceolate, under 1 in. Pedicels 1 to 4 lines
long.
Leaves ovate, flat (¼ to 1 in.). Perianth glabrous, with
long points to the segments . . . . 50. *P. Cunninghamii.*
Leaves lanceolate to almost ovate, flat (⅓ to 1 in.) Perianth
pubescent, with moderate points . . . . 51. *P. myrtilloides.*
Leaves ovate (2 to 3 lines) to lanceolate (3 to 6 lines). Pe-
rianth glabrous, without points . . . . 52. *P. oxycoccoides.*
Leaves narrow-linear . . . . 53. *P. nutans.*
Ovary glabrous Leaves linear (broad or narrow), concave or
grooved above or nearly flat without any prominent mid-
rib.
Leaves mostly 1 to 1½ in. long.
Leaves oblong-linear or linear-lanceolate, 1½ to 2 lines
broad. Pedicels short and thick . . . . 54. *P. angulata.*
Leaves narrow-linear. Pedicels slender . . . . . . 55. *P. virgata.*
Leaves mostly ½ to ¾ in. long.
Leaves narrow-linear, not pungent. Ovary 1-ovulate . . 56. *P. chamæpeuce.*
Leaves narrow-linear or subulate, pungent-pointed . . . 57. *P. juniperina.*
Leaves filiform, not pungent.
Ovary 2-ovulate. Perianth-segments without points . . 58. *P. tenuifolia.*
Ovary 1-ovulate. Perianth-segments with subulate points 59. *P. acerosa.*

SECT. 1. PYCNOSTYLIS, Meissn.—Style short, often as thick as the
ovary, incurved or hooked at the end, burying the stigma in a cavity of
the upper perianth-segment below the anther.

1. **P. hakeæformis,** *Meissn. in DC. Prod.* xiv. 330. A very rigid
shrub, the young shoots and inflorescence softly pubescent or villous.
Leaves terete, very spreading, mostly recurved at the end, grooved
underneath, thick and rigid, 1 to 2 in. long. Pedicels 2 to 3 lines long,
softly villous, crowded into a terminal or subterminal raceme, with most
of the floral leaves reduced to small bracts. Perianth glabrous, not
above 5 lines long, the upper segment saccate below the anther, which
is quite adnate and almost or quite sterile, the other anthers free except at
the base, the connective produced into a thick obtuse appendage. Ovary
glabrous, contracted at the base into a thick stipes, and tapering into a
short thick style curved and hooked at the end, with a lateral stigma
buried in the cavity of the upper perianth-segment. Fruit not seen.

W. Australia, *Drummond, 4th coll. n.* 275.

2. **P. teretifolia,** *R. Br. in Trans. Linn. Soc.* x. 160, *Prod.* 372. A
bushy shrub of several ft., the young shoots and inflorescence ferrugi-

nous-tomentose or shortly villous, the adult foliage glabrous. Leaves terete, rather slender but rigid, not continuously grooved, although sometimes irregularly so owing to the shrivelling in drying, 1½ to 3 in. long but mostly about 2 in., the floral ones similar or a few of them much reduced in size. Pedicels 1 to 2 lines long, solitary in the axils but sometimes crowded at the base or at the end of a shoot with the lower floral leaves abortive. Perianth declinate, ferruginous-pubescent, about ½ in. long, the upper segment very concave but not saccate. Anthers all perfect, the connective produced into a long point. Ovary scarcely contracted at the base, tapering into short thick style, curved against the upper perianth-segment below the anthers, ovules 2. Drupe obliquely ovoid-oblong, ½ in. long or more.—Meissn. in DC. Prod. xiv. 329; *P. scoparia,* Meissn. l.c.

W. Australia. Lucky Bay, *R. Brown;* King George's Sound towards Cape Riche and Salt river, *Harvey, Baxter, Drummond, 4th coll. n.* 276.

Var.? *amblyanthera.* Appendage to the anthers short and obtuse, the specimens showing no other difference, yet perhaps a distinct species.—Murchison river, *Oldfield.*

3. **P. saccata,** *R. Br. Prot. Nov.* 12. An erect shrub of 2 to 6 ft., the young shoots and inflorescence pubescent or villous, the adult foliage glabrous. Leaves linear-terete, sometimes almost filiform, 2 to 4 in. long or even longer, more or less distinctly grooved underneath or the groove doubled by the prominence of the midrib between the recurved or thickened margins, but the groove always very narrow and sometimes very faint. Pedicels 2 to 3 lines long, mostly at the ends or below the ends of the branches and crowded into racemes with the floral leaves much reduced or abortive, rarely all axillary. Perianth very oblique, pubescent, about ½ in. long, the upper segment saccate below the anther. Anthers all perfect, the connective produced into a rather long point. Ovary contracted into a rather short thick style curved into the cavity of the upper perianth-segment. Ovules 2. Fruit obovoid, rather shorter and thicker than in *P. teretifolia.*—Meissn. in DC. Prod. xiv. 329; *P. Fraseri,* R. Br. Prot. Nov. 12, not of Meissn.; *P. macrostachya,* Lindl. Swan Riv. App. 35; Meissn. in Pl. Preiss. i. 531, and in DC. Prod. xiv. 330.

W. Australia. West coast, *Baudin's Expedition;* Swan river, *Fraser, Drummond, 1st coll. n.* 598, *Preiss, n.* 730 ; Cape Naturaliste, *Collie ;* Donnelly river, *T. C. Carey.* The specimens of *P. Fraseri,* both in Brown's and in Hooker's herbarium, have lost all their flowers, but in other respects correspond entirely with the *P. saccata,* evidently a common plant about Swan river.

4. **P. Saundersiana,** *Kipp. ; Meissn. in Hook. Kew Journ.* vii. 72, *and in DC. Prod.* 330. Branches virgate, pubescent or villous, the adult foliage glabrous. Leaves linear, 3 to 6 in. long, flat but thick, rigid and very narrow, with a double groove on each surface formed by the prominent midrib and marginal or submarginal nerves. Pedicels 3 to 4 lines long, villous, all axillary or crowded at the base of the shoots with the lower floral leaves abortive. Perianth glabrous or slightly pubescent with appressed hairs, about ½ in. long, the upper segment saccate

below the anther. Anthers all perfect, the connective produced into an obtuse appendage. Ovary broad, scarcely contracted at the base, tapering into a cylindrical style, hooked at the end under the upper anthers. Ovules 2. Young fruit obliquely ovoid-globular.

W. Australia, *Drummond, 5th coll. suppl. n.* 4.

5. **P. comata,** *Meissn. in Hook. Kew Journ.* vii. 71, *and in DC. Prod.* xiv. 330. A low shrub with erect branches softly pubescent as well as the young shoots, the adult foliage glabrous. Leaves linear-cuneate or oblanceolate, 1½ to 3 in. long, obtuse with a small point, contracted into a short petiole, of a pale green, thick and rigid, with nerve-like margins and a prominent midrib, and when broad with several very obscure oblique almost longitudinal veins. Pedicels 2 to 6 lines long, villous, axillary or forming a raceme at the base of the shoots with the lower floral leaves reduced to small bracts. Perianth shortly pubescent, rather above ½ in. long, very oblique with a short incurved point, the upper segment saccate below the anther. Anthers all perfect, the connective produced beyond the cells. Ovary glabrous, contracted into a short thick stipes, thickened upwards and incurved, with an oblique stigma buried in the cavity of the upper perianth-segment. Ovules 2.

W. Australia. Murchison river, *Oldfield;* near Dundagaran, *Drummond, 6th coll. n.* 178.

6. **P. brachystylis,** *F. Muell. Fragm.* vi. 221. An erect shrub of several ft., the branches and young leaves softly tomentose, the adult foliage glabrous. Leaves linear-cuneate, 1 to 2 in. long, obtuse with a small callous point, contracted into a short petiole, thick and rigid, with the midrib and margins prominent underneath. Pedicels axillary, 2 to 4 lines long. Perianth oblique, with a short recurved point, more or less silky-hairy, about 5 lines long, the upper segment concave below the anthers but not saccate. Anthers all perfect, rather long, the connective produced into a short obtuse appendage. Ovary glabrous, contracted into a short thick stipes, tapering into a short thick incurved style concealing the stigma under the upper anthers. Ovules 2.

W. Australia. Murchison river, *Oldfield.*

7. **P. falcata,** *R. Br. in Trans. Linn. Soc.* x. 162, *Prod.* 373. Usually a small tree, glabrous or the young shoots minutely tomentose-pubescent. Leaves linear or lanceolate, falcate, 4 to 8 in. long, and very variable in width, obtuse or acuminate, contracted into a petiole, the midrib prominent, the margins usually nerve-like, the lateral veins obscure or fine and very oblique. Pedicels slender, ¼ to ½ in. long, glabrous as well as the perianth or very minutely hoary-pubescent, sometimes all axillary, but more frequently forming a long raceme with the lower floral leaves reduced to bracts and growing out at the end into a leafy shoot. Ovary glabrous, of nearly uniform thickness with the short stipes and incurved style, the stigma oblique under the upper anther. Anthers all perfect, but the upper one usually smaller, the connective produced into a long

or short point.—Meissn. in DC. Prod. xiv. 331; *P. mimosoides*, A. Cunn. Herb.

N. Australia. Islands of the Gulf of Carpentaria, *R. Brown;* M'Adam Range, *F. Mueller;* Escape Cliff, *Hulls;* Victoria river, *Bynoe;* Cygnet Bay, N.W. coast, *A. Cunningham* (a narrow-leaved form, with the pedicels only remaining, but apparently rather this species than *P. longifolia*).
Queensland. Cape York, *Daemel;* Dayman's island, Endeavour Straits, *W. Hill;* Endeavour river, *A. Cunningham, W. Hill;* Æstuary of the Burdekin, *Fitzalan;* Kennedy district, *Daintree;* Edgecombe and Rockingham Bays, *Dallachy;* Cape and Bowen rivers, *Bowman.*

SECT. 2. ACRANTHERA.—Style elongated beyond the anthers, with a terminal stigma. Anthers all perfect, the connective produced into an appendage beyond the cells.

8. **P. trinervis,** *Meissn. in DC. Prod.* xiv. 332. A shrub with the young shoots silky-pubescent or villous, the adult foliage glabrous. Leaves oblong-lanceolate or linear-spathulate, obtuse with a callous point, narrowed into a short petiole, thick and rigid, 3-nerved but the lateral nerves often almost marginal and less conspicuous on the upper than on the under surface, 1 to 2 in. long. Flowers several together clustered in the axils, sessile or on very short thick pedicels, each within a small villous bract. Perianth densely villous with ferruginous hairs, very obtuse, about 5 lines long. Anthers with very short obtuse appendages to the connective. Ovary densely villous, contracted into a very short stipes; style straight, rather thick but elongated, with a terminal stigma; ovule solitary.

W. Australia, *Drummond, 5th coll. suppl. n.* 5.

9. **P. tortifolia,** *Meissn. in DC. Prod.* xiv. 331. Stems in our specimens several from a woody rhizome, branching, not above 1 ft. high, the branches and young shoots silky or hoary-pubescent, the foliage at length glabrous or nearly so. Leaves oblong- or linear-lanceolate, rarely almost obovate, mucronate, contracted into a short petiole, spirally twisted at least when dry, coriaceous, 1-nerved on the upper surface, mostly 3-nerved underneath, under 1 in. or a few of the larger ones 1½ in. long. Flowers solitary in the axils on very short pedicels. Perianth about ½ in. long, silky-ferruginous with appressed hairs. Anthers with rather long narrow appendages to the connective. Ovary densely villous, on a glabrous stipes, tapering into a long straight glabrous style, with a terminal stigma; ovule solitary.

W. Australia, *Drummond, n.* 169, *4th coll. n.* 272.

10. **P. angustiflora,** *Benth.* A shrub of about 1 ft., the erect branches minutely hoary-tomentose, the foliage glabrous. Leaves very narrow linear or terete as in *P. microcarpa,* but usually more rigid, rarely above 3 in. long, and the thicker midrib and more prominent margins give them a more striate or doubly grooved appearance. Flowers usually 2 or 3 together in the axils, on pedicels of 1 to 2 lines. Perianth fully ½ in. long, narrower than in *P. microcarpa,* silky-villous with ful-

vous hairs.   Anthers with rather long points or appendages to the connective.   Ovary shortly stipitate, densely villous; style straight, elongated ; ovule solitary.—*P. Fraseri*, Meissn. in Pl. Preiss. i. 532, and in
DC. Prod. xiv. 334, not of R. Br.

W. Australia. Swan river, *Drummond, 1st coll. n.* 597, *Preiss, n.* 729.

Var ? *pedicellaris.* Pedicels 3 to 6 lines long. Flowers much less villous or
sprinkled only with a few hairs. Ovary not thickened, and the ovule abortive in the
flowers examined.—Murchison river, *Oldfield.*

11. **P. rudis,** *Meissn. in DC. Prod.* xiv. 333. A shrub apparently
spreading or procumbent, the younger parts densely hirsute with soft
spreading hairs, the older foliage glabrous or nearly so. Leaves crowded,
linear-subulate, nearly terete, grooved without any prominent midrib,
acute but not pungent, the longer ones on the main stem sometimes 2
in. but mostly about 1 in. long. Pedicels axillary, solitary, 1 to 2 lines
long.   Perianth about $\frac{1}{2}$ in. long, hirsute with a few long spreading
hairs.   Anthers with long points or appendages to the connective.
Ovary on a short narrow stipes, densely hirsute with long hairs, tapering into a long hirsute style with a terminal stigma. Ovule solitary.

W. Australia, *Drummond, 4th coll. n.* 273.

12. **P. microcarpa,** *R. Br. in Trans. Linn. Soc.* x. 160, *Prod.* 372.
An erect broom-like shrub of 3 to 5 ft., the young shoots villous with
fine appressed hairs, the adult foliage glabrous. Leaves terete and
more or less grooved underneath, or very narrow linear and flat with
a scarcely prominent midrib, acute, 2 to 4 in. long. Flowers axillary,
usually in clusters of 2 or 3, on very short villous pedicels rarely exceeding 1 line. Bracts ovate or lanceolate, villous, sometimes as long
as the pedicels. Perianth villous with appressed hairs, 4 to 5 lines
long. Anthers inserted nearly at the base of the perianth-segments,
the connective produced into a short broad appendage. Ovary glabrous, very shortly stipitate, tapering into a thick style recurved at the
end with an oblique stigma, but longer than the anthers ; ovules 2.
Drupe small, nearly globular.—Meissn. in DC. Prod. xiv. 334.

W. Australia. King George's Sound and adjoining districts, *R. Brown, A. Cun-*
ningham, Drummond, 3rd coll. n. 260, *Preiss, n.* 728, and many others.

13. **P. sulcata,** *Meissn. in DC. Prod.* xiv. 333. A much-branched
shrub, apparently divaricate or procumbent, glabrous or with a slight
pubescence on the branches. Leaves crowded, spreading, subulate,
rigid and pungent-pointed, doubly grooved by the prominent margins
and midrib, mostly $\frac{3}{4}$ to 1 in. long. Pedicels axillary, solitary, 1 to 3
lines long, glabrous. Perianth glabrous, acute, about 4 lines long,
rather attenuate towards the base. Anthers with long narrow appendages or points to the connective. Ovary glabrous, contracted into a
short stipes, and tapering into an elongated angular straight style with
a terminal stigma ; ovules 2.

W. Australia, *Drummond, n.* 168, *4th coll. n.* 274.

c c 2

14. **P. acicularis,** *F. Muell. Fragm.* vi. 220. A rigid shrub of about 1 ft. (*Oldfield*), the branches shortly villous, the foliage glabrous or sprinkled with a few short rigid hairs. Leaves crowded, spreading, linear-subulate, rigid and pungent-pointed, doubly grooved as in *P. sulcata,* but rarely exceeding ½ in. in length. Pedicels axillary, solitary, glabrous, 3 to 4 lines long. Perianth fully 5 lines long, quite glabrous, contracted above the broad base, the segments very acute. Anthers with the connective produced into very short thick and obtuse appendages. Ovary glabrous, contracted into a short stipes and tapering into an elongated straight style.

W. **Australia.** Sandy plains, Murchison river, *Oldfield.* In the only ovary I examined I only found one ovule, but I may have overlooked a second abortive one, and the specimens were too few to sacrifice more flowers.

15. **P. scabrella,** *Meissn. in Hook. Kew Journ.* vii. 72, and in *DC. Prod.* xiv. 333. A rigid shrub with erect branches minutely pubescent when young. Leaves crowded, incurved, linear-terete, thick and rigid, deeply grooved underneath but without any prominent midrib, obtuse or callous-pointed, glabrous or minutely scabrous when young, ½ to 1 in. long. Flowers solitary in the axils and sessile. Perianth about 4 lines long, villous with appressed hairs, contracted at the base. Anthers with small globular tips to the connective. Ovary on a slender stipes, broad and glabrous; style folded immediately above the ovary, then erect and filiform with a terminal stigma. Ovules 2.

W. **Australia.** Between Moore and Murchison rivers, *Drummond, 6th coll. n.* 177.

16. **P. dillwynioides,** *Meissn. in DC. Prod.* xiv. 333. A bushy glabrous shrub. Leaves crowded, narrow-linear or terete, grooved underneath but without any prominent midrib, obtuse or scarcely acute, rarely above ½ in. long, quite smooth and glabrous. Pedicels solitary in the axils, very short. Perianth glabrous, about 5 lines long, slightly contracted above the base. Anthers with very short obtuse appendages to the connective. Ovary shortly stipitate, glabrous, tapering into a long style with a terminal stigma; ovules 2. Drupe broad and very oblique, 4 to 5 lines diameter.

W. **Australia,** *Drummond, 5th coll. n.* 403 ; Fitzgerald river, *Maxwell.*

17. **P. striata,** *R. Br. Prot. Nov.* 13. Quite glabrous or the branches silky-pubescent. Leaves linear, obtuse or mucronate-acute, contracted into a short petiole, flat but thick, striate with 3 to 5 prominent closely approximate longitudinal nerves, glabrous or minutely papillose, 1 to 1½ in. long. Pedicels solitary in the axils, 2 to 3 lines long. Perianth glabrous or slightly pubescent, 4 to 5 lines long. Anthers with the connective produced into an appendage usually rather long but variable. Ovary glabrous, slightly contracted into the stipes, tapering into a straight filiform style with a terminal stigma. Ovule solitary. Drupe ovoid, oblique, 4 to 5 lines long.—Meissn. in DC. Prod. xiv. 332 ; *P. striolata,* Meissn. l.c.

W. **Australia.** King George's Sound or to the eastward, *Baxter, Drummond, 5th coll. suppl. n.* 6 ; near W. Mount Barren, *Maxwell.*

18. **P. quinquenervis,** *Hook. Ic. Pl. t.* 425. A shrub of 5 to 7 ft., the branches slightly pubescent. Leaves broadly linear-spathulate or oblong-lanceolate, mucronate, contracted into a short petiole, rigid, with 5 prominent longitudinal nerves, glabrous or minutely papillose, 1 to 1½ in. long. Pedicels solitary in the axils, ¼ to ½ in. long, glabrous or slightly pubescent. Perianth nearly ½ in. long, glabrous or sprinkled with a very few small hairs. Anthers with rather short appendages to the connective. Ovary slightly contracted into a short stipes; style straight, with a broad terminal stigma. Ovule solitary.—Meissn. in Pl. Pr. i. 532, and in DC. Prod. xiv. 332.

W. Australia. Swan river, *Drummond,* 1st coll. n. 596; York district, *Preiss, n.* 531, 731. The species scarcely differs from *P. striata,* except in the broader leaves and rather larger flowers, and perhaps in stature.

19. **P. rufiflora,** *Meissn. in Hook. Kew Journ.* vii. 72, *and in DC. Prod.* xiv. 332. A shrub at first sight closely resembling *P. trinervis,* but very different in the structure of the flower, the young branches slightly pubescent, the foliage glabrous. Leaves oblong-linear or slightly spathulate, obtuse with a minute point, contracted into a very short petiole, thick, 1-nerved with thickened nerve-like margins or rarely with intramarginal nerves on the under side, 1 to 2 in. long. Flowers sessile or nearly so, solitary or 2 or 3 together in the axils. Perianth about 4 lines long, very densely ferruginous-villous. Anthers not reaching above half the length of the segments, with short gland-like appendages to the connective. Ovary abruptly stipitate, broad, glabrous; style elongated, slender, with a terminal stigma. Ovules 2.

W. Australia. Between Moore and Murchison rivers, *Drummond,* 6th coll. n. 176.

20. **P. scabra,** *R. Br. in Trans. Linn. Soc.* x. 162, *Prod.* 373. Apparently a bushy shrub, the branches pubescent. Leaves crowded, linear-lanceolate or more or less spathulate but never so much so as in some Eastern species, obtuse or mucronate-acute, contracted at the base, ½ to 1 in. long, 1-nerved, scabrous with small papillæ or crystalline or opaque asperities. Flowers solitary in the axils on short pedicels. Perianth glabrous, or sprinkled with a few hairs, about 5 lines long. Anthers with the connective produced into rather long appendages. Ovary stipitate, glabrous, with a straight filiform style and stigma.—Meissn. in DC. Prod. xiv. 337; *P. flexifolia,* R. Br. in Trans. Linn. Soc. x. 162, Prod. 372, not of Lodd. and others; *P. spathulata,* R. Br. ll. cc. 162 and 373, not of Lodd. and others.

W. Australia. Lucky Bay, *R. Brown.* The only specimens that I have seen of these three supposed species are the almost single and very unsatisfactory ones in Brown's Herbarium, with scarcely any flowers. They are all from the same locality, and the differences between them appear to me to be very slight, no more than what we constantly observe between different specimens of other species. In *P. flexifolia* the leaves are rather smaller and narrower than in *P. scabra,* and there are fewer hairs on the perianth; in *P. spathulata* the leaves are rather larger and broader, with similar slightly bairy perianths. All three agree in the habit and other characters which distinguish them from any that I have seen in other collections. Meissner places them in a wrong section, but was misled by Loddiges' figures representing garden plants falsely named, his *P. flexifolia* being *P. nutans*, and *P. spathulata* most probably *P. rigida*, both of them eastern species of the section *Amblyanthera*.

21. **P. graminea,** *R. Br. in Trans. Linn. Soc.* x. 164, *Prod.* 374. A dwarf shrub or undershrub, with short procumbent stems, glabrous except a few appressed hairs sprinkled on the young shoots and inflorescence. Leaves numerous, erect, narrow-linear or very rarely linear-lanceolate, acute, flat, but the midrib and sometimes the margins prominent underneath, mostly 6 to 8 in. long, or even more. Flowers small, distant, in slender one-sided leafless racemes, terminal or in the upper axils, not half so long as the leaves. Bracts subulate. Pedicels solitary within each bract, ⅓ to 1 line long. Perianth about 2 lines long, glabrous or sprinkled with a few appressed transparent hairs. Anthers rather short, the connective produced into a long slender appendage. Ovary glabrous, globular, on a short stipes. Style thick but elongated, angular, tapering upwards, with a terminal oblique stigma. Ovules 2.—Meissn. in Pl. Preiss. i. 533, and in DC. Prod. xiv. 331.

W. Australia. King George's Sound, *R. Brown, Baxter, Drummond, Preiss, n.* 725; Wilson's Inlet, *Oldfield.* This is the only species with a truly racemose inflorescence, without the rhachis growing out into a leafy shoot.

SECT. 3. AMBLYANTHERA.—Style elongated beyond the anthers, with a terminal stigma. Anthers all perfect, the connective not produced beyond the cells.

22. **P. longifolia,** *R. Br. in Trans. Linn. Soc.* x. 164, *Prod.* 374. A tall shrub or small tree of 10 to 20 ft., glabrous or the young branches minutely silky-pubescent. Leaves linear or linear-lanceolate, 5 to 8 in. long, callous-pointed or obtuse, narrowed into a short petiole, not thick, finely veined, the midrib slightly prominent, the lateral veins almost longitudinal and scarcely conspicuous. Flowers solitary in the axils but owing to the abortion of the lower or of nearly all the floral leaves forming often terminal or subterminal racemes. Pedicels 3 to 6 lines long. Perianth more or less ferruginous with minute silky hairs, 5 to 6 lines long. Anther-connective not produced beyond the cells. Ovary glabrous, on a stipes distinctly articulate above the base; style elongated with a terminal stigma; ovules 2. Drupe broad and oblique, about 5 lines diameter.—Meissn. in Pl. Preiss. i. 533, and in DC. Prod. xiv. 343; *P. Drummondii,* Lindl. Swan Riv. App. 35.

W. Australia. King George's Sound, *R. Brown* and others, and thence to Swan river, *Drummond, 1st coll. and n.* 89, *Preiss, n.* 724, 732; Vasse river, *Mrs. Molloy;* Cape Naturaliste, *Collie.*

23. **P. articulata,** *R. Br. in Trans. Linn. Soc.* x. 164, *Prod.* 374. A tall shrub or small tree, closely resembling *P. longifolia* and intermediate as it were between that species and *P. elliptica.* Leaves much broader than in *P. longifolia* and not falcate, oblong-lanceolate or oblanceolate, 3 to 6 in. long, with very oblique almost longitudinal anastomosing lateral veins. Inflorescence as in *P. longifolia* often racemose from the

abortion or reduction of the floral leaves. Flowers usually larger than in that species, the perianth fully ½ in. long. Anthers and pistil the same, but the articulation of the stipes of the ovary rather more prominent, being often above the level of the hypogynous glands.—Meissn. in DC. Prod. xiv. 342.

W. Australia. King George's Sound, *R. Brown, Baxter, Harvey, Drummond, n. 88, 96, Oldfield, F. Mueller;* Vasse river, *Oldfield.*

24. **P. elliptica,** *R. Br. in Trans. Linn. Soc.* x. 164, *Prod.* 373. A shrub or tree of 10 to 20 ft., quite glabrous. Leaves from broadly obovate to elliptical or broadly lanceolate, obtuse with a callous point or almost acute when narrow, contracted into a short petiole, not thick, penniveined with oblique anastomosing veins, 1½ to near 3 in. long. Pedicels 1 to 2 lines long, mostly axillary and solitary but sometimes forming racemes by the abortion or reduction of the floral leaves as in the two preceding species. Perianth glabrous, 4 to 5 lines long. Anther-connective not produced beyond the cells. Ovary glabrous, on a stipes articulate near the base; style elongated, with a terminal stigma. Ovules 2.—Meissn. in DC. Prod. xiv. 341; *P. laureola,* Lindl. Swan Riv. App. 35; Meissn. in Pl. Preiss. i. 532, and in DC. Prod. xiv. 341; Hook. Ic. Pl. t. 426.

W. Australia. King George's Sound and adjoining districts, *R. Brown, Baxter, A. Cunningham,* and many others, and thence to Swan river, *Drummond, 1st coll., 4th coll. n. 277, Preiss, n.* 726, *Clarke.*

The articulation of the stipes, so conspicuous in the three preceding Western species, is also observable in most of the following Eastern species, but it is much less marked and usually at the very base of the stipes, not forming a ring above the base.

25. **P. ferruginea,** *Sm. Exot. Bot.* ii. 47, *t.* 83. A tall shrub, the branches and young shoots slightly ferruginous or hoary-tomentose, the adult foliage glabrous. Leaves ovate oblong-elliptical or almost lanceolate, acute or obtuse with a callous point, contracted at the base and sometimes shortly petiolate, 2 to 3 or rarely 4 in. long, the midrib prominent underneath and obscurely or distinctly penniveined. Flowers either solitary in the axils or more frequently in dense clusters of 4 to 6, either terminal or pedunculate in the upper axils, with small bracts (reduced floral leaves) under each flower. Pedicels very short or scarcely any. Perianth densely ferruginous-pubescent, about 6 lines long. Anther-connective not produced beyond the cells. Ovary densely villous, on a short glabrous pedicel; style elongated, with a terminal stigma; ovules 2.—R. Br. in Trans. Linn. Soc. x. 163, Prod. 373; Meissn. in DC. Prod. xiv. 343 (except as to F. Mueller's specimens); *P. laurina,* Pers. Syn. i. 118.

N. S. Wales. Port Jackson and Blue Mountains, *R. Brown, Sieber, n.* 58, *A. Cunningham* and others. The Victorian plant referred here by Meissner having the ovary quite glabrous, is much nearer to *P. lanceolata,* and is described below as *P. confertiflora.*

26. **P. media,** *R. Br. Prot. Nov.* 16. A tall erect shrub, the young branches ferruginous-pubescent. Leaves lanceolate or almost elliptical,

sometimes oblique or slightly falcate, acuminate or acute, contracted
into a short petiole, 2 to 4 in. long, flat thin and glabrous, very obliquely veined. Pedicels axillary, 1 to 2 lines long, ferruginous or
glabrous, solitary or very rarely irregularly clustered. Perianth fully
5 lines long, pubescent with short appressed hairs, the segments tipped
with short subulate points. Anther-connective not produced beyond
the cells. Ovary more or less silky-hirsute but usually much less so
than in *P. ferruginea* and *P. cornifolia* and the hairs sometimes almost
disappearing, the stipes glabrous; style elongated, with a terminal
stigma; ovules 2.—Meissn. in DC. Prod. xiv. 342.

Queensland. Brisbane river, Moreton Bay, *Fraser, W. Hill, F. Mueller.*
N. S. Wales. Hastings and Clarence rivers, *Beckler.*

Some specimens come near to some varieties of *P. lanceolata*, but independently of
the hairs of the ovary, *P. media* has usually thinner, more veined leaves, larger pedicels, and more prominent subulate tips to the perianth-segments.

27. **P. cornifolia,** *A. Cunn.; R. Br. Prot. Nov.* 16. A tall erect
shrub, the branches pubescent, the young leaves ciliate on the margins
and sometimes pubescent, the adult foliage glabrous. Leaves from
broadly obovate or ovate to elliptical-oblong or even broadly lanceolate,
acute when narrow or obtuse when broad, usually mucronate, contracted
into a very short petiole, mostly 1 to 2 in. long, flat, rather rigid, the
midrib slightly prominent and sometimes obscurely and very obliquely
veined. Pedicels solitary in the axils or clustered on a very short
axillary branch with the floral leaves reduced or abortive, sometimes
very short, rarely 2 to 3 lines long. Perianth shortly silky-pubescent,
5 to 6 lines long, without points to the segments. Anther-connective
not produced beyond the cells. Ovary densely silky-villous, on a very
short glabrous stipes; style elongated, with a terminal stigma; ovules
2.—Meissn. in DC. Prod. xiv. 341; *P. tinifolia*, A. Cunn. Herb.

Queensland. Logan and Brisbane rivers, Moreton Bay and island, *Fraser, A.
Cunningham, F. Mueller,* and others.
N. S. Wales. Macleay, Clarence, and Richmond rivers, *Beckler, C. Moore;*
New England, *C. Stuart.*

The foliage of this species is very variable, but the leaves are usually shorter, broader,
and less veined than in the preceding species, and the perianth has neither the ferruginous hairs of *P. ferruginea* nor the subulate tips of *P. media;* the hairs of the ovary
are constantly dense.

28. **P. marginata,** *R. Br. Prot. Nov.* 16. A shrub with pubescent
branches. Leaves ovate obovate or broadly elliptical-oblong, mucronate, contracted at the base but scarcely petiolate, ¾ to 1½ in. long,
flat, coriaceous, glabrous and often shining, the midrib prominent
underneath, more or less distinctly penniveined and sometimes 3-
nerved. Pedicels exceedingly short, pubescent, solitary or clustered
with much reduced or abortive floral leaves. Perianth slightly pubescent, about 5 lines long, the segments tipped with horn-like dorsal
points. Anther-connective not produced beyond the cells. Ovary
densely hirsute, on a short glabrous stipes; style elongated, with a

terminal stigma; ovules 2.—Meissn. in DC. Prod. xiv. 341; *P. obcordata*, A. Cunn. Herb.

N. S. Wales. Barren rocky hills north of Bathurst and on the Cujeegong river, *A. Cunningham.* R. Brown describes the anthers as silky on the back, which I have not found to be the case in the flowers I examined. Meissner distinguishes two varieties with oval and obovate leaves, both however may be observed on the same specimens.

29. **P. sericea,** *A. Cunn.; R. Br. Prot. Nov.* 14. An erect shrub resembling at first sight some forms of *P. rigida*, but the indumentum softer; branches densely ferruginous-pubescent or villous. Leaves obovate oblong or oblanceolate, mucronate, but otherwise obtuse, contracted at the base but scarcely petiolate, ¾ to 1½ in. long, flat, pubescent on both sides and in the broader leaved forms densely silky, penniveined. Flowers solitary in the axils but a few of the floral leaves sometimes much reduced. Pedicels villous, at first very short, 2 to 3 lines long when in fruit. Perianth pubescent, about 5 lines long, the segments obtuse, without points. Anther-connective not produced beyond the cells. Ovary densely villous, on a glabrous stipes; style elongated, villous; ovules 2.—Meissn. in DC. Prod. xiv. 342.

N. S. Wales. Liverpool plains and neighbouring ranges, *A. Cunningham, Leichhardt, C. Moore* (with the leaves nearly all obovate); barren country south-west of Lachlan river, *A. Cunningham;* New England, *C. Stuart;* Clarence river, *Beckler* (all with the leaves mostly narrow).

The broad-leaved more silky form, constituting Meissner's var. β, or *P. velutina,* A. Cunn. MS., appears to be the one that R. Brown had chiefly in view, the narrow-leaved specimens come very near to *P Mitchellii,* and I should have kept up the two at least as distinct varieties, but that some of the latter have a few broadly obovate leaves on the principal stems, and some broad-leaved specimens have narrow leaves on the lateral branches. The specimens are none of them in very good flower.

30. **P. Mitchellii,** *Meissn. in Hook. Kew Journ.* vii. 73, *and in DC. Prod.* xiv. 342. An erect shrub with pubescent or villous branches. Leaves oblong-linear spathulate or rarely almost obovate, obtuse with a small point or acute, narrowed at the base, 1 to 2 in. long, pubescent when young, at length glabrous or nearly so. Pedicels 3 to 4 lines long, erect or recurved, axillary or forming terminal or subterminal racemes by the reduction of the floral leaves to small bracts. Perianth villous, 4 to 5 lines long, the segments without points. Anther-connective not produced beyond the cells. Ovary villous, on a glabrous stipes; style elongated; ovules 2.

Queensland. Burnett river, *Haly;* Brisbane river, Moreton Bay, *F. Mueller.* (I have not seen Mitchell's typical specimen from the interior.)
N. S. Wales. New England, *C. Stuart.*

I do not feel very confident in having properly distinguished this species from *P. sericea.* Possibly some of the narrow-leaved specimens which I have referred to the latter may belong rather to *P. Mitchellii,* or *P. Mitchellii* itself may be only a variety of *P. sericea.* The series of specimens is not complete enough to determine these points.

31. **P. fastigiata,** *R. Br. Prot. Nov.* 13. A shrub with rather slender branches shortly pubescent. Leaves rather crowded, incurved, narrow-linear, scarcely acute, attenuate at the base, the margins re-

curved or revolute, $\frac{3}{4}$ to $1\frac{1}{2}$ in. long, more or less scabrous-pubescent. Pedicels slender, solitary or clustered in the axils (the floral leaves abortive), 1 to 2 lines long. Perianth minutely pubescent or hirsute, 3 to $3\frac{1}{2}$ lines long. Anther-connective not produced beyond the cells. Ovary densely hirsute, on a glabrous stipes; style elongated; ovules 2. —Meissn. in DC. Prod. xiv. 337.

N. S. Wales. In the interior, *Fraser;* Robinson Ranges and Dogwood Creek, *Leichhardt.* Resembles sometimes *P. curvifolia,* but with smaller flowers and hirsute ovaries.

32. **P. hirsuta,** *Pers. Syn.* i. 118. A spreading shrub of 2 or 3 ft., the whole plant pubescent or hirsute, or the foliage at length scabrous only or rarely nearly glabrous. Leaves sessile, broadly linear-lanceolate or oblong, obtuse or scarcely acute, with revolute margins, rarely above $\frac{1}{2}$ in. long. Flowers sessile or on pedicels rarely exceeding 1 line, solitary in the axils but usually several crowded at or near the ends of the branches and sometimes the lower floral leaves abortive. Perianth densely hirsute with spreading hairs, about 5 lines long. Anther-connective not produced beyond the cells. Ovary densely hirsute on a glabrous stipes; style elongated; ovule solitary. Drupe obliquely ellipsoid, $\frac{1}{2}$ to $\frac{3}{4}$ in. long.—R. Br. in Trans. Linn. Soc. x. 161, Prod. 372, Prot. Nov. 13; Rudge in Trans. Linn. Soc. x. 291, t. 16; Meissn. in DC. Prod. xiv. 337; Lodd. Bot. Cab. t. 327.

N. S. Wales. Port Jackson to the Blue Mountains, *R. Brown, Sieber, n.* 56, and many others. However variable the leaves may be in breadth, they scarcely supply the means of distinguishing marked varieties as proposed by Meissner. In our specimens of Sieber's *P. arida,* Pl. Exs. *n.* 55, they are not broader than in those of his *n.* 56, but rather scabrous than hirsute. In A. Cunningham's *P. aspera,* from between Hunter's and Richmond rivers, they are very scabrous, short, and rather broad; in R. Brown's typical specimens they are narrow and very hispid.

33. **P. chamæpitys,** *A. Cunn. in Field, N. S. Wales,* 329. A prostrate or divaricate shrub, quite glabrous or the young branches slightly pubescent. Leaves crowded, narrow-linear, acute and rigid but scarcely pungent-pointed, channelled underneath with revolute margins, not scabrous, mostly about $\frac{1}{2}$ in. long. Flowers very shortly pedicellate, solitary within each floral leaf but crowded into short heads or clusters at or below the ends of the branches, with reduced floral leaves. Perianth 4 to 5 lines long, glabrous or sprinkled with a few appressed hairs. Anther-connective not produced beyond the cells. Ovary densely hirsute, on a glabrous stipes; style elongated; ovule solitary.—R. Br. Prot. Nov. 13; Meissn. in DC. Prod. xiv. 335 ; *P. gnidioides,* Sieb. in Spreng. Syst. Cur. Post. 45, and in Roem. and Schult. Syst. iii. Mant. 269.

N. S. Wales. Blue Mountains, *A. Cunningham, Sieber, n.* 53.

34. **P. arborea,** *F. Muell. Fragm.* v. 37, vi. 221. A tree of about 30 ft., the branches and sometimes the young shoots hoary-pubescent or ferruginous. Leaves oblong-lanceolate, obtuse or scarcely pointed, contracted into a petiole, glabrous above, minutely pubescent under-

neath, mostly 2 to 4 in. long, flat or the margins slightly recurved. Flowers solitary in the axils but the floral leaves sometimes reduced to small bracts or quite abortive at the base of the young shoots. Pedicels pubescent, 1 to 2 lines long. Perianth larger than in any species except *P. Gunnii*, 8 to 9 lines long, silky-pubescent, the segments with rather thick conical dorsal tips. Anther-connective not produced beyond the cells. Ovary stipitate, glabrous; style elongated; ovules 2.

Victoria. Moist shady woods on the upper Tyers, Tarwin, Latrobe, Tangil, and Yarra rivers, and in the beech woods of the Lower regions of the Baw-Baw Mountains, *F. Mueller.*

35. **P. salicina,** *Pers. Syn.* i. 118. A tall shrub, the young branches minutely pubescent, the foliage glabrous. Leaves oblong-lanceolate, more or less oblique or falcate, obtuse or mucronate-acute, contracted into a short petiole, mostly 4 to 8 in. long, flat, not very thick, more or less distinctly 3-nerved and obliquely veined. Flowers at the base of the shoots axillary or below the leaves, the floral leaves mostly abortive or much reduced. Pedicels 2 to 4 lines long, glabrous or pubescent. Perianth glabrous or slightly pubescent, 5 to 6 lines long, the segments without terminal points. Anther-connective not produced beyond the cells. Ovary stipitate, glabrous; style elongated; ovules 2.—R. Br. in Trans. Linn. Soc. x. 163, Prod. 373; Meissn. in DC. Prod. xiv. 343.

N. S. Wales. Port Jackson, *R. Brown, Sieber, n.* 60, and many others; Hastings river, *Beckler;* Port Macquarrie, *Tozer.*

Var.? *Muelleri.* Quite glabrous. Leaves broader, less falcate or straight, thicker, with nerve-like margins, and mostly 2 to 4 lines long. Pedicels shorter. Perianth not seen, the specimens all in fruit only.—Twofold Bay and Genoa river, *F. Mueller.*

36. **P. prostrata,** *R. Br. in Trans. Linn. Soc.* x. 163, *Prod.* 373. A trailing prostrate or low and diffuse shrub, extending sometimes to 2 or 3 ft., glabrous or the young shoots slightly pubescent. Leaves broadly elliptical oblong or almost lanceolate, acute or mucronate, contracted into a very short petiole, 1 to 1½ in. long, not so rigid as in *P. cornifolia*, scarcely veined except the slightly prominent midrib. Pedicels short, axillary, solitary or clustered on reduced axillary shoots. Perianth glabrous, 4 to 5 lines long, the segments without terminal points. Anther-connective not produced beyond the cells. Ovary stipitate, glabrous; style elongated; ovules 2.—Meissn. in DC. Prod. xiv. 342; *P. daphnoïdes*, A. Cunn.; R. Br. Prot. Nov. 15; Meissn. in DC. Prod. xiv. 339.

Queensland. Sandy Cape, Hervey Bay, *R. Brown.*
N. S. Wales. Hunter's river, *A Cunningham;* New England near Armidale and head of Macleay river, *C. Stuart;* Mount Mitchell, *Beckler.*

The specimens are several of them very unsatisfactory, but all described as prostrate. The foliage is nearly that of some short leaved specimens of *P. cornifolia*, but the ovary is glabrous; the leaves also resemble those of *P. revoluta*, but are flatter and green on both sides.

37. **P. lanceolata,** *Andr. Bot. Rep. t.* 74. An erect shrub of several ft., the branches and young shoots hoary-pubescent, the adult foliage

glabrous or nearly so. Leaves lanceolate or oblong-lanceolate, mucro-nate-acute, much contracted into a short petiole, mostly 1½ to 2½ in. long, flat, the midrib slightly prominent, the margins scarcely nerve-like, otherwise veinless. Pedicels exceedingly short, solitary or 2 together, pubescent, rarely 1 line long. Perianth about 5 lines long, pubescent with very short appressed hairs, the segments without ter-minal points. Anther-connective not produced beyond the cells. Ovary glabrous, contracted into a short stipes ; style elongated ; ovules 2.—R. Br. in Trans. Linn. Soc. x. 162, Prod. 373 ; Meissn. in DC. Prod. xiv. 340 ; Lodd. Bot. Cab. t. 25 ; *P. ligustrina,* Knight, Prot. 100 ; *P. glaucescens,* Sieb. in Roem. and Schult. Syst. iii. Mant. 271.

N. S. Wales. Port Jackson, *R. Brown, Sieber, n.* 47 *and* 57, and many others : New England, *C. Stuart.*

Var. ? *lævis.* The whole plant glabrous. Leaves rather longer and thinner, broad or narrow. Pedicels not quite so short.—Clarence and Macleay rivers, *Beckler;* Port Jackson ? (from garden specimens.)

To the above variety may probably be referred *Linkia lævis,* Cav. Ic. iv. 61, t. 389 (very badly figured and described), *P. latifolia,* Andr. Bot. Rep. t. 280, and perhaps also of Lodd. Bot. Cab. t. 1509. *P. attenuata,* R. Br. Prot. Nov. 16, Meissn. in DC. Prod. xiv. 342, from Moreton Bay, *Fraser,* is described from specimens in fruit only, and may be either a variety of *P. media* or this variety of *P. lanceolata,* which connects in some measure the two species as to foliage and length of pedicels, neglecting the hairiness of the ovary in *P. media.* The exact discrimination of *P. lanceolata* and several allied species is often very difficult.

38. **P. confertiflora,** *Benth.* An erect shrub of 1 to 3 ft., the branches and inflorescence slightly ferruginous-pubescent, otherwise glabrous. Leaves broadly lanceolate or ovate-elliptical, acute, con-tracted at the base into a very short petiole, 2 to 3 in. long, flat, the midrib prominent underneath, more or less distinctly veined, and some-times 3-nerved. Flowers 6 to 10 together in short dense leafless axillary racemes or clusters, the rhachis at length growing out to 3 or 4 lines, ferruginous-pubescent as well as the very short pedicels, the racemes or clusters sometimes shortly pedunculate, with a pair of small bracts. Perianth about 5 lines long, villous with appressed ferruginous or silky hairs, the segments without terminal points. Anther-connec-tive not produced beyond the cells. Ovary very shortly stipitate, quite glabrous ; style elongated ; ovules 2. Drupe small, ovoid.

Victoria. Scrubby declivities of the Stringy-bark ranges towards Gipps' Land, Mitta-Mitta and Genoa rivers, *F. Mueller.* These specimens were referred by F. Mueller and Meissner (in DC. Prod. xiv. 343) to *P. ferruginea,* which they resemble at first sight, but the ovary is perfectly glabrous, and the affinity appears to me to be much greater with *P. lanceolata.*

39. **P. lucida,** *R. Br. in Trans. Linn. Soc.* x. 161, *Prod.* 372. A tall shrub or small tree of 12 to 16 ft. with a lamellose bark, the young branches tomentose or shortly pubescent, the foliage glabrous. Leaves linear-lanceolate, mostly acute, shortly contracted at the base, 2 to 4 in. long, flat or with recurved margins, rather thin, obscurely and finely veined. Pedicels 1 to 2 lines long, more slender and less pubescent than in *P. lanceolata,* all axillary or forming racemes at the base of the

branches owing to the abortion or reduction of the lower floral leaves. Perianth slightly pubescent, 4 or 5 lines long, the segments with minute terminal points or without any. Anther-connective not produced beyond the cells. Ovary glabrous, on a very short stipes; style elongated; ovules 2.—Meissn. in DC. Prod. xiv. 339.

N, S. Wales. Nepean river, *Bauer ;* Port Jackson to the Blue Mountains, *Woolls, A. Cunningham ;* Sydney woods, *Macarthur, Paris Exhibition,* 1855, *n.* 150 *and* 225, from Port Jackson, and n. 11 from Illawarra.

Var.? *latifolia.* Leaves rather broader and thinner. Perianth rather longer, with more evident points to the segments.—Clarence river, *Beckler.*

40. **P. linearis,** *Andr. Bot. Rep. t. 77.* A tall shrub or small tree of 10 to 20 ft., the young branches pubescent or villous, the adult foliage usually glabrous. Leaves rather crowded, linear, acute or almost obtuse, contracted at the base, 1 to 2 in. long, ¾ to 1½ lines broad, obscurely veined. Pedicels solitary, 1 to 3 lines long. Perianth about 5 lines long, more or less pubescent with short appressed hairs. Anther-connective not produced beyond the cells. Ovary glabrous, stipitate; style elongated; ovules 2. Drupe ovoid.—R. Br. in Trans. Linn. Soc. x. 161, Prod. 372; Meissn. in DC. Prod. xiv. 335; Vent. Jard. Malm. t. 32; Bot. Mag. t. 760; *P. angustifolia* Knight, Prot. 99; *P. pinifolia,* Sieb. Pl. Exs. ; *P. filifolia,* Dietr.; Roem. and Schult. Syst. iii. 401; *P. pruinosa,* A. Cunn.; Steud. Nom. Bot. ed. 2? (the specimens not in flower); *Pentadactylon angustifolium,* Gærtn. f. Fr. iii. 219, t. 220 ; *Persoonia pentadactylon* Steud. Nom. Bot. ed. 2.

Queensland. Stradbrooke island, *Fraser.*
N. S. Wales. Port Jackson to the Blue Mountains, *R. Brown, Sieber, n.* 50, and many others; northward to Hastings and Macleay rivers, *Beckler ;* southward to Illawarra, *Shepherd, A. Cunningham, Macarthur, Paris Exhibition, n.* 115, and Twofold Bay, *F. Mueller.*
Victoria. Genoa and Snowy rivers and Nangatta mountains, *F. Mueller.*

Var. *sericea.* Silky-pubescent at the time of flowering.—Shoalhaven river, *C. Moore.* The fruit of this species is said to be one of those most known under the name of "Geebung." The plant varies much in the breadth of the leaves. R. Brown describes the bark as smooth, and the leaves of his specimens are all very narrow; F. Mueller and others have sent with their specimens, mostly with broader leaves, a lamellose bark like that of *P. lucida.* The real distinction between the two species requires therefore further elucidation from the observation of living specimens.

In referring Gærtner's *Pentadactylon* to this species, I have been guided by a specimen in the Banksian Herbarium named *Persoonia angustifolia* in the same handwriting as the name of *Persoonia ferruginea,* there given to a specimen of the latter plant from the same collection (Mr. Burton), which was evidently the one which Gærtner also described under the latter name. The lobed embryo he figures has also been observed by F. Mueller in *P. chamæpeuce.*

41. **P. pinifolia,** *R. Br. in Trans. Linn. Soc.* x. 160, *Prod.* 372. A shrub attaining 8 to 10 ft., with virgate branches "often pendulous," pubescent as well as the young leaves, the adult foliage glabrous. Leaves crowded, erect or incurved, linear-filiform, acute, with recurved margins, channelled underneath, 1 to 2 in. long, the floral ones not exceeding the flowers but otherwise similar. Flowers on very short pedicels or almost sessile, solitary within the floral leaves but owing to their reduced size

forming a dense terminal raceme of 2 or 3 in. or even twice that length.
Perianth slightly silky-pubescent, 4 to 5 lines long, the segments rather
obtuse. Anther-connective not produced beyond the cells. Ovary glabrous, stipitate; style elongated; ovule solitary. Fruit ovoid, scarcely
oblique, about ½ in. long.—Rudge in Trans. Linn. Soc. x. 290, t. 16;
Meissn. in DC. Prod. xiv. 334.

N. S. Wales. Port Jackson to the Blue Mountains, *R. Brown,* and many others.
Sydney woods, Paris Exhibition, 1855, *M'Arthur, n.* 216.

42. **P. Caleyi,** *R. Br. Prot. Nov.* 13. A shrub of several ft., the
branches tomentose-pubescent. Leaves narrow-linear, acute or almost
obtuse, 1 to 2 in. long, with slightly recurved margins, veinless, glabrous above, pale and often minutely pubescent underneath. Flowers
axillary, solitary (or rarely 2 together?), sessile or on exceedingly short
pedicels. Perianth slender, slightly tomentose or nearly glabrous, about
5 lines long, the segments tipped with fine points. Anther-connective
not produced beyond the cells. Ovary glabrous, on a short slender
stipes.—Meissn. in DC. Prod. xiv. 335.

N. S. Wales. Jarvis Bay, *Caley.*
Victoria? Wilson's Promontory, *Baxter,* but possibly some mistake in this station.
I have seen this plant only in R. Brown's herbarium, where there are several good
specimens. It is evidently allied to *P. ledifolia* and *P. revoluta,* but with the very
narrow leaves of *P. linearis,* and distinguished from all three by the long points of the
perianth-segments.

43. **P. ledifolia,** *A. Cunn.; Meissn. in DC. Prod.* xiv. 339. A tall
erect bushy shrub, the branches shortly ferruginous-villous. Leaves
oblong-linear or nearly lanceolate, obtuse, very shortly contracted at
the base, 1 to 2 in. long, the margins recurved, glabrous and smooth
above, pale or hoary and often minutely pubescent underneath. Pedicels
axillary, exceedingly short, villous. Perianth about 5 lines long, slightly
villous, with spreading hairs, the segments without terminal points.
Anther-connective not produced beyond the cells. Ovary glabrous, on
a very short stipes; style elongated; ovules 2. Drupe obliquely ovoid.

N. S. Wales. Illawarra, *A. Cunningham, Shepherd.*

44. **P. revoluta,** *Sieb. in Roem. and Schult. Syst.* iii. *Mant.* 272. An
erect or spreading shrub of 2 to 4 ft., the young shoots silky, the adult
foliage glabrous or nearly so. Leaves in the typical form obovate-oblong
to oblong-lanceolate, very obtuse, shortly contracted at the base, but
not distinctly petiolate, ¾ to 1 in. long, with recurved or revolute margins, coriaceous, glabrous and smooth above, pale or white and long retaining a slight pubescence underneath. Flowers solitary, erect on very
short pedicels. Perianth silky-pubescent, about 5 lines long, the segments tipped with short points. Ovary glabrous, stipitate; style elongated; ovules 2.—R. Br. Prot. Nov. 14; Meissn. in DC. Prod. xiv. 339.

N. S. Wales. Blue Mountains, *Sieber, n.* 48, *A. Cunningham, Woolls, Miss
Atkinson;* Berrima, *M'Arthur.*

Var. *angustifolia.* Leaves mostly oblong-linear.—Blue Mountains, *Miss Atkinson;*
Argyle County, *M'Arthur.*

This species as well as the *P. prostrata* was designated by A. Cunningham by the name of *P. daphnoides.*

45. **P. Gunnii,** *Hook. f. in Hook. Lond. Journ.* vi. 283, *Fl. Tasm.* i. 321. An erect bushy shrub of 6 to 10 ft., the branches and young shoots more or less hoary or silky-tomentose. Leaves rather crowded, cuneate-oblong oblong-linear or linear-spathulate, ¾ to 1½ in. long, obtuse, flat, thick, nerveless, contracted into a short petiole, quite glabrous in the typical form. Flowers larger than in any species except *P. arborea,* on short thick tomentose pedicels, all solitary in the axils, but sometimes clustered towards the ends of the branches. Perianth hoary-pubescent or nearly glabrous, 8 to 9 lines long. Anther-connective not produced beyond the cells. Ovary stipitate, glabrous, style elongated; ovules 2.— Meissn. in DC. Prod. xiv. 340.

Tasmania. Mountain regions, Lake St. Clair, May-Day plains, *Gunn;* Port Davy, *Milligan;* Mount Lapeyrouse, *C. Stuart.*

Var. *angustifolia.* More silky; leaves narrower and longer; perianth more pubescent.—Macquarrie Harbour, *Milligan.* Some of the Mount Lapeyrouse specimens closely connect the two extreme forms.

46. **P. mollis,** *R. Br. in Trans. Linn. Soc.* x. 161, *Prod.* 372. A tall erect shrub or small tree, the young branches ferruginous-villous. Leaves lanceolate or linear-lanceolate, acute, shortly contracted at the base, mostly 1½ to 2½ in. long, the margins recurved, glabrous above when young and more or less veined, paler pubescent and veinless underneath or rarely when old quite glabrous. Flowers sessile or on very short villous pedicels, the floral leaves sometimes almost opposite, or in whorls of 3. Perianth densely villous, 5 to 6 lines long or even rather longer. Anther-connective not produced beyond the cells. Ovary stipitate, glabrous; style elongated; ovules 2, but one of them often abortive at a very early stage. Drupe small, oblique.—Meissn. in DC. Prod. xiv. 339.

N. S. Wales. Blue Mountains, *R. Brown, Sieber, n.* 54, *A.* and *R. Cunningham, Miss Atkinson.* In some herbaria specimens of *P. hirsuta* are designated by A. Cunningham under the name of *P. mollis.*

47. **P. rigida,** *R. Br. Prot. Nov.* 14. A bushy shrub of 3 or 4 ft., the branches softly tomentose-villous and often ferruginous. Leaves from obovate to linear-spathulate, mucronate, much contracted at the base but scarcely petiolate, ¾ to 1½ in. long, the margins usually recurved or replicate, softly or scabrous-pubescent at least when young, and more so and the midrib more conspicuous on the upper than on the under surface, the old leaves rarely glabrous. Flowers almost sessile and solitary in the axils, but sometimes several crowded at the base of axillary shoots with one or two of the floral leaves reduced to small bracts. Perianth densely villous, 5 to nearly 6 lines long. Anther-connective not produced beyond the cells. Ovary stipitate, glabrous; style elongated; ovules 2. Fruit broad, about ½ in. long.—Meissn. in DC. Prod. xiv. 337; *P. spathulata,* Sieb. in Roem. and Schult. Syst. iii.

271, and probably also Lodd. Bot. Cab. t. 1199, and therefore of Meissn. in DC. l.c. 338 partly, but not of R. Br.

N. S. Wales. Blue Mountains, *Caley, A. Cunningham;* Macquarrie and Lachlan rivers, *A. Cunningham.*

Victoria. Buffalo range, Mount Alexander, Forest Creek, Broken and King rivers, *F. Mueller.*

Var.? *microphylla.* The whole plant very scabrous. Leaves broadly spathulate, about ⅓ in. long including the long winged petiole. Perianth slightly scabrous or hispid.—Blue Mountains, *Caley;* Castlereagh, *C. Moore;* also in *Leichhardt's* collection.

The species has sometimes some resemblance to some varieties of *P. sericea,* but readily distinguished by the glabrous ovary.

48. **P. curvifolia,** *R. Br. Prot. Nov.* 13.   A shrub of 1 to 3 ft., the branches softly pubescent. Leaves rather crowded, usually incurved, narrow-linear with revolute margins so as to be almost terete with a deep groove underneath, acute, very scabrous, ¾ to 1½ in. long. Flowers solitary in the axils on very short pedicels or almost sessile, crowded at or near the ends of the branches and a few of the floral leaves very small or abortive, forming a very short terminal or subterminal raceme or spike. Perianth 5 or 6 lines long, slightly pubescent. Anther-connective not produced beyond the cells. Ovary glabrous, stipitate; style elongated; ovules 2.—Meissn. in DC. Prod. xiv. 337; *P. abietina,* A. Cunn.; Meissn. l.c. 336.

N. S. Wales. Harvey's range, west of Wellington valley, *A. Cunningham;* St. George's range, *Fraser;* Castlereagh river, *C. Moore.* Resembles some specimens of *P. fastigiata,* under which name it also occurs in A. Cunningham's collections, but is readily distinguished by the glabrous ovary.

49. **P. oblongata,** *A. Cunn.; R. Br. Prot. Nov.* 14.   Glabrous or the branches very slightly pubescent. Leaves ovate-lanceolate or lanceolate, mucronate-acute, rounded or contracted at the base, nearly sessile, mostly about 1½ in. long, flat, not thick, scarcely veined besides the midrib. Pedicels axillary, slender, recurved, ¼ to ½ in. long or even longer, usually glabrous. Perianth about 5 lines long, much constricted above the base, glabrous, the segments without subulate points. Ovary stipitate, quite glabrous or rarely sprinkled with very few hairs. —Meissn. in DC. Prod. xiv. 341 ; *P. planifolia,* A. Cunn. Herb.

N. S. Wales. Towards Hunter's river, *A. Cunningham;* Blue Mountains, *Woolls.*

50. **P. Cunninghamii,** *R. Br. Prot. Nov.* 15.   An erect or spreading low shrub, closely allied to *P. myrtilloides* and perhaps a variety, the branches sparingly pubescent or hirsute. Leaves numerous, ovate, mucronate-acute, ½ to 1 in. long, the larger ones often very broad, coriaceous, flat, 1-nerved. Pedicels all axillary, glabrous, slender, 2 to 4 lines long. Perianth 4 to 5 lines long, glabrous, the segments tipped with long dorsal subulate points. Anther-connective not produced beyond the cells. Ovary stipitate, glabrous; style elongated; ovules 2.— Meissn. in DC. Prod. xiv. 342.

N. S. Wales. Country north of Cujeegong river, *A. Cunningham;* sandy ridges, Liverpool plains, *C. Moore.* A specimen without flowers from Mudgee, *Woolls,* may possibly belong also to this species.

51. **P. myrtilloides,** *Sieb. in Roem. and Schult. Syst.* iii. *Mant.* 272. A much-branched spreading shrub of 4 ft. (*Fraser*), the branches and sometimes the young shoots pubescent, the adult foliage glabrous or nearly so. Leaves in the typical form oblong-lanceolate, mucronate, contracted at the base and sometimes shortly petiolate, ¾ to above 1 in. long, rigid, the nerve-like margin slightly recurved, the midrib scarcely prominent, otherwise flat and veinless. Pedicels axillary, pubescent, spreading, 1 to 2 lines long. Perianth 4 to 5 lines long, pubescent, the segments tipped with dorsal horn-like points. Anther-connective not produced beyond the cells. Ovary stipitate, glabrous; style elongated; ovules 2.—R. Br. Prot. Nov. 14; Meissn. in DC. Prod. xiv. 339; *P. oleifolia,* A. Cunn. Herb.

N. S. Wales. Blue Mountains, *A. Cunningham, Fraser, Sieber, n.* 52, and others.

Var. *brevifolia.* Leaves ½ to ¾ in. long, varying from lanceolate to almost ovate, the floral ones sometimes reduced to small bracts. Pedicels pubescent. Perianth not seen.

Victoria. Upper Genoa river and Nangatta mountains up to 4000 ft. elevation, *F. Mueller,* the specimens all in fruit only.

52. **P. oxycoccoides,** *Sieb. in Spreng. Syst. Cur. Post* 45, *and in Roem. and Schult. Syst.* iii. *Mant.* 270. A much-branched shrub, sometimes low and bushy or even procumbent, sometimes said to be very tall, more or less scabrous-pubescent or quite glabrous. Leaves very shortly petiolate, orbicular ovate elliptical or oblong-lanceolate, obtuse, rigid, with recurved margins, veinless except the midrib, 2 to 3 lines long when broad, twice that length when narrow. Pedicels axillary, spreading, 1 to 2 lines long. Perianth glabrous, about 4 lines long, the segments without terminal points. Anther-connective not produced beyond the cells. Ovary stipitate, glabrous; style elongated; ovule solitary in all the ovaries opened.—R. Br. Prot. Nov. 15; Meissn. in DC. Prod. xiv. 338; *P. thymifolia,* A. Cunn.; R. Br. l.c.; Meissn. l.c.; *P. microphylla,* R. Br. l.c.; Meissn. l.c.

N. S. Wales. Blue Mountains, *A. Cunningham, Sieber, n.* 49, *Woolls;* Argyle County, *A. Cunningham;* Berrima, *M'Arthur;* southern districts, *C. Moore;* Shoalhaven, *Rietmann.*

The majority of the numerous specimens I have seen are in fruit only, and in that state the three supposed species do not appear to me to be distinguishable even as marked varieties, notwithstanding the great differences in the size and shape of the leaves. Some very imperfect specimens from Darling Downs, *Lau,* may possibly belong to the same species.

Var.? *longifolia.* "A small tree." Leaves lanceolate, acute, about 1 in. long.— New England, *C. Stuart,* and perhaps the same from the Blue Mountains in Herb. F. Mueller, where both are referred to *P. oxycoccoides.* This determination is, however, very doubtful, and there are no flowers on either specimen.

53. **P. nutans,** *R. Br. in Trans. Linn. Soc.* x. 162, *Prod.* 373. An erect glabrous shrub, usually low and bushy with numerous slender branches, the young shoots rarely minutely hoary-pubescent. Leaves narrow-linear, acute, contracted at the base, ¾ to above 1 in. long, flat or the margins slightly recurved, the midrib prominent underneath. Pedicels axillary, solitary or rarely 2 together, filiform, 3 to 5 lines

long, very spreading or at length reflexed. Perianth glabrous, about 4 lines long, the segments tipped with fine points usually short. Anther-connective not produced beyond the cells. Ovary on a rather long stipes, glabrous; style elongated, often but not always very flexuose at the base; ovules 2.—Meissn. in DC. Prod. xiv. 335, *P. linearis,* Sieb. Pl. Exs.; *P. flexifolia,* Lodd. Bot. Cab. t. 922, and consequently Meissn. in DC. Prod. xiv. 337 in part, not of R. Br.

N. S. Wales. Port Jackson to the Blue Mountains, *R. Brown, Sieber, n.* 46, *and Fl. Mixt. n.* 472, and many others.

Var. *apiculata.* Perianth rather longer, the points of the segments long and fine.— *P. apiculata,* Meissn. in Hook. Kew Journ. vii. 73, and in DC. Prod. xiv. 335.—Liverpool road near Sydney, with the typical form, *A. Cunningham.* The bending of the style above the ovary and the greater or less prominence of the tips of the segments are both very variable in the numerous specimens before me of *P. nutans.*

54.? **P. angulata,** *R. Br. Prot. Nov.* 14. An erect shrub, glabrous or nearly so. Leaves crowded, oblong-linear or linear-lanceolate, acute or with a short callous point, contracted at the base, 1 to 1½ in. long, thick, nerveless, concave. Pedicels solitary, about 1 line long. Flowers and fruit unknown.—Meissn. in DC. Prod. xiv. 339.

N. S. Wales. Blue Mountains, *A. Cunningham.* Evidently distinct from all other species, and most probably a *Persoonia,* but it must remain doubtful till flowering specimens have been examined.

55. **P. virgata,** *R. Br. in Trans. Linn. Soc.* x. 161, *Prod.* 372. A large shrub, with numerous slender branches more or less pubescent, the adult foliage glabrous or nearly so. Leaves narrow-linear, acuminate, contracted at the base, 1 to 1½ in. long, flat concave or grooved above, convex underneath without any prominent midrib. Pedicels slender, glabrous or minutely pubescent, about ¼ in. long. Perianth glabrous, 4 to 4½ lines long, the segments tipped with minute points. Anther-connective not produced beyond the cells. Ovary glabrous, on a rather short and thick stipes; style elongated; ovules 2.—Meissn. in DC. Prod. xiv. 338; *P. linariifolia,* A. Cunn. Herb.; *P. tenuifolia,* Meissn. l.c. 334, not of R. Br.

Queensland. Sandy Cape, Hervey Bay, *R. Brown;* sandy shores of Stradbrooke Island, *A. Cunningham, Fraser;* Moreton Island, *M'Gillivray, F. Mueller;* Pine river, *Fitzalan.*

N. S. Wales. Cape Byron, *C. Moore.*

This species resembles in some respects *P. nutans,* but the flowers are more erect and the leaves grooved on the upper and not on the under side, the margins rather incurved than recurved.

56. **P. chamæpeuce,** *Lhotsky; Meissn. in DC. Prod.* xiv. 336. A low decumbent or erect shrub, quite glabrous or the young shoots sparingly pubescent. Leaves spreading, linear, acute but not pungent, contracted at the base, rigid, flat or slightly concave, without any prominent midrib, ½ to 1 in. long. Pedicels axillary, solitary, rarely above 2 lines long. Perianth glabrous, acuminate, about 5 lines long. Anther-connective not produced beyond the cells. Ovary stipitate, glabrous; style elongated; ovules solitary in the ovaries examined.—

P. suffruticosa, F. Muell. 1st Gen. Rep. 17 ; *P. Caleyi,* F. Muell. Pl. Vict. ii. t. 69, not of R. Br.

N. S. Wales. Barren rocky hills near Bathurst, *A. Cunningham, Woolls.*
Victoria. Hardinge and Buffalo ranges, mountains on Macalister river, between Loddon and Creswick Creeks, between Broken and Ovens rivers, *F. Mueller.*

57. **P. juniperina,** *Labill. Pl. Nov. Holl.* i. 33, *t.* 45. A bushy or divaricate shrub, sometimes low and spreading, more frequently 5 or 6 ft. high or even much taller, the young shoots silky-pubescent, the older foliage glabrous or nearly so. Leaves sessile, narrow-linear, rigid and pungent-pointed, flat or concave, the midrib prominent underneath but not always very distinct, ½ to 1 in. long. Flowers axillary, solitary, on very short pedicels. Perianth shortly silky-pubescent or nearly glabrous, about 4 lines long, the segments without subulate points. Anther-connective not produced beyond the cells. Ovary stipitate, glabrous ; style elongated ; ovules 2.—R. Br. in. Trans. Linn. Soc. x. 160, Prod. 372 ; Meissn. in DC. Prod. xiv. 336 ; Hook. f. Fl. Tasm. i. 321 ; *P. surrecta,* F. Muell. in Adelaide Deutsch. Zeit. 1851, (Meissn.)

Victoria. Port Phillip, *R. Brown ;* from the Glenelg, *Robertson,* and Melbourne, *Adamson,* to Genoa river, *F. Mueller ;* Dandenong ranges and Grampians, *F. Mueller ;* Wimmera, *Dallachy.*
Tasmania, *Labillardière;* Port Dalrymple, *R. Brown ;* very common, ascending to 3200 ft., *J. D. Hooker.*
S. Australia. Mount Lofty range, *F. Mueller ;* Tattiara country, *Woods.*

58. **P. tenuifolia,** *R. Br. Prot. Nov.* 12, *but not of Meissn.* A shrub with slender slightly pubescent branches. Leaves crowded, filiform, more or less spreading, acute but not pungent, nearly terete, grooved above, glabrous, ½ to ¾ in. long. Pedicels axillary, rarely 1 line long. Perianth glabrous, 4 to 5 lines long, the segments without subulate points. Anther-connective not produced beyond the cells. Ovary stipitate, glabrous ; style elongated ; ovules 2. Fruit broad.

Queensland. Logan and Brisbane rivers, *Fraser ;* Glasshouses and Moreton Bay *F. Mueller.*
N. S. Wales. New England, *C. Stuart ;* Biroa, *Leichhardt.*

59. **P. acerosa,** *Sieb. in Roem. and Schult. Syst.* iii. *Mant.* 269. A shrub of 2 to 6 ft., with slender virgate branches, the whole plant quite glabrous or a few hairs on the young shoots, the specimens usually drying very black. Leaves crowded, more erect than in *P. tenuifolia,* filiform, acute but not pungent, nearly terete, grooved above, ½ to ¾ in. long. Flowers axillary, on very short erect pedicels, scattered or forming a long leafy raceme. Perianth glabrous, 4 to 5 lines long, the segments tipped with long subulate points. Anther-connective not produced beyond the cells. Ovary stipitate, glabrous ; style elongated ; ovules solitary in all the ovaries examined. Fruit ovoid.—R. Br. Prot. Nov. 13 ; Meissn. in DC. Prod. xiv. 335 ; *P. pallida,* Grah. in Edinb. New Phil. Journ. 1828-9, 177 ; Meissn. l.c. 334.

N. S. Wales. Blue Mountains, *Sieber, n.* 59, *Fraser, Backhouse, Miss Atkinson, Woolls;* Illawarra, *Shepherd.* I find this species as well as *P. curvifolia* sometimes designated as *P. abietina,* A. Cunn.

SUBORDER 2. FOLLICULARES.—Fruit dehiscent, follicular or 2-valved or rarely drupaceous and indehiscent. Flowers usually in pairs, with a single bract to each pair, or rarely the inflorescence anomalous. Ovules 2 or more, collateral in each pair.

TRIBE 5. GREVILLEEÆ.—Ovules 2 or 4, all collateral. Seeds without any intervening substances or separated only by a thin lamina or mealy substance. Flowers in racemes or rarely in umbels or clusters, with deciduous or abortive bracts or rarely surrounded by an involucre of imbricate bracts.

13. HELICIA, Lour.

Flowers hermaphrodite. Perianth regular, the tube slender, the laminæ small, the segments all much revolute when separating. Anthers on short filaments inserted a little below the laminæ, the connective produced into a short appendage. Hypogynous glands equal, distinct or united in a ring or cup round the ovary. Ovary sessile, with a long straight style, slightly thickened at the end with a terminal stigma; ovules 2, ascending, laterally attached near the base. Fruit hard, nearly globular, indehiscent (without any fleshy exocarp?). Seeds either solitary and globular or two together and hemispherical; testa veined or rugose; cotyledons thick and fleshy.—Trees or tall shrubs. Leaves alternate, entire or toothed. Flowers pedicellate in pairs, in terminal or axillary simple racemes, the pedicels of each pair often more or less connate. Bracts very deciduous (or sometimes none?).

The genus is spread over tropical Asia extending northwards to Japan. The Australian species appear to be all endemic, although one of them closely resembles one of the most widely dispersed of the Asiatic ones. The young fruits I have seen appear to be fleshy externally, the few ripe ones are detached, woody and smooth, but the exocarp may be deciduous as in *Macadamia ternifolia*.

Leaves mostly entire. Flowers glabrous.
Perianth 7 to 8 lines long. Hypogynous glands oblong or
    obovoid, quite distinct . . . . 1. *H. præalta.*
Perianth slender, about ½ in. long. Hypogynous glands broad,
    truncate but distinct . . . . 2. *H. australasica.*
Perianth slender, about 4 lines long. Hypogynous glands more
    or less connate in a truncate ring or cup . . . . 3. *H. glabriflora.*
Leaves mostly serrate. Perianth small, densely ferruginous-villous
    as well as the whole inflorescence. . . . . 4. *H. ferruginea.*

1. **H. præalta,** *F. Muell. Fragm.* iii. 37. A moderate sized or sometimes lofty tree attaining 100 ft., glabrous except the inflorescence which is often minutely tomentose. Leaves lanceolate, usually narrow, obtuse or acuminate, contracted into a petiole, quite entire in all the specimens seen, only 3 or 4 in. long in a few specimens, mostly 6 to 10 in. in others, coriaceous, often shining, penniveined and reticulate, the veins fine. Racemes axillary or lateral, 3 to 6 in. long, the rhachis rigid. Pedicels 2 to 3 lines long, united to above the middle. Perianth 7 to 8 lines long. Hypogynous glands quite distinct and narrow. Ovary

glabrous. Fruit smooth and hard, above 1 in. diameter, but not seen quite ripe. Seed (nearly full-grown) with a deeply rugose testa.

Queensland. Scrubs near Brisbane, *W. Hill.*
N. S. Wales. Clarence river, *Beckler, C. Moore;* Richmond river, *C. Moore.*

2. **H. australasica,** *F. Muell. in Hook. Kew Journ.* ix. 27. A small tree, quite glabrous. Leaves oval-elliptical, obtuse or scarcely acuminate, entire or irregularly toothed, contracted into a very short petiole, glabrous and veined on both sides, 4 to 8 in. long. Racemes axillary or lateral, shorter than the leaves or rarely exceeding the shorter upper ones, quite glabrous. Flowers in pairs on an exceedingly short common pedicel so as to appear almost sessile on the rhachis. Perianth slender, glabrous, about ½ in. long. Hypogynous glands broad and truncate so as apparently to form a ring or cup, but really free. Ovary villous. Fruit not seen.

N. Australia. Towards Macadam range, *F. Mueller;* Port Darwin, *Schultz.*
The species is very nearly allied to the common Asiatic *H. robusta.*

3. **H. glabriflora,** *F. Muell. Fragm.* ii. 91. A small tree quite glabrous. Leaves ovate-elliptical, obtuse or obtusely acuminate, entire or very rarely toothed, contracted into a short petiole, 2 to 3 in. long, coriaceous with the veins less conspicuous than in *H. australasica.* Racemes terminal axillary or lateral, very slender, glabrous, about as long as the leaves, the rhachis almost filiform. Pedicels free or shortly united at the base, ½ to 1 line long. Perianth very slender, glabrous, about 4 lines long. Hypogynous glands more or less connate in a truncate ring or cup. Fruit only seen young.—*H. conjunctiflora,* F. Muell. Fragm. v. 38.

N. S. Wales. Camden Haven, *C. Moore;* Leycester Creek, Richmond river, *Beckler.*

4. **H. ferruginea,** *F. Muell. Fragm.* iii. 37. A moderate-sized tree, the branches and inflorescence densely villous with ferruginous or fulvous hairs, which often persist on the principal veins of the underside of the leaves. Leaves shortly petiolate, ovate-elliptical or oblong, acuminate, serrate, contracted or rounded at the base, 3 to 4 in. long in some specimens, twice that size in others, the veins very prominent underneath, the primary ones sometimes numerous and regular, in others fewer and more unequal, the minor reticulations also very variable. Racemes terminal or axillary, rather dense, shorter than the leaves. Flowers small, in pairs on a very short common pedicel. Perianth slender, densely rufous-villous, 2½ to 3 lines long. Hypogynous glands short and broad, irregular, 2 of them sometimes united. Fruit not seen ripe.

Queensland. Rockingham Bay, *Dallachy.*
N. S. Wales. Tweed, Richmond and Clarence rivers, *C. Moore.*

14. MACADAMIA, F. Muell.

Flowers hermaphrodite. Perianth regular or slightly irregular, the tube opening earlier on the under side and the segments, at least the lower ones, less revolute than in *Helicia.* Anthers on short filaments, inserted a little below the laminæ, the connective produced into a gland or very short appendage. Hypogynous glands equal, distinct or united in a ring or cup round the ovary. Ovary sessile, with a long straight style, ovoid or clavate at the end, with a small terminal stigma; ovules 2, descending, laterally attached at or near the top. Fruit globular, indehiscent, with a hard thick putamen and rather thin fleshy exocarp. Seeds either solitary and globular or 2 and hemispherical; testa membranous; cotyledons thick and fleshy.—Trees or tall shrubs. Leaves verticillate, entire or serrate. Flowers pedicellate in pairs, in terminal or axillary simple racemes, the pedicels not connate. Bracts very deciduous.

The genus is endemic in Australia. It is, as observed by F. Mueller, closely allied to *Helicia,* but the verticillate leaves, constantly free pedicels, slightly oblique flowers, descending ovules and more drupaceous fruits, may justify the retaining it as distinct.

Perianths about 8 lines long, in short dense racemes. Hypogynous
glands ovoid, distinct . . . . 1. *M. Youngiana.*
Perianths 2 to 3 lines long, in slender racemes. Hypogynous
glands broad, truncate, united in a cup or ring.
Leaves in whorls of 3 or 4. Racemes long . . . . . . 2. *M. ternifolia.*
Leaves in whorls of 5 to 7. Racemes much shorter than the
leaves . . . . 3. *M. verticillata.*

1. **M. Youngiana,** *F. Muell.* A shrub of 8 to 10 ft., the young branches and inflorescence ferruginous-pubescent. Leaves shortly petiolate, in whorls of 3 or 4, oblong-elliptical, acute or acuminate, entire or with a few small teeth, rounded or contracted at the base, 2 to 4 in. long, glabrous above, silky underneath when young. Racemes terminal, rather dense, shorter than the leaves. Pedicels ferruginous-pubescent, rarely above 2 lines long. Perianth pubescent, about 8 lines long, the three lower segments remaining longer coherent and less revolute than the upper one. Anther-connectives produced into a short obtuse appendage. Hypogynous glands oblong, quite distinct. Ovary villous; style-end ovoid.—*Helicia Youngiana,* F. Muell. Fragm. iv. 84.

N. S. Wales. Head of the Clarence river, *C. Moore;* Richmond river, *C. Moore, Fawcett, Henderson.*

2. **M. ternifolia,** *F. Muell. in Trans. Phil. Inst. Vict.* ii. 72, *with a plate.* A small tree with a very dense foliage, glabrous or the young branches and inflorescence minutely pubescent. Leaves sessile or nearly so, in whorls of 3 or 4, oblong or lanceolate, acute, serrate with fine or prickly teeth, glabrous and shining, from a few in. to above 1 ft. long. Racemes often as long as the leaves, with numerous small flowers, the pairs often clustered or almost verticillate. Pedicels at first very

short and not above 2 lines when in fruit. Perianth minutely pubes
cent or glabrous, nearly 3 lines long. Hypogynous glands united in a
ring. Ovary villous; style-end clavate. Fruit with a 2-valved fleshy
exocarp; the putamen globular, smooth and shining, thick and woody,
often above 1 in. diameter.—*Helicia ternifolia,* F. Muell. Fragm. ii. 91,
vi. 191.

Queensland. Pine river and Moreton Bay, *W. Hill;* Dawson and Burnett rivers,
Leichhardt (with the leaves less toothed and the flowers rather larger).
N. S. Wales. Clarence and Richmond rivers, *C. Moore* (leaves rather small).

3. **M. verticillata,** *F. Muell.* Young shoots and inflorescence
slightly hoary or rusty-tomentose. Leaves in whorls of 5 or 6, oblong-
lanceolate, obtuse with a small callous point or almost acute, coarsely
toothed, contracted into a very short petiole, 3 to 5 in. long, coriaceous,
much reticulate. Racemes much shorter than the leaves, with numerous small crowded flowers. Bracts on the very young spikes broad,
villous, falling off long before the flowers expand. Pedicels filiform,
scarcely 2 lines long, hirsute. Perianth glabrous, about 1½ lines long.
Hypogynous glands united in a ring or cup. Ovary densely villous;
style-end clavate.—*Helicia verticillata,* F. Muell. Fragm. vi. 191.

Queensland or **N. S. Wales,** *Leichhardt,* the precise station not given (*Herb.*
F. Mueller).

15. XYLOMELUM, Sm.

Flowers partially polygamous. Perianth regular, nearly cylindrical
in the bud, the segments revolute, dilated at the end into short concave laminæ. Anthers apparently perfect in all the flowers, on short
filaments inserted a little below the laminæ, the connective produced
beyond the cells in a short obtuse or gland-like appendage. Hypogynous glands 4, small. Ovary in the fertile flowers shortly stipitate
or almost sessile, tapering into a filiform style clavate at the end, with
a terminal stigma either small and scarcely prominent or large thick
and pulvinate; ovules 2 laterally attached below the middle; in the
sterile flowers the ovary is abortive, but the style is clavate at the end
though without any stigmatic surface. Fruit large, ovoid or tapering
above the middle, very thick and woody, tardily opening along the
upper side or in 2 valves. Seed flat, obliquely ovate, with a long terminal oblique or falcate wing.—Trees or tall shrubs. Leaves opposite,
entire or prickly-toothed. Flowers in opposite dense spikes, axillary
or at first forming a terminal cluster, becoming lateral by the elonga-
tion of the branch. Bracts small, at first imbricate, but falling off
long before the flowering. Flowers sessile in pairs within each bract,
the lower ones of the spike usually perfect, the upper ones with abortive ovaries.

The genus is endemic in Australia, where it is widely spread, the fruits generally
known under the name of *wooden pears.*

Leaves veined, those at least of the barren branches prickly-
toothed. Perianths ferruginous-villous, 5 to 6 lines long.
Leaves of the flowering branches usually entire. Eastern
species . . . . 1. *X. pyriforme.*

Leaves of the flowering branches usually prickly-toothed.
　Western species . . . . 2. *X. occidentale*.
Leaves all quite entire.　Perianth silky, under 4 lines long.
Leaves lanceolate, often falcate, veined.　Fruits nearly glabrous,
　narrow.　Eastern species . . . . 3. *X. salicinum*.
Leaves linear or linear-lanceolate, thick and veinless.　Fruits
　closely tomentose.　Western species . . . . 4. *X. angustifolium*.

1. **X. pyriforme,** *Knight, Prot.* 105.　A tree of moderate size, the
young shoots ferruginous-villous or tomentose but soon becoming glabrous, the spikes remaining densely tomentose-villous.　Leaves of the
flowering branches usually entire, lanceolate or ovate-lanceolate, very
acute, 4 to 6 in. long and tapering into a rather long petiole, those of
flowerless branches or of younger plants often sinuate and prickly-
toothed and attaining 8 in. with short petioles, all at length coriaceous
and shining.　Spikes very dense, 2 to 3 in. long, usually clustered 3 to
6 together and at first appearing terminal, but soon lateral by the
growing out of the shoots.　Bracts woolly-villous, orbicular, 1½ to 2
lines diameter.　Perianth about 5 lines long.　Style-end clavate, but
smaller and shorter than in *X. occidentale*.　Fruit 2½ to 3 in. long and
above 1 in. diameter near the base, somewhat tapering above the
middle.—R. Br. in Trans. Linn. Soc. x. 189, Prod. 387, Prot. Nov.
31; Meissn. in DC. Prod. xiv. 422; Endl. Iconogr. t. 47, 48; Reichb.
Ic. Exot. t. 90; *Banksia pyriformis*, Gærtn. Fr. i. 220, t. 47; Sm. in
White, Voy. 224, t. 21; *Hakea pyriformis*, Cav. Anal. Hist. Nat. i. 217,
Ic. vi. 25, t. 536; *Conchium pyriforme*, Willd. Enum. Hort. Berol. 141.

N. S. Wales.　Port Jackson, *R. Brown, Sieber, n.* 53, and many others.

2. **X. occidentale,** *R. Br Prot. Nov.* 31.　An irregular shrub or tree
of 12 to 25 ft., the young shoots and inflorescence densely ferruginous
or hoary-tomentose, the older leaves glabrous.　Leaves petiolate, ovate
elliptical or oblong, irregularly marked by a few coarse undulate prickly
teeth, 3 to 5 in. long.　Spikes in the upper axils, 3 to 5 in. long,
flowering from the base and forming a large terminal densely branched
panicle with the floral leaves much reduced, or sometimes the leafy
branch growing out leaving a few pairs of spikes at the base.　Bracts
very broad, concave, truncate, 1 to 2 lines diameter, very deciduous.
Perianth 5 to 6 lines long, softly tomentose-villous as well as the bracts
and rhachis.　Anther-connectives produced into an oblong appendage.
Ovary very densely villous.　Style-end clavate.　Fruit 2 to 3 in. long
and about 1 in. diameter near the base, somewhat tapering above the
middle, but very obtuse.—Meissn. in Pl. Preiss. i. 580, and in DC. Prod.
xiv. 423; Hook. Ic. Pl. t. 446.

W. Australia.　King George's Sound, *M'Lean;* Geographe Bay, *Fraser;* Swan
river, *Drummond,* 1*st coll. n.* 616, *Preiss, n.* 754; Vasse river, *Oldfield.*　I have great
doubts whether this species be sufficiently distinct from *X. pyriforme*.

3. **X. salicinum,** *A. Cunn. in R. Br. Prot. Nov.* 31.　A small tree,
glabrous except the inflorescence or the young shoots minutely hoary-
pubescent.　Leaves lanceolate, broad or narrow, obtuse, falcate, entire,
contracted into a slender petiole, 4 to 8 in. long, of a pale green above,

usually glaucous or whitish underneath. Spikes 1½ to 2 in. long, the rhachis and flowers tomentose-pubescent and whitish, sometimes silvery. Bracts small and broad, imbricate in the very young spike, but very deciduous. Perianth 3 to 3½ lines long. Ovary villous. Style shortly clavate, and in the fertile flowers capped by a thick broad pulvinate stigma, which I have not observed in any other species. Fruit 2½ to 3 in. long, not so broad as in *X. pyriforme* and tapering into a thick beak recurved at the end.—*X. pyriforme,* var. *salicinum,* R. Br. Prot. Nov. 31; Meissn. in DC. Prod. xiv. 423; *Helicia Scottiana,* F. Muell. Fragm. iv. 107; *Xylomelum Scottianum,* F. Muell. Fragm. v. 174, 215.

Queensland. Moreton Bay, *A. Cunningham;* Dogwood Creek, *Leichhardt;* Rockingham Bay, *Dallachy;* Darling Downs, *Lau.*

4. **X. angustifolium,** *Kipp.; Meissn. in DC. Prod.* xiv. 423. A shrub of 6 to 8 ft., with erect virgate branches, the inflorescence and often the very young shoots silky-pubescent, the adult foliage glabrous. Leaves linear or linear-lanceolate, 4 to 6 in. long, tapering into a fine rigid point when perfect, or sometimes obtuse even when young, contracted into a petiole, thick but flat, veinless except the midrib. Spikes rather loose, shorter than the leaves, flowering from near the base. Bracts broad, silky-villous, imbricate in the very young spikes but falling off very early. Perianth silky, about 3½ lines long. Anther-connectives tipped with a small gland. Style slender, slightly clavate at the end, the thickened portion covered with a short transparent pubescence in some flowers, glabrous in others, but in all those examined I found the anthers perfect and the ovary abortive, the perfect flowers may possibly therefore have the thick stigma of *X. salicinum.* Fruit ovoid, oblique, about 2½ in. long, slightly contracted towards the end, covered with a very close but dense tomentum.

W. Australia, *Drummond, 5th coll. suppl. n. 7; Forrest?* (in herb. F. Mueller); Ironstone range, Murchison river, *Oldfield.*

16. **CARNARVONIA,** F. Muell.

Flowers hermaphrodite. Perianth regular, nearly cylindrical in the bud, the segments free or nearly so, recurved in the upper portion, without distinct laminæ. Anthers all perfect, linear, sessile below the middle of the perianth, the connective produced beyond the cells. No hypogynous glands. Ovary sessile or nearly so, tapering into an erect style with a small terminal stigma; ovules 2, laterally attached below the middle to a short funicle. Fruit a hard incurved follicle. Seeds compressed, produced at the upper end into a long wing.—A tree with compound leaves. Flowers small, in axillary simple or compound irregular racemes.

The genus is limited to a single species, endemic in Australia, with the flowers nearly of *Persoonia,* but with the fruit of *Hakea.* It is also closely allied in character to *Orites,* but with a very different habit and inflorescence.

1. **C. araliæfolia,** *F. Muell. Fragm.* vi. 81, *t.* 55, 56. A small or moderate sized tree, the young leaves slightly pubescent underneath, the inflorescence usually pubescent, the older leaves glabrous. Leaves compound, with 3 to 5 petiolulate leaflets digitate at the extremity of the petiole, or 1 to 3 of the leaflets replaced by pinnæ, each with 2 or 3 petiolulate leaflets not digitate, the leaflets from broadly obovate and very obtuse to elliptical oblong or lanceolate and acute, entire or remotely toothed, tapering at the base, 3 to 5 in. long, the whole leaf from 6 or 8 in. to twice that length. Racemes very irregular, simple or more frequently compound, much shorter than the leaves with small deciduous trifoliolate bracts under the branches, and a narrow entire one under each flower, or under a cluster of 3 to 6 flowers terminating the peduncle or branches. Pedicels softly hirsute, 2 to 3 lines long, glabrous and twice as long when in fruit. Perianth about 2 lines long, densely hirsute with soft hairs. Ovary glabrous. Follicle much incurved, acuminate, 1½ in. long. Wing of the seed twice as long as the seed itself, the raphe much within the margin.

Queensland. Rockingham Bay, *Dallachy.*

17. ORITES, R. Br.

(Oritina, *R. Br.*)

Flowers hermaphrodite. Perianth regular, nearly cylindrical in the bud, the segments free or nearly so, dilated at the end into short usually concave laminæ. Anthers all perfect, enclosed in the perianth-laminæ in the bud, but with short filaments inserted below the laminæ, the cells adnate to the slender connective which is not produced beyond them and is often scarcely conspicuous. Hypogynous glands linear, obtuse. Ovary sessile, with a terminal filiform straight style, scarcely thickened at the end, obtuse, with a small terminal stigma; ovules 2, amphitropous, laterally attached at or below the middle. Fruit an obliquely acute coriaceous follicle, more or less boat-shaped, the dorsal suture curved, the ventral one nearly straight. Seed compressed, with a terminal oblique or falcate wing, sometimes decurrent along the margins.—Shrubs or trees. Leaves alternate, more or less petiolate, entire toothed or rarely (in the same species) lobed. Flowers small, in terminal or axillary spikes, sessile or nearly so, in pairs within each bract. Bracts concave, imbricate in the very young spike but falling off long before the flowers expand.

The genus is endemic in Australia, and exclusively eastern.

Spikes all axillary. Leaves lanceolate, often toothed or divided on
the barren branches.
Tall tree. Leaves mostly above 4 in. Branches inflorescence
and ovary glabrous . . . . 1. *O. excelsa.*
Shrub. Leaves mostly under 3 in. Branches rhachis and ovary
villous . . . . 2. *O. diversifolia.*
Spikes terminal, rarely also in the upper axils. Leaves small and
crowded.
Leaves flat, ovate, all toothed . . . . 3. *O. Milligani.*

Leaves flat, oblong or lanceolate, obtuse, all entire . . . .   4. *O. lancifolia.*
Leaves linear with revolute margins, obtuse or scarcely acute .   5. *O. revoluta.*
Leaves terete, grooved above, mucronate or pungent. Seeds surrounded by a narrow wing (the wing terminal and long in all
  other species) . . . .   6. *O. acicularis.*

1. **O. excelsa,** *R. Br. Prot. Nov.* 32. A handsome tree of 40 to 60
ft., usually quite glabrous. Leaves on the flowering branches lanceolate, obtuse or acute, tapering into a rather long petiole, entire or
slightly toothed, 4 to 6 in. long, flat, reticulate, shining above, glaucous
underneath; those of the barren branches often larger, toothed or
deeply divided into 3 or 5 lanceolate toothed lobes. Spikes axillary,
interrupted, shorter than the leaves, usually glabrous, the flowers in
distant pairs. Bracts at first ovate acute and imbricate, but falling off
at a very early stage. Perianth glabrous, about 3 lines long. Filaments broad, attached to about the middle of the claws. Ovary glabrous; style short. Follicle acuminate, about 1 in. long. Seed flat,
the nucleus about 4 lines long, with a terminal wing at least as long.
—Meissn. in DC. Prod. xiv. 423.

N. S. Wales. Deep shaded forests at the sources of the Hastings river, *A. Cunningham;* Macquarrie river, *Fraser;* Tweed, Richmond and Clarence rivers,
C. Moore.

2. **O. diversifolia,** *R. Br. in Trans. Linn. Soc.* x. 190, Prod. 388
A shrub of 3 to 4 ft., with erect tomentose villous branches. Leaves
lanceolate, with a callous point, contracted into a petiole of 1 to 2 or
even 3 lines, those of the floral branches usually 1½ to 2 in. long, entire
or with a few teeth towards the end, coriaceous, with recurved margins,
smooth above, glaucous or slightly ferruginous underneath, glabrous or
sprinkled with a few short hairs; those of the barren branches often
twice as long and irregularly toothed above the middle or almost to
the base and with more conspicuous veins. Spikes axillary, rather
dense and about as long as the leaves or sometimes longer and looser,
the rhachis ferruginous-tomentose. Perianth glabrous, about 2 lines
long, not very slender, the laminæ scarcely broader than the claws.
Ovary villous; style short in some flowers, elongated in others. Follicle acute, ¾ in. long.—Meissn. in DC. Prod. xiv. 424; Hook. f. Fl.
Tasm. i. 326.

Tasmania. Near the summit of Mount Wellington, *R. Brown, J. D. Hooker,* and
others; Mount Field East, *F. Mueller.*

3. **O. Milligani,** *Meissn. in DC. Prod.* xiv. 424. A rigid densely
bushy shrub of 1½ to 3 ft., glabrous and glaucous except the inflorescence. Leaves shortly petiolate, ovate, coarsely toothed, thick and
rigid, penniveined but the veins not very prominent underneath and
inconspicuous or slightly impressed above, ½ to 1 in. long. Spikes
terminal, 1 to 1½ in. long. Bracts ovate, concave, rigid, ciliate or quite
glabrous, very deciduous. Perianth glabrous, slender, nearly 3 lines
long. Filaments very short, almost immediately under the laminæ.
Ovary ferruginous-villous.—Hook. f. Fl. Tasm. i. 326.

Tasmania. Mount Sorrel, Macquarrie harbour, at an elevation of 4000 to 5000 ft.,
Milligan.

4. **O. lancifolia,** *F. Muell. in Trans. Phil. Soc. Vict.* i. 108. A handsome glabrous shrub. Leaves shortly petiolate, crowded, oblong or oblong-lanceolate, obtuse, flat, with nerve-like or slightly recurved margins, mostly ¾ to 1 in. long, coriaceous, prominently reticulate underneath, obscurely so or quite smooth and shining above. Spikes terminal or terminating short axillary shoots, or rarely also in the upper axils without leaves at their base, exceeding the leaves and sometimes 2 in. long, the rhachis ferruginous. Bracts villous, very deciduous. Perianth 2 to 2½ lines long. Filaments narrow, inserted more than ½ line below the short concave laminæ. Ovary villous. Fruit about ¾ in. long. Seed with a broad oblique terminal wing.

Victoria. Rocky summits of the Australian Alps at an elevation of 5000 to 6000 ft., *F. Mueller.* Meissner, in DC. Prod. xiv. 423, reduces this plant to a variety of *O. excelsa,* but the small entire leaves, the inflorescence, the villous ovary, and other characters, appear to me to be constant.

5. **O. revoluta,** *R. Br. in Trans. Linn. Soc.* x. 190, *Prod.* 388. A bushy shrub of 4 to 6 ft., the branches hoary or ferruginous-pubescent. Leaves rather crowded, sessile or shortly petiolate, linear, obtuse or scarcely acute, the margins revolute, thick, rigid, glabrous and smooth above, the under surface slightly tomentose but usually concealed, ½ to ¾ in. or rarely 1 in. long. Spikes terminal, sessile, mostly above 1 in. long, the rhachis ferruginous. Bracts ovate or oblong, villous outside, very deciduous. Perianth about 2½ lines long. Filaments narrow and short. Ovary densely villous. Follicle ½ to ¾ in. long, silky-villous and usually ferruginous. Seed with a broad falcate terminal wing.—Meissn. in DC. Prod. xiv. 424; Hook. f. Fl. Tasm. i. 326; A. Rich. Sert. Astrol. t. 25.

Tasmania. Mount Wellington, *R. Brown ;* abundant on all the mountain ranges above 3000 ft. elevation, *J. D. Hooker.*

6. **O. acicularis,** *R. Br. Prot. Nov.* 32. A bushy shrub of 4 or 5 ft. or sometimes only half that size, the foliage of a yellowish sickly green, quite glabrous. Leaves crowded, terete with a very narrow groove on the upper side, mucronate-acute and often pungent, contracted into a slender petiole, rigid and smooth, ¾ to 1½ in. long. Spikes terminal, usually shorter than the leaves, the rhachis ferruginous-pubescent. Perianth glabrous, scarcely 2 lines long, not so slender as in *O. revoluta,* the segments more tardily revolute and the laminæ more concave than in that species. Filaments very short and broad, inserted immediately below the laminæ. Ovary villous. Follicle about ½ in. long, glabrous or nearly so. Seed oblong, flat, surrounded by a narrow wing rather broader at the upper end, like that of many *Grevilleæ.*—Meissn. in DC. Prod. xiv. 424; Hook. f. Fl. Tasm. i. 326; A. Rich. Sert. Astrol. t. 25; *Oritina acicularis,* R. Br. in Trans. Linn. Soc. x. 224.

Tasmania. Mount Wellington, *R. Brown ;* abundant on all the mountains at an elevation of 3000 to 4000 ft., *J. D. Hooker.* In its seeds this species approaches *Grevillea,* but the hypogynous glands and the distinct though short filaments are those of *Orites.* It differs from the other species of *Orites* as *Grevillea* does from *Hakea,* but the affinity in all other respects is so close with *O. revoluta* that botanists have all followed Brown in suppressing the genus he had at first proposed for it.

18. LAMBERTIA, Sm.

Flowers hermaphrodite. Perianth regular or nearly so, the tube
elongated, often dilated upwards and slightly incurved, the lobes narrow,
spirally revolute, the two lower ones sometimes more deeply separated.
Anthers all perfect, inserted on the lobes and revolute with them, the
connective shortly produced beyond the cells. Hypogynous scales
either flat at least as long as the ovary and free or connate, or in one
species wanting. Ovary very small, densely covered with long hairs,
with 2 pendulous ovules. Style filiform sometimes slightly thickened
and grooved on a level with the anthers; stigma small, terminal or
shortly decurrent on the upper side. Fruit a short hard truncate sessile
follicle, the lower (dorsal) margin produced into a thick horn,
and often a horn also on the upper angle of each valve. Seeds where
known flat, bordered by a narrow margin.—Shrubs. Leaves mostly in
whorls of 3, rarely of 4, or sometimes scattered at the base of luxuriant
shoots, entire or with spinescent teeth. Flowers red or yellow, usually
long, solitary or 7 together sessile within an involucre of imbricate coloured bracts; the inner bracts long and narrow the outer ones short and
broad, the involucres sessile and terminal or axillary.

The genus is endemic in extratropical Australia, the species all Western except *L.
formosa.* The species with uniflorous involucres have the inflorescence, perianth, stamens
and style very nearly of *Adenanthos,* but are readily distinguished by the whorled
leaves, biovulate ovary and follicular fruit; the perianth-segments sometimes remain
closed as in that genus after the style has emerged a little lower down from the slit in
the perianth-tube, but when open they are much more revolute than in *Adenanthos.*

Involucres 1-flowered, ½ in. long or under. Leaves entire.
　Leaves ovate, obovate or oblong, mucronate . . . . 1. *L. uniflora.*
　Leaves linear . . . . 2. *L. rariflora.*
Involucres 7-flowered, terminal or in the upper axils. Leaves entire.
　Leaves very obtuse, usually small.
　　Leaves obovate to linear, flat or nearly so. Bracts rather
　　　obtuse . . . . 3. *L. inermis.*
　　Leaves linear with revolute margins. Bracts with subulate
　　　points . . . . 4. *L. ericifolia.*
　Leaves mucronate, mostly linear and 1 to 2 in. long.
　　Inner bracts of the involucre half as long as the perianth.
　　　Western species . . . . 5. *L. multiflora.*
　　Inner bracts about as long as the perianth. Eastern species 6. *L. formosa.*
Involucres 7 flowered, all axillary. Leaves pungent-pointed and
　often prickly-toothed.
　Perianth 1¼ to 1½ in. long, the tube dilated upwards and incurved
　　(as in all the preceding species) . . . . 7. *L. echinata.*
　Perianth 7 to 8 lines long, the tube slender and straight . . . 8. *L. ilicifolia.*

1. **L. uniflora,** *R. Br. in Trans. Linn. Soc.* x. 188, *Prod.* 386. An
irregularly branched shrub, sometimes low and diffuse, sometimes erect
and 6 to 10 ft. high, glabrous or the young branches pubescent. Leaves
crowded about the flowers, often in distant clusters in the lower part of
the branches, very shortly petiolate, from ovate and under ½ in. to
broadly oblong and above 1 in. long, mucronate acute and sometimes
almost pungent or rounded at both ends, flat, smooth or reticulate, the

midrib prominent underneath. Involucres 1-flowered, he bracts very narrow and acute, almost scarious, the inner ones about $\frac{1}{2}$ in. long. Perianth nearly $1\frac{1}{2}$ in. long, dilated upwards and incurved, 2 of the lobes more united and less deeply revolute than the 2 others, the laminæ short, without terminal appendages. Anther-connectives produced into oblong tips. Hypogynous scales free. Follicle smooth, the dorsal suture acuminate, the valves rounded.—Meissn. in Pl. Preiss. i. 578, ii. 263, and in DC. Prod. xiv. 420.

W. Australia. King George's Sound and adjoining districts, *R. Brown, Baxter, Drummond, 3rd coll. n.* 261, 262 (*in some herbaria*), *Preiss, n.* 762, and many others.

2. **L. rariflora,** *Meissn. in Pl. Preiss.* ii. 263, *and in DC. Prod.* xiv. 420. An erect shrub, the young branches pubescent and often hirsute with long fine spreading hairs, the adult foliage glabrous. Leaves very shortly petiolate or almost sessile, linear or the floral ones lanceolate at the base, mucronate, rather thinner and the veins more prominent than in *L. multiflora,* mostly 1 to 2 in. long, but a few exceeding 3 in. Involucre sessile, 1-flowered, the inner bracts 3 to 4 lines long and acute, the outer ones gradually shorter broader and more obtuse. Perianth at least $1\frac{1}{4}$ in. long, pubescent dilated and incurved towards the middle, the laminæ without appendages. Anther-connectives tipped with exceedingly short gland-like ends. Hypogynous scales free. Follicle smooth.

W. Australia, *Drummond, 2nd coll. n.* 312.

3. **L. inermis,** *R. Br. in Trans. Linn. Soc.* x. 188, *Prod.* 387. A shrub of 6 to 10 ft., the branches minutely tomentose or silky-pubescent. Leaves from obovate or oblong-spathulate to linear, obtuse, contracted into a short petiole, $\frac{1}{2}$ to $\frac{3}{4}$ in. long, flat or slightly convex or concave, glabrous and smooth above, minutely silky pubescent and often ferruginous underneath. Involucres 7-flowered, terminal, solitary or rarely 2 together, the bracts more obtuse than in the other species, the inner ones $\frac{1}{2}$ to $\frac{3}{4}$ in. long. Perianth red according to some, yellow according to others, about $1\frac{1}{2}$ in. long, dilated and incurved in the middle, the lobes narrow with short pubescent tips, all nearly equally revolute. Anther-connectives produced into appendages of $\frac{1}{2}$ line. Hypogynous scales free. Follicle smooth.—Meissn. in Pl. Preiss. i. 578, ii. 263, and in DC. Prod. xiv. 420, *L. Drummondii,* Gardn. in Field. Sert. t. 22.

W. Australia. Lucky Bay, *R. Brown, Baxter;* King George's Sound to the Stirling Range, Cape Riche, and towards Swan river, *Drummond, 1st coll. n.* 594, and n. 87, *Preiss, n.* 763, 764, *Harvey, Oldfield, Roe, F. Mueller;* eastward to Middle Mount Barren, *Maxwell.*

4. **L. ericifolia,** *R. Br. Prot. Nov.* 30. A shrub of 6 to 10 ft., with virgate branches, the young shoots silky-pubescent, at length glabrous. Leaves linear, obtuse, with closely revolute margins, sessile or contracted into a very short petiole, usually about $\frac{1}{2}$ in. long, but $\frac{3}{4}$ in. on luxuriant sterile branches. Involucres 7-flowered, terminal, solitary or clustered 2 or 3 together, the bracts almost scarious ciliate and pubes-

cent or nearly glabrous, the inner ones subulate-acuminate and above ½ in. long, the outer ones short lanceolate and acute. Perianth above 1½ in. long, much dilated and incurved in the middle, the narrow laminæ cohering late round the style, viscid, with short hood-shaped tips, the segments at length revolute. Anther-connectives produced into oblong appendages. Hypogynous scales lanceolate, acute, more or less connate (2 connate and 2 free in the flowers examined). Follicle smooth. —Meissn. in Pl. Preiss. ii. 263, and in D.C. Prod. xiv. 420.

W. Australia, *Baxter, Drummond, 3rd coll. n.* 264; Stirling range, Salt river, and Cape Riche, *Maxwell;* at the base of Stirling range, *F. Mueller.*

5. **L. multiflora,** *Lindl. Swan Riv. App.* 32. A shrub of 3 or 4 ft., quite glabrous or the young shoots minutely pubescent, the flowering branches often acutely angular, the older ones terete. Leaves sessile, linear or the floral ones sometimes cordate-lanceolate, 1 to 2 in. long, mucronate, rigid, the midrib prominent underneath, the transverse veins chiefly conspicuous on the floral leaves. Involucres terminal, all 7-flowered (rarely fewer-flowered by abortion?), but owing to 2 or 3 heads being usually closely clustered together they have been described as 14- to 21-flowered, the inner bracts linear, fringed at the end and fully ½ in. long, the outer ones gradually shorter broader and entire. Perianth about 1½ in. long, slightly dilated and incurved above the middle, the laminæ tipped with small hood-shaped appendages. Anther-connectives shortly produced beyond the cells. Hypogynous scales free, oblong-lanceolate. Ovary very densely hirsute. Follicle smooth, the valves terminating in lanceolate points.—Meissn. in Pl. Preiss. i. 579, ii. 264 and in DC. Prod. xiv. 421; Field. Sert. t. 23.

W. Australia. Swan river, *Drummond, 1st coll. n.* 595 (*2nd coll.?*) *n.* 136, *Preiss, n.* 766; between Moore and Murchison rivers, *Drummond, 6th coll. n.* 198.

6. **L. formosa,** *Sm. in Trans. Linn. Soc.* iv. 214, *t.* 20. A tall shrub, glabrous or with a slight pubescence on the young shoots and here and there a few spreading hairs. Leaves linear or slightly linear-cuneate, rarely linear-lanceolate, mucronate with a fine pungent point, the margins recurved, contracted into a very short petiole, rigid, shining above, pale or almost ferruginous underneath, with a prominent midrib, varying from scarcely 1 in. to above 2 in. long. Involucres terminal, usually solitary, 7-flowered (or fewer-flowered by abortion?), the inner bracts narrow, silky-pubescent outside, 1½ to 2 in. long, the outer ones short and ovate. Perianth 1½ to 2 in. long, glabrous outside, dilated in the middle, the segments bearded inside below the anthers, the laminæ with pubescent tips. Anther-connectives produced into minute appendages. Hypogynous scales united in a truncate tube or cup surrounding the ovary. Follicle smooth, glabrous or villous.—R. Br. in Trans. Linn. Soc. x. 188, Prod. 387, Prot. Nov. 30; Meissn. in DC. Prod. xiv. 421; Cav. Anal. Hist. Nat. i. 233, t. 15, Ic. vi. 34, t. 547; Lodd. Bot. Cab. t. 80; Bot. Reg. t. 528; Andr. Bot. Rep. t. 69; *Protea nectarina,* Wendl. Sert. Hann. 5, t. 21.

W. Australia. Port Jackson, *R. Brown, Sieber, n.* 24, and many others.

7. **L. echinata,** *R. Br. in Trans. Linn. Soc.* x. 189, *Prod.* 387, *Prot. Nov.* 31. A shrub of 8 to 10 ft. with rigid stout or virgate branches usually pubescent or hirsute. Leaves more or less cuneate, ¾ to above 1 in. long, dilated at the end and truncate toothed or lobed, the teeth or lobes undulate and pungent-pointed, tapering into a narrow or broad base, sessile or petiolate, glabrous, rigid, the midrib and principal veins prominent, the whole foliage in shape texture and arrangement much resembling that of some *Gastrolobia.* Involucres 7-flowered, axillary, sessile, the bracts glabrous and acute, the inner ones narrow and 1 in. long or more, the outer ones gradually smaller and broader. Perianth yellow, 1¼ in. long, dilated and incurved above the middle, the segments nearly equally revolute. Anther-connectives tipped with small almost gland-like points. Hypogynous scales rather short, obtuse, free. Follicle more or less echinate with short thick prickles.—Meissn. in Pl. Preiss. i. 579, and in DC. Prod. xiv. 421.

W. Australia. Lucky Bay, *R. Brown, Baxter;* King George's Sound and adjoining districts, *A. Cunningham, Drummond, 3rd coll.* 263, *Preiss, n.* 761, and many others.

L. propinqua, R. Br. Prot. Nov. 30, Meissn. in DC. Prod. xiv. 420, from King George's Sound or to the eastward, *Baxter,* appears to me to be scarcely even a variety of *L. echinata,* but merely slender branches with the leaves less toothed or entire and truncate.

8. **L. ilicifolia,** *Hook. Ic. Pl. t.* 553. A bushy shrub, glabrous and somewhat glaucous or the branches slightly pubescent, readily distinguished from all other species by the small flowers. Leaves in the typical form cuneate, acuminate, pungent-pointed, entire or with 1 or 2 prickly teeth on each side and 1 in. long, in other specimens ovate, pungent-pointed, entire or with 1 or 2 lateral teeth, rounded at the base and under ½ in. long, all thick rigid and scarcely veined besides the midrib. Involucres 7-flowered, all axillary, the bracts glabrous, the inner ones linear, about 5 lines long, the outer ones gradually smaller and broader. Perianth 7 to 8 lines long, with a few long hairs on the upper portion, the tube slender, the segments equally revolute with short obtuse laminæ. Anther-connectives very shortly produced beyond the cells. Hypogynous scales entirely wanting in all the flowers examined. Style more or less bearded. Follicle unknown.—Meissn. in Pl. Preiss. i. 580, ii. 264 and in DC. Prod. xiv. 422.

W. Australia, *Drummond, 3rd coll. n.* 262, *Preiss, n.* 766.

19. ADENOSTEPHANUS, Kl.

Flowers hermaphrodite. Perianth somewhat irregular, the tube slender, slightly incurved, the limb obliquely globular, the segments separating to the base, unequally revolute. Anthers all perfect, ovate, sessile in the base of the concave laminæ, the connective not produced beyond the cells. Hypogynous glands short and thick, free or more or less united, all 4 equal or 2 shorter or deficient. Ovary sessile or shortly stipitate, with 2 pendulous orthotropous ovules; style filiform, shortly clavate at the end, with a small stigma in the centre of a lateral convex disk. Fruit unknown.—Trees or shrubs. Leaves usually pinnate, with petiolulate entire or toothed leaflets. Flowers rather small, pedicellate in pairs in terminal or rarely axillary racemes, the pedicels often more or less connate. Bracts very deciduous (or sometimes none ?).

The genus as far as hitherto known is chiefly Brasilian, with one New-Caledonian species, besides the Australian one which is endemic.

1. **A. Bleasdalii,** *Benth.* A small but beautiful tree (*Dallachy*), the branches petioles and inflorescence minutely ferruginous-tomentose. Leaflets 3 to 17, petiolulate, ovate to oblong-lanceolate, acutely acuminate, somewhat undulate and irregularly mucronate-serrate, tapering or cuneate at the base, 2 to 5 in. long, penniveined and reticulate, silky-villous on both sides when very young, but glabrous and green when full-grown, rather more shining above than below, the common rhachis varying from 1 or 2 in. to above 1 ft. in length, and often irregularly winged at least between the upper leaflets, the terminal leaflet always developed and sometimes larger than the others. Racemes 1 to 2 in. long, usually several in a short panicle or solitary in the upper axils. Pedicels about 1 line long, those of each pair completely united into a single one with the two flowers obliquely sessile at or near the end. Perianth about 3½ lines long. Torus oblique, with one broad 2-lobed gland on the lower side sometimes almost divided into 2. Ovary glabrous, continuous with a short stipes at least as thick as the ovary.— *Grevillea Bleasdalii* or *Bleasdalea cupanioides*, F. Muell. Fragm. v. 90.

Queensland. Rockingham Bay, *Dallachy.* The structure and position of the flowers are in every respect those of some Brasilian *Adenostephani*, and especially of *A. organensis*, Endl., except as to the hypogynous glands, of which the two upper ones are absent, but in *A. organensis* I find them very unequal, two sometimes much smaller than the two others. The pendulous orthotropous ovules at once distinguish this plant both from *Grevillea*, of which it has in some respects the perianth and anthers, and from *Helicia*, of which it has the inflorescence. Whether *Adenostephanus* itself with the New Caledonian *Kermadecia* should or should not be reunited with *Rhopala* as sections, is a question the determination of which would require a careful re-examination of all the American species. The differences in the obliquity of the torus, in the hypogynous glands, and in the style-end are not greater than those which separate different sections of *Grevillea*, and the united genus would be at once a natural and to all appearance a definitely characterised one.

20. GREVILLEA, R. Br.

(Lysanthe *and* Stylurus, *Salisb.*, Anadenia, *R. Br.*, Manglesia, *Endl*, Strangea *and* Molloya, *Meissn.*)

Flowers hermaphrodite. Perianth irregular or regular, the tube revolute or curved under the limb or straight and slender, the limb globular or rarely ovoid, usually oblique, the laminæ usually cohering long after the tube has opened. Anthers all perfect, ovate, sessile in the base of the concave laminæ, the connective not produced beyond the cells. Hypogynous glands united in a single semi-annular or semicircular gland occupying the upper (often the shortest) side of the torus or rarely completely annular surrounding the ovary, or altogether wanting. Ovary stipitate or rarely sessile, with 2 amphitropous ovules laterally attached about the middle ; style filiform or somewhat dilated, usually long and protruding from the slit on the lower side of the perianth tube before the summit is set free from the limb, ultimately straightened and erect or in a few species of *Lissostyles* and *Conogyne* remaining hooked, more or less dilated at the end into a straight oblique or lateral cone or disk bearing the small stigma in the centre of the disk or at the summit of the cone. Fruit a follicle, usually oblique with the ventral suture curved, either coriaceous and opening along the upper margin, or rarely woody and opening almost or quite in two valves. Seeds 1 or 2, flat orbicular or oblong, bordered all round by a membranous wing, or narrowly winged at the end or outer margin only or entirely wingless.—Shrubs or trees. Leaves alternate, very diversified in shape. Flowers in pairs along the rhachis of a short and umbel-like or elongated raceme, rarely reduced to a single pair ; the racemes either terminal or also axillary, rarely all axillary. The indumentum usually consists of closely appressed hairs attached by the centre, rarely of erect or spreading hairs, and then usually forked at the base or clustered.

With the exception of three or four New Caledonian species the genus is limited to Australia. In the distributing the numerous species of this beautiful genus into sections, I have been unable in all respects to follow Brown, and still less Meissner, especially as to the foliage or seeds, for the former is far too variable to serve for much beyond specific distinction, and the seeds are unknown in a large number of the species. The following sectional characters are derived chiefly from the inflorescence and flowers, and if less absolute than could have been wished, are the best I could devise. The first eight sections constitute Meissner's subgenus *Eugrevillea*, with the perianth revolute under the limb, and the stigmatic disk oblique or lateral, usually flat or convex, the tenth and eleventh proposed as subgenera by Meissner, have the perianth straight, and the stigmatic disk replaced by a straight cone, the intermediate ninth section, *Conogyne*, and most of the species of the sixth, with the perianth recurved or revolute at the top, as in the preceding ones, have the straight stigmatic cone of *Anadenia* and *Manglesia*. A very few of the species of the eighth section, *Lissostyles*, have also the stigmatic cone shaped as in the last sections, but very oblique. The absence of the hypogynous gland, one of the chief characters on which the genus *Anadenia* was founded, occurs also in a few species belonging to other sections of true Grevilleas. The peculiar style of *Manglesia* passes into that of *Anadenia* through *G. acrobotrya*, and *G. didymobotrya* closely connects *Conogyne* with *Anadenia*.

SECT. 1. **Eugrevillea.**—*Racemes secund, and elongated, or few-flowered. Perianth-tube dilated below the middle and usually opening on the lower side, revolute under the limb. Torus small, straight or slightly oblique. Stigmatic disk lateral.*

SERIES 1. **Leiogynæ.**—*Ovary glabrous, stipitate. Torus sometimes oblique, but with the gland side the longest. Species all Western, one also in S. Australia.*

Racemes sessile or nearly so on leafy branches. Leaves linear,
    undivided, obtuse.
Racemes loose, several flowered. Stipes of the ovary thick
    and flattened.
    Leaves narrow-linear . . . . 1. *G. pinaster.*
    Leaves oblong-linear . . . . 2. *G. obtusifolia.*

Racemes mostly reduced to a single pair of flowers. Leaves
very narrow-linear . . . . 3. *G. sparsiflora.*
(See also sect. *Lissostyles,* ser. *Puniceæ.*)
Racemes sessile, few-flowered. Leaves lobed or divided. Styles
very long.
Leaves with 3 broad triangular pungent-pointed lobes . . 4. *G. macrostylis.*
Leaves with 3 or 5 narrow-linear divaricate rigid pungent-
pointed segments . . . . 5. *G. tripartita.*
Racemes 2- or 4-flowered, usually several on a short leafless
flexuose peduncle or branch.
Leaves once or twice 3 lobed with broad rigid lobes. Rhachis
very flat . . . . 6. *G. platypoda.*
Leaves once or twice ternately divided into narrow rigid
divaricate pungent-pointed segments . . . . 7. *G. patentiloba.*
Leaves regularly pinnate with narrow rigid but not pungent
segments . . . . 8. *G. pectinata.*
Racemes loose but short, several on long leafless peduncles or
branches.
Leaves regularly pinnate with long narrow-linear rigid but
not pungent segments . . . . 9. *G. plurijuga,*
Leaves simple, narrow-linear, often very long . . . . . 10. *G. nudiflora.*
Racemes rather dense, many-flowered. Leaves pinnate with
narrow-linear or filiform segments.
Leaves simply pinnate . . . . 11. *G. stenomera.*
Leaf-segments mostly again divided . . . . 12. *G. Thelemanniana.*

SERIES 2. **Hebegynæ.**—*Ovary sessile or scarcely stipitate, densely villous. Torus
sometimes but very rarely slightly oblique, with the gland side the shortest.*

Leaves obtuse or mucronate, not pungent.
Leaves narrow-linear, rigid, mostly undivided, 1 to 2 in.
long. Racemes short . . . . 13. *G. concinna.*
Leaves mostly pinnate with narrow-linear rigid segments
doubly grooved underneath.
Leaf-segments 3 to 9, under 2 in. long. Racemes silky,
1½ to 3 in. long.
Leaves glabrous when full-grown . . . . 14. *G. Hookeriana.*
Leaves silky on both sides . . . . 15. *G. Baxteri.*
Leaf-segments few or leaves entire, 4 to 8 in. long.
Racemes villous.
Racemes 2 to 4 in. long. Perianth ¼ in. . . . . . 16. *G. pterosperma.*
Racemes dense, 3 to 6 in. long. Perianth ½ in. long . 17. *G. eriostachya.*
Leaf-segments numerous, regular, under 2 in. long.
Racemes densely villous, 2 to 4 in. . . . . 18. *G. thyrsoides.*
Leaves pinnate with linear or lanceolate segments, glabrous
above, silky underneath. Perianth villous.
Leaf-segments very narrow, 4 to 8 in long, often divided 19. *G. chrysodendron.*
Leaf-segments few, linear-lanceolate, 2 to 4 in. long . . 20. *G. Banksii.*
Leaf-segments numerous, regular, linear-oblong, ¾ to 1½ in.
long . . . . 21. *G. Caleyi.*
Leaves entire or pinnately-toothed or lobed, lanceolate to
ovate.
Leaves long, lanceolate. Perianth silky.
Tall and erect. Racemes dense. Perianth-tube narrow . 22. *G. asplenifolia.*
Prostrate. Racemes loose. Perianth-tube broad, ex-
panding into an orbicular disk . . . . 23. *G. cirsiifolia.*
Leaves ovate or oblong, obtuse, entire, 2 to 5 in. long . 24. *G. laurifolia.*
Leaves oblong or lanceolate, acute, 6 to 10 in. long, entire
or broadly pinnatifid . . . . 25. *G. Barklyana.*

Leaves or leaf-lobes pungent-pointed (the first five species
all Eastern).
Leaves ovate or cuneate, with prickly teeth or lobes.
Prostrate. Leaves ovate, undulate, shortly prickly-toothed.
Torus slightly oblique . . , ... 26. *G. repens.*
Erect or spreading. Leaves ovate oblong or cuneate,
irregularly lobed. Torus straight.
Leaves usually villous, pinnately many-toothed or lobed.
Stipes of the ovary very short . . . . 27. *G. aquifolium.*
Leaves glabrous above or nearly so, silky underneath,
cuneate with few lobes. Stipes of the ovary as long
as the ovary . . . . 28. *G. ilicifolia.*
Leaves glabrous, deeply pinnatifid, with broad prickly lobes.
Leaf-lobes oblong or ovate, entire or rarely 2- or 3-lobed 29. *G. Gaudichaudii.*
Leaf-lobes mostly cuneate, very rigid, 2- or 3-lobed.
Racemes dense . . . . 30. *G. acanthifolia.*
Leaf-lobes mostly again pinnatifid. Racemes loose.
Western species . . . . 31. *G. bipinnatifida.*
Leaves once or twice ternately divided into linear rigid
divaricate segments. Western species.
Racemes dense. Flowers numerous, about ¼ in. long,
nearly sessile . . . . 32. *G. armigera.*
Racemes loose. Flowers nearly ½ in. long, on slender
pedicels . . . . 33. *G. asparagoides.*

SECT. 2. **Ptychocarpa.**—*Racemes short, often umbel-like. Perianth-tube dilated
below the middle and usually opening on the lower side, revolute under the limb.
Torus small, straight or nearly so. Ovary sessile or very shortly stipitate, densely
villous or rarely with only a tuft of hairs at the base. Stigmatic disk lateral. Leaves
entire. Species all Eastern.*

Perianth densely villous.
Leaves ovate or oblong, mostly obtuse, the upper surface
glabrous or minutely scabrous . . . . 34. *G. floribunda.*
Leaves ovate to oblong-lanceolate, mucronate-acute, the
upper surface scabrous-dotted . . . . 35. *G. cinerea.*
(See also 39, *G. arenaria.*)
Leaves oblong to linear, obtuse, pubescent or villous on both
sides . . . . 36. *G. alpina.*
Perianth sprinkled or silky with appressed hairs.
Leaves oblong or lanceolate, obtuse or with a small callous
point.
Leaves glabrous or scabrous above, mostly narrow.
Perianth segments acuminate or acute . . . . . . 37. *G. montana.*
Perianth-segments obtuse . . . . 38. *G. obtusiflora.*
Leaves silky or minutely pubescent above, mostly rather
broadly-oblong or cuneate. Perianth-segments acumi-
nate, sometimes villous . . . . 39. *G. arenaria.*
Leaves ovate or lanceolate, acutely acuminate or mucronate
with a fine point. Perianth-segments obtuse . . . . 40. *G. mucronulata.*
Perianth quite glabrous.
Leaves ovate or lanceolate, nearly flat, glabrous as well as
the branches, or slightly pubescent . . . . 41. *G. Baueri.*
Leaves linear, obtuse, much revolute, villous or hirsute as
well as the branches . . . . 42. *G. lanigera.*
Leaves linear or linear-lanceolate, mostly acute and revo-
lute, the upper surface glabrous, scabrous or slightly
hirsute.
Ovary villous. Spreading or diffuse shrub usually villous 43. *G. ericifolia.*

Ovary glabrous except a tuft of hairs at the base.
Spreading shrub with linear-subulate leaves, mostly
　under 1 in. long . . . . 44. *G. divaricata.*
Erect shrub with erect leaves mostly above 1 in. long　45. *G. rosmarinifolia.*

SECT. 3. **Plagiopoda.**—*Racemes various. Perianth-tube dilated below the middle and usually opening on the lower si!e, revolute under the limb. Torus very oblique, the gland side the shortest. Ovary villous except in a few axillary-flowered species. Stigmatic disk very oblique or lateral.*

Racemes terminal, erect, sometimes secund. Ovary densely
　villons. Style very long. Eastern species (except *G.
　Wilsoni*).
Leaves oval-elliptical or oblong-lanceolate, large, entire . . 46. *G. Goodii.*
Leaves mostly above 6 in. long, simple or pinnate with
　narrow-lanceolate obliquely penniveined lobes . . . . 47. *G. venusta.*
Leaves linear, above 6 in. long, simple or pinnate with long-
　linear lobes.
　Racemes oblong, glabrous except the ovary . . . . . 48. *G. longistyla.*
　Racemes usually paniculate, viscid-villous . . . . . 49. *G. juncifolia.*
Leaves ternately divided into narrow-linear rigid divaricate
　pungent-pointed segments. Western species . . . . 50. *G. Wilsoni.*
　(See also 73. *G. Huegelii*, with a glabrous ovary.)
Racemes terminal, short, umbel-like. Style very long. Leaves
linear or linear-lanceolate, acute, entire. Eastern species . 51. *G. lavandulacea.*
Racemes short, few-flowered, sessile, terminal and in the upper
　· axils. Style short. Western species (except *G. aspera*).
Leaves ovate, rigid, sinuate and prickly-toothed . . . . 52. *G. insignis.*
　(See also 71. *G. Cunninghamii*, with a glabrous ovary).
Leaves entire, narrow or rarely ovate.
　Branches sparingly or shortly pubescent.
　　Leaves ovate or lanceolate, ½ to ¾ in. long or oblong
　　　and longer, obtuse . . . . 53. *G. Brownii.*
　　Leaves linear or lanceolate, ½ to 2 in. long. Flowers
　　　small.
　　　Leaves smooth or minutely scabrous above . . . 54. *G. fasciculata.*
　　　Leaves veined and very scabrous above . . . . 55. *G. aspera.*
　　Leaves linear or lanceolate, 2 to 4 in. long . . . . 56. *G. brachystylis.*
　Branches densely and softly villous. Perianth-tube broadly
　　saccate at the base . . . . 57. *G. saccata.*
　Branches hirsute with long fine hairs. Perianth small,
　　not saccate . . . . 58. *G. Drummondii.*
Racemes reduced to 1 or 2 pairs of flowers mostly axillary.
　Torus sometimes less oblique. Leaves entire.
Leaves narrow-linear. Ovary villous. Style long.
　Leaves angular-terete, ½ to ¾ in. long . . . . 59. *G. disjuncta.*
　Leaves convex and smooth above, channelled under-
　　neath, ¾ to 1½ in. long . . . . 60　*G. haplantha.*
Leaves narrow-linear. Ovary villous. Style short . . . 61. *G. pinifolia.*
Leaves linear-subulate, pungent-pointed. Ovary glabrous.
　Style long . . . . 62. *G. acuaria.*
Leaves ovate or orbicular, small, flat. Ovary glabrous. Style
　long . . . . ·. . . . 63. *G. singuliflora.*
Leaves linear-cuneate or oblong, 1-nerved. Ovary glabrous.
　Fruit small . . . . 64. *G. pauciflora.*
Leaves linear or linear-cuneate, thick, nerveless.
　Fruit 1 to 1½ in. long. Flowers unknown. Eastern species 65. *G. Strangea.*
　Fruit 2 to 2½ in. long. Ovary villous. Style short. Pe-
　　rianth unknown. Western species . . . . 66. *G. cynanchicarpa.*

SECT. 4. **Calothyrsus.**—*Racemes secund, usually many-flowered. Perianth-tube more or less dilated below the middle and usually opening on the lower side, revolute under the limb. Torus oblique, the gland-side the shortest. Ovary glabrous, stipitate. Species all tropical except G.* quercifolia *and G.* Huegelii.

Leaves undivided, ovate or lanceolate, angular or prickly-
   toothed.
   Leaves petiolate or tapering at the base.
     Leaves glabrous, mostly sinuate-toothed.
      Racemes mostly terminal on long peduncles. Perianth
        bearded inside with very short hairs . . . . . . 67. *G. quercifolia.*
      Racemes axillary, shortly pedunculate. Perianth
        densely bearded inside with erect hairs . . . . . 68. *G. angulata.*
     Leaves silky-pubescent at least when young, mostly an-
      gular. Racemes axillary. Perianth bearded inside
      with spreading or reflexed hairs.
      Perianth slightly dilated at the base as in *G. angulata* 69. *G. Wickhami.*
      Perianth much-dilated at the base as in *G. Cunning-*
       hamii . . . . 70. *G. agrifolia.*
Leaves sessile, deeply cordate with large stem-clasping
   auricles, sinuate and prickly-toothed . . . . 71. *G. Cunninghamii.*
Leaves sessile, deeply and regularly pinnatifid with rigid
   pungent pointed lobes . . . . 72. *G. pungens.*
Leaves once or twice divided into short linear rigid divaricate
   segments. Racemes very short with long flowers. Desert
   species . . . . 73. *G. Huegelii.*
Leaves not toothed, entire or divided into long narrow seg-
   ments. Racemes usually paniculate.
   Leaves undivided, broadly falcate, longitudinally reticulate 74. *G. dimidiata.*
   Leaves mostly pinnate, the segments oblong-lanceolate,
     longitudinally reticulate . . . . 75. *G. heliosperma.*
   Leaves mostly pinnate, the segments oblong-lanceolate or
     linear, penniveined with numerous oblique parallel pri-
     mary veins . . . . 76. *G. refracta.*
   Leaf-segments numerous, linear, obscurely veined above,
     1-nerved underneath. Racemes long. Perianth above
     ½ in. long . . . . 77. *G. Dryandri.*
   Leaf-segments not numerous, linear or lanceolate, obscurely
     veined above, 1 nerved underneath. Racemes dense.
     Perianth under ¼ in. long . . . . 78. *G polystachya.*
Leaves mostly bipinnatifid with lanceolate lobes or segments 79. *G. robusta.*

SECT. 5. **Cycladenia.**—*Racemes many-flowered, paniculate (scarcely secund?) Perianth nearly of* Cycloptera, *but larger. Torus straight. Hypogynous gland annular (deficient on the lower side in all other sections). Ovary glabrous, stipitate. Western species.*

Leaves divided into short rigid linear divaricate pungent-
   pointed segments . . . . 80. *G. annulifera.*
Leaves divided into numerous very long narrow-linear segments . . . . 81. *G. leucopteris.*

SECT. 6. **Cycloptera.**—*Racemes dense, usually paniculate. Flowers small. Perianth-tube narrow, recurved or reflexed under the limb. Torus straight. Ovary glabrous, stipitate. Fruit usually broad. Seeds winged all round. Tropical or subtropical species.*

Leaves longitudinally veined.
   Leaves mostly pinnate.
     Leaf-segments linear, very long and narrow . . . . . 82. *G. leucadendron.*
     Leaf-segments linear-cuneate, obtuse, under 5 in. long . 83. *G. pyramidalis.*
   Leaves undivided, very long, with 9 to 13 closely parallel
     veins . . . . 84. *G. striata.*

Leaves undivided, falcate, longitudinally reticulate and irregularly several-veined . . . . 85. *G. mimosoides.*
Leaves large, penniveined, ovate-lanceolate or oblong.
Leaves entire or deeply pinnatifid, rather thin, glabrous
above, silvery-silky underneath . . . . 86. *G. Hillii.*
Leaves rather thick, entire, minutely pubescent on both
sides . . . . · ·· . . . . . 87. *G. gibbosa.*

SECT. 7. **Eriostylis.**—*Racemes umbel-like, sessile or nearly so. Flowers small,
villous. Perianth-tube narrow or rather broad, revolute under the limb. Torus
straight. Ovary shortly stipitate, villous as well as the style; stigmatic disk lateral.
Leaves entire.*

Hairy style produced into an appendage beyond the stigmatic
disk. Eastern species.
Leaves ovate to broadly lanceolate. Stigmatic disk orbicular,
the hairy appendage reflexed . . . . 88. *G. buxifolia.*
Leaves lanceolate or linear. Stigmatic disk oblong, the
hairy appendage erect . . . . 89. *G. phylicoides.*
Hairy style not produced (or obscurely so in *G. sphacelata*),
the disk orbicular or oval. Western species (except *G.
sphacelata*).
Leaves lanceolate or linear, scabrous-punctate. Young
branches closely silky. Stigmatic disk oval.
Stipes of the ovary much longer than the gland. Stig-
matic disk not projecting beyond the hairy style . . 90. *G. sphacelata.*
Stipes of the ovary very short. Stigmatic disk with a
free glabrous margin . . . . 91. *G. occidentalis.*
Young branches villous. Stigmatic disk thick, with an
incurved turbinate glabrous back . . . . 94. *G. oxystigma.*
Leaves narrow-linear, rarely lanceolate, smooth.
Leaves linear-terete, pungent-pointed. Stigmatic disk
oval, flat . . . . 92. *G. acerosa.*
Leaves linear, not pungent. Stigmatic disk orbicular,
flat . . . . 93. *G. umbellulata.*
Leaves linear or lanceolate, not pungent. Stigmatic disk
thick with an incurved turbinate glabrous back . . 94. *G. oxystigma.*
Hairy style not produced beyond the base of the stigmatic
disk which terminates in an oblong involute appendage.
Leaves mostly lanceolate, smooth. Stigmatic disk glabrous
on the back . . . . 95. *G. Candolleana.*
Leaves mostly linear, scabrous-punctate. Stigmatic disk
tomentose on the back . . . . 96. *G. scabra.*

SECT. 8. **Lissostylis.**—*Racemes short and dense (except* G.Victoriæ *and* G. trachytheca). *Perianth-tube narrow, revolute or recurved under the limb. Torus straight.
Ovary glabrous, stipitate. Stigmatic disk (or cone in the last two species) very oblique
or lateral.*

SERIES 1. **Puniceæ.**—*Flowers not numerous or in a loose raceme, the perianth
about ½ in. long. Leaves entire. Eastern species.*

Leaves penniveined, ovate to broadly lanceolate. Racemes
loose. Style not very long.
Leaves oval or ovate-oblong (1½ to 2½ in.), veinless above,
penniveined underneath . . . . 97. *G. Miqueliana.*
Leaves obovate or oval (¾ to 1¼ in.), veinless above, 1-nerved
underneath . . . . 98. *G. brevifolia.*
Leaves lanceolate (2 to 4 in.), penniveined above, 1-nerved
underneath · . . . . 99. *G. Victoriæ.*

Leaves penniveined, mostly lanceolate, the lateral nerves if
    present close to the margin. Style long.
    Leaves mostly under 1½ in. long. Racemes mostly terminal 100. *G. punicea.*
    Leaves mostly 2 to 4 in. long. Racemes mostly axillary . 101. *G. oleoides.*
Leaves narrow, rigid, pungent-pointed.
    Leaves linear or lanceolate, mostly 3 nerved . . . . . 102. *G. trinervis.*
    Leaves linear-subulate, mostly 1-nerved . . . . 103. *G. juniperina.*
    (See also *Eugrevillea,* ser. *Leiogynæ.*)

SERIES 2. **Sericeæ.**—*Flowers numerous in a short dense raceme. Perianth-tube
under 4 lines long. Leaves entire. Fruit usually smooth. Eastern species.*

Leaves oblong-lanceolate or almost linear, obtuse or with a
    small point, silky underneath.
    Flowers silky-pubescent. Leaves rarely 1½ in. long . . . 104. *G. sericea.*
    Flowers ferruginous-villous. Leaves mostly 1½ to 2 in. long 105. *G. capitellata.*
Leaves oblong-lanceolate or linear, flat, green on both sides . 106. *G. leiophylla.*
Leaves mostly linear, very acute.
    Leaves with the midrib very prominent underneath, doubly
        grooved when narrow.
    Leaves open underneath between the midrib and margin.
        Perianth-tube 2½ to 3 lines long . . . . 107. *G. linearis.*
    Leaves very rigid, doubly grooved underneath.
        Perianth tube 2½ to 3 lines long . . . . 108. *G. confertifolia.*
        Perianth-tube not 2 lines long . . . . 109. *G. parviflora.*
    Leaves (under 1 in.) with the midrib not prominent under-
        neath, singly grooved when narrow. Perianth-tube about
        2 lines long . . . . 110. *G. australis.*

SERIES 3. **Occidentales.**—*Flowers numerous in a dense raceme or head Perianth-
tube under 4 lines long. Leaves entire or divided. Fruit usually (but not always)
rugose or tuberculate. Western species.*

Racemes short. Bracts none or minute and falling off early.
    Stigmatic disk flat or convex.
    Leaves rather thick, obtuse or mucronate, oblong-cuneate
        or linear, entire or divided.
        Racemes ovate, on short peduncles. Perianth-tube about
            2 lines long
            Leaves entire or rarely 2- or 3-toothed when broad . 111. *G. commutata.*
            Leaves mostly pinnate with narrow linear segments . 112. *G. pinnatisecta.*
        Racemes globular, on filiform peduncles. Perianth-tube
            about 1 line long . . . . 113. *G. argyrophylla.*
    Leaves rather thick, narrow-linear, all entire.
        Leaves doubly grooved underneath. No hypogynous
            gland. Fruit smooth . . . . 114. *G. brachystachya.*
        Leaves long, 1-nerved underneath, concave and nerveless
            above. Gland pulvinate . . . . 115. *G. Endlicheriana.*
    Leaves not very thick, varying from broadly cuneate and
        acutely toothed or lobed to narrow-linear and very
        acute.
        Leaves silky-pubescent underneath . . . . 116. *G. manglesioides.*
        Leaves glabrous on both sides . . . . 117. *G. diversifolia.*
    Leaves linear terete, singly or doubly grooved.
        Leaves slender, entire or rarely 2- or 3-lobed. Flowers
            very small.
            Leaves 4 to 6 in. long. Racemes axillary and ter-
                minal . . . . 118. *G. filifolia.*
            Leaves 1 to 2 in. long. Racemes axillary . . . 119. *G. hakeoides.*
        Leaves ternately divided into rigid divaricate pungent-
            pointed segments . . . . 120. *G. teretifolia.*

Racemes short. Bracts membranous, broad, imbricate in the
    young racemes, persisting nearly to the flowering. Stig-
    matic disk flat or convex.
Racemes on long terminal leafless simple or branched pe-
    duncles.
    Leaves large, glaucous, undulate, deeply pinnatifid, with
        obovate or oblong lobes . . . . 121. *G. eryngioides.*
    Leaves narrow-linear, rigid, entire or deeply trifid, doubly
        grooved underneath. . . . . 122. *G. bracteosa.*
Racemes dense, terminal. Stigmatic disk conical in the centre
    or replaced by an oblique cone.
    Racemes short, sessile. Leaves short, crowded, with 3 or 5
        narrow-linear segments . . . . 123. *G. crithmifolia.*
    Racemes elongated, cylindrical. Leaves narrow linear,
        entire or 3-lobed . . . . 124. *G. trachytheca.*

SECT. 9. **Conogyne.**—*Racemes dense or. rarely slender, short or cylindrical.
Flowers small. Perianth-tube slender, recurved under the limb. Torus straight.
Ovary stipitate. Style filiform, with an erect stigmatic cone.*

Hypogynous gland none or very obscure. Racemes short or
    rarely elongated and loose or cylindrical.
    Ovary villous. Leaves ternately divided. Eastern species.
        Leaf-segments narrow-linear, rigid, pungent-pointed . . 125. *G. triternata.*
        Leaf-segments oblong-cuneate or lanceolate, prickly-
            toothed . . . . 126. *G. ramosissima.*
    Ovary glabrous. Western species except *G. nematophylla.*
        Leaves toothed or pinnatifid.
            Leaves ovate, prickly-toothed, glabrous, glaucous .. . 127. *G. monticola.*
            Leaves mostly linear-cuneate or lanceolate, toothed or
                pinnatifid, silky underneath . . . . 128. *G. Muelleri.*
        Leaves cuneate or linear, 3-fid or 3-toothed.
            Racemes short and sessile. Entire base of the leaf
                short and broad or linear . . . . 129. *G. trifida.*
            Racemes oblong-cylindrical, pedunculate. Entire base
                of the leaf long and cuneate. Leaf very glaucous . 130. *G. synapheæ.*
        Leaves mostly pinnate with pinnatifid or pinnate pinnæ.
            Ultimate leaf-segments short and broad, rhachis
                flexuose. Racemes oblong-cylindrical, compact . . 131. *G. flexuosa.*
            Leaf-segments narrow, rhachis and stems very slender.
                Racemes elongated, loose . . . . 132. *G. leptobotrya.*
        Leaves twice or thrice ternately divided into linear pun-
            gent-pointed segments.
            Leaf-segments short. Racemes short and sessile . . 133. *G. brevicuspis.*
            Leaf-segments long. Racemes cylindrical elongated
                and loose . . . . 134. *G. intricata.*
        Leaves filiform, entire. Racemes cylindrical, paniculate . 135. *G. didymobotrya.*
Hypogynous gland semiannular. Racemes cylindrical, nar-
    row, in a terminal leafless panicle.
    Leaves flat, oblong or lanceolate. Ovary stipes very short 136. *G. polybotrya.*
    Leaves linear-terete, very long. Ovary stipes long . . . 137. *G. nematophylla.*

SECT. 10. **Anadenia.**—*Racemes dense, short or cylindrical. Flowers small.
Perianth-tube slender, straight, limb erect. Torus straight. Style filiform or dilated
upwards, not contracted under the erect stigmatic cone. Western species except* G.
anethifolia.

Hypogynous gland semiannular. Leaves linear-terete, ter-
    nately divided.
    Leaf-segments divaricate, under 1 in. long, pungent-pointed.
        Racemes short, sessile. Style dilated and flattened . . 138. *G. anethifolia.*

Racemes cylindrical, spike-like, sessile, the rhachis densely
villous. Style long, filiform . . . . 139. *G. paradoxa.*
Leaf-segments erect, above 1 in. long, not pungent. Ra-
cemes spike-like, pedunculate along a common leafless
peduncle . . . . 140. *G. petrophiloides.*
No hypogynous gland.
Leaves flat, pinnate, with 3-lobed or pinnatifid pinnæ.
Racemes rather short. Ovary stipitate.
Pinnæ 3 or 5. Perianth 2½ lines long . . . . . . 141. *G. tenuiflora.*
Pinnæ 7 to 11. Perianth 1½ lines long . . . . . . 142. *G. pulchella.*
Leaves entire or toothed at the end. Racemes cylindrical.
Ovary nearly sessile.
Leaves narrow-cuneate, 3-lobed or 3-toothed at the end.
Ovary villous . . . . 143. *G. rudis.*
Leaves obovate-oblong or lanceolate, entire. Ovary
glabrous.
Leaves glabrous . . . . 144. *G. Shuttleworthiana*
Leaves more or less pubescent . . . . 145. *G. integrifolia.*
Leaves narrow-linear. Ovary glabrous.
Fruit 3 or 4 times as long as broad . . . . . . 146. *G. stenocarpa.*

SECT. 11. **Manglesia.**—*Racemes short, dense, axillary. Flowers small. Perianth-
tube straight, slender or fusiform; limb erect. Torus straight. Ovary glabrous, stipitate. Style turgid in the middle or fusiform, constricted under the erect stigmatic
cone. Western species.*

Hypogynous gland none. Style fusiform. Stem-leaves
broadly cuneate; floral leaves with 3 linear pungent seg-
ments . . . . 147. *G. acrobotrya.*
Hypogynous gland semiannular. Style turgid in the middle.
Capsule very rugose. Leaves mostly above 1 in. long.
Leaves quite glabrous. Racemes branching.
Leaves broad, once or twice trifid with short lobes . 148. *G. glabrata.*
Leaves narrow, with 3 lanceolate lobes . . . . . 149. *G. ornithopoda.*
Leaves mostly biternate with terete pungent-pointed
segments. Fruit erect . . . . 150. *G. paniculata.*
Leaves more or less hoary, at least when young. Racemes
simple.
Leaves mostly biternate with narrow pungent-pointed
segments. Fruit transverse . . . . 151. *G. biternata.*
Leaves linear-cuneate, simple or trifid, villous under-
neath . . . . 152. *G. triloba.*
Capsule smooth. Leaves mostly under 1 in. long, with
pungent-pointed lobes or segments or teeth.
Leaves broad, stem-clasping with large auricles, prickly-
toothed . . . . 153. *G. amplexans.*
Leaf lobes more or less dilated, showing the under surface 154. *G. vestita.*
Leaf-segments narrow-linear, very rigid, doubly grooved
underneath . . . . 155. *G. tridentifera.*
Leaf-segments slender, terete, mostly 1-grooved . . . . 156. *G. erinacea.*

G. berberifolia, podocarpifolia and *trifurcata,* Sweet, and *G. Flindersii* and *mucro-
nifolia,* A. Cunn., included in Steud. Nom. Bot. ed. 2, are garden names of unpublished
species, which, if genuine *Grevilleæ,* must be the same as some of those here described.

SECT. 1. EUGREVILLEA.—Racemes secund and elongated or few-
flowered, rarely reduced to 1 or 2 pairs of flowers, usually terminal.
Perianth-tube usually dilated below the middle, and opening on the
lower side, the segments otherwise long-cohering, revolute above the

middle. Torus small, straight or slightly oblique. Stigmatic disk flat
or convex, lateral.

SERIES 1. LEIOGYNÆ.—Ovary glabrous, stipitate. Torus sometimes oblique but with the gland-side the longest (not the shortest as
in *Plagiopoda* and *Calothyrsus*).

This series differs from *Lissostylis* in the more secund inflorescence and in the shape
of the perianth.

1. G. pinaster, *Meism. in Hook. Kew Journ.* vii. 76, *and in D C. Prod.*
xiv. 367. A bushy shrub attaining 3 or 4 ft. the young branches
tomentose, the adult foliage glabrous. Leaves linear, usually very
narrow and doubly grooved underneath, the lower ones on the young
plants sometimes broader linear-lanceolate flat 3-nerved and slightly
silky underneath, all obtuse or with a small callous point, 1 to 2 in.
long. Racemes spreading, rather loose, secund, 1 to 1½ in. long, the
rhachis minutely pubescent. Pedicels filiform, 1 to 2 lines long.
Perianth quite glabrous outside or sprinkled with few hairs, bearded
inside to below the middle with short hairs, the tube 3 to 4 lines long,
dilated at the base, attenuate and revolute under the globular limb.
Hypogynous gland thick, rather broad, semicircular. Ovary glabrous,
on a long stipes; style long, clavate under the broad very oblique or
lateral stigmatic disk. Fruit glabrous, nearly smooth, rather narrow,
the stipes dilated upwards and flattened. Seeds with a narrow wing
on the outer edge.

W. Australia. Murchison river, *Oldfield, Drummond, 6th coll. n.* 182.

Var. *brevifolia.* Leaves all under 1 in. long and the racemes short ; in one specimen a few of the larger leaves divided into 3 segments.—Murchison river, *Oldfield.*

Var. *hirtella.* Leaves of the preceding variety, but more or less hirsute with short
fine spreading hairs often clustered.—Champion Bay, *Walcott;* a single specimen in
herb. F. Mueller.

2. G. obtusifolia, *Meissn. in D C. Prod.* xiv. 356. A much-branched
shrub, apparently spreading or procumbent, the young branches
slightly pubescent with appressed hairs. Leaves oblong-linear or
linear-cuneate, obtuse, with recurved or revolute margins, contracted
into a short petiole, glabrous and smooth above, silky-ferruginous
underneath, with a prominent midrib. Racemes short, secund, rather
loose, the rhachis and pedicels glabrous or nearly so. Pedicels slender,
1 to 2 lines long. Perianth glabrous outside, bearded inside to below
the middle with very short hairs, the tube fully 3 lines long, broad at
the base, attenuate and revolute under the obliquely globular limb.
Hypogynous gland thick, broad, semicircular. Ovary glabrous, on a
long stipes. Style long, thickened at the end under the very oblique
or lateral stigmatic disk. Fruit above ½ in. long, nearly smooth, the
stipes thickened upwards and flattened.

W. Australia, *Drummond, 4th coll n.* 278, *also n.* 10 *and* 34. The species is
very near to and perhaps a variety of *G. pinaster.*

3. **G. sparsiflora,** *F. Muell. Fragm.* vi. 206. A bushy shrub of about 3 ft., with erect branches, the young shoots minutely silky-pubescent, the adult foliage nearly glabrous. Leaves rather crowded, erect, very narrow-linear or almost terete, obtuse or with a small callous point, ¾ to 1 in long, doubly grooved underneath by the thickened margin and midrib. Pedicels in pairs or even solitary, axillary and terminal, 2 to 3 lines long. Perianth glabrous or nearly so, the tube nearly 4 lines long, rather narrow, attenuate and revolute under the globular limb, densely bearded inside about the middle with reflexed hairs. Torus small, straight. Ovary stipitate, glabrous; style long, with an orbicular lateral stigmatic disk.

W. Australia. Sand flats near Eyre's Relief, Cape Arid, *Maxwell.*

4. **G. macrostylis,** *F. Muell. Fragm.* i. 137. A shrub of 4 to 6 ft., very near *G. tripartita,* with the same inflorescence flowers and fruit, but a different foliage. Leaves on short petioles, cuneate at the base, more or less deeply divided into 3 broad triangular or lanceolate pungent-pointed lobes, nearly glabrous and more or less veined above, silvery-silky underneath, the whole leaf usually about 1 in. long and broad. Flowers few in umbel-like axillary or terminal racemes, more or less secund. Perianth above ½ in. long, entirely as in *G. tripartita* as well as the hypogynous gland and pistil.

W. Australia. Eyre's Relief and East Mount Barren, *Maxwell.* In one specimen the leaves are much narrower and deeply 3-fid, or a few of them linear-lanceolate and entire.

5. **G. tripartita,** *Meissn. in DC. Prod.* xiv. 373. An erect shrub of 3 to 5 ft., the branches tomentose. Leaves pinnate, with 3 or 5 linear divaricate very rigid and pungent-pointed segments, flat but thick, doubly grooved underneath, glabrous or slightly silky, the whole leaf 1 to 1½ in. long, the common petiole short. Racemes sessile, few-flowered, terminal or in the upper axils. Pedicels pubescent, 4 to 5 lines long. Perianth slightly pubescent outside and minutely so inside about the middle, the tube rather broad, 7 to 8 lines long, attenuate and revolute under the very oblique usually tomentose limb. Torus slightly oblique, the gland-side uppermost. Hypogynous gland broad, thick, obliquely semicircular. Ovary glabrous on a short stipes; style very long, scarcely thickened under the large lateral stigmatic disk. Fruit hard, 6 to 8 lines long, smooth or with a few prominent tubercles.

W. Australia, *Drummond,* 4th coll. n. 285, *Roe;* sandy ridges, Phillips river, *Maxwell.*

6. **G. platypoda,** *F. Muell. Fragm.* vi. 205. A shrub with stout minutely tomentose branches, very angular when young. Leaves shortly petiolate, deeply pinnatifid, with 3 or 5 broadly cuneate mostly 3-lobed segments, all short and pungent-pointed, the whole leaf 1 to 2 in. long and broad, firmly coriaceous, glabrous and shining above, minutely silky underneath, the primary veins prominent. Racemes 2-

to 4-flowered, very shortly pedunculate, several together in terminal or lateral raceme-like panicles of 2 or 3 in., the common rhachis broadly fasciate and flexuose. Pedicels 2 to 3 lines long. Perianth slightly pubescent outside, shortly bearded inside above the middle, the tube 4 to 5 lines long, slightly dilated below the middle, much revolute and attenuate under the globular limb. Torus nearly straight. Gland very prominent, obliquely semicupular. Ovary glabrous, shortly stipitate; style not very long, shortly thickened under the broadly oval lateral stigmatic disk.

W. Australia. Stirling range, *F. Mueller*, a single specimen in herb. F. Mueller. The dilatation of the rhachis of inflorescence may possibly be abnormal.

7. **G. patentiloba,** *F. Muell. Fragm.* i. 137. A spreading shrub of about 4 ft., glabrous or the young shoots minutely silky-pubescent. Leaves mostly twice pinnatifid, with 3 to 7 primary pinnæ, the lower ones with 3 to 5 segments and sometimes some of these again divided, and a few leaves with only 3 to 5 segments altogether, the segments all linear, often short, rigid, divaricate, pungent-pointed, thick but flat, smooth above, doubly grooved underneath, the whole leaf under 2 in. diameter. Racemes 2- to 4-flowered, on very short peduncles but often rather numerous in a raceme-like panicle with a common minutely pubescent rhachis of 1½ to 3 in. Pedicels 1 to 2 lines long. Perianth pubescent outside and in, the tube broad, about 4 lines long, revolute above the middle and much constricted under the obliquely globular limb. Torus straight or oblique with the gland-side the longest. Gland prominent, oblique. Ovary glabrous, shortly stipitate; style long, slightly thickened under the oblique almost lateral stigmatic disk.

W. Australia, *Drummond;* Phillips Range, *Maxwell.*

8. **G. pectinata,** *R. Br. Prot. Nov.* 23. A low spreading or procumbent shrub, rarely above 2 ft. high, the young shoots minutely silky-pubescent, the adult foliage glabrous, of a pale colour. Leaves pinnate, with 9 to 11 segments, all approximate and parallel, narrow linear, thick but flat, obtuse or with a small callous point, doubly grooved underneath or on both sides by the prominent margins and midrib, the lower ones of each leaf ½ to ¾ in. long and regularly diminishing to the end. Racemes very short and loose, simple or rarely branched. Pedicels 2 to 3 lines long. Perianth slightly hoary-pubescent or silky outside, very sparingly bearded inside, the tube ½ in. long, not very broad at the base, tapering into a revolute neck under the globular limb. Torus nearly straight. Hypogynous gland very prominent, thin, erect, semicupular. Ovary glabrous, on a rather short stipes; style very long, slightly clavate under the lateral stigmatic disk, fruit nearly globular, 4 to 5 lines long, prominently rugose. Meissn. in DC. Prod. xiv. 372; *G. etenophylla,* Meissn. l.c.

W. Australia, *Drummond, 5th coll. n.* 407; between Lucky Bay and Cape Arid, *Baxter;* East Mount Barren and Phillips Range, *Maxwell.*

9. **G. plurijuga,** *F. Muell. Fragm.* iv. 84. A spreading shrub attaining 5 or 6 ft. in height and 12 ft. diameter, quite glabrous or the young shoots minutely tomentose. Leaves simply pinnate, with 9 to 21 linear-terete rigid but rather slender mucronate segments, mostly $\frac{3}{4}$ to $1\frac{1}{2}$ in. long, singly or doubly grooved underneath. Flowering branches almost leafless or with simple leaves at the base of the racemes. Racemes loose, secund, 2 to 4 in. long. Pedicels slender, 3 to 4 lines long, glabrous as well as the rhachis. Perianth glabrous but apparently viscid outside, shortly bearded inside at about the middle, the tube 5 to 6 lines long, rather broad in the lower part, attenuate above the middle and much revolute under the obliquely globular limb. Torus straight. Gland semicircular, slightly prominent. Ovary glabrous, very shortly stipitate; style very long, the stigmatic disk lateral. Fruit obliquely ovoid, hard, 7 to 8 lines long.

W. Australia. Sand flats, Point Malcolm, *Maxwell.*

10. **G. nudiflora,** *Meissn. in DC. Prod.* xiv. 366. A diffuse prostrate or trailing shrub extending sometimes to several feet, glabrous or the young shoots minutely silky-pubescent. Leaves rather crowded at the base of the branches, undivided, very narrow-linear, rigid but not pungent, doubly grooved underneath by the prominent midrib and margins, varying from under 2 in. to fully 6 in. long. Flowering branches long and leafless, often compressed. Racemes loose but few-flowered, on short distant simple or branched peduncles, the subtending leaves reduced to small scales or entirely deficient. Pedicels 2 to 3 lines long. Perianth red, glabrous outside, bearded inside to below the middle with short hairs, the tube 4 or 5 lines long, rather broad, attenuate and revolute under the globular limb. Torus straight. Gland prominent, semi-annular. Ovary glabrous, rather shortly stipitate; style long, slightly thickened under the very oblique or lateral stigmatic disk. Fruit broad, about $\frac{1}{2}$ in. long, smooth or slightly chagrined.—*G. pedunculosa,* F. Muell. Fragm. i. 135.

W. Australia, *Drummond, 5th coll. n.* 406; Upper Kalgan river, *Oldfield, F. Mueller;* Phillips ranges, Salt and Fitzgerald rivers, *Maxwell.*

11. **G. stenomera,** *F. Muell. Fragm.* iv. 85. A spreading shrub of 4 or 5 ft., the young branches hoary or silvery with a minute tomentum, the adult foliage glabrous. Leaves pinnate, with narrow-linear segments twice as long and not quite so regular as in *G. pectinata,* mostly 1 to 2 in. long, obtuse or mucronate, doubly grooved underneath. Racemes solitary or several in a terminal panicle, spreading, loose, secund, $1\frac{1}{2}$ to 2 in. long, the rhachis and pedicels minutely pubescent or nearly glabrous. Pedicels about 2 lines long in flower, twice as long in fruit. Perianth nearly glabrous outside, bearded inside above the middle, the tube about 3 lines long, dilated in the lower part, attenuate and revolute under the globular limb. Torus straight. Gland broad, thick, semicircular. Ovary glabrous, on a long

stipes. Style long, slightly thickened under the lateral stigmatic disk. Fruit oblong, smooth or slightly rugose, 6 to 8 lines long.

W. Australia. Murchison river and near Bunbury, *Oldfield.* This may prove to be a variety of *G. Thelemanniana*, with less divided leaves and a looser inflorescence.

12. **G. Thelemanniana,** *Endl. Nov. Stirp. Dec.* 6. A spreading shrub of 3 to 5 ft., the young branches softly tomentose the foliage glabrous or very slightly silky, of a pale or glaucous hue, not unlike that of some *Artemisiæ*. Leaves pinnate with the lower pinnæ usually again divided, the segments rather numerous, linear, terete, slender, not pungent, singly or doubly grooved underneath, the whole leaf 1 to 2 in. long. Racemes terminal, spreading, secund, rather dense, 1 to 1½ in. long, the rhachis tomentose. Flowers pink with green tips. Pedicels 1 to 2 lines long. Perianth sprinkled outside with a few appressed hairs, bearded inside above the middle with short hairs, the tube 3 to 3½ lines long, somewhat dilated below the middle, attenuate and revolute under the globular limb. Torus straight. Gland broad, semi-orbicular, thick but flat or obscurely 3-lobed. Ovary glabrous, on a long stipes; style long, thickened under the oblique or lateral stigmatic disk. Fruit smooth, 5 or 6 lines long.—Meissn. in DC. Prod. xiv. 372; Baill. Hist. Pl. ii. 389, f. 216; *G. Preissii*, Meissn. in Pl. Preiss. i. 543, ii. 253 and in DC. Prod. xiv. 371; Bot. Mag. i. 5837.

S. Australia? Murray Desert near Lake Alexandrina, *Wurth*, a single specimen in herb. F. Mueller. Can it be a cultivated one?

W. Australia. Swan river, *Drummond, n.* 69, *1st coll. n.* 637, *Preiss, n.* 709, and others; between Swan river and King George's Sound, *Harvey;* King George's Sound, *Fraser.*

I have not seen any typical specimens of the plant originally described by Endlicher from Baron Huegel's garden, but his character agrees well with the common Swan river specimens. Baillon's figure above quoted represents well the foliage and inflorescence, but the enlarged figures 217 and 218 differ both from Endlicher's description and from our specimens in the shape of the perianth, its dense pubescence, and in the very short stipes of the ovary.

SERIES 2. HEBEGYNÆ.—Ovary sessile or scarcely stipitate, densely villous. Torus sometimes but very rarely slightly oblique with the gland-side the shortest.

This series differs from *Ptychocarpa* chiefly in the oblong or elongated secund racemes, in the perianth glabrous inside as well as the style, and in the leaves not so constantly entire as in *Ptychocarpa*. The absence or prominence of ribs on the fruit is a character very rarely appreciable. The seeds are in some species more winged than in *Ptychocarpa*, but that appears to be no more than a specific distinction, and can very rarely be ascertained from herbarium specimens.

13. **G. concinna,** *R. Br. in Trans. Linn. Soc.* x. 172, *Prod.* 377, *and Prot. Nov.* 18. An erect bushy shrub of several feet, the young branches tomentose hoary or ferruginous. Leaves mostly entire, linear or linear-lanceolate, very shortly mucronate, 1 to 2 in. long, smooth above and glabrous when old, silky-pubescent underneath, but when narrow the under surface concealed by the revolute margins and thick

midrib, and occasionally on young plants a few leaves deeply lobed. Racemes terminal, shortly pedunculate, dense, secund, $\frac{1}{2}$ to 1 in. long in the typical form. Pedicels very short, silky as well as the rhachis. Perianth silky outside, glabrous inside, the tube $3\frac{1}{2}$ to 4 lines long, rather broad at the base, much attenuate and revolute under the globular limb. Torus straight. Gland broad, depressed, semilunar. Ovary shortly stipitate, densely villous; style long, glabrous, slightly thickened under the broad oblique stigmatic disk. Fruit acuminate, about $\frac{1}{2}$ in. long, obscurely ribbed.—Meissn. in DC. Prod. xiv. 367; Sweet, Fl. Austral. t. 7; Bot. Reg. t. 1383; *G. Lemanniana*, Meissn. l.c. 366.

W. Australia, *Drummond, 5th coll. n.* 405; Lucky Bay, *R. Brown, Baxter;* Gardiner, Fitzgerald and Phillips ranges, West Mount Barren, Bremer Bay, *Maxwell.*

Var. *racemosa.* Racemes longer and more erect.—*G. Hewardiana*, Meissn. in DC. Prod. xiv. 366, and *G. coccinea*, Meissn. l.c. 367.—*Drummond, 5th coll. n.* 404; Mount Manypeak, *Preiss, n.* 711; Gardiner river, *Maxwell.*

14. **G. Hookeriana,** *Meissn. in Pl. Preiss.* i. 546, *and in D C. Prod.* xiv. 374. An erect shrub of several feet, the branches tomentose. Leaves pinnate, with 3 to 9 very narrow-linear segments, rigid but not pungent, doubly grooved underneath by the revolute margins and prominent midrib, glabrous when full grown, the whole leaf 6 to 8 in. long in some specimens with distant segments of 1 to 2 in., the leaf in other specimens 1 to 2 in. with segments of $\frac{1}{2}$ to 1 in. Racemes spike-like, dense, erect, secund, $1\frac{1}{2}$ to 3 in. long, the rhachis tomentose-villous. Pedicels scarcely any. Perianth silky-villous outside, glabrous inside, the tube about 4 lines long, slightly dilated below the middle, attenuate and revolute below the globular limb. Torus straight. Gland broad, horizontally spreading, semiorbicular. Ovary densely villous, contracted at the base but scarcely stipitate; style long, glabrous, the stigmatic disk oblique. Fruit obtusely angular, shortly acuminate, about $\frac{3}{4}$ in. long.—*G. tetragonoloba*, Meissn. in DC. Prod. xiv. 374.

W. Australia, *Drummond, 1st coll. n.* 633, *4th coll. n.* 282; Gardiner river and Doubtful Island Bay, *Oldfield.*

Drummond's specimens 6th coll. n. 184, referred here by Meissner, although much resembling *G. Hookeriana* in foliage, have very different flowers and constitute the *G. pinnatisecta.*

15 ? **G. Baxteri,** *R. Br. Prot. Nov.* 22. Leaves pinnate, silky on both sides, the segments about 1 in. long, narrow-linear, mucronate. Racemes erect. Perianth and pistil silky. Stigmatic disk dilated, convex, nearly vertical.—Meissn. in DC. Prod. xiv. 372.

W. Australia. Cape Arid, *Baxter.* I have not seen this species, the parcel of R. Brown's collection containing the original specimen could not be found, and I have not met with it in any other set of Baxter's plants. It appears to be very close to *G. Hookeriana,* and perhaps one of its forms, but differing in the silky leaves and vertical stigmatic disk.

16. **G. pterosperma,** *F. Muell. in Trans. Phil. Soc. Vict.* i. 22, *and in Hook. Kew Journ.* viii. 208. A shrub of several feet with numerous

erect branches, silky-tomentose when young. Leaves very narrow-linear, erect, entire or rarely divided into 2 or 3 segments, rigid but not pungent, 3 to 6 in. long, doubly grooved underneath, sprinkled with small appressed silky hairs or at length glabrous. Racemes terminal, secund, rather loose, 2 to 4 in. long, the flowers numerous. Pedicels rarely 1 line long, silky-tomentose as well as the rhachis. Bracts membranous and imbricate on the young racemes but falling away very early. Perianth silky-villous outside, glabrous inside, the tube nearly 3 lines long, somewhat dilated below the middle, narrow and revolute under the globular limb. Torus straight. Gland semi-annular, broader and flatter in the western than in the eastern specimens. Ovary distinctly stipitate, villous with long hairs ; style glabrous, the stigmatic disk oblique. Fruit nearly globular, densely tomentose, about ½ in. diameter. Seed-wing rather broad, especially on the outer margin.—Meissn. in DC. Prod. xiv. 384 ; *G. sericostachya,* Meissn. l.c. (previously named but without diagnosis in Hook. Kew Journ. iv. 186.)

N. S. Wales. Near the junction of the Murrumbidgee and the Murray rivers, *F. Mueller ;* between the Lachlan and Darling rivers, *Burkitt.*
Victoria. Wimmera and Murray Desert, *Dallachy.*
S. Australia. Cooper's Creek, *Howitt's Expedition.*
W. Australia, *Drummond,* 5th coll. suppl. n. 10, also n. 70.

17. **G. eriostachya,** *Lindl. Swan Riv. App.* 36. A stout erect shrub of 3 to 6 ft., the young branches silky-hoary or tomentose. Leaves very narrow-linear, occasionally undivided but mostly pinnate with 3 to 5 long distant segments, rigid but not pungent, doubly grooved underneath, glabrous or very minutely pubescent, 4 to 8 in. long. Racemes terminal, erect, dense, secund, 3 to 6 in. long on short thick tomentose-villous peduncles, and sometimes several on a long leafless branch, but often the floral branches leafy, the whole raceme densely tomentose-villous. Pedicels very short. Bracts membranous, villous, imbricate on the young raceme but falling off very early. Perianth silky-villous outside, glabrous inside, the tube about ½ in. long, slightly dilated at the base, revolute under the oblique ovoid-globular limb. Torus straight. Gland broad, flat, semi-lunar. Ovary sessile, densely villous ; style long, glabrous, slightly clavate under the oblique stigmatic disk. Fruit thick but flattened, oblique, broad, ¾ in. long. Seeds broadly winged all round.—Meissn. in Pl. Preiss. i. 545, and in DC. Prod. xiv. 383.

W. Australia, *Drummond,* 1st coll. n. 636, 2nd coll. n. 328, also n. 73 ; Murchison river, *Oldfield ;* Champion Bay, *Walcott.*

G. pityophylla, F. Muell. Fragm. vi. 208, described from a mere fragment in Drummond's collection, appears to me to be this species with the leaves entire, the margins more revolute showing only a single groove underneath.

18. **G. thyrsoides,** *Meissn. in Hook. Kew Journ.* vii. 77, *and in DC. Prod.* xiv. 375. Stems apparently decumbent, leafy at the base, slightly tomentose or silky-pubescent. Leaves pinnate, with 6 to 14 pairs of very narrow linear segments, rigid but not pungent, more or

less scabrous-punctate, doubly grooved underneath, the whole leaf
2 to 4 in. long, the segments ¾ to 1½ in. Flowering stems virgate,
leafless, often above 1 ft. long, bearing at the end 1 to 3 shortly
pedunculate racemes and several abortive ones lower down. Perfect
racemes secund, dense, 2 to 4 in. long, the rhachis tomentose-villous.
Pedicels very short. Bracts ovate or lanceolate, imbricate on the
young raceme but falling off very early. Perianth rather loosely
silky-villous outside, glabrous inside, the tube about 4 lines long,
broad and almost gibbous at the base, narrow from the middle, revolute under the globular limb. Torus straight. Gland broad, horizontal, semi-lunar. Ovary densely villous on a distinct stipes; style
long, more or less bearded, shortly thickened under the oblique
stigmatic disk. Fruit unknown.

W. Australia. Between Dundagaran and Smith river, *Drummond, 6th coll. n.*
1 83.

19. **G. chrysodendron,** *R. Br. in Trans. Linn. Soc.* x. 176, *Prod.*
379. A tree of 15 to 20 ft., the young branches tomentose. Leaves
pinnate with numerous very narrow linear segments of 4 to 8 in. the
lower ones sometimes forked, the rhachis angular, 6 in. to 1 ft. long,
the segments silky underneath, becoming glabrous above, rarely rather
broader and veined, the margins revolute. Racemes erect, terminal
or in the upper axils, rather dense, secund, 3 to 5 in. long. Flowers
yellow. Pedicels 1 to 2 lines long, tomentose-villous as well as the
rhachis. Perianth pubescent or villous outside, glabrous inside, the
tube not broad, scarcely 3 lines long, slightly contracted and much
revolute under the globular limb. Torus straight. Gland almost or
quite divided into two, broad and short. Ovary sessile, villous; style
long, the stigmatic disk oblique or lateral, with a prominent central
umbo. Fruit obliquely ovate-oblong, compressed, about ¾ in. long.
Seedwing surrounding the nucleus, but narrow.—Meissn. in DC. Prod.
xiv. 383; *G. pteridifolia* Knight, Prot. 121; *G. Mitchellii,* Hook. in
Mitch. Trop. Austral. 265; Meissn. l.c.

N. Australia. Islands of the Gulf of Carpentaria, *R. Brown;* Victoria river,
Bynoe, F. Mueller; Port Essington, *Armstrong;* Port Darwin, *Schulz;* Melville
island, *Fraser,* and other points of the N. coast, *A. Cunningham,* and others.
Queensland. Endeavour river, *R. Brown;* Rockingham Bay, *Dallachy;* Mistake
Creek, *Fitzalan;* Cape river, *Bowman;* Brigalow scrub on the Belyando, *Mitchell.*
Meissner distinguishes two species amongst Mitchell's specimens, differing chiefly in
the breadth of the perianth-tube and in the degree of obliquity of the stigmatic disk,
differences which however I have failed to appreciate in the specimens quoted.

20. **G. Banksii,** *R. Br. in Trans. Linn. Soc.* x. 176, *Prod.* 379. A
tall shrub or slender tree of 15 to 20 ft. the branches and inflorescence
softly ferruginous-tomentose. Leaves deeply pinnatifid or pinnate,
with 3 to 11 broadly linear or lanceolate segments, obtuse or mucronate,
with recurved margins, 2 to 4 in. long, glabrous above, silky-ferruginous
underneath, the midrib alone prominent or obscurely penniveined, the
whole leaf 4 to 8 in. long and here and there a smaller leaf undivided.
Racemes terminal, erect, dense, secund, 2 to 4 in. long, solitary or 2 or 3

on a terminal leafless peduncle. Flowers red. Pedicels 3 to 4 lines long, tomentose as well as the rhachis. Perianth tomentose outside, glabrous inside, the tube not very broad, 6 or 7 lines long, contracted and revolute under the limb. Torus straight or nearly so. Gland prominent, semiannular, more or less lobed or jagged. Ovary sessile, densely villous; style long and glabrous, clavate under the very oblique or lateral convex stigmatic disk. Fruit obliquely ovate, compressed, almost acute, about 1 in. long.—Meissn. in DC. Prod. xiv. 375; Bauer, Illustr. t. 9.

Queensland. Broad Sound, Keppel and Shoalwater Bays, *R. Brown;* open barren hills, upper Brisbane river, *A. Cunningham; Rockhampton, Thozet;* head of Cape river. *Bowman;* Wide Bay, *Bidwill;* Keppel Bay, *O'Shanesy;* mouth of Fitzroy river, *C. Haynes;* Facing Island, *W. Hill.*

21. **G. Caleyi,** *R. Br. Prot. Nov.* 22. A slender shrub of 5 or 6 ft., the branches petioles and inflorescence densely villous with soft spreading ferruginous hairs. Leaves deeply pinnatifid or pinnate with numerous (above 20) oblong-linear divaricate segments, obtuse or mucronate, with recurved margins, glabrous above, softly villous underneath, ¾ to 1½ in. long but very regular on the same leaf, the whole leaf 3 to 6 in. long. Racemes terminal or in the upper axils, erect, rather dense, secund, shortly pedunculate, 1½ to 2 in. long. Pedicels 1 to 2 lines long. Perianth pubescent or villous outside, glabrous inside, the tube about 3 lines long, slightly dilated at the base, contracted and revolute under the ovoid limb. Torus straight. Gland semicircular, not very prominent. Ovary shortly stipitate, villous; style long, glabrous, shortly thickened under the oblique umbonate stigmatic disk. Fruit broadly falcate, slightly compressed, ¾ in. long, villous but the concave edge marked with longitudinal glabrous lines.—Meissn. in DC. Prod. xiv. 375; Bot. Mag. t. 3133; *G. blechnifolia,* A. Cunn. Herb.

N. S. Wales. Port Jackson, *Caley, A. Cunningham.*

22. **G. asplenifolia,** *Knight Prot.* 120. A tall shrub or small slender tree of 12 to 15 ft., the branches minutely silky-pubescent when very young. Leaves lanceolate or linear-lanceolate, mucronate-acute, entire acutely toothed or pinnatifid with short broad acute lobes, contracted into a short petiole, 4 to 10 in. long, glabrous and more or less distinctly penniveined above, silky-silvery or fulvous underneath, the midrib alone prominent. Racemes sessile or shortly pedunculate, terminal or in the upper axils, secund, 1 to 2 in. long. Pedicels scarcely 1 line long, minutely tomentose as well as the rhachis. Perianth silky-pubescent outside, glabrous inside, the tube narrow, 4 or 5 lines long, revolute under the obliquely globular limb. Torus straight or nearly so. Gland semiannular, not very prominent. Ovary shortly stipitate, villous; style long, glabrous; stigmatic disk oblique, convex.—R. Br. in Trans. Linn. Soc. x. 175, Prod. 379; Meissn. in DC. Prod. xiv. 376; *G. longifolia,* R. Br. Prot. Nov. 22; Meissn. l.c.

N. S. Wales. Port Jackson to the Blue Mountains, *Caley, A. Cunningham, Fraser,* and others; Sydney woods, Paris Exhibition, 1855, *M'Arthur, n.* 181.

436   CIV. PROTEACEÆ.   [*Grevillea.*

23. **G. cirsiifolia,** Meissn. in Pl. Preiss. ii. 253, and in DC. Prod. xiv.
376. Stems prostrate, not much branched, silky-tomentose. Leaves
linear or lanceolate, 3 to 6 in. long, entire remotely toothed or pinnatifid, the teeth or lobes short and falcate or rarely longer and lanceolate, glabrous above when full grown, silky underneath. Racemes
lateral or axillary, loose, secund, shortly pedunculate, 2 to, 3 in. long,
the rhachis pedicels and perianths silky and often fulvous. Bracts
small, often persistent. Pedicels 1½ to 3 lines long. Perianth glabrous
inside, the tube scarcely dilated at the base, revolute above the middle
and the 2 lower segments there dilated into broad semiorbicular horizontally spreading appendages, forming a broad disk entirely concealing the revolute globular limb. Torus straight. Gland obsolete.
Ovary sessile or nearly so, densely villous with long fulvous hairs;
style glabrous, thick but flattened; stigmatic disk lateral, thick, the
stigma on a prominent central point. Fruit very oblique, ovoid, about
4 lines long.

W. Australia, *Drummond, 3rd coll. n.* 267. The curious form of the perianth is
quite anomalous in the genus.

24. **G. laurifolia,** Sieb. in Roem. and Schult. Syst. iii., Mant. 279 and in
Spreng. Syst. Cur. Post. 45. A procumbent or trailing shrub, the young
branches minutely silky-tomentose. Leaves petiolate, ovate oblong or
broadly lanceolate, obtuse or mucronate, entire, rounded or cuneate at
the base, 2 to 5 in. long, glabrous above, closely silky underneath,
the primary veins nearly parallel and arching into an intramarginal
nerve but not quite so regular as in *G. Goodii,* and the reticulate veinlets scarcely conspicuous. Racemes terminal or lateral, shortly pedunculate, secund, rather dense, 1 to 1½ or rarely 2 in. long. Pedicels 1
to 2 lines long, closely ferruginous-tomentose as well as the rhachis.
Perianth slightly hairy outside, glabrous inside, the tube obliquely
dilated at the base, about 4 lines long, attenuate and revolute under
the globular limb. Torus straight. Gland prominent, semiannular.
Ovary stipitate, villous; style long, glabrous; stigmatic disk oblique,
umbonate.—R. Br. Prot. Nov. 17; Meissn. in DC. Prod. xiv. 352;
G. humifusa, A. Cunn. Herb.

N. S. Wales. Blue Mountains, *Caley, Sieber, n.* 26, *A. Cunningham, Fraser,
Woolls.* In habit and foliage this species resembles *G. Goodii,* but the flowers are very
different.

25. **G. Barklyana,** F. Muell. (ined. ?). A shrub, probably tall, the
young branches hoary-tomentose or ferruginous-silky. Leaves either
oblong-lanceolate entire and 4 to 8 in. long, or pinnatifid with 3 to 7
triangular or broadly lanceolate lobes often above 1 in. long and the
whole leaf 6 to 10 in. long, penniveined, glabrous above, ferruginous
or hoary-tomentose underneath. Racemes nearly sessile, dense, secund,
2 to 3 in. long, terminal or at length lateral. Pedicels exceedingly short,
tomentose as well as the rhachis. Perianth pubescent outside, glabrous
inside, the tube scarcely dilated at the base, about 3 lines long, revolute
under the globular limb. Torus straight. Gland semiannular, scarcely

prominent. Ovary very shortly stipitate, villous; style long, glabrous, the stigmatic disk slightly oblique, convex. Fruit acuminate, not $\frac{1}{2}$ in. long.

Victoria. Ranges on the upper Tarwan and Bunyip rivers, *F. Mueller.* I have been unable to discover where F. Mueller has published this species.

26. **G. repens,** *F. Muell. in Linnæa,* xxvi. 355. A prostrate shrub, spreading to a great extent, the young branches slightly pubescent. Leaves glabrous or sprinkled underneath with appressed hairs, very shortly petiolate, from broadly ovate to oval-oblong, cordate truncate or cuneate at the base, bordered by short prickly teeth, the margins often undulate but not recurved, penniveined with the primary veins prominent underneath, $\frac{3}{4}$ to $1\frac{1}{2}$ in. long when broad or twice as long when narrow. Racemes terminal or on short axillary branches, shortly pedunculate, secund, 1 to 2 in. long. Pedicels 1 to 2 lines long. Perianth silky-pubescent outside, glabrous inside, the tube 3 to $3\frac{1}{2}$ lines long, somewhat dilated below the middle, revolute under the globular limb. Torus slightly oblique. Gland semicircular, thick and rather broad. Ovary villous, on a stipes as long as itself; style long, glabrous, the stigmatic disk very oblique, with a central umbo or small cone.—Meissn. in DC. Prod. xiv. 377.

Victoria. Goulburn ranges, Watts and Loddon rivers, *F. Mueller.*

27. **G. aquifolium,** *Lindl. in Mitch. Three Exped.* ii. 178. A shrub of several ft., the branches more or less tomentose or villous. Leaves petiolate, ovate ovate-lanceolate or oblong, undulate and prickly-toothed or pinnatifid with short pungent-pointed lobes, cuneate or truncate at the base, 1 to 3 in. long, rigid and veined, sometimes nearly glabrous but more frequently pubescent above and silky or softly villous underneath, and often ferruginous. Racemes terminal or on short axillary branches, nearly sessile, dense, secund, 1 to 2 in. long. Pedicels very short, villous as well as the rhachis. Perianth villous outside, glabrous inside, the tube about 4 lines long, dilated below the middle, attenuate and revolute under the globular limb. Torus nearly straight ; gland semiannular. Ovary densely villous on a very short stipes ; style long, glabrous ; stigmatic disk slightly oblique with a central umbo.—Meissn. in DC. Prod. xiv. 378 ; *G. variabilis* Lindl. in Mitch. Three Exped. ii. 179 ; Meissn. l.c. ; *G. induta,* F. Muell. First Gen. Rep. 17.

Victoria. Grampians, *Mitchell, F. Mueller ;* Wimmera, *Dallachy ;* near Bridgewater Bay, *Robertson ;* Portland, *Allitt.*

28. **G. ilicifolia,** *R. Br. Prot. Nov.* 21. A large spreading shrub attaining 6 ft. or more though often much smaller, the branches more or less silky or hoary-pubescent. Leaves in the typical form cuneate, undulate prickly-toothed and lobed at the end, with a long tapering base, the whole leaf 1 to 2 in. long, but sometimes longer and deeply pinnatifid with narrow lobes, more rarely pinnatifid with short lobes from near the base almost as in *G. aquifolium,* or as broad as long and once or twice 3-lobed, the lobes or teeth always rigid and pungent-pointed, glabrous and veined above, more or less silky underneath. Ra-

cemes terminal, secund, 1 to 2 in. long, the rhachis and pedicels silky-pubescent or villous. Pedicels about 1 line long. Bracts sometimes persistent. Perianth villous outside, glabrous inside, the tube about 4 lines long, revolute under the limb. Torus nearly straight. Ovary stipitate, villous; style long, glabrous; stigmatic disk oblique. Fruit oblique, acuminate, about ½ in. long.—Meissn. in DC. Prod. xiv. 377; *Anadenia ilicifolia*, R. Br. in Trans. Linn. Soc. x. 167, Prod. 375; *G. Behrii*, Schlecht. in Linnæa, xx. 585.

Victoria, *Harvey ;* Forest Creek, Mount Corong, Station Peak, *F. Mueller.*
S. Australia. Port Lincoln, *R. Brown ;* Kangaroo Island, *Baxter, Waterhouse :* Spencer's Gulf, *F. Mueller.*

Var. *lobata.* Leaves with lanceolate or rarely linear lobes, and often again lobed.— *G. lobata,* F. Muell. in Trans. Phil. Soc. Vict. i. 22, and in Hook. Kew Journ. viii. 207; Meissn. in DC. Prod. xiv. 379. *G. dumetorum,* Meissn. l.c. 378.—N. W. Victoria, *L. Morton, Dallachy,* and others ; Grampians, *Mitchell ;* Murray Desert, *F. Mueller ;* Tattiara country, *Woods.*

29. **G. Gaudichaudii,** *R. Br. in Gaudich. Freyc. Voy. Bot.* 443. *t.* 46 ; *Prot. Nov.* 22. An erect shrub with the habit inflorescence and flowers of *G. acanthifolia,* of which it may be a variety with less divided and less prickly leaves. Branches slightly silky-pubescent when young, but soon becoming glabrous. Leaves deeply pinnatifid, the lobes oblong or ovate, all entire and pungent-pointed, or the terminal one or sometimes the lateral ones also cuneate with 2 or 3 pungent-pointed teeth or short secundary lobes, the whole leaf 2 to 4 in. long and the lower lobes sometimes above 1 in., glabrous above, more or less distinctly penniveined, with the primary veins confluent in an intra-marginal nerve, paler underneath and often sprinkled with appressed hairs. Racemes terminal, secund, silky-villous, entirely as in *G. acanthifolia,* with the same perianth, nearly sessile densely villous ovary, long glabrous style and oblique stigmatic disk.—Meissn. in DC. Prod. xiv. 377.

N. S. Wales. Blue Mountains, *Gaudichaud, A. Cunningham, Fraser.*

30. **G. acanthifolia,** *A. Cunn. in Field. N. S. Wales,* 328 *with a plate.* An erect or straggling shrub of several ft., glabrous except the inflorescence. Leaves deeply pinnatifid; lobes or segments usually 9 to 15 but sometimes more or fewer, either cuneate and 3- to 5-lobed or the upper ones lanceolate and entire, all rigid, pungent-pointed, green on both sides, the whole leaf 1½ to 3 in. long, the lobes or segments in some specimens all under ½ in., in others ½ to ¾ in. long. Racemes terminal or in the upper axils, sessile or shortly pedunculate, dense, secund, 2 to 4 in. long. Pedicels exceedingly short, densely villous as well as the rhachis. Bracts broad, villous, membranous, imbricate on the young spike and sometimes persisting till the flowers expand. Perianth pink, silky-villous outside, glabrous inside, the tube 4 to 5 in. long, slightly dilated below the middle, contracted and revolute under the globular limb. Torus nearly straight. Gland semi-annular, not very prominent. Ovary nearly sessile, densely villous with long silky hairs; style long, glabrous; stigmatic disk oblique, convex.—R. Br. Prot. Nov. 22;

Meissn. in DC. Prod. xiv. 377; Bot. Mag. t. 2807; Lodd. Bot. Cab. t. 1153; Lindl. and Paxt. Fl. Gard. iii. 103, f. 281.

N. S. Wales. Blue Mountains, *A. Cunningham, Fraser, Sieber, n.* 32, and others.

Var. *stenomera,* F. Muell. Prostrate; leaf-lobes linear-lanceolate.—Head of Macleay river, *C. Moore.*

31. **G. bipinnatifida,** *R. Br. Prot. Nov.* 23. A diffuse or prostrate shrub of 3 or 4 ft., the branches tomentose-pubescent with appressed hairs. Leaves broad, deeply pinnatifid or pinnate; lobes or segments 9 to 21, either oblong or cuneate or again pinnatifid with triangular or lanceolate pungent-pointed lobes, the whole leaf usually 3 to 4 in. long and 2 to 3 in. broad, rather rigid, the upper surface glabrous and reticulate, the lower sprinkled with a few hairs or glabrous, the primary veins alone prominent. Racemes loose, secund, usually 2 to 4 in. long, solitary or several in a terminal panicle, the rhachis ferruginous-tomentose. Pedicels 3 to 5 lines long. Perianth red, silky-pubescent outside, glabrous inside, the tube 7 to 8 lines long, dilated and somewhat gibbous below the middle, attenuate and revolute under the globular limb. Torus straight. Gland obovate or orbicular, convex, horizontally spreading. Ovary sessile or nearly so, shortly villous; style very long, glabrous, slightly clavate under the broad oblique stigmatic disk.—Meissn. in Pl. Preiss. i. 541, and in DC. Prod. xiv. 376.

W. Australia. Swan river, *Fraser, Drummond,* 1st coll. *n.* 632, *Preiss, n.* 707, 708; Harvey and Blackwood rivers, *Oldfield.*—In some specimens the raceme is much elongated and very loose, but not constituting a distinct variety.

32. **G. armigera,** *Meissn. in DC. Prod.* xiv. 373. A stout shrub the branches softly tomentose, the foliage scabrous-punctate. Leaves once twice or three times divided into narrow-linear rigid divaricate pungent-pointed segments, doubly grooved underneath, rather thicker and broader than in *G. asparagoides,* the whole leaf 1 to 2 in. diameter. Racemes terminal, sessile, dense, secund, about 2 in. long, the rhachis tomentose. Pedicels exceedingly short or scarcely any. Perianth silky-villous outside, glabrous inside, the tube about 3 lines long, much dilated and almost gibbous below the middle, attenuate and revolute under the globular limb. Torus straight. Gland broadly ovate, spreading. Ovary nearly sessile, densely villous; style long, glabrous, slightly thickened under the oblique stigmatic disk. Young fruit globular.

W. Australia, *Drummond* (2nd coll.?), *n.* 164, 4th coll. *n.* 284; Plantagenet and Stirling Ranges, *Maxwell.*

33. **G. asparagoides,** *Meissn. in DC. Prod.* xiv. 373. A divaricately branched or prostrate intricate shrub of several ft., the branches tomentose, the foliage minutely pubescent or glabrous. Leaves once twice or three times ternately divided into rigid divaricate pungent-pointed segments, very narrow linear and doubly grooved underneath, the whole leaf 1 to 2 in. diameter. Racemes terminal, loose, secund, rarely above 1 in. long, the rhachis and pedicels shortly hirsute and glandular-viscid.

Pedicels about 2 lines long. Perianth pubescent or hirsute outside, glabrous or nearly so inside, the tube 4 to 5 lines long, dilated or somewhat gibbous at the base, attenuate and revolute under the globular limb. Torus straight. Gland horizontal, convex, semi-annular, not broad, sometimes 3-crenate. Ovary sessile or nearly so, villous; style long, glabrous, slightly clavate under the oblique stigmatic disk. Fruit ovoid, acuminate, 7 to 8 lines long.

W. Australia, *Drummond*, (2nd *coll.?*) *n.* 165, *4th coll. n.* 283; Salt river and Phillips Range, *Maxwell*.

SECT. 2. PTYCHOCARPA.—Racemes short, often umbel-like and not at all or scarcely secund. Perianth-tube dilated below the middle and usually opening on the lower side, the segments otherwise long-cohering, attenuate and revolute above the middle. Torus small, straight or nearly so. Style hirsute ciliate or tomentose. Ovary sessile or very shortly stipitate, densely villous or rarely glabrous except a tuft of hairs at the base on the upper side. Stigmatic disk lateral. Leaves entire with revolute margins.

This section differs from the *Hebegynæ* series of *Eugrevillea* chiefly in the inflorescence, in the perianth more or less bearded or hirsute inside as well as the style, and in the leaves which appear never to break out into teeth or lobes as they do constantly or occasionally in all the species of that series. The section is usually distinguished by the ribbed fruit, but the ribs are often very obscure, and quite disappear in several species otherwise inseparable from the group.

34. **G. floribunda,** *R. Br. Prot. Nov.* 19. An erect or spreading shrub " not exceeding 5 ft." the branches ferruginous-tomentose. Leaves nearly sessile, oval or oblong, obtuse or with a small callous point, the margins recurved or revolute, $\frac{3}{4}$ to nearly 2 in. long, villous when young, minutely scabrous above and silky-tomentose underneath when full-grown, sometimes faintly penniveined. Racemes terminal, sessile or shortly pedunculate and often 1 in. long, the rhachis and flowers very densely villous with ferruginous hairs. Pedicels 1 to 2 lines long. Perianth bearded inside about the middle, the tube from about 3 lines to above 4 lines long, broad and gibbous at the base, attenuate and much revolute above the middle, the limb globular and obtuse. Torus nearly straight. Gland horizontal, broad, not very prominent. Ovary sessile, densely villous; style not very long, villous, thick; stigmatic disk lateral. Fruit about $\frac{1}{2}$ in. long, slightly ribbed.—Meissn. in DC. Prod. xiv. 361; *G. sphacelata*, A. Cunn. Herb.; *G. autumnalis*, Lhotzk. MSS. (*Meissner*); *G. chrysophæa*, F. Muell. First Gen. Rep. 17; Meissn. in Linnæa, xxvi. 357, and in DC. Prod. xiv. 361; *G. ferruginea*, Grah. in Maund, Botanist, t. 153, not of Sieber.

N. S. Wales. Goulburn and Hunter rivers, *A. Cunningham, Fraser;* ravines near Mount Owen and Mount Clift, *Mitchell*.
Victoria. Avon, Macalister and Latrobe rivers, *Stieglitz;* Station Peak, *F. Mueller*; Geelong, *Dallachy*.

35. **G. cinerea,** *R. Br. in Trans. Linn. Soc.* x. 173, *Prod.* 378. A tall shrub, the branches tomentose-villous. Leaves obovate or

broadly oblong in the typical form, mucronate-acute, the margins recurved, contracted into a very short petiole, ¾ to above 1 in. long, scabrous-punctate and more or less distinctly veined above, densely silky-tomentose and hoary or ferruginous underneath. Racemes terminating short leafy branches, umbel-like with few rather large flowers. Pedicels 1 to 2 lines long. Perianth densely villous outside, the shorter segments scantily bearded inside below the middle, the tube nearly 6 lines long, not very broad, contracted and revolute under the oblique obtuse limb. Torus nearly straight. Gland semi-annular. Ovary sessile, villous; style long, more or less ciliate, channelled at the base; stigmatic disk lateral.—Meissn. in DC. Prod. xiv. 358, partly.

N. S. Wales. Port Jackson, *R. Brown;* on the road to Illawarra, *A. Cunningham;* Blue Mountains, *Fraser.*

Var. *angustifolia.* Leaves mostly narrow-oblong.—*G. attenuata,* A. Cunn. Herb. —Towards Hunter's river, *A. Cunningham.*

As observed by Meissner this species is near to *G. mucronulata,* but appears to me to be constantly distinct in the densely villous perianth. The specimens therefore specially described by Meissner under the name of *G. cinerea* would belong to the true *G. mucronulata.* Meissner was enabled to take only a very cursory glance over R. Brown's own set of *Proteaceæ,* and not sufficient to verify with precision any critical species.

Lysanthe stylosa, Knight, Prot. 117 (*Grevillea stylosa,* Steud. Nom. Bot. ed. 2), is probably either this species or *G. montana.*

36. **G. alpina,** *Lindl. in Mitch. Three Exped.* ii. 179. A much-branched shrub, erect spreading or diffuse, densely tomentose or villous with spreading hairs. Leaves rather crowded, sessile or nearly so, oval, oblong-lanceolate or almost linear, obtuse or with a small point, the margins revolute, all under ½ in. long in some specimens, but sometimes attaining 1 in., hirsute or rarely scabrous only above, silky-villous underneath. Racemes very short, terminal, sessile. Pedicels 2 to 4 lines long, pubescent or villous as well as the rhachis. Perianth villous outside, bearded inside to below the middle, the tube from under 4 lines to above 5 lines long, broad and obliquely gibbous at the base on the upper side, attenuate and much revolute above the middle, the limb ovoid-globular, obtuse or very shortly acuminate. Torus nearly straight. Gland very prominent projecting almost horizontally into the gibbosity of the perianth. Ovary sessile, densely villous; style densely villous; stigmatic disk lateral, slightly umbonate. Fruit about ½ in. long.—Meissn. in DC. Prod. xiv. 360; *G. oreophila* and *G. Dallachiana,* F. Muell. First Gen. Rep. 17; *G. alpestris,* Meissn. l.c. 361; Bot. Mag. t. 5007.

Victoria. Mount William in the Grampians, *Mitchell;* Mount Disappointment, Buffalo ranges, Upper Yarra and Ovens ranges, *F. Mueller.*

The variations in the foliage and indumentum do not sufficiently correspond with those in the size of the perianth ór in the obtuseness of its limb to admit of the establishing two distinct forms as proposed.

37. **G. montana,** *R. Br. in Trans. Linn. Soc.* x. 172, *Prod.* 378. A spreading shrub, the branches densely tomentose or villous with

spreading hairs. Leaves shortly petiolate, oblong or lanceolate, obtuse with a small callous point, the margins recurved, contracted at the base, ¾ to 1½ in. long, glabrous scabrous or slightly hairy above and often veined, densely silky-tomentose and usually ferruginous underneath. Racemes terminal, sessile, short and umbel-like, few-flowered. Perianth as large as in *G. cinerea* but nearly glabrous outside, the tube not very broad, fully ½ in. long, revolute and attenuate under the ovoid acuminate limb. Torus nearly straight. Gland semi-annular, slightly prominent. Ovary sessile, densely villous; style long, more or less tomentose and hirsute with short hairs.—Meissn. in DC. Prod. xiv. 358; *G. ferruginea*, Sieb. in Roem. and Schult. Syst. iii. Mant. 280; R. Br. Prot. Nov. 19; Meissn. l.c. 359.

N. S. Wales. Blue Mountains, *R. Brown, Sieber, n.* 27 ; ̄on the Bulga road, *A. Cunningham;* Bent's Basin, *Woolls;* Harper's Hill, Hunter's river, *Backhouse* (with shorter points to the perianth-limb); Illawarra, *Shepherd* (no perianths to the specimens and the determination doubtful). The degree of acumination of the perianth-limb is variable in this as in *G. arenaria*, from which this species differs chiefly in the indumentum of its various parts and in its larger flowers.

38. **G. obtusiflora,** *R. Br. Prot. Nov.* 19. A spreading shrub, the branches tomentose. Leaves oblong or lanceolate, obtuse or with a small often recurved point, the margins revolute, contracted at the base and sometimes shortly petiolate, ½ to 1 in. long, the upper surface at first pubescent but becoming glabrous, the under side silky-tomentose. Racemes terminal, short, sessile, the rhachis and pedicels pubescent. Perianth sprinkled with appressed hairs, the tube not very broad, much revolute above the middle, the limb very obliquely globular, obtuse. Torus nearly straight. Gland semi-annular, slightly prominent. Ovary sessile, densely villous; style hirsute, not very long.—Meissn. in DC. Prod. xiv. 359.

Queensland ? Wide Bay, *Bidwill*, but possibly some mistake in the station.
N. S. Wales. Brushy hills, North of Bathurst, *A. Cunningham.*—The species appears to be very close to *P. montana*, but with a very obtuse perianth-limb. The specimens seen are however not good.

39. **G. arenaria,** *R. Br. in Trans. Linn. Soc.* x. 172, *Prod.* 378. An erect shrub of about 6 ft., the branches densely tomentose. Leaves shortly petiolate, obovate-oblong to narrow oblong, obtuse with a very small point, the margins recurved, ¾ to 1½ in. long, minutely hoary-tomentose and scarcely veined on the upper side, densely tomentose and often ferruginous underneath. Racemes terminal, short, umbel-like and few-flowered, mostly reflexed. Pedicels 1 to 2 lines long. Perianth densely tomentose or pubescent outside in the typical form, bearded inside to below the middle, the tube about 5 lines long, rather broad at the base, much revolute from the middle upwards, contracted under the ovoid acuminate limb. Torus nearly straight. Gland semi-annular, slightly prominent. Ovary sessile, villous; style long, tomentose but not hirsute as in several of the allied species; stigmatic disk lateral. Fruit fully ½ in. long.—Meissn. in DC. Prod. xiv. 358; Bot. Mag. t. 3285 ; *Lysanthe cana*, Knight, Prot. 117.

N. S. Wales. Nepean river, *R. Brown;* near Goulburn, *Backhouse;* Shoalhaven gullies, *C. Moore;* Sidmouth Valley and Lachlan river, *Woolls.*

Var. *canescens.* Perianth more villous, the points to the laminæ longer.—*G. cinerea,* A. Cunn. in Field, N. S. Wales, 329, not of R. Br.; *G. canescens,* R. Br. Prot. Nov. 18; Meissn. in DC. Prod. xiv. 359; Bot. Mag. t. 3185.—North of Bathurst, *A. Cunningham, Fraser;* Macquarrie river, *Bowman.*

The species differs from *G montana* chiefly in habit and indumentum.

40. **G. mucronulata,** *R. Br. in Trans. Linn. Soc.* x. 173, *Prod.* 378. A large shrub with rather slender hirsute branches. Leaves shortly petiolate, ovate in the typical form and either rounded at the end with a fine point in the centre, or tapering into a fine point, flat or with recurved margins, mostly about ½ in. long, scabrous and obscurely or distinctly veined above, silky-tomentose and sometimes ferruginous underneath, with the midrib alone prominent. Racemes short, loose, few-flowered, on slender terminal or axillary peduncles often longer than the leaves. Pedicels 1 to 2 lines long, silky-pubescent or hirsute as well as the rhachis. Perianth sprinkled with appressed hairs outside, the shorter segments slightly bearded inside below the middle, the tube about 5 lines long, rather broad, gibbous at the base, revolute above the middle and attenuate under the very oblique obtuse limb. Torus nearly straight. Gland semi-annular. Ovary sessile, villous; style long, more or less ciliate; stigmatic disk lateral, large.—Meissn. in DC. Prod. xiv. 357; Sweet Fl. Austral. t. 38; *Lysanthe podalyriæfolia,* Knight, Prot. 117; *G. podalyriæfolia,* Sw. in Steud. Nom. Bot. ed. 2; *G. cinerea,* Lodd. Bot. Cab. t. 857; Meissn. l.c. partly not of R. Br.; *G. myrtacea,* Sieb. in Roem. and Schult. Syst. iii. Mant. 280, and in Spreng. Syst. Cur. Post. 46; *G. acuminata,* Sw. Fl. Austral. t. 55.

N. S. Wales. Port Jackson to the Blue Mountains, *R. Brown, Sieber, n.* 39, and others. The typical form in Brown's herbarium and from Hunter's river, *A. Cunningham,* and represented in Sweet's plate, n. 38, has most of the leaves rounded at the ends and shortly mucronate, with only a few of the leaves acuminate and tapering in a fine point; in Sieber's and other specimens, including those represented in Sweet's plate 55, the majority are thus acuminate, but not so narrow as in Brown's typical *G. acuminata.*

Var. *angustifolia.* Leaves mostly lanceolate-acuminate with a fine point,—*G. acuminata,* R. Br. in Trans. Linn. Soc. x. 173, Prod. 378; Meissn. in DC. Prod xiv. 358, as to R. Brown's plant, but not as to A. Cunningham's, which is a var. of *G. cinerea.*—Hunter's river, *R. Brown.*

The six preceding species appear very much to run into each other.

41. **G. Baueri,** *R. Br. in Trans. Linn. Soc.* x. 173, *Prod.* 378, *Prot. Nov.* 19. A bushy shrub attaining several ft. the branches more or less pubescent. Leaves rather crowded, sessile, oblong, narrow or broad, obtuse, scarcely contracted at the base, the margins recurved, ½ to 1 in. long, glabrous and 1-nerved on both sides or rarely sprinkled with a few short hairs, obscurely reticulate above, paler and sometimes penniveined underneath. Racemes very short and umbel-like, sessile on short leafy branches. Pedicels slender, glabrous, about 3 lines long. Perianth greenish yellow or tinged with red, glabrous outside,

bearded inside at or below the middle with reflexed hairs, the tube about 4 lines long, broad in the lower part, much revolute and attenuate under the very oblique limb. Torus slightly oblique. Gland broad, thick, semicircular. Ovary sessile; densely hirsute; style long, ciliate, rather thick, channelled on the upper side; stigmatic disk lateral.—Meissn. in DC. Prod. xiv. 357; *G. pubescens,* Hook. Exot. Fl. t. 216; Lodd. Bot. Cab. t. 1229; *G. daphnoides,* Sieb. in Roem. and Schult. Syst. iii. Mant. 281; *G. myrtillifolia,* A. Cunn. Herb.

N. S. Wales. Port Jackson to the Blue Mountains. *R. Brown, Sieber, n.* 25, *and Fl. Mixt. n.* 478; near Bathurst, *Fraser;* Camden and Argyle Counties, *A. Cunningham.*

Var. *pubescens.* Leaves usually narrow, pubescent above, silky underneath.— Shoalhaven, *Woolls.* These specimens are much more densely pubescent than the garden ones on which *G. pubescens* was founded, which are sprinkled only with a few hairs.

42. **G. lanigera,** *A. Cunn. in R. Br. Prot. Nov.* 20. Branches densely tomentose or villous. Leaves crowded, sessile or nearly so, linear, obtuse, with revolute margins, mostly about ½ in. long, rather thick, pubescent above, silky-tomentose underneath. Racemes terminal, short but rather loose, sessile or shortly pedunculate, quite glabrous. Pedicels 1 to 2 lines long. Perianth glabrous outside, the shorter segments bearded inside about the middle, the tube about 4 lines long, broad and somewhat gibbous at the base, much revolute and attenuate under the very oblique limb. Torus straight. Ovary almost sessile, densely villous; style long, ciliate; stigmatic disk lateral. Fruit 5 or 6 lines long.—Meissn. in DC. Prod. xiv. 363.

N. S. Wales. Camden and Argyle Counties, Lachlan and Murrumbidgee rivers, *A. Cunningham;* Nangas, *M'Arthur;* Gabo Island, *Maplestone.*
Victoria. Mitta-Mitta and Wilson's Promontory, *F. Mueller;* near Albury, *Beattie.*

43. **G. ericifolia,** *R. Br. Prot. Nov.* 20. A low shrub, spreading or diffuse (or sometimes erect and taller?), the branches pubescent or tomentose-villous. Leaves sessile, linear or lanceolate, mucronate-acute, with revolute margins, ½ to 1 in. long, silky-pubescent or villous when young, becoming glabrous above when full-grown. Racemes terminal, short but rather loose and often shortly pedunculate, quite glabrous. Pedicels slender, 2 or 3 lines long. Perianth glabrous outside, densely bearded inside below the middle, the tube 3 to 3½ lines long, broad and gibbous at the base, much revolute and attenuate under the very oblique limb. Torus nearly straight. Gland broad, thick, semicircular. Ovary sessile, densely villous especially on the upper side; style long, thick, more or less ciliate; stigmatic disk lateral.—Meissn. in DC. Prod. xiv. 365; *G. Latrobei,* Meissn. in. Pl. Preiss. i. 539 and in DC. Prod. xiv. 364.

N. S. Wales. Lachlan river, *A. Cunningham, Mitchell;* Limestone Creek west from Bathurst, *A. Cunningham.*

Victoria. Near Melbourne, *Adamson, F. Mueller ;* Plenty Creek, Genoa river, *F. Mueller.*

Var. *scabrella.* Leaves more scabrous. Perianth rather larger. — *G. scabrella,* Meissn. in DC. Prod. xiv. 365.—Near Nangas, *M'Arthur.*

44. **G. divaricata,** *R. Br. Prot. Nov.* 20. A bushy shrub more slender and spreading than *G. rosmarinifolia* and sometimes low and diffuse, the branches more or less pubescent or hirsute. Leaves linear-subulate, from under ½ in. to above ¾ in. long, glabrous and smooth or more or less hirsute or scabrous. Racemes short sessile and glabrous, and perianth entirely as in *G. rosmarinifolia,* and the ovary as in that species sessile and glabrous as well as the style except a small tuft of hairs at the base on the upper side.—Meissn. in DC. Prod. xiv. 364 ; *G. nutans,* Meissn. l.c. (with rather longer leaves).

N. S. Wales. Lachlan and Cujeegong rivers, and Euryalean scrub and forest land north of Bathurst, *A. Cunningham, Mitchell.*
Victoria. Forest and Darebin Creeks, *F. Mueller ;* Skipton, *Whan ;* Wimmera, *Dallachy.*

G. glabella, R. Br. Prot. Nov. 20 ; Meissn. in DC. Prod. xiv. 364, is a slight variety with finer, more crowded and more erect leaves, but several specimens are intermediate. The species is very near both to *G. ericifolia* and *G. rosmarinifolia.* The three might indeed be regarded as varieties of a single one. *G. lavandulacea* sometimes approaches them in habit, but has a differently shaped perianth and the torus always oblique,

45. **G. rosmarinifolia,** *A. Cunn. in Field, N. S. Wales,* 328. An erect shrub of 5 or 6 ft., the branches virgate, closely tomentose. Leaves mostly erect, rather crowded, sessile, linear-subulate or the larger ones linear-lanceolate, mucronate-acute, the margins much revolute, ½ to 1½ in. long, scabrous-pubescent or glabrous above, the under surface, when exposed, silky-pubescent. Racemes short, dense, sessile, terminal but often appearing lateral from the shortness of the flowering branches, the rhachis quite glabrous. Pedicels 1 to 2 lines long. Perianth glabrous outside, densely bearded inside below the middle, the tube about 3½ lines long, broad and gibbous at the base, much revolute and attenuate under the very oblique limb. Torus nearly straight. Gland broad, thick, semicircular. Ovary sessile, glabrous as well as the style except a tuft of hairs at the base on the upper side ; stigmatic disk lateral. Fruit rather narrow, incurved, 6 or 7 lines long, not distinctly ribbed.—R. Br. Prot. Nov. 20 ; Meissn. in DC. Prod. xiv. 363 ; Sweet, Fl. Austral. t. 30 ; Lodd. Bot. Cab. t. 1479 ; *G. riparia,* Sieb. in Roem. and Schult. Syst. iii. Mant. 278.

N. S. Wales, *Sieber, n.* 33 ; Cox's river, *A. Cunningham, Fraser ;* Sidmouth valley, *Woolls.*

The dilatation of the torus or summit of the pedicel in a ring outside the perianth, supposed to be characteristic of this species, is variable in degree both in *G. rosmarini-folia* and in *G. divaricata,* and is not always absolutely wanting in *G. ericifolia.*

SECT. 3. PLAGIOPODA.—Racemes various, erect and secund in the first few species, short and few-flowered in the others. Perianth-tube dilated below the middle and usually opening on the lower side, revo-

lute under the limb. Torus very oblique, the gland side the shortest.
Ovary villous except in a few of the axillary-flowered species. Stigmatic disk very oblique or lateral.

The oblique torus, which is the chief character of this and the following section, is
perhaps a somewhat artificial one, but is usually well marked. The first five of the
following species differ from the hebegynous *Eugrevilleæ* chiefly in the torus, the remainder of the present section correspond rather with *Ptychocarpa.*

46. **G. Goodii,** *R. Br. in Trans. Linn. Soc.* x. 174, *Prod.* 379. Stems
prostrate, diffuse (or sometimes erect ?), the young branches minutely
tomentose. Leaves petiolate, oval-elliptical to oblong-lanceolate, obtuse, rounded or cuneate at the base, 3 to 8 in. long, glabrous or the
under surface minutely silky, of a pale colour, penniveined with numerous primary veins uniting in an intramarginal nerve and minor
reticulations conspicuous on both sides. Racemes terminal, solitary or
2 or 3 together, pedunculate, secund, the rhachis 1½ to 2 in. long.
Pedicels 2 to 3 lines long, minutely tomentose as well as the rhachis.
Perianth 6 to 8 lines long, nearly glabrous outside, bearded inside with
reflexed hairs, the tube obliquely dilated at the base, attenuate above
the middle and revolute under the obliquely globular depressed limb.
Torus very oblique, linear, about 3 lines long. Gland horseshoe-
shaped, slightly prominent. Ovary villous with long hairs, stipitate on
the upper margin of the torus; style very long, more or less ciliate;
stigmatic disk broad, lateral.—Meissn. in DC. Prod. xiv. 351; Guillem.
Ic. Pl. Austral. t. 16.

N. Australia. North Coast, *R. Brown;* Port Essington, *Armstrong;* Point
Pearce and Newcastle Range, *F. Mueller.*

Queensland. Sandstone country, head of Cape and Flinders rivers, *Bowman.*

The foliage bears some resemblance to that of *G. laurifolia,* but the flowers are totally
different.

47. **G. venusta,** *R. Br. in Trans. Linn. Soc.* x. 175, *Prod.* 379. A
tall shrub or small tree, the young branches and inflorescence ferruginous-tomentose. Leaves simple or deeply pinnatifid, with 3 to 7 long
narrow-lanceolate lobes, the whole leaf 4 to 8 in. long, glabrous and
penniveined with numerous oblique parallel veins as in *G. refracta,* but
not so close and often confluent in an intramarginal nerve, minutely
silky-tomentose underneath with ferruginous veins. Racemes terminal,
rather loose but short like those of *G. Wilsoni.* Perianth sprinkled
with appressed hairs, the tube about 5 lines long, very obliquely dilated
at the base, attenuate and revolute above the middle. Torus very
oblique, narrow, 2 lines long. Gland horseshoe-shaped. Ovary densely
villous, stipitate on the upper margin of the torus; style long, nearly
glabrous; stigmatic disk lateral.—Meissn. in DC. Prod. xiv. 351;
Guillem. Ic. Pl. Austral. t. 11.

Queensland. Shoalwater Bay, *R. Brown.* I have not seen this species anywhere
except in R. Brown's Herbarium.

48. **G. longistyla,** *Hook. in Mitch. Trop. Austr.* 343. An erect
shrub of 7 or 8 ft., the young branches minutely tomentose. Leaves

linear, 6 to 10 in. long, from very narrow to above 2 lines broad, entire
or deeply divided into 3 to 5 segments, glabrous above, the margins
recurved, silky-pubescent and silvery or fulvous underneath, with a
prominent midrib. Racemes erect, shortly pedunculate, terminal or
in the upper axils, rather loose, secund, 1 to 2 in. long. Pedicels 2 to
4 in. long, tomentose-pubescent and apparently viscid as well as the
rhachis. Torus very oblique. Gland large, disk-shaped or almost
horseshoe-shaped. Ovary densely but shortly villous, on a short
stipes at the upper end of the torus; style very long, glabrous,
thickened under the broad lateral stigmatic disk. Fruit hard, semi-
globular, about ½ in. long. Seed scarcely winged.—Meissn. in DC.
Prod. xiv. 351 ; *G. neglecta,* R. Br. App. Sturt Exped. 24 ; Meissn. l.c.

Queensland. Sandstone ranges near Mount Pluto and the Pyramids, *Mitchell ;*
Burnett ranges, *F. Mueller ;* Boyd's river, *Leichhardt ;* Flinders river, *Sutherland.*

The pinnate-leaved specimens on which *G. neglecta* was founded, cannot be distinguished as a variety from the simple-leaved ones, as both forms occur frequently on the
same branch.

49. **G. juncifolia,** *Hook. in Mitch. Trop. Austr.* 341. A tall erect
shrub, the branches softly tomentose. Leaves very narrow-linear, 6 to
10 in. long, entire or here and there divided into two or three similar
segments, rigid and rather thick, doubly grooved underneath and sometimes obscurely so above, glabrous or minutely pubescent. Racemes
rather loose, secund, 3 to 6 in. long, usually several together in a terminal leafy panicle. Pedicels about ½ in. long, viscid-pubescent as
well as the rhachis. Perianth yellow, slightly pubescent and apparently viscid outside, glabrous inside, the tube broad at the base, 4 or 5
lines long, much attenuate and revolute under the obliquely globular
limb, the lobes with a horn-like dorsal appendage. Torus oblique.
Gland broad, semicircular. Ovary villous, almost sessile on the upper
margin of the torus; style very long, slightly thickened upwards;
stigmatic disk lateral, convex or umbonate. Fruit very oblique, almost
transverse, nearly 1 in. long. Seed broadly winged all round.—Meissn.
in DC. Prod. xiv. 351 ; *G. Sturtii,* R. Br. App. Sturt Exped. 23 ;
Meissn. l. c. 383, from the character given.

Queensland. Near Mount Pluto, *Mitchell.*
S. Australia. Near Central Mount Stuart, *M'Douall Stuart ;* scrub near Forster's
range, *Herb. F. Mueller* (collector not named).

This and the preceding species approach in habit *G. chrysodendron* and *G. Banksii,*
but are at once distinguished by the oblique torus.

50. **G. Wilsoni,** *A. Cunn. in Wils. Voy.* 273. An erect shrub of 3
to 5 ft., glabrous and somewhat glaucous. Leaves twice or three times
or rarely only once ternately divided into narrow-linear or subulate
rigid divaricate pungent-pointed segments ½ to 1 in. long, doubly
grooved underneath, the common petiole usually very short. Racemes
loose, erect, often branched, the rhachis ½ to 1 in., the pedicels about
¾ in. long, all glabrous. Perianth glabrous outside, very shortly
bearded inside below the middle, the tube about 6 or 7 lines long,

slightly dilated and gibbous at the base, attenuate and revolute under the oblique depressed-globular limb. Torus very oblique. Gland large and horseshoe-shaped but scarcely prominent. Ovary very villous, on a short stipes at the upper end of the torus; style very long, glabrous or villous in the lower portion, scarcely thickened under the lateral stigmatic disk.—Meissn. in DC. Prod. xiv. 373; *G. Lindleyana*, Meissn. in Pl. Preiss. i. 542.

W. Australia. Swan river, *Wilson, Drummond, 1st coll. n.* 631, *Preiss, n.* 692, *Harvey ;* Canning river, *Oldfield.*

51. **G. lavandulacea,** *Schlecht. Linnæa,* xx. 586. A low shrub, the branches more or less tomentose or silky-pubescent. Leaves sessile, entire, oblong-linear or lanceolate, mucronate-acute, the margins recurved or closely revolute, pubescent or scabrous above, silky-tomentose underneath with the midrib scarcely prominent, ½ to near 1 in. long. Racemes terminal, sessile, very short and almost umbel-like. Pedicels 1 to 2 lines long, tomentose as well as the rhachis. Perianth red, more or less silky-pubescent outside, at least on the limb, the shorter segments bearded inside about the middle with reflexed hairs, the tube fully 4 lines long, rather broad, gibbous at the base, revolute and attenuate under the very oblique limb. Torus very oblique. Gland broad, thick, semicircular. Ovary villous-tomentose, shortly stipitate on the upper margin of the torus; style long, glabrous or hirsute in the lower portion, thickened under the very oblique stigmatic disk.—Meissn. in DC. Prod. xiv. 362; Lem. Illustr. Hortic. t. 61; *G. rosea,* Lindl. in Paxt. Fl. Gard. ii. 91, t. 56; Meissn. l.c.; *G. ramulosa,* F. Muell.; Meissn. l.c. 362 (with very narrow leaves).

Victoria. N.W. districts of the Colony, *F. Mueller, Dallachy ;* Glenelg river, *Robertson, F. Mueller.*

S. Australia, *Behr ;* St. Vincent's Gulf, *Blandowski ;* Encounter Bay, *Whittaker;* Lofty ranges, Lake Torrens, *F. Mueller.*

Var. *sericea.* Leaves very narrow, closely revolute, silky-hairy on both sides.— Mount Barker, *F. Mueller;* Wimmera, *Dallachy.*

The species is sometimes confounded with *G. ericifolia,* from which it may be at once distinguished by the oblique torus and pubescent perianth. The honey exuded from the hypogynous gland is in this species and a few others copiously secreted also from a foveola at the base of the ovary.

52. **G. insignis,** *Kipp.; Meissn. in DC. Prod.* xiv. 379. A rigid shrub, glabrous and glaucous in all its parts except the ovary. Leaves petiolate, broadly ovate, undulate sinuate and prickly-toothed, truncate or scarcely cuneate at the base, 1½ to 2½ in. long, rigidly coriaceous and veined on both sides. Racemes short and loose, solitary in the upper axils but often crowded into a short terminal panicle with a few small floral leaves, the rhachis of each raceme rarely above ½ in. long, the pedicels 3 to 4 lines. Perianth glabrous outside, very shortly bearded inside, the tube 4 or 5 lines long, broad below the middle, slightly contracted and revolute under the very oblique limb. Torus oblique and concave lined by the scarcely prominent gland. Ovary villous on a short stipes at the upper margin of the torus. Style not

very long, slightly villous, clavate at the end with an obovate lateral disk or scarcely prominent flat or concave face, round the small lateral stigma.

W. Australia, *Drummond, 5th coll. suppl. n.* 12.

53. **G. Brownii,** *Meissn. in Pl. Preiss.* i. 537, *and in DC. Prod.* xiv. 370. A prostrate diffuse or spreading shrub of 2 or 3 ft., the branches slightly pubescent. Leaves sessile, ovate or lanceolate, ½ to ¾ in. long, passing into narrower leaves twice as long, mostly with a small callous point, the margins recurved, glabrous and smooth or scabrous-punctate above, silky-tomentose or white underneath. Racemes umbel-like, few-flowered sessile and terminal. Perianth pubescent outside, bearded inside with a transverse line of reflexed hairs, the tube broad and almost saccate at the base, about 3 lines long below the curve, slightly contracted and revolute under the limb. Torus very oblique. Gland very prominent, almost horizontal. Ovary shortly stipitate on the upper margin of the torus, densely villous; style villous, not very long, with a thick oblique or lateral stigmatic disk. Fruit 5 or 6 lines long.

W. Australia. King George's Sound or adjoining districts, *Preiss, n.* 719, *Drummond, n.* 22; Kalgan river, *Oldfield* (with narrow leaves); Mount Barker, Upper Hay river and Perongerup, *F. Mueller;* Perongerup and Phillips ranges, *Maxwell.*

G. depauperata, R. Br. Prot. Nov. 21; Meissn. in DC. Prod. xiv. 370, from King George's Sound, *Baxter,* appears to me to be a variety or state of this species with the racemes reduced to 1 or 2 flowers, and *G. Brownii* itself (of which I have not seen Preiss's typical specimen) may be a variety only of the common *G. fasciculata.*

54. **G. fasciculata,** *R. Br. Prot. Nov.* 20. A shrub, low and prostrate in the typical form but sometimes erect bushy and attaining 3 or 4 ft., the young shoots slightly pubescent with appressed hairs. Leaves sessile or very shortly petiolate, linear-lanceolate or the lower ones oblong-elliptical, obtuse or with a callous point, the margins revolute, ½ to 1 in. long or rather more when narrow, scabrous-punctate above, the under surface silky-tomentose but usually concealed. Racemes umbel-like, few-flowered, sessile, axillary or terminal. Pedicels 1 to 2 lines long. Perianth red, clothed or sprinkled with appressed hairs outside, bearded inside with a transverse line of reflexed hairs, the tube broad, 3 to 4 lines long, saccate at the base, contracted and revolute under the globular limb. Torus very oblique. Gland broad, truncate. Ovary villous, shortly stipitate on the upper margin of the torus; style not very long, villous with short appressed hairs; stigmatic disk very oblique or lateral, broad and thick, often concave. Fruit about 5 lines long.—Meissn. in DC. Prod. xiv. 369; *G. Meissneriana,* F. Muell. in Linnæa xxvi. 357, Meissn. l.c. 360; *G. aspera* var. *linearis,* Meissn. in Pl. Preiss. i. 537.

W. Australia. King George's Sound and adjoining districts, *Baxter, Drummond, Preiss, n.* 712 *and* 718, *Harvey,* and others; eastward to E. Mount Barren, *Maxwell.*

The species varies much in the size of the flowers, the breadth of the leaves and even in the degree of obliquity of the torus, but I have been unable among the numerous specimens seen to mark out any distinct narrow-leaved small-flowered variety as represented by Preiss's, n. 712 (*G. Meissneriana*).

55. **G. aspera,** *R. Br. in Trans. Linn. Soc.* x. 172, *Prod.* 377. A shrub of 5 or 6 ft., "with pendulous branches," densely tomentose. Leaves sessile or very shortly petiolate, linear or lanceolate, obtuse or acute but always mucronate, with revolute margins, 1 to 2 in. long, scabrous-pubescent and very obliquely penniveined above, silky-tomentose underneath with the midrib alone prominent. Racemes loose but short, shortly pedunculate, solitary or 2 or 3 together at the ends of the branches and usually reflexed. Pedicels 1 to 2 lines long, tomentose-pubescent as well as the rhachis. Perianth silky or villous outside, bearded inside above the middle, the tube about 3 lines long, rather broad and very oblique at the base, shortly contracted and incurved under the very oblique depressed-globular limb. Torus very oblique. Gland horseshoe-shaped. Ovary glabrous, stipitate at the upper margin of the torus; style short, thick, dilated under the large concave lateral stigmatic disk.—Meissn. in DC. Prod. xiv. 360.

S. Australia. Port Lincoln, *R. Brown, Trevor;* Gawler Ranges, *Sullivan.*
W. Australia ? Some of Baxter's and other specimens from the coast to the East of King George's Sound, may belong to this species but are not in flower.

Notwithstanding the glabrous ovary this species appears too nearly allied to the two preceding and to some of the following ones to be removed from the section.

56. **G. brachystylis,** *Meissn. in Pl. Preiss.* i. 538, ii. 252, *and in DC. Prod.* xiv. 350. A loosely branched shrub of 1 or 2 ft., the young shoots silky or ferruginous with short hairs. Leaves linear or lanceolate, shortly contracted at the base, the margins recurved or revolute, 2 to 4 in. long, glabrous above with a prominent midrib, ferruginous-tomentose or silky underneath. Racemes umbel-like, few-flowered, sessile, terminal or axillary. Pedicels 1 to 2 lines long. Perianth ferruginous-villous outside, bearded inside with erect (not reflexed hairs) the tube very oblique and adnate at the base, erect for about 2 lines then revolute, the limb of the lower (longer) segments very broad almost constricted into 2 concave lobes the inner one containing the anther, the outer larger one empty, the limb of the smaller upper segments normal. Torus very oblique. Gland disk-like but small. Ovary densely villous, shortly stipitate on the upper margin of the torus; style rather short, villous; stigmatic disk broad, produced into a broadly oblong appendage pubescent on the back, the stigma in the centre of the broader lower part. Fruit about ½ in. long.

W. Australia, *Drummond, 2nd coll. n.* 322 ; Sussex district, *Preiss, n.* 717.

57. **G. saccata,** *Benth.* Apparently procumbent or very spreading, the branches and young leaves softly villous, almost woolly, the older foliage rarely glabrous. Leaves sessile, linear or lanceolate, with a callous point, the margins revolute, ¾ to 1½ or sometimes 2 in. long, smooth or sparingly scabrous-punctate when the hairs wear off.

Racemes umbel-like, few-flowered, terminal or in the upper axils. Pedicels 2 or 3 lines long. Perianth more or less pubescent outside, the tube very oblique, nearly 4 lines long and about as broad, the upper side dilated saccate and separated from the remainder by two longitudinal densely hairy ribs inside, the whole tube contracted and revolute at the top with a ring of reflexed hairs inside, the lower larger laminæ of the limb dilated on the outer side but not constricted as in *G. brachystylis.* Torus very oblique. Gland broad, disk-like but scarcely prominent. Ovary villous, nearly sessile near the upper margin of the torus; style pubescent; stigmatic disk lateral.

W. Australia, *Drummond.* The habit is nearly that of *G. Drummondii,* but the perianth is very different.

58. **G. Drummondii,** *Meissn. in Pl. Preiss.* i. 536, ii. 252, *and in D C. Prod.* xiv. 350. Stems apparently diffuse or procumbent, the branches tomentose and hirsute with long fine spreading hairs. Leaves sessile, rather crowded, oblong lanceolate or linear, obtuse or mucronate, the margins recurved, ¾ to 1 or rarely 1½ in. long, sprinkled and ciliate with long fine hairs when young, scabrous-punctate above when the hairs wear off or nearly smooth, pale and glabrous or sometimes silky-tomentose underneath. Racemes umbel-like, sessile, terminal or on very short axillary tufts, the flowers smaller than in the allied species. Pedicels rarely above 1 line long. Perianth glabrous or hirsute with fine hairs, bearded inside near the top with reflexed hairs, the tube rather above 2 lines long, not saccate, contracted and recurved under the oblique limb. Torus oblique. Ovary villous, on a rather long stipes on the upper margin of the torus; style short, with a large lateral stigmatic disk.

W. Australia, *Drummond, 3rd coll. n.* 327, *4th coll. n.* 335; near Mandurah, *Clarke.*

59. **G. disjuncta,** *F. Muell. Fragm.* vi. 206. An erect shrub of about 2 ft. the young branches hoary or silky with appressed hairs, the foliage glabrous. Leaves rather crowded, erect, linear-terete, scarcely mucronate, with several prominent longitudinal ribs or angles and singly grooved, ½ to ¾ in. long. Pedicels axillary, solitary or in pairs, nearly glabrous, 2 to 3 lines long. Perianth pubescent outside with appressed hairs, densely bearded inside with reflexed hairs, the tube 3 to 4 lines long, broad and slightly gibbous on the upper side at the base, the gibbosity glabrous inside, contracted and revolute under the obliquely globular limb. Torus oblique. Gland broad, semiannular, scarcely prominent. Ovary densely villous, nearly sessile on the upper margin of the torus. Style very long, nearly glabrous; stigmatic disk lateral.

W. Australia. Salt river and rocky ranges east of Stirling river, *Maxwell.* An imperfect specimen from *Drummond* in Herb. F. Mueller may also belong to this species.

60. **G. haplantha,** *F. Muell. Herb.* Branches tomentose, the foliage minutely pubescent or at length glabrous. Leaves sessile, narrow-

linear, thick and rigid, mucronate, ¾ to 1½ in. long, smooth and nerveless on the convex upper side, broadly channelled underneath. Pedicels
axillary or lateral, in small sessile clusters sometimes reduced to a
single one, 1 to 3 lines long, woolly-pubescent. Perianth pubescent
outside, bearded inside to below the middle with reflexed hairs, the
tube about 4 lines long, rather broad and slightly gibbous at the base,
attenuate and revolute under the globular limb. Torus very oblique.
Gland broad, semiannular or almost disk-shaped, slightly prominent.
Ovary densely villous, nearly sessile on the upper margin of the torus ;
style long, pubescent or villous ; stigmatic disk orbicular, lateral.

W. Australia, *Drummond* (a single specimen in Herb. F. Mueller); East Mount
Barren, *Maxwell.*

61. **G. pinifolia,** *Meissn. in DC. Prod.* xiv. 350. Erect and shrubby,
the young shoots silky-tomentose. Leaves linear-terete, grooved underneath, slender but rigid, erect and rather crowded, minutely pointed,
glabrous and smooth when full-grown, 1 to nearly 2 in. long. Racemes
umbel-like, few-flowered, axillary and sessile. Pedicels scarcely 1 line
long. Perianth ferruginous-villous outside, very sparingly hairy inside,
the tube rather broad, scarcely 2 lines long, contracted and recurved at
the top only. Torus oblique. Gland truncate or emarginate. Ovary
densely villous, very shortly stipitate at the upper margin of the torus ;
style short, nearly glabrous ; stigmatic disk lateral, broad and thick.
Fruit small.

W. Australia, *Drummond* (*2nd coll.?*) n. 161, *4th coll.* n. 281.

62. **G. acuaria,** *F. Muell. Herb.* Branches divaricate, sparingly
pubescent as well as the foliage with minute appressed hairs or nearly
glabrous. Leaves divaricate, linear-terete, slender but rigid, pungent-
pointed, slightly grooved, ½ to nearly 1 in. long. Pedicels solitary or
clustered few together, axillary or terminal, 2 to 3 lines long. Perianth
glabrous outside, bearded inside to below the middle, with short reflexed hairs, the tube nearly 4 lines long, rather narrow but open on
the lower side only, attenuate and recurved under the globular limb.
Torus very oblique. Gland broad and flat, slightly prominent. Ovary
glabrous, stipitate on the upper margin of the torus. Style long, stigmatic disk lateral, orbicular.

W. Australia, *Drummond.* This is referred to by F. Mueller, Fragm. vi. 207,
as a variety of *G. sparsiflora,* which it much resembles in inflorescence and flowers ; but,
besides the foliage, it differs in the very oblique torus. It is only known from a small
specimen which is however abundantly in flower.

63. **G. singuliflora,** *F. Muell. Fragm.* vi. 92. A densely branched
glabrous shrub, probably small. Leaves sessile or very shortly petiolate,
broadly ovate or orbicular, very obtuse, flat or undulate, with a nerve-
like margin, 4 to 6 lines long, faintly penniveined on both sides.
Pedicels in pairs in the upper axils, filiform, glabrous, 3 to 4 lines
long. Perianth glabrous outside, slightly pubescent inside about the
middle, the tube gibbous at the base on the upper side, about 4 lines

long, revolute under the globular limb. Torus very oblique, linear, about 2 lines long. Gland small and horseshoe-shaped at the lower end. Ovary glabrous on a long stipes at the upper end of the torus; style rather long; stigmatic disk lateral.

Queensland. Dogwood Creek, *Leichhardt.*

64. **G. pauciflora,** *R. Br. in Trans. Linn. Soc.* x. 171, *Prod.* 377. An erect bushy shrub, the branches silky or hoary-tomentose. Leaves linear or oblong, usually cuneate, very obtuse or with a small callous point, with nerve-like sometimes recurved margins tapering to the base and sometimes shortly petiolate, ¾ to 1½ in. long, glabrous above when full grown, silky underneath or at length nearly glabrous, the midrib alone prominent. Pedicels in pairs or in very short racemes of 2 or 3 pairs, axillary or terminal, 1 to 2 lines long. Perianth red, sprinkled with a few small hairs outside, densely bearded inside by a ring of reflexed hairs above the middle, the tube about 3 lines long, dilated below the middle but almost acute not gibbous at the base, attenuate above the middle and recurved under the globular limb. Torus oblique. Gland semiannular. Ovary glabrous on a short thick stipes; style short; stigmatic disk lateral. Fruit about ¾ in. long.—Meissn. in DC. Prod. xiv. 356; *G. oligantha,* F. Muell. Fragm. vi. 206.

S. Australia. Port Lincoln, *R. Brown;* Marble Ranges, *Wilhelmi;* Spencer's Gulf, *Warburton.*

W. Australia. Lucky Bay, *R. Brown;* S.W. end of Russel Range and Phillips river, *Maxwell.*

The specimens are many of them very unsatisfactory.

65. **G. Strangea,** *Benth.* A small erect glabrous shrub. Leaves rather crowded, erect, linear or linear-cuneate, obtuse or mucronate, contracted at the base, 1½ to 2½ in. long, thick and veinless. Flowers unknown. Fruits solitary on lateral recurved pedicels of 3 or 4 lines, stipitate on the torus, quite glabrous, fusiform, slightly compressed, thick and hard, 1 to 1½ in. long, opening along the upper suture as in all genuine Grevilleas. Seed probably like that of *G. cynanchicarpa,* but only known from a single separate membranous outer coating in Herb. F. Mueller, similar to that of *G. cynanchicarpa.—Strangea linearis,* Meissn. in Hook. Kew Journ. vii. 66, and in DC. Prod. xiv. 348.

Queensland. Wide Bay, *Bidwill;* swamps near Durval, *Leichhardt;* Sandy ridge, Cape Byron, *C. Moore.*

As far as the characters are derived from the fruit and foliage, the only parts known, this species appears to me to be even sectionally inseparable from *G. cynanchicarpa;* but both require further investigation from more perfect specimens.

66. **G. cynanchicarpa,** *Meissn. in Hook. Kew Journ.* vii. 75. A spreading but stout and rigid shrub of 3 or 4 ft., the young branches tomentose, the adult foliage glabrous. Leaves rather crowded, erect, linear, acute or mucronate, thick and rigid but flat or slightly concave, veinless, contracted at the base but scarcely petiolate, 2 to 4 in. long in some specimens, 3 to 6 in. in other. Pedicels solitary in the axils,

about 3 lines long. Perianth unknown. Torus very oblique. Gland prominent, thick, obliquely semiannular. Ovary tomentose-villous, stipitate on the upper margin of the torus, with 2 laterally attached ovules as in all other *Grevilleæ*; style short and thick; stigmatic disk large and lateral. Fruit 2 to 2½ in. long, rather narrow, hard, tapering at both ends, obtusely ribbed, opening along the upper suture. Seed flat, oblong, 1½ to 2 in. long, thin but not distinctly winged, slightly thickened about the hilum which is very near the base, the outer membranous coating separating from the inner and opening in two valves. Embryo of the shape of the seed and equally distant from both ends, the radicle exceedingly short at the lower end.—*Molloya cynanchicarpa*, Meissn. in DC. Prod. xiv. 348.

W. Australia. Moore river and sand plain north of Diamond river, *Drummond, 6th coll. n.* 190; Cockleshell gully, Murchison river, *Oldfield.*

This has been proposed as a distinct genus on the supposition that the seed was winged at the upper end like that of a *Hakea*, and that the ovule was solitary. That proves however to be a mistake. The ovary, fruit, and seed are precisely those of several true *Grevilleæ*, except as to what appears to be the outer coating of the seed, which in this species and in *G. Strangea* separates itself from the inner in a manner not observed in other species, but which, in the absence of any other character, can scarcely justify the generic separation of these plants so long as the seeds of so many allied species remain unknown.

Sect. 4. Calothyrsus.—Racemes secund, usually terminal and many-flowered. Perianth-tube more or less dilated below the middle and usually opening on the lower side, revolute under the limb. Torus very oblique, the gland-side the shortest. Ovary glabrous, stipitate. Species all tropical, except *G. Huegelii.*

This section comprises all the species with a very oblique torus and glabrous ovary excepting a very few with almost solitary axillary flowers, which I thought better placed in *Plagiopoda* with others of the same exceptional inflorescence.

67. **G. quercifolia,** *R. Br. Prot. Nov.* 23. An undershrub or shrub of 1 to 3 ft., glabrous and glaucous like *G. Synapheæ* but the branches terete or nearly so. Leaves ovate or oblong, sinuate-pinnatifid with short broad mucronate or pungent-pointed lobes, cuneate or rarely truncate at the base and very shortly petiolate, mostly 3 to 4 in. sometimes 5 in. long, or those of long lateral branches much smaller, all prominently veined and often undulate. Racemes dense, ½ to 1 in. long, pedunculate in the upper axils or 3 or 4 on a long terminal leafless peduncle. Bracts broad, very deciduous or rarely more persistent. Pedicels 1 to 2 lines long. Perianth purple, glabrous outside, pubescent inside about the middle, very much revolute in the bud, slightly dilated at the base, abruptly bent down above the ovary when expanded, the limb very obliquely ovoid. Torus very oblique. Gland obsolete. Ovary glabrous, on a long stipes on the upper margin; style flattened; stigmatic disk large and lateral.—Meissn. in Pl. Preiss. i. 551, and in DC. Prod. xiv. 390; *G. brachyantha*, Lindl. Swan Riv. App. 31.

W. Australia. Swan river, *Fraser, Drummond, 1st coll. n.* 619, *Preiss, n.* 693;
Vasse river, *Oldfield;* Cape Leeuwin, *Lay and Collie* (with long loose racemes).

Var. *angustifolia.* Leaves lanceolate, pinnatifid, 3 to 6 in. long.—W. Australia,
Drummond; Mount Barker, *Oldfield, F. Mueller;* Donelly and Blackwood rivers,
T. C. Carey.

68. **G. angulata,** *R. Br. Prot. Nov.* 24. A shrub of 3 to 6 ft.,
quite glabrous and usually glaucous, or the young shoots scarcely
pubescent. Leaves petiolate, ovate or oblong, cuneate at the base,
more or less undulate and prickly-toothed, 1 to 2 in. long or rather
more when narrow, reticulate on both sides. Racemes dense, secund,
axillary and terminal, on peduncles of ½ to 1 in., the rhachis 6 to 8
lines long. Pedicels slender, 1½ to 2 lines long. Perianth glabrous
or pubescent with short appressed hairs outside, densely bearded inside
with erect hairs, the tube nearly 4 lines long, not much dilated at the
base, attenuate and much revolute under the globular limb. Torus
oblique. Gland prominent, horseshoe-shaped. Ovary glabrous, stipitate on the upper margin of the torus; style rather long; stigmatic
disk ovate, lateral. Fruit very obtuse with the base of the style
lateral, 4 to 6 lines long.—Meissn. in DC. Prod. xiv. 380; *G. ilicifolia,*
A. Cunn. Herb. not of R. Br.

N. Australia. Sims's Island, *A. Cunningham;* Victoria river, *Bynoe, F. Mueller;*
Fitzmaurice river, *F. Mueller.*

Var.? *lancifolia,* F. Muell. Leaves oblong-lanceolate, 2 to 3 in. long.—Stony ranges,
Central Australia, *Herb. F. Mueller* (collector not named).

69. **G. Wickhami,** *Meissn. in DC. Prod.* xiv. 380. A shrub of 4 to
6 ft. or a small tree, the young branches and foliage minutely silky-
pubescent, the older leaves nearly glabrous. Leaves petiolate, ovate,
angular or sinuate, with prickly-pointed angles or teeth, cuneate at the
base, 1 to 1½ in. long, reticulate as in *G. angulata,* or thicker with the
reticulations less conspicuous. Racemes secund, ¾ to 2 in. long, on
short axillary peduncles or terminating short branches. Pedicels 1 to
2 lines long, glabrous as well as the rhachis. Perianth glabrous outside, bearded inside about the middle, the tube scarcely 3 lines long,
broad and very oblique at the base, contracted and much revolute
under the globular limb. Torus oblique. Gland horseshoe-shaped,
large but not very prominent. Ovary glabrous, stipitate on the upper
margin of the torus; style glabrous or minutely papillose-pubescent,
with a large lateral stigmatic disk. Fruit very obtuse, 4 to 5 lines
long.

N. Australia. Usborne's Harbour, N.W. coast, *Wickham;* Roebuck Bay, *Martin;*
King's Sound and Collier Bay, *Chapman;* Port Darwin, *Schultz;* Gulf of Carpentaria,
F. Mueller.

70. **G. agrifolia,** *A. Cunn. in R. Br. Prot. Nov.* 24. A shrub of
strong growth, the young shoots minutely silky-tomentose, often
ferruginous and the old foliage scarcely glabrous. Leaves petiolate,
obovate-cuneate, more or less undulate and angular or sinuate with
prickly-pointed angles or teeth, tapering from the middle downwards,

1½ to 3 in. long, more obliquely penniveined than in *G. Wickhami,* the reticulations prominent on both sides. Racemes rather dense, secund, 1 to 1½ in. long, on short peduncles, axillary or terminating short branches. Pedicels 1 to 2 lines long, glabrous as well as the rhachis. Perianth glabrous outside, densely bearded inside with reflexed hairs, the tube 3 or 4 lines long, scarcely dilated at the base, much revolute under the globular limb. Torus oblique. Gland large, horseshoe-shaped. Ovary glabrous, on a slender stipes, at the upper margin of the torus; style long, slender, glabrous; stigmatic disk lateral. Fruit obliquely globular, 7 or 8 lines diameter.—Meissn. in DC. Prod. xiv. 380.

N. Australia. Cape Pond, Sims's, Lacrosse, and Goulburn Islands, *A. Cunningham;* Nichol Bay, *F. Gregory's Expedition;* Gulf of Carpentaria, *F. Mueller;* in the interior, *M'Douall Stuart's Expedition.*

71. **G. Cunninghamii,** *R. Br. Prot. Nov.* 23. A glabrous and more or less glaucous shrub attaining 10 ft. Leaves sessile or nearly so, ovate, deeply cordate with broad stem-clasping auricles, undulate and prickly-toothed, 1 to 2 in. long and broad, prominently but finely reticulate on both sides. Racemes axillary, short and few-flowered, on slender peduncles of about ½ in., the pedicels 1 to 2 lines long, all quite glabrous. Perianth glabrous outside, sparingly bearded inside about the middle, the tube scarcely 3 lines long, broad and very oblique below the middle, contracted and much revolute under the globular limb. Torus oblique. Gland horseshoe-shaped, large but slightly prominent. Ovary glabrous, shortly stipitate near the upper margin of the torus; style not very long, broad and flattened, stigmatic disk oval-oblong, lateral. Fruit oblique, very obtuse, 4 to 5 lines long.—Meissn. in DC. Prod. xiv. 379; *G. carduifolia,* A. Cunn. Herb.

N. Australia. Montague Sound, N.W. coast, *A. Cunningham.*

72. **G. pungens,** *R. Br. in Trans. Linn. Soc.* x. 175, *Prod.* 379. An erect shrub of 2 to 5 ft., the branches and foliage silky-pubescent. Leaves sessile, deeply and regularly pinnatifid, with 11 to 21 lanceolate rigid pungent-pointed lobes, the lower ones often 2-fid or 3-fid, much veined above, more densely silky underneath with the midribs alone prominent, the whole leaf 1 to 2 in. or even longer, variable in the breadth and depth of the lobes. Racemes terminal, secund, solitary or 2 together, shortly pedunculate, 2 to 3 in. long, quite glabrous. Pedicels 1 to 2 lines long. Perianth glabrous inside and out, the tube about 4 lines long, somewhat dilated below the middle, attenuate upwards and revolute under the very obliquely globular limb. Torus oblique. Gland prominent, half cup-shaped, truncate or 2-lobed. Ovary glabrous on a short stipes; style not very long, slightly thickened under the lateral stigmatic disk.—Meissn. in DC. Prod. xiv. 372.

N. Australia. W. coast of the Gulf of Carpentaria, *R. Brown, Leichhardt;* Maria island, *Gulliver.*

73. **G. Huegellii,** *Meissn. in Pl. Preiss.* i. 543, *and in DC. Prod.* xiv. 372. An erect and spreading or procumbent rigid shrub attaining sometimes several ft., the young branches slightly tomentose. Leaves pinnate with 3 to 7 segments, all entire or the lower ones again divided, linear, rigid, pungent-pointed, thick but flat, glabrous and smooth above, doubly grooved underneath, and more or less silky in the grooves. Racemes very short and few-flowered, sessile, solitary or 2 or 3 together at the ends of the branches or at the old nodes, the rhachis closely tomentose. Pedicels 2 or 3 lines long. Perianth-tube rather narrow, nearly straight, 7 to 8 lines long, scarcely contracted under the slightly recurved limb. Torus very oblique and elongated. Gland adnate, scarcely prominent. Ovary glabrous, on a long stipes inserted near the upper margin of the torus; style not very long, shortly clavate under the oblique convex stigmatic disk.—*G. rigidissima,* F. Muell. Pl. Vict. ii. t. 71; Meissn. in Linnæa xxvi. 356, and in DC. Prod. xiv. 350.

N. S. Wales. Darling and Murray desert, *Mitchell, Dallachy, Mrs. Ford,* and others.
Victoria. N.W. district of the colony, *L. Morton.*
S. Australia. Murray river to St. Vincent's and Spencer's Gulfs, *F. Mueller;* Gawler river, *Weidenbach.*
W. Australia. Swan river, *Drummond,* 1st coll. n. 634; York district, *Preiss, n.* 691.

74. **G. dimidiata,** *F. Muell. Fragm.* iii. 146. A tree (?), quite glabrous with a glaucous foliage resembling in some respects *G. mimosoides,* but at once distinguished by the broad leaves and large differently shaped flowers. Leaves falcate, 6 to 10 in. long and 1 to 3 in. broad in the middle in the few specimens seen, tapering into a short petiole, flat, rather thick, with numerous almost longitudinal veins and reticulations, not very prominent but equally visible on both sides. Racemes rather loose, 2 to 4 in. long, pedunculate and collected several together in a loose terminal panicle. Pedicels 1 to 1½ lines long. Perianth quite glabrous as well as the whole inflorescence, the tube about 4 lines long, revolute under the globular limb. Torus very oblique, concave. Gland adnate, scarcely prominent. Ovary glabrous, on a long stipes on the upper part of the torus; style long, flattened; stigmatic disk oblique, shortly conical in the centre.

N. Australia. Careening Bay, *A. Cunningham* (leaves only); Victoria river, *F. Mueller;* Roper river, *M'Douall Stuart's Expedition.*

75. **G. heliosperma,** *R. Br. in Trans. Linn. Soc.* x. 176, *Prod.* 380. A small slender tree, the young shoots minutely silvery or fulvous-pubescent, the adult foliage glabrous and more or less glaucous. Leaves once or twice pinnate, the segments not very numerous, oblong-lanceolate, obtuse, flat, 3 to 4 in. long, tapering at the base and often petiolulate, triplinerved or penniveined with few almost longitudinal primary veins and numerous almost longitudinal reticulations conspicuous on both sides, the whole leaf 6 in. to 1 ft. long. Racemes very loose, secund, 2 to 4 in. long, terminal or lateral, often branched but shorter

than the leaves. Perianth glabrous outside as well as the whole inflorescence, bearded inside below the middle, the tube fully 4 lines long,
dilated towards the base, somewhat contracted upwards and much revolute under the obliquely globular limb. Torus very oblique and narrow, 3 lines long. Gland adnate, scarcely prominent. Ovary glabrous,
stipitate at the upper end of the torus; style very long; stigmatic disk
very oblique or lateral, very convex. Fruit nearly globular, oblique, 1
in. diameter, with very thick hard valves. Seed broadly winged all
round.—Meissn. in DC. Prod. xiv. 380.

N. Australia. N. coast, *R. Brown;* Port Raffles, *A. Cunningham;* Melville
island, *Fraser;* Point Pearce, M'Adam Range, Roper river, *F. Mueller;* Port Darwin,
Schultz; Caledon Bay and Liverpool river, *Gull.*

76. **G. refracta,** *R. Br. in Trans. Linn. Soc.* x. 176, *Prod.* 380. A
tall shrub or small tree, the young branches tomentose. Leaves mostly
pinnate with 3 to 11 segments 2 to 5 in. long, linear-lanceolate or the
terminal one broader, or sometimes reduced to a single oblong-cuneate
leaf, the segments acute or obtuse when broad, tapering at the base and
sometimes petiolulate, the margins usually recurved, nearly glabrous
above and penniveined with numerous very oblique and nearly parallel
primary veins, densely silky-pubescent underneath with the midrib alone
prominent. Racemes short, secund, nearly sessile, usually several together in a small sessile terminal panicle, the rhachis of each raceme
rarely above 1 in. long, the pedicels 2 to 3 lines, hoary-tomentose.
Perianth silky-pubescent outside, bearded inside about the middle, the
tube 4 to 5 lines long, dilated towards the base, contracted upwards and
much revolute under the obliquely globular limb. Torus very oblique.
Gland very prominent, half cup-shaped, 2-lobed. Ovary glabrous,
stipitate on the upper margin of the torus; style thick, not very long;
stigmatic disk lateral. Fruit very hard, nearly globular, about 1 in.
diameter. Seed broadly winged.—Meissn. in DC. Prod. xiv. 382.

N. Australia. Islands of the Gulf of Carpentaria, *R. Brown;* Cambridge Gulf,
N.W. coast, *A. Cunningham;* Cygnet Bay, *Wickham;* Victoria river, *Bynoe, F.
Mueller;* Sea Range and Fitzmaurice river, *F. Mueller;* Short's Range, Newcastle
water, Rilliart's springs, *M'Douall Stuart's Expedition.*

Var. *ceratophylla.* Leaf-segments often rather broader and several of the leaves undivided, the primary veins more conspicuous underneath.— *G. ceratophylla,* R. Br. in
Trans. Linn. Soc. x. 177, Prod. 380; Meissn. in DC. Prod. xiv. 382 ; *G. heterophylla,*
A. Cunn. in R. Br. Prot. Nov. 24 ; Meissn. l.c. 381.—Islands of the Gulf of Carpentaria, *R. Brown;* N.W. coast, *A. Cunningham.*

Var. *velutina,* Meissn. Segments of the leaves all broad and densely silky-ferruginous on both sides.— *G. velutina,* A. Cunn. Herb.—Greville island, Regent's river, *A.
Cunningham;* Glenelg river, *Martin.*

77. **G. Dryandri,** *R. Br. in Trans. Linn. Soc.* x. 175, *Prod.* 379. A
tall shrub, the branches minutely hoary-tomentose or quite glabrous.
Leaves pinnate, with numerous narrow-linear rather rigid mucronate
segments, the lower ones 2 to 4 in. long, the upper ones gradually
smaller, all with recurved or revolute margins, glabrous above and more
or less marked with very oblique or longitudinal veins, silky-pubescent

underneath with the midrib prominent. Racemes loose and secund, glabrous and glaucous, from 4 or 5 in. to nearly 1 ft. long, and often several on a long stout terminal peduncle. Pedicels 3 to 5 lines long. Perianth white, glabrous outside, slightly bearded inside, the tube 7 to 9 lines long, dilated at the base, attenuate from the middle and revolute under the obliquely globular limb, the laminæ of the longer segments bearing a longitudinal dorsal keel-like appendage. Torus oblique but not very much so. Gland prominent, semiannular, often 2 lobed. Ovary glabrous, on a long stipes; style very long; stigmatic disk lateral.— Meissn. in DC. Prod. xiv. 374; *G. rigens,* A. Cunn.; Meissn. l.c.; *G. callipteris,* Meissn. l.c. 375.

N. Australia. Islands of the Gulf of Carpentaria, *R. Brown, Henne;* Upper Victoria river, *F. Mueller;* Goulburn island, *A. Cunningham;* Port Essington, *Armstrong.*

Queensland. Cape Flinders, *A. Cunningham.*

78. **G. polystachya,** *R. Br. in Trans. Linn. Soc.* x. 177, *Prod.* 380. A tall shrub or small tree, attaining about 30 ft., the branches minutely silky-pubescent or hoary when young. Leaves linear or linear-lanceolate and undivided, or dilated upwards and irregularly divided into 2 to 6 long linear-lanceolate segments, the whole leaf 6 to 10 in. long, acuminate, tapering into a petiole, glabrous above, more or less silky-pubescent underneath, with the midrib and often longitudinal lateral veins prominent. Racemes rather dense, secund, 3 to 4 in. long, glabrous, usually several in a short terminal panicle. Pedicels 1 to 2 lines long. Perianth white, glabrous outside, shortly bearded inside, the tube about 4 lines long, not much dilated at the base, narrow and revolute under the globular limb. Torus oblique but not very much so. Gland semicircular. Ovary glabrous, stipitate; style rather long, shortly thickened under the very oblique stigmatic disk. Fruit woody, obliquely orbicular, ¾ to 1 in. diameter. Seeds broadly winged.—Meissn. in DC. Prod. xiv. 384; *G. parallela,* Knight, Prot. 121; *G. polybotrya,* F. Muell. in Hook. Kew Journ. ix. 23; Meissn. l.c. 698, not of Meissn. l.c. 386.

N. Australia. Macadam Range and S. Alligator river, *F. Mueller.*

Queensland. Shoalbay passage, *R. Brown;* Gilbert river, *F. Mueller;* Edgecombe and Rockingham Bays, *Dallachy;* Port Denison, *Fitzalan;* Broad Sound, Flinders and Bowen rivers, *Bowman;* Liverpool river, *Gulliver;* Port Mackay, *Nernst.*

Var. *hebestachya.* Racemes minutely hoary-pubescent. Flowers rather smaller.— Cape York, *Daemel;* Dayman's island, *W. Hill.* Some of the Rockingham Bay specimens are intermediate between these and the typical form.

G. angustata, R. Br. Prot. Nov. 24; Meissn. in DC. Prod. xiv. 384, described from specimens in leaf only from Cape Cleveland, *A. Cunningham,* is probably only a very narrow leaved form of *G. polystachya.*

79. **G. robusta,** *A. Cunn. in R. Br. Prot. Nov.* 24. A tree sometimes small and slender, sometimes robust and 80 to 100 ft. high, the young branches hoary or ferruginous-tomentose. Leaves pinnate with about 11 to 21 pinnatifid pinnæ, the secondary lobes or segments entire or again lobed, lanceolate or rarely linear, often above 1 in. long, the

margins recurved, glabrous above or sprinkled with appressed hairs and
obscurely veined; silky underneath, the whole leaf 6 to 8 in. long and
nearly as broad. Racemes secund, 3 or 4 in. long, solitary or several
together on very short leafless branches on the old wood. Pedicels
slender, about ½ in. long, glabrous as well as the rhachis. Perianth
glabrous outside and in, the tube nearly 3 lines long, scarcely dilated
at the base, revolute under the ovoid limb. Torus slightly oblique.
Gland prominent, semiannular. Ovary glabrous, stipitate; style long,
the stigmatic disk somewhat oblique with a central cone. Fruit broad,
very oblique, 8 or 9 lines long. Seed winged all round.—Meissn. in
DC. Prod. xiv. 381; Bot. Mag. t. 3184; *G. umbratica*, A. Cunn.;
Meissn. l.c.

Queensland. Brisbane river, Moreton Bay, *A. Cunningham, Fraser.*
N. S. Wales. From the Richmond to the Tweed rivers, *C. Moore;* Sydney
woods, Paris Exhibition, 1855, *Macarthur, n.* 159, *C. Moore, n.* 88; Clarence and
Richmond brushes, London Exhibition, 1862, *C. Moore, n.* 108.

Under cultivation the leaf is sometimes a foot long, almost tripinnate, with numerous
pinnæ and narrow acute segments.

SECT. 5. CYCLADENIA.—Racemes many-flowered (scarcely secund?)
several in a terminal panicle. Perianth-tube narrow, recurved or reflexed under the limb. Torus straight. Gland annular, surrounding
the stipes of the ovary. Ovary glabrous; style long with a lateral
stigmatic disk.

The inflorescence is nearly that of the tropical species of *Calothyrsus*, but the flowers
are more crowded and apparently not secund. The perianth is nearly that of *Cycloptera*, but longer, and the section is readily distinguished from all others by the regular
annular hypogynous gland.

80. **G. annulifera,** *F. Muell. Fragm.* iv. 85. A shrub of 6 to 8 ft.,
quite glabrous and more or less glaucous. Leaves pinnate; segments
5 or 7, narrow-linear, divaricate, rigid, pungent-pointed, ¾ to 1 in.
long, the margins revolute, smooth above, doubly grooved underneath.
Racemes loose, many-flowered, 2 to 4 in. long, shortly pedunculate
and usually several together in a terminal panicle. Pedicels 2 to 3
lines long, glabrous as well as the rhachis. Perianth white, glabrous
outside, slightly bearded inside near the base, the segments equally
separating from the base and scarcely dilated, much revolute under the
obliquely globular limb, about ½ in. long if unrolled. Torus straight.
Gland annular, rather thick but not very prominent. Ovary glabrous,
on a long stipes free in the centre of the gland; style very long, with
an oblique or lateral stigmatic disk.

W. Australia. Murchison river, *Oldfield.*

81. **G. leucopteris,** *Meissn. in Hook. Kew Journ.* vii. 76, *and in DC.
Prod.* xiv. 382. A shrub of 4 to 8 ft., the branches and petioles hoary
or ferruginous with a close but soft and dense tomentum. Leaves pinnate; segments numerous, narrow-linear, 4 to 10 in. long, of a pale
colour, convex and smooth above, more or less tomentose and doubly

grooved underneath, the whole leaf often above 1 ft. long. Racemes loose, many-flowered, 3 to 4 in. long, shortly pedunculate in a terminal panicle often raised upon branches of 3 or 4 ft. either leafless or with a few simple leaves or bracts, or rarely the panicle close upon the pinnate leaves ; the common peduncle tomentose up to the racemes, but the rhachis and pedicels glabrous. Bracts broad, membranous, villous, imbricate on the young racemes but very early deciduous. Pedicels rarely above 2 lines long. Perianth glabrous outside, densely bearded inside near the base, the segments equally separating and slightly dilated at the base, revolute under the globular limb, fully 5 lines long. Torus straight. Gland annular, thick but not very prominent. Ovary glabrous, on a long stipes in the centre of the gland ; style very long, slightly thickened under the lateral stigmatic disk. Fruit broad, very oblique, nearly 1 in. long.—*G. segmentosa,* F. Muell. Fragm. iii. 145, iv. 176.

W. Australia. Murchison river, *Oldfield, Drummond, 6th coll. n.* 188.

SECT. 6. CYCLOPTERA.—Flowers small in dense terminal racemes usually paniculate. Perianth-tube narrow, recurved or reflexed under the limb. Torus straight. Ovary glabrous, stipitate. Fruit usually broad. Seed winged all round.

This section has the flowers of *Lissostylis* with the general habit more of *Calothyrsus,* and all the species except *G. Hillii* have the stigma on an erect cone as in *Conogyne.*

82. **G. leucadendron,** *A. Cunn. in R. Br. Prot. Nov.* 25. A tall shrub or small tree, the foliage silky or sprinkled with small appressed hairs or rarely almost glabrous and usually glaucous. Leaves mostly pinnate, with 3 to 11 long linear flat segments, sometimes scarcely broader than thick, sometimes 2 or 3 lines broad, with about 3 longitudinal veins simple or anastomosing and prominent on both sides, the segments varying from under 6 in. to 1½ ft. in length, and occasionally again divided or rarely the whole leaf simple. Flowers very small, in dense erect racemes of 2 or 3 in., shortly pedunculate and usually several together in a terminal leafless panicle shorter than the leaves. Pedicels ½ to 1 line long. Perianth glabrous as well as the whole inflorescence, the tube narrow, about 2 lines long, reflexed under the globular limb. Torus small. Gland horseshoe-shaped, not very prominent. Ovary glabrous, on a long stipes ; style not very long, rather thick, the stigmatic cone nearly straight. Fruit broad, oblique, compressed, about ¾ to 1 in. long. Seed winged all round.—Meissn. in DC. Prod. xiv. 382 ; *G. obliqua,* R. Br. Prot. Nov. 25 ; Meissn. l.c. ; *G. longiloba,* F. Muell. Fragm. i. 136.

N. Australia. Cambridge Gulf and Enderby island, Dampier's Archipelago, *A. Cunningham ;* Sea range, Victoria river, and sources of the Roper, Wickham, and Alligator rivers, Gulf of Carpentaria, *F. Mueller.*

Cunningham's specimens have leaves and fruits only, those of *G. leucadendron* from Cambridge Gulf with narrower leaf-segments than those of *G. obliqua* from Enderby island ; F. Mueller's are in good flower and closely connect the two forms, the leaf-segments varying in breadth from 1 to 3 lines. After as careful a comparison as the specimens admit of I have no doubt but that all belong to one species.

83. **G. pyramidalis,** *A. Cunn. in R. Br. Prot. Nov.* 25. A tall shrub or small tree, the adult foliage glabrous and glaucous. Leaves once or twice pinnate, the segments not numerous, linear-cuneate or oblanceolate, obtuse, 3 to 5 in. long, tapering at the base and often petiolulate, flat, rather thick, longitudinally but irregularly veined on both sides. Flowers very small, in paniculate glabrous racemes like those of *G. leucadendron,* and agreeing precisely in structure and proportions with that species, of which *G. pyramidalis* may prove to be a short-leaved variety.—Meissn. in DC. Prod. xiv. 381.

N. Australia. Regent's river, N.W. coast, *A. Cunningham.*

84. **G. striata,** *R. Br. in Trans. Linn. Soc.* x. 177, *Prod.* 380. A small or large tree, the branches closely tomentose, the foliage minutely and sometimes sparingly silky-pubescent. Leaves undivided, linear or linear-lanceolate, 6 to 18 in. long, often curved, 2 to 5 lines broad, obscurely veined above, striate underneath, with 9 to 13 raised parallel nerves, separated by intervals much narrower than the nerves themselves. Flowers small, in slender spike-like erect racemes of 2 or 3 in., shortly pedunculate and usually several together in a leafless panicle shorter than the leaves, the rhachis tomentose. Pedicels scarcely 1 line long. Perianth silky-pubescent outside, glabrous inside, the tube about 2 lines long, narrow, revolute under the globular limb. Torus small. Gland semiannular, prominent. Ovary glabrous, on a slender stipes; style not very long, the stigmatic cone straight. Fruit broad, very oblique, compressed, about ¾ in. long.—Meissn. in DC. Prod. xiv. 385; *G. lineata,* R. Br. App. Sturt. Exped. 24; Meissn. l.c.

N. Australia. Victoria river, *F. Mueller;* islands of the Gulf of Carpentaria, *R. Brown.*

Queensland. Wide Bay, *Bidwill;* Port Denison, *Fitzalan;* Kennedy district, *Daintree;* Flinders and Dawson rivers, *Sutherland;* in the interior, *Mitchell.*

N. S. Wales. Darling desert, *Victorian Expedition;* Bogan river, *C. Stuart.*

S. Australia. Cooper's Creek, *Howitt's Expedition.*

85. **G. mimosoides,** *R. Br. in Trans. Linn. Soc.* x. 177, *Prod.* 380, *Prot. Nov.* 25. A tree quite glabrous, but the foliage glaucous. Leaves undivided, lanceolate, falcate, 6 to 10 in. long, varying from under ½ in. to above 1 in. in breadth, obtuse or with a callous point, tapering into a short petiole, flat, with several sometimes many longitudinal veins or nerves and very oblique almost longitudinal veinlets visible on both sides, but not very prominent. Flowers small, " pinkish white" in slender glabrous racemes of 3 or 4 in. shortly pedunculate and usually several in a terminal leafless panicle. Pedicels ½ to 1 in. long. Perianth glabrous, the tube narrow, about 2 lines long, revolute under the globular limb. Torus small. Disk semiannular, scarcely prominent or very obscure. Ovary glabrous, on a long stipes; style long, with a short nearly straight stigmatic cone or conical disk. Fruit broad, very oblique, somewhat compressed, fully 1 in. long. Seed-wing narrow, coriaceous.—Meissn. in DC. Prod. xiv. 385.

N. Australia. Careening Bay, N.W. coast, *A. Cunningham;* Victoria river,

M'Adam Range, Fitzmaurice river, *F. Mueller;* islands of the Gulf of Carpentaria, *R. Brown, Henne.*

The leaves closely resemble those of some of the tropical phyllodinous *Acaciæ;* when broad they are also very nearly those of *G. dimidiata,* which has however very different flowers.

86. **G. Hilliana,** *F. Muell. in Trans. Phil. Inst. Vict.* ii. 72. A large tree, the young branches minutely tomentose. Leaves petiolate, either entire obovate-oblong or elliptical, very obtuse, tapering at the base and 6 to 8 in. long, or still longer and deeply divided at the end into 2 or 3 diverging lobes, or deeply pinnatifid with 5 to 7 oblong or lanceolate lobes of several inches, the whole leaf then sometimes above 1 ft. long, glabrous above penniveined and reticulate with the primary veins confluent in an intramarginal nerve, more or less silvery-silky underneath. Flowers small and very numerous in dense cylindrical racemes of 4 to 8 in., on short axillary shoots accompanied often by 1 or 2 smaller racemes. Pedicels about 1 line long, minutely pubescent as well as the rhachis. Perianth minutely silky outside, glabrous or scarcely pubescent inside, the tube slender, about 3 lines long, revolute under the globular limb. Torus straight. Gland semiannular, not very prominent. Ovary glabrous, stipitate ; style long and slender, the stigmatic disk lateral. Fruit slightly compressed, nearly 1 in. long. Seed rather narrowly winged all round.

Queensland. Brisbane river, Moreton Bay, *W. Hill, F. Mueller;* Rockingham Bay, *Dallachy.*

N. S. Wales. From the Clarence to the Tweed river, *C. Moore.*

87. **G. gibbosa,** *R. Br. in Trans. Linn. Soc.* x. 177, *Prod.* 380. A small or large tree sometimes reduced to a tall shrub, the branches and foliage softly tomentose-pubescent with very short hairs silky on the young shoots and persisting on both sides of the adult leaves. Leaves entire, ovate ovate-lanceolate or oblong-elliptical, obtuse or almost acute, tapering into a short petiole, 3 to 4 or rarely 5 in. long, penniveined with rather numerous oblique primary veins confluent in an intramarginal nerve. Flowers small, in dense spike-like racemes of 3 to 6 in., shortly pedunculate and usually 3 together at the ends of the branches. Pedicels 1 to 1½ lines long, pubescent as well as the rhachis. Perianth sprinkled or clothed with appressed hairs outside, glabrous inside, the tube slender, about 2 lines long, revolute under the globular limb. Torus small. Gland very prominent, semi-cupular, truncate or 2-lobed. Ovary glabrous, shortly stipitate ; style long, filiform, the stigmatic cone straight or nearly so. Fruit obliquely globular, 1 to 1½ in. diameter, opening in 2 very hard thick hemispherical valves, enclosing 1 or 2 flat (broadly winged ?) seeds.— Meissn. in DC. Prod. xiv. 385 ; *G. glauca,* Knight, Prot. 121.

Queensland. Endeavour river, *Banks and Solander, W. Hill;* Cape York, *M'Gillivray, Daemel;* Albany island, *F. Mueller, W. Hill;* Suttor, Cape, and Burdekin rivers, *Leichhardt, F. Mueller, Bowman,* and others.

SECT. 7. ERIOSTYLIS. R. Br.—Racemes umbel-like, sessile or nearly so. Flowers small, villous. Perianth-tube revolute under the limb. Torus straight. Ovary shortly stipitate, villous as well as the style; stigmatic disk lateral. Leaves all entire.

The section is in many respects allied to the series *Hebegynæ* of *Eugrevillea*, but readily distinguished by the foliage from all the species of that series except *G. concinna*, which has a glabrous style.

88. **G. buxifolia,** *R. Br. in Trans. Linn. Soc.* x. 174, *Prod.* 379. A bushy shrub of 4 to 6 ft., the branches ferruginous-pubescent and villous. Leaves rather crowded, sessile, from ovate to oblong or almost lanceolate, obtuse or acute, the margins recurved, $\frac{1}{4}$ to $\frac{3}{4}$ in. but mostly about $\frac{1}{2}$ in. long, minutely scabrous veined and often shining above, ferruginous or silky-pubescent underneath. Racemes terminal, sessile, umbel-like. Pedicels 3 to 6 lines long, villous as well as the rhachis. Perianth villous outside, bearded inside with reflexed hairs, the tube nearly 3 lines long, rather broad, reflexed under the globular limb. Torus straight. Gland semiannular. Ovary almost sessile, densely villous; style long, villous, clavate at the end, with a broad thick lateral stigmatic disk, the villous back produced beyond the disk into a spreading or reflexed appendage.—Meissn. in DC. Prod. xiv. 369; Bot. Reg. t. 443; Lodd. Bot. Cab. t. 1562; *Embothrium buxifolium*, Sm. Spec. Bot. N. Holl. 29. t. 10; Andr. Bot. Rep. t. 218; *Embothrium genianthum*, Cav. Ic. iv. 60. t. 387; *Stylurus buxifolia* and *S. collina*, Knight, Prot. 115, 116.

N. S. Wales. Port Jackson, *R. Brown, Sieber, n.* 37, *Fl. Mixt. n.* 477, and many others.

89. **G. phylicoides,** *R. Br. in Trans. Linn. Soc.* x. 174, *Prod.* 379. A shrub of 3 or 4 ft., the branches ferruginous-villous. Leaves sessile or very shortly petiolate, linear-lanceolate or lanceolate, acute, mucronate, mostly $\frac{3}{4}$ to 1 in. long, the margins revolute, very scabrous above, pubescent or villous underneath. Racemes umbel-like, terminal, sessile or very shortly pedunculate. Pedicels 1 to 3 lines long, tomentose as well as the rhachis. Perianth densely villous outside, bearded inside with reflexed hairs, the tube about 2 lines long reflexed under the globular limb. Torus straight. Gland very prominent, semiannular, crenate. Ovary villous, shortly stipitate; stigmatic disk lateral, thick, oblong, the villous back produced beyond the disk into a rather long erect horn.—Meissn. in DC. Prod. xiv. 369.

N. S. Wales. Grose river, *R. Brown;* Blue Mountains, *Fraser, Sieber, n.* 29, and others; Clarence river, *Beckler.*

90. **G. sphacelata,** *R. Br. in Trans. Linn. Soc.* x. 174, *Prod.* 378. A shrub of several feet, resembling *G. phylicoides,* but the indumentum of the branches and underside of the leaves closely appressed. Leaves sessile or very shortly petiolate, linear-lanceolate, obtuse with a callous point, the margins revolute, $\frac{3}{4}$ to 1 in. long, scabrous-punctate on the upper side. Racemes umbel-like, terminal and usually sessile. Pedicels 1 to 2 lines long, pubescent as well as the rhachis. Perianth villous outside, bearded inside with reflexed hairs shorter than in *G.*

phylicoides, the tube scarcely 2 lines long, revolute under the ovoid limb. Torus straight. Gland very prominent, semiannular. Ovary villous, stipitate ; style villous ; stigmatic disk lateral, oval, thick, with a raised border, the villous back scarcely produced beyond the disk.—Meissn. in DC. Prod. xiv. 369.

N. S. Wales. Cook's river and George's river, *R. Brown;* Liverpool, *Leichhardt;* Illawarra, *A. Cunningham, Shepherd;* near Appin, *Backhouse;* Ashfield, *Herb. F. Mueller.*

91. **G. occidentalis,** *R. Br. in Trans. Linn. Soc.* x. 173, *Prod.* 378. A loosely branched shrub of 3 or 4 ft., the branches silky-tomentose with appressed hairs. Leaves sessile or nearly so, lanceolate or almost linear, with short callous points, the margins revolute, scabrous-punctate above, silky underneath, ½ to 1 in. long. Racemes umbel-like, terminal, shortly pedunculate or rarely sessile. Pedicels 2 to 4 lines long, shortly villous. Perianth villous outside, bearded inside in the upper half with reflexed hairs, the tube rather broad, fully 2 lines long, revolute under the globular limb. Torus straight. Gland semi-annular. Ovary villous, very shortly stipitate ; style villous; stigmatic disk lateral, broadly orbicular, the villous dorsal centre not reaching to the margin of the disk. Fruit ovoid-oblong, about ½ in. long.—Meissn. in Pl. Preiss. i. 539, ii. 252, and in DC. Prod. xiv. 370.

W. Australia. King George's Sound and adjoining districts, *R. Brown, Drummond, 2nd coll. n.* 270, *Preiss, n.* 713, and several others.

92. **G. acerosa,** *F. Muell. Fragm.* i. 136. A shrub of about 3 ft. the branches rather slender but rigid, the young shoots silky or ferruginous. Leaves sessile, linear-terete, grooved underneath, rigid and pungent-pointed, ½ to ¾ or rarely 1 in. long, glabrous and smooth when full grown. Racemes umbel-like, small, sessile, terminal or in axillary leafy tufts. Pedicels filiform, 1 to 2 lines long, villous. Perianth densely villous, the tube much revolute, the upper shorter segments scarcely above 1 line long, the limb globular. Torus straight. Gland prominent, semiannular. Ovary densely villous, very shortly stipitate ; style villous; stigmatic disk lateral, oval, flat or convex, the small stigma prominent in the centre. Fruit 4 or 5 lines long.

W. Australia, *Drummond, n.* 126; Salt river and Cape Knob, *Maxwell.*

93. **G. umbellulata,** *Meissn. in Pl. Preiss.* ii. 252, *and in DC. Prod.* xiv. 371. A shrub with rather slender virgate branches, the young shoots slightly silky-hairy, the adult foliage glabrous. Leaves narrow-linear or rarely linear-lanceolate, acute, with revolute margins, not scabrous, 1½ to 3 in. long, a few floral ones much smaller. Racemes umbel-like, numerous, sessile, axillary and terminal. Pedicels 1 to 2 lines long, densely villous as well as the rhachis. Perianth densely villous outside, bearded inside, the tube scarcely 1½ lines long, revolute under the globular limb. Torus straight. Gland very short, semiannular. Ovary villous, nearly sessile ; style villous and filiform

to the end; stigmatic disk lateral, orbicular, flat with a raised border, the stigma sessile or nearly so in the centre.

W. Australia, *Drummond, 2nd coll. n.* 324; Port Gregory, *Oldfield.*—Very near *G. oxystigma,* but the differences in the style appear to be constant.

94. **G. oxystigma,** *Meissn. in Pl. Preiss.* i. 540, *and in DC. Prod.* xiv. 370. A shrub of 1 to 4 ft., the branches silky-pubescent or glabrous. Leaves of the main stems sometimes lanceolate and 1 to 2 in. long, those of the flowering branches linear or linear-lanceolate and $\frac{1}{2}$ to 1 in. long, in other specimens all crowded appressed and under $\frac{1}{2}$ in., all with revolute margins, glabrous and smooth above, the concealed under surface often silky. Racemes umbel-like, numerous, terminal or on very short axillary branches. Pedicels 2 to 4 lines long, villous as well as the rhachis. Perianth villous, the tube about $1\frac{1}{2}$ lines long, revolute under the globular limb. Torus straight. Gland short, semiannular. Ovary villous, very shortly stipitate; style villous, filiform, the obliquely clavate glabrous end forming a thick lateral orbicular disk, with the stigma raised on a prominent point in the centre. Fruit 4 or 5 lines long.—*Hakea pilulifera,* Lindl. Swan Riv. App. 36.

W. Australia. Swan river to King George's Sound, *Drummond, 1st coll. n.* 629, *Preiss, n.* 710, 714, 715, 716; Gordon and Kalgan rivers, *Oldfield;* base of Stirling Range, *F. Mueller.*

Var.? *villosa.* Branches densely villous; adult leaves scabrous-punctate.—Between Swan river and King George's Sound, *Harvey;* near Belgarup, *Oldfield.*

95. **G. Candolleana,** *Meissn. in Pl. Preiss.* i. 541, *and in DC. Prod.* xiv. 371. Apparently a low shrub, with erect and virgate or ascending and loosely branched stems, the branches more or less tomentose. Leaves sessile or nearly so, oblong-lanceolate or almost linear, mucronate, with recurved or revolute margins, $\frac{3}{4}$ to $1\frac{1}{2}$ in. long or smaller on the side branches, glabrous and smooth above with a prominent midrib, white-tomentose underneath. Racemes umbel-like, sessile, terminal and in the upper axils. Pedicels $\frac{1}{4}$ to $\frac{1}{2}$ in. long, densely villous. Perianth densely villous with soft hairs, the tube much revolute, the lower segments about 2 lines, the upper scarcely 1 line long under the globular limb. Torus straight. Gland small. Ovary villous on a short stipes; style villous, elongated; stigmatic disk lateral, produced at the end into a narrow-oblong involute appendage glabrous on the back, the stigma sessile on the face near the base of the disk. Fruit oblique, about $\frac{1}{2}$ in. long.

W. Australia. Swan river, *Drummond, 1st coll. n.* 628, *Oldfield;* also *Preiss, n.* 2625 (*Meissner*).

96. **G. scabra,** *Meissn. in Pl. Preiss.* i. 541, *and in DC. Prod.* xiv. 371. A shrub, apparently more rigid and erect than *G. Candolleana,* the young leaves and branches hirsute with scattered spreading hairs. Leaves sessile and often clustered in the axils, linear or rarely lanceolate, mucronate, the margins closely revolute, $\frac{3}{4}$ to above 1 in. long, very scabrous above after the hairs have worn off, the under surface

silky-tomentose but usually concealed. Racemes umbel-like, sessile, numerous, terminal or in the axillary tufts. Pedicels about ½ in. long, villous. Perianth villous and revolute as in *G. Candolleana*, but larger, the lower segments with a claw of 3 lines and the concave lamina 1½ lines broad, the upper segments much smaller. Torus small. Gland scarcely prominent. Ovary villous, on a short stipes. Style villous, elongated ; stigmatic disk lateral, shortly produced at the base below its insertion and at the end forming an oblong involute appendage like that of *G. Candolleana*, but more or less tomentose on the back.

W. Australia. Swan river, *Drummond, 1st coll. n.* 627.

SECT. 8. LISSOSTYLIS, R. Br.—Racemes short and dense (except in *G. Victoriæ* and *G. trachytheca*). Perianth-tube narrow, revolute or recurved under the limb. Torus straight. Ovary glabrous, stipitate. Stigmatic disk (or in 2 species stigmatic cone) very oblique or lateral.

SERIES 1. PUNICEÆ.—Flowers not numerous or loosely racemose, the perianth about ½ in. long. Leaves entire. Species all Eastern.

The species of this series approach the *Eugrevilleæ* of the series *Leiogynæ*, but the perianth, although not so small as in the rest of the section, has a narrower tube, scarcely dilated below the middle, and the segments more equally separating than in *Eugrevillea*. The racemes are also scarcely if at all secund.

97. **G. Miqueliana**, *F. Muell. in Trans. Vict. Inst.* 1855, 132, *and in Hook. Kew Journ.* viii. 206. An erect shrub, the young branches loosely tomentose pubescent or villous. Leaves shortly petiolate, ovate or oval-oblong, obtuse, with or without a small callous point, the margins slightly recurved, 1½ to 2½ in. long, glabrous or minutely scabrous and veinless above, tomentose-pubescent penniveined and more or less reticulate underneath. Racemes short, dense, on very short recurved terminal peduncles. Pedicels 1 to 1½ lines long, densely pubescent as well as the rhachis. Perianth loosely pubescent or villous outside, densely bearded inside about the middle, the tube narrow, 7 or 8 lines long. Torus straight. Gland semiannular. Ovary glabrous, shortly stipitate ; style not much exceeding the perianth, with a large lateral stigmatic disk. Fruit ¾ in. long.—Meissn in DC. Prod. xiv. 352.

Victoria. Upper valley of the Avon, Mount Baw-Baw, summits and higher regions of Mount Useful, sources of the Macalister river, &c., *F. Mueller.*

98. **G. brevifolia**, *F. Muell. Herb.* A low bushy shrub, the branches hoary-tomentose. Leaves obovate or oval, obtuse, contracted into a very short petiole, the margins slightly recurved, ¾ to 1¼ in. long, glabrous shining and veinless above, hoary or silvery-tomentose with the midrib slightly prominent underneath. Racemes very short, terminal, recurved. Pedicels 1 to 1½ lines long, ferruginous-silky as well as the rhachis. Perianth of *G. punicea* but rather larger, ferruginous-silky outside, densely bearded inside a little below the middle, the tube fully 7 lines long, slightly dilated towards the base, incurved under the obliquely globular limb. Torus straight. Gland semiannular, scarcely

prominent. Ovary glabrous, shortly stipitate ; style not much exceeding the perianth, with a large almost lateral stigmatic disk.

Victoria. Mount Tambo at an elevation of 5000 ft., *F. Mueller.* Included by F. Mueller (Trans. Phil. Soc. Vict. i. 108, and Hook. Kew Journ. viii. 205) in *G. Victoriæ,* but it appears to me that both in foliage and inflorescence it is nearer to *G. Miqueliana,* but distinct from both.

99. **G. Victoriæ,** *F. Muell. in Trans. Phil. Soc. Vict.* i. 107, *and in Hook. Kew Journ.* viii. 205. An erect handsome shrub of 8 to 12 ft., the branches softly tomentose. Leaves petiolate, lanceolate, broad or narrow, mostly acute, with recurved margins, 2 to 4 in. long, glabrous, often shining and prominently penniveined above, silky-pubescent underneath with the midrib alone prominent. Flowers in rather loose terminal racemes on short recurved peduncles often branched, the rhachis of each raceme ½ to 1 in. long, tomentose. Pedicels 1 to 2 lines long. Perianth ferruginous-silky outside, bearded inside rather below the middle, the tube 7 or 8 lines long, slightly dilated below the middle, incurved under the very oblique globular limb. Torus straight. Gland semiannular. Ovary glabrous, stipitate ; style not much exceeding the perianth, with a lateral stigmatic disk. Fruit rather narrow, about ¾ in. long.—Meissn. in DC. Prod. xiv. 353.

Victoria. Mount Aberdeen, Mount Latrobe, Buffalo Range, *F. Mueller.*

Var. ? *leptoneura.* Leaf-veins much less prominent, the leaves obtuse, and the perianth slender as in *G. punicea,* but with the shorter style of *G. Victoriæ.*—Sources of the Genoa river, *F. Mueller.*

100. **G. punicea,** *R. Br. in Trans. Linn. Soc.* x. 169, *Prod.* 376. An erect shrub, the young branches silky-tomentose. Leaves shortly petiolate, oblong elliptical or almost oval, obtuse with a small callous point, the margins recurved, all under 1 in. long in some specimens, a few exceeding 2 in. in others, glabrous often shining and obscurely penniveined above and frequently with a prominent marginal or inter-marginal nerve, silvery-silky or ferruginous underneath with the midrib alone prominent. Racemes very short, rather dense, almost sessile at the ends of the branches, very spreading or recurved. Pedicels 1 to 2 lines long, tomentose as well as the rhachis. Perianth slightly silky outside, densely bearded inside above the middle, the tube narrow, 5 or 6 lines long, revolute under the obliquely globular limb. Torus straight. Gland semiannular, not very prominent. Ovary glabrous, stipitate but scarcely thicker than the stipes and the long style ; stigmatic disk very oblique or lateral. Fruit ½ in long.—Meissn. in DC. Prod. xiv. 354 ; Bot. Reg. t. 1319 ; Lodd. Bot. Cab. t. 1357 ; Reichb. Icon. Exot. t. 105 ; *Lysanthe speciosa,* Knight, Prot. 118.

N. S. Wales. Port Jackson to the Blue Mountains, *R. Brown, Sieber, n.* 31, and several others.

101. **G. oleoides,** *Sieb. in Roem. and Schult. Syst.* iii. *Mant.* 277. An erect shrub, closely allied to *G. punicea* and probably a variety only, with the same indumentum venation of leaves flowers and fruit, but the leaves are longer and narrower, 2 to 4 in. long and linear or lanceolate, and

the racemes are mostly sessile in the axils of the leaves or terminating very short axillary branches.—R. Br. Prot. Nov. 17 ; Meissn. in DC. Prod. xiv. 353 ; Reichb. Icon. Exot. t. 104, *G. Seymouriæ*, Sweet ; Meissn. l.c. 354, (partly).

N. S. Wales. Blue Mountains ? *Sieber, n.* 35 ; George's river, *Macarthur, n.* 214 ; near Appin, *Backhouse ;* Illawarra, *A. Cunningham, Fraser.*

Var. *dimorpha* Leaves rather more rigid, lanceolate in some specimens, linear in others, racemes more constantly axillary and flowers sometimes but not always rather smaller.—*G. dimorpha*, F. Muell. in Trans. Phil. Soc. Vict. i. 21, and in Hook. Kew Journ. viii. 206 ; Meissn. in DC. Prod. xiv. 353.

Victoria. Grampians, *Mitchell, Wilhelmi, Robertson, F. Mueller.*

102. **G. trinervis,** *R. Br. Prot. Nov.* 18. A low rigid spreading shrub with the habit of and closely allied to *G. juniperina,* the branches softly tomentose. Leaves lanceolate or linear, rigid and pungent-pointed, with recurved or revolute margins, under 1 in. long, glabrous above and prominently 3 nerved, but the lateral nerves sometimes close to the margin, silky-tomentose underneath. Racemes short, dense, terminal. Pedicels very short or rarely nearly 2 lines long. Perianth silky-pubescent outside, densely bearded inside about the middle, the tube about ½ in. long, slightly dilated below the middle, revolute under the globular limb. Torus nearly straight. Gland semiannular, slightly prominent. Ovary glabrous, stipitate ; style rather long, with a broad lateral stigmatic disk umbonate in the centre.—Meissn. in DC. Prod. xiv. 363.

N. S. Wales. Argyle County and near Bathurst, *Fraser, Macarthur, Woolls ;* Campden, *Leichhardt ;* Clarence river *Beckler ;* heads of Hastings and Macleay rivers, *C. Moore.*

103. **G. juniperina,** *R. Br. in Trans. Linn. Soc.* x. 171, *Prod.* 377. An erect and bushy or spreading shrub, the branches softly tomentose or villous. Leaves very spreading, linear-subulate, rigid and pungent-pointed, with revolute margins, ½ to ¾ or rarely 1 in. long, glabrous and 1-nerved or rarely 2-nerved above, silky-pubescent underneath. Racemes very short, almost umbel-like, sessile, terminal. Pedicels mostly ½ to 1 line long. Perianth silky-pubescent outside, bearded inside about the middle, usually pale yellow and green but often more or less tinged with red, the tube 5 to 6 lines long, slightly dilated below the middle, revolute under the globular limb. Torus nearly straight. Gland semi-annular, slightly prominent. Ovary glabrous, stipitate ; style rather long, with a broad almost lateral stigmatic disk. Fruit about 4 lines long.—Meissn. in DC. Prod. xiv. 363 ; Guillem. Ic. Pl. Austral. t. 8 ; Bot. Reg. t. 1089 ; Lodd. Bot. Cab. t. 1003.

N. S. Wales. Port Jackson to the Blue Mountains, *R. Brown, Sieber, n.* 34, and others (all with the perianth more or less tinged with red).

Var. *sulphurea.* Perianth without any or scarcely any red tint.—*G. sulphurea*, A. Cunn. in Field, N. S. Wales, 329 ; R. Br., Prot. Nov. 17 ; Meissn. in DC. Prod. xiv. 362 ; Lodd. Bot. Cab. t. 1723 ; *G. aciphylla*, Sieb. Pl. Exs. ; *G. acicularis*, Roem. and Schult. Syst. iii. Mant. 278 ; *G. acifolia*, Spreng. Syst. Cur. Post. 46.—Blue Mountains ? *Sieber, n.* 28 ; Cox's river, *A. Cunningham, Fraser ;* Berrima, *Macarthur ;* on the Murrumbidgee, *Woolls.*

The pink and the yellow-flowered specimens are not distinguishable when dry, and the red tinge appears to be very variable.

SERIES 2. SERICEÆ.—Flowers numerous in short dense racemes. Perianth-tube under 4 lines long. Leaves entire. Species all Eastern.

104. **G. sericea,** *R. Br. in Trans. Linn. Soc.* x. 170, *Prod.* 376. An erect spreading or diffuse shrub, the branches rather slender, silky-pubescent and often angular when young. Leaves very shortly petio-late, oblong-lanceolate or almost linear, mucronate, with recurved margins, ½ to 1 in. long, or twice as long when narrow, glabrous or sparingly silky above and more or less distinctly penniveined, closely silky-tomentose underneath with the midrib alone prominent. Racemes very dense, rather short, on short terminal peduncles. Pedicels ½ to 1½ lines long. Perianth more or less pink, silky-pubescent outside, densely bearded inside about the middle, the tube 2 to nearly 4 lines long, slightly dilated below the middle, revolute under the small globular limb. Torus straight. Gland semiannular, entire or 2-lobed. Ovary glabrous, stipitate; style rather long, slender, with a large lateral stigmatic disk   Fruit about ½ in. long.—Meissn. in DC. Prod. xiv. 354; Lodd. Bot. Cab. t. 880; Reichb. Ic. et Descr. Pl. t. 76; *Embothrium sericeum,* Sm. Specim. Bot. Nov. Holl. 25, t. 9; Andr. Bot. Rep. t. 100; *Embothrium cytisoides* Cav. Ic. iv. 60, t. 386; *Grevillea dubia,* R. Br. in Trans. Linn. Soc. x. 169, Prod. 376; Meissn. l.c.; Bot. Mag. t. 3798; *Lysanthe sericea* and *L. cytisifolia,* Knight, Prot. 118, 119.

N. S. Wales. Port Jackson to the Blue Mountains, *R. Brown, Sieber, n.* 38, and many others.

Var. *diffusa.* Leaves narrow, branches scarcely angular. — *G. diffusa,* Sieb. in Roem. and Schult. Syst. iii. Mant. 279, and in Spreng. Syst. Cur. Post. 46; R. Br. Prot. Nov. 17; Meissn. in DC. Prod. xiv. 355.—Port Jackson or Blue Mountains, *Sieber, n.* 36. This seems to me scarcely to form a distinct variety. *G. planifolia,* Lodd. Bot. Cab. t. 1737, referred by Meissner to *G. Seymouriæ,* seems to me from the figure to be a narrow-leaved specimen of *G. sericea,* with darker-coloured flowers.

The perianth in this species is rather less slender than in the following ones, and connects them with *G. punicea.*

105. **G. capitellata,** *Meissn. in DC. Prod.* xiv. 356. A low spreading shrub, the young branches densely pubescent or villous. Leaves very shortly petiolate, oblong-lanceolate or almost linear, obtuse or with a callous point, the margins revolute, 1 to 2 in. long, glabrous above and distantly penniveined, silky-tomentose or villous underneath. Racemes short, very dense, shortly pedunculate, terminal. Pedicels very short, villous as well as the rhachis. Perianth densely villous outside, bearded inside about the middle, the tube slender, 2 to 2½ lines long, recurved only under the globular limb. Torus straight. Gland semiannular. Ovary glabrous, stipitate; style not very long, with a large lateral stigmatic disk.

N. S. Wales. Illawarra, *A. Cunningham, Shepherd.*—Very near the var. *diffusa* of *G. sericea,* but more villous, the leaves longer and more veined, and the perianth smaller and more slender.

106. **G. leiophylla,** *F. Muell. Herb.* Stems in all the specimens seen erect from a thick rhizome, simple or branched, scarcely above 1 ft. high, the whole plant except the inflorescence glabrous or sprinkled with a few rare appressed hairs. Leaves linear or linear-lanceolate, mucronate-acute, shortly contracted at the base, 1 to near 2 in. long, with recurved or revolute margins or quite flat, green on both sides, veinless except the prominent midrib. Inflorescence of *P. linearis,* the flowers rather smaller but of the same structure, and with the same dense tuft of hairs in the perianth-tube.

Queensland. Glasshouse ranges, Moreton Bay, *F. Mueller,* and probably from the same neighbourhood, *Leichhardt.* I have been unable to ascertain whether F. Mueller has published this species, or whether he subsequently considered it as an outlying variety of *G. linearis.*

107. **G. linearis,** *R. Br. in Trans. Linn. Soc.* x. 170, *Prod.* 376. An erect or spreading shrub attaining 5 or 6 ft., the branches and young shoots minutely silky-pubescent. Leaves linear or rarely linear-lanceolate, acute, with revolute margins, contracted at the base and sometimes shortly petiolate, 1 to 1½ in. long and spreading in some specimens, in others more rigid narrow and 2 to 3 in. long, glabrous above, silky-tomentose underneath, the midrib always prominent underneath and sometimes on both sides. Flowers small, in short dense somewhat secund racemes, sessile or pedunculate at the ends of the branches or in the upper axils. Pedicels 1 to 2 lines long, silky-tomentose as well as the rhachis. Perianth silky-pubescent outside, densely bearded inside about the middle, the tube slender, 2½ to near 3 lines long, revolute under the globular limb. Torus straight. Gland semiannular, scarcely prominent. Ovary glabrous, stipitate; style filiform, but little exceeding the perianth, with a very oblique or lateral stigmatic disk.—Meissn. in DC. Prod. xiv. 355; Bot. Mag. t. 2661; Lodd. Bot. Cab. t. 50, 858; Reichb. Ic. et Descr. Pl. t. 76; *Embothrium lineare,* Andr. Bot. Rep. t. 272; *E. linearifolium,* Cav. Ic. iv. 59, t. 386; *Lysanthe linariæfolia,* Knight, Prot. 119; *Grevillea riparia,* R. Br. in Trans. Linn. Soc. x. 170, Prod. 377; Meissn. l.c. 355; *G. stricta,* R. Br. ll. cc.; Meissn. l.c. 356.

N. S. Wales. Port Jackson to the Blue Mountains, *R. Brown, Sieber, n.* 30, and many others. R. Brown's three species, judging from his own specimens only, differ considerably in the length, breadth, more or less revolute margins and rigidity of the leaves, but in the numerous other specimens before me pass so gradually one into the other that I have been unable to sort them into distinct varieties. Endlicher's figure of *G. riparia,* Iconogr. t. 33, represents the flowers rather larger and more secund than I have usually found them. The few short hairs near the summit of the style commented upon by Meissner, occur occasionally in this and in several of the allied species.

108. **G. confertifolia,** *F. Muell. in Trans. Phil. Soc. Vict.* i. 22, *and in Hook. Kew Journ.* viii. 207. A rigid spreading shrub, the young shoots minutely silky-pubescent. Leaves often crowded, narrow-linear, rigid, mucronate and often pungent-pointed, with revolute margins, ¾ to 1½ in. long, more or less distinctly 3-nerved above, the midrib very

prominent underneath. Racemes short, dense, somewhat secund, sessile, terminal. Pedicels about 2 lines long, silky-pubescent as well as the rhachis. Perianth silky-pubescent outside, bearded inside about the middle with a tuft of reflexed hairs, the tube slender, 2 to 2½ lines long, revolute under the globular limb. Torus straight. Gland semi-annular, slightly prominent. Ovary glabrous, stipitate; style not much exceeding the perianth, with a very oblique or lateral stigmatic disk.—Meissn. in DC. Prod. xiv. 368.

Victoria. Summits of Mount William in the Grampians, *Wilhelmi*, *F. Mueller.*— Very near *G. linearis* and *G. parviflora*, with the flowers of the former and the leaves nearly of the latter.

109. **G. parviflora,** *R. Br. in Trans. Linn. Soc.* x. 171, *Prod.* 377. An erect bushy shrub of 3 to 6 ft., the branches and young shoots minutely pubescent, the foliage glabrous or sprinkled with a few silky hairs. Leaves very narrow linear, acute, with revolute margins, ¾ to 1½ in. long, doubly grooved underneath by the prominence of the midrib and margins. Flowers small, in very short umbel-like racemes shortly pedunculate and mostly terminal. Pedicels 1½ to 2 lines long, silky-pubescent as well as the rhachis. Perianth silky-pubescent outside, very minutely or scarcely bearded inside, the tube slender, about 1½ lines long, revolute under the globular limb. Torus straight. Gland scarcely prominent. Ovary glabrous, shortly stipitate. Style filiform, scarcely exceeding the perianth; stigmatic disk very oblique or lateral. Fruit ¼ to ½ in. long.—Meissn. in DC. Prod. xiv. 367 ; *G. micrantha*, Meissn. in Linnæa, xxvi. 358, and in DC. l.c.

N. S. Wales. Blue Mountains, *R. Brown ;* Illawarra, *Shepherd ;* Camden, *Leichhardt.*

Victoria. Mitta-Mitta, upper branches of the Genoa river, *F. Mueller ;* Skipton and Creswick, *Whan ;* Portland, *Robertson.*

Var. *acuaria*, F. Muell. Leaves shorter and more rigid.

S. Australia. Kangaroo Island, *Waterhouse.*

110. **G. australis,** *R. Br. in Trans. Linn. Soc.* x. 171, *Prod.* 377. A much-branched shrub sometimes erect and 3 or 4 ft. high, sometimes very spreading and under 1 ft., or prostrate and clinging to rocks, the branches and young shoots minutely pubescent. Leaves linear or rarely oblong, with rigid often pungent points, the margins either closely re-volute concealing the under surface or recurved only, contracted at the base but scarcely petiolate, from under ½ in. to nearly 1 in. long, gla-brous above, the under surface when exposed more or less silky-tomen-tose. Flowers small, in short umbel-like racemes, sessile or shortly pedunculate, terminal or in the upper axils. Pedicels ½ to 1 line long in flower, rarely 2 lines long in fruit, tomentose as well as the rhachis. Perianth silky-pubescent outside, shortly bearded inside, the tube slender, scarcely exceeding 2 lines, recurved under the globular limb. Torus straight. Gland scarcely prominent. Ovary glabrous, shortly stipitate ; style filiform, scarcely exceeding the perianth, thickened under the very oblique stigmatic disk. Fruit ellipsoid or ovoid, ¼ to ½

in. long.—Meissn. in DC. Prod. xiv. 359; Hook. f. Fl. Tasm. i. 322; *G. tenuifolia*, R. Br. in Trans. Linn. Soc. x. 171, Prod. 377.

Victoria. Sources of the Yarra, Baw-baw, Haidinger and Cobra ranges, Mount Wellington, ascending to 6000 ft., *F. Mueller;* Portland, *Allitt.*

Tasmania. Derwent river and Port Dalrymple, *R.Brown;* abundant throughout the colony, ascending to 4000 ft., *J. D. Hooker.*

J. D. Hooker has carefully distinguished seven Tasmanian varieties according to the erect, spreading or prostrate habit, and the length and breadth of the leaves, the latter character depending however often on the degree in which the margins have become revolute in drying. The most remarkable variety is the *planifolia*, Hook. f., in which the leaves are ½ to ¾ in. long and 1 to 2 lines broad, with the midrib prominent above, although as in all other forms scarcely conspicuous underneath. *G. Stuartii*, Meissn. in Linnæa xxvi. 357, and in DC. Prod. xiv. 355, appears to me to be a very luxuriant form of the variety *planifolia*, with the leaves 1¼ in. long, like those of some specimens of *G. linearis*, but with smaller flowers, and the midrib not prominent underneath.

SERIES 3. OCCIDENTALES.—Flowers numerous in dense racemes or heads. Perianth-tube under 4 lines long. Leaves entire or divided. Fruit usually (but not always) rugose or tuberculate. Western species.

111. **G. commutata,** *F. Muell. Fragm.* vi. 207. A spreading shrub of 4 to 12 ft. the young shoots minutely silky-tomentose or ferruginous. Leaves linear, sometimes rather broad or linear-cuneate, entire or very rarely 2- or 3-toothed, with revolute margins, 1½ to 3 in. long, glabrous and smooth above, silky or ferruginous underneath. Racemes dense, short or oblong, on peduncles of ½ to 1 in., usually several together in a short terminal panicle. Pedicels 1 to 1½ lines long, pubescent as well as the rhachis. Perianth silky-pubescent outside, bearded inside with a ring of hairs about the middle, the tube slender, about 2 lines long, revolute under the globular limb. Torus straight. Gland scarcely prominent. Ovary glabrous, on a short stipes; style filiform; stigmatic disk very oblique or lateral. Fruit rugose.

W. Australia. Murchison river, *Oldfield.*

112. **G. pinnatisecta,** *F. Muell.* Young shoots silky-pubescent. Leaves variously divided into narrow-linear segments, very rarely linear-lanceolate and entire, glabrous and smooth above, the midrib and revolute margins prominent underneath, silky-pubescent between them when broad enough to expose the under surface, the whole leaf from under 2 in. to nearly 6 in. long. Racemes short and dense, on short terminal often branched peduncles, the rhachis tomentose. Pedicels 1 to 2 lines long, silky-pubescent. Perianth silky-pubescent outside, bearded inside with a ring of hairs about the middle, the tube slender, not 2 lines long, revolute under the small globular limb. Torus straight. Gland prominent, semiannular. Ovary glabrous, on a rather long stipes; style filiform, with a very oblique or lateral stigmatic disk. Fruit oblong, slightly rugose, 4 to 5 lines long.

W. Australia. Between Moore and Murchison rivers, *Drummond, 6th coll. n.* 184. Meissner, who had only seen the fruiting specimens, included them in his *G. Hookeriana*, which differs however widely in the perianth, the villous ovary, &c. F. Mueller, Fragm. vi. 208, thinks they may form a variety only of *G. commutata.*

113. **G. argyrophylla,** *Meissn. in Hook. Kew Journ.* vii. 75, *and in D C. Prod.* xiv. 357. A shrub, probably tall, nearly allied to *G. diversifolia,* but with a more silvery aspect, more obtuse leaves and more globular racemes. Leaves lanceolate or oblong-cuneate, obtuse or with a small callous point, entire or shortly and obtusely 2-lobed, tapering into a short petiole, 1 to 1½ in. long, glabrous and veined above, more or less silvery-silky underneath. Racemes numerous, small, nearly globular, on filiform axillary peduncles. Flowers small, crowded on the very short pubescent rhachis. Pedicels scarcely 1 line long. Perianth pubescent with appressed hairs, the tube slender, about 1 line long, revolute under the globular limb. Torus straight. Gland broad, semicircular. Ovary glabrous, shortly stipitate ; style filiform, with an orbicular oblique stigmatic disk. Fruit about 4 lines long, more or less rugose.

W. Australia. Murchison river, *Oldfield, Drummond, 6th coll. n.* 179.

114. **G. brachystachya,** *Meissn. in Pl. Preiss.* ii. 254, *and in DC. Prod.* xiv. 366. Branches virgate, the young ones hoary-pubescent and often angular, the foliage glabrous or sprinkled with minute appressed hairs. Leaves erect, narrow-linear, thick and rigid, smooth above, doubly grooved underneath by the very prominent midrib and revolute margins, 2 to 4 in. long. Racemes very short and dense, terminal, almost sessile, the rhachis villous. Pedicels hirsute, about 2 lines long. Perianth more or less pubescent, especially the limb, with spreading glandular hairs, the tube slender, scarcely 3 lines long, recurved under the globular limb. Torus straight, without any gland. Ovary glabrous, on a rather long stipes; style filiform with a lateral orbicular stigmatic disk. Fruit ¾ in. long, smooth.

W. Australia, *Drummond, 2nd coll. n.* 319 *;* Murchison river, *Oldfield.*

115. **G. Endlicheriana,** *Meissn. in Pl. Preiss.* i. 546, *and in DC. Prod.* xiv. 356. An erect shrub attaining 7 ft., with virgate branches, silky when young, but soon glabrous. Leaves linear or linear-lanceolate, the point often recurved, tapering into a short petiole, the lower ones 2 to 4 in. long, those on the side shoots and the floral ones much reduced and often distant, flat or with recurved margins, silvery-silky on both sides, the midrib slightly prominent above, inconspicuous underneath, the upper leaves sometimes almost terete. Racemes dense, oblong, ½ to ¾ in. long, shortly pedunculate, terminal or in the axils of short or almost abortive floral leaves. Bracts imbricate on the young buds but falling away very early. Pedicels under 1 line long. Perianth glabrous outside, slightly villous inside, the tube about 2 lines long, slender as in others of this section but opening only on the lower side as in *Eugrevillea,* revolute under the globular limb. Torus straight. Gland semiannular. Ovary glabrous, stipitate ; style long, filiform, with an oblique orbicular stigmatic disk.

W. Australia. Swan river, *Drummond, 1st coll. n.* 630, *Preiss, n.* 698, *Oldfield.*

116. G. manglesioides, *Meissn. in Pl. Preiss.* i. 547, ii. 255, *and in DC. Prod.* xiv. 368. A loosely-branched shrub attaining 8 ft., the young shoots silky-ferruginous. Leaves either entire and from lanceolate to obovate-oblong, or cuneate and 3- or 5-toothed or lobed at the end, the teeth or lobes mucronate, contracted into a short petiole or nearly sessile, 1 to 1½ in. long and sometimes nearly as broad when lobed, glabrous and veined above, more or less silky or ferruginous underneath. Racemes terminal or on short axillary shoots, pedunculate or nearly sessile, the flowers crowded on a tomentose rhachis of about ½ in. and usually secund. Pedicels slender, about 1 line long. Perianth nearly glabrous outside, bearded inside above the middle, the tube slender, about 2 lines long, revolute under the small globular limb. Torus straight. Gland semiannular. Ovary glabrous, stipitate; style filiform, with an oblique or lateral orbicular stigmatic disk. Fruit about ½ in. long, slightly tuberculate-rugose.

W. Australia, *Drummond,* 2nd coll. *n.* 317, 318, *and suppl. n.* 11; Vasse river, *Preiss, n.* 720 ; Blackwood river, *Walcott.*

Var.? *angustissima.* Leaves narrow-linear, 2 in. long or more, ferruginous or silky underneath. Pedicels rather longer. Flowers the same.—W. Australia, *Drummond.*

117. G. diversifolia, *Meissn. in Pl. Preiss.* i. 547, ii. 255, *DC. Prod.* xiv. 368. A tall shrub attaining sometimes 12 ft., quite glabrous or with a very slight pubescence on the young shoots and inflorescence. Leaves linear or linear-lanceolate, mucronate-acute, the margins revolute, contracted into a short petiole or nearly sessile, 1½ to 2½ in. long, entire or with 2 or 3 divaricate lobes at the end, glabrous on both sides, 1- or 3-nerved. Racemes very short and dense, nearly sessile or on slender peduncles, mostly axillary, the rhachis nearly glabrous and rarely 2 lines long. Pedicels 1 to 2 lines long. Perianth slightly silky-pubescent outside, bearded inside below the middle, the tube slender, scarcely above 1 line long, revolute under the globular limb. Torus straight. Gland small, semiannular. Ovary glabrous, very shortly stipitate; style filiform, with an orbicular very oblique or lateral stigmatic disk. Fruit above ½ in. long, smooth or slightly tuberculate.

W. Australia, *Drummond,* 2nd coll. *n.* 316, suppl. *n.* 55, 56 ; Vasse river, *Preiss, n.* 697 ; Stirling ranges, *Maxwell.*

Var.? *rigida,* Meissn. Leaves shorter and more rigid, slightly silky underneath. Flowers not seen. Fruit very rugose. Perhaps a distinct species.—W. Australia, *Drummond,* 1st coll. 4th coll. *n.* 286.

118. G. filifolia, *Meissn. in Pl. Preiss.* i. 547, *and in DC. Prod.* xiv. 365. Branches slender, at length glabrous. Leaves semiterete, filiform, 4 to 6 in. long, not rigid, hooked at the end, silky, nerveless, obscurely grooved above or nearly flat. Racemes terminal or axillary, pedunculate, short and dense. Pedicels 2 lines long, glabrous. Perianth about 2 lines long, glabrous outside, bearded inside. Gland prominent, semiannular. Ovary glabrous, stipitate; style filiform, with an oval lateral stigmatic disk.

W. Australia, *Preiss, n.* 699. I have not seen this species. From the above character taken from Meissner it appears to differ from *G. hakeoides* chiefly in its much longer leaves.

119. **G. hakeoides,** *Meissn. in Pl. Preiss.* ii. 252, *and in DC. Prod.* xiv. 365. A bushy shrub with the habit of the slender terete-leaved *Hakeæ,* the young shoots minutely silky-pubescent, the older foliage glabrous. Leaves linear-terete, rigid but slender, slightly grooved, not pungent, 1 to 2 in. long. Flowers very small, in very short dense racemes either terminal or in the upper axils, the rhachis pubescent. Pedicels 1 to 1½ lines long. Perianth glabrous inside and out, scarcely 1½ lines long, the tube slender, revolute under the globular limb. Torus straight. Gland semiannular, prominent. Ovary glabrous, shortly stipitate; style filiform, with a very oblique or lateral orbicular stigmatic disk. Fruit 3 to 4 lines long, rugose.

W. Australia, *Drummond, 2nd coll. n.* 325, 326.

120. **G. teretifolia,** *Meissn. in Pl. Preiss.* ii. 255, *and in DC. Prod.* xiv. 373. A rigid shrub, quite glabrous or the young branches minutely pubescent. Leaves linear-terete, mostly 3-fid with the lateral branches again trifid or bifid, the segments rigid, pungent-pointed, singly or doubly grooved, from 3 or 4 lines to 1 in. long, the common petiole about as long as the branches or segments. Racemes very short and dense, terminal or on short axillary shoots, sessile or nearly so, the whole inflorescence as well as the flowers glabrous or sprinkled with a few appressed hairs. Pedicels 1 to 1½ lines long. Perianth scarcely above 2 lines long, shortly bearded inside above the middle, the tube slender, revolute under the obliquely globular limb. Torus straight. Gland semiannular. Ovary glabrous, on a rather long stipes; style filiform, with an orbicular lateral stigmatic disk. Fruit (if correctly matched) 3 to 4 lines long, rugose.

W. Australia, *Drummond, 3rd coll. n.* 271 (in flower), *Baxter* (in fruit). The foliage is that of *G. triternata* and of *G. anethifolia,* but the flowers are very different from those of either species. Baxter's fruiting specimen appears to correspond with the one which Brown referred to *G. anethifolia,* but in that there was probably some mistake, as Baxter did not collect in the interior of N. S. Wales.

121. **G. eryngioides,** *Benth.* A rigid stout undershrub or shrub, the foliage and leafy part of the stem glabrous and very glaucous. Leaves deeply pinnatifid, with few obovate or oblong lobes broadly decurrent on the rhachis and to the base of the petiole, all obtuse or with a callous point, undulate and penniveined, the whole leaf usually 3 to 5 in. long. Upper part of the branches erect, leafless, sometimes above 1 ft. long, bearing several oblong or cylindrical dense racemes of ¾ to 1 in., on peduncles of several inches, usually tomentose-pubescent as well as the rhachis. Bracts broadly ovate, membranous and coloured, falling off shortly before the flowers expand. Pedicels about 1 line long. Perianth glabrous inside and out, the tube slender, about 2 lines long, revolute under the large very oblique limb. Torus straight. Gland semiannular, but scarcely prominent. Ovary glabrous, on a long

stipes; style filiform, somewhat thickened under the orbicular lateral stigmatic disk.

W. Australia, *Drummond, n.* 16. The foliage of this species bears but little resemblance to that of any other Proteaceous plant.

122. **G. bracteosa,** *Meissn. in Pl. Preiss.* ii. 254, *and in DC. Prod.* xiv. 366. Branches elongated, pubescent with short appressed hairs when young, the adult foliage glabrous. Leaves distant or here and there crowded, very narrow-linear, entire or deeply divided into 3 segments, rigid but not pungent, doubly grooved underneath, 1 to 3 in. long. Racemes very short or oblong, dense, pedunculate, terminal or along almost leafless flowering branches. Bracts large, membranous, coloured, but falling off before the flowers expand. Pedicels 1 to 2 lines long. Perianth glabrous, 3 lines long but very much revolute, the upper segments much shorter. Torus straight. Gland slightly prominent, semiannular. Ovary glabrous, on a long stipes; style filiform, slightly thickened under the very oblique or lateral stigmatic disk.

W. Australia, *Drummond, 3rd coll. n.* 269.

123. **G. crithmifolia,** *R. Br. Prot. Nov.* 23. A bushy shrub of 1 to 4 ft., the branches softly tomentose-pubescent, the foliage sprinkled with a few hairs or at length glabrous. Leaves rather crowded, narrow-linear, pinnately divided into 3 to 5 segments on a short common petiole or here and there entire, obtuse or scarcely acute, thick but flat, doubly grooved underneath, the whole leaf usually under 1 in. long. Racemes very short and dense, sessile, terminal or on very short axillary branches, surrounded usually by a few imbricate very deciduous bracts as in *Hakea,* the rhachis villous. Pedicels glabrous, 2 to 4 lines long. Perianth glabrous, scarcely 2 lines long, the tube slender, revolute under the globular limb. Torus straight. Gland more than semi-annular with a small free one at the back of the stipes. Ovary glabrous, on a long stipes; style filiform; stigmatic disk very oblique, conical in the centre. Fruit ovoid, tubercular and muricate, ½ line long.— *Meissn. in Pl. Preiss.* i. 544 and in DC. Prod. xiv. 387; *G. Sternbergiana,* Hortul. (Meissn.).

W. Australia. Swan river, *Fraser, Drummond, 1st coll. n.* 625, *Preiss, n.* 599, 690, and others.

124. **G . trachytheca,** *F. Muell. Fragm.* vi. 207. A tall shrub attaining 8 to 10 ft., the young branches densely clothed with a soft fulvous tomentum. Leaves sessile, narrow-linear, mucronate-acute, entire or 3-lobed, ¾ to 1½ in. long, rather thick, smooth above, doubly grooved underneath. Racemes cylindrical, terminal, not dense, 1½ to 3 in. long, the linear bracts sometimes persisting till the flowers are nearly out, the rhachis softly villous. Pedicels filiform, 1 to 1½ lines long. Flowers small and numerous. Perianth glabrous, the tube slender, about 1 line long, much revolute under the globular limb. Torus straight, without any gland. Ovary glabrous, very shortly stipitate; style long, filiform, with a very oblique stigmatic cone some-

times expanding into a disk on the lower side. Fruit very oblique, muricate, 4 to 5 lines long.

W. Australia. Murchison river, *Oldfield.* This and the preceding species connect *Lissostylis* with *Conogyne.*

SECT. 9. CONOGYNE.—Racemes dense or rarely slender, short or cylindrical. Perianth-tube slender, recurved under the limb. Torus straight. Ovary stipitate. Style filiform, with an erect stigmatic cone.

125. **G. triternata,** *R. Br. Prot. Nov.* 21. An erect bushy shrub of several ft., the young shoots and inflorescence ferruginous or silky-pubescent, the adult foliage glabrous or nearly so. Leaves twice or thrice ternately divided into narrow-linear rigid pungent-pointed divaricate segments doubly grooved underneath, the whole leaf 1½ to 3 in. long and broad. Racemes terminal or in the upper axils, cylindrical, rather dense, sessile and shorter than the leaves. Pedicels ½ to nearly 1 line long. Perianth strigose-pubescent, the tube slender, under 2 lines long, recurved under the globular limb. Torus straight. Gland semiannular but scarcely prominent. Ovary hirsute, contracted into a short stipes, tapering into a rather thick style; stigmatic cone erect, surrounded by a prominent margin. Young fruit like that of *G. ramosissima,* from which this species differs chiefly in its narrow leaf-segments.—Meissn. in DC. Prod. xiv. 387 ; *Anadenia triternata,* A. Cunn. Herb.

N. S. Wales. Lachlan, Cujeegong, and Hunter's rivers, *A. Cunningham, Fraser;* Namoi river, *C. Moore;* Medway, *Miss Atkinson;* Berrima and Castlereagh, *Woolls.*

126. **G. ramosissima,** *Meissn. in DC. Prod.* xiv. 388. A bushy or spreading shrub of about 2 ft., the branches inflorescence and under side of the leaves clothed with a soft ferruginous almost silky pubescence. Leaves mostly twice trifid or the primary segments pinnate, with oblong-cuneate and 3-toothed or lanceolate and entire segments, all confluent and decurrent on the petiole, the teeth or lobes pungent-pointed, the margins recurved, the upper surface glabrous often shining and veined, the midrib alone prominent underneath, the whole leaf 1½ to 2 in. long in short bushy specimens, twice that in luxuriant ones. Racemes cylindrical, not very dense, 1 to 1½ in. long, terminal, mostly sessile and shorter than the leaves. Pedicels scarcely ½ line long. Perianth villous, recurved, scarcely 2 lines long. Torus straight, without any gland. Ovary hirsute, contracted into a short stipes and tapering into a rather thick style; stigmatic cone erect, surrounded by a prominent margin. Fruit curved, obliquely beaked.—*Anadenia Caleyi,* R. Br. Prot. Nov. 16.

N. S. Wales. Barren hills north of Bathurst, Liverpool plains, Cujeegong river, *A. Cunningham;* Reedy Creek, *C. Moore;* Goulburn, *Backhouse;* Macquarrie river, *Fraser.*

127. **G. monticola,** *Meissn. in Pl. Preiss.* ii. 259, *and in DC. Prod.* xiv. 390. A shrub of about 2 ft., glabrous and glaucous, with the

branches often angular, but less so than in *G. synapheæ*. Leavès broadly ovate, undulate and prickly toothed, cuneate at the base but scarcely petiolate, 1½ to 2 in. long or rather more, rigid and strongly veined. Racemes pedunculate in the upper axils or 2 or 3 in a short terminal panicle, ½ to ¾ in. long. Pedicels filiform, 1 to 2 lines long. Perianth glabrous, the tube slender, about 2 lines long, revolute under the globular limb. Torus straight, without any gland. Ovary glabrous, stipitate ; style rather thick ; stigmatic cone erect, bordered by a prominent margin.—*Anadenia aquifolium*, Lindl. Swan Riv. App. 31 ; *G. aquifolium*, Meissn. in Pl. Preiss. i. 551, not of Lindl.

W. Australia. Swan river, *Drummond, 1st coll.*

128. **G. Muelleri,** *Benth.* An undershrub or low shrub with erect branches of about 1 ft., rather slender and more or less hoary-pubescent. Lower leaves cuneate or oblong, 1 to 1½ in. long, toothed at the end ; upper ones mostly linear-cuneate or lanceolate, acuminate, with 1 or 2 lateral lanceolate lobes or teeth on each side, contracted into a long narrow base, the whole leaf 2 to 4 in. long, the floral ones usually linear, 1 to 2 in. long, entire or with 2 or 3 rigid divaricate linear lobes at the end ; all the leaves with revolute margins, prominent midribs, glabrous and smooth above, minutely silky-pubescent underneath. Racemes very short and dense, almost globular, sessile in the upper axils or terminal. Pedicels filiform, about 2 lines long. Perianth glabrous, slender, revolute, about 2 lines long, with a globular limb. Torus straight, without any gland. Ovary glabrous, on a filiform stipes ; style shortly thickened at the base ; stigmatic cone short, erect, with a very prominent margin.

W. Australia. Summit of Stirling range, *F. Mueller.*

129. **G. trifida,** *Meissn. in Pl. Preiss.* i. 553, *and in DC. Prod.* xiv. 389. An erect shrub of 2 or 3 ft., the branches slightly hoary. Leaves mostly cuneate, rather broad and shortly 3-lobed or 3-toothed, or narrower with the lobes again 3-toothed or rarely pinnately 5-lobed, all the teeth rigid and pungent-pointed, the margins recurved, the midribs prominent, the upper surface glabrous, the under often silky, the whole leaf ½ to 1 in. long when broad or 1½ in. when narrow ; or in some specimens the lower undivided portion of the leaf narrow-linear, 1½ in. long, with 2 or 3 short divaricate lobes at the end or some of the upper ones quite entire. Racemes axillary, sessile, dense, shorter than the leaves or scarcely exceeding them, the rhachis pubescent. Pedicels 1 to 2 lines long, filiform, glabrous. Perianth glabrous, "pale yellow," slender, revolute, under 2 lines long, the limb globular, prominently 4-ribbed. Torus straight, without any gland. Ovary glabrous, on a short stipes. Style filiform ; stigmatic cone short, erect, with a broadly prominent margin. Fruit nearly smooth.—*Anadenia trifida*, R. Br. in Trans. Linn. Soc. x. 167, Prod. 375 ; Prot. Nov. 16.

W. Australia. King George's Sound and adjoining districts, *R. Brown, Baxter, A. Cunningham, Preiss, n.* 701, and others.

130. **G. synaphéæ,** *R. Br. Prot. Nov.* 23. An undershrub of 1 to 2 ft., quite glabrous and glaucous or the young shoots minutely pubescent, the branches acutely angular. Leaves on a long linear-cuneate base deeply 3-lobed, the lobes cuneate and shortly 3-lobed 3-toothed or here and there lanceolate and entire, or the central one again 3-toothed, the lobes all rather broad, flat. mostly pungent-pointed, the whole leaf 2 to 4 in. long. Racemes dense, ½ to 1 in. long, pedunculate in the upper axils or 3 or 4 forming a terminal panicle. Pedicels filiform, ½ to 1 in. long. Perianth glabrous, slender, revolute, nearly 2 lines long, the limb globular. Torus straight, without any gland. Ovary glabrous, stipitate ; style rather thick and flat ; stigmatic cone short, erect, with a prominent margin. Fruit tuberculate-rugose, curved, about ½ in. long.—Meissn. in Pl. Preiss. i. 552, ii. 259, and in DC. Prod. xiv. 390; *Anadenia gracilis*, Lindl. Swan Riv. App. 31.

W. Australia. Swan river, *Fraser, Drummond, 1st coll., 2nd coll. n.* 313, *Preiss, n.* 702 (partly) 706.

131. **G. flexuosa,** *Meissn. in Pl. Preiss.* i. 553, *and in DC. Prod.* xiv. 389. Glabrous and glaucous with elongated branches. Leaves pinnate, the common rhachis flexuose terete or dilated and concave, 3 to 6 in. long or even more ; segments 8 to 12 or more, distant, deeply pinnatifid, confluent or petiolate, with lanceolate or triangular divaricate lobes, rigid and pungent-pointed. Racemes cylindrical, dense, 1 to 1½ in. long, pedunculate in the upper axils or 2 or 3 in a terminal panicle. Pedicels 1 to 2 lines long. Perianth glabrous, slender, revolute, above 2 lines long ; limb broadly globular. Torus straight, without any gland. Ovary glabrous, on a rather long stipes ; style filiform ; stigmatic cone short, erect, with a prominent margin.—*Anadenia flexuosa*, Lindl. Swan Riv. App. 31.

W. Australia, *Drummond, 1st. coll. n.* 613.

Var.? *pauciloba.* Leaves reduced to 3 linear lanceolate segments, or to 5 with the lower pair trifid at the base.—Darling range, *Oldfield*, fragmentary specimens possibly taken from a side-branch of the normal form, or from some abnormal specimen of *G. synapheæ.*

132. **G. leptobotrya,** *Meissn. in Pl. Preiss.* ii. 256, *and in DC. Prod.* xiv. 388. A slender diffuse or procumbent shrub, quite glabrous or sprinkled with minute appressed hairs. Leaves on long slender petioles, mostly twice pinnate, with linear or linear-lanceolate rigid acute divaricate segments, with recurved margins, smooth above, with the midrib prominent underneath, the whole leaf often 2 to 4 in. long and broad, on a common petiole at least as long ; or sometimes the lower leaves or those of the slender branches simply pinnate with narrow cuneate segments deeply divided into 3 lanceolate lobes, or pinnatifid with 5 lobes. Racemes terminal or in the upper axils, very slender, simple or branched. Pedicels filiform, 1½ to 4 lines long, the pairs often distant. Perianth sprinkled with appressed hairs, slender, revolute, scarcely 1½ lines long, the limb small, globular. Torus straight, without any gland. Ovary glabrous, stipitate ; style filiform ; stigmatic cone short, with a promi-

nent margin. Fruit oblique, smooth or obscurely rugose, about ½ in. long.

W. Australia, *Drummond,* 3rd *coll. n.* 268 ; hills on the Gordon river, *Maxwell.*

133. **G. brevicuspis,** *Meissn. in Pl. Preiss.* ii. *256, and in DC. Prod.* xiv. 388. A shrub with rather slender divaricate or flexuose branches, quite glabrous. Leaves numerous, with short petioles, mostly twice divided into narrow-linear rigid divaricate pungent-pointed segments, with revolute margins, the whole leaf 1 to 1½ or rarely 2 in. long. Racemes short, dense, quite glabrous, sessile in the axils. Pedicels filiform, 1 to 2 lines long. Perianth glabrous, slender, revolute, about 1½ lines long, the limb globular, 4-ribbed. Torus straight, without any gland. Ovary glabrous, on a slender stipes; style long, filiform, shortly thickened under the stigmatic cone, which is surrounded by a very prominent margin. Fruit very oblique, smooth, about 4 lines long.

W. Australia, *Drummond,* 2nd *coll. n.* 321 ; Murchison river, *Oldfield.*

134. **G. intricata,** *Meissn. in Hook. Kew Journ.* vii. 74, *and in DC. Prod.* xiv. 387. A shrub of 2 or 3 ft. (or sometimes 6 to 10 ft. ?), with slender branches, glabrous or the young shoots and inflorescence sprinkled with appressed hairs. Leaves long and slender, once twice or three times ternately divided into linear-subulate almost terete rigid acute segments, singly or doubly grooved, often above 1 in. long on a common petiole of 2 in. or more. Racemes slender, pedunculate, 1 to 2 in. long and sometimes branched, terminal or lateral. Pedicels filiform, 2 to 3 lines long. Perianth glabrous, slender, revolute, scarcely 1½ lines long, the limb globular. Torus straight, without any gland. Ovary glabrous, on a slender stipes; style filiform, stigmatic cone bordered by a prominent margin. Fruit 5 or 6 lines long, slightly rugose.

W. Australia. Murchison river and Champion Bay, *Oldfield, Drummond,* 6th *coll. n.* 189.

135. **G. didymobotrya,** *Meissn. in DC. Prod.* xiv. 386. A shrub of 3 or 4 ft. (*Oldfield*) or tree of 10 to 15 ft. (*Herb. F. Mueller*), minutely hoary or silvery-pubescent or at length glabrous. Leaves linear-terete, slender but rigid, acute or mucronate, mostly 2 to 4 in. long, finely striate and sometimes channelled underneath. Racemes shortly pedunculate, cylindrical, dense, rather narrow, 1½ to 2 in. long, terminal or in the upper axils, forming a terminal corymbose leafy panicle, the rhachis pubescent. Pedicels ½ to ¾ line long. Perianth sprinkled with appressed hairs, slender, revolute, scarcely 1½ lines long, the limb globular. Torus straight, without any gland. Ovary glabrous, nearly sessile; style filiform; stigmatic cone short, surrounded by a broad margin and sometimes slightly oblique. Fruit incurved, semiobcordate, about 3 lines long.—*Anadenia filifolia,* Endl. in Ann. Wien. Mus. ii. 209, and Nov. Stirp. Dec. 88, from the description given, and consequently *G. capillaris,* Meissn. in DC. Prod. xiv. 386.

W. Australia, *Drummond, n.* 163, 166, 4*th coll. n.* 280 ; Murchison river, *Old-field ;* Irwin river, *Herb. F. Mueller,* collector not named.

Anadenia Roei, Endl. Gen. Pl. Suppl. iv. 83, as yet undescribed, is believed by Meissner to be the same as the above *A. filifolia.*

136. **G. polybotrya,** *Meissn. in DC. Prod.* xiv. 386. An erect probably tall shrub, the branches and inflorescence tomentose-pubescent, the young foliage sprinkled with silvery shining hairs, glabrous but pale when full grown. Leaves entire, oblong and obtuse or oblong-lanceolate and acute, contracted into a short petiole, $\frac{3}{4}$ to $1\frac{1}{2}$ in. long, flat and rather thick, veinless or the midrib and sometimes 2 lateral veins slightly prominent. Racemes or spikes shortly pedunculate, cylindrical, dense, $1\frac{1}{2}$ to 2 in. long, numerous in a large, dense leafless terminal panicle. Flowers nearly sessile. Perianth sprinkled with appressed hairs, slender, revolute, fully 2 lines long, the limb globular. Torus straight. Gland prominent, semiannular. Ovary glabrous or slightly viscid, very shortly stipitate ; style long and filiform ; stigmatic cone rather short, with a prominent margin. Fruit about $\frac{1}{2}$ in. long, compressed, apparently viscid.—*G. Martinii,* F. Muell. Fragm. iv. 129, t. 32 ; *G. polybotrya,* F. Muell. Fragm. vi. 208, but not the one described in Hook. Kew Journ. ix. 23.

N. Australia. Glenelg river, N.W. coast, *Martin.*
W. Australia, *Drummond, n.* 90, 162, 4*th coll. n.* 279.

137. **G. nematophylla,** *F. Muell. Fragm.* i. 136. A shrub of 5 or 6 ft., minutely pubescent or glabrous. Leaves entire, linear-terete, slender but rigid, acute, 3 to 6 in. long, obscurely striate or channelled. Racemes cylindrical, pedunculate, rather loose, $1\frac{1}{2}$ to $2\frac{1}{2}$ in. long, several in a terminal panicle shorter than or scarcely exceeding the leaves, the rhachis glabrous. Pedicels scarcely $\frac{1}{2}$ line long. Perianth glabrous or sprinkled with appressed hairs, slender, revolute, about 2 lines long, the limb globular. Torus straight. Gland prominent, semiannular. Ovary glabrous, on a long stipes ; style filiform ; stigmatic cone slightly oblique.

N. S. Wales. Mount Murchison, *Dallachy.*

SECT. 10. ANADENIA.—Racemes dense, short or cylindrical. Flowers small. Perianth-tube slender, straight ; limb erect. Torus straight. Style filiform or dilated upwards, not contracted under the erect stigmatic cone.

138. **G. anethifolia,** *R. Br. Prot. Nov.* 21. A shrub with the foliage of *G. triternata,* but with very different flowers, the branches shortly pubescent, the foliage glabrous. Leaves once twice or three times divided into narrow-linear almost terete rigid divaricate pungent-pointed segments, singly or doubly grooved, the whole leaf under 2 in. long. Racemes short, dense, axillary or terminal. Pedicels glabrous, 1 to 2 lines long. Perianth glabrous, slender, straight or nearly so, $2\frac{1}{2}$ lines long, the limb globular. Torus straight. Gland prominent,

semiannular. Ovary glabrous, on a long slender stipes; style much dilated and flattened except at the base; stigmatic cone surrounded by a prominent margin.—Meissn. in DC. Prod. xiv. 387; *Anadenia anethifolia,* A. Cunn. Herb.

N. S. Wales. Rare in the barren flat country near Peel's Range, *A. Cunningham.*

139. **G. paradoxa,** *F. Muell. Fragm.* vi. 246. A stout rigid bushy shrub, the young shoots ferruginous or silky-pubescent, the adult foliage glabrous. Leaves very intricate, once twice or three times divided into linear-terete rigid divaricate pungent-pointed segments, rarely above ½ in. long and not grooved, the whole leaf under 2 in. Racemes or spikes cylindrical, thick and very dense, 1½ to 2 in. long, sessile amongst the last leaves, the rhachis densely hirsute. Pedicels ½ to 1 line long. Perianth glabrous, straight or slightly curved, slender, 2½ to 3 lines long, the limb ovoid, erect. Torus straight. Gland prominent, semiannular. Ovary sessile or nearly so, glabrous or slightly glandular-pubescent. Style very long, filiform; stigmatic cone very narrow. Fruit ovoid, acute, tomentose, 2 to 3 lines long.

W. Australia, *Drummond, 5th coll. suppl. n.* 11.

140. **G. petrophiloides,** *Meissn. in Pl. Preiss.* ii. 257, *and in DC. Prod.* xiv. 387. A shrub of 4 or 5 ft., with rigid erect branches, the whole plant glabrous and glaucous or the young shoots very slightly ferruginous-pubescent. Leaves twice or three times ternately or pinnately divided into linear-terete erect segments, sometimes very slender and 1 to 2 in. long, sometimes shorter more rigid and angular, the whole leaf 4 to 8 in. long. Racemes very dense and spike-like, 2 to 4 in. long, solitary and terminal or several along a terminal stout leafless common peduncle often above 1 ft. long. Pedicels ½ to 1 line long. Perianth glabrous, slender, straight, 4 to 4½ lines long, the limb narrow-oblong, erect. Torus straight. Gland semiannular. Ovary glabrous, stipitate; style very long and filiform; stigmatic cone very narrow. Fruit nearly globular, with an oblique point, scarcely 3 lines long.

W. Australia, *Drummond, n. 83, 3rd coll. n. 300, 5th coll. Suppl. n.* 8; Murchison river, *Oldfield.*

141. **G. tenuiflora,** *Meissn. in Pl. Preiss.* i. 554, *and in DC. Prod.* xiv. 389. A bushy shrub, the branches and foliage more or less pubescent with short often glandular hairs. Leaves pinnately divided into 5 or the upper ones into 3 segments or lobes, the segments broadly cuneate 3- or 5-lobed, the lobes triangular and entire or broadly 3-lobed, all rigid and shortly pungent-pointed, often shining above and opaque underneath, but retaining a minute pubescence on both sides, the whole leaf 1 to 2 in. long and broad. Racemes terminal, dense, sessile or shortly pedunculate, 1 to 1½ in. long, the rhachis pubescent. Bracts small, ovate, very deciduous. Pedicels filiform, 1½ lines long. Perianth nearly glabrous, slender, straight

2½ lines long, the limb ovoid, 4-angled. Torus straight, without any gland. Ovary on a rather long stipes, sprinkled with few glandular hairs; style filiform, slightly thickened and pubescent towards the end; stigmatic cone very narrow, with a slightly projecting margin round the base.—*Anadenia tenuiflora,* Lindl. Swan Riv. App. 31.

W. Australia. Swan river, *Drummond, 1st coll., Preiss, n.* 703.

142. **G. pulchella,** *Meissn. in Pl. Preiss.* i. 553, *and in DC. Prod.* xiv. 389. A rather slender divaricate undershrub or shrub of 1 to 2 ft., scabrous-pubescent and often glandular, or the foliage at length glabrous. Leaves pinnate; segments 7 to 11, cuneate, trifid or 3-toothed, distinct or the upper ones confluent and more entire, the lobes or teeth triangular or lanceolate, acute or pungent-pointed, the margins revolute, the whole leaf 1½ to 3 in. long. Racemes dense, usually glabrous, ½ to 1 in. long, terminal or in the upper axils, on short slender peduncles. Pedicels scarcely 1 line long. Perianth white, glabrous, slender, straight, about 1½ lines long, the limb ovoid-fusiform. Torus straight, without any gland. Ovary stipitate, sprinkled with glandular hairs; style filiform, thickened under the narrow stigmatic cone. Fruit 3 to 4 lines long, the valves very open and revolute when ripe.—*Anadenia pulchella,* R. Br. in Trans. Linn. Soc. x. 167, Prod. 374.

W. Australia. King George's Sound, *R. Brown, Baxter, A. Cunningham, Preiss, n.* 700, and many others.

143. **G. rudis,** *Meissn. in Hook. Kew Journ.* vii. 73, *and in DC. Prod.* xiv. 390. An erect shrub or undershrub, the branches and foliage very scabrous and more or less hirsute with long spreading hairs. Leaves in the lower part of the branches cuneate, dilated and shortly once or twice ternately lobed or broadly toothed at the end, narrowed to the base but not petiolate, thick, rigid, penniveined, 1 to 2 in. long, the upper leaves distant, sessile, lanceolate, entire, ¼ to ½ in. long. Racemes cylindrical, dense, 1 to 1½ in. long, hirsute and glandular-viscid, terminal or accompanied by one or two smaller ones lower down the branch. Bracts acuminate and comose on the very young raceme. Pedicels about 1 line long. Perianth hir ute, slender, straight, 2 lines long, the limb ovoid. Torus straight, without any gland. Ovary almost sessile, hirsute; style long, filiform but thickened at the end under the narrow stigmatic cone.

W. Australia. Between Moore and Murchison rivers, *Drummond, 6th coll. n.* 180.

144. **G. Shuttleworthiana,** *Meissn. in Pl. Preiss.* ii. 258, *and in DC. Prod.* xiv. 386. An erect shrub perfectly glabrous or with a very slight hoary pubescence on the branches. Leaves obovate or cuneate, undulate, mucronate or acuminate, entire, contracted into a distinct petiole, ¾ to 1 in. or rarely 1½ in. long, rigid and obliquely veined on both sides. Racemes cylindrical, rather dense, pedunculate, 1 to 2 in. long, terminal or in the upper axils forming sometimes a terminal

leafy panicle, the rhachis and flowers glabrous.   Perianth slender, straight, under 2 lines long, the limb nearly globular.   Torus straight, without any gland.   Ovary glabrous, nearly sessile ; style filiform, with a narrow stigmatic cone.   Fruit small, semi-obcordate, falcate, the lower edge or back dilated.

W. Australia, *Drummond, 2nd coll. n.* 299, *3rd coll. n.* 266.

145. **G. integrifolia,** *Meissn. in DC. Prod.* xiv. 385.   An erect shrub, the branches and especially the foliage more or less silvery or silky-pubescent.   Leaves in the typical form oblong lanceolate or oblong-cuneate, obtuse with a callous point or acute, contracted into a petiole, 1 to 1½ in. long, very obliquely veined and often 3-nerved. Racemes pedunculate, cylindrical, rather loose, 1½ to 2 in. long, several in a terminal leafy panicle, the rhachis slightly pubescent. Pedicels ½ to 1 line long.   Perianth glabrous, slender, straight, scarcely 2 lines long, the limb ovoid.   Torus straight, without any gland.   Ovary almost sessile, glabrous ; style filiform, the stigmatic cone almost linear, marked by a slightly projecting rim at the base. Fruit not seen ripe, when young more like that of *G. Shuttleworthiana* than of *G. stenocarpa.*—*Anadenia integrifolia,* Endl. in Ann. Wien. Mus. ii. 209, and Nov. Stirp., Dec. 88.

W. Australia, *Drummond, n.* 157, *5th coll. suppl. n.* 6.

Var. *obovata.*   Leaves mostly obovate, ¼ to ¾ in. long, but on some specimens throwing off branches with leaves of the typical form.—*G. biformis,* Meissn. in Pl. Preiss. ii. 258, and in DC. Prod. xiv. 386, as to the barren specimens.—W. Australia, *Roe, Drummond.*

146. **G. stenocarpa,** *F. Muell. Herb.*   An erect bushy shrub of 3 or 4 ft., more or less silvery or hoary-tomentose, the older foliage rarely quite glabrous.   Leaves narrow-linear, thick but more or less flattened, acute or obtuse, striate with 3 to 5 closely approximate longitudinal nerves, 2 to 4 in. long or in a few specimens only 1 to 2 in., slightly attenuate and almost terete at the base.   Racemes shortly pedunculate, rather loose, 1½ to 2½ in. long, several in a terminal leafy panicle.   Pedicels very short.   Perianth glabrous, slender, straight, scarcely 2 lines long, the limb ovoid.   Torus straight, without any gland.   Ovary glabrous, nearly sessile ; style filiform ; stigmatic cone very narrow, with a projecting rim at the base.   Fruit (only known in very few specimens), very narrow, obliquely clavate, 5 to 8 lines long. —*G. biformis,* Meissn. in Pl. Preiss. ii. 258, and in DC. Prod. xiv. 386, as to the flowering specimens.

W. Australia, *Drummond, 3rd coll. n.* 265, *6th coll.' n.* 181 ; Murchison river, *Oldfield.*   In some herbaria, Drummond's specimens n. 265 are associated with others without flowers of *G. integrifolia,* and had evidently been so transmitted by Drummond, in other cases the corresponding leafy specimens had been correctly sent with the flowering ones of *G. integrifolia.*   I have therefore been unable to retain Meissner's specific name founded upon the supposed dimorphous foliage.

Sect. 11. Manglesia.—Racemes short, dense, axillary.   Flowers small.   Perianth-tube straight, slender or fusiform, the limb erect.

Torus straight. Ovary glabrous, stipitate. Style turgid in the middle or fusiform, constricted under the erect stigmatic cone.

147. **G. acrobotrya,** *Meissn. in Hook. Kew Journ.* vii. 74, *and in DC. Prod.* xiv. 391. Branches rigidly virgate, hoary-pubescent as well as the foliage. Stem-leaves nearly sessile, broadly cuneate or fan-shaped or almost rhomboidal, ¾ to 1 in. long, coarsely toothed at the end with mucronate or prickly teeth, glabrous above, almost silky underneath with very prominent primary veins, the upper ones passing into the smaller floral leaves, deeply divided into 3 or more linear rigid pungent-pointed lobes. Racemes dense, sessile, scarcely above ½ in. long, all axillary, either distant or crowded towards the ends of the branches. Pedicels filiform, 2 to 4 lines long. Perianth glabrous, about 2 lines long, the tube rather thick and scarcely longer than the globular limb. Torus straight. Gland obsolete. Ovary glabrous, stipitate; style thick and fusiform with a small terminal stigma. Fruit very oblique, about 4 lines long.

W. Australia. Between Moore and Murchison rivers, *Drummond, 6th coll. n.* 185. The style in this species is anomalous, but nearer to that of *Manglesia* than of *Anadenia.*

148. **G. glabrata,** *Meissn. in Pl. Preiss.* i. 549, ii. 255, *and in DC. Prod.* xiv. 391. An erect shrub of 5 or 6 ft., perfectly glabrous and more slender than the allied species. Leaves broadly cuneate, shortly and broadly 3-lobed, the lobes acute with fine pungent points, contracted into a petiole, flat, with prominent primary veins, the whole leaf 1 to 1½ in. long. Racemes axillary, as long as or longer than the leaves, the upper ones forming a terminal panicle; rhachis slender. Pedicels filiform, 3 to 6 lines long. Perianth glabrous, straight, about 1½ to 2 lines long, the tube much longer than the globular limb. Torus straight. Gland semiannular. Ovary glabrous, on a long stipes; style contracted above the ovary, then thick and fusiform and again contracted under the stigmatic cone.—Baill. Hist. Pl. ii. 390, f. 219 to 222; *Manglesia glabrata,* Lindl. Swan Riv. App. 37; *M. cuneata,* Endl. Nov. Stirp. Dec. 25; *Anadenia Manglesii,* Grah. in Hook. Ic. Pl. t. 337; *Grevillea Manglesii,* Hortul. (*Meissn.*)

W. Australia. Swan river, *Drummond, 1st coll. n.* 621; *Preiss, n.* 695.

149. **G. ornithopoda,** *Meissn. in Pl. Preiss.* ii. 256, *and in DC. Prod.* xiv. 391. A perfectly glabrous shrub with rather slender branches. Leaves cuneate, tapering into a long narrow base or winged petiole, divided into 3 lanceolate acutely acuminate lobes of ½ to 1 in., the undivided part 1½ to 2½ in. long, the three primary veins prominent underneath. Racemes axillary, simple or branched, rarely exceeding the leaves, the rhachis slender and glabrous. Pedicels filiform, 3 to 6 lines long. Perianth glabrous, nearly 2 lines long, the tube fusiform, not twice the length of the globular limb. Torus straight. Gland semiannular. Ovary glabrous, on a long stipes;

style shortly contracted at the base, then turgid and slightly contracted under the stigmatic cone. Fruit very oblique, rugose.

W. Australia, *Drummond, 2nd coll. n.* 314.

150. **G. paniculata,** *Meissn. in Pl. Preiss.* i. 550, *and in DC. Prod.* xiv. 392. A shrub of 6 to 8 ft., glabrous and often glaucous or the young shoots slightly hoary with a minute tomentum. Leaves linear-terete, slender but rigid, pungent-pointed, more or less grooved, once or twice divided into 3 rarely 2 segments ½ to 1 in. long, the undivided base about as long. Racemes axillary, shorter or rather longer than the leaves, simple or branched. Pedicels 2 to 4 lines long. Perianth glabrous, under 2 lines long, the tube not twice as long as the globular limb. Torus straight. Gland semiannular. Ovary glabrous, on a long stipes; style contracted at the base, then dilated and again contracted under the stigmatic cone. Fruit ½ in. long, oblique, but the stipes not so lateral as in the allied species, very rugose.

N. Australia, *Drummond, n.* 105, *Preiss, n.* 617, *a* and *b;* Fitzgerald Flats, *Maxwell.*

151. **G. biternata,** *Meissn. in Pl. Preiss.* i. 549, ii. 256, *and in DC. Prod.* xiv. 392. Branches softly tomentose-pubescent, the young shoots ferruginous. Leaves very narrow-linear, pungent-pointed, doubly grooved underneath, mostly once or twice divided into 3 or sometimes 2 segments of 1 to 1½ in., the undivided base as long or rather shorter and in some specimens a few leaves quite entire. Racemes axillary, sessile, shorter than the leaves. Perianth glabrous, under 2 lines long, the tube not twice the length of the globular limb. Torus straight. Gland semiannular. Ovary glabrous, on a long stipes; style shortly constricted at the base then dilated and again contracted under the stigmatic cone. Fruit rugose as in *G. triloba,* of which this may prove to be a very narrow-leaved variety.

W. Australia, *Drummond, n.* 188, *1st coll. n.* 624, *2nd coll. n.* 315, 323 ; north of Cape Paisley, Phillips and Thomas rivers, *Maxwell.*

Var. *leptostachya.* A shrub of 10 to 15 ft., more glabrous than the typical form, with slender racemes, about 1 in. long.—Champion Bay, *Walcott.*

152. **G. triloba,** *Meissn. in Hook.Kew Journ.* vii. 74, *and in DC. Prod.* xiv. 388. Branches rather stout, softly ferruginous-villous or hoary. Leaves with a linear-cuneate base of about 1 in., usually divided into 3 linear-lanceolate divaricate lobes of ½ to 1½ in., mucronate or sometimes pungent-pointed, with revolute margins, glabrous above when full-grown and obliquely veined, softly pubescent or villous underneath ; a few of the lower leaves sometimes entire and oblong-lanceolate. Racemes axillary, dense, sessile, shorter than the leaves, the rhachis villous. Pedicels filiform, 3 to 5 lines long. Perianth glabrous, nearly 2 lines long, the fusiform tube much longer than the globular limb. Torus straight. Gland semiannular. Ovary glabrous, on a long stipes; style contracted at the base, then dilated into a swelling of the

shape of the ovary and again contracted under the stigmatic cone. Fruit tuberculate-rugose, about 4 lines long.

W. Australia. Murchison river, *Oldfield, Drummond, 6th coll. n.* 187.

153. **G. amplexans,** *F. Muell. Herb.* Quite glabrous and glaucous. Leaves sessile, nearly orbicular, deeply cordate, clasping the stem by broad auricles closed or overlapping behind the stem, veined on both sides, the principal veins produced into short pungent points. Racemes axillary or terminal, loose and somewhat branched in the only specimen seen, about 1 in. long. Pedicels 2 to 3 lines long. Perianth glabrous, straight, about 2 lines long, the tube not twice as long as the globular limb. Torus straight. Gland semiannular. Ovary glabrous, stipitate; style contracted at the base, then dilated and again contracted under the stigmatic cone. Fruit smooth, about 4 lines long.

W. Australia. Northern districts, *Herb. F. Mueller,* the collector not named.

154. **G. vestita,** *Meissn. in Pl. Preiss.* i. 548, ii. 255, *and in DC. Prod.* xiv. 391. An erect bushy shrub of 6 to 9 ft., the young shoots ferruginous-villous. Leaves cuneate, broad or narrow, tapering towards the very narrow base, $\frac{3}{4}$ to $1\frac{1}{2}$ in. long, more or less deeply 3- or rarely 5-lobed at the end, the lobes broad mucronate and often pungent, the margins recurved, glabrous above when old and veined, pubescent or villous underneath; a few of the lower leaves often entire and oblanceolate. Racemes axillary, dense, scarcely exceeding the leaves, the rhachis pubescent or villous. Pedicels $\frac{1}{4}$ to $\frac{1}{2}$ in. long. Perianth glabrous, nearly 2 lines long, the tube much longer than the globular limb. Torus straight. Gland semiannular. Ovary glabrous, on a long stipes; style contracted at the base, then thickened and again contracted under the stigmatic cone. Fruit very oblique, quite smooth, about 4 lines broad.—*Manglesia vestita,* Endl. Nov. Stirp. Dec. 26.

W. Australia. Swan river, *Drummond, n.* 65, 67, *1st coll. n.* 620, *Preiss, n.* 72 ; King George's Sound, Williams and Murray rivers, *Oldfield* (all with broad leaves); *Drummond, n.* 29, 64, *1st coll. n.* 622, *2nd coll. n.* 320 (with linear-lanceolate leaf-lobes).

Var. *stenogyne.* Style and stigmatic cone much more slender, approaching the style of *G. acrobotrya.*—W. Australia, *Drummond.*

155. **G. tridentifera,** *Meissn. in Pl. Preiss.* i. 547, *and in DC. Prod.* xiv. 392. A rigid shrub said to be quite glabrous in the typical form which I have not seen, the branches densely tomentose-villous in our specimens. Leaves on very short petioles, divided into 2 or 3 narrow-linear rigid pungent-pointed segments of about $\frac{1}{2}$ in., entire or again 2-lobed, convex above, doubly grooved underneath. Racemes axillary, sessile, very short, the rhachis tomentose. Pedicels filiform, 3 or 4 lines long. Perianth glabrous, about $1\frac{1}{2}$ lines long, the tube not twice as long as the globular limb. Torus straight. Gland semiannular. Ovary glabrous, on a long stipes; style with a rather long filiform base, then dilated and again contracted under the rather broad stigmatic cone. Fruit smooth, at least when young.—*Manglesia tridentifera,* Endl. Nov. Stirp. Dec. 25 (*Meissn.*).

W. Australia, *Drummond, 1st coll. n.* 623.

156. **G. erinacea,** *Meissn. in Hook. Kew Journ.* vii. 74, *and in DC. Prod.* xiv. 392. Branches hoary-tomentose or ferruginous when young. Leaves crowded, deeply and ternately once or twice divided into slender but rigid and pungent-pointed segments, linear-terete and singly grooved or slightly flattened and doubly grooved underneath, the whole leaf under 1 in. long including the short petiole. Racemes axillary, sessile, short and dense, the rhachis slightly tomentose. Pedicels filiform, 3 to 4 lines long. Perianth glabrous, 1½ lines long, the tube not twice as long as the globular limb. Torus straight. Gland semiannular, slightly prominent. Ovary glabrous, stipitate; style filiform at the base, then turgid and again contracted under the stigmatic cone. Fruit very oblique, perfectly smooth, 4 to 5 lines broad.

W. Australia. Between Moore and Murchison rivers, *Drummond, 6th coll.* n. 186.

21. HAKEA, Schrad.

(Conchium, *Sm.*)

Flowers hermaphrodite. Perianth irregular or rarely regular, the tube revolute or curved under the limb or rarely straight, the limb globular or rarely ovoid, often oblique, the laminæ often cohering long after the tube has opened. Anthers all perfect, sessile in the base of the concave laminæ, the connective not produced beyond the cells. Hypogynous glands united in a single semiannular or semicircular rarely disk-shaped gland occupying the upper side of the torus, in some species very small. Ovary stipitate but usually very shortly so, with 2 amphitropous ovules laterally attached about the middle; style either long and protruding from the slit of the perianth before the summit is set free from the limb as in *Grevillea,* or not exceeding the perianth, more or less dilated at the end into a straight or oblique or lateral cone or disk, bearing the small stigma in the centre of the disk or at the summit of the cone. Fruit a hard usually woody capsule opening in 2 valves. Seeds 2, compressed and collateral, the testa produced at the upper end into a broad membranous wing usually longer than the nucleus and more or less decurrent down the upper or both margins and sometimes completely surrounding the nucleus, the nucleus itself flat and smooth on the inner face (next the other seed), convex on the outer face and usually rugose or muricate, the protuberances fitting into corresponding cavities in the valve; each seed with its wing sometimes covering the whole inner surface of the valve, more frequently placed near the upper margin and covering about half only or rather more, the remainder of the valve a hard woody mass.—Shrubs or rarely small trees. Leaves alternate, very diversified in shape, flat or terete, the margins rarely recurved and the two surfaces usually similar and equally veined. Flowers in pairs along the rhachis of a short and dense raceme or cluster or rarely in a longer raceme; the clusters or racemes sessile in the axils or rarely also terminal or in a very few species all terminal. Indumentum as in *Grevillea*, consisting of closely appressed hairs attached by the centre, rarely of erect or spreading hairs.

The genus is limited to Australia. As will be perceived on comparing the above character with that of *Grevillea*, there is no one organ in which the two genera are absolutely distinct excepting the seed-wing, and even that, although essentially terminal in *Hakea* and annular when present in *Grevillea*, is exceptional in *Hakea platysperma*, for instance, where the wing is almost of equal breadth all round the nucleus ; but even there the texture and venation of the wing is that of *Hakea*, not of *Grevillea*, and the two genera are with few exceptions so natural, that there are very few species that would not at once be referred to their right genus even without the fruit, especially as the wing of the seed can generally be traced in the ovule immediately after fecundation. The racemes are generally axillary and reduced to sessile clusters in *Hakea*, terminal and loose in *Grevillea*, but they are loose and elongated in the section *Grevilleoides* of *Hakea*, and terminal in *H. ruscifolia* and a few others ; whilst the section *Manglesia* of *Grevillea* as well as *G. hakeoides* and a few others have the inflorescence of *Hakea*. The so called involucres or imbricate bud-scales enveloping the nascent inflorescences of *Hakea*, appear to be wanting in *Grevillea*, but they are also deficient in the section *Grevilleoides* of *Hakea*, and are always so deciduous as to be generally absent from flowering specimens. The same variations of form in the perianth occur in the two genera, but in *Hakea* the hairs or beard inside the tube of many *Grevilleæ* are always wanting and the ovary is always glabrous. *Hakea* has also the various modifications of the pistil of *Grevillea*, except the turgid style of the section *Manglesia*. The fruit is in general totally different in the two genera, and yet that of *Grevillea gibbosa* is certainly a near approach to that of *Hakea platysperma*.

The determination of the species of *Hakea* generally requires the presence both of flowers and fruit. Species, especially amongst the terete-leaved ones, with scarcely distinguishable flowers and leaves, have sometimes very different fruits, whilst closely similar fruits have occasionally very different flowers and leaves. I have found the flowers chiefly available for sectional, the fruits for specific distinction. The dorsal protuberances on the fruit-valves of some species usually called *spurs*, appear to me to be more appropriately termed *horns*, as they occur always close to the apex not to the base of the valve.

Sect. 1. **Grevilleoides.**—*Flowers in oblong or cylindrical or rarely short racemes, without any involucre. Perianth much revolute. Stigmatic disk oblique or lateral, flat or broadly conical. Tropical or subtropical species.*

Leaves terete, usually very long.
  Racemes as well as the whole plant quite glabrous . . . .1. *H. chordophylla.*
  Racemes pubescent or villous.
    Leaves mostly above 1 ft. long. Racemes 3 to 6 in. long.
      Perianth-tube 4 to 5 lines.
      Torus very oblique . . . . 2. *H. Cunninghamii.*
      Torus scarcely oblique . . . . 3. *H. lorea.*
    Leaves slender, mostly about ½ ft. long. Racemes 1 to 2
      in. Perianth-tube under 3 lines . . . . 4. *H. Fraseri.*
Leaves flat, linear, usually long.
  Seed-wing not at all or scarcely decurrent along the nucleus.
    Racemes 3 to 6 in. long. Perianth villous, 5 to 6 lines
      long . . . . 5. *H. macrocarpa.*
    Racemes under ½ in. long, pedunculate. Perianth silky,
      under 2 lines . . . . 6. *H. arborescens.*
  Seed-wing decurrent on both sides round the base of the
    nucleus . . . . 7. *H. stenophylla.*
Leaves flat, 3-nerved, oblong or lanceolate. Racemes glabrous,
  about 2 in. long. Perianth 3 lines . . . . 8. *H. trineura.*

Sect. 2. **Euhakea.**—*Racemes usually short or reduced to sessile clusters, enclosed before their development in an involucre or bud of imbricate scales. Perianth revolute,* *at least under the limb. Stigmatic disk oblique or lateral, flat or slightly convex, without any cone (except in* H. rugosa *and* H. rostrata).

Series 1. **Obliquæ.**—*Perianth pubescent. Torus oblique, the ovary at the shortest margin, the remainder occupied by a large very concave adnate gland. Leaves entire. Species all Western.*

Leaves flat, tapering at the base, obscurely several-nerved.
    Leaves 4 to 8 in. long. Fruit curved with a broad, rather
      long, closely inflexed beak . . . . 9. *H. cyclocarpa.*
    Leaves under 3 in. Fruit large and thick, nearly smooth,
      with a very small inflexed beak . . . . 10. *H. crassifolia.*
    Leaves 2 to 4 in. long. Fruit very large and thick, covered
      with large conical tubercles . . . . 11. *H. pandanicarpa.*
    Leaves narrow, under 2 in. Fruit unknown   . . . . 12. *H. Roei.*
Leaves linear-terete.
    Leaves erect, with short erect or curved points. Fruit rather
      large, very broad, with a small inflexed beak . . . . . 13. *H. adnata.*
    Leaves spreading, pungent-pointed. Fruit twice as long as
      broad, with a short nearly straight beak . . . . . . 14. *H. obliqua.*

Series 2. **Pubiflorae.**—*Perianth pubescent. Torus straight or rarely oblique, the gland thick or semiannular. Leaves entire, toothed or divided.*

Leaves flat. Fruit-valves without dorsal appendages (except
    in the two doubtful species).
    Leaves thick, tapering at the base, obscurely several-veined
      (unless very narrow). Western species.
      Leaves (mostly 3 in. or more), oblong-spathulate or lanceo-
        late, obtuse, entire. Fruit large and thick   . . . . 15. *H. Hookeriana.*
      Leaves (1 to 2 in.) oblong linear or linear-lanceolate, entire.
        Fruit large and thick . . . . 16. *H. incrassata.*
    Leaves fan-shaped, toothed at the end.
      Leaves truncate at the toothed end. Perianth-tube
        narrow, under 3 lines long . . . . 17. *H. flabellifolia.*
      Leaves rounded at the toothed end. Perianth-tube
        broad, above 3 lines long.
        Leaf-teeth short and callous. Seed-wing decurrent
          on both margins of the nucleus . . . . 18. *H. Brownii.*
        Leaf-teeth prickly. Seed-wing decurrent only on the
          upper margin of the nucleus . . . . 19. *H. Baxteri.*
    Leaves from broadly cuneate to long and narrow, mostly
      lobed or variously divided . . . . 20. *H. ceratophylla.*
Leaves not so thick, lanceolate, obscurely penniveined.
    Fruit much longer than broad.
    Leaves 1 to 2 in. long. Perianth densely villous. Wes-
      tern species . . . . 21. *H. lasiantha.*
    Leaves 3 to 5 in. long. Perianth silky. Eastern species . 22. *H. eriantha.*
Leaves thick and veinless or obscurely penniveined, but
      flowers unknown and therefore the affinities doubtful.
      Fruit-valves with dorsal appendages.
      Leaves obovate-oblong (1½ to 3 in.). Fruit above 2 in.
        long, 1½ in. broad . . . . 23. *H. megalosperma.*
      Leaves narrow-oblong (1 to 2 in.). Fruit ¾ in. long,
        under ½ in. broad . . . . 24. *H. clavata.*
Leaves mixed linear-terete and flat, or some or all linear-terete
    and divided. Western species.
    Leaves narrow-linear or rarely terete, entire or pinnate,
      grooved underneath. Fruit large, with a long straight
      beak . . . . 25. *H. orthorrhyncha.*

Leaves some linear-terete and some linear and flat, all entire,
   not grooved. Perianth loosely villous. Fruit rather large,
   with a short inflexed beak . . . . 26. *H. Candolleana.*
Leaves all or almost all terete and divided. Perianth villous.
   Fruit narrow.
   Leaves mostly 1½ to 3 in., with 3 segments ; a few often
      entire and flat. Stigmatic disk broad . . . . . . 27. *H. trifurcata.*
   Leaves mostly 1 in. with 3 or 5 segments. Stigmatic disk
      oblong-linear . . . . 28. *H. erinacea.*
Leaves all entire, linear-terete.
   Western species.
     Leaves thick, 3 to 5 in. long. Perianth above 4 lines.
      Fruit globular, smooth, 1½ to 2 in. diameter. Seeds
      winged all round . . . . 29. *H. platysperma.*
     Leaves slender, 1 to 3 in. Perianth 1 line. Fruit com-
      pressed, rugose, 1 in. diameter. Seeds winged all round 30. *H. brachyptera.*
     Leaves slender, 1 to 3 in. Perianth small. Fruit under
      1 in., smooth, thick, broadly beaked. Seed-wing decur-
      rent on one side . . . . 31. *H. Kippistiana.*
     Leaves thick, short. Perianth 2 lines. Fruit narrow, two-
      horned Seed-wing decurrent on one side . . . . . 32. *H. Preissii.*
   Eastern species.
     Fruit-beak straight or obsolete.
      Fruit narrow, with an obliquely transverse crestlike
        rugose prominence below the beak . . . . . . 33. *H. pugioniformis.*
      Fruit with a broad gibbous base and scarcely distinct
        broad beak, without any crest . . . . 34. *H. Pampliniana.*
      Fruit ovoid, nearly smooth, with a broad smooth dark
        line down each suture . . . . 35. *H. vittata.*
     Fruit recurved at the base, then incurved with an in-
      flexed beak.
      Fruit rugose. Stigmatic disk with a central cone.
        Fruit above 1 in. long, ¾ in. broad . . . . 36. *H. rostrata.*
        Fruit ½ to ¾ in. long, under ½ in. broad . . . . . 37. *H. rugosa.*
      Fruit smooth. Stigmatic disk flat . . . . . . 38. *H. epiglottis*

SERIES 3. **Glabrifloræ.**—*Perianth glabrous.* Torus straight or slightly oblique,
the gland semiannular or none.

Leaves flat, 1-nerved or obscurely penniveined. Western
   species, except *H. saligna.*
   Leaves prickly-toothed or lobed, stem-clasping at the base.
     Leaves 4 to 8 in. long, gradually expanded at the base
      into large prickly-toothed auricles . . . . 39. *H. amplexicaulis.*
     Leaves 1¼ to 3 in., more or less obovate or oblong-cu-
      neate, usually contracted near the base . . . . . 40. *H. glabella.*
     Leaves 1½ to 3 in., narrow, dilated at the end, with 3 to
      5 prickly teeth or lobes, auriculate at the base . . . 41. *H. auriculata.*
   Leaves distinctly petiolate, obovate, undulate and prickly-
     toothed . . . . 42. *H. cristata.*
   Leaves linear-lanceolate, sessile, prickly-toothed or entire . 43. *H. linearis.*
   Leaves 3 to 4 in., oblong-cuneate, entire. Umbels axillary,
     pedunculate . . . . 44. *H. stenocarpoides.*
   Leaves small, petiolate, entire, with a fine point. Branches
     hirsute with long hairs. Inflorescence terminal . . . 45. *H. ruscifolia.*
   Leaves lanceolate, obtuse, entire. Eastern species . . . 46. *H. saligna.*
Leaves terete. Eastern species.
   Style at least twice as long as the perianth.
     Leaves undivided. Fruit 2 horned. Seed-wing scarcely
      decurrent . . . . 47. *H. verrucosa.*

Leaves divided. Fruit scarcely horned. Seed-wing de-
current all round the nucleus . . . . 48. *H. purpurea.*
Style not twice as long as the perianth.
Pedicels pubescent or hirsute, perianth alone glabrous.
  Branches densely villous. Fruit as broad as long, with
  a short incurved beak . . . . 49. *H. gibbosa.*
  Branches glabrous or scarcely pubescent.
    Perianth 1 line long.
      Fruit as broad as long, straight and obtuse.
      Leaves rather stout . . . . 50. *H. propinqua.*
      Fruit narrower than long, with a slightly incurved
      beak. Leaves slender . . . . 51. *H. nodosa.*
    Perianth 2 lines long or more . . . . 52. *H. acicularis.*
Pedicels glabrous as well as the flower.
  Flowers in short racemes, the peduncle and rhachis ½ to
  ¾ in. long . . . . 53. *H. leucoptera.*
  Flowers in sessile clusters.
    Leaves all terete. Fruit 1¼ in. long and nearly as
    broad. Seed-wing decurrent all round the nucleus 54. *H. cycloptera.*
    Leaves some terete and some flat. Fruit ½ in. long
    ¼ in. broad. Seed-wing not decurrent . . . . 55. *H. microcarpa.*
Leaves terete. Western species. Pedicels always glabrous.
  Leaves 3 to 5 in., rigid, thick, spreading or recurved. Peri-
  anth 3 lines long . . . . 56. *H. recurva.*
  Leaves ¾ to 1¼ in. long, terete and smooth.
    Branches tomentose. Perianth 2 lines. Racemes or
    clusters mostly terminal. Seed-wing decurrent all round
    the nucleus . . . . 57. *H. circumalata.*
    Quite glabrous. Racemes or clusters mostly axillary.
    Perianth near 3 lines. Fruit about ½ in. long. Seed-
    wing decurrent on one side of the nucleus . . . . 58. *H. commutata.*
    Perianth 2 lines. Fruit 1½ to 2 in. long, very thick
    and broad. Seed-wing decurrent all round the
    nucleus . . . . 59. *H. strumosa.*

SECT. 3. **Conogynoides.**—*Racemes usually short or reduced to sessile clusters, rarely elongated, enclosed before their development in an involucre or bud of imbricate scales. Perianth glabrous, revolute under the limb. Stigmatic cone erect or scarcely oblique.*

SERIES 1. **Longistylæ.**—*Leaves flat, entire, oblong-lanceolate or narrow. Style at least twice as long as the perianth. Stigmatic cone long and narrow. Western species, one also central.*

Leaves (6 to 8 in.) very finely many-nerved. Racemes oblong,
1 to 3 in. long . . . . 60. *H. multilineata.*
Leaves (4 to 6 in.) 3- or 5-nerved, often falcate, on long petioles.
Racemes globular . . . . 61. *H. laurina.*
Leaves (under 3 in.) 3-nerved, sessile . . . . 62. *H. obtusa.*
Leaves (4 to 7 in.) very thick, narrow, very prominently tripli-
nerved, tapering at the base . . . . 63. *H. cinerea.*
Leaves (under 3 in.) thick, 1-nerved, tapering to the base.
Flower-clusters almost verticillate . . . . 64. *H. corymbosa.*

SERIES 2. **Petiolares.**—*Leaves broad, triplinerved and reticulate (the veins prominent or obscure), tapering into a distinct petiole. Style not twice as long as the perianth. Western species.*

Leaves prickly-toothed, prominently-veined . . . . . . 65. *H. undulata.*

Leaves quite entire.
　　Leaves prominently veined. Perianth near 3 lines long.
　　　　Fruit with a rather long straight beak.
　　　　Leaves acuminate. Petiole long. Seed-wing decurrent on
　　　　　　both margins . . . . 66. *H. petiolaris.*
　　　　Leaves scarcely acuminate. Petiole short. Seed-wing de-
　　　　　　current on one margin only . . . . 67. *H. neurophylla.*
　　Leaves obscurely veined. Perianth about 2 lines. Fruit
　　　　with a short slightly curved beak . . . . 68. *H. loranthifolia.*

　　Series 3. **Sessiles.**—*Leaves broad to oblong-lanceolate, 3- or more-nerved and reticulate, sessile or nearly so. Western species.*

Leaves (of the flowering stems) broadly orbicular-cordate, entire
　　or prickly-toothed . . . . 69. *H. cucullata.*
Leaves ovate to ovate-lanceolate, acute or acuminate, entire.
　　Leaves glabrous, often cordate, shortly acuminate. Fruit 1 in.
　　　　long, including the long beak . . . . 70. *H. ferruginea.*
　　Leaves usually pubescent, acutely acuminate, never cordate.
　　　　Fruit ½ in. long, not distinctly beaked . . . . . . 71. *H. smilacifolia.*
Leaves obtuse or with a small callous point.
　　Leaves broadly oval or elliptical, several-nerved . . . . 72. *H. elliptica.*
　　Leaves oblong-elliptical to lanceolate, 3-nerved . . . . . 73. *H. ambigua.*

　　Series 4. **Nervosæ.**—*Leaves lanceolate or linear, prominently 3- or more-nerved
(rarely 2- or 1-nerved when very narrow), smooth between the nerves.*

Eastern species. Seed-wing decurrent on one side of the nucleus.
　　Leaves 4 to 6 in. long, falcate, 5- to 9-nerved. Perianth 2 lines
　　　　long . . . . 74. *H. plurinervia.*
　　Leaves lanceolate or linear-lanceolate, usually 3-nerved.
　　　　Perianth 1 line long.
　　　　Leaves usually lanceolate, obtuse or acute, not pungent.
　　　　　　Pedicels silky-pubescent . . . . 75. *H. dactyloides.*
　　　　Leaves usually linear-lanceolate or linear, pungent-pointed.
　　　　　　Pedicels glabrous . . . . 76. *H. ulicina.*
Western species. Seed-wing decurrent all round the nucleus.
　　Leaves linear lanceolate, 3-nerved, under 3 in. long . . . 77. *H. falcata.*
　　Leaves linear, very thick and silky, 4 to 8 in. long . . . 78. *H. pycnoneura.*

　　Series 5. **Uninerves.**—*Leaves narrow or small, flat, 1-nerved with nerve-like
margins. Fruit-valves without dorsal appendages. Western species.*

Leaves linear, very thick and silky, 4 to 8 in. long . . . . 78. *H. pycnoneura.*
Leaves linear or lanceolate, 1½ to 4 in. long. Fruit 1 to 1½ in.
　　long, 2 to 3 lines broad . . . . 79. *H. stenocarpa.*
Leaves lanceolate, ¾ to 1½ in. long. Fruit ¾ in. long, 3 to 4
　　lines broad . . . . 80. *H. marginata.*
Leaves mostly ovate, ½ to ¾ in. long, with a long point . . . 81. *H. myrtoides.*
Leaves linear, pungent-pointed, under ½ in. long, the keel or
　　midrib very prominent underneath . . . . 82. *H. costata.*

　　Series 6. **Enerves.**—*Leaves mostly flat, obscurely penniveined, the midrib not
prominent, entire prickly-toothed or lobed. Fruit-valves with dorsal horns or protuberances. Western species.*

Leaves oblong-lanceolate, quite entire or very rarely with 1 or
　　2 minute teeth . . . . 83. *H. oleifolia.*
Leaves lanceolate, acute, mostly prickly-toothed . . . . . 84. *H. florida.*
Leaves more or less cuneate or pinnatifid with prickly teeth or
　　lobes, thick and here and there almost terete . . . . . 85. *H. varia.*

SERIES 7. **Teretifoliæ.**—*Leaves linear-terete entire, usually angular or striate (rarely a few flat ones intermixed). Fruit-valves without dorsal appendages. Western species except* H. flexilis.

Leaves sulcate, occasionally flat.
  Fruit ovoid ; beak straight. Flowers in close clusters . . 86. *H. sulcata.*
  Fruit small, ovoid-globular, with an inflexed beak. Flowers
    in short dense racemes . . . . 87. *H. Meissneriana.*
Leaves very finely striate. Fruit-beak straight . . . . . 88. *H. subsulcata.*
Leaves angular or terete and smooth.
  Fruit muricate with fringed tubercles. Western species . 89. *H. Lehmanniana.*
  Fruit smooth or rugose. · Eastern species . . . . 90. *H. flexilis.*

SECT. 4. **Manglesioides.**—*Racemes short or reduced to sessile clusters, enclosed before their development in an involucre or bud of imbricate scales. Perianth glabrous, straight, the limb erect in the bud. Stigmatic cone erect. Western species.*

Leaves obovate oblong or lanceolate, entire or prickly-toothed . 91. *H. nitida.*
Leaves terete (under 2 in.), smooth, undivided, pungent-pointed 92. *H. Oldfieldii.*
Leaves terete (3 to 4 in.), simply pinnate or also undivided,
  stout . . . . 93. *H. suaveolens.*
Leaves terete (1 to 1½ in.), twice or simply pinnate with few
  segments.
  Rhachis of the clusters under ¼ in. long. Fruits ½ to ¾ in.
    long, with short dorsal horns . . . . 94. *H. lissocarpa.*
  Rhachis of the racemes ¼ to ½ in. long. Fruits ¾ to 1 in.
    long, with long dorsal horns . . . . 95. *H. bipinnatifida.*

H. carduifolia, Lodd., *H. echinata,* Mackay, *H. Lamberti,* Sweet, *H. latifolia,* Lodd., and *H. subulata,* Cunn., entered in Steud. Nom. Bot. ed. 2, are garden names without descriptions, belonging probably to some of the species here enumerated.
    H. longifolia and *H. tenuifolia,* Dum. Cours. in Roem. and Schult. Syst. iii. 425, are also garden plants described only as to their foliage, and quite insufficiently for recognition.

SECT. 1. GREVILLEOIDES.—Flowers in oblong or cylindrical or rarely short racemes, without any involucre as far as known. Perianth much revolute, opening early on the lower side. Stigmatic disk oblique or lateral, flat or broadly conical. Species all tropical or subtropical.

1. **H. chordophylla,** *F. Muell. in Hook. Kew Journ.* ix. 23. A tree perfectly glabrous and more or less glaucous. Leaves terete, smooth, mostly from ½ to 1 ft. long. Racemes from the old wood, loosely cylindrical, 3 to 4 in. long, quite glabrous. Pedicels about ¼ in. long. Perianth glabrous, the tube 4 to 5 lines long, much dilated and oblique at the base, revolute under the limb. Torus very oblique. Gland large, horseshoe-shaped. Ovary stipitate ; style long ; stigmatic disk broad, oblique, not convex. Fruit obliquely lanceolate, about 1½ in. long and ½ in. broad near the base, tapering into a short slightly incurved beak. Seed-wing not decurrent along the nucleus.—Meissn. in DC. Prod. xiv. 699.

N. Australia. Sturt's Creek, *F. Mueller* (the perianth-tube scarcely 4 lines long) ; Kekwick springs, *Waterhouse* (the perianth-tube fully 5 lines long).

2. **H. Cunninghamii,** *R. Br. Prot. Nov.* 26. A small tree of 12 to 16 ft. Leaves terete, rigid, mostly above 1 ft. long, but usually not so long as in *H. lorea.* Racemes lateral on the old wood, loosely cylin-

drical, 3 to 6 in. long, the rhachis pedicels and perianths clothed with appressed silky hairs. Pedicels 3 to 5 lines long. Perianth-tube 4 to 5 lines long, somewhat dilated and very oblique at the base, attenuate and revolute under the limb. Torus very oblique. Gland large, horseshoe-shaped. Ovary on a long stipes. Style very long, slightly clavate under the oblique convex or broadly conical stigmatic disk. Fruit obliquely ovate-lanceolate, 1¼ to 1½ in. long and ¾ in. thick. Seed-wing not decurrent along the nucleus or only very shortly so on the upper margin.—Meissn. in DC. Prod. xiv. 394; *H. longifolia*, A. Cunn. Herb.; F. Muell. Fragm. vi. 190.

N. Australia. Bay of Rest, N.W. coast, *A. Cunningham;* Nichol Bay, *F. Gregory's Expedition;* Victoria river, *F. Mueller.*

3. **H. lorea,** *R. Br. Prot. Nov.* 25. A tall shrub or tree attaining 20 ft. Leaves terete, smooth, often above 2 ft. long and rarely under 1 ft., very rarely (on barren branches? or young plants?) a few once or even twice forked or trifid. Racemes cylindrical, in the upper axils, sometimes forked or in a terminal cluster, more dense than in *H. Cunninghamii,* from under 3 in. to fully 6 in. long, the rhachis pedicels and perianths densely pubescent with shorter hairs much less appressed than in *H. Cunninghamii.* Perianth-tube nearly 4 lines long, slightly dilated below the middle, revolute upwards. Torus oblique but less so than in *H. Cunninghamii.* Gland large, horseshoe-shaped. Ovary stipitate; style long, with a very oblique broadly conical stigmatic disk. —Meissn. in DC. Prod. xiv. 394; F. Muell. Fragm. vi. 189; *Grevillea lorea,* R. Br. in Trans. Linn. Soc. x. 177, Prod. 380.

N. Australia. Attack Creek, *M'Douall Stuart's Expedition.*
Queensland. Shoalwater Bay, *R. Brown;* Port Denison, *Fitzalan;* Rockhampton, *Thozet;* Cape river and Nerkool Creek, *Bowman;* Dyngie, *Miss Ross;* also in *Leichhardt's* collection.

Several of the above-quoted specimens are not in flower, and are therefore in some measure doubtful.

4. **H. Fraseri,** *R. Br. Prot. Nov.* 26. A tall shrub, the branches much more slender than in *H. lorea,* of which it may possibly be a variety. Leaves much more slender, 4 to 8 in. long. Racemes only 1 to 2 in. long. Pedicels 2 to 3 lines. Flowers of *H. lorea,* but smaller, the perianth-tube not above 3 lines long. Fruit unknown in both species.—Meissn. in DC. Prod. xiv. 394.

N. S. Wales. Hastings river, *Fraser,* and probably from the same neighbourhood, *Herb. F. Mueller,* apparently from *Leichhardt.*

5. **H. macrocarpa,** *A. Cunn. in R. Br. Prot. Nov.* 30. A tree of 15 to 20 ft. of robust growth with a rugged bark. Leaves linear-lanceolate, 6 to 8 in. long, 2 to 5 lines broad, mostly obtuse, tapering at the base, thick, minutely silky-pubescent on both sides, the veins longitudinal, slightly anastomosing, scarcely prominent. Racemes loosely cylindrical, 3 to 6 in. long, tomentose-pubescent as in *H. lorea.* Pedicels 2 to 5 lines long. Perianth-tube about 5 lines, oblique, dilated

below the middle, revolute upwards. Torus oblique. Gland large,
semicircular. Ovary shortly stipitate, style rather thick, not long;
stigmatic disk oblique, broadly conical. Fruit ovate-lanceolate, 1½ to
1¾ in. long, 7 to 8 lines broad.—Meissn. in DC. Prod. xiv. 411; F.
Muell. Fragm. vi. 191; *Grevillea Alphonsiana,* F. Muell. in Hook. Kew
Journ. ix. 22; Meissn. l.c. 699.

N. Australia. Arid shores of Cygnet Bay, N.W. coast, *A. Cunningham;* Roebuck Bay and Glenelg river, *Martin;* remotest parts of Sturt's Creek, *F. Mueller.*
W. Australia. Three hundred miles up the Murchison river, *Walcott.*

6. **H. arborescens,** *R. Br. in Trans. Linn. Soc.* x. 187, *Prod.*
386. A tall shrub or small tree, the young shoots silky. Leaves
linear linear-lanceolate or sometimes rather broader and slightly
falcate, obtuse or acuminate, contracted at the base, longitudinally but
obscurely several-veined, minutely silky-pubescent on both sides, 3 to
6 in. long. Flowers small, in short dense almost globular racemes or
clusters on lateral peduncles of ¼ to ½ in., often on the old wood. Pedicels filiform, 1 to 2 lines long. Perianth silky-pubescent, the tube
slender, scarcely 1½ lines long, revolute under the limb. Torus straight.
Gland semiannular, not very prominent. Ovary shortly stipitate. Stigmatic disk nearly straight with a short broad cone. Fruit nearly
straight, 1½ to nearly 2 in. long, 1 in. broad, very shortly beaked. Seed-
wing decurrent along the upper margin of the nucleus to the base.—
Meissn. in DC. Prod. xiv. 410.

N. Australia. Islands of the Gulf of Carpentaria, *R. Brown, Henne;* Copeland
island, *A. Cunningham;* Victoria river, *F. Mueller;* Port Essington, *Armstrong.*
Queensland. Rockingham Bay, *Dallachy;* Mount Elliott, *Fitzalan.*

7? **H. stenophylla,** *A. Cunn.; Meissn. in DC. Prod.* xiv. 417. A
small tree of 12 to 15 ft., glabrous or the young shoots minutely hoary.
Leaves very narrow-linear, entire, flat, 2 to 4 in. long, rather thick,
veinless or with a scarcely prominent midrib. Flowers unknown.
Fruit rather above 1 in. long, ¾ in. thick, smooth with a very short
conical straight beak. Seed-wing broad, more or less decurrent all
round the nucleus.

N. Australia. Arid sands, Bay of Rest, N.W. coast, *A. Cunningham.*—Appears
allied to *H. arborescens,* but the affinities must be very uncertain until the flowers are
known.

8. **H. trineura,** *F. Muell. Fragm.* iii. 146. A tree?, the branches
and foliage very minutely hoary-tomentose. Leaves oblong-cuneate or
lanceolate, obtuse, contracted into a short petiole, 3 to 5 in. long, rather
thick, triplinerved and very obliquely almost longitudinally veined, the
margins nerve-like, not recurved, the veins equally conspicuous on
both sides. Racemes axillary, about 2 in. long. Pedicels ½ to 1 in.
long. Perianth glabrous as well as the whole inflorescence, the tube
fully 3 lines long, revolute under the globular limb. Torus straight.
Gland thick, semicircular. Ovary shortly stipitate; style long; stigmatic disk oblique, broad, conical in the centre. Fruit 1½ in. long, ¾
in. thick, smooth, slightly incurved, scarcely beaked. Seed-wing shortly

decurrent on the upper margin of the nucleus.—*Grevillea trineura*, F. Muell. l.c.

Queensland.   Broad Sound, *Bowman;* Rockhampton, *Thozet.*

SECT. 2. EUHAKEA.—Racemes usually short or reduced to sessile clusters, enclosed before their development in an involucre or bud of imbricate scales, falling off very early. Perianth revolute either from the middle or close under the limb. Stigmatic disk oblique or lateral, flat or slightly convex, very rarely (in *H. rugosa* and *H. rostrata*) with a central cone, and those species differing from *Conogynoides* in the pubescent perianth and short cone.

SERIES 1. OBLIQUÆ.—Perianth pubescent. Torus oblique, the ovary inserted at the shortest margin, the remainder occupied by a large very concave adnate gland.

9. **H. cyclocarpa,** *Lindl. Swan Riv. App.* 36. A shrub of 5 or 6 ft., the branches and foliage glabrous in our specimens. Leaves oblong-lanceolate, acute or obtuse, but not rounded at the end, tapering from the middle into a short petiole, 4 to 8 in. long, thick, entire, with obscure longitudinal veins, sometimes nearly 1½ in. broad in the middle. Flowers in axillary clusters or very short racemes. Pedicels silky-pubescent, 2 to 3 lines long. Perianth silky-pubescent, about 5 lines long, reflexed under the globular limb. Torus oblique, the greater portion occupied by a large very concave gland. Ovary shortly stipitate at the lower margin; style not long; stigmatic disk oblique. Fruit 1½ in. long, 1 in. broad, compressed, recurved at the base, then incurved, with a broad inflexed beak, the valves with a dorsal truncate protuberance at the top. Seed-wing narrowly decurrent on the upper margin.—Meissn. in Pl. Preiss. i. 573, and in DC. Prod. xiv. 415.

W. Australia.   Swan river to King George's Sound, *Drummond, n.* 108, 279, 1st *coll. n.* 609, *Preiss, n.* 576, *Harvey.*—The shape of the fruit is near that of *H. rostrata* and its allies, the flowers and foliage very different.

10. **H. crassifolia,** *Meissn. in Pl. Preiss.* i. *570, and in DC. Prod.* xiv. 412. A tall shrub, attaining 12 to 15 ft., the branches closely tomentose, the adult foliage glabrous. Leaves mostly oblong, broad or narrow, rounded at the end with a small callous point, shortly contracted at the base, 1½ to 3 in. long, very thick, with obscure longitudinal veins. Racemes short, shortly pedunculate, axillary, the rhachis and peduncle together 3 or 4 lines long. Pedicels silky-pubescent, about 3 lines long. Perianth silky-pubescent, the tube fully 5 lines long, revolute under the ovoid almost acuminate limb. Torus oblique, the upper portion occupied by a large concave gland. Ovary nearly sessile on the lower margin; style not long, thickened under the oblique stigmatic disk. Fruit nearly 1½ in. long and broad, neither rugose nor cristate, yet not smooth. Seed-wing decurrent down both sides round the base of the nucleus.

W. Australia, *Drummond, 4th coll. n.* 293; sand plains, Kalgan river, *Oldfield;* Stirling Range to E. Mount Barren, *Maxwell.*

11. **H. pandanicarpa,** *R. Br. Prot. Nov.* 29. An erect shrub, rarely over 6 ft. high, the young shoots hoary or silky, the adult foliage glabrous. Leaves narrowly oblong-cuneate, obtuse with a callous point, tapering into a short petiole and often shortly decurrent on the branch, 2 to 4 in. long, thick and veinless or with a few obscure longitudinal veins. Racemes short, axillary, silky-pubescent, the rhachis often ½ in. long, the pedicels 3 to 4 lines. Perianth silky-pubescent, the tube about 5 lines long, revolute under the ovoid limb. Torus very oblique, the upper portion occupied by a large concave gland. Ovary nearly sessile, very spreading from the lower margin of the torus. Style not long; stigmatic disk oblong, almost lateral. Fruit the largest of the genus, ovoid-globular, 2 to 2½ in. diameter, covered with large conical protuberances resembling the drupes of a Pandanus fruit but very hard. Seed-wing surrounding the nucleus, but broader above than below.— Meissn. in DC. Prod. xiv. 412 ; Hook. Ic. Pl. t. 434.

W. Australia. Towards Cape Arid, *Baxter ;* gravelly coast hills, Stokes Inlet to Cape Le Grand, Cape Arid, Cape Paisley, and Russell Range, *Maxwell.*

12. **H. Roei,** *Benth.* Branches minutely hoary. Leaves lanceolate or oblong-linear, mucronate, sometimes falcate, tapering into a short petiole, 1½ to 2½ in. long, thick, veinless or obscurely marked with a few longitudinal veins. Flowers few in sessile axillary clusters. Pedicels 2 to 3 lines long, silky-ferruginous as well as the flowers. Perianth-tube 4 or 5 lines long, recurved under the globular limb. Torus very oblique, the upper portion occupied by a large concave gland. Ovary shortly stipitate from the lower margin of the torus ; style not long, slightly thickened under the broad lateral stigmatic disk.

W. Australia, *J. S. Roe.* The foliage is nearly that of *H. incrassata,* the flowers very different, the fruit unknown.

13. **H. adnata,** *R. Br. Prot. Nov.* 26. A rigid bushy shrub of 3 or 4 ft., the young branches hoary, the adult foliage glabrous. Leaves terete, smooth, with a straight or incurved point, attenuate at the base, mostly 2 to 3 in. long. Flowers few together in axillary clusters on silky pedicels of 1 to 2 lines. Perianth silky-pubescent, the tube 3 to 3½ lines long, revolute under the globular limb. Torus oblique, but not so much so as in *H. obliqua,* the greater portion occupied by the large concave gland. Ovary almost sessile at the lower margin of the torus ; style not very long, dilated at the end into a large oblique concave disk with the stigma prominent in the centre. Fruit 1 to 1½ in. long and 1 in. broad, very oblique, much compressed and smooth when nearly ripe, very thick and sometimes rugose when perfect, with a small conical inflexed beak, without dorsal appendages to the valves. Seed-wing as broad as the capsule, decurrent along the upper margin and very narrowly so along the lower margin of the nucleus.—Meissn. in DC. Prod. xiv. 396 ; *H. lativalvis,* F. Muell. Fragm. vi. 219.

W. Australia. South coast, *Baxter ;* sand plains north of Cape Arid, Esperance Bay and Russell Range, *Maxwell.*

14. **H. obliqua,** *R. Br. in Trans. Linn. Soc.* x. 180, *Prod.* 382. A spreading shrub of 2 or 3 ft., the branches minutely hoary-pubescent, the foliage glabrous. Leaves terete, smooth, rigid, mucronate, 1 to 2 in. long, thick and divaricate in some specimens, twice as long, thinner and more erect in others. Flowers in sessile axillary clusters. Pedicels 1 to 1½ lines long. Perianth silky, the tube 3 lines long or rather more, revolute under the ovoid shortly acuminate limb. Torus very oblique, the greater part occupied by a large concave gland. Ovary shortly stipitate at the lower margin of the torus; style not very long with a broad lateral stigmatic disk umbonate in the centre. Capsule recurved at the base, about 1 in. long and ½ in. broad, rugose, with a short straight conical beak, without dorsal appendages. Seed-wing decurrent along both margins round the base of the nucleus.—Meissn. in DC. Prod. xiv. 395.

W. Australia. Lucky Bay, *R. Brown;* between Swan river and King George's Sound, *Drummond, 2nd coll. n.* 329, 330 ; scrubs north of Stirling Range, *F. Mueller;* Cape Paisley, Cape Le Grand and Orleans Bay, *Maxwell.*

SERIES 2. PUBIFLORÆ.—Perianth pubescent. Torus straight or if oblique the gland, thick or semiannular, inserted on the lowest side.

15. **H. Hookeriana,** *Meissn. in DC. Prod.* xiv. 412. A tree of 15 to 20 ft., the young branches minutely silky-pubescent, the foliage glabrous. Leaves oblong-spathulate or oblanceolate, very obtuse, entire, tapering into a short petiole, 3 to 4 in. long, ½ to 1 in. wide, very thick, with faintly visible longitudinal veins. Flowers small, in sessile axillary clusters. Pedicels silvery-silky, not 1 line long. Perianth pubescent with small appressed reddish hairs, the tube about 2 lines long, rather broad, reflexed under the ovoid-globular limb. Torus straight. Gland semiannular. Ovary nearly sessile; style not long; stigmatic disk oblique. Fruit very thick, almost didymous, oblique, 2 to 2½ in. long and nearly 2 in. broad, broadly furrowed along the upper suture, rugose with irregular raised lines, the beak very short or obtuse. Seed-wing scarcely decurrent.

W. Australia. *Drummond, 5th coll. n.* 413 ; E. Mount Barren Range, *Maxwell.* —The contrast of the white pedicels and reddish flowers is very evident in the dried specimens.

16. **H. incrassata,** *R. Br. Prot. Nov.* 29. A shrub of 2 or 3 ft., with tomentose branches. Leaves oblong-linear or linear-lanceolate, acute but scarcely pungent, entire, contracted into a very short petiole, 1 to 2 or rarely 3 in. long, thick and veinless or obscurely 3-nerved. Flowers very small, in sessile axillary clusters. Pedicels about 2 lines long, hirsute. Perianth hirsute, the tube scarcely 1½ lines long, much revolute, the upper segments short. Torus small. Gland rather broad. Ovary very shortly stipitate; style short, with a large lateral stigmatic disk. Fruit globular, 1 to near 1½ in. diameter, smooth with a few rough blotches, slightly furrowed at the sutures, with a very short inflexed beak on the upper side. Seed-wing broadly decurrent down

both sides of the nucleus.—Meissn. in DC. Prod. xiv. 411 ; Hook. Ic.
Pl. t. 442 ; *H. leucadendron,* Meissn. in Pl. Preiss. i. 572, and in DC.
Prod. xiv. 411.

W. Australia. Swan river, *Fraser, J. S. Roe, Drummond, n.* 100, *1st coll. n.*
603, *Preiss, n.* 568, 578 ; Belgarup, *Oldfield.*

17. **H. flabellifolia,** *Meissn. in Hook. Kew Journ.* vii. 116, *and in DC.
Prod.* xiv. 409. An erect shrub, with minutely hoary branches or quite
glabrous. Leaves cuneate but not so broad as in *H. Baxteri* and rather
truncate than rounded at the toothed end, tapering into a long
petiole, 1½ to 2 in. long, ¾ to 1 in. broad at the end, very thick and
obscurely veined. Flowers in axillary clusters, much smaller than in
H. Baxteri. Pedicels not 1 line long. Perianth silky-pubescent, the
tube under 3 lines long, revolute under the globular limb. Torus
nearly straight. Gland thick, not very large. Ovary very shortly
stipitate ; style not very long, with an oblique stigmatic disk. Fruit
unknown.

W. Australia. Between Moore and Murchison rivers, *Drummond, 6th. coll. n.*
196.

18. **H. Brownii,** *Meissn. in Pl. Preiss.* i. 569, *and* ii. 261, *and in DC.
Prod.* xiv. 409. Very closely allied to *H. Baxteri,* and perhaps a variety.
Leaves rather thicker, the veins less conspicuous and the teeth shorter
and less prickly. Flowers quite the same. Fruit rather broader than
in *H. Baxteri,* the very small beak more oblique and the seed-wing
decurrent down both margins of the nucleus.

W. Australia, *Drummond, 4th coll. n.* 296, *Preiss, n.* 552.

19. **H. Baxteri,** *R. Br. Prot. Nov.* 28. An erect shrub of 6 to 8 ft.,
glabrous or the young shoots and branches minutely hoary-pubescent.
Leaves very broadly cuneate fan-shaped or almost reniform, the
broad rounded end undulate and shortly prickly-toothed, contracted
into a distinct but broad petiole, shortly decurrent on the stem, thick,
with obscure radiating branching veins, 1½ to 2½ in. broad. Flowers
in sessile axillary clusters. Pedicels rarely above 1 line long. Perianth
ferruginous-pubescent, the tube broad, 3 to 3½ lines long, contracted
above the middle and revolute under the ovoid limb. Torus oblique.
Gland thick, semicircular. Ovary very shortly stipitate, spreading
from the base ; style long ; stigmatic disk oblique. Fruit about 1½ in.
long, 1 in. broad, rugose, shortly beaked. Seed-wing decurrent down
the upper margin of the nucleus to the base, but not along the lower
margin.—Meissn. in Pl. Preiss. i. 569, and in DC. Prod. xiv. 409 ;
Hook. Ic. Pl. t. 439.

W. Australia. King George's Sound or to the eastward, *Baxter, Drummond, n.*
161, *4th coll. n.* 295 ; Cape Riche, *Preiss, n.* 553, *Maxwell ;* Stirling Range, *F. Mueller.*

20. **H. ceratophylla,** *R. Br. in Trans. Linn. Soc.* x. 184, *Prod.* 384.
An erect or spreading shrub of 2 to 5 ft., the young shoots slightly
silky-pubescent, the adult foliage glabrous. Leaves remarkably diver-

sified in form, usually narrow, more or less cuneate, tapering into a
long base, irregularly and deeply divided in the upper half into 3 unequal coarsely toothed lobes, but a few leaves sometimes quite entire
and linear or linear-lanceolate, or the whole leaf narrow and pinnately
divided into short and broad or long and narrow divaricate lobes, or the
3 lobes long, narrow, and toothed, or all 3 broadly cuneate and 3- or 5-
lobed, the whole leaf 2 to 4 in. long, or longer when very narrow,
thick and irregularly almost longitudinally veined, the veins obscure or
equally prominent on both sides, the lobes or teeth pointed and often
pungent. Flowers in small sessile clusters, axillary or at the old nodes.
Pedicels 1 to 2 lines long, silky. Perianth ferruginous-pubescent, the
tube 2 to 2½ lines long, revolute under the globular limb. Torus
oblique. Gland large, very prominent, concave. Ovary very shortly
stipitate and spreading from the lower margin of the torus; style not
very long, with a broad oblique stigmatic disk. Fruit nearly 1 in. long,
about ½ in. broad. Seed-wing decurrent on the upper margin only of
the nucleus.—Meissn. in Pl. Preiss. i. 569, and in DC. Prod. xiv. 410;
Reichb. Ic. et Descr. Pl. t. 24; *Conchium ceratophyllum*, Sm. in Trans.
Linn. Soc. ix. 124; *Hakea acanthophylla*, Link. Enum. Hort. Berol. i.
118; *H. laciniosa*, F. Muell. Fragm. iv. 49.

W. Australia. King George's Sound and adjoining districts, *R. Brown, Drummond, n.* 16, *2nd coll. n.* 334, *Preiss, n.* 579, 580, and many others.

Var *elongata.* Leaves narrow-linear, 4 to 6 in. long, entire or with a few distant linear
lobes, but readily distinguished from those of *H. orthorrhyncha* by the midrib if present,
prominent on both sides and not grooved underneath.—*Drummond,* (*3rd coll. ?*) *n.* 297.

21. **H. lasiantha,** *R. Br. Prot. Nov.* 29. A shrub attaining 6 to 8 ft.,
the branches densely tomentose-villous. Leaves oblong lanceolate or
oval-elliptical, with a short rigid callous point, contracted into a short
petiole, 1 to 1½ or sometimes 2 in. long, thick, entire, nerveless or very
obscurely and almost longitudinally penniveined, ferruginous-silky
when young, at length glabrous. Flowers not numerous, in axillary
clusters. Pedicels villous, 1 to 2 lines long. Perianth densely villous
with spreading hairs, the tube 3 to 4 lines long, recurved under the limb.
Torus small. Gland prominent. Ovary nearly sessile; style short,
with a large lateral stigmatic disk. Fruit about 1 in. long and 4 lines
broad, somewhat incurved, smooth, with a short beak. Seed-wing
decurrent on the upper side only to below the nucleus.—Meissn. in Pl.
Preiss. i. 571, and in DC. Prod. xiv. 411.

W. Australia. King George's Sound or adjoining districts, *Baxter, Drummond,
n.* 96, and others; Arthur river, *Oldfield;* Stirling range and Gordon river, *Maxwell.*

Var. *angustifolia.* Leaves lanceolate, 1½ to 3 in. long, approaching those of *H.
eriantha.* Fruit 1¼ in. long, 3 to 4 lines broad.—W. Australia, *Drummond, n.* 21.

22. **H. eriantha,** *R. Br. Prot. Nov.* 29. A tall shrub or small tree,
the young shoots silky-pubescent, the adult foliage glabrous. Leaves
lanceolate, acuminate, acute or with a callous point, tapering into a
short petiole, veinless except the scarcely prominent midrib, of a pale
colour like those of *H. saligna* but rather thicker, 3 to 5 in. long.

Flowers in axillary clusters, not very numerous. Pedicels silky-villous, 1 to 2 lines long. Perianth silky, the tube about 3 lines long, reflexed under the almost acute limb. Torus small. Gland prominent, semi-annular. Ovary shortly stipitate; style not very long, with a large lateral stigmatic disk. Fruit about 1 in. long and under ½ in. broad, slightly incurved, rather smooth, with a short incurved or straight beak, very obscure when the fruit is quite ripe. Seed-wing very shortly decurrent on the upper side only of the nucleus.—Meissn. in DC. Prod. xiv. 417.

N. S. Wales. Mount Lindsay, *Fraser ;* Hastings river, *Beckler ;* New England, *C. Stuart ;* Twofold Bay, *L. Morton.*
Victoria. Tambo, Mitta-Mitta to Genoa river in Gipps' Land, *F. Mueller.*

23 ? **H. megalosperma,** *Meissn. in Hook. Kew Journ.* vii. 117, *and in DC. Prod.* xiv. 417. A tall shrub quite glabrous, the foliage glaucous. Leaves obovate-oblong to narrow oblong, very obtuse, tapering into a short petiole, 1½ to 3 in. long, very thick, veinless or obscurely penni-veined with the very faint primary veins very oblique. Flowers unknown. Fruit 2 to 2½ in. long, 1½ in. broad, smooth, scarcely beaked, the valves with dorsal ridges near the end forming prominent truncate appendages. Seed-wing broadly decurrent on both sides round the base of the nucleus.

W. Australia. Mount Lesueur, between Moore and Murchison rivers, *Drummond, 6th coll. n.* 154.

The affinities of this and the following species must remain very uncertain until the flowers shall be known.

24 ? **H. clavata,** *Labill. Pl. Nov. Holl.* i. 31, *t.* 41. A rigid stout spreading shrub of 3 or 4 ft., our specimens quite glabrous. Leaves narrow-oblong, obtuse with a short very rigid point, contracted at the base, 1 to 2 in. long, very thick and veinless. Flowers unknown: Fruit about ¾ in. long, 4 or 5 lines broad, obtuse, not rugose, the valves with dorsal conical horns at the end. Seed-wing narrowly decurrent at least on the upper side of the nucleus.—R. Br. in Trans. Linn. Soc. x. 187, Prod. 386; Meissn. in DC. Prod. xiv. 417; *Conchium clavatum,* Willd. Enum. Hort. Berol. 141.

W. Australia. King George's Sound or to the eastward, *Labillardière, R. Brown, Baxter ;* Cape Arid, *Maxwell.*

25. **H. orthorrhyncha,** *F. Muell. Fragm.* vi. 214. A shrub of 2 or 3 ft., the young branches minutely hoary, the foliage glabrous. Leaves in the typical form narrow-linear, entire or pinnately divided into 3 to 5 segments, mucronate, 3 to 6 in. long, thick but flat, with the margins so closely revolute as to leave only a very narrow depressed line between them on the under side, sometimes contracted into a short terete petiole. Flowers in small sessile clusters chiefly on the old wood. Pedicels 1 to 1½ lines long. Perianth silky-pubescent, the tube shortly dilated almost gibbous at the base, the longer segments about 5 lines long, much revolute above the middle. Torus straight. Gland very

prominent, semiannular. Ovary shortly stipitate; style very long, with a lateral stigmatic disk. Fruit 1½ in. long or rather more, ¾ in. broad, smooth, tapering into a rather long straight beak. Seed-wing broadly decurrent on the upper side of the nucleus, less so on the lower side.

W. Australia. Sandy plains, Murchison river, *Oldfield*.

Var. *filiformis*, F. Muell. Leaves narrower, more often divided, sometimes terete, less distinctly and sometimes not at all grooved.—W. Australia, *Drummond*, in fruit only, but probably the same species.

26. **H. Candolleana,** *Meissn. in Pl. Preiss.* ii. 262, *and in DC. Prod.* xiv. 397. Branches virgate, pubescent when young, the foliage glabrous. Leaves some and sometimes nearly all narrow-linear, obtuse or with a callous point, tapering into a short petiole, thick but flat with very obtuse thickened margins, the midrib more or less prominent underneath; others or sometimes nearly all terete or slightly flattened and nerveless, mostly 1 to 2 in. long. Flowers few, in sessile axillary clusters. Pedicels villous, about 1 line long. Perianth villous with spreading hairs, 1½ to near 2 lines long but very much revolute with the upper segments short. Torus straight. Gland prominent, semi-annular. Ovary contracted into a very short stipes; style not long, with a broad lateral stigmatic disk. Fruit 1 to 1¼ in. long, 7 to 10 lines broad, incurved, nearly smooth, with a short conical inflexed beak. Seed-wing decurrent on both margins of the nucleus.—*H. falcata, var.* Meissn. in Pl. Preiss. i. 572.

W. Australia, *Drummond, n.* 99, 1st *coll. n.* 605, 2nd *coll. n.* 331 ; Canning river, *Preiss, n.* 603 ; Hampden, *Clarke*.

Var. *campylorrhyncha*, F. Muell. Leaves nearly all flat and 1½ lines broad. —Murchison river,*'Oldfield*.

The S. Australian *H. flexilis*, R. Br., to which Meissner refers some of Drummond's specimens, is a very different plant in flowers and fruit, although somewhat similar in foliage.

27. **H. trifurcata,** *R. Br. in Trans. Linn. Soc.* x. 183, *Prod.* 383. A much-branched shrub, attaining 8 or 10 ft., the young shoots silky-pubescent, the adult foliage glabrous. Leaves mostly terete, slender but rigid, divided into 3 divaricate segments as long as the entire base or undivided, acute and usually pungent-pointed, smooth or slightly grooved, the whole leaf 1½ to 3 in. long; mixed with these are usually a few flat oval oblong obtuse or sometimes lanceolate and acute entire leaves ¾ to 1½ in. long. Flowers not very numerous, in sessile axillary clusters. Pedicels silky, about 2 lines long. Perianth silky-villous or hirsute, the tube straight, 3 lines long, the limb slightly recurved. Torus small. Gland very prominent, truncate, half cup-shaped. Ovary almost sessile ; style not long, with a large lateral stigmatic disk. Fruit ¾ in. long or rather less, ¼ in. broad, slightly incurved, obtuse, smooth. Seed-wing not decurrent.—Meissn. in Pl. Preiss. i. 558, and in DC. Prod. xiv. 404 ; *Conchium trifurcatum*, Sm. in Trans. Linn. Soc. ix. 122 ; *H. mixta* and *H. tricruris*, Lindl. Swan Riv. App. 35.

W. Australia. King George's Sound, *R. Brown, Baxter, A. Cunningham,* and others; Swan river to King George's Sound, *Drummond, n.* 102, 103, *1st coll. n.* 626, *Preiss, n.* 619 ; Murchison river, *Oldfield.*

H. Boucheana, Kunth. Ind. Sem. Hort. Berol. 1844, in Linnæa, xviii. 499 ; Meissn. in DC. Prod. xiv. 404, said to have been raised from Tasmanian seeds, is nevertheless probably the Western *H. trifurcata,* the only difference relied upon being in the flat leaves described as lanceolate and .pungent-pointed, but that occurs here and there in several of the King George's Sound specimens.

28. **H. erinacea,** *Meissn. in Pl. Preiss.* i. 559, *and in DC. Prod.* xiv. 404. An erect shrub of 2 or 3 ft., the branches tomentose or villous, the foliage glabrous. Leaves divided into 3 or 5 terete rigid pungent-pointed segments either entire or again bifid or trifid, usually shorter than the entire part and the whole leaf rarely much above 1 in. long. Flowers in small sessile axillary clusters. Pedicels pubescent, about 3 lines long. Perianth densely but shortly villous, the tube about 3 lines long, slightly recurved under the broadly conical limb. Torus straight. Gland very prominent, semiannular. Ovary almost sessile ; style rather short, with a lateral oblong-linear stigmatic disk about 1 line long. Fruit 7 to 8 lines long, 2 to 3 lines broad, smooth, falcate or abruptly curved in the middle into a beak almost as broad as the thicker base. Seed-wing not decurrent.

W. Australia. Swan river, *Drummond, n.* 107, *1st coll. n.* 601, *Preiss, n.* 601.

Var. *longiflora.* Perianth-tube 4 lines, the limb and the stigmatic disk 2 lines long. —Swan river, *Drummond.*

29. **H. platysperma,** *Hook. Ic. Pl. t.* 433. Branches minutely hoary-pubescent. Leaves terete, smooth, thick, rigid, pungent-pointed, somewhat attenuate at the base, mostly erect, 3 to 5 in. long. Flowers in axillary sessile clusters. Pedicels minutely silky-pubescent, 2 to 3 lines long. Perianth minutely silky-pubescent, the tube shortly dilated at the base, much revolute, above 4 lines long, the upper segments much shorter. Torus straight. Gland broad, concave. Ovary stipitate ; style long, with a long narrow lateral stigmatic disk. Fruit when perfect quite globular, fully 2 in. diameter, very hard. Seed-wing covering the whole inner face of the valves, nearly equally broad all round the nucleus, which is more muricate on the outer face than in any other species.—Meissn. in. Pl. Preiss. i. 555, ii. 259, and in DC. Prod. xiv. 394.

W. Australia, *Drummond,* 2*nd coll. n.* 329, 4*th coll. n.* 287, *Preiss, n.* 551 ; Stirling Range, *Maxwell.*

30. **H. brachyptera,** *Meissn. in DC. Prod.* xiv. 396. A low intricately-branched shrub, the young shoots silky, the older foliage glabrous. Leaves terete, slender, divaricate, acute, scarcely attenuate at the base, 1 to 3 in. long. Flowers very small, in sessile axillary clusters. Pedicels silky, about 1 line long. Perianth silky-pubescent, revolute under the globular limb, the shorter segments scarcely 1 line long. Torus straight. Gland semiannular. Ovary nearly sessile ; style not long, with a large oblique stigmatic disk. Fruit compressed, very rugose, 1 in. long and nearly as broad, with a very short lateral

beak sometimes scarcely prominent. Seed-wing decurrent along both sides round the base of the nucleus.

W. Australia, *Drummond, 4th coll. n.* 291.

31. **H. Kippistiana,** *Meissn. in Hook. Kew Journ.* vii. 115, *and in D C. Prod.* xiv. 402. Branches virgate, nearly glabrous, but the young shoots more or less silky-pubescent. Leaves terete, slender, smooth, mucronate, attenuate at the base, 1½ to 3 in. long. Flowers small, in shortly pedunculate axillary clusters, but not seen fully out, the rhachis pedicels and perianths sprinkled with appressed hairs. Perianth-tube slender, reflexed under the globular limb. Gland prominent, semicupular. Ovary shortly stipitate; style (not long?) with an oblique broad stigmatic disk. Fruit ¾ to 1 in. long, nearly ½ in. broad, smooth, with a broad obtuse compressed beak, the valves sometimes with a slight dorsal gibbosity near the end. Seed-wing decurrent on the upper side only.

W. Australia, *Drummond, 5th coll. suppl. n.* 14.—The fruit, like that of *H. Pampliniana* and *H. commutata,* is shaped like that of *H. leucoptera,* with which F. Mueller, Fragm. vi. 219, unites *H. Kippistiana,* but the flowers are different and the fruit quite smooth.

32. **H. Preissii,** *Meissn. in Pl. Preiss.* i. 557, *and in D C. Prod.* xiv. 399. A tall shrub or tree attaining 30 ft., with very rigid branches, our specimens quite glabrous, even the young shoots. Leaves terete, smooth, very thick and rigid, pungent-pointed, in some specimens all under 1 in. in others 1 to 1½ in. long. Flowers in axillary clusters, the rhachis ½ to 1 line long, often persistent on the old branches. Pedicels silky-pubescent, 2 to 3 lines long. Perianth silky-pubescent, the tube slightly dilated towards the base, scarcely 2 lines long, much revolute upwards. Torus straight. Gland large, semicupular. Ovary very shortly stipitate; style slightly clavate under the broad lateral stigmatic disk. Fruit about ¾ in. long, 3 or 4 lines broad, transversely truncate at the end with the horizontally conical dorsal horns of the valves, which are sometimes very prominent, sometimes obscure.

W. Australia. Swan river, *Drummond, n.* 190, *1st coll., Preiss, n.* 617 *b ;* Darling ranges, *Oldfield ;* Jarramup, *Maxwell.*

In herb. F. Mueller there is a specimen in fruit, with longer leaves, apparently of this species, but with the fruit rather longer though not broader. We have also specimens in leaf only of what may be the same species, from Sharks Bay, *Milne.*

33. **H. pugioniformis,** *Cav. Ann. Hist. Nat.* i. 213, t. 11, *Ic.* vi. 24, *t.* 533. A shrub usually of 2 to 4 ft., but sometimes twice as tall, the branches and foliage glabrous or very minutely silky-pubescent. Leaves terete, smooth, rigid with a short pungent point, from under 1 in. to near 2 in. long. Flowers few, in axillary sessile clusters. Pedicels 1 to 2 lines long, pubescent as well as the perianth with appressed or spreading hairs. Perianth-tube slender, 2 to 3 lines long, reflexed under the ovoid limb. Torus small. Gland prominent,

semiannular or semicupular. Ovary shortly stipitate; style long, with a nearly straight convex or very broadly conical disk. Fruit lanceolate, acuminate, about 1 in. long and ¼ in. broad, rugose outside about the middle with an obliquely transverse prominent crest. Seed-wing scarcely decurrent on the upper side of the nucleus.—R. Br. in Trans. Linn. Soc. x. 178, Prod. 381 ; Meissn. in DC. Prod. xiv. 398 ; Hook. f. Fl. Tasm. i. 324 ; Lodd. Bot. Cab. t. 353; Reichb. Ic. et Descr. Pl. t. 23.—*Conchium pugioniforme,* Sm. in Trans. Linn. Soc. ix. 122; *Conchium longifolium,* Sm. l.c. 121 ; *Lambertia teretifolia,* Gærtn. f. Fr. iii. t. 217; *Banksia teretifolia,* Salisb. Prod. 51 ; *Hakea glabra,* Schrad. Sert. Hann. 27. t. 17 ; *H. glauca* and *H. parilis,* Knight, Prot. 106.

N. S. Wales. Port Jackson to the Blue Mountains, *R. Brown, Sieber, n.* 13, and others ; Hastings river, *Beckler.*

Victoria. Grampians, low scrubby hills towards Mount Ararat, between Melbourne and the Dandenong ranges and in Gipps' Land, *F. Mueller.*

Tasmania. Port Dalrymple and Derwent river, *R. Brown;* common in many parts of the island in poor soils, *J. D. Hooker.*

The Tasmanian and Victorian specimens have generally but not constantly rather larger flowers, with more spreading hairs than those from N. S. Wales.

34. H. Pampliniana, *Kipp. ; Meissn. in DC. Prod.* xiv. 395. Young branches silky-tomentose, the adult foliage glabrous. Leaves terete, slender, finely pointed, slightly attenuate at the base, 1½ to 3 in. long. Flowers in axillary clusters mostly on short peduncles. Pedicels about 2 lines long. Perianth silky-pubescent, the tube slender, about 3 lines long, reflexed under the globular limb. Torus small. Gland prominent, semiannular. Ovary shortly stipitate ; style not very long, with an oblique disk scarcely umbonate in the centre. Fruit 1 in. long, nearly ¾ in. broad, with a short broad straight beak, shaped like the fruit of *H. leucoptera* but smooth, the valves thickened at the end but scarcely horned. Seed-wing shortly decurrent on the upper margin of the nucleus.

Queensland ? Curriwillighie, *Dalton* (the specimen incomplete).

N. S. Wales. Castlereagh river, *Woolls.*

Victoria ? Murray desert, *Herb. F. Mueller.*

S. Australia. Spencer's Gulf, *Herb. F. Mueller ;* Streaky Bay, *Babbage;* Encounter Bay, *Whittaker ;* Tattiara Country, *Woods.*

35 ? H. vittata, *R. Br. in Trans. Linn. Soc.* x. 182, *Prod.* 383. Young shoots minutely ferruginous or hoary, the adult foliage glabrous. Leaves terete, slender, rigid, finely almost pungent-pointed, not attenuate at the base, 1½ to 3 in. long. Flowers not seen except some loose remains which appear to have been like those of *H. Pampliniana,* the perianth silky-pubescent, the torus small, the style rather long with an orbicular lateral stigmatic disk. Fruit ovoid, nearly ¾ in. long, fully 4 lines broad, spotted or slightly verrucose, scarcely beaked, marked with a broad smooth dark line down each suture, the valves with a short dorsal horn near the end. Seed-wing decurrent along the upper margin only of the nucleus.—Meissn. in DC. Prod. xiv. 401.

S. Australia. Port Lincoln, *R. Brown.* The foliage is that of *H. Pampliniana,* and if the fragments of flowers above-mentioned really belong to the specimens, it is possible that some of the flowering specimens referred to *H. Pampliniana* may belong rather to *H. vittata.* The fruits of the two as far as known are very different.

36. **H. rostrata,** *F. Muell.; Meissn. in Linnæa* xxvi. 259, *and in .DC. Prod.* xiv. 396. An erect shrub of several feet, glabrous except the inflorescence. Leaves terete, smooth, rigid, mucronate, 1½ to 4 in. long. Flowers in sessile axillary clusters. Perianth silky-pubescent, the tube about 3 lines long, much revolute under the globular limb. Torus small. Gland usually crenate or shortly 2- or 3-lobed. Ovary shortly stipitate; style not very long, the stigmatic disk with a prominent cone as in *H. rugosa.* Fruit recurved at the base, incurved from the middle, with a closely inflexed conical beak of 3 or 4 lines, the whole fruit 1 to 1½ in. long and nearly ¾ in. broad, more or less rugose but not cristate.

Victoria. Grampians, *Mitchell*; Mount Sturgeon, *Robertson;* Glenelg river and Mount Abrupt, *F. Mueller.*

S. Australia. Lofty range and Kangaroo Island, *F. Mueller;* St. Vincent's Gulf, *Blandowski.*

37. **H. rugosa,** *R. Br. in Trans. Linn. Soc.* x. 179, *Prod.* 381. A spreading or prostrate shrub, glabrous except the inflorescence or the young shoots minutely silky. Leaves terete, divaricate, smooth, rigid, pungent-pointed, ¾ to 1½ or rarely 2 in. long. Flowers small, in axillary clusters. Pedicels silky-pubescent, short. Perianth silky-pubescent, the tube about 2 lines long, recurved under the limb. Torus straight. Gland semiannular, truncate or crenate. Ovary shortly stipitate; style not very long; stigmatic disk oblique, with a prominent cone. Fruit about ¾ in. long, under ½ in. broad, recurved at the base, incurved above the middle, very rugose or cristate, with an inflexed conical beak. Seed-wing not decurrent.—Meissn. in DC. Prod. xiv. 397.

Victoria. Macalister river, towards its source, *F. Mueller;* Glenelg river, *Robertson.*

S. Australia. Port Lincoln, *R. Brown;* Murray river, Lofty range, *F. Mueller;* Bethanie, *Oswald;* Kangaroo Island, *Waterhouse.*

The cone on the stigmatic disk of this species and of *H. rostrata,* and in a less degree on that of *H. pugioniformis,* connects them with the section *Conogynoides,* but the pubescent flowers and general affinity with *H. epiglottis* place them rather in *Euhakea.*

38. **H. epiglottis,** *Labill. Pl. Nov. Holl.* i. 30, t. 40. An erect shrub attaining 7 or 8 ft., glabrous except the inflorescence and the silky-pubescent young shoots. Leaves terete, smooth, rigid, mucronate, 1½ to 3 in. long. Flowers small, in axillary clusters. Perianth silky-pubescent, the tube near 2 lines long, recurved under the limb. Torus straight. Gland semiannular. Ovary shortly stipitate; style not very long; stigmatic disk oblique, without the prominent cone of *H. rugosa.* Fruit 7 to 9 lines long, 3 to 4 lines broad, rugose but not cristate, recurved at the base, incurved above the middle, with a short conical incurved beak. Seed-wing not decurrent.—R. Br. in Trans. Linn. Soc. x. 179, Prod. 382; Meissn. in DC. Prod. xiv. 395 ; Hook. f. Fl. Tasm. i. 324; *Conchium*

epiglottis, Willd. Enum. Hort. Berol. 141; *Conchium teretifolium,* Gærtn.
f. Fr. iii. 217, t. 219; *Hakea Milligani,* Meissn. in DC. Prod. xiv. 395.

Tasmania. Port Dalrymple and Derwent river, *R. Brown;* common in various
parts of the island, ascending to 3000 feet, *J. D. Hooker.*

SERIES 2. GLABRIFLORÆ.—Perianth glabrous. Torus straight or
slightly oblique, the gland semiannular or none.

39. **H. amplexicaulis,** *R. Br. in Trans. Linn. Soc.* x. 184, *Prod.* 384.
An erect shrub attaining 10 to 12 ft., our specimens glabrous. Leaves
sessile, ovate-oblong or lanceolate, undulate sinuate and prickly-toothed,
deeply cordate and clasping the stem with rounded prickly-toothed
auricles, the whole leaf 3 to 8 in. long, rigid, more or less glaucous,
penniveined but the veins scarcely prominent. Flowers white, numerous,
in short axillary pedunculate clusters or racemes, the peduncles 1 to 2
lines long and glabrous, the rhachis about as long and villous. Pedicels
glabrous, filiform, 4 to 6 lines long. Perianth glabrous, the tube about
3 lines long, much revolute under the globular limb. Torus somewhat
oblique. Gland very prominent, semicupular. Ovary shortly stipitate;
style long, clavate under the oblique convex disk. Fruit about 1 in.
long and ½ in. broad or rather larger, smooth or slightly muricate, shortly
acuminate. Seed-wing shortly decurrent, especially on the upper margin of the nucleus.—Meissn. in Pl. Preiss. i. 565, and in DC. Prod. xiv.
407; *H. triformis,* Lindl. Swan Riv. App. 36.

W. Australia. King George's Sound, *R. Brown, Oldfield, F. Mueller;* and
thence towards Swan river, *Drummond, 1st coll. n.* 610, *Preiss, n.* 548.

40. **H. glabella,** *R. Br. Prot. Nov.* 28. An erect spreading or diffuse shrub described sometimes as growing into a small tree of 12 to
15 ft., sometimes as low and prostrate, quite glabrous or the branches
more or less villous. Leaves sessile, obovate to oblong-cuneate, entire or
more frequently sinuate and prickly-toothed, usually contracted below
the middle, dilated at the base, deeply cordate and embracing the stem
with rounded and entire or angular and prickly-toothed auricles, the
whole leaf 1½ to 3 in. long. Flowers rather smaller than in *H. amplexicaulis,* in axillary clusters or short racemes, the rhachis 1 to 3 lines long
and quite glabrous as well as the pedicels and perianths. Pedicels 2 to
4 lines long. Perianth-tube 2 to 2½ lines long, reflexed under the limb.
Torus nearly straight. Gland prominent, semiannular. Ovary shortly
stipitate; style long, with an oblique stigmatic disk. Fruit 1 to 1¼ in.
long, 6 to 7 lines broad, smooth or sparingly muricate, with a broad,
obtuse slightly incurved beak. Seed-wing decurrent along the upper
margin of the nucleus.—Meissn. in Pl. Preiss. i. 564, and in DC. Prod.
xiv. 407; *H. denticulata,* R. Br. Prot. Nov. 28; *H. prostrata,* R. Br. in
Trans. Linn. Soc. x. 184, Prod. 384; Meissn. in Pl. Preiss. i. 565, and
in DC. Prod. xiv. 407.

W. Australia. King George's Sound and adjoining districts, *R. Brown, Baxter,
Drummond, 3rd coll. n.* 278, *Preiss, n.* 539, 542, and others; Swan river, *Fraser,
Drummond, 1st coll. n.* 612, *Preiss, n.* 538; Murchison river, *Oldfield.*

H. glabella and *H. prostrata* are usually distinguished by the former being a tall glabrous Swan river shrub, and the latter prostrate with villous stems from King George's Sound, but some of Oldfield's specimens with villous branches are described as attaining 12 to 15 ft., and some of F. Mueller's glabrous ones from Stirling Range as 10 to 12 ft. high, whilst some of Drummond's glabrous Swan river ones are evidently as prostrate as Brown's typical *H. prostrata.* The foliage is equally variable in all, and the flowers and fruits the same as far as known.

41. **H. auriculata,** *Meissn. in Hook. Kew Journ.* vii. 116, *and in DC. Prod.* xiv. 406. An erect shrub of 2 ft. or more, glabrous or the branches pubescent. Leaves usually cuneate, broad and truncate or sinuate and prickly-toothed at the end, tapering into a long narrow lower portion, sessile, and again dilated at the base into stem-clasping auricles more or less angular or prickly-toothed, but sometimes the upper end divided into 3 narrow pungent-pointed lobes or the whole leaf from a broad stem-clasping base linear-lanceolate pungent-pointed and entire or with a pair of divaricate lobes, the leaf varying from 1½ to near 3 in. long. Flowers in axillary clusters. Pedicels 2 or 3 lines long, glabrous as well as the rhachis. Perianth glabrous, the tube about 2 lines long, reflexed under the globular limb. Torus nearly straight. Gland small, semiannular. Ovary nearly sessile; style rather long, thickened under the oblique convex stigmatic disk. Fruit about ¾ in. long and ½ in. broad, muricate with long prickles, the valves with a thick dorsal horn near the end. Seed-wing surrounding the nucleus and occupying nearly the whole breadth of the valves.

W. Australia. Murchison river, *Oldfield, Drummond, 6th coll. n.* 197.

Var. *spathulata.* Leaves with the long winged base half stem-clasping, but scarcely dilated in the majority of leaves, although here and there showing small prickly-toothed auricles.—*H. attenuata,* Meissn. in DC. Prod. xiv. 406, as to Drummond's specimens, not of R. Brown.—Swan river, *Drummond, 1st coll. n.* 615.

Some of the forms of *H. varia* have the foliage almost of some forms of *H. auriculata,* but the flowers and fruits are very different.

42. **H. cristata,** *R. Br. Prot. Nov.* 28. An erect shrub, attaining 6 to 8 ft., our specimens quite glabrous. Leaves from broadly obovate to oblong, sinuate and prickly-toothed, tapering into a short petiole, 1½ to 3 in. long, rigid, glaucous, obscurely penniveined. Flowers small, in short axillary racemes, the villous rhachis 2 to 4 lines long. Pedicels filiform, glabrous, 2 to 3 lines long. Perianth glabrous, the tube about 1½ lines long, revolute under the globular limb. Torus slightly oblique. Gland small but prominent, erect, obovate, truncate. Ovary shortly stipitate; stigmatic disk broad, oblique, slightly convex. Fruit about 1½ in. long and 1 in. broad, with a broad slightly incurved beak; each valve bearing usually next the sutures an irregular longitudinal rigid toothed wing or crest sometimes decurrent along both sutures nearly to the base, sometimes along the upper suture only or almost obsolete. Seed-wing narrowly decurrent down both margins of the nucleus, which is less lacunose on the outer face than in most species.—Meissn. in Pl. Preiss. i. 564, and in DC. Prod. xiv. 406; Hook. Ic. Pl. t. 443.

W. Australia. Swan river, *Fraser, Drummond, 1st coll. n.* 614, *Preiss, n.* 546.

43. **H. linearis,** *R. Br. in Trans. Linn. Soc.* x. 183, *Prod.* 384. An erect bushy bright green glabrous shrub. Leaves sessile, linear-lanceolate, pungent-pointed, entire or bordered by a few small prickly teeth, 1 to 1½ or rarely 2 in. long, thick and rigid, veinless except the scarcely prominent midrib. Flowers small, in axillary clusters or short racemes, the rhachis 1 to 2 lines long, quite glabrous. Perianth glabrous, white, the tube slender, 2¼ to 3 lines long, revolute under the globular limb. Torus small. Gland small but prominent, truncate or 2-lobed. Ovary stipitate; style not very long, with an oblique orbicular stigmatic disk. Fruit (not seen attached) 1 in. long, about ¼ in. broad, on a recurved stipes, slightly incurved, smooth, with a short conical beak, the valves with conical dorsal protuberances or short horns near the end. Seed-wing shortly decurrent on the upper margin only of the nucleus.—Meissn. in Pl. Preiss. i. 562, and in DC. Prod. xiv. 405; Sweet. Fl. Austral. t. 43; Bot. Reg. t. 1489.

W. Australia. King George's Sound, *R. Brown, Fraser, Milne, Maxwell, Drummond,* 2nd coll. n. 335; Canning river, *Oldfield.*

44. ? **H. stenocarpoides,** *F. Muell. Herb.* Apparently tall and quite glabrous. Leaves oblong-cuneate, obtuse, tapering to the base, entire, 3 to 4 in. long, thick, obscurely triplinerved with the addition sometimes of a few very oblique veins. Flowers about 6 or 8 together, umbellate on a common axillary peduncle of about ½ in., the pedicels scarcely ¼ in. long, all as well as the flowers quite glabrous. Perianth about ½ in. long, with a short broad oblique almost gibbous base, then narrow, revolute under the globular limb. Torus small, without any gland. Ovary on a stipes of nearly 3 lines much thickened at the base; style scarcely 2 lines long, with a large orbicular lateral stigmatic disk. Fruit unknown.

W. Australia, *Drummond* (5th coll.?) suppl. n. 15. The inflorescence is quite anomalous, and the genus must remain uncertain until the fruit shall have been observed. There are certainly only 2 ovules in the ovary, collaterally attached, and the plant has much more the aspect of a *Hakea* than of a *Grevillea.*

45. **H. ruscifolia,** *Labill. Pl. Nov. Holl.* i. 30, t. 39. An erect bushy shrub of 6 to 8 ft., the branches and young shoots hirsute with long fine spreading hairs, intermingled with a fulvous tomentum. Leaves on long petioles when small, the larger ones almost sessile, ovate oblong or lanceolate, pungent-pointed, all under ½ in. in some specimens, in others narrower and near 1 in. long, veinless and scabrous-pubescent or glabrous above, tomentose or at length nearly glabrous underneath with the midrib prominent. Flowers small, white, in dense clusters terminating short leafy branches. Pedicels glabrous, filiform, 2 to 4 lines long. Perianth glabrous, the tube about 1½ lines long, reflexed under the limb. Torus small. Gland prominent, semiannular. Ovary nearly sessile; style short, the stigmatic disk broad, slightly convex. Fruit ½ to ¾ in. long, ¼ in. broad, scarcely beaked. Seed-wing decurrent only on the upper margin of the nucleus.—R. Br. in Trans.

Linn. Soc. x. 186, Prod. 385 ; Meissn. in Pl. Preiss. i. 576, and in DC. Prod. xiv. 419 ; *Conchium ruscifolium,* Willd. Enum. Hort. Berol. 141.

W. Australia. King George's Sound and adjoining districts, *Labillardiere, R. Brown,* and many others, and thence to Swan river, *Fraser, Drummond,* 1*st coll.,* 3*rd coll. n.* 276, *Preiss, n.* 611, and others. This species also stands alone without any immediate affinity with any other.

46. **H. saligna,** *Knight, Prot.* 108. A tall bushy shrub, quite glabrous or the young shoots slightly silky. Leaves usually lanceolate, obtuse or with a short callous point, tapering into a short petiole, 3 to 6 in. long, but sometimes oblong-elliptical and 2 to 4 in. long, of a pale colour, veinless or obscurely and obliquely penniveined. Flowers small, in dense axillary clusters, the very short rhachis hirsute. Pedicels glabrous, filiform, about 3 lines long. Perianth glabrous, the tube scarcely 2 lines long, much revolute under the globular limb. Torus small. Gland small. Ovary nearly sessile ; style long, with a large lateral convex stigmatic disk. Fruit about 1 in. long, $\frac{1}{2}$ to $\frac{3}{4}$ in. broad, with a short incurved beak, more or less rugose and sometimes covered with large very prominent tubercles. Seed-wing shortly decurrent along the upper margin of the nucleus.—R. Br. in Trans. Linn. Soc. x. 185, Prod. 385 ; Meissn. in DC. Prod. xiv. 416 ; Sweet, Fl. Austral. t. 27 ; *Embothrium salignum,* Andr. Bot. Rep. t. 215 ; *Conchium salignum,* Sm. in Trans. Linn. Soc. ix. 124 ; *Embothrium salicifolium,* Vent. Jard. Cels. t. 8 ; *Conchium salicifolium,* Gærtn. f. Fr. iii. 217 ; *Hakea mimosoides* A. Cunn. ; Meissn. in DC. Prod. xiv. 416 ; *H. florulenta,* Meissn. in Hook. Kew Journ. vii. 116, and in DC. Prod. xiv. 416.

Queensland. Araucaria ranges, *Leichhardt;* Brisbane river, Moreton Bay, *A. Cunningham, F. Mueller.*

N. S. Wales. Port Jackson to the Blue Mountains, *R. Brown,* and many others; Argyle County, *Fraser ;* New England, *C. Stuart.*

The Queensland specimens, to which the name of *H. mimosoides* specially applies, have longer leaves than most of the N. S. Wales ones, and F. Mueller's have the fruits narrower and less tuberculate, but several N. S. Wales ones have the same long leaves, with the fruits unknown. Some of C. Stuart's New England specimens have the leaves shorter and broader than usual. It will require, however, much more complete specimens to establish any definite varieties.

47. **H. verrucosa,** *F. Muell. Fragm.* v. 25, vi. 218. A handsome shrub of several ft., the branches closely pubescent. Leaves terete, smooth, rigid, mucronate, $1\frac{1}{2}$ to 3 in. long. Flowers red, in short racemes, mostly terminating short leafy branches, the pubescent or villous rhachis 2 to 4 lines long. Pedicels glabrous, slender, 1 to 2 lines long. Perianth glabrous, the tube 4 to 5 lines long, narrow, opening on the lower side only, revolute under the ovoid-globular limb. Torus rather oblique. Gland broad, semicircular, scarcely prominent. Ovary very shortly stipitate ; style long, with an oblique almost lateral stigmatic disk. Fruit above 1 in. long, about $\frac{1}{2}$ in. broad, recurved at the base, the valves with a conical dorsal protuberance near the end. Seed-wing decurrent about half way down the upper margin of the nucleus.

W. or **E. Australia?** A very distinct species, although allied to *H. purpurea*, cultivated in the Melbourne Botanic Garden as West Australian, but suspected by F. Mueller to be of eastern origin.

48. **H. purpurea,** *Hook. in Mitch. Trop. Austr.* 348. A hard rigid bushy shrub of several ft., glabrous except a few silky hairs on the very young shoots. Leaves terete, smooth, once or twice bifid or trifid, rigid and pungent-pointed, the whole leaf usually 1½ to 2 in. long, the divided portion about as long as the simple base. Flowers "crimson," in sessile or shortly pedunculate axillary umbels. Pedicels glabrous, filiform, 3 or 4 lines long. Perianth glabrous, the tube about 5 lines long, dilated below the middle, open early along the under side, attenuate and revolute under the limb. Torus straight, rather broad. Gland scarcely prominent. Ovary shortly stipitate; style long, with an oblique almost lateral stigmatic disk. Fruit 1¼ to 1½ in. long, ¾ in. broad, nearly straight, scarcely beaked. Seed-wing very broad, decurrent down both margins and round the base of the nucleus.—Meissn. in Linnæa xxvi. 358, and in DC. Prod. xiv. 404; *Grevillea trisecta,* F. Muell. First Gen. Rep. 17 (name only).

N. S. Wales. Warrego river, *Mitchell;* Darling Desert, *Nielson* (specimens in leaf only).
Victoria. N.W. interior of the Colony, *F. Mueller* (the specimens seen all cultivated).

49. **H. gibbosa,** *Cav. Anal. Hist. Nat.* i. 215, *Ic.* vi. 24, *t.* 534. A shrub of several ft., the branches and young leaves hirsute with spreading hairs, the older foliage sometimes glabrous. Leaves terete, entire, smooth, rigid, pungent-pointed, 1 to 3 in. long. Flowers in sessile axillary clusters. Pedicels short, densely villous. Perianth glabrous, the tube about 3 lines long, revolute under the globular limb. Ovary contracted into a very short stipes; style not long, with an oblique stigmatic disk. Fruit ovoid-globular, oblique, about 1 in. diameter, rugose, with a very short thick obtuse oblique or incurved beak, the valves with small dorsal horns near the end. Seed-wing narrowly decurrent down both margins of the nucleus.—R. Br. in Trans. Linn Soc. x. 181, Prod. 382; Meissn. in DC. Prod. xiv. 401; *Banksia gibbosa,* Sm. in White Voy. 224, t. 22, f. 2; *Conchium gibbosum* Sm. in Trans. Linn. Soc. ix. 119; *Conchium sphæroideum,* Sm. l.c. 120; *Conchium cornutum,* Gærtn. f. Fr. iii. 216, t. 219; *Hakea pubescens,* Schrad. Sert. Hannov. 27; *Conchium pubescens.* Willd. Enum. Hort. Berol. 141; *H. pinifolia,* Salisb. Prod. 51; *H. lanigera,* Ten. Fl. Nap. i. 22, t. 6.

N. S. Wales. Port Jackson, *R. Brown, Sieber, n.* 14, and others.

50. **H. propinqua,** *A. Cunn. in Field, N. S. Wales,* 327. A bushy shrub, the adult foliage glabrous, the branches scarcely pubescent. Leaves crowded, terete, smooth, mucronate, rather thick, shortly attenuate at the base, mostly 1 to 1½ in. long. Flowers very small, in little axillary clusters. Pedicels hirsute, scarcely 1 line long. Perianth glabrous, revolute, about 1 line long. Torus straight. Gland small.

514 CIV. PROTEACEÆ.

Ovary nearly sessile; style not long, with a large oblique stigmatic disk. Fruit above 1 in. long and nearly as broad, very rugose, with large prominent obtuse tubercles, straight, scarcely beaked. Seed-wing decurrent along the upper margin of the nucleus.—Meissn. in DC. Prod. xiv. 397 ; *H. pachyphylla*, Sieb. in Spreng. Syst. Cur. Post. 46, and in Roem. and Schult. Syst. iii. Mant. 282 ; R. Br. Prot. Nov. 26.

N. S. Wales. Blue Mountains, *A. Cunningham, Sieber, n.* 11 ; *Backhouse.*

51. **H. nodosa,** *R. Br. in Trans. Linn. Soc.* x. 179, *Prod.* 382. A shrub of 2 to 6 ft., quite glabrous or the young shoots minutely silky-pubescent, the branches rather slender. Leaves rather crowded, terete and slender or rather broader and slightly compressed, mucronate, smooth, slightly attenuate at the base, 1 to 1½ or rarely 2 in. long. Flowers minute, in axillary clusters. Pedicels scarcely 1 line long, slightly silky. Perianth glabrous, scarcely 1 line long, revolute under the limb. Torus straight or nearly so. Gland prominent, erect, semi-annular. Ovary very shortly stipitate; style not long, with a large very oblique stigmatic disk. Fruit ¾ to near 1 in. long, ½ in. broad, either verrucose with a broad obtuse smooth but otherwise scarcely distinct beak, or the whole fruit smooth. Seed-wing decurrent down the upper margin to below the nucleus.—Meissn. in DC. Prod. xiv. 397 as to Brown's plant only ; *H. flexilis*, R. Br. in Trans. Linn. Soc. x. 180, Prod. 382 ; Meissn. l.c. 396, also as to Brown's plant only ; *H. semiplana*, F. Muell. ; Meissn. in Linnæa xxvi. 359, and in DC. Prod. xiv. 397.

Victoria. Port Phillip, *R. Brown, Adamson;* heaths near Bridgewater Bay and Portland, *Robertson;* marshy pastures from Dandenong Creek to Gipps' Land, *F. Mueller ;* Grampians? *Mitchell.*

52. **H. acicularis,** *R. Br. in Trans. Linn. Soc.* x. 181, *Prod.* 383. A tall shrub or small bushy tree, glabrous except the inflorescence, or the young branches silky and the foliage rarely minutely pubescent. Leaves terete, smooth, rigid, pungent-pointed, 1 to 2 or rarely near 3 in. long, not attenuate at the base. Flowers in sessile axillary clusters. Pedicels silky-pubescent, 1 to 2 lines long. Perianth glabrous, the tube about 2 lines long in the typical form, revolute under the limb. Torus straight. Gland prominent, semiannular. Ovary contracted into a very short stipes; style not long, with an oblique stigmatic disk. Fruit usually about 1 in. long and ½ to ¾ in. broad, very thick and rugose, with a short obtuse smooth and straight beak. Seed-wing decurrent along the upper margin only of the nucleus.—Meissn. in DC. Prod. xiv. 400; Endl. Iconogr. t. 24; Reichb. Ic. et Descr. Pl. t. 24; *Conchium aciculare*, Vent. Jard. Malm. t. 111 ; Sm. in Trans. Linn. Soc. ix. 121 ; *Hakea sericea*, Schrad. Sert. Hannov. 27 ; *Conchium compressum*, Sm. in Trans. Linn. Soc. ix. 121; *Banksia tenuifolia*, Salisb. Prod. 50 (Sm.) ; *H. decurrens*, R. Br. Prot. Nov. 27; Meissn. in DC. Prod. xiv. 401.

N. S. Wales. Port Jackson to the Blue Mountains, *R. Brown, Sieber, n.* 10, *and Fl. Mixt n.* 481, and many others; Liverpool Plains, *A. Cunningham;* Twofold Bay. *F. Mueller.*

Var. *lissosperma.* Leaves usually more rigid and stouter, sometimes 3 to 4 in. long, but sometimes like those of the typical form. Perianth about 3 lines long. Fruit broader, scarcely beaked. Nucleus of the seed less rugose but rarely quite smooth on the outer face. All these characters, however, occur occasionally in N. S. Wales specimens, or are scarcely marked in southern ones.—*H. lissosperma,* R. Br., in Trans. Linn. Soc. x. 180, Prod. 382; Meissn. in DC. Prod. xiv. 401; *H. acicularis* and *H. lissosperma,* Hook. f. Fl. Tasm. i. 325; *H. brachyrrhyncha,* F. Muell., First Gen. Rep. 17 (name only); *H. obliqua,* Lodd. Bot. Cab. t. 1682? not of R. Br.

Victoria. Wilson's Promontory, Macalister river, and ranges near Stieglitz, *F. Mueller.*

Tasmania. Derwent river, *R. Brown;* Cape Barren and Flinders Islands, Bass's Straits, *Gunn, Milligan;* common in subalpine stations from 2000 to 4000 ft. elevation, *J. D. Hooker.*

53. **H. leucoptera,** *R. Br. in Trans. Linn. Soc.* x. 180, *Prod.* 382. A shrub with rather slender virgate branches, minutely hoary-pubescent. Leaves terete, smooth, mucronate with fine straight rigid points, more or less attenuate at the base, 1½ to 3 in. long. Flowers small, in short racemes or clusters pedunculate in the axils or rarely terminating short leafy branches, the peduncle and rhachis minutely silky-pubescent, ½ to ¾ in. long. Pedicels glabrous, 2 to 2½ lines long. Perianth glabrous, the tube about 2½ lines long, slightly dilated below the middle, revolute under the limb. Torus slightly oblique. Gland semiannular. Ovary stipitate; style not long, with a very oblique almost lateral stigmatic disk. Fruit about 1 in. long, ½ to ¾ in. broad, often somewhat verrucose, with a short conical beak, the valves without any or with scarcely prominent dorsal protuberances at the end. Seed-wing usually more or less decurrent along the upper margin only of the nucleus.—Meissn. in DC. Prod. xiv. 396; F. Muell. Fragm. vi. 219 (but not all the synonyms adduced); *H. leucocephala,* Dietr. Syn. Pl. i. 531 (by a misprint); *H. virgata* R. Br. Prot. Nov. 26; Meissn. in DC. Prod. xiv. 395; *H. tephrosperma,* R. Br. l.c.; Meissn. l.c. 402; *H. longicuspis,* Hook. in Mitch. Trop. Austr. 397; Meissn. l.c. 395; *H. stricta,* F. Muell.; Meissn. in Linnæa xxvi. 360 and l.c. 400.

Queensland. Armadilla, *Barton.*
N. S. Wales. Field's and Harrington's Plains, Lachlan river, *A. Cunningham, Fraser;* Plains near the Gwydir, *Mitchell;* Lachlan and Darling rivers to the Barrier range, *Victorian and other Expeditions;* Mount Murchison, *Bonney.*
Victoria. Murray Desert, *F. Mueller;* N.W. districts, *L. Morton.*
S. Australia. Head of Spencer's Gulf, *R. Brown;* Cooper's Creek, *Murray.*

In some specimens of Fraser's the fruit appears much narrower, but is evidently not fully ripe.

54. **H. cycloptera,** *R. Br. in Trans. Linn. Soc.* x. 182, *Prod.* 383. Branches virgate, quite glabrous. Leaves terete, smooth, rigid, pungent-pointed, not contracted at the base, 3 to 5 in. long. Flowers in sessile axillary clusters, the whole inflorescence quite glabrous. Pedicels 1 to 2 lines long. Perianth glabrous, the tube not 2 lines long, revolute under the limb. Torus straight. Gland small. Ovary on a rather long stipes; style not long, with an oblique stigmatic disk. Fruit nearly 1½ in. long and above 1 in. broad, rugose, obtuse, the valves with dorsal horns near the end. Seed-wing broad, decurrent along

L L 2

both margins and round the base of the nucleus.—Meissn. in DC.
Prod. xiv. 402.

S. Australia. Port Lincoln, *R. Brown*, *Wilhelmi;* the former specimens in fruit
only, the latter in flower with detached fruits.

55. **H. microcarpa,** *R. Br. in Trans. Linn. Soc.* x. 182, *Prod.* 383.
A shrub varying from 2 or 3 ft. to twice that height, quite glabrous or
with a very minute pubescence on the young branches and foliage.
Leaves mostly terete and smooth, slender or thick, from 1 in. in some
specimens to 4 in. long in others, but sometimes the lower ones or in
other specimens nearly or quite all more or less compressed or channelled above, or quite flat and linear-lanceolate with the midrib and
margins prominent underneath. Flowers in axillary clusters. Perianth
glabrous as well as the pedicels, the tube usually about 2 lines long
but variable in size, revolute under the limb. Torus straight. Gland
semiannular. Ovary shortly stipitate; style not long, with a broad
somewhat oblique stigmatic disk. Fruit $\frac{1}{2}$ to nearly $\frac{3}{4}$ in. long, about
$\frac{1}{4}$ in. broad, oblique, smooth or slightly rugose, the valves with short
dorsal horns near the end sometimes reduced to small protuberances or
almost obsolete.—Meissn. in DC. Prod. xiv. 400; Hook. f. Fl. Tasm.
i. 324; Bot. Reg. t. 475; Lodd. Bot. Cab. t. 219; *H. patula*, R. Br.
Prot. Nov. 27; Meissn. l.c. 401; *H. bifrons*, Meissn. l.c. 400.

N. S. Wales. Macquarrie river, *A. Cunningham;* near Bathurst, *Fraser;* Berrima, *Woolls;* New England, *C. Stuart;* Clarence river, *Beckler.*
Victoria. Elephant plains, *Robertson, F. Mueller;* Ovens, King and Upper Genoa
rivers, ascending the Australian Alps to 5000 or 6000 ft., *F. Mueller.*
Tasmania. Port Dalrymple, *R. Brown;* common especially on gravelly banks of
rivers, ascending to 3000 ft., *J. D. Hooker.*

The flat or channelled leaves, very rare in Tasmanian specimens, are common in many
of the N.S. Wales and Victorian ones, and sometimes to the exclusion of the terete
leaves. It is probable therefore that *H. Mitchellii*, Meissn. in DC. Prod. xiv. 398, of
which I have seen no authentic specimen, should be included in *H. microcarpa.*

56. **H. recurva,** *Meissn. in DC. Prod.* xiv. 394. A very stout rigid
shrub, the branches and foliage glabrous. Leaves terete, smooth, very
thick and rigid, pungent-pointed, very spreading or recurved, 3 or 4
in. long. Flowers numerous in sessile axillary clusters or dense
racemes, with a villous rhachis of 1 to 2 lines. Pedicels glabrous, filiform, 4 or 5 lines long. Perianth glabrous, the tube about 3 lines
long, slightly dilated below the middle, much revolute upwards. Torus
straight. Gland very prominent, semicupular. Ovary contracted into
a very short stipes; style not very long, with an oblique convex
stigmatic disk.

W. Australia, *Drummond, 4th coll. n.* 288, *6th coll. n.* 160; Murchison river,
Oldfield.

57. **H. circumalata,** *Meissn. in Hook. Kew Journ.* vii. 114, *and in
DC. Prod.* xiv. 402. A bushy shrub, the young branches tomentose.
Leaves terete, smooth, rigid, rather thick, pungent-pointed, $\frac{3}{4}$ to above
1 in. long, rarely contracted at the base, at first pubescent, at length

nearly glabrous. Flowers in terminal sessile clusters. Pedicels 1 to 2
lines long, sparingly pubescent. Perianth glabrous, the tube about 2
lines long, much revolute above the middle, the upper segments much
shorter. Torus small. Gland semiannular, not very prominent.
Ovary contracted into a thick stipes ; style rather long, with a broad
oblique stigmatic disk. Fruit about ¾ in. long, ½ in. broad, very
rugose or nearly smooth, very shortly beaked, the valves with short or
long dorsal protuberances or horns near the end. Seed-wing decurrent along both margins round the base of the nucleus.

W. Australia, *Drummond, 4th coll. n.* 290, *6th coll. n.* 192 ; Murchison river,
Oldfield.

58. **H. commutata,** *F. Muell. Fragm.* v. 26. Glabrous in every
part even the young shoots and rhachis of the inflorescence. Leaves
terete, thick, smooth, mucronate, attenuate at the base, ½ to 1 in. long
on the flowering branches, sometimes twice as long on the main stems.
Flowers in shortly pedunculate clusters, terminal or lateral on the old
wood, the rhachis and peduncle together 1 to 3 lines, the pedicels 2 to
3 lines long. Perianth about 3 lines long, much revolute, the upper
segments short. Torus straight. Gland small. Ovary almost sessile ;
style rather long, slightly thickened under the very oblique almost
lateral stigmatic disk. Fruit (if correctly matched) ¾ in. long, about
5 lines broad, smooth, with a short broad very obtuse beak, straight or
slightly incurved. Seed-wing decurrent on the upper margin only of
the nucleus.—*H. nodosa,* Meissn. in Pl. Preiss. i. 555, and in DC. Prod.
xiv. 397 as to the western specimens, not of R. Br.

W. Australia, *Drummond, 5th coll. n.* 412 (in flower), *n.* 41, *and 5th coll. suppl.*
n. 13 (in fruit).

59. **H. strumosa,** *Meissn. in DC. Prod.* xiv. 402. A shrub of 2 to
6 ft., quite glabrous even the inflorescence. Leaves terete, smooth,
rigid, mucronate, scarcely contracted at the base, 1½ to 3 in. long.
Flowers in sessile axillary clusters. Pedicels 2 to 3 lines long. Perianth glabrous, the tube about 2 lines long, revolute under the globular
limb. Torus small. Gland very prominent, erect, oblong. Ovary
nearly sessile ; style rather long, with an oblique stigmatic disk.
Fruit 1½ to 2 in. long, 1 to 1½ in. broad, very thick and gibbous, with
a very small lateral beak often almost obsolete. Seed-wing very
broadly decurrent along both margins round the base of the nucleus.

W. Australia, *Drummond, 4th coll. n.* 289 ; poor ridges from Mount Bland to
Esperance Bay, *Maxwell.*

SECT. 3. CONOGYNOIDES.—Racemes usually short or reduced to
sessile clusters, rarely elongated, enclosed before their development in
an involucre or bud of imbricate scales. Perianth glabrous, recurved
or revolute under the limb. Stigmatic cone erect or scarcely oblique.

The flowers are difficult to distinguish from those of the section *Conogyne* of *Grevillea,* but the inflorescence and fruits are those of *Hakea.*

SERIES 1. LONGISTYLÆ.—Leaves flat, entire, oblong-lanceolate or narrow. Style at least twice as long as the perianth. Stigmatic cone long and narrow.

60. **H. multilineata,** *Meissn. in Pl. Preiss.* ii. 261, *and in DC. Prod.* xiv. 410. A tree or tall shrub. Leaves linear-lanceolate to oblong-cuneate, very obtuse, tapering towards the base, 6 to 8 in. long when narrow, 3 to 5 in. when broad, flat, thick, minutely pubescent, striate with numerous parallel not prominent nerves. Racemes axillary, sessile, very dense and spike-like, 1½ to above 3 in. long, the rhachis and flowers quite glabrous in the typical form. Perianth-tube 2 to 3 lines long, slender, reflexed under the oblong limb. Torus rather oblique. Gland broad, flat, semicircular. Ovary very shortly stipitate ; style long, filiform, with a long narrow somewhat oblique stigmatic cone. Fruit ovoid-globular, with a very short beak, about ½ in. diameter.

S. Australia. Gawler ranges, *Sullivan.*
W. Australia, *Drummond, 3rd coll. n.* 275, *5th coll. suppl. n.* 18.

Var. *grammatophylla.* Rhachis of the raceme densely tomentose. I can perceive no other difference, but the specimens are not satisfactory.— *Grevillea grammatophylla,* F. Muell. Fragm. v. 25 ; *Hakea grammatophylla,* F. Muell. Fragm. vi. 214.—Central Mount Stuart, *M'Douall Stuart's Expedition.*

H. Francisiana, F. Muell. Fragm. i. 20, from Spencer's Gulf, *Francis,* of which I find no specimen among F. Mueller's collections, is probably, from the character given, not different from *H. multilineata,* which differs widely from all other species in the venation of the leaves.

61. **H. laurina,** *R. Br. Prot. Nov.* 29. A shrub of 10 ft. or more or a small tree attaining 30 ft., the branches minutely hoary-tomentose or glabrous as well as the foliage. Leaves narrowly elliptical-oblong or oblong-lanceolate, often shortly acuminate but obtuse, tapering into a long petiole, mostly 4 to 6 in. long, triplinerved besides the thick nerve-like margins and sometimes with an additional longitudinal vein between the nerves. Flowers " crimson " in large dense globular clusters sessile in the axils, the globular rhachis densely villous. Pedicels glabrous, about 2 lines long. Perianth glabrous, the tube scarcely above 3 lines long, reflexed under the oblong limb. Torus very oblique. Gland large, disk-shaped. Ovary shortly stipitate on the upper margin of the torus ; style long, with a long narrow stigmatic cone. Fruit 1 to 1¼ in. long and ¾ in. broad or rather larger, more or less cristate along the upper suture, with a very short small beak. Seed-wing decurrent along both margins round the base of the nucleus.—Meissn. in DC. Prod. xiv. 411 ; *H. eucalyptoides,* Meissn. in Pl. Preiss. i. 573, ii. 262, and in DC. l.c. 413 ; F. Muell. Fragm. iv. 130.

W. Australia, *Drummond, 3rd coll. n.* 274, *4th coll. n.* 294 ; between Lucky Bay and Cape Arid, *Baxter ;* towards Cape Riche, *Preiss, n.* 565 ; between Perth and King George's Sound, *Harvey ;* Stirling, Fitzgerald, &c. ranges all the way to Cape Arid, *Maxwell.*

62. **H. obtusa,** *Meissn. in DC. Prod.* xiv. 411. A spreading shrub of 2 or 3 ft., the young shoots silky-tomentose, the adult foliage glabrous. Leaves oblong-lanceolate, obtuse or with a callous point contracted into a very short petiole or almost sessile, 1½ to 2½ in. long, prominently 3-nerved. Flowers (red ?) in dense clusters on the old wood, the villous rhachis 1 to 1½ lines long. Pedicels glabrous, about 2 lines long. Perianth glabrous, the tube about 2 lines long, reflexed under the limb. Torus very oblique and narrow, the gland horseshoe-shaped occupying the lower portion. Ovary nearly sessile at the upper end of the torus; style very long with a long narrow stigmatic cone. Fruit nearly 1 in. long, ½ to ¾ in. broad, with a short straight beak; sometimes several fruits clustered together and almost connate.

W. Australia, *Drummond, 5th coll. n.* 409 ; coast hills near E. Mount Barren, *Maxwell.*

63. **H. cinerea,** *R. Br. in Trans. Linn. Soc.* x. 186, *Prod.* 385. A stout shrub of 5 or 6 ft., the branches densely tomentose. Leaves linear-cuneate or oblanceolate, obtuse, tapering to the base but scarcely petiolate, 4 to 7 in. long, very thick and rigid, of an ashy grey colour and minutely scabrous, very prominently triplinerved, with nerve-like margins. Flowers numerous in globular axillary clusters, the villous rhachis very short. Perianth glabrous, the tube very slender, about ½ in. long, reflexed under the oblong limb. Torus oblique. Gland not very prominent, but almost surrounding the sessile ovary. Style long, with a long narrow stigmatic cone. Fruit small, erect, ½ to ¾ in. long including the rather long straight beak, 4 to 5 lines broad, smooth or tubercular-rugose. Seed-wing decurrent on the upper margin only of the nucleus.—Meissn. in DC. Prod. xiv. 414; *H. canescens,* Link. Enum. Hort. Berol. i. 118; *H. tricostata,* Hook. Ic. Pl. t. 435.

W. Australia. Lucky Bay, *R. Brown, Baxter ;* Esperance Bay to Cape Arid, *Maxwell.*

64. **H. corymbosa,** *R. Br. Prot. Nov.* 28. A much-branched flat-topped shrub of 1 to 2 ft., the branches tomentose, the foliage minutely silky-pubescent or at length glabrous. Leaves linear or linear-cuneate, mucronate, tapering to the base but scarcely petiolate, 1½ to 3 in. long, very thick, 1-nerved underneath, almost nerveless above, the margins thick. Flowers in axillary clusters, the floral leaves sometimes crowded in false-whorls, the bud-scales or involucre sometimes persisting till the flowers expand. Flowers pale yellow, not very numerous. Pedicels glabrous, 1 to 2 lines long. Perianth glabrous, the tube narrow, fully 6 lines long, reflexed only under the oblong limb. Torus oblique. Gland almost disk-shaped. Ovary shortly stipitate on the upper end of the torus; style long, with a long narrow stigmatic cone. Fruit ¾ to almost 1 in. long and almost as thick, with a very short straight beak. Seed-wing decurrent along both margins and round

the base of the nucleus but very narrow on the upper margin.—
Meissn. in Pl. Preiss. i. 574, and in DC. Prod. xiv. 418.

W. Australia. King George's Sound or to the eastward, *Baxter ;* Stirling Range,
F. Mueller ; S.W. Bay, Kalgan and Tone rivers, *Oldfield ;* Phillips ranges, *Maxwell.*

SERIES 2. PETIOLARES.—Leaves broad, triplinerved and reticulate,
the veins prominent or obscure, tapering into a distinct petiole. Style
not twice as long as the perianth.

65. **H. undulata,** *R. Br. in Trans. Linn. Soc.* x. 185, *Prod.* 384. An
erect shrub of 6 to 8 ft., the young shoots ferruginous-tomentose or
villous, the adult foliage glabrous. Leaves obovate ovate oblong or
rarely lanceolate, more or less undulate and prickly-toothed, tapering
into a petiole, 2 to 3 or rarely 4 in. long, rigid, triplinerved or quintuplinerved and reticulate. Flowers small, in axillary clusters, the rhachis
villous, sometimes 1 line long. Pedicels glabrous, 1 to 2 lines long.
Perianth glabrous, the tube not $1\frac{1}{2}$ lines long, revolute under the limb.
Torus small, slightly oblique. Gland very small. Ovary shortly stipitate ; style not long, with a narrow stigmatic cone. Fruit recurved
at the base, 1 to $1\frac{1}{4}$ in. long, $\frac{1}{2}$ to $\frac{3}{4}$ in. broad, somewhat rugose, distinctly beaked. Seed-wing decurrent on the upper margin of the
nucleus to the base.—Meissn. in Pl. Preiss. i. 566, and in DC. Prod.
xiv. 407; Hook. Ic. Pl. t. 447; *Anadenia hakeoides,* Lindl. Swan Riv.
App. 30.

W. Australia. King George's Sound, *R. Brown, Fraser ;* thence to Swan river,
Drummond, n. 92, *1st coll. n.* 613 *;* Green Mountain, *Preiss, n.* 560; Tone and Canning
rivers, *Oldfield.*

66. **H. petiolaris,** *Meissn. in Pl. Preiss.* i. 577, *and in DC. Prod.*
xiv. 413. A tall shrub, attaining 6 to 8 ft., glabrous and glaucous, the
young shoots silky. Leaves orbicular or ovate, shortly acuminate,
entire, contracted into a rather long petiole, $1\frac{1}{2}$ to 4 in. long and sometimes quite as broad, thick and rigid, with nerve-like margins, more or
less distinctly triplinerved or sometimes 5-plinerved and reticulate.
Flowers very numerous in axillary clusters or racemes, the thick villous
rhachis sometimes 3 or 4 lines long. Pedicels glabrous, 3 or 4 lines
long. Perianth glabrous, the tube slender, nearly 3 lines long, revolute under the ovoid limb. Torus oblique. Gland thick, flat, semiorbicular. Ovary stipitate on the upper margin of the torus; style
rather long, with a narrow stigmatic cone. Fruit 1 to $1\frac{1}{4}$ in. long, 6 to
8 lines broad, nearly smooth, with a rather narrow straight beak.
Seed-wing rather broadly decurrent along both margins of the nucleus.
—*H. crassinervia,* Meissn. in Pl. Preiss. i. 578, and in DC. Prod. xiv.
413.

W. Australia, *Drummond, n.* 95, *1st coll. n.* 607; rocks of Mount Currie and
Mount Hardy, *Preiss, n.* 557, 559. The relative prominence and length of 2 or more
of the principal primary veins is very variable, even in different leaves of the same
specimen.

67. **H. neurophylla,** *Meissn. in Hook. Kew Journ.* vii. 117, *and in DC. Prod.* xiv. 413. Young shoots minutely hoary, the adult foliage glabrous and glaucous. Leaves ovate-elliptical to oblong-lanceolate, shortly acuminate or rarely almost obtuse, contracted into a broad but distinct petiole, 2 to 3 in. long, very thick and rigid, with nerve-like margins, irregularly triplinerved or quintuplinerved, with few anastomosing veins. Flowers in axillary clusters, the rhachis villous, about 1 line long. Pedicels glabrous, about 2 lines long, revolute under the ovoid-globular limb. Torus small, oblique. Gland small. Ovary sessile or nearly so; style not very long, with a straight stigmatic cone. Fruit about 1 in. long, $\frac{3}{4}$ in. thick, with a nearly straight conical beak. Seed-wing rather broadly decurrent on the upper margin only of the nucleus.

W. Australia. Between Moore and Murchison rivers, *Drummond, 6th coll. n.* 195. Possibly a variety only of *H. petiolaris.*

68. **H. loranthifolia,** *Meissn. in Pl. Preiss.* i. 574, *and in DC. Prod.* xiv. 411. A shrub of 6 ft., all our specimens quite glabrous. Leaves obovate to elliptical-oblong, acuminate with a rigid point, tapering into a petiole, 1½ to 2 in. long, very rigid, obscurely triplinerved with sometimes 2 or 3 additional longitudinal primary veins but very faint. Flowers small, in axillary clusters or short racemes, with a pubescent rhachis of $\frac{1}{2}$ to 1 line. Pedicels glabrous, under 1 line long. Perianth glabrous, the tube under 2 lines long, reflexed under the limb. Torus small. Gland prominent and thick but small. Ovary contracted into a short stipes; style not long, with a straight stigmatic cone. Fruit about $\frac{3}{4}$ in. long, $\frac{1}{2}$ in. thick, with a very short slightly curved beak. Seed-wing decurrent on the upper margin only of the nucleus.

W. Australia, *Drummond, 1st coll. n.* 606; near York, *Preiss, n.* 567.

SERIES 3. SESSILES.—Leaves from very broad to oblong-lanceolate, 3- or more-nerved and reticulate, sessile or nearly so. Style not twice as long as the perianth.

69. **H. cucullata,** *R. Br. Prot. Nov.* 30. An erect stout shrub, attaining sometimes 12 to 14 ft., the branches softly tomentose and often villous with spreading hairs, the young shoots entirely tomentose or villous, the adult foliage glabrous or minutely scabrous. Leaves on the flowering branches sessile, orbicular or reniform, spreading and concave, 2 to 4 in. diameter, rigid, entire crenate sinuate or bordered by short prickly teeth, several-nerved and reticulate on both sides; those of the young plant or barren shoots oblong or elliptical, penniveined and prickly-toothed. Flowers clustered in the axils, almost concealed at the base of the leaves. Pedicels not exceeding 1 line. Perianth glabrous, the tube 4 or 5 lines long, revolute under the ovoid-oblong limb. Torus oblique. Gland large, horseshoe-shaped. Ovary nearly sessile; style long with a long narrow stigmatic cone. Fruit oblique, about 1 in. long and $\frac{3}{4}$ in. broad, more or less rugose, the beak

very short. Seed-wing decurrent along both margins round the base of the nucleus.—Meissn. in DC. Prod. xiv. 408; Hook. Ic. Pl. t. 441; Bot. Mag. t. 4528; copied into Lem. Fl. Jard. t. 45; *H. conchifolia*, Hook. Ic. Pl. t. 432; Meissn l.c., *H. Victoriæ*, Drumm. in Bot. Mag. lxxiv. Comp. 2; Meissn. l.c. 409.

W. Australia. Mount Gardner, *Baxter*; King George's Sound or towards Swan river, *Drummond*, 1*st coll n.* 611, West Mount Barren, *Drummond*, 4*th coll. n.* 300; near Wuljenup and Mount Manypeak, *Preiss, n.* 537; Kalgan river, *Oldfield;* north of Stirling Range, *F. Mueller.*

The difficulty of reducing specimens of this fine plant to herbarium size is the cause of our inability to ascertain whether there really is any sufficient character to distinguish the three supposed species. I can find no difference in the flowers or fruits, and the leaves appear to vary, quite entire or denticulate. The West Mount Barren specimens are described by Drummond as forming erect almost simple stems of 12 ft. or more, with closely packed leaves showing distinctly each year's growth of 6 to 9 inches, the lower or earlier ones very large, the upper ones gradually diminishing, and all with richly-coloured veins, the young ones whitish yellow, the intermediate ones orange, turning to a deep crimson. No such colours are described in the more common Stirling Range plant, and Drummond's dried specimens have entirely lost that which they had.

70. **H. ferruginea,** *Sweet, Fl. Austral. t.* 45. A slender shrub of 3 or 4 ft., the young branches tomentose-pubescent. Leaves glabrous or villous, sessile, from cordate-ovate to ovate-lanceolate, shortly acuminate, with a callous point, entire or with slightly sinuate or undulate margins, $1\frac{1}{2}$ to 3 in. long, flat, several-nerved and reticulate on both sides. Flowers small, in axillary clusters. Pedicels 1 to $1\frac{1}{2}$ lines long. Perianth glabrous, about 3 lines long, much revolute, the upper segments scarcely 2 lines long, the limb ovoid. Torus oblique. Gland small, semiorbicular. Ovary shortly stipitate; style not very long, with a narrow stigmatic cone. Fruit nearly 1 in. long and $\frac{3}{4}$ in. broad, with a long narrow beak. Seed-wing decurrent down the upper margin only of the nucleus.—Bot. Mag. t. 3424; *H. repanda*, R. Br. Prot. Nov. 30; Meissn. in Pl. Preiss. i. 568, ii. 261, and in DC. Prod. xiv. 408; Lodd. Bot. Cab. t. 1750.

W. Australia. King George's Sound and adjoining districts, *Baxter, Cunningham, Drummond, 3rd coll. n.* 279; Mount Wuljenup, *Preiss, n.* 547; Mount Barker, *Oldfield;* Stirling Range, *F. Mueller;* W. Mount Barren, *Maxwell.*

71. **H. smilacifolia,** *Meissn. in Pl. Preiss.* i. 567, *and in DC. Prod.* xiv. 408. An erect shrub, attaining 3 or 4 ft., difficult to distinguish from *H. ferruginea* without the fruit. Branches tomentose-villous and often hirsute with spreading hairs. Leaves sessile, broadly ovate, acutely acuminate, rounded or cuneate at the base and never cordate, entire or slightly sinuate, 1 to 2 in. long, usually pubescent, several-nerved and reticulate. Flowers small, in axillary clusters, resembling those of *H. ferruginea*, but not seen in a very good state. Fruit very smooth, about $\frac{1}{2}$ in. long and $\frac{1}{4}$ in. broad, not so distinctly beaked as in *H. ferruginea*, and without the thick woody protuberance of most Hakea fruits.

W. Australia. *Drummond, n.* 97; Quangen plains, *Preiss, n.* 535; Cujong, *Oldfield.*

72. **H. elliptica,** *R. Br. in Trans. Linn. Soc.* x. 187, *Prod.* 386. A shrub attaining 6 to 8 ft., the branches and young shoots ferruginous-tomentose, the adult foliage glabrous or minutely pubescent. Leaves sessile or nearly so, broadly oval or elliptical, obtuse but usually with a small callous point, 1½ to 3 in. long, many-nerved and reticulate on both sides. Flowers numerous, in axillary clusters on very short racemes, the villous rhachis rarely above 1 line long. Pedicels glabrous, 2 to 4 lines long. Perianth glabrous, the tube slender, about 2 lines long, revolute under the ovoid limb. Torus oblique. Gland prominent, flat, semiorbicular. Ovary stipitate on the upper margin of the torus; style filiform, with an erect stigmatic cone. Fruit about 1 in. long and nearly ¾ in. broad, usually smooth, with a small oblique beak. Seed-wing decurrent on the upper margin about halfway down the nucleus.—Meissn. in Pl. Preiss. i. 568, and in DC. Prod. xiv. 412; *Conchium ellipticum,* Sm. in Trans. Linn. Soc. ix. 123.

W. Australia. King George's Sound and adjoining districts, *R. Brown, A. Cunningham, Preiss, n.* 558, and many others.

73. **H. ambigua,** *Meissn. in Pl. Preiss.* ii. 260, *and in DC. Prod.* xiv. 415. Branches tomentose or almost villous. Leaves from narrow-lanceolate to oblong-elliptical, obtuse or with a small callous point, contracted at the base but sessile or nearly so, minutely pubescent or glabrous when full grown, prominently triplinerved and sometimes reticulate between the nerves, 1½ to near 3 in. long. Flowers only seen in a rather imperfect state but apparently like those of *H. elliptica.* Fruit ¾ to near 1 in. long, ½ in. broad, with a rather long nearly straight conical beak. Seed-wing decurrent on the upper margin only of the nucleus.—*H. trinervis,* Meissn. in DC. Prod. xiv. 414.

W. Australia, *Drummond, 2nd coll. n.* 277, *5th coll. n.* 408. F. Mueller considers this as a narrow-leaved variety of *H. elliptica,* but the leaves have only 3 principal nerves even when broad, and the fruit has a much longer straight beak.

SERIES 4. NERVOSÆ.—Leaves lanceolate or linear, prominently 3- or more nerved (rarely 2- or 1-nerved when very narrow) smooth between the nerves.

74. **H. plurinervia,** *F. Muell. Herb.* A shrub of 6 or 7 ft., the young branches loosely tomentose. Leaves lanceolate, falcate, obtuse and often oblique at the end, with a small callous point, tapering into a very short petiole or almost sessile, 4 to 6 in. long and ½ to ¾ in. broad, with about 7 longitudinal nerves prominent on both sides. Flowers small and numerous in axillary clusters, the villous rhachis rarely above 1 line long. Pedicels glabrous, 2 or 3 lines long. Perianth glabrous, the tube scarcely above 2 lines long, slender, revolute under the ovoid-globular limb. Torus small, oblique. Gland scarcely prominent. Ovary shortly stipitate; style not long, with an erect stigmatic cone. Fruit above 1 in. long and nearly ¾ in. broad, more or less falcate, with an incurved conical beak. Seed-wing narrowly but unequally decurrent along both sides of the nucleus.

Queensland. Rockingham Bay, *Dallachy.*

75. **H. dactyloides,** *Cav. Anal. Hist. Nat.* i. 215, t. 12; *Ic.* vi. 25, t. 535. A tall shrub with erect branches, the young shoots usually silky, the adult foliage rarely retaining more or less of pubescence, usually quite glabrous. Leaves from linear-lanceolate to oblong-lanceolate, acute or scarcely obtuse, tapering into a short petiole, falcate oblique or straight, 2 to 4 in. long, rigid, prominently triplinerved, smooth between the nerves or rarely in the broader leaves a few irregular veins forming almost 1 or 2 additional longitudinal nerves. Flowers very small and numerous in axillary clusters or short racemes, the villous rhachis 1 to 1½ lines long. Pedicels silky-hairy, 1 to 1½ lines long. Perianth glabrous, the tube about 1 line long, revolute under the globular limb. Torus small, nearly straight. Gland small. Ovary shortly stipitate; style short, with an erect stigmatic cone. Fruit ¾ to 1 in. long, ½ to ¾ in. thick, smooth or slightly rugose, with a very small straight beak. Seed-wing narrowly decurrent along the upper margin only of the nucleus.—R. Br. in Trans. Linn. Soc. x. 186, Prod. 385; Meissn. in DC. Prod. xiv. 415; Bot. Mag. t. 3760; *Banksia dactyloides,* Gærtn. Fr. i. 221, t. 47; *Conchium dactyloides,* Vent. Jard. Malm. t. 110; Sm. in Trans. Linn. Soc. ix. 123; *Banksia oleifolia,* Salisb. Prod. 51; *Conchium nervosum,* Sm. in Willd. Enum. Hort. Berol. 141; *Hakea nervosa,* Knight, Prot. 108; *H. ferruginea,* Lodd. Bot. Cab. t. 1501? not of Sw.

N. S. Wales. Port Jackson to the Blue Mountains, *R. Brown, Sieber, n.* 12, and many others; Clarence river, *Beckler;* New England, *C. Stuart,* some of the latter specimens with more rigid almost pungent narrow leaves, approaching those of *H. ulicina,* but with silky pedicels.

76. **H. ulicina,** *R. Br. Prot. Nov.* 29. Very near the narrow-leaved forms of *H. dactyloides,* but the leaves all linear acute and pungent-pointed, 4 to 8 in. long in the typical form, more spreading rather broader and 1 to 2 in. long in some specimens, very rarely 2 lines wide, prominently 1- to 3-nerved underneath, the nerves less conspicuous and sometimes obsolete on the upper surface. Flowers still smaller than in *H. dactyloides,* the pedicels always glabrous as well as the perianth. Fruit rarely above ½ in. long, with a short straight beak. Seed as in *H. dactyloides.*—Meissn. in DC. Prod. xiv. 415; *H. angustifolia,* Hortul. (Meissn.).

N. S. Wales. Twofold Bay, *Baxter, F. Mueller.*
Victoria. Glenelg river, *Robertson ;* Grampians, *Mitchell, F. Mueller;* Port Phillip, Mount Sturgeon and Mount Abrupt, *F. Mueller.*
Tasmania. Flinders island, *Milligan.*

Var. *carinata,* F. Muell. Leaves mostly 1-nerved underneath, nerveless above, but in some specimens normal 3-nerved leaves mixed with the others.—*H. carinata,* F. Muell.; Meissn. in Linnæa xxvi. 360, and in DC. Prod. xiv. 418.

S. Australia. Mount Lofty and Bugle Range, *F. Mueller ;* near Adelaide, *Whittaker, Blandowski;* Encounter Bay, *Whittaker ;* Tattiara country, *Woods.*

77. **H. falcata,** *R. Br. Prot. Nov.* 29. A tall shrub, closely and minutely tomentose, becoming at length glabrous. Leaves lanceolate or almost linear, acute or with a callous point, straight or somewhat fal-

cate, contracted into a very short petiole, 1½ to 2½ in. long, more or
less prominently triplinerved. Flowers small and numerous in axillary
clusters. Pedicels glabrous, about 1 line long. Perianth glabrous,
slender, nearly 2 lines long, revolute under the globular limb. Torus
small, oblique. Gland rather thick. Ovary sessile on the upper margin
of the torus; style not very long, with a narrow stigmatic cone. Fruit
from under ¾ to nearly 1 in. long and about ½ in. broad, with a small
more or less incurved beak. Seed-wing decurrent along both margins
round the base of the nucleus.—Meissn. in Pl. Preiss. i. 572 (partly),
ii. 262, and in DC. Prod. xiv. 414.

W. Australia. King George's Sound or to the eastward, *Baxter, Drummond,*
2nd coll. n. 333, *and* (*5th coll. ?*) *suppl. n.* 15, 16. The leaf-veins are sometimes as
prominent as in *H. dactyloides,* from which *H. falcata* is then only to be distinguished
by the flowers not quite so small and by the seed-wing decurrent all round the nucleus.
Sometimes even on the same specimen the leaves are thicker and the veins obscure
almost as in *H. incrassata.* The fruit varies much in size.

78. **H. pycnoneura,** *Meissn. in Hook. Kew Journ.* vii. 117, *and in*
DC. Prod. xiv. 414. A straggling shrub of several ft., the branches and
foliage minutely silvery-silky. Leaves linear, obtuse or with a callous
point, tapering at the base but scarcely petiolate, 4 to 8 in. long, very
thick and rigid, usually flexuose, the margins and midrib prominent
and sometimes 1 or 2 additional short longitudinal veins. Flowers
purple, in short very dense axillary racemes, the thick ovoid villous
rhachis 3 to 4 lines long. Pedicels glabrous, 2 or 3 lines long. Perianth glabrous, the tube 2 lines long, reflexed under the ovoid-oblong
limb. Torus small. Gland semiannular, not very prominent. Ovary
scarcely stipitate; style short, with a long stigmatic cone. Fruit usually
¾ in. long, about ½ in. broad, with a short straight beak; in some
specimens the whole fruit longer. Seed-wing decurrent along both
margins round the base of the nucleus.

W. Australia. Murchison river, *Oldfield, Drummond, 6th coll. n.* 193.

SERIES 5. UNINERVES.—Leaves narrow or small, flat, 1-nerved with
nerve-like margins. Fruit-valves without dorsal appendages.

79. **H. stenocarpa,** *R. Br. Prot. Nov.* 29. A glabrous bushy shrub of
several ft. Leaves linear or linear-lanceolate, acute or with a callous
point, contracted at the base but scarcely petiolate, 1½ to 4 in. long,
with a prominent midrib and nerve-like margins as in *H. marginata.*
Flowers very small, in axillary clusters, the rhachis very short and villous. Pedicels glabrous, not 1 line long. Perianth glabrous, scarcely
1½ lines long, revolute under the limb. Torus and gland small. Ovary
almost sessile; style not long, with a straight stigmatic cone. Fruit 1
to 1½ in. long, 2 to 3 lines broad at the base, tapering into a long
slightly incurved beak. Seed-wing narrowly decurrent along both
margins of the nucleus.—Meissn. in Pl. Preiss. i. 575, and in DC. Prod.
xiv. 417; Hook. Ic. Pl. t. 444.

W. Australia. Swan river, *Fraser, Drummond, 1st coll., Preiss, n.* 574.

80. **H. marginata,** *R. Br. in Trans. Linn. Soc.* x. 185, *Prod.* 385. A bushy shrub, attaining 6 to 8 ft., the young shoots minutely silky-pubescent, the adult foliage glabrous or nearly so. Leaves lanceolate, acute, pungent-pointed, shortly contracted at the base but scarcely petiolate, ¾ to 1½ in. long, rather thick and rigid, with the midrib and nerve-like margins prominent on both sides, otherwise veinless. Flowers small, in axillary clusters. Pedicels glabrous, scarcely 1 line long. Perianth glabrous, the tube nearly 2 lines long, revolute under the ovoid limb. Torus oblique. Gland rather large, almost stipitate. Ovary very shortly stipitate on the upper margin of the torus ; style not long, with a straight stigmatic cone. Fruit about ¾ in. long, 3 to 4 lines broad, with a rather long straight beak. Seed-wing decurrent along the upper margin only of the nucleus.—Meissn. in Pl. Preiss. i. 575, and in DC. Prod. xiv. 418.

W. Australia. Lucky Bay, *R. Brown, Baxter;* Hotham river, *Oldfield;* between King George's Sound and Swan river, *Harvey, Drummond,* 1*st coll. n.* 604.

81. **H. myrtoides,** *Meissn. in Pl. Preiss.* i. 577, *and in DC. Prod.* xiv. 418. A spreading shrub of 2 or 3 ft., the branches rather loosely villous at length glabrous. Leaves sessile, ovate, usually broad, mucronate with a rigid or pungent point, ½ to ¾ in. long, rather shining, with a prominent midrib and nerve-like margins, faintly penniveined and reticulate. Flowers " pink," in axillary clusters, the rhachis very short and villous. Pedicels glabrous, 2 to 3 lines long. Perianth glabrous, the tube about 2 lines long, reflexed under the globular limb. Torus nearly straight. Gland broad. Ovary almost sessile ; style long, with a long narrow stigmatic cone. Fruit scarcely ½ in. long, about ¼ in. broad, with a rather long incurved beak. Seed-wing broadly decurrent along both margins round the base of the nucleus.—Bot. Mag. t. 4643, copied into Lem. Fl. Jard. t. 272.

W. Australia. Swan river, *Fraser, Drummond, n.* 96, 1*st coll. n.* 608, *Preiss, n.* 534, *Oldfield.*

82. **H. costata,** *Meissn. in Pl. Preiss.* i. 575, *and in DC. Prod.* xiv. 418. An erect shrub attaining 3 or 4 ft., the branches loosely tomentose or villous. Leaves crowded, those of the flowering branches linear, pungent-pointed, rigid, under ½ in. long, with thickened margins and a very prominent keel or midrib, the lower ones sometimes oblong, flat, 2 lines broad and tapering into a short petiole, 1-nerved but otherwise veinless. Flowers small, in numerous axillary clusters. Pedicels glabrous, under 1 line long. Perianth white, glabrous, the tube about 1½ lines long, reflexed under the globular limb. Torus small. Gland very small. Ovary sessile or nearly so ; style long and slender, with a long narrow stigmatic cone. Fruit under ½ in. long, 3 or 4 lines broad, shortly beaked. Seed-wing broadly decurrent along both margins of the nucleus.

W. Australia, *Drummond, n.* 17, *and* 2*nd coll. n.* 332 ; Quangen plains, *Preiss, n.* 532, 533 (*Meissn.*) ; near Yatheroo and Toodyay, *Oldfield.*

SERIES 6. ENERVES.—Leaves mostly flat, obscurely penniveined, the midrib not prominent, entire prickly-toothed or lobed. Fruit-valves with dorsal horns or protuberances.

83. **H. oleifolia,** *R. Br. in Trans. Linn. Soc.* x. 185, *Prod.* 385. A tall shrub or small tree of 15 to 20 ft., the branches and young shoots ferruginous or silky-tomentose, the adult foliage glabrous or nearly so. Leaves oblong-lanceolate or cuneate-oblong, rounded at the end, with a fine minute point, contracted into a short petiole, 1 to above 2 in. long, rather thick, pale coloured, penniveined but the midrib usually scarcely prominent and the veins immersed or inconspicuous. Flowers in dense axillary clusters or racemes, the villous rhachis 1 to 2 lines long. Pedicels glabrous, about 3 lines long. Perianth glabrous, the tube not 2 lines long, reflexed under the globular limb. Torus small, oblique. Gland prominent but small. Ovary very shortly stipitate ; style not very long, with an erect stigmatic cone. Fruit ¾ to 1 in. long, ½ to ¾ in. broad, rugose, scarcely beaked, the valves with dorsal conical horns near the end, sometimes wanting on one valve. Seed-wing shortly decurrent along the upper margin of the nucleus.—Meissn. in Pl. Preiss. i. 571, and in DC. Prod. xiv. 416 ; *Conchium oleifolium,* Sm. in Trans. Linn. Soc. ix. 124 ; *Hakea ligustrina,* Knight, Prot. 108.

W. Australia. King George's Sound and adjoining districts, *R. Brown, Drummond, n.* 14, *and 5th coll. n.* 410, *Preiss, n.* 554, *Oldfield, Maxwell, F. Mueller.*

84. **H. florida,** *R. Br. in Trans. Linn. Soc.* x. 183, *Prod.* 384. An erect rigid shrub, attaining 5 or 6 ft., the branches and young leaves pubescent or villous, the adult foliage glabrous. Leaves sessile or nearly so, lanceolate or linear-lanceolate, very acute and pungent-pointed, bordered by a few prickly teeth or small lobes, one pair usually close to the cuneate base, 1 to 1½ or rarely 2 in. long, thick and veinless above, the midrib alone prominent underneath. Flowers very small, in axillary clusters, the villous rhachis very short. Pedicels glabrous, about 2 lines long. Perianth glabrous, the tube slender, scarcely 1½ lines long, revolute under the globular limb. Torus small. Gland prominent, truncate. Ovary stipitate ; style not very long, with a nearly straight rather broad stigmatic cone. Fruit above 1 in. long, nearly ¾ in. thick, scarcely beaked, the valves with very short dorsal protuberances near the end.—Meissn. in Pl. Preiss. i. 562, and in DC. Prod. xiv. 405 ; Bot. Mag. t. 2579.

W. Australia. King George's Sound, *R. Brown, Preiss, n.* 584, *Oldfield ;* Champion Bay, *Bower ;* Tulbinup ranges, *Maxwell.* —The leaves are sometimes like those of *H. linearis,* but the species is at once distinguished by the pubescent branches and rhachis and by the stigmatic cone.

85. **H. varia,** *R. Br in Trans. Linn. Soc.* x. 183, *Prod.* 383. A bushy or scrubby shrub, erect and 6 to 8 ft. high or sometimes spreading and diffuse, the branches tomentose and hirsute with spreading hairs or nearly glabrous. Leaves silky when young, glabrous when full grown, lanceolate linear-oblong or cuneate, with 2 or 3 short prickly lobes at the end or sinuate and prickly-toothed to below

the middle, or pinnatifid with few narrow or broad prickly-pointed lobes, rarely quite entire and linear, from under 1 in. long in some specimens to near 2 in. in others, thick and obscurely penni-veined or veinless, always tapering at the base. Flowers small, in clusters terminating short leafy branches or sessile in the upper axils, the villous rhachis rarely 1 line long. Pedicels glabrous, about 2 lines long or sometimes longer. Perianth glabrous, not 2 lines long, re-curved under the globular limb. Torus small. Ovary shortly stipitate; style not long, with a straight stigmatic cone. Fruit ¾ in. long, ⅓ in. broad, the valves with dorsal horns near the end. Seed-wing narrowly decurrent along the upper margin of the nucleus.—Meissn. in Pl. Preiss. i. 561, and in DC. Prod. xiv. 405; *H. attenuata* and *H. ilicifolia*, R. Br. in Trans. Linn. Soc. x. 183, 184; Meissn. in Pl. Preiss. i. 563 and in DC. Prod. xiv. 406; *H. tuberculata*, R. Br. Prot. Nov. 28; Meissn. ll. cc. 561 and 405; *H. lasiocarpha*, R. Br. Prot. Nov. 27; Meissn. ll. cc. 561 and 403 (as to Baxter's and Drummond's specimens); *H. heterophylla* and *H. intermedia*, Hook. Ic. Pl. t. 437, 445.

W. Australia. King George's Sound and Lucky Bay, *R. Brown;* eastward to Cape Paisley, *Maxwell;* from the same districts and towards Swan river, *Drummond,* n. 173, 197, *1st coll.* n. 615, 617, *4th coll.* n. 299, *Preiss,* n. 593, 600, *Oldfield, F. Mueller.*

The several supposed species here united are distinguished chiefly by the foliage which is truly protean, and specimens might be selected to represent several types so marked in their aspect that I should have retained them as distinct species were it not that other specimens occur combining the different forms of leaves on one stem, if not on the same branch. *H. ilicifolia* has generally rather large and broadly pinnately-toothed leaves, only shortly narrowed at the base; *H. attenuata* has them broad and toothed or shortly lobed at the end, tapering into a long narrow base; *H. varia* proper rather long leaves from linear almost terete and entire to deeply pinnatifid; *H. tuber-culata*, small, usually crowded leaves, cuneate or linear, and toothed, pinnatifid, serrate or entire, in a few specimens scarcely exceeding ½ in.; all these forms apparently dis-tinct in luxuriant specimens, but variously mixed in stunted scrubby ones. The fruit is frequently tuberculate or muricate but sometimes smooth in *H. tuberculata*, smooth or rarely muricate in *H. varia* proper. *H. lasiocarpha*, Br., is founded on a stunted spe-cimen, little more than a fragment, with deeply-divided narrow leaves, and the scaly buds larger than usual, but showing no other difference.

SERIES 7. TERETIFOLIA.—Leaves linear-terete, entire, rarely a few of the lower ones flat. Fruit-valves without any dorsal appendages.

86. **H. sulcata,** *R. Br. in Trans. Linn. Soc.* x. 180, *Prod.* 382, *Prot. Nov.* 27. An erect shrub attaining 5 or 6 ft., but often low, the young shoots silky, the adult foliage glabrous. Leaves linear-terete, angular and furrowed, rigid, mucronate, sometimes pungent-pointed, from under 2 in. to above 4 in. long, in the typical form. Flowers small, in dense axillary clusters, the small rhachis densely villous. Pedicels glabrous, ½ to 1 line long. Perianth glabrous, varying from 1½ to 3 lines, the tube slender, reflexed under the rather large limb. Torus and gland small. Ovary sessile; style rather long, with a long stig-matic cone. Fruit ovoid, about ½ in. long or rather longer, with a short narrow-conical straight beak. Seed-wing decurrent along the

upper margin of the nucleus.—Meissn. in Pl. Preiss. i. 556 and in DC. Prod. xiv. 399.

W. Australia. Lucky Bay, *R. Brown;* Swan river, *Drummond, 1st coll. n.* 599, *Preiss, n.* 608; Hill river, *Oldfield;* towards King George's Sound, *Drummond, 5th coll. n.* 411.

Var. *scoparia.* Branches and bud scales more pubescent. Leaves mostly longer, sometimes 8 in., less pointed, but occasionally short on some branches. Perianth sometimes larger sometimes rather smaller than the average typical size.—*H. scoparia,* Meissn. in Pl. Preiss. i. 556, and in DC. Prod. xiv. 399 ; Bot. Mag. t. 4644, copied into Lem. Fl. Jard. t. 376.—Swan river, *Drummond, 1st coll. n.* 600.

Var. *Gilbertii.* Leaves short, slender, with rather long pungent points. Fruit rather smaller.—*H. Gilbertii,* Kipp. in Hook. Kew Journ. vii. 115 ; Meissn. in DC. Prod. xiv. 399.—Swan river, *Gilbert.*

87. **H. Meissneriana,** *Kipp. in Hook. Kew Journ.* vii. 114. A tall erect shrub, the young shoots silky, the adult foliage glabrous. Leaves linear-terete, rigid, obscurely or more distinctly angular and striate, obtuse or mucronate, mostly 2 to 4 in. long. Flowers small, in dense axillary racemes or clusters, the villous rhachis 2 to 4 lines long. Pedicels glabrous, scarcely ½ line long. Perianth glabrous, the tube about 1½ lines long, revolute under the globular limb. Torus oblique. Gland small. Ovary contracted into a very short stipes; style not long, with a nearly straight stigmatic cone. Fruit ovoid, somewhat incurved, about ½ in. long, smooth or nearly so, with a short conical beak. Seed-wing narrowly decurrent on both margins of the nucleus.— Meissn. in DC. Prod. xiv. 399.

W. Australia, *Drummond, 3rd coll. n.* 272, *5th coll. suppl. n.* 16, *6th coll. n.* 191.

88 ? **H. subsulcata,** *Meissn. in Pl. Preiss.* i. 555, *and in DC. Prod.* xiv. 398. A shrub of 6 or 7 ft., the young shoots silky-pubescent, the adult foliage usually glabrous. Leaves terete, mucronate, finely striate, 2 to 4 in. long. Flowers not seen. Fruit densely clustered, ovoid, erect, smooth, ½ to ¾ in. long, 3 or 4 lines broad, tapering into a conical erect beak. Seed-wing decurrent along the upper margin of the nucleus but scarcely reaching the base.

W. Australia, *Drummond, 5th coll. suppl. n.* 15, *Preiss, n.* 607.

89. **H. Lehmanniana,** *Meissn. in Pl. Preiss.* i. 557, *and in DC. Prod.* xiv. 398. A bushy shrub attaining from 2 to 4 ft., glabrous or the branches minutely pubescent. Leaves linear-terete 3-angled or channelled above, not attenuate at the base, rigid, mucronulate, from under 1 in. to nearly 3 in. long. Flowers in very dense axillary clusters, the villous rhachis very short. Pedicels glabrous, about 2 lines long. Perianth glabrous, the tube scarcely 2 lines long, revolute under the limb. Torus rather oblique. Gland large, semiannular. Ovary shortly stipitate ; style long, with a rather long straight stigmatic cone. Fruit about 1 in. long and above ½ in. broad, more or less muricate with fringed tubercles or branching prickles, the small conical beak rather oblique. Seed-wing decurrent down both margins to the base of the nucleus.

W. Australia, *Drummond, 3rd coll. n* 273; Gordon river, *Preiss, n.* 604; between Swan river and King George's Sound, *Harvey;* Salt, Gordon, Franklin, and Tone rivers, *Maxwell.*

90. **H. flexilis,** *F. Muell. in Linnæa* xxvi. 359, *not of R. Br.* A tall shrub or small tree of about 20 ft., the branches and foliage quite glabrous. Leaves linear-terete, very spreading, mostly angular or slightly compressed, acute and sometimes pungent-pointed, not attenuate at the base, mostly $1\frac{1}{2}$ to 3 in. long. Flowers small, in axillary clusters, the hirsute rhachis very short. Pedicels glabrous, $\frac{1}{2}$ to 1 line long. Perianth glabrous, the tube scarcely 2 lines long, revolute under the globular limb. Ovary nearly sessile; style not long, with a straight stigmatic cone. Fruit $\frac{1}{2}$ to $\frac{3}{4}$ in. long, 4 or 5 lines broad, smooth or irregularly rugose, with a short conical straight or slightly inflexed beak. Seed-wing decurrent along the upper margin and sometimes partially also along the lower margin of the nucleus.—Meissn. in DC. Prod. xiv. 396 (*H. flexibilis* by a misprint) under *H. flexilis,* Br.

Victoria. N.W. districts of the Colony, *L. Morton.*
S. Australia. Murray Desert and Lake Hindmarsh, *F. Mueller ;* near Adelaide, *Herb. Hooker;* Kangaroo island, *F. Mueller, Waterhouse.*

SECT. 4. MANGLESIOIDES.—Racemes short or reduced to sessile clusters, enclosed before their development in an involucre or bud of imbricate scales. Perianth glabrous, slender, straight, the limb erect in the bud. Stigmatic cone erect.

The flowers in this section are quite those of the section *Manglesia* of *Grevillea,* except that the style is filiform.

91. **H. nitida,** *R. Br. in Trans. Linn. Soc.* x. 184, *Prod.* 384. A dense shrub of 6 to 8 ft., the branches and foliage quite glabrous. Leaves obovate oblong or rarely lanceolate, sometimes quite entire and obtuse with a small pungent point, sometimes acute pungent-pointed and irregularly bordered by a few prickly teeth or lobes, tapering at the base but scarcely petiolate, $1\frac{1}{2}$ to 3 or even 4 in. long, thick and veinless or obscurely and very obliquely penniveined. Flowers small and numerous in axillary racemes, the rhachis rigid, tomentose, $\frac{1}{2}$ to $\frac{3}{4}$ in. long. Pedicels glabrous, filiform 2 to 3 lines long. Perianth glabrous, straight, the tube about $1\frac{1}{2}$ lines long, the limb globular, erect in the bud. Torus small. Gland small but prominent. Ovary nearly sessile; style short with an erect broad stigmatic cone. Fruit 1 in. long or rather more, $\frac{3}{4}$ in. broad, scarcely beaked, with a conical horn near the end of one or both the valves. Seed-wing decurrent along the upper margin of the nucleus and sometimes narrowly so along the lower margin.—Meissn. in DC. Prod. xiv. 406; Bot. Mag. t. 2246; *H. pycnobotrys,* F. Muell. Fragm. v. 72.

W. Australia. Lucky Bay, *R. Brown;* probably from the same district, *Drummond, 4th coll. n.* 298; S.W. Bay, *Oldfield;* Gardner and Phillips rivers, E. Mount Barren, Esperance Bay, *Maxwell.*

92. **H. Oldfieldii,** *Benth.* Glabrous in all its parts. Leaves terete, smooth, pungent-pointed, $\frac{3}{4}$ to near 2 in. long, all undivided. Flowers small, in axillary racemes or clusters, the rhachis 1 to 2 lines long,

quite glabrous as well as the flowers. Pedicels filiform, at first short, 3 or 4 lines long when the flowers are fully out. Perianth slender, straight, scarcely 2 lines long, with a globular limb. Torus small. Gland prominent, obovate. Ovary shortly stipitate; style not long, with an erect stigmatic cone. Fruit not seen.

W. Australia. Champion Bay, *Oldfield;* in the interior, *J. S. Roe.*

93. **H. suaveolens,** *R. Br. in Trans. Linn. Soc.* x. 182, *Prod.* 383. An erect shrub of 5 or 6 ft., the young shoots silky-pubescent, the adult foliage glabrous. Leaves terete, erect, a few of them undivided grooved above and 3 or 4 in. long, but mostly pinnate with few or many erect segments of 1 or 2 in., all rigid and pungent-pointed. Flowers small, in dense racemes in the upper axils, the rigid pubescent rhachis ½ to ¾ in. long. Pedicels glabrous, 2 or 3 lines long. Perianth glabrous, straight; the tube about 1½ lines long, the limb globular, erect. Torus small. Gland small but prominent. Ovary shortly stipitate; style short, with an erect stigmatic cone. Fruit ¾ to 1 in. long, ½ to ¾ in. thick, smooth and almost shining but marked with warts, with a very small incurved almost lateral beak, the valves with very small dorsal protuberances near the end. Seed-wing decurrent on the outer side only.—Meissn. in Pl. Preiss. i. 558, and in DC. Prod. xiv. 403; *H. pectinata,* Colla, Hort. Rip. App. 2. 320, t. 11.

W. Australia. Middle Island, *R. Brown;* King George's Sound or adjoining districts, *Labillardière (Meissn.), A. Cunningham, Drummond, n. 93, Preiss, n.* 605, *Collie, Oldfield, F. Mueller;* eastward to Eyre's range, Cape Le Grand and Cape Arid, *Maxwell.*

Conchium drupaceum, Gærtn. f. Fr. iii. 217, t. 219 (*Hakea drupacea,* Roem. and Schult. Syst. iii. 426), which Gærtner had from Labillardière, is probably this species.

94. **H. lissocarpha,** *R. Br. Prot. Nov.* 27. A densely branched rigid shrub of 2 or 3 ft., the branches and young shoots more or less tomentose or hirsute, the foliage minutely scabrous-punctate after the hairs have worn off or rarely glabrous and smooth. Leaves pinnately divided into 3 to 7 terete rigid pungent-pointed segments, all entire or some of them forked, sometimes very short and thick, sometimes longer and slender, the rhachis often somewhat flattened and grooved above, the whole leaf 1 to 1½ in. long. Flowers small, in dense almost sessile axillary clusters, the thick villous rhachis 1 to 1½ lines long. Pedicels glabrous, about 3 lines long. Perianth glabrous, scarcely 1½ lines long, straight, the globular limb erect in the bud. Torus small. Gland semiannular. Ovary shortly stipitate; style short, with an erect stigmatic cone. Fruit ½ to ¾ in. long, 3 to 4 lines broad, scarcely beaked, with small dorsal protuberances near the end of the valves, sometimes obsolete. Seed-wing decurrent along the upper margin of the nucleus.—Meissn. in Pl. Preiss. i. 559, and in DC. Prod. xiv. 403; *H. intricata,* R. Br. Prot. Nov. 27; Meissn. in DC. Prod. xiv. 404; *H. petrophiloides,* Hortul. (Meissn.)

W. Australia. Swan river, *Fraser;* S. coast, *Baxter;* from Swan river to King

George's Sound and Cape Riche, *Drummond, n.* 106, 172, 1*st coll. n.* 602, 4*th coll. n.* 292, *Preiss, n.* 598, *Harvey ;* Stirling range, *Oldfield, Maxwell.*

The leaves when not much divided resemble those of some forms of *H. varia,* the flowers are very different.

95. **H. bipinnatifida,** *R. Br. Prot. Nov.* 28. A bushy shrub of 2 or 3 ft., the branches and foliage glabrous. Leaves terete, once or twice ternately divided or pinnate with the lower segments forked or trifid, the segments all slender, usually divaricate, mucronate-acute, ½ to ¾ in. long, smooth or singly grooved, the whole leaf under 2 in. long. Flowers small, in short dense almost sessile racemes, axillary or terminating short leafy branches, the villous rhachis ¼ to ½ in. long. Pedicels glabrous, 1 to 3 lines long. Perianth glabrous, scarcely 1¼ lines long, straight, the limb erect in the bud. Torus straight. Gland prominent, semiannular. Ovary shortly stipitate; style short, with an erect stigmatic cone. Fruit nearly 1 in. long, 3 to 4 lines broad, smooth, tapering at both ends, the valves with a prominent dorsal horn near the end. Seed-wing decurrent along the upper margin of the nucleus.—Meissn. in Pl. Preiss. i. 560, and in DC. Prod. xiv. 403.

W. Australia. W. coast, *Baudin's Expedition;* Swan river, *Fraser, Drummond, n.* 17, 22, 104, *Preiss, Oldfield.*

22. BUCKINGHAMIA, F. Muell.

Flowers hermaphrodite. Perianth irregular, the tube slender, revolute under the globular limb. Anthers all perfect, sessile in the base of the concave laminæ, the connective not produced beyond the cells. Hypogynous glands united in a single semiannular truncate and crenulate gland. Ovary shortly stipitate, with 4 collateral amphitropous ovules attached about the middle; style filiform, with an oblique almost lateral disk at the end, with the small stigma in its centre. Fruit a compressed follicle, opening along the upper suture. Seeds very flat and thin, surrounded by a narrow wing-like margin.—Tree. Leaves undivided, penniveined. Flowers small, pedicellate in pairs in terminal racemes. Bracts none or very deciduous.

The genus is limited to a single species endemic in tropical Australia and closely allied to the *Grevilleæ* of the section *Cycloptera,* differing only in the number of ovules and seeds.

1. **B. celsissima,** *F. Muell. Fragm.* vi. 248. A tree attaining 60 ft. or more, the young branches and inflorescence minutely hoary-tomentose. Leaves petiolate, elliptical-oblong, acute or obtuse, tapering at the base, 3 to 5 in. long, dark green and apparently glabrous above but sprinkled with minute hairs only visible under a lens, glaucous or almost silvery underneath, covered with the same grevillioid hairs. Racemes 4 to 8 in. long, somewhat secund, the flowers crowded nearly from the base. Pedicels slender, about ¼ in. long. Perianth silvery, 3 to 4 lines long. Anthers broad, with a thick rather

broad connective. Ovary glabrous. Follicle broadly and obliquely ovate, about 1 in. long, with a short incurved point. Seed broadly obovate.

Queensland. Rockingham Bay, *Dallachy.*

23. DARLINGIA, F. Muell.

Flowers hermaphrodite. Perianth regular, the tube slender, straight, the limb globular, erect. Anthers all perfect, sessile in the base of the concave laminæ, the connective produced beyond the cell into a minute gland-like appendage. Hypogynous glands 4, globular. Ovary sessile, with 4 collateral amphitropous ovules attached about the middle; style filiform, with an ovoid-fusiform end and a small terminal stigma. Fruit a compressed follicle, opening along the upper suture. Seeds very flat and thin, surrounded by a wing-like margin.—Tree. Leaves entire or pinnatifid, penniveined. Flowers sessile in pairs in terminal paniculate racemes. Bracts none or minute and deciduous.

The genus is limited to a single species endemic in tropical Australia. Like *Buckinghamia* it is closely allied to *Grevillea*, differing in the number of ovules and seeds, and distinguished from *Buckinghamia*, like the sections *Anadenia* and *Manglesia* from *Eugrevillea*, chiefly in the straight perianth. The sessile flowers and minute appendages to the anthers remove it also in a slight degree from both genera.

1. **D. spectatissima,** *F. Muell. Fragm.* v. 152. A tree, quite glabrous or the inflorescence minutely ferruginous-pubescent. Leaves oblong or oblanceolate, obtuse or acute, entire or deeply 3-lobed or pinnatifid with 5 to 7 long lanceolate acute lobes, tapering into a rather long petiole, the whole leaf 8 or 9 in. to 1½ ft. long, penniveined with rather numerous almost parallel primary veins. Racemes in the upper axils 4 to 8 in. long, forming a terminal panicle shorter than the leaves, with numerous flowers. Perianth glabrous, ¾ in. long, the tube slender, the laminæ tipped with small dorsal obtuse appendages, distinct in the bud. Ovary villous; style long. Follicle 1½-to nearly 2 in. long, nearly 1 in. broad, recurved. Seeds oblong, as long as the follicle.—*Helicia Darlingiana,* F. Muell. Fragm. v. 24; *Knightia Darlingii,* F. Muell. l.c. 152.

Queensland. Rockingham Bay, *Dallachy.*

TRIBE 6. EMBOTHRIEÆ.—Ovules several, imbricate in 2 rows. Seeds usually separated by thin laminæ or a mealy substance, (possibly the outer coating of the seeds detached and united as in *Banksia*).

24. TELOPEA, R. Br.

(Hylogyne, *Salisb.*)

Flowers hermaphrodite. Perianth irregular, the tube open early on the under-side, tapering and recurved under the limb, the laminæ oblique, broad. Anthers broad, sessile at the base of the laminæ, the connective not produced beyond the cells. Hypogynous glands united in a short very oblique nearly complete ring. Ovary contracted into a long stipes and tapering into a long style, clavate at the end, with a lateral stigma; ovules several, imbricate upwards in 2 rows, laterally attached near the base. Fruit a recurved coriaceous follicle. Seeds flat, terminating in a nearly straight or oblique membranous wing.— Tall shrubs. Leaves alternate, entire or toothed. Flowers pedicellate in pairs, in very dense globular or ovoid terminal racemes, surrounded by an involucre of imbricate coloured bracts, the bracts within the raceme small. Perianths as well as the whole inflorescence red.

The genus is endemic in Australia. It is allied in many respects to *Hakea*, differing chiefly in the number of ovules and seeds, and in habit.

Leaves prominently veined, mostly toothed. Involucre 2 to 3 in.
  long . . . . 1. *T. speciosissima.*
Leaves scarcely veined, mostly entire. Involucre under 1 in.
  long.
  Involucre glabrous . . . . 2. *T. oreades.*
  Involucre silky-ferruginous . . . . 3. *T. truncata.*

1. **T. speciosissima,** *R. Br. in Trans. Linn. Soc.* x. 198, *Prod.* 388. A stout erect glabrous shrub of 6 to 8 ft. Leaves cuneate-oblong or almost obovate, 5 to 10 in. long, mostly toothed in the upper part, tapering into a rather long petiole, coriaceous, penniveined with the midrib prominent, a few rarely quite entire. Flowers crimson, in a dense ovoid or globular head or raceme of about 3 in. diameter. Involucral bracts coloured, ovate-lanceolate, the inner ones 2 to 3 in. long, the outer ones few and small, surrounded by a dense tuft of floral leaves like the stem ones but smaller and more entire. Bracts under the pairs of flowers very short. Pedicels thick, recurved, $\frac{1}{4}$ to $\frac{1}{2}$ in. long. Perianth glabrous nearly 1 in. long. Ovules 12 to 16. Fruit recurved, 3 to 4 in. long. Seeds 10 to 20, the nucleus broad, obliquely quadrate, the wing obliquely truncate, $\frac{1}{4}$ to above $\frac{1}{2}$ in. long.—Meissn. in DC. Prod. xiv. 446; *Embothrium speciosissimum*, Sm. Specim. Bot. Nov. Holl. i. 19, t. 7; Bot. Mag. t. 1128; *E. spathulatum*, Cav. Ic. iv. 60, t. 388; *E. speciosum*, Salisb. Parad. Lond. t. 111; *Hylogyne speciosa*, Knight, Prot. 126.

N. S. Wales. Port Jackson to the Blue Mountains, *R. Brown, Sieber, n.* 22, and many others, known by the name of *Warratau* or *Waratah.*

2. **T. oreades,** *F. Muell. Fragm.* ii. 170. A shrub with the habit of *T. speciosissima*, the branches slightly ferruginous-pubescent, the foliage glabrous. Leaves obovate-oblong or almost lanceolate, acute or obtuse, 4 to 8 in. long, tapering into a long petiole, entire or rarely with a few teeth at the end, usually glaucous underneath, the veins scarcely conspicuous except the midrib. Racemes short broad and dense as in *T. speciosissima*, but the glabrous involucre in one specimen coloured and obtuse with the inner bracts 1 in. long, in the other specimens all herbaceous rigid mucronate and the inner ones scarcely $\frac{1}{2}$ in.

long. Flowers of *T. speciosissima.* Fruit 3 in. long, besides the stipes and persistent style.

Victoria. Nangatta mountains and Canus river, Gipps' Land, *F. Mueller.*

3. **T. truncata,** *R. Br. in Trans. Linn. Soc.* x. 198, *Prod.* 389. A stout shrub of 6 to 8 ft., the young branches ferruginous-pubescent or villous, the foliage glabrous. Leaves mostly oblong-cuneate, but varying from oblong-linear to almost obovate, obtuse or with a small callous point, tapering into a short petiole, 2 to 3 or rarely 4 in. long, thick, the veins often impressed above and scarcely conspicuous underneath, the margins often recurved. Racemes short and dense, about 2 in. diameter. Involucral bracts ovate, clothed with appressed hairs, the inner ones ¾ in. long, the outer ones shorter, more acuminate and sometimes with herbaceous tips. Pedicels glabrous, about ½ in. long. Perianth under 1 in. long, the broad part shorter in proportion than in *T. speciosissima,* tapering into a recurved neck at least as long. Fruit about 2 in. long, besides the persistent style. Seeds about 16.—Meissn. in DC. Prod. xiv. 446; Hook. f. Fl. Tasm. i. 327; *Embothrium truncatum,* Labill. Pl. Nov. Holl. i. 32, t. 44; *Hylogyne australis,* Knight, Prot. 127.

Tasmania. Mount Wellington, *R. Brown ;* abundant in cool humid mountainous regions at an elevation of 2000 to 4000 ft., *J. D. Hooker.*

25. LOMATIA, R. Br.

(Tricondylus, *Salisb.*)

Flowers hermaphrodite. Perianth irregular, the tube oblique, open along the lower side, tapering at the top, the limb ovoid-globular, recurved, the laminæ long cohering. Anthers ovate, sessile in the concave laminæ. Hypogynous glands 3, broad and truncate, the fourth upper one deficient. Ovary on a long stipes, tapering into a long style dilated at the top into a flat oblique disk stigmatic in the centre ; ovules several, laterally attached below the middle, amphitropous, imbricate upwards in 2 rows. Follicle coriaceous, opening almost flat. Seeds imbricate upwards, with a broad terminal nearly straight wing, surrounded by the marginal raphe.—Shrubs or trees. Leaves alternate, entire, toothed or pinnately divided, very variable on the same individual. Flowers pedicellate in pairs, in terminal or axillary simple or slightly branched racemes. Bracts under each pair usually small narrow and very deciduous or often entirely wanting. Perianths white or pale yellow, sometimes assuming at length a reddish tint.

The genus is also represented in the mountains of extratropical South America, but the Australian species appear to be all endemic. The structure and proportions of the parts of the flower and fruit are remarkably uniform in the Australian species, leaving little for their distinction besides the foliage which is eminently variable. The thin fragile pellicle or powdery substance interposed between the seeds in this genus and in *Telopea,* appears to be an epidermal production of the seed itself, but its real nature can scarcely be ascertained without observing it in a fresh state both before and after the maturity of the seed.

Leaves pinnate with ovate petiolulate segments . . . . . .   1. *L. fraxinifolia.*
Leaves undivided or once or twice pinnate, with sessile or decur-
   rent segments usually reticulate and toothed.
   Leaves mostly undivided, ovate to lanceolate, acutely toothed,
     rarely pinnate . . . .   2. *L. ilicifolia.*
   Leaves mostly undivided, linear-lanceolate, with callous ser-
     ratures . . . .   3. *L. longifolia.*
   Leaves mostly once twice or thrice pinnate . . . .   4. *L. silaifolia.*
Leaves narrow, undivided, pinnatifid or pinnate, otherwise entire
   or rarely toothed at the end, rather thick and veinless.
   Leaves mostly pinnate, glabrous or nearly so. Racemes long
     and loose . . . .   5. *L. tinctoria.*
   Leaves mostly undivided, closely and densely tomentose
     underneath. Racemes short and dense . . . .   6. *L. polymorpha.*

1. **L. fraxinifolia,** *F. Muell. Herb.* A tall shrub or small tree, the branches and foliage glabrous and drying black, the inflorescence slightly ferruginous-tomentose. Leaves mostly pinnate; segments 3 to 7, ovate or ovate-lanceolate, acuminate, coarsely-toothed, contracted into a distinct petiolule, 2 to 3 or rarely 4 in. long, coriaceous and shining above, the veins not very conspicuous; occasionally the lower leaves are undivided, or on luxuriant shoots one or two of the segments are again divided. Racemes 6 to 8 in. long, solitary or several in a broad terminal panicle. Pedicels 3 or 4 lines long. Perianth glabrous, 4 to 5 lines long, the limb ovoid. Fruit only seen young.

Queensland. Rockingham Bay, *Dallachy.*

2. **L. ilicifolia,** *R. Br. in Trans. Linn. Soc.* x. 200, *Prod.* 390, *Prot. Nov.* 33. An erect branching shrub of several ft., growing out sometimes into a small tree, quite glabrous or the young shoots and inflorescence more or less ferruginous-pubescent. Leaves petiolate, ovate oblong or lanceolate, irregularly prickly-toothed or lobed, varying from 2 or 3 in. in some specimens, to twice that size in others, glabrous above and more or less reticulate, closely and shortly silky-pubescent underneath; the upper leaves often small and distant, and on barren shoots the leaves sometimes pinnate with numerous small sessile or decurrent lanceolate toothed segments. Racemes long and loose, simple or slightly branched. Pedicels ¼ to ½ in. long. Perianth glabrous or pubescent with small appressed hairs, the tube 3 to 3½ lines long. Fruit 1 to 1½ in. long.—Meissn. in DC. *Prod.* xiv. 447; Bot. Mag. t. 4023; *Embothrium ilicifolium,* Poir. Dict. Suppl. ii. 551; *L. Fraseri,* R. Br. Prot. Nov. 34; Meissn. l.c.

N. S. Wales. Wombat Brush, *Fraser, A. Cunningham;* Berrima, *Woolls;* New England, *C. Stuart;* Clarence river, *Beckler, Lennans;* Mount Lindsay, *W. Hill;* snowy mountains at the head of Macleay and Bellinger rivers, *C. Moore;* southward to Twofold Bay, *A. Cunningham, F. Mueller.*

Victoria. Port Phillip *R. Brown;* Wilson's Promontory, *Baxter;* Dandenong ranges, Mount Disappointment, Seeler's Cove, Bunip Creek, Mount Aberdeen, Delatite Mountains, &c., *F. Mueller.*

The northern specimens are generally more ferruginous-pubescent than the southern ones; some from New England have the leaves all small and ovate; in those from Clarence river they are frequently pinnate, and in one instance some are pinnatifid with few lobes, and others pinnate with many segments on the same specimen; in Fraser's specimens they vary from slightly toothed to deeply pinnatifid.

3. **L. longifolia,** *R. Br. in Trans. Linn. Soc.* x. 200, *Prod.* 390. An erect shrub of 8 to 10 ft., glabrous or with a slight ferruginous pubescence on the young shoots and inflorescence. Leaves linear-lanceolate or rarely oblong-lanceolate, acuminate, bordered by distant serratures, tapering into a short petiole, mostly 4 to 8 in. long, not very prominently veined. Racemes axillary and terminal, shorter than the leaves or rarely longer. Perianth glabrous, 4 to 5 lines long. Fruit about 1 in. long.—Meissn. in DC. Prod. xiv. 447; Bot. Reg. t. 442; *Embothrium myricoides,* Gærtn. f. Fr. iii. 215, t. 218; *E. longifolium,* Poir. Dict. Suppl. ii. 551; *Tricondylus myricæfolius,* Knight, Prot. 122; *L. angustifolia,* Schnitzl. Ic. ii. 113? (name and fruit only).

N. S. Wales. Port Jackson to the Blue Mountains, *R. Brown, Sieber, n.* 16, *Fl. Mixt. n.* 473, and many others; Sydney woods, Paris Exhibition, 1855, *M'Arthur, n.*177; Argyle County, *Backhouse;* Twofold Bay, *L. Morton.*

Victoria. King river, Mitta-Mitta and Buffalo ranges, *F. Mueller.*

Var. *arborescens.* A small tree of 20 to 25 ft., with rather longer, more terminal racemes and flowers, smaller or shorter slender pedicels.—Sydney woods, Paris Exhibition, 1855, *M'Arthur, n.* 219.

4. **L. silaifolia,** *R. Br. in Trans. Linn. Soc.* x. 199, *Prod.* 389, *Prot. Nov.* 33. A shrub of 2 or 3 ft., glabrous or the young shoots and inflorescence minutely pubescent. Leaves mostly twice or thrice pinnate, rarely simply pinnate; segments sessile and decurrent, linear or lanceolate, usually deeply and sharply toothed, narrow or broad, long or short, the whole leaf usually 4 to 8 in. long and broad, or the lower ones larger, the reticulations obscure or prominent. Racemes terminal, long and loose, simple or branched, the flowers larger than in the preceding species. Pedicels ¼ to ½ in., perianth 7 to 8 lines long.—Meissn. in DC. Prod. xiv. 448; Bot. Mag. t. 1272; *Embothrium silaifolium,* Sm. Specim. Bot. Nov. Holl. 23, t. 8; *E. herbaceum,* Cav. Ic. iv. 58; t. 384; *E. crithmifolium,* Sm. (Steud.); *Tricondylus silaifolius,* Knight, Prot. 122.

Queensland. Brisbane river, Moreton Bay, *F. Mueller.*

N. S. Wales. Port Jackson to the Blue Mountains, *R. Brown, Sieber, n.* 15, and others; New England, *C. Stuart, C. Moore ;* Hastings river, *Fraser.*

Var. *induta,* F. Muell. Leaves silky-pubescent underneath, passing into the cut-leaved forms of *L. ilicifolia.*—Brisbane river, Moreton Bay, *Leichhardt, F. Mueller.*

Some specimens from Hastings river, *Beckler,* with simply pinnate leaves and toothed segments may be a variety either of *L. silaifolia* or *L. ilicifolia.*

5. **L. tinctoria,** *R. Br. in Trans. Linn. Soc.* x. 199, *Prod.* 389. A small shrub, rarely exceeding 2 ft., and increasing by subterraneous runners so as to form large patches, glabrous or the young shoots inflorescence and underside of the leaves silky-pubescent. Leaves pinnate bipinnate or rarely undivided; segments linear, obtuse, entire or lobed, varying from under ½ in. to above 1 in. long, but rather regular in the same leaf, scarcely veined besides the midrib. Racemes terminal or in the upper axils, pedunculate, loose, 4 to 8 in. long. Pedicels

about ½ in., perianth about 5 lines long. Fruit from ½ to near 1 in. long.—Meissn. in DC. Prod. xiv. 448 ; Hook. f. Fl. Tasm. i. 328; Bot. Mag. t. 4110; *Embothrium tinctorium,* Labill. Pl. Nov. Holl. i. 31, t. 43; *Tricondylus tinctorius,* Knight, Prot. 122.

Tasmania. Port Dalrymple and Derwent river, *R. Brown;* abundant in sandy soil, ascending to 3000 ft., *J. D. Hooker.*

6. **L. polymorpha,** *R. Br. in Trans. Linn. Soc.* x. 200, *Prod.* 389. A tall slender shrub, the branches and inflorescence ferruginous or silky-tomentose. Leaves mostly oblong-linear or lanceolate, obtuse or acute, entire, tapering into a petiole, 1 or 2 in. long, rarely more or less pinnatifid, thick, smooth and veinless above, densely but closely tomentose underneath, the midrib prominent, the margins often nerve-like or recurved. Racemes terminal, short and dense. Pedicels 3 to 6 lines long. Perianth pubescent, ½ in. long.—Meissn. in DC. Prod. xiv. 448; Hook. f. Fl. Tasm. i. 327; *Embothrium tinctorium, var.* Labill. Pl. Nov. Holl. i. 31, t. 42.

Tasmania. Port de l'Esperance and Mount Wellington, *R. Brown;* abundant in the western and central alpine districts, *J. D. Hooker.*

26. CARDWELLIA, F. Muell.

Flowers hermaphrodite. Perianth somewhat irregular, the tube open along the lower side, tapering at the top and recurved under the obliquely globular limb. Anthers ovate, sessile in the concave laminæ. Hypogynous glands 4. Ovary contracted into a short stipes; style elongated, dilated at the top into a lateral disk stigmatic in the centre; ovules several, laterally attached near the top and imbricate downwards in 2 rows. Fruit thick and woody, opening at length into a broad follicle. Seeds very flat, oblong, surrounded by a wing-like margin.— A tree. Leaves alternate, abruptly pinnate. Flowers in terminal racemes, in pairs, with the very short pedicels united. Bracts not seen.

The genus is limited to a single species endemic in Australia.

1. **C. sublimis,** *F. Muell. Fragm.* v. 24, 38, 73, and 152. A tree of 80 to 90 ft., the young branches and inflorescence minutely hoary-tomentose, the adult foliage glabrous. Leaves above 1 ft. long ; leaflets 4 to 10, opposite or alternate, all on rather long petiolules, ovate or oblong, obtuse, coriaceous, veined, green above, pale glaucous or fulvous underneath, 3 to 8 in. long. Racemes several in a terminal panicle, sometimes shorter sometimes longer than the leaves. Pedicels exceedingly short. Perianth hoary-tomentose, the tube about ½ in. long. Hypogynous glands globular. Ovules 12 to 16. Fruit when unripe with the aspect of that of a *Xylomelum,* but one old one in Herb. F. Mueller has opened out into a broad orbicular follicle, 3 in. diameter. Seed about 3 in. long, ¾ in. broad.

Queensland. Mountains about Rockingham Bay, *Dallachy.*

27. STENOCARPUS, R. Br.

(Agnostus, *A. Cunn.*)

Flowers hermaphrodite. Perianth slightly irregular, the tube opening along the lower side, the limb nearly globular and recurved, the segments at length separating. Anthers broad, sessile within the concave laminæ, the connective not produced beyond the cells. Hypogynous glands united in a short semiannular disk or cup or almost obsolete. Ovary stipitate, tapering into a long style dilated at the top into a flat oblique disk, stigmatic in the centre; ovules several, laterally attached at or near the top, imbricate downwards in 2 rows. Fruit a follicle, usually narrow, coriaceous; seeds produced at the lower end into a membranous wing.—Trees. Leaves alternate or scattered, entire or deeply pinnatifid with few lobes. Peduncles terminal or in the upper axils, sometimes several in an umbel or short raceme, each bearing an umbel of pedicellate red or yellow flowers. Bracts none or falling off at a very early stage.

The genus extends to New Caledonia, the Australian species are however all endemic.

Leaves 6 in. to 1 ft. long. Perianths above 1 in. long, the
  pedicels radiating in a single row round the disk-like end of
  the peduncle . . . . 1. *S. sinuatus.*
Leaves under 6 in. Perianths ½ in. long or less, the pedicels
  irregularly crowded on the summit of the peduncle.
  Ovary usually pubescent . . . . 2. *S salignus.*
  Ovary quite glabrous . . . . 3. *S. Cunninghamii.*

1. S. sinuatus, *Endl. Gen. Pl. Suppl.* iv. 88. A tree sometimes described as small and slender, sometimes said to attain 60 to 100 ft., glabrous or the inflorescence minutely tomentose. Leaves petiolate, either undivided oblong-lanceolate and 6 to 8 in. long, or pinnatifid and above 1 ft. long, with 1 to 4 oblong lobes on each side, mostly obtuse, quite glabrous but reddish underneath, penniveined and minutely reticulate. Peduncles terminal, either 2 or more together in a general umbel, or several at some distance forming a short broad raceme, each peduncle 2 to 4 in. long, and bearing an umbel of 12 to 20 bright red flowers, the pedicels about ½ in. long, radiating in a single row round the disk-like dilated summit of the peduncle. Perianth tube 1 in. long or rather more, straight, tapering upwards, the limb recurved, globular, about 2 lines diameter. Ovary densely pubescent, on a glabrous stipes, with a rather thick glabrous style. Ovules 12 to 14.—Meissn. in DC. Prod. xiv. 451; *Agnostus sinuatus,* A. Cunn. in Loud. Hort. Brit. 580; *Stenocarpus Cunninghamii,* Hook. Bot. Mag. t. 4263 (copied into Fl. des. Serres. iii. 189, t. 7) not of R. Br.; Paxt. Mag. xiv. i. with a fig.

Queensland. Brisbane river, Moreton Bay, *A. Cunningham, W. Hill;* Araucaria ranges, *Leichhardt;* Queensland woods, London Exhibition, 1862, *W. Hill, n.* 17.
N. S. Wales. Richmond river, *C. Moore, Fawcett;* Tweed river, *C. Moore.*

2. S. salignus, *R. Br. in Trans. Linn. Soc.* x. 202, *Prod.* 391. A moderate-sized tree, glabrous or the inflorescence minutely pubescent.

Leaves in the typical form ovate-lanceolate or elliptical, acute acuminate or rarely obtuse, tapering into a short petiole, 2 to 4 in. long, varying from penniveined to triplinerved (the lower primary veins scarcely longer or much longer and thicker than the others), but the veins usually indistinct slightly prominent or almost immersed, a few leaves on young trees or barren branches larger and pinnatifid. Peduncles slender, terminal or in the upper axils, usually shorter than the leaves, bearing a single umbel of 10 to 20 flowers or in luxuriant specimens as many as 30 flowers. Pedicels ¼ to ½ in. long, irregularly crowded on the summit of the peduncles. Perianth usually under ½ in. long. Ovary slightly silky-pubescent or nearly glabrous. Ovules 6 to 8, not so closely imbricate nor so narrow and compressed as in *S. sinuatus.*— Meissn. in DC. Prod. 451; Bot. Reg. t. 441; *Hakea rubricaulis,* Colla, Hort. Ripul. App. i. 114, t. 3; *Embothrium rubricaule,* Giord. Obs. 1837 (Meissn.); *Stenocarpus acacioides,* F. Muell. Fragm. i. 135.

Queensland. Warwick, *Nernst.*

N. S. Wales. Grose river, *R. Brown;* Blue Mountains, *A. and R. Cunningham;* Tweed river, *C. Moore;* Illawarra, *A. Cunningham, Shepherd;* Sydney woods, Paris Exhibition, 1855, *M'Arthur, n.* 187. Known under the name of " Silky Oak."

Var. *Moorei.* Leaves broader and usually more distinctly tripli- or quintupli-nerved, the ovary minutely pubescent.—*S. Moorei,* F. Muell. Fragm. i. 134, v. 154.—Rockingham Bay, *Dallachy;* Mount Lindsay, *W. Hill;* Illawarra, *C. Moore;* Mount Warming, *C. Moore* (with a few leaves deeply pinnatifid with 3 or 5 long narrow lobes).

Var. *concolor.* Leaves more prominently tripli- or rarely quintupli-nerved, the reticulations also more distinct. Flowers rather larger. Ovary glabrous or nearly so.—*S. concolor,* F. Muell. Fragm. iii. 147, v. 154.—Broad Sound and near Maryborough, *Bowman.*

3. **S. Cunninghamii,** *R. Br. Prot. Nov.* 34. A tall bushy shrub or small tree, glabrous or the inflorescence slightly pubescent, the specimens closely resembling those of *S. salignus* in which the leaves are rather narrow, thick and obscurely veined. Leaves oblong-lanceolate, obtuse or acuminate, varying in breadth, about 2 to 4 in. long, tapering into a short petiole, faintly tripli- or quintupli-nerved, the smaller veins rarely visible. Flowers precisely as in *S. salignus,* except that the ovary appears to be constantly quite glabrous.—Meissn. in DC. Prod. xiv. 451.

N. Australia. Vansittart's Bay, N W. coast, *A. Cunningham* (with small flowers and a slightly pubescent inflorescence); sources of the Roper river, *F. Mueller* (with small flowers and a nearly glabrous inflorescence); Liverpool river, *Cadell's Expedition* (with rather larger flowers and the inflorescence quite glabrous). The whole should probably be considered as varieties of *S. salignus.*

TRIBE 7. BANKSIEÆ.—Ovules 2, collateral. Seeds separated by a hard or membranous, usually bifid, sometimes double plate, rarely wanting. Flowers in dense spikes or *cones* with closely imbricate persistent bracts within or below the spike.

The singular plate intervening between the two seeds in this tribe has been explained by Brown to consist of the outer coating of one side of each seed, separating from the inner coatings as they advance towards maturity, the two becoming usually consolidated opposite the nuclei, remaining distinct opposite the seed-wings. This plate is entirely free from the walls of the pericarp, except at the point of attachment of the seed, forming a portion of the latter, not of the former, and has therefore no title to the name of a dissepiment, real or spurious, still given to it in systematic works, even in the Prodromus.

28. BANKSIA, Linn. f.

Flowers hermaphrodite. Perianth regular or nearly so, straight or curved, the slender tube opening equally or along the lower side only, the limb ovoid oblong or linear, the laminæ remaining long coherent, or rarely separating as the tube opens. Anthers narrow, sessile in the concave laminæ, the connective thick, usually very shortly produced beyond the cells. Hypogynous scales 4, very thin and membranous (rarely deficient?). Ovary very small and sessile; style usually longer than the perianth, rigid, curved and protruding from the slit in the perianth-tube until the end is set free by the separation of the laminæ, and then either straightened or remaining hooked or curved, rarely straight from the first and not exceeding the perianth; the stigmatic end on a level with the anthers, of a different texture but smooth, or striate and furrowed, continuous with the style or with a prominent rim at the base, the real stigma small and terminal; ovules 2, collaterally attached about the middle. Fruit a compressed capsule, opening at the broad end (or rather outer margin, for the scar of the style is lateral) in two hard often woody horizontal valves. Seeds usually 2, compressed, with a terminal membranous wing broad and rounded like the valves, the seeds separated by a plate of the same shape (the consolidated outer integuments of the inner side of the two seeds) free from the ripe seeds, simple (completely consolidated) between the nuclei, double (remaining distinct) between the wings.—Trees or shrubs. Leaves alternate or rarely verticillate or nearly so, usually narrow, entire toothed pinnatifid or pinnate, with numerous (rarely few) short teeth lobes or segments, the primary veins numerous and transverse, rarely inconspicuous or irregular and the minute reticulations numerous on the under surface, with a minute tomentum rarely wanting in the areolæ, and sometimes white and covering the whole under surface, the upper surface almost always glabrous and smooth. Flowers sessile in pairs, in dense terminal cylindrical oblong or globular spikes, either terminal and sessile above the last leaves or rarely lateral or on short lateral branches; each pair of flowers subtended by one bract and two lateral rather smaller bracteoles, both bracts and bracteoles densely woolly-villous on the sides, the tips glabrous tomentose or villous, either clavate and obtuse or truncate, or shortly acuminate, always densely imbricate in parallel spiral or rarely vertical lines. Perianth-tube very slender and entire within the bracts, ultimately splitting beyond them. In fruit the bracts and bracteoles become consolidated with the rhachis into a thick woody cone, either covered with the withered remains of the perianths amongst which the capsules are entirely concealed, or, where the flowers are wholly deciduous, the valves of the capsules protrude more or less beyond the bracts, the lower indehiscent portion containing the nuclei of the seeds remaining imbedded among the bracts. The proportion of perfect capsules is usually very small in relation to the number of flowers, of which there are often from 500 to above 1000 in the same spike.

The genus is endemic in Australia, and the greater number of species are Western, two only of the Eastern species penetrate into the tropics, besides one which is exclusively tropical, if it be really more than a variety of the most widely diffused of the Eastern species.

Sect. 1. **Oncostylis.**—*Leaves linear or rarely lanceolate, with revolute margins or nearly flat but very white underneath, entire denticulate or pinnate with small nume-rous regular segments. Style remaining hooked after the perianth-limb has opened, the stigmatic end very small.*

Perianth-tube villous, less than half as long as the style, the limb glabrous. Leaves small, entire. Western species.
    Perianth-tube about 4 lines long. Leaves incurved or erect, ¼ to
        ½ in. long . . . . 1. *B. pulchella.*
    Perianth-tube about 3 lines long. Leaves spreading or reflexed,
        not exceeding ¼ in. . . . . 2. *B. Meissneri.*
Perianth-tube more than half as long as the style, silky as well as the limb.
    Leaves linear with closely revolute entire margins and not trun-
        cate at the end. Western species.
        Leaves mostly short. Perianth under 1 in. long. Bracts
            with glabrous tips . . . . 3. *B. nutans.*
        Leaves mostly long. Perianth above 1 in. long. Bracts en
            tirely woolly-villous . . . . 4. *B. sphœrocarpa.*
    Leaves linear, truncate or notched at the end and often denticu-
        late, especially near the end.
        Western species, leaves long.
            Leaves (2 to 4 in.) very narrow, with closely revolute entire
                margins. Bracts villous to the end . . . . 5. *B. tricuspis.*
            Leaves (2 to 4 in.) with revolute or recurved margins, entire
                or denticulate towards the end. Bracts with glabrous tips 6. *B. occidentalis.*
            Leaves (4 to 8 in.) more open, showing the tomentose under
                surface. Bracts tomentose at the end . . . . . . 7. *B. littoralis.*
        Eastern species.
            Leaves (about ½ in.) very narrow with closely revolute entire
                margins . . . . 8. *B. ericifolia.*
            Leaves (1½ to 3 in.) narrow-linear with closely revolute
                entire or denticulate margins . . . . 9. *B. spinulosa.*
            Leaves (1½ to 3 in.) linear, more open, showing the white
                under surface, denticulate to the base or rarely entire . 10. *B. collina.*
Leaves mostly verticillate, oblong-lanceolate or broadly linear,
    entire or rarely toothed at the end, white underneath.
    Western species . . . . 11. *B. verticillata.*
Leaves pinnate with numerous small regular contiguous but
    distinct segments.
    Leaf-segments broad, triangular. Spikes small, globular or
        ovoid . . . . 12. *B. dryandroides.*
    Leaf-segments narrow, falcate. Spikes large, oblong or
        cylindrical . . . . 13. *B. Brownii.*

Sect. 2. **Cyrtostylis.**—*Leaves flat or undulate, the margins not revolute, toothed, pinnatifid or pinnate. Style arched or nearly straight and turned upwards after flowering, not hooked, the stigmatic end small, not striate. Western species.*

Perianth obtuse or acute, not aristate.
    Leaves narrow, regularly serrate, usually white underneath.
        Spikes narrow. Perianth glabrous, under ¾ in. long . . . 14. *B. attenuata.*

Spikes broad. Perianth 1 in. long, the tube villous, the limb
  at length glabrous . . . . 15. *B. media.*
Leaves large, on long petioles, irregularly toothed or lobed.
  Tree. Outer bracts short . . . . 16. *B. Solandri.*
Low prostrate shrubs. Outer bracts linear-subulate.
  Leaves closely surrounding the spike and not along the pros-
    trate stem . . . . 17. *B. Goodii.*
  Leaves erect along the prostrate stem, white underneath,
    none round the spike . . . . 18. *B. petiolaris.*
Leaves large, on long petioles, deeply and irregularly pinnatifid.
  Low prostrate shrubs. Spikes oblong.
  Perianth 1 in. long, the limb hirsute with loose usually persis-
    tent hairs . . . . 19. *B. repens.*
  Perianth scarcely ¾ in. long, the limb clothed with intricate
    loose ferruginous very deciduous hairs . . . . 20. *B. prostrata.*
Leaves large, pinnate, with triangular distinct but contiguous seg-
  ments. Spikes cylindrical . . . . 21. *B. grandis.*
Perianth acuminate with long awn-like points. Leaves nearly
  sessile, not very large, irregularly toothed or lobed.
  Spikes 3 to 4 in. long, rather narrow. Bracts with glabrous tips 22. *B. quercifolia.*
  Spikes 4 to 8 in. long, very thick. Bracts villous at the end . 23. *B. Baueri.*

SECT. 3. **Eubanksia.**—*Leaves linear-lanceolate, oblong or cuneate, with recurved or revolute, entire or dentate margins, white underneath. Style at first curved, straight and very spreading or reflexed after the perianth-limb has opened, the stigmatic end small, not striate. Eastern or tropical species.*

Leaves (mostly 1 to 2 in.) entire or rarely toothed, reticulate underneath, without any or with few and irregular primary transverse veins . . . . 24. *B. marginata.*
Leaves (mostly 3 to 6 in.) entire or rarely toothed, with transverse
  primary veins underneath, usually numerous but not much more
  prominent than the reticulations and white like them . . . . 25. *B. integrifolia.*
Leaves (mostly 4 to 8 in.) broad, coarsely toothed, the transverse
  primary veins prominent underneath and not so white as the
  reticulations . . . . 26. *B. dentata.*

(*B. latifolia,* has nearly the flowers and style of *Eubanksia,* but flat leaves not white underneath).

SECT. 4. **Orthostylis.**—*Leaves flat or undulate (irregularly in* B. Caleyi *and* B. coccinea), *serrate, pinnatifid or pinnate, with short lobes or segments. Perianth usually straight. Style, after the perianth-limb has opened, curved upwards near the base, then straight and erect, the stigmatic end prominently angled and furrowed or striate.*

Eastern species.
  Leaves 2 to 3 in. long, broad, irregularly toothed. Style end of
    Eubanksia . . . . 27. *B. latifolia.*
  Leaves 3 to 6 in. long, ¾ to 1 in. broad, regularly serrate. Style-
    end thickened at the base.
    Style-end cylindrical . . . . 28. *B. serrata.*
    Style-end ovoid, very short . . . . 29. *B. æmula.*
  Leaves 2 to 4 in. long, ½ to ¾ in. broad, regularly serrate. Style-
    end oblong . . . . 30. *B. ornata.*
Western species.
  Perianth villous.
    Leaves ¾ in. broad or more, very shortly sinuate, toothed.
      Leaves 1½ to 2½ in. long, very broad, often cordate. Peri-
        anths, before opening, in double-straight rows alternating
        with double rows of styles . . . . 31. *B. coccinea.*

Leaves 1½ to 2½ in. long. Spikes long, the perianths and
   styles alternating in single rows . . . . 32. *B. sceptrum.*
Leaves 6 in. to 1 ft. long . . . . 33. *B. Menziesii.*
Leaves under ½ in. broad, regularly serrate, the veins inconspicuous underneath.
Leaves 2 to 4 in. long. Spikes globular. Style-end small
   and slender . . . . 34. *B. lævigata.*
Leaves 4 to 8 in. long. Spikes oblong. Style-end long,
   with a thickened base . . . . 35. *B. Hookeriana.*
Leaves deeply and regularly serrate or lobed, the transverse
   veins connivent in each lobe. Style-end stipitate above
   its thickened base.
Leaf-lobes short and broad, not reaching halfway to the
   midrib . . . . 36. *B. prionotes.*
Leaf-lobes triangular, acuminate, reaching more than half-
   way to the midrib . . . . 37. *B. Victoriæ.*
Leaves pinnate with contiguous broad acute segments. Style
   hairy.
Spikes oblong. Perianth-limb obtuse. Leaves often 1 ft.
   long . . . . 38. *B. speciosa.*
Spikes globular. Perianth-limb acute. Leaves under 6 in.
   long . . . . 39. *B. Baxteri.*
Perianth glabrous.
Leaves 1 to 1½ in. long, oblong, truncate, sinuate-toothed.
   Spikes oblong cylindrical . . . . 40. *B. marcescens.*
Leaves 1½ to 3 in. long, obovate-oblong, toothed. Spikes large,
   nearly globular . . . . 41. *B. Lemanniana.*
Leaves 3 to 6 in. long, narrow, sinuate and prickly-toothed.
   Spikes nearly globular. Perianth-limb half as long as the
   tube . . . . 42. *B. Caleyi.*
Leaves 2 to 4 in. long, regularly serrate. Perianth-limb not
   half as long as the tube . . . . 43. *B. Lindleyana.*
Leaves 6 in. to 1 ft. long or more, with numerous regular
   triangular lobes or segments.
Leaves lobed only . . . . 44. *B. elegans.*
Leaves divided to the midrib . . . . 45. *B. Candolleana.*

SECT. 5. **Isostylis.** — *Spikes reduced to depressed-globular heads. Perianths
straight, the limb opening as soon as the tube and style straight as in most* Dryandræ,
but the outer bracts few as in Banksiæ.
Leaves 1 to 3 in. long, obovate-oblong or cuneate, undulate and
   prickly-toothed . . . . 46. *B. ilicifolia.*

B. Huegelii, Br., *B. longifolia,* Desf., *B. mimosoides,* Don, *B. rubra,* Don, and *B.
virens,* Don, are names only of plants which, if true *Banksiæ,* belong probably to some
of the species above enumerated.

SECT. 1. ONCOSTYLIS.—Leaves linear or rarely lanceolate, with
revolute margins or nearly flat but very white underneath, entire denticulate or pinnate with small numerous regular segments. Style
remaining hooked after the perianth-limb has opened, the stigmatic
end very small and not distinctly furrowed.

1. **B. pulchella,** *R. Br. in Trans. Linn. Soc.* x. 202, *Prod.* 391. A
shrub with villous or tomentose branches. Leaves crowded, erect or
incurved, linear, sometimes very narrow or almost terete, obtuse or
almost acute, the margins entire and closely revolute, narrowly grooved
or more broadly channelled underneath, the midrib not prominent, ¼

to ½ in. long. Spikes ovoid-globular, the rhachis 1 to 1½ in. long.
Bracts villous. Perianth-tube densely villous, about 4 lines long, the
limb acute, glabrous. Ovary villous. Style nearly 1 in. long, remaining hooked, with a very small broad stigmatic end. Fruiting cone
globular, about 2 in. diameter, the capsules usually very numerous and
closely packed, very flat, projecting but slightly, the margin becoming
glabrous, nearly 1 in. broad when perfect.—Meissn. in Pl. Preiss. ii.
264, and in DC. Prod. xiv. 452.

W. Australia. Lucky Bay, *R. Brown,* and probably from the same neighbourhood, *Baxter, Drummond, n.* 24, *and 2nd coll. n.* 338.

2. **B. Meissneri,** *Lehm.* ; *Meissn. in Pl. Preiss.* i. 582, *and in DC.
Prod.* xiv. 452. A spreading shrub of 2 or 3 ft., or sometimes low
and straggling, the branches slightly hoary. Leaves linear, rather
crowded, very spreading or reflexed, obtuse or scarcely acute, with
revolute margins, singly grooved or channelled underneath, not above
¼ in. long and thicker than in *B. pulchella.* Spikes ovoid and flowers
smaller than in that species. Perianth-tube loosely villous, scarcely
3 lines long, the glabrous limb very small. Ovary glabrous? Style
about ¾ in. long, remaining hooked with the small depressed stigmatic
end of *B. pulchella.* Fruiting cone not seen.

W. Australia. Between Swan river and King George's Sound, *Drummond, n.*
109, *2nd coll. n.* 282, *Preiss, n.* 488, *Harvey;* near Arthur, *Oldfield;* Beaufort and
Gordon plains, *Maxwell;* and with more erect leaves, Phillips river to Esperance Bay,
Maxwell. F. Mueller thinks that this is a variety only of *B. pulchella,* with small
thick spreading leaves. The ovary appeared to me to be glabrous, but that character
may require further confirmation. Both species differ from all other *Banksiæ* in their
small perianth, very short in proportion to the style.

3. **B. nutans,** *R. Br. in Trans. Linn. Soc.* x. 203, *Prod.* 391. A
shrub, glabrous or nearly so except the inflorescence. Leaves crowded,
very narrow-linear, almost terete, very shortly mucronate, the margins
closely revolute and entire, singly grooved underneath, ½ to 1 in. long.
Spikes globular or shortly oblong, erect or nodding, the rhachis from
under 1 to near 2 in. long. Bracts with small glabrous tips. Perianth-
tube ¾ in. long, silky-villous as well as the limb. Ovary glabrous.
Style remaining hooked, with a short thick stigmatic end not distinctly
furrowed. Fruiting cone globular, 2 to 4 in. diameter ; capsules very
thick and scarcely protruding, the end in some specimens above 1 in.
broad and nearly 1 in. thick, smooth and at first raised along the
suture, at length depressed the thick almost turgid backs of the valves
very rugose ; in some specimens the capsules smaller and smoother,
but perhaps not full-grown.—Meissn. in Pl. Preiss. i. 581, and in DC.
Prod. xiv. 453 ; F. Muell. Fragm. iv. 108.

W. Australia. Lucky Bay, *R. Brown ;* King George's Sound or adjoining districts, *Baxter, Drummond, n.* 168, *3rd coll. n.* 281, *Oldfield, Maxwell.* Meissner describes the capsules as somewhat tomentose all over and not turgid on the top; but he
had probably either a mismatched fruit or a distinct variety from any I have seen, for I
have always found the capsules perfectly glabrous, and more deserving the character
of turgid at the top than any other species.

546	CIV. PROTEACEÆ.	[*Banksia.*

4. **B. sphærocarpa,** *R. Br. in Trans. Linn. Soc.* x. 203, *Prod.* 391. A shrub of 3 or 4 ft., minutely silvery or hoary-tomentose. Leaves linear, obtuse or scarcely mucronate, with closely revolute entire margins, under 1 in. long in the typical specimens, in others 2 to 3 in. long. Spikes globular or nearly so, 2 to 3 in. diameter. Perianth silky, varying from a little above 1 in. to fully 1½ in. long, the limb narrow, obtuse. Style longer than the perianth, hooked, with a small cylindrical stigmatic end. Fruiting cone globular, dense; capsules slightly prominent, glabrous, thick, with a prominent ridge at the suture, nearly 1 in. broad when perfect.—Meissn. in Pl. Preiss. i. 581, and in DC. Prod. xiv. 452; *B. pinifolia,* Meissn. in DC. Prod. xiv. 453.

W. Australia. King George's Sound, *R. Brown, Baxter,* and others, and thence to Swan river, *Drummond, n.* 99, 100, *1st coll. n.* 648, 649, *2nd coll. n.* 336, *Preiss, n.* 486, 487, 494, 497, and others; Murchison river, *Oldfield;* between Moore and Murchison rivers, *Drummond, 6th coll. n.* 199.

Some of the northern specimens, which constitute the *B. pinifolia,* have larger flower-heads and flowers and longer leaves, and a fruit of Drummond's which, from his notes, may belong to this *B. pinifolia* is also much larger, with more prominent and thinner capsules. Other specimens from the same district have precisely the flowers of the common form. In some specimens the bracts have conical tomentose tips, in others they are quite flat. It is possible therefore that two species may be here confounded, but the specimens are insufficient for their distinction.

Var. *glabrescens,* Meissn. Flower-heads and flowers smaller, not so villous, the fulvous hairs of the bracts not so prominent.—W. Australia, *Drummond, 2nd coll. n.* 337.

Var. *latifolia,* F. Muell. Leaves short, 1 to 1¼ lines broad. Flowers large, silky-villous with long rather loose hairs.—Perongerup Range, *Maxwell.*

5. **B. tricuspis,** *Meissn. in Hook. Kew Journ.* vii. 118, *and in DC. Prod.* xiv. 453. Branches rather slender, glabrous or very slightly hoary. Leaves narrow-linear, truncate or almost notched, with a small callous point, the margins entire and closely revolute, 2 to 4 in. long. Spikes oblong-cylindrical, 5 to 6 in. long. Bracts obtuse, fulvous-villous. Perianths silky-villous but all withered and revolute in our specimens. Style above 1½ in. long, hooked, with a very small ovoid stigmatic end. Fruiting cone with very closely imbricate obtuse bracts; capsules very prominent, not thick, becoming glabrous, 9 to 10 lines broad.

W. Australia. Mount Lesueur and Gardner's Range, *Drummond, 6th coll. n.* 205.

6. **B. occidentalis,** *R. Br. in Trans. Linn. Soc.* x. 204, *Prod.* 392. An erect shrub of 4 or 5 ft., the branches glabrous or minutely hoary. Leaves linear, truncate notched or 3-toothed at the end, otherwise entire or with a few small teeth towards the end, the margins recurved only, showing the white under surface and prominent midrib, 2 to 4 in. long. Spikes from ovoid and 3 in. to cylindrical and twice as long. Bracts with small glabrous tips. Perianth silky-villous, about ¾ in. long, the limb narrow. Ovary villous; style about 1 in. long, hooked, the stigmatic end scarcely distinct. Fruiting cone tomentose with the closely packed bracts; capsules prominent, not very thick, rounded, tomentose-villous, becoming glabrous at the suture, about ¾ in. broad.

—Meissn. in Pl. Preiss. i. 582, and in DC. Prod. xiv. 454, Bot. Mag. t. 3535; Lindl. and Paxt. Mag. i. t. 35, copied into Flora des Serres vi. 636, and into Lem. Fl. Jard. t. 119.

W. Australia. King George's Sound and adjoining districts, *R. Brown, Baxter, Drummond, 3rd coll. n.* 283, *Preiss, n.* 491, and others.

7. **B. littoralis,** *R. Br. in Trans. Linn. Soc.* x. 204, *Prod.* 392. A tree of 20 to 40 ft., the branches closely tomentose. Leaves scattered or irregularly whorled, linear, broadly and distantly serrate or rarely entire, tapering into a petiole, 4 to 8 in. long, the margins recurved or nearly flat, the under surface hoary-tomentose or white. Spikes oblong or cylindrical, 6 to 10 in. long. Bracts truncate and tomentose at the end. Perianth silky, nearly 1 in. long. Style rather longer than the perianth, remaining hooked, with a very small ovoid stigmatic end. Fruiting cones tomentose with the closely packed bracts after the perianths have fallen away; capsules shortly protruding, rounded, not thick, tomentose, ½ to ¾ in. broad.—Meissn. in Pl. Preiss. i. 583, and in DC. Prod. xiv. 454.

W. Australia. King George's Sound and adjoining districts, *R. Brown, Fraser, Drummond, n.* 109, *1st coll. n.* 647, *Preiss, n.* 479, 496, *Oldfield, Maxwell, F. Mueller.* Very near in many respects to the eastern *B. collina*, but at once distinguished by the long leaves.

8. **B. ericifolia,** *Linn. f. Suppl.* 127. A tall shrub or small tree of 12 to 14 ft., glabrous except the inflorescence. Leaves crowded, narrow-linear, truncate or notched at the end and sometimes with an intermediate point, otherwise entire with closely revolute margins, rarely exceeding ½ in. Spikes cylindrical, 6 to 10 in. long. Bracts with broad shortly acuminate silky-pubescent tips. Perianth yellow, silky, the tube about ¾ in. long, the limb ovoid. Style about 1 in. long, hooked, with a very short thick stigmatic end. Fruiting cones long and cylindrical. Capsules scarcely protruding, villous but often becoming glabrous, the flat top ¾ to 1 in. broad and 4 or 5 lines thick.— R. Br. in Trans. Linn. Soc. x. 203, Prod. 391; Meissn. in DC. Prod. xiv. 453; Cav. Ic. vi. t. 538; Andr. Bot. Rep. t. 156; Bot. Mag. t. 738; Baill. Hist. Pl. ii. 393, f. 227 to 229.

N. S. Wales. Port Jackson, *R. Brown, Sieber, n.* 7, and many others; Hastings river, *Beckler.*

9. **B. spinulosa,** *Sm. Specim. Bot. N. Holl.* 13, *t.* 4. A tall shrub, glabrous or the young branches minutely pubescent. Leaves narrow-linear, notched at the end with a prominent point in the notch and often bordered towards the end with 2 or 3 small teeth on each side, otherwise entire, with revolute margins and the midrib prominent underneath, 1½ to 3 in. long. Spikes ovoid and 2 to 3 in. long, or rarely cylindrical and twice as long. Bracts with broad shortly acuminate silky-pubescent tips. Flowers yellow, larger than in *B. ericifolia.* Perianth silky, the tube nearly 1 in. long. Style 1¼ to 1½ in. long, often purple, with a very short stigmatic end not thicker than

the style. Fruiting cone cylindrical. Capsules scarcely protruding, glabrous, thick, smooth.—R. Br. in Trans. Linn. Soc. x. 203, Prod. 392; Meissn. in DC. Prod. xiv. 453; Cav. Ic. t. 537; Andr. Bot. Rep. t. 457; *B. denticulata*, Dum. Cours. (Meissn.).

N. S. Wales. Port Jackson, *R. Brown, Sieber, n.* 1, *Woolls,* and many others; near Richmond, *Wilhelmi;* southward to Twofold Bay, *F. Mueller.*

10. **B. collina,** *R. Br. in Trans. Linn. Soc.* x. *204, Prod.* 392. A tall erect shrub attaining 8 to 12 ft., the young branches tomentose or villous. Leaves linear, much broader than in *B. spinulosa,* and always showing the white under surface, the margins only slightly recurved, more or less denticulate or rarely quite entire, 1½ to 3 in. long. Spikes oblong or cylindrical, 3 to 6 in. long. Bracts with broad flat or scarcely acuminate ends. Perianths silky, the tube above 1 in. long, the limb narrow-ovoid. Style longer than the perianth, hooked, with a very small stigmatic end. Fruiting cone cylindrical like that of *B. ericifolia* or longer. Capsules thick and scarcely protruding as in that species but quite glabrous.—Meissn. in DC. Prod. xiv. 454; *B. ledifolia,* A. Cunn. Herb.; *B. Cunninghamii,* Sieb. in Spreng. Syst. Cur. Post. 47, and in Roem. and Schult. Syst. iii. Mant. 289; R. Br. Prot. Nov. 35; Meissn. in DC. Prod. xiv. 454; Reich. Iconogr. Exot. t. 81; *B. littoralis,* Lindl. Bot. Reg. t. 1363, Grah. in Bot. Mag. t. 3060, not of R. Br.; *B. prionophylla,* F. Muell. 1st Gen. Rep. 17; *B. marginata* var. *macrostachya,* Hort. Petrop.

Queensland. Glasshouses, Moreton Bay, *C. Moore.*
N. S. Wales. Hunter's river, *Caley;* Blue Mountains? *Sieber, n.* 6; western descent of the Blue Mountains, *A. Cunningham;* New England, *C. Stuart;* Richmond, Clarence and Hastings rivers, *Beckler;* Sydney woods, Paris Exhibition, 1855, *M'Arthur, n.* 215.
Victoria. Wilson's Promontory, *Baxter;* Sealer's Cove and towards Mount Ararat, *F. Mueller;* Upper Yarra river, *C. Walter.*

When the leaves are small and rather broad, they are somewhat like those of *B. marginata,* but the species is readily distinguished by the large flowers, hooked style and thick capsules.

11. **B. verticillata,** *R. Br. in Trans. Linn. Soc.* x. *207, Prod.* 394. A small tree, the young branches tomentose and sometimes villous. Leaves in whorls of 4 to 6 sometimes irregular or broken on luxuriant branches, shortly petiolate, oblong-lanceolate or broadly linear, with recurved margins, white underneath, those of the flowering stems 1½ to 3 in. long, obtuse, entire or slightly toothed, but in some specimens without flowers (from young trees?) longer, narrower and more or less serrate. Spikes oblong-cylindrical, 4 to 8 in. long. Bracts truncate or very shortly acuminate with woolly-villous ends. Perianth yellow, silky, nearly 1 in. long. Style scarcely longer, hooked, with a very small stigmatic end. Fruiting cones long and narrow, the perianths deciduous leaving the closely packed bracts in hoary areolæ, with a more glabrous centre, or with slightly protruding flat capsules, ½ to ¾ in. broad, the valves not thickened.—Meissn. in Pl. Preiss. i. 583, and in DC. Prod. xiv. 457; Hook. Exot. Fl. t. 96.

W. Australia. King George's Sound, *R. Brown, Baxter, Drummond, n.* 167 (with smaller flowers), 4*th coll. n.* 304, *Preiss, n.* 493, 495 (the latter a barren specimen with denticulate leaves).

12. **B. dryandroides,** *Baxt. in Sw. Fl. Austral. t.* 56.   A shrub of 2 or 3 ft., with very spreading tomentose branches.   Leaves sessile, 3 to 6 in. long, flexuose, divided nearly or quite to the midrib into numerous contiguous triangular lobes or segments, the largest of which are 3 to 4 lines long and broad, thick, with revolute margins, white or ferruginous-tomentose underneath.   Spikes globular or rarely ovoid, about 1½ in. diameter, shortly pedunculate, more lateral than in most species.   Perianth-tube silky-villous, about ½ in. long, the limb hirsute with longer deciduous hairs, about 1 line long, acute.   Style scarcely longer than the perianth, remaining hooked, with a very small almost capitate stigmatic end.   Fruiting cone globular, about 2 in. diameter.   Capsules protruding, rounded at the end, rather flat, ¾ in. broad, at first villous, at length glabrous.—R. Br. Prot. Nov. 36; Meissn. in Pl. Preiss. i. 588, and in. DC. Prod. xiv. 465.

W. Australia. Towards Cape Riche, *Drummond,* 3rd *coll. n.* 287, *Preiss, n.* 490, *Maxwell;* Mount Gardner, *Baxter;* sand plains, Kalgan river, *Oldfield.*

13. **B. Brownii,** *Baxt. in R. Br. Prot. Nov.* 37.   A small tree of 10 to 20 ft.   Leaves very shortly petiolate, 3 to 5 in. long, divided to the midrib into very numerous lanceolate falcate regular segments, the largest scarcely above 3 lines long, with recurved margins, white under-neath.   Spikes oblong-cylindrical, very thick, 6 to 8 in. long.   Perianth silky-villous, about 1 in. long, the limb small narrow and acute.   Style longer than the perianth, hooked, with a very small stigmatic end.   Fruiting cone oblong or cylindrical, thick.   Capsules protruding, rounded, not thick, shortly villous or at length glabrous, about ¾ in. broad.—Meissn. in Pl. Preiss. i. 588, and in DC. Prod. xiv. 465.

W. Australia. Towards Cape Riche, *Baxter, Preiss, n.* 478, *Drummond,* 5th *coll. n.* 415.

SECT. 2. CYRTOSTYLIS.—Leaves flat or undulate, the margins not revolute, toothed pinnatifid or pinnate.   Style arched or nearly straight and turned upwards or curved, but not hooked after flowering, the stigmatic end small, not furrowed.

The foliage is that of *Orthostylis,* but the style less rigid and erect, and the stigmatic end that of *Oncostylis* and *Eubanksia.*

14. **B. attenuata,** *R. Br. in Trans. Linn. Soc.* x. 209, *Prod.* 395. A tree of 40 ft. with tomentose branches.   Leaves linear or oblanceo-late, serrate, tapering into a short petiole, 3 to 6 in. long, 3 to 5 lines broad towards the end, rather thick, flat, hoary-tomentose underneath with transverse veins and reticulations.   Spikes cylindrical, 4 to 8 in. long.   Bracts densely hirsute at the end.   Perianth glabrous, the tube about ½ in., the limb 2 lines long, obtuse.   Style remaining arched but not hooked, with a small slender stigmatic end.   Fruiting cone thick. Capsule scarcely protruding from the remains of the flowers, villous,

above 1 in. broad and ½ in. thick, showing the scar or even the base of the style on the right-hand margin.—Meissn. in Pl. Preiss, ii. 264, and in DC. Prod. xiv. 458; F. Muell. Fragm. vii. 55; *B. cylindrostachya*, Lindl. Swan Riv. App. 34; Meissn. in Pl. Preiss. i. 583, and in DC. Prod. xiv. 455.

W. Australia. King George's Sound, *R. Brown;* Stirling Range, *F. Mueller;* thence to Swan river, *Fraser, Drummond, 1st coll. suppl. n.* 114, *3rd coll. n.* 286, *Preiss, n.* 475; Serpentine and Murchison rivers, *Oldfield.*

15. **B. media,** *R. Br. Prot. Nov.* 35. A tall shrub or small tree, the branches hoary-tomentose. Leaves lanceolate-cuneate, truncate, serrate, tapering into a short petiole, 2 to 3 in. long in some specimens, twice as long in others, ½ to ¾ in. broad, flat, tomentose underneath with parallel transverse veins and reticulate between them. Spikes oblong or cylindrical, 3 to 6 in. long. Bracts hirsute at the end. Perianth about 1 in. long, the tube shortly silky-pubescent, the limb at first pubescent but soon becoming glabrous. Fruiting cone thick. Capsules immersed in the persistent remains of the flowers, nearly glabrous.— Meissn. in DC. Prod. xiv. 457; Bot. Mag. t. 3120.

W. Australia. Lucky Bay, Point Malcolm, to Cape Arid, *Baxter;* interior from Cape Riche, Gardner, Fitzgerald and Phillips Ranges, and away to the eastward, *Maxwell.*

16. **B. Solandri,** *R. Br. Prot. Nov.* 36. A tree, with tomentose branches. Leaves on rather long petioles, oblong, truncate, more or less divided into irregular triangular lobes very rarely reaching the midrib, 6 to 8 in. long, 1½ to 4 in. broad, flat, very rigid, the under surface pale and sometimes white, with numerous prominent transverse veins and conspicuous reticulations. Spikes oblong or cylindrical, 3 to 8 in. long. Perianths very slender, scarcely 1 in. long, the tube loosely silky-hairy, the limb narrow, acute, glabrous or with a very few long fine hairs. Style remaining curved but not hooked, with a small very short stigmatic end. Fruiting cone ovoid or oblong, 2 in. diameter. Capsules quite glabrous, thick with a slightly prominent acute ridge at the suture.—Meissn. in DC. Prod. xiv. 463; *B. Hookeri*, Drumm. in Bot. Mag. lxxiv. Comp. 1.

W. Australia. Mountains near King George's Sound, *Baxter;* summit of Mongerup, *Drummond, 4th coll. n.* 305; Perongerup ranges and sand plains, Kalgan river, *Oldfield.*

17. **B. Goodii,** *R. Br. Prot. Nov.* 36. Stems short, woolly or tomentose, and apparently prostrate as in the three following species, but without leaves excepting close under the inflorescence. Leaves on long petioles, ½ to 1 ft. long, 1 to 3 in. broad, sinuate and irregularly toothed or lobed but the lobes rarely reaching half way to the midrib and usually very short, very rigid, the under surface tomentose but the tomentum deciduous and never white, the primary transverse veins prominent. Spikes oblong-cylindrical, 3 or 4 in. long, closely surrounded by the floral leaves and a few subulate plumose outer bracts. Perianth-tube not 1 in. long, loosely villous, the limb narrow, acute, at first bearded

with long hairs but soon glabrous. Style remaining curved but not hooked, with a very small stigmatic end.—Meissn. in DC. Prod. xiv. 463; *B. barbigera,* Meissn. in Pl. Preiss. ii. 264, and in DC. Prod. xiv. 463.

W. Australia. King George's Sound or to the eastward, *Baxter, Drummond, 3rd coll. n.* 290.

18. **B. petiolaris,** *F. Muell. Fragm.* iv. 109. Stems short, prostrate, thick and tomentose. Leaves erect, on long petioles, above 1 ft. long, truncate, sinuate with short callous teeth, tapering at the base, about 1 in. broad towards the top, flat or undulate, thick, the veins concealed on the under surface by a white tomentum. Spike erect as in *B. repens,* cylindrical, 5 in. long in the specimen before me. Perianth about ¾ in. long, the tube loosely pubescent, the limb nearly 2 lines long, obtuse, bearing longer more deciduous hairs. Style remaining curved, with a very small stigmatic end.

W. Australia. Sand plains, Cape Le Grand to Cape Arid, *Maxwell* (a single specimen in Herb. F. Mueller). Possibly a variety of *B. repens,* as suggested by F. Mueller, Fragm. vii. 58.

19. **B. repens,** *Labill. Voy.* i. 411, *t.* 23. Stems short, prostrate, thick, densely tomentose or woolly. Leaves erect, on long petioles, often a foot long, deeply and irregularly pinnatifid, the lobes varying from lanceolate or falcate entire and 1 to 1½ in. long to oblong-lanceolate or somewhat cuneate entire lobed or pinnatifid and 1 to 4 in. long, or to short broad and almost triangular, all thick and rigid, flat or undulate, the transverse veins prominent underneath and sometimes also on the upper surface. Spikes turned up at the end of the stems, not closely surrounded by leaves, oblong or cylindrical, 3 to 4 in. long. Perianths about 1 in. long, the tube pubescent with short crisped hairs, the limb recurved, nearly 2 lines long, obtuse, villous with much longer crisped hairs sometimes deciduous. Style remaining curved but not hooked, with a very small stigmatic end.—R. Br. in Trans. Linn. Soc. x. 211, Prod. 396; Meissn. in Pl. Preiss. i. 586, and in DC. Prod. xiv. 462; *B. polypodifolia,* Knight, Prot. 113; *B. blechnifolia,* F. Muell. Fragm. iv. 108; *B. pinnatisecta,* F. Muell. Fragm. vii. 58 (name only).

W. Australia. King George's Sound or adjoining districts, *Labillardière, Baxter, Drummond, 3rd coll. n.* 291, *Oldfield;* sandy plains from Stirling Range to Young river, *Maxwell.*

20. **B. prostrata,** *R. Br. Prot. Nov.* 36. Stems prostrate, tomentose. Leaves erect, on long petioles, often above 1 ft. long and 1 to 1½ in. broad, divided about half way to the midrib into broad ovate or triangular mostly obtuse lobes, thick flat and rigid, the transverse veins scarcely prominent even on the under surface. Spikes turned up at the ends of the stems as in *B. repens,* not closely surrounded by leaves, oblong or cylindrical, rarely above 3 in. long. Perianth scarcely above ¾ in. long, the tube loosely hirsute, the limb recurved, narrow, obtuse, at first densely bearded with long crisped and intricate ferruginous

woolly hairs, but soon becoming glabrous. Style remaining curved but not hooked, with a minute stigmatic end. Capsules slightly prominent, tomentose-villous, thick, 1 in. broad.—Meissn in Pl. Preiss. i. 587, and in DC. Prod. xiv. 462; Bot. Reg. t. 1572.

W. Australia. Sand plains, King George's Sound and neighbouring districts, *Baxter, Drummond, 3rd coll. n.* 289, *Preiss, n.* 480, and several others.

21. **B. grandis,** *Willd. Spec. Pl.* i. 535. A tree attaining about 40 ft., the branches tomentose. Leaves often 1 ft. long or more, divided to the midrib into ovate-triangular contiguous segments, the larger ones 1½ to 2 in. long and 1 in. broad at the base, the lower ones gradually smaller, all flat, with several primary transverse veins impressed above, prominent underneath, the under surface pale, reticulate, tomentose in the areolæ. Spike cylindrical, 8 to 12 in. long. Perianths above 1 in. long, the tube loosely villous, the limb glabrous, obtuse, scarcely 1½ lines long. Style long, remaining curved but not hooked, with a small oblong stigmatic end. "Capsules glabrous, 6 to 8 lines broad."—R. Br. in Trans. Linn. Soc. x. 210, Prod. 396; Meissn. in Pl. Preiss. i. 587, and in DC. Prod. xiv. 464.

W. Australia. King George's Sound, *R. Brown, Oldfield, F. Mueller;* Cape Riche, *Preiss, n.* 474, 492; Swan river, *Drummond, 1st coll., Oldfield.* The foliage is nearly that of *B. Baxteri,* the spikes and flowers very different.

22. **B. quercifolia,** *R. Br. in Trans. Linn. Soc.* x. 210, *Prod.* 396. An erect shrub of 5 or 6 ft., the branches and foliage glabrous. Leaves sessile or nearly so, oblong-cuneate, truncate, deeply and irregularly prickly-toothed or pinnatifid, tapering to the base, 2 to 4 in. long, flat or undulate, the transverse veins and reticulations more or less conspicuous underneath. Spikes oblong-cylindrical, dense but rather narrow, 3 to 4 in. long. Bracts with very short glabrous tips. Perianth-tube about ½ in. long, ferruginous-villous, the limb narrow, reflexed, 2½ lines long with an awn-like point at least as long, pubescent with shorter hairs than the tube. Style about ¾ in. long, remaining curved, with a small very narrow stigmatic end. Capsules rounded, thick, glabrous or slightly tomentose, ¾ in. broad.—Meissn. in Pl. Preiss. i. 585, and in DC. Prod. xiv. 462; Bot. Reg. t. 1430.

W. Australia. King George's Sound, *R. Brown, Baxter, Harvey, Preiss, n.* 489, *Oldfield,* and others. The foliage is nearly that of *B. Caleyi.*

Var. *integrifolia,* F. Muell. Fragm. vii. 57. Leaves cuneate, truncate, with a small central pungent point, entire or minutely 2- or 3-toothed. Capsules very thick, 1 in. broad.—East Mount Barren and Tulbinup, *Maxwell.*

23. **B. Baueri,** *R. Br. Prot. Nov.* 35. Probably arborescent, the branches tomentose or nearly glabrous. Leaves oblong-cuneate or almost lanceolate, truncate, sinuate-toothed, very shortly petiolate, mostly 3 to 4 in., sometimes 5 in. long, flat, the transverse veins prominent underneath and the reticulations conspicuous, scarcely tomentose. Spikes very thick and dense, globular or oblong, 6 to 8 in. long. Bracts densely villous at the end. Perianth-tube pubescent, the limb densely

villous, narrow, abruptly reflexed, about 3 lines long, ending in a plumose awn-like point of ½ in. or more. Style remaining curved, with
a narrow acute stigmatic end. Capsules concealed among the dense
perianth-remains, very thick, glabrous, smooth, 1¼ to 1½ in. broad.—
Meissn. in DC. Prod. xiv. 460; F. Muell. Fragm. iv. 107.

W. Australia. King George's Sound or the neighbouring districts, *Baxter,
Drummond, 4th coll. n.* 303. The long fine points to the perianth-laminæ forming
awn like ends to the limb before it opens, are quite peculiar to this and the preceding
species.

SECT. 3. EUBANKSIA.—Leaves linear-lanceolate oblong or cuneate,
with recurved or revolute entire or dentate margins, white underneath.
Style at first curved, straight and very spreading or reflexed after the
perianth-limb has opened, the stigmatic end very small, not furrowed.

The three species here included, divided into many more by R. Brown, Meissner and
others, are so closely allied and so frequently connected by intermediates, that they
might almost be considered as varieties of a single one.

24. **B. marginata,** *Cav. Anal. Hist. Nat.* i. 227, *t.* 13. *Ic.* vi. 29, *t.*
544. Usually a bushy shrub of 10 to 15 ft., growing out sometimes
into a tree of considerable size or sometimes low and straggling or
depressed, the branches tomentose or villous. Leaves of the flowering
branches very shortly petiolate, oblong-lanceolate or broadly linear,
obtuse or retuse, usually entire, with recurved margins, 1 to 2 in. long,
in some flowerless branches or even on some flowering specimens some
or all rather larger and more or less serrate with short rigid or prickly
teeth, all very white underneath, minutely reticulate, without any or
with very few of the transverse veins of *B. integrifolia.* Spikes oblong-
cylindrical, 2 to 3 or rarely near 4 in. long, or in the dwarf varieties
sometimes nearly globular and small. Bracts tomentose at the end.
Perianths silky, 7 to 8 lines long. Style straightening after the
perianth-laminæ have separated, and usually very spreading or
reflexed, with a small slender stigmatic end. Fruiting cone oblong-
cylindrical; capsules prominent above the closely packed bracts, flat,
not thick, rounded, ½ in. broad, at first pubescent but the hairs wearing
off.—R. Br. in Trans. Linn. Soc. x. 204, Prod. 392, Meissn. in DC.
Prod. xiv. 455; Bot. Mag. t. 1947; *B. microstachya,* Cav. Anal. Hist.
Nat. i. 224, Ic. vi. 28, t. 541 (specimens with serrate leaves);
B. marginata, Lodd. Bot. Cab. t. 61, and *B. oblongifolia,* Lodd.
Bot. Cab. t. 241, not of others (both with serrate leaves); *B.
australis,* R. Br. in Trans. Linn. Soc. x. 206; Prod. 393; Meissn. in
DC. Prod. xiv. 456; Hook. f. Fl. Tasm. i. 329; Bot. Reg. t. 787; *B.
depressa, B. patula* and *B. insularis,* R. Br. in Trans. Linn. Soc. x. 205,
206, Prod. 393; Meissn. l.c. 456; *B. Gunnii,* Meissn. l.c.

N. S. Wales. Port Jackson, *R. Brown, Sieber, n.* 8, and others; Berrima and
Mudgee, *Woolls.*
Victoria. Port Phillip, *R. Brown;* Wanganatta and Dandenong, *F. Mueller;*
Melbourne, *Adamson;* Glenelg river, *Robertson.*

Tasmania. Port Dalrymple, Derwent river, and King's Island, *R. Brown.* Abundant throughout the island, ascending to 3000 ft., *J. D. Hooker.*

S. Australia. Port Lincoln, *R. Brown; Boston Point, Wilhelmi;* near Adelaide, *Whittaker, Blandowski;* Mount Barker and Cook's Creek, *Neumann;* Kangaroo Island, *Waterhouse.*

It appears from R. Brown's labels that he had originally referred all his southern specimens to *B. marginata,* and the characters upon which he afterwards thought he could distinguish four southern species, fail so completely when applied to the large number of specimens we now possess that I have felt obliged to return to his original views. As a whole the species differs from *B. integrifolia* generally in the smaller leaves and flowers and in the leaves reticulate only without transverse veins. In some specimens however some of the leaves show a few of these veins, especially when toothed there is often one entering into each tooth.

B. præmorsa, Dum. Cours., *B. ferrea,* Vent., and *B. hypoleuca.* Hoffmsg., are names of garden plants which have been referred by Meissner and others to this species. *B. marcescens,* Bonpl. Jard. Malm. 116, t. 48, appears to me to represent the toothed-leaved state of *B. marginata,* and not the true *B. marcescens,* Br.

25. B. integrifolia, *Linn. F. Suppl.* 127. A tree attaining sometimes a considerable size, the young branches closely tomentose. Leaves scattered, sometimes irregularly verticillate, oblong cuneate or lanceolate, quite entire or irregularly toothed, tapering into a short petiole, 3 to 4 in. long in some specimens, twice that length in others, especially the northern ones, ½ to near 1 in. broad, white underneath, with numerous transverse veins and reticulations not very prominent; the young shoots are also sometimes tomentose or villous with richly coloured fulvous almost woolly hairs persisting on the under side till the leaves are nearly full grown. Spikes oblong or cylindrical, 3 to 6 in. long. Bracts tomentose at the end. Perianth usually about 1 in. long, silky. Style straightening after the perianth-laminæ have separated and usually very spreading or reflexed as in *B. marginata.* Fruiting cone oblong, cylindrical, the capsules prominent and not thick, as in that species.—R. Br. in Trans. Linn. Soc. x. 206, Prod. 393; Meissn. in DC. Prod. xiv. 456; Cav. Ic. vi. t. 546; Bot. Mag. t. 2770; *B. spicata,* Gærtn. Fr. i. 221, t. 48; *B. oleifolia,* Cav. Anal. Hist. Nat. i. 228, t. 14, Ic. vi. 30, t. 545; *B. macrophylla,* Link. Enum. Hort. Berol. i. 116; *B. compar,* R. Br. in Trans. Linn. Soc. x. 207, Prod. 393; Meissn. in DC. Prod. xiv. 457.

Queensland. Keppel Bay, *R. Brown, O'Shanesy;* Brisbane river, Moreton Bay, *A. Cunningham, F. Mueller,* and others; Condamine river, *Leichhardt;* Mount Archer, *Bowman;* Rockhampton and Rockingham Bay, *Dallachy.*—The greater number of these northern specimens have remarkably long leaves, sometimes 8 to 10 in long and ¾ in. wide, and constitute the *B. compar,* Br. They have also usually rather larger flowers, but neither character is at all constant, and R. Brown had himself at first referred his specimens to *B. integrifolia.*

N. S. Wales. Port Jackson, *R. Brown, Sieber, n.* 4, and many others; northward to Hastings river, *Beckler;* Richmond river, *Fawcett;* New England, *C. Stuart;* Mount Lindsay, *W. Hill;* southward to Twofold Bay, *F. Mueller.*

Victoria. Sealer's Cove, Port Phillip, Brighton, *F. Mueller.*

Var. *paludosa.* Flowers scarcely larger than in *B. marginata,* the perianth 7 to 8 lines long, but the leaves of one of the common short-leaved forms of *B. integrifolia.*— *B. paludosa,* R. Br. in Trans. Linn. Soc. x. 207; Prod. 394; Meissn. in DC. Prod. xiv. 457; Bot. Reg. t. 697; Lodd. Bot. Cab. t. 392.— Port Jackson, *R. Brown, Sieber, n.* 5. Distributed also from the Botanical Garden, St. Petersburgh as *B. integrifolia.*

B. oblongifolia, Cav. Anal. Hist. Nat. i. 225, Ic. vi. t. 542; R. Br. in Trans. Linn.
Soc. x. 208, Prod. 394; Meissn. in DC. Prod. xiv. 461, appears to be referrible to *B.
integrifolia,* the specimens of *Sieber,* n. 5, and from Mount Lindsay, *Fraser,* have rather
more coriaceous leaves than usual with the transverse veins more prominent, approaching in some degree *B. dentata,* but not otherwise distinguishable from the typical *B.
integrifolia. B. glauca,* and *B. salicifolia,* Cav. Anal. Hist. Nat. i. 230, 231, Ic. vi. 31,
B. asplenifolia, Salisb. Prod. 51, *B. cuneifolia* and *B. reticulata,* Hoffmsg. in Roem.
and Schult. Syst. iii. Mant. 379; Meissn. in DC. Prod. xiv. 466, *Hakea pubescens,*
Hort. Cels. in Steud. Nom. Bot. ed. 2, are garden plants which appear to have been
correctly referred to *B. integrifolia,* although several of them have been described only
as to their foliage.

26. **B. dentata,** *Linn. F. Suppl.* 127. A small tree of 15 to 20 ft.
closely allied to *B. oblongifolia.* Leaves shortly petiolate, cuneate-
oblong, 4 to 8 in. long, 1 to 2 in. broad, irregularly toothed, the
margins slightly recurved, white underneath with the primary transverse veins more prominent than in *B. integrifolia* and not so white.
Spikes oblong or cylindrical, usually larger than in *B. integrifolia* but
the flowers in all other respects as well as the fruits entirely those of
B. integrifolia. Styles about 1½ in. long, becoming straight, with a
small narrow stigmatic end.—R. Br. in Trans. Linn. Soc. x. 210, Prod.
396; Meissn. in DC. Prod. xiv. 462; F. Muell. Fragm. vii. 57.

N. Australia. Islands of the Gulf of Carpentaria, *R. Brown;* Port Hurd, opposite Melville Island, *A. Cunningham;* Point Pearce, *F. Mueller;* Glenelg river, N.W.
coast, *Martin.*

Queensland. Endeavour river, *Banks and Solander, A. Cunningham.*

SECT. 4. ORTHOSTYLIS.—Leaves flat or undulate, regularly or
rarely irregularly serrate pinnatifid or pinnate, with short lobes or segments. Perianth straight or the limb rarely reflexed. Style after the
perianth limb has opened curved upwards at the base only, then straight
rigid and erect, the stigmatic end prominently angled and furrowed or
striate.

The foliage is that of *Cyrtostylis,* but the regular rigid erect often almost imbricate
styles give the cones after the flowers have opened a different aspect, and the stigmatic
ends of the styles are well marked. A few species have the styles elegantly curved
before they are set free from the perianth-limb, and *B. latifolia* in its flowers and styles
is almost intermediate between *Eubanksia* and *Orthostylis.*

27. **B. latifolia,** *R. Br. in Trans. Linn. Soc.* x. 208, *Prod.* 394. A
low but stout shrub, the branches densely tomentose. Leaves shortly
petiolate, obovate-oblong, often truncate, irregularly serrate with short
usually prickly teeth, contracted at the base, 4 to 8 in. long, 1½ to 3 in.
broad, flat, minutely tomentose but not white underneath, with prominent transverse veins and reticulations. Spikes oblong-cylindrical,
3 to 5 in. long. Perianth slender, about 1 in. long, the tube shortly
silky-pubescent, the limb glabrous, narrow, acute, scarcely 2 lines long.
Style becoming straight and spreading as in *Eubanksia,* with a very
small stigmatic end. Fruiting cones large and thick; capsules villous,
not thick, protruding, about 6 or 7 lines diameter.—Meissn. in DC.
Prod. xiv. 460; Bot. Mag. t. 2406; *B. robur,* Cav. Anal. Hist. Nat. i.

226, Ic. vi. 29, t. 543; *B. uncigera* and *B. dilleniæfolia*, Knight, Prot. 112, 113; *B. fagifolia*, Hoffmsg.; Roem. and Schult. Syst. iii. Mant. 379 (Meissn.).

Queensland. Moreton Bay, *W. Hill, F. Mueller.*
N. S. Wales. Marshes about Port Jackson, *R. Brown, A. Cunningham, Leichhardt;* Hastings river, *Beckler.*

28. **B. serrata,** *Linn. f. Suppl.* 126. A tree, the young shoots tomentose or villous and sometimes densely so with richly coloured ferruginous very deciduous hairs. Leaves oblong-lanceolate, acute or truncate, regularly and deeply serrate, tapering into a petiole, 3 to 6 in. long, ½ to 1 in. wide, coriaceous, flat, hoary or rarely white underneath, with parallel transverse veins. Spikes oblong-cylindrical or rarely globular, 3 to 6 in. long, very thick. Perianth shortly silky, the tube above 1 in. long, the laminæ narrow, acuminate, nearly 3 lines long, the silky hairs longer than those of the tube. Style at length straight, with a cylindrical somewhat furrowed stigmatic end, about ½ line long and thickened at the base. Capsules very prominent, tomentose, thick and hard, obliquely rounded or ovate, above 1 in. broad.— R. Br. in Trans. Linn. Soc. x. 209, Prod. 395; Sm. in White, Voy. 223, t. 18 to 20; Meissn. in. DC. Prod. xiv. 461; F. Muell. Fragm. vii. 56; Andr. Bot. Rep. t. 82; *B. conchifera*, Gærtn. Fr. i. 221, t. 48; *B. mitis*, Knight, Prot. 112; *B. dentata*, Wendl. Hort. Herrenh. t. 8; *B. media*, Hook. f. Fl. Tasm. i. 329, not of R. Br.

N. S. Wales. Botany Bay, *Banks and Solander*; Port Jackson, *A. Cunningham,* also according to Meissner, *Sieber, n.* 2, partly.
Victoria. Port Albert, *F. Mueller* (I have not seen the specimens).
Tasmania. N. coast on two hills called the Sisters, between Rocky and Table Capes, *Backhouse, Gunn.*

The plant figured by Cavanilles as *B. serrata* appears to be rather *B. æmula;* Baillon's figure, Hist. Pl. ii. 394, f. 230, is most probably taken from *B. attenuata.*

29. **B. æmula,** *R. Br. in Trans. Linn. Soc.* x. 210. *Prod.* 395. A shrub very closely allied to *B. serrata* and difficult to distinguish from it except by the stigmatic end of the style which is very much shorter and ovoid. The flowers are also said to be of a yellowish green without the bluish grey tinge of *B. serrata.* The spikes are usually not so thick, the foliage precisely the same. Capsules at least as large as in *B. serrata,* the tomentum easily wearing off.—Meissn. in DC. Prod. xiv. 461; Bot. Mag. t. 2671; Bot. Reg. t. 688; *B. serrata*, Cav. Ic. vi. 27, t. 540, not of Linn. f.; *B. serratifolia*, Salisb. Prod. 51 or *B. serræfolia*, Knight, Prot. 112 (*R. Br.*); *B. elatior*, R. Br. in Trans. Linn. Soc. x. 209, Prod. 395; Meissn. in DC. Prod. xiv. 458; *B. undulata*, Lindl. Bot. Reg. t. 1316.

Queensland. Sandy Cape, *R. Brown;* Stradbrooke Island, Moreton Bay, *A. Cunningham.* I have not seen Brown's own specimens of *B. elatior*, which have been mislaid, but there seems no doubt that Cunningham was right in his identification.
N. S. Wales. Port Jackson, *R. Brown, Sieber, n.* 2 (our specimens at least), and others; Hastings river, *Beckler;* Twofold Bay, *L. Morton?* (leaves only).
Victoria. Gipps' Land, *F. Mueller.*

30. **B. ornata,** *F. Muell. Meissn. in Linnæa* xxvi. 352, *and in DC. Prod.* xiv. 460. A shrub of 5 or 6 ft., the branches densely hirsute. Leaves oblong-cuneate, mostly truncate, regularly serrate, tapering into a short petiole, 2 to 4 in. long, ½ to ¾ in. broad, flat, the transverse veins prominent underneath. Spikes globular or oblong-ovoid, 2 to 4 in. long. Bracts obtuse, villous. Perianth slender, villous with spreading hairs, 1 to 1¼ in. long, the limb narrow, 3 lines long. Style curved upwards from the base, then becoming straight, stigmatic end narrow, furrowed. Fruiting cone ovoid; capsules prominent, very thick, tomentose-villous, fully ¾ in. broad.—F. Muell. Fragm. vii. 56.

Victoria. N.W. districts, *L. Morton;* Wimmera, *Dallachy.*

S. Australia. Encounter Bay, *Whittaker;* Onkaparinga river and towards Guichin Bay, *F. Mueller.*

31. **B. coccinea,** *R. Br. in Trans. Linn. Soc.* x. 207, *Prod.* 394. An erect shrub attaining 12 to 15 ft., the branches densely tomentose, with a few long spreading hairs often intermixed. Leaves sessile or very shortly petiolate, from broadly oblong or obovate to almost orbicular or broader than long, truncate or retuse, often cordate at the base, bordered by small irregular prickly teeth, 1½ to 2½ in. long, flat, rigid, prominently penniveined and reticulate underneath. Spikes globular, about 2 in. diameter, the flowers regularly imbricate in vertical (not spiral) rows, the tubes of those of each pair opening inwards for the emission of the style of which the end is retained in the reflexed limb, the spike thus long remaining elegantly striped by double rows of arched richly coloured red styles alternating with double rows of villous perianths. Each perianth about 1 in. long with a limb of about 2 lines. When at length liberated the style straightens; bearing a stigmatic end of about ¾ line, furrowed, with a prominent rim round its base. Fruiting cone after the fall of the perianths ovoid, 1 to 1½ in. diameter, tomentose-villous; capsules very small thin and scarcely protruding, 4 or rarely 5 lines broad.—Meissn. in Pl. Preiss. i. 585, and in DC. Prod. xiv. 459; Bauer, Illustr. t. 3.

W. Australia. King George's Sound and adjoining districts, *R. Brown, Drummond, 3rd coll. n.* 284, *Preiss, n.* 481, and many others.

32. **B. sceptrum,** *Meissn. in Hook. Kew Journ.* vii. 120, *and in DC. Prod.* xiv. 459. A tall shrub or small tree of 10 to 15 ft., with thick closely tomentose branches. Leaves petiolate, oblong truncate, shortly sinuate-toothed, 1½ to 2½ in. long, flat, rigid, transversely veined and reticulate underneath. Spike oblong-cylindrical, thick and dense, 6 to 8 in. long, the curved styles protruding before the perianth-limb opens, alternating in single rows with the perianths Perianth silky-villous, the tube ½ in., the obtuse limb 4 or 5 lines long. Style after it is set free from the perianth straight or flexuose, much longer than the perianth, with a thick furrowed stigmatic end of 1½ to 2 lines. Capsules prominent, very thick, variegated and hirsute, often 1 in. broad.

W. Australia. Hutt river, *Drummond, 6th coll. n.* 206; Murchison river, *Oldfield.*

33. **B. Menziesii,** *R. Br. Prot. Nov.* 36. A tree of 30 to 40 ft., the branches thick and tomentose. Leaves shortly petiolate, 6 in. to 1 ft. long, ¾ to 1 in. wide, truncate, bordered by short broad teeth, more or less ferruginous-tomentose underneath with numerous parallel transverse veins. Spikes thick, oblong, 4 to 5 in. long. Bracts with broad obtuse tomentose pale coloured ends surrounded by the deeply coloured woolly hairs of the sides, marking the spike both in bud and after the perianths have fallen with a lozenge-shaped pattern in numerous spiral rows. Perianth-tube about 1 in. long, silky-pubescent, the limb erect, villous with longer hairs, about 3 lines long. Style incurved at the base, then erect and straight, with a furrowed stigmatic end about 1 to 1½ lines long. Capsules very prominent, oblique, thick, tomentose.—Meissn. in Pl. Preiss. i. 584, and in DC. Prod. xiv. 459.

W. Australia. Swan river, *Collie, Drummond,* 1st coll., *Preiss, n.* 477; Murchison river, *Oldfield.*

34. **B. lævigata,** *Meissn. in DC. Prod.* xiv. 458. A shrub? with tomentose branches. Leaves linear-cuneate, truncate, serrate, contracted into a short petiole, 2 to 4 in. long, thick, flat, with the transverse veins very fine and slightly impressed underneath. Spikes globular, resembling those of *B. ornata,* 2 to 3 in. diameter. Perianths incurved at the base, erect, hirsute with spreading hairs, scarcely 1 in. long, the narrow limb about 1½ lines long. Style slender, incurved, with a small narrow slightly furrowed stigmatic end. Fruiting cone globular, about 3 in. diameter; capsules slightly prominent, rounded, thick, villous, about ½ in. broad.

W. Australia. Between Swan river and Cape Riche, *Drummond,* 5th coll. n. 414, or in some herbaria, 415; East Mount Barren, *Maxwell.*

35. **B. Hookeriana,** *Meissn. in Hook. Kew Journ.* vii. 119, *and in DC. Prod.* xiv. 458. A shrub of 5 or 6 ft., with densely tomentose branches. Leaves linear-cuneate, 4 to 8 in. long, 4 to 5 lines broad near the end, tapering into a short petiole, divided nearly half-way to the midrib into numerous broadly triangular teeth or lobes, minutely tomentose underneath, the veins inconspicuous. Spikes oblong, very thick, 4 to 5 in. long. Perianth curved upwards, nearly 1½ in. long, the limb about 3 lines long, densely hirsute with long spreading hairs. Style rigid, incurved at the base, then erect and straight, with a slender furrowed stigmatic end.

W. Australia. Between Tea-tree swamp and Irwin river, *Drummond,* 6th coll. n. 202.

36. **B. prionotes,** *Lindl. Swan Riv. App.* 34. A tree of about 30 ft., with thick tomentose branches. Leaves 8 in. to above 1 ft. long, ½ to 1 in. broad, truncate, pinnatifid with numerous rather regular lobes not reaching half-way to the midrib, broader than long, rounded, flat, with short rigid but not pungent points, the transverse veins numerous and fine, visible underneath and converging at the apex of each lobe. Spikes thick, oblong, 3 to 5 in. long. Perianth incurved

and erect, the tube nearly 1 in. long, villous, the limb 3 lines long, very densely villous with spreading hairs. Style rigid, incurved at the base, then erect, with a narrow furrowed stigmatic end of 1 to 1¼ lines. Fruiting cones after the fall of the perianth-remains showing the prominent conical tomentose ends of the bracts; capsules prominent, rounded, rather thick, tomentose or shortly villous, about ¾ in. broad, the lateral base of the style more or less prominent.—Meissn. in Pl. Preiss. i. 584, and in DC. Prod. xiv. 459.

W. Australia. Between Swan river and King George's Sound, *Drummond,* 1*st coll.,* 3*rd coll. n.* 288, *Preiss, n.* 476, *Harvey ;* Upper Gardner river, *Hassell ;* Murchison river, *Oldfield.*

37. **B. Victoriæ,** *Meissn. in Hook. Kew Journ.* vii. 119, *and in D C. Prod.* xiv. 464. A shrub of 12 to 15 ft., nearly allied to *B. prionotes* but the branches more hirsute, the leaves divided more than halfway to the midrib into broad triangular acute or acuminate lobes, the larger ones fully ½ in. long and broad, and the loose ferruginous wool more persistent although ultimately deciduous. Spike of *B. prionotes,* but the outer bracts at the base above ½ inch long and plumose with long hairs. Perianth rather longer than in *B. prionotes,* much more villous, especially the limb. Style the same. Capsules more prominent, 1 in. broad, densely villous with purple hairs.—Bot. Mag. t. 4906; *B. speciosa,* Lindl. Bot. Reg. t. 1728, not of R. Br.

W. Australia. Hutt river, *Drummond, 6th coll. n.* 203 ; Baker's Well, *Oldfield.*

38. **B. speciosa,** *R. Br. in Trans. Linn. Soc.* x. 210, *Prod.* 396. A tall shrub, with thick tomentose branches. Leaves shortly petiolate, 8 in. to above 1 ft. long, divided to the midrib into numerous contiguous rounded or triangular shortly acuminate segments, the larger ones ¾ in. broad at the base and nearly as long, diminishing towards each end of the leaf, flat, rigid, retaining more or less of a white tomentum underneath, with numerous transverse converging veins. Spikes very thick, oblong, 4 to 5 in. long. Perianths incurved upwards, hirsute, the tube about 1 in., the obtuse hirsute limb about 2½ lines long. Style incurved at the base, erect, rigid, hairy; stigmatic end stipitate and furrowed.—Meissn. in DC. Prod. xiv. 464 ; Bot. Mag. t. 3052 (the leaves not quite correct) ; *B. grandidentata,* Dum. Cours (Meissn).

W. Australia. Lucky Bay, *R. Brown, Baxter.*

39. **B. Baxteri,** *R. Br. Prot. Nov.* 36. A tall shrub, the branches glabrous or hirsute under the spikes with long fine spreading hairs. Leaves mostly 3 to 4 in. long, divided to the middle into ovate-triangular acute contiguous segments, the larger ones 1 in. long and ¾ in. broad at the base but mostly smaller, flat, rigid, pale or whitish underneath with several fine and faint transverse converging veins. Spikes globular, 2 to 3 in. diameter, the outer linear bracts plumose with long fine hairs. Perianths hirsute with long fine hairs, 1½ in. long, the limb narrow, acute or acuminate, about 4 lines long. Style incurved at the base, erect, thick and rigid, densely hairy, the stig-

560 CIV. PROTEACEÆ. [*Banksia.*

matic end narrow, acute, furrowed. Capsules prominent, very thick
and woody, 1½ in. broad.—Meissn. in Pl. Preiss. i. 587, and in DC.
Prod. xiv. 464.

W. Australia. King George's Sound or adjoining districts, *Baxter, Drummond,*
4th coll. n. 306, *Preiss, n.* 485, *Harvey;* flat sandy plains from Stirling Range to Salt
river, *Maxwell.*

40. **B. marcescens,** *R. Br. in Trans. Linn. Soc.* x. 208, *Prod.* 395. A
shrub of 5 or 6 ft. the branches tomentose. Leaves petiolate, oblong,
truncate, serrate, almost obtuse at the base, 1 to 1½ in. long and about
½ in. broad, flat, minutely tomentose underneath with faint transverse
veins and reticulations. Spikes oblong or cylindrical, dense, 3 to 10 in.
long, like those of *B. media.* Bracts tomentose at the end. Perianth
purple, glabrous, scarcely 1 in. long, the limb narrow, obtuse, about 2
lines long. Style erect, about as long as the perianth, the stigmatic
end short and sulcate. Capsules usually buried in the persistent remains of the flowers, rather thick, rounded, about ¾ in. broad, quite
glabrous and shining but chagrined with raised dots or tubercles.—
Meissn. in Pl. Preiss. i. 586, and in DC. Prod. xiv. 461; Sw. Fl.
Austral. t. 14; Bot. Mag. t. 2803; *B. præmorsa,* Andr. Bot. Rep. t. 258;
B. asplenifolia, Knight, Prot: 113, not of Salisb. (*R. Br.*).

W. Australia. King George's Sound, *Menzies, Baxter, Drummond, 3rd coll. n.*
285, *Preiss, n.* 484.

Bonpland's figure and description of *B. marcescens,* Jard. Malm. 116, t. 48, appear to
me to represent rather one of the garden varieties of *B. marginata.* I have not seen
Preiss's specimens above quoted from Meissner.

41. **B. Lemanniana,** *Meissn. in DC. Prod.* xiv. 462. Branches tomentose or shortly villous. Leaves petiolate, obovate-oblong, less truncate than in most species, almost regularly toothed, cuneate at the base,
1½ to 3 in. long, flat, loosely tomentose underneath when young, the
transverse veins and reticulations visible but not prominent. Spikes
globular or shortly oblong, very thick, 3 to 4 in. long. Perianths glabrous, above 1 in. long, the narrow obtuse limb about 4 lines. Style
slightly curved, erect, the stigmatic end long narrow and furrowed.

W. Australia, *Drummond, 4th coll. n.* 302.

42. **B. Caleyi,** *R. Br. Prot. Nov.* 35. A low shrub, the branches
tomentose. Leaves oblong-lanceolate or narrow-cuneate, usually truncate, sinuate and broadly prickly-toothed or almost pinnatifid, tapering
into a short petiole, 3 to 6 in. long, flat or undulate, green on both
sides, finely and not prominently transversely veined and reticulate underneath. Spikes ovoid-oblong or globular, 2 to 3 in. long. Bracts
obtuse, densely villous. Perianths nearly 1 in. long, quite glabrous or
with a minute and scanty pubescence on the tube, the limb very angular and obtuse, about 4 lines long. Style incurved, erect, the stigmatic end long narrow and furrowed, with a projecting rim at the base.
Meissn. in DC. Prod. xiv. 462.

W. Australia, *Baxter, Drummond, 4th coll. n.* 301.

43. **B. Lindleyana,** *Meissn. in Hook. Kew Journ.* vii. 120, *and in DC. Prod.* xiv. 455. A shrub of 3 or 4 ft., differing slightly from *B. Caleyi* in the narrower serrate leaves and the flowers usually larger. Young shoots tomentose and villous, leafy branches hoary or almost glabrous. Leaves linear-lanceolate, serrate, tapering into a short petiole, 2 to 4 in. long, flat, slightly tomentose, reticulate and pitted underneath. Spikes very thick, ovoid-globular, about 4 in. long. Bracts woolly-tomentose with short obtuse points prominent above the bracteoles. Perianth glabrous, the tube nearly 1 in. long, the obtuse angular limb 3 to 4 lines. Style incurved, erect, the stigmatic end long, narrow and furrowed.

W. Australia. Murchison river, *Oldfield, Drummond, 6th coll. n.* 204.

44. **B. elegans,** *Meissn. in Hook. Kew Journ.* vii. 119, *and in DC. Prod.* xiv. 465. A small tree, the specimens at first sight closely resembling those of *B. Candolleana,* the leaves of the same size, with numerous broad pungent-pointed lobes, but divided only a little more than half way to the midrib, and the under surface pale or whitish with a minute tomentum, which almost conceals the veins, the smaller reticulations quite inconspicuous. Spikes globular, larger and more dense than in *B. Candolleana.* Perianth straight, fully 1 in. long, the tube minutely pubescent, the limb narrow, glabrous, fully 2 lines long. Style curved, erect, the stigmatic end fusiform and furrowed.

W. Australia. Valley of the Lakes, Hill river, *Drummond, 6th coll. n.* 200.

45. **B. Candolleana,** *Meissn. in Hook. Kew Journ.* vii. 118, *and in DC. Prod.* xiv. 465. A shrub with a creeping underground trunk and erect leafy stems of 1 to 2 ft., the flowering ones often short with few leaves, all minutely tomentose or glabrous. Leaves shortly petiolate, often 1 ft. long or more, divided to the midrib into numerous broad ovate-triangular contiguous segments, the larger ones scarcely above 4 lines long and broad, all pungent-pointed, flat, rigid, strongly veined and reticulate underneath. Spikes ovoid-globular, not surrounded by leaves, about 1½ in. long without the perianths, which are not so dense as in most species, straight, about 1 in. long, the tube slender, minutely pubescent or glabrous, the limb oblong, glabrous, striate, about 2 lines long. Style curved, erect; stigmatic end fusiform, sulcate. Capsules very prominent, hard, thick, tomentose, the projecting portion 2 in. long and 1¼ in. broad, with a small lateral conical beak or persistent base of the style.—F. Muell. Fragm. vii. 58.

W. Australia. Dundagaran and Hill river, *Drummond, 6th coll. n.* 201.

Sect. 5. Isostylis.—Spikes reduced to depressed globular heads. Perianth-limb opening as soon as the limb, the style straight, not longer than the perianth, with a small stigmatic end.

46. **B. ilicifolia,** *R. Br. in Trans. Linn. Soc.* x. 211, *Prod.* 396, *Prot. Nov.* 37. A tree attaining from 20 to 40 ft., or sometimes remaining shrubby and 8 to 10 ft. high, the branches tomentose and often hirsute

with a few long spreading hairs. Leaves shortly petiolate, oval-oblong obovate or cuneate, truncate, undulate and irregularly prickly-toothed or lobed, 1 to 3 in. long, green on both sides, veined and reticulate underneath, but the veins rarely prominent. Spikes terminal, depressed-globular, sessile amongst the floral leaves, the rhachis with the closely packed villous bracts about ½ in. diameter   Perianths erect, straight, the tube shortly silky-pubescent, 1 to 1¼ in. long, the limb obtuse, glabrous or nearly so, not 2 lines long. Style not longer than the perianth, erect, straight, glabrous, with a small scarcely distinct stigmatic end. Fruiting cone very small. Capsules usually 1 or 2 only, very prominent, obliquely ovoid, thick, tomentose, the projecting portion ½ to ¾ in. long, with a scarcely prominent lateral beak or scar indicating the base of the style.—Meissn. in Pl. Preiss. i. 589 and in DC. Prod. xiv. 466 ; *B. aquifolium,* Lindl. Swan Riv. App. 34.

W. Australia. King George's Sound and the neighbouring districts, *R. Brown, Baxter, A. Cunningham, Oldfield, F. Mueller ;* Swan river, *Drummond,* 1st *coll., Preiss,* n. 482.

The specimens at first sight closely resemble those of some forms of *Dryandra floribunda,* to which I find them referred in several herbaria, as also by F. Mueller, Fragm. vi. 92, and vii. 50.

Var. *integrifolia.* Leaves obovate, entire or scarcely toothed.—Swan river, *Preiss,* n. 482 (some specimens).

29. DRYANDRA, Br.

(Hemiclidia, *Br.,* Josephia, *Salisb.*)

Flowers hermaphrodite. Perianth regular or nearly so, usually straight, the tube slender, the limb oblong or linear, the laminæ separating as the tube opens, or rarely remaining long coherent as in *Banksia,* and the limb thus sometimes reflexed before opening, the tube separating into the four claws to below the middle, the base of the tube remaining entire. Anthers narrow, sessile in the concave laminæ, the connective thick, usually very shortly produced beyond the cells. Hypogynous scales 4, very narrow, thin and membranous (rarely deficient ?), usually accompanied by a few long hairs. Ovary very small and sessile ; style straight and scarcely exceeding the perianth, or longer, curved and protruding from a slit in the perianth-tube until the end is set free by the separation of the laminæ and then straightened ; the stigmatic end, on a level with the anthers, of a different texture, smooth or striate and furrowed, continuous with the style or thickened at the base into a slightly prominent rim, the real stigma small and terminal ; ovules 2 (usually or always ?), collaterally attached at or near the top. Fruit a compressed capsule, opening at the dilated end (or outer margin) in two coriaceous or rarely almost woody broad valves. Seeds 2, or 1 by abortion, compressed, with a terminal membranous wing broad and rounded like the valves, the seeds either separated by a plate simple between the nuclei, double between the valves, as in *Banksia,* but not so thick, or the outer integuments of the 2 seeds remain distinct from each other but separated from the seeds forming two membranous plates between the seeds, or remaining attached to the nucleus or to the whole seed leaving the seeds separate, each with a double or single wing.—Shrubs, often low or flowering near the base. Leaves alternate, very rarely entire, usually either sinuate and prickly-toothed, or pinnatifid or pinnate with numerous small regular lobes or segments, usually smooth and veinless on the upper surface, white-tomentose or marked with parallel transverse veins underneath. Flowers sessile, in pairs, in dense terminal or lateral heads in an involucre of numerous imbricate scale-like bracts and usually surrounded by a ring of floral leaves similar to the stem leaves ; receptacle flat or convex, densely villous or woolly, with narrow-linear villous or woolly bracts or paleæ subtending each pair of flowers, sometimes very small or deficient at least in the centre of the head. Perianth usually yellow, the short entire base glabrous or villous towards the divided part, the remainder of the tube or claws usually pubescent or villous, the limb occasionally, the whole perianth very rarely, glabrous. Ovary almost always hairy. Capsules usually villous, but the hairs very readily rubbing off, and in some species apparently glabrous from the first.

The genus is endemic in West Australia. It is readily distinguished from *Banksia* by the involucre, by the flat or nearly flat receptacle, and by the fruit; but the structure of the flowers is so uniform that it is very difficult to establish any definite sections. The differences in the foliage correspond but very little with those in inflorescence, and both are variable in some species. Meissner has founded his groups on the former, I have preferred the inflorescence, which appears to me more characteristic. With regard to the sections founded upon the differences in the so-called dissepiment of the capsule (the plate intervening between the seeds), I have adopted them upon the supposition that these differences are constant, but the seeds remain to be examined in a considerable number of species. If it should prove that these species, here arranged according to their apparent affinity with those whose seeds are known, have been misplaced, all practical utility in these sections will be lost, and some other principle of division must be sought for, although no good one has as yet suggested itself.

SECT. 1. **Eudryandra.**—*Outer integuments of the inner faces of the two seeds united in a bifid plate separating from them. Involucres various, the bracts narrow or very rarely rather broad.*

SERIES 1. **Armatæ.**—*Flower-heads usually large, mostly terminal, enclosed in floral leaves longer than the flowers. Involucres broad. Perianths above 1 in. long. Leaves with prickly teeth or lobes.*

Involucre (2 in.) as long as the flowers. Leaves obovate-oblong,
    deeply prickly-toothed, not white underneath . . . . . . 1. *D. quercifolia.*
Involucre about half as long as the flowers.
    Leaves obovate or oblong-cuneate, prickly-toothed.
        Leaves white underneath . . . . 2. *D. præmorsa.*
        Leaves green on both sides . . . . 3. *D cuneata.*
    Leaves pinnatifid, with flat pungent-pointed lobes.
        Perianth-limb glabrous. Fruit 1-seeded . . . . 4. *D. falcata.*
        Perianth-limb more or less hairy. Fruit 2-seeded . . . . 5. *D. armata.*
    Leaves divided to the midrib or nearly so into small rigid segments with revolute margins.
        Leaves 6 in. to 1 ft. long, the lobes lanceolate or triangular . 6. *D. longifolia.*
        Leaves 2 to 4 in. long, the segments linear, distant. . . . 7. *D. Fraseri.*

o o 2

Series 2. **Floribundæ.**—*Flower-heads small, mostly terminal, the floral leaves either shorter than the flowers or few and spreading. Involucres broad. Perianths under 1 in. long.*

Leaves obovate or cuneate, prickly-toothed, flat.   Perianth silky-
   pubescent . . . . 8. *D. floribunda.*
Leaves lanceolate, prickly-toothed or semipinnatifid, flat. Perianth
   silky-hairy . . . . 9. *D. carduacea.*
Leaves linear with revolute margins, entire or with few prickly
   teeth.   Perianth glabrous . . . . 10. *D. carlinoides.*
Leaves pinnate with numerous small segments, the margins re-
   volute.
   Leaf-segments narrow, distant.   Perianth-limb glabrous . . 11. *D. polycephala.*
   Leaf-segments short, approximate.   Perianth-limb narrow,
     densely villous . . . . 12. *D. Kippistiana.*

Series 3. **Concinnæ.**—*Flower-heads small, broad, axillary, the bracts narrow, the floral leaves usually spreading. Leaves flat or nearly so, tomentose underneath, pinnatifid, with short lobes.*

Leaves narrow, the lobes small and distant . . . . 13. *D. squarrosa.*
  (See also 33, *D. patens*, with the flower-heads of the *Concinnæ*
  but the foliage of the *Obvallatæ.*)
Leaf-lobes contiguous, ovate-triangular, mucronate-acute.
Leaf-lobes reaching about halfway to the midrib.
   Involucral bracts acute, ciliate . . . . 14. *D. serra.*
   Involucral bracts obtuse, tomentose . . . . 15. *D. concinna.*
Leaf-lobes divided nearly to the midrib . . . . 16. *D. foliolata.*

Series 4. **Formosæ.**—*Flower-heads large, broad, terminal or axillary. Involucral bracts broad, villous. Leaves flat or nearly so, with numerous contiguous triangular lobes or segments, tomentose underneath, acute but not pungent-pointed.*

Leaf-lobes scarcely reaching above halfway to the midrib.   Flower-
  heads mostly terminal . . . . 17. *D. stupposa.*
Leaf-lobes deep but not reaching the midrib.   Flower-heads mostly
  lateral.
   Styles nearly 2 in. long . . . . 18. *D. nobilis.*
   Styles under 1½ in. long . . . . 19. *D mucronulata.*
Leaves divided to the midrib.
   Leaf-segments 2 to 4 lines long.   Flower-heads mostly ter-
    minal . . . . 20. *D. formosa.*
   Leaf-segments under 2 lines long.   Flower-heads mostly lateral 21. *D. Baxteri.*

Series 5. **Niveæ.**—*Flowering stems from a creeping trunk very short, with one or few ovoid flower-heads surrounded by long floral leaves. Leaves pinnate with numerous rigid segments white underneath except in* D. nana.

Leaf-segments contiguous, triangular or falcate, 1 to 3 lines long 22. *D. nivea.*
Leaf-segments separated by broad sinuses, linear, 2 to 4 lines long.
   Style under 2 in. long ; stigmatic end narrow . . . . . 23. *D. arctotidis.*
   Style about 3 in. long ; stigmatic end large, ovoid . . . . 24. *D. nana.*
Leaf segments linear, ¼ to above 1 in. long, some of them again
   lobed . . . . 25. *D. Preissii.*
  (See also 30, *D. vestita*, which has sometimes dwarf flower-
  ing-stems.)

Series 6. **Obvallatæ.**—*Flower-heads axillary, ovoid or small, enveloped in long floral leaves. Leaves either pinnate with very small rigid segments or more frequently pinnatifid with very rigid pungent-pointed lobes.*

Leaves pinnate with numerous decurrent segments, under 2 lines
  long, the margins revolute.

Involucral bracts numerous, with long plumose points.
Perianth about ¾ in. long . . . . 26. *D. sclerophylla.*
Involucral bracts few besides the leafy ones. Perianth
nearly 1 in. long . . . . 27. *D. pulchella.*
Leaves pinnatifid with pungent-pointed lobes.
Involucral bracts with long plumose-hairy points, or some
of them leafy.
Leaf-lobes triangular, approximate, white underneath . 28. *D. plumosa.*
Leaf-lobes linear or lanceolate, usually distant . . . 29. *D. seneciifolia.*
Involucral bracts numerous, narrow, tomentose or villous,
but not plumose.
Involucre narrow, 1 in. long. Leaf-lobes nearly flat.
Leaf-lobes about as long as the broad rhachis . . . 30. *D. vestita.*
Leaf-lobes much longer than the narrow rhachis . . 31. *D. cirsioides.*
Involucre campanulate or broadly ovoid, under ¾ in.
long. Leaf-lobes distant, with revolute margins,
white underneath.
Perianth-limb glabrous. Involucre broad, ½ in. dia-
meter.
Bracts with acute, usually recurved tips. Floral
leaves appressed . . . . 32. *D. Hewardiana.*
Bracts obtuse, appressed. Floral leaves spreading 33. *D. patens.*
Perianth-limb hairy. Involucre ovoid, ¾ in. long, the
bracts appressed or inflexed . . . . 34. *D. conferta.*
Involucral bracts hirsute, the inner bracts above 1 in. long,
the upper half reflexed and deciduous . . . . . . 35. *D. horrida.*
Involucres glabrous or nearly so, the bracts rather broad
and closely appressed.
Leaves 2 or 3 in. long, with linear or lanceolate lobes not
distant. Involucre ¾ in. long . . . . 36. *D. serratuloides.*
Leaves 6 in. to above 1 ft. long, very narrow, with small
distant lobes. Involucre above 1 in. long . . . . 37. *D. comosa.*

SERIES 7. **Gymnocephalæ.**—*Flower-heads lateral, on very short scaly peduncles
without floral leaves outside the involucre. Involucral bracts very numerous and
narrow, a few of them leaf-like in one species.*

Involucral bracts all very narrow, acute and dry.
Leaves (2 to 4 in.) pinnate with numerous very small seg-
ments with revolute margins and white underneath.
Involucre 1 in. long . . . . 38. *D. Shuttleworthiana.*
Leaves (3 to 5 in.) narrow and entire. Involucre 2 in. long 39. *D. speciosa.*
Several of the outer involucral bracts leaf-like. Leaves under
2 in. long, linear-cuneate, mostly 3-toothed . . . . . 40. *D. tridentata.*

SECT. 2. **Aphragmia.**—*Outer integuments of the two seeds not connate or readily
separable from each other (seeds without any or with a double plate between them).
Involucre large, with numerous broad bracts.*

Involucres broad, lateral below the leafy branches, the bracts
black, glabrous or minutely ciliate.
Leaves very narrow, entire, or with few or very numerous
short not pungent-pointed segments . . . . 41. *D. tenuifolia.*
Leaves under ¼ in. broad, pinnatifid with distant triangular pungent-pointed lobes . . . . 42. *D. proteoides.*
Leaves above ½ in. broad, pinnatifid with broadly triangular rigid acute lobes . . . . 43. *D. runcinata.*
Involucres ovoid, terminating very short ascending stems,
with a few leaves below them.
Leaf-lobes broadly triangular, rigid, acute . . . . . . 43. *D. runcinata.*

Leaf-lobes short, very numerous, regular and obtuse. In-
  volucre 2 in. long, glabrous and black . . . . . . 44. *D. obtusa.*
Leaf-lobes linear, often again divided. Involucre 3 in.
  long, pale-coloured, tomentose when young . . . . . . 45. *D. bipinnatifida.*
Involucres terminal, broad, villous, surrounded by long floral
  leaves.
  Leaf-segments linear or narrow-lanceolate . . . . 46. *D. pteridifolia.*
  Leaf-segments ovate lanceolate or triangular . . . . . 47. *D. calophylla.*

SECT. 1. EUDRYANDRA, Meissn.—Outer integuments of the inner
faces of the two seeds united in a bifid plate separating from them.
Involucres various.

See below, the observations under Sect. 2.

SERIES 1. ARMATÆ.—Flower-heads usually large, mostly terminal,
enclosed in floral leaves longer than the flowers. Involucres broad.
Perianths above 1 in. long. Stigmatic end of the style slender, often
scarcely distinct. Leaves with prickly teeth or lobes.

This series differs from the *Formosæ* chiefly in the foliage.

1. **D. quercifolia,** *Meissn. in DC. Prod.* xiv. 467. Branches stout,
tomentose or villous. Leaves obovate-oblong or oblong-cuneate, undulate and deeply prickly-toothed or lobed, contracted into a short petiole,
3 to 4 in. long, flat, very rigid, veined and reticulate underneath but
quite glabrous. Flower-heads terminal, very large, surrounded by
floral leaves longer than the flowers. Involucre hemispherical or
nearly globular, nearly 2 in. long, densely villous, the outer bracts
subulate-acuminate, the inner ones linear or linear-lanceolate. Perianth about as long as the involucre, hoary-tomentose above the short
glabrous base, the remainder silky-villous, the limb narrow, 3 lines
long. Style longer than the perianth, the stigmatic end long slender
and furrowed. Capsule obovate-falcate, fully ½ in. broad.—F. Muell.
Fragm. vii. 50.

W. Australia, *Drummond, 4th coll. n.* 307.

2. **D. præmorsa,** *Meissn. in Pl. Preiss.* ii. 265, *and in DC. Prod.* xiv.
467. Branches tomentose and sometimes hispid with spreading hairs.
Leaves obovate or oblong-cuneate, truncate, undulate, coarsely prickly-
toothed or lobed, 1½ to 3 in. long, white underneath, with prominent
transverse veins. Flower-heads terminal, surrounded by floral leaves
at least as long as the flowers. Involucre broad, the outer bracts
broadly lanceolate and tomentose, the inner ones narrow and acute, about
half as long as the flowers. Perianth above 1 in. long, silky-villous,
the limb 2 lines long, villous with longer hairs than those of the tube.
Style longer than the perianth, with a distinctly sulcate stigmatic end
of about 1 line. Capsule obovate-falcate, rather above ½ in. long.

W. Australia, *Drummond, n.* 26, 125, *2nd. coll. n.* 339, *5th coll. n.* 422.

3. **D. cuneata,** *R. Br. in Trans. Linn. Soc.* x. 212, *Prod.* 397. A
tall shrub, the branches rather thick, tomentose and often hispid with
long spreading hairs. Leaves shortly petiolate, from obovate to oblong-

cuneate, undulate and deeply prickly-toothed or almost entire, tapering
at the base, 1½ to 2½ or rarely 3 in. long, penniveined and reticulate
but not white underneath. Flower-heads terminal, closely surrounded
by floral leaves longer than the flowers. Involucre broad, about ½ in.
long, silky-tomentose, the outer bracts lanceolate and some of them
almost leafy, the inner ones very narrow, passing into the filiform paleæ.
Perianth about 1¼ in. long, hirsute with fine hairs, short on the tube
rather longer on the limb, the limb narrow, acute, 2½ lines long. Style
nearly 1½ in. long, the stigmatic end slender, obscurely furrowed. Capsule broadly rounded, about ½ in. diameter.—Meissn. in Pl. Preiss. i.
590, and in DC. Prod. xiv. 468.

W. Australia. King George's Sound or adjoining districts, *R. Brown, Baxter,
Drummond, n.* 175, *3rd coll. n.* 292, *Maxwell.*

4. **D. falcata,** *R. Br. in Trans. Linn. Soc.* x. 213, *Prod.* 397. A
shrub of 4 or 5 ft., the young branches usually tomentose and hirsute
with spreading hairs. Leaves more or less cuneate, pinnatifid or deeply
toothed with lanceolate pungent-pointed teeth or lobes, tapering at the
base but almost sessile, mostly 2 to 3 in. long, flat or undulate, very
rigid and not white underneath. Flower-heads terminal, closely surrounded by floral leaves longer than the flowers. Involucres broadly
ovoid or almost globular, 7 to 8 lines long, the outer bracts linear-lanceolate and tomentose, the inner ones narrow-linear. Perianth 1¼ to 1½
in. long, the tube woolly-tomentose above the glabrous base, the limb
glabrous. Style scarcely exceeding the perianth, the stigmatic end
slender and not very distinct. Capsule "1-seeded by abortion, the
abortive ovule forming a wing-like appendage to the interseminal plate."
—*Hemiclidia Baxteri,* R. Br. Prot. Nov. 40 ; Meissn. in Pl. Preiss. i.
601, and in DC. Prod. xiv. 482; Bot. Reg. t. 1455.

W. Australia. King George's Sound or to the eastward, *R. Brown, Baxter,
Drummond, 4th coll. n.* 321 ; near Cape Riche, *Preiss, n.* 527.

I have not succeeded in finding any capsules in any of our sets of Baxter's or of
Drummond's specimens, but as far as I can understand the characters given, the difference in the fruit upon which the genus *Hemiclidia* was founded is merely the result of
the abortion of one ovule, which occurs occasionally or perhaps constantly in one or two
other species of *Dryandra.* The foliage and inflorescence of *D. falcata* are precisely
those of *D. armata,* from which I am unable to distinguish flowering specimens except
by the glabrous perianth-limb.

5. **D. armata,** *R. Br. in Trans. Linn. Soc.* x. 212, *Prod.* 397. A
much-branched shrub of 2 to 4 ft., the young branches tomentose.
Leaves 2 to 3 in. long, deeply pinnatifid with lanceolate or triangular
pungent-pointed lobes, very rigid, flat or undulate, veined reticulate
and sometimes slightly tomentose underneath. Flower-heads terminal,
closely surrounded by floral leaves longer than the flowers. Involucre
broadly ovoid or almost globular, about ¾ in. long ; the bracts at first
villous at length becoming glabrous, the outer ones broad, the inner
narrow. Perianth above 1 in. long, more or less villous, the limb narrow, obtuse, becoming glabrous at the end but not entirely so as in that

species. Style exceeding the perianth, with a very narrow furrowed stigmatic end of about 1½ lines. Capsule "ripening both seeds imbedded normally in the interseminal plate."—Meissn. in Pl. Preiss. i. 590, and in DC. Prod. xiv. 468; Bot. Mag. t. 3236; *D. favosa,* Lindl. Swan Riv. App. 33.

W. Australia. King George's Sound or neighbouring districts, *R. Brown, Baxter, Drummond, n.* 1, *and 5th coll. n.* 421; Swan river, *Preiss, n.* 519; Blackwood river and Toodyay, *Oldfield;* Mount Melville and sources of the Kalgan river, *F. Mueller;* summit of Cape Arid, *Maxwell.* I have not seen ripe capsules of this species.

6. **D. longifolia,** *R. Br. in Trans. Linn. Soc.* x. 215, *Prod.* 398. A tall shrub, with tomentose branches. Leaves narrow, 6 in. to 1 ft. long, pinnatifid with lanceolate or triangular rigid acute lobes, contiguous or distant, 2 to 3 lines long or longer when narrow, the undivided rhachis 1 to 2 lines broad, the margins revolute, the under surface hoary or white. Flower-heads large, terminating short branches, surrounded by long floral leaves. Involucre broad, 1½ in. long, the outer bracts with a short broad base and subulate recurved points, the inner ones linear-lanceolate and shortly acuminate but variable in breadth. Perianth silky-pubescent, 1½ in. long, the limb hirsute with a few longer hairs, narrow, 2½ lines long. Style shortly exceeding the perianth, the stigmatic end scarcely distinct, slightly angular.—Meissn. in DC. Prod. xiv. 477; Bot. Mag. t. 1582; Sweet, Fl. Austral. t. 3; Paxt. Mag. iii. 171, with a fig.

W. Australia. Lucky Bay (?), *R. Brown, Baxter;* summit of Cape Arid, *Maxwell.*

7. **D. Fraseri,** *R. Br. Prot. Nov.* 39. An erect shrub of 2 or 3 ft., the young branches tomentose. Leaves narrow, 2 to 4 in. long, divided to the midrib into rather distant linear segments rigid and pungent-pointed, divaricate or recurved, 3 to 4 lines long, the margins revolute and narrowly decurrent to near the next segments. Flower-heads rather large and terminal or a few smaller ones on short axillary branches, all closely surrounded by floral leaves longer than the flowers. Involucre ¾ to 1 in. long, tomentose, the outer bracts broad at the base, tapering into long slender hairy points, the innermost linear. Perianth slightly silky except the glabrous base, 1¼ in. long, the limb narrow, about 2 lines long. Style exceeding the perianth, curved, the stigmatic end not thickened and only distinguishable by a somewhat darkened colour.—Meissn. in Pl. Preiss. i. 596, and in DC. Prod. xiv. 476.

W. Australia. Swan river, *Fraser, Drummond, n.* 129, *and 1st coll. n.* 642; York district, *Preiss, n.* 517; Dundagaran and Port Gregory, *Oldfield.*

SERIES 2. FLORIBUNDÆ.—Flower-heads small, mostly terminal, the floral leaves either shorter than the flowers or few and spreading, leaving the flowers more exposed than in any other series. Involucres broad. Perianths under 1 in. long. Stigmatic end of the style small, but thickened and distinct. Leaves with prickly or rigid teeth or lobes.

8. **D. floribunda,** *R. Br. in Trans. Linn. Soc.* x. 212, *Prod.* 397. A bushy shrub of 4 to 8 ft., the young shoots more or less silky-hairy. Leaves sessile or nearly so, obovate to cuneate, more or less undulate and prickly-toothed, especially towards the end, otherwise flat, neither prominently veined nor white underneath, all under 1 in. in some specimens, 2 in. long or even more in others. Flower-heads terminal, usually numerous, closely surrounded by floral leaves not exceeding the flowers. Involucre campanulate, under ½ in. long, pubescent; bracts not very acute, the outer ones lanceolate, the inner very narrow. Perianth not quite 1 in. long, the tube silky-pubescent above the glabrous base, the limb obtuse, almost glabrous. Style thickened and bulbous-like above the base, scarcely exceeding the perianth, the stigmatic end short, slightly clavate. Capsule obovate-falcate, ½ in. long in some specimens, smaller in others.—Meissn. in Pl. Preiss. i. 589, and in DC. Prod. xiv. 468; *Josephia sessilis*, Knight, Prot. 110.

W. Australia. King George's Sound, *R. Brown, Baxter*, and thence to Swan river, *Fraser, Drummond, n.* 118, *1st coll. n* 638, 639, *2nd coll. n.* 344, *Preiss, n.* 520, 521, *Oldfield ;* Champion Bay, *Oldfield.*

Var. *major.* Branches more tomentose and hairy. Leaves 2 to 2½ in. long, more frequently cordate ; flowers larger.—Bot. Mag. t. 1581.—Cape Naturalist, *Oldfield.*

The arborescent form mentioned by F. Mueller, Fragm. vi. 92, and vii. 50, is *Banksia ilicifolia.*

9. **D. carduacea,** *Lindl. Swan Riv. App.* 33. A tall shrub attaining sometimes 12 ft., the young branches slightly tomentose or glabrous. Leaves mostly sessile, linear-cuneate or lanceolate, undulate, deeply prickly-toothed or pinnatifid with pungent-pointed lobes, 1 to 2 or rarely 3 in. long, hoary or whitish underneath, but the margins not revolute. Flower-heads rather small, terminal, the floral leaves not exceeding the flowers. Involucre campanulate, about ½ in. long, the bracts very numerous, lanceolate or linear, with recurved tips. Perianth under 1 in. long, the limb about 1 line long, silky-hairy as well as the tube. Style scarcely exceeding the perianth, with a small slightly thickened stigmatic end. Capsule rounded, about 5 lines long and broad, 1-seeded by abortion in the one examined.—Meissn. in Pl. Preiss. i. 591, and in DC. Prod. xiv. 469; Bot. Mag. t. 4317.

W. Australia. Swan river, *Drummond, 1st coll., Preiss, n.* 516; Williams river and Toodyay, *Oldfield.* Some of Drummond's specimens belong to a form with longer and less prickly leaves and rather larger flower-heads, with the involucral bracts less squarrose, approaching in some respects *D. falcata* and *D. armata*, but with the habit and shorter floral leaves of the *Floribundæ.*

10. **D. carlinoides,** *Meissn. in Pl. Preiss.* ii. 267, *and in DC. Prod.* xiv. 479. An erect shrub, with the branches often almost verticillate round the old flower-heads (proceeding from the axils of some of the leafy bracts). Leaves linear or lanceolate, rigid and pungent-pointed, entire or with 1 or 2 prickly teeth on each side near the end, the margins revolute, tapering at the base, ¾ to 1 in. long, hoary or white underneath. Flower-heads terminal, usually numerous. Involucre

hemispherical or nearly globular, ¾ to 1 in. diameter, more or less
villous, with a few outer leafy bracts longer than the flowers, but
spreading and not enclosed in floral leaves, mostly dilated at the base
and passing into the imbricate bracts, which are very numerous, lanceolate with long narrow points. Paleæ plumose with long woolly
hairs. Perianths glabrous, about 7 lines long, the limb narrow, mucronate, 1½ lines long. Style rather longer than the perianth, the stigmatic end short, slightly thickened and angular. Capsules scarcely
above ¼ in. long.

W. Australia, *Drummond, 2nd coll. n.* 345.

11. **D. polycephala,** *Benth.* Branches rather slender, glabrous or
nearly so. Leaves narrow, divided to the midrib into small rather
distant segments, the lower leaves 3 to 6 in. long with short broad
obtuse segments, those of the flowering branches 1 to 2 in. long, very
spreading or recurved, with narrow acute segments of 1 to 2 lines; all
the segments very rigid, with recurved margins decurrent along the
rhachis to the next segment. Flower-heads small, numerous, terminating lateral branches or crowded at the end of the principal ones, the
floral leaves few and spreading. Involucre broadly campanulate, 3 to 4
lines long, the bracts numerous, narrow, with subulate usually recurved
points. Perianth-tube about ½ in. long, silky-villous except the minute
glabrous base, the limb glabrous, about 1 line long. Style longer than
the perianth, with a small but distinct clavate stigmatic end. Capsule
broadly obovate, not 3 lines long.—*D. squarrosa,* Meissn. in Pl. Preiss.
ii. 266, and in DC. Prod. xiv. 474, not of R. Br.

W. Australia, *Drummond, 1st coll., 2nd coll. n.* 342.

12. **D. Kippistiana,** *Meissn. in Hook. Kew Journ.* vii. 122, *and in
DC. Prod.* xiv. 473. An erect shrub of 2 or 3 ft., the branches loosely
hoary-tomentose, the young shoots often hairy. Leaves narrow, 1½ to
4 in. long, pinnate; segments divided to the midrib, numerous,
obliquely triangular, obtuse or acute, 1 to 2 lines long, the margins
revolute, decurrent along the rhachis, but shortly so the segments
being much closer than in *B. polycephala,* usually white underneath.
Flower heads scarcely larger than in *D. polycephala,* terminal with a few
also on very short axillary branches, the floral leaves few and spreading.
Involucre broadly campanulate, under ½ in. long, the bracts not
numerous, broad and tomentose at the base, tapering into fine points
ciliate with long hairs. Perianth-tube nearly ½ in. long, loosely hairy
above the glabrous base, the limb narrow, above 1 line long, densely
villous with longer hairs. Style longer than the perianth, with a small
but distinct dark-coloured obtuse stigmatic end.—*D. foliolata,* Meissn.
in Pl. Preiss. ii. 266, not of R. Br.

W. Australia, *Drummond, 4th coll. n.* 343; near Dundagaran, *Oldfield.*

SERIES 3. CONCINNÆ.—Flower-heads small, broad, axillary, the
bracts narrow, the floral leaves usually spreading. Perianth under ¾
in. long. Stigmatic end of the style small but thickened and distinct.

Leaves flat or nearly so, tomentose underneath, semipinnatifid with short acute mucronate or rarely pungent-pointed lobes.

This series has the flower heads of the *Floribundæ* but axillary, with the leaves of the *Plumosæ* but less deeply divided.

13. **D. squarrosa,** *R. Br. Prot. Nov.* 38. A shrub with rather slender branches, at first tomentose but soon becoming glabrous. Leaves narrow, the lower ones 4 to 8 in. long, those of the flowering branches usually about half that length, notched, prickly-toothed or pinnatifid, with short pungent-pointed or angular rather distant teeth or lobes rarely reaching half-way to the midrib, the entire centre of the leaf of a uniform breadth of $1\frac{1}{2}$ to $2\frac{1}{2}$ lines, the whole leaf flat or undulate, hoary or tomentose underneath. Flower-heads small, often numerous, mostly axillary surrounded by a few spreading floral leaves. Involucre broadly campanulate, under $\frac{1}{2}$ in. long, the bracts numerous, narrow, acute or with subulate often recurved points. Perianths silky-villous, about 7 lines long, the limb about 1 line long, villous with longer hairs. Style longer than the perianth, with a small slightly thickened stigmatic end.—Meissn. in DC. Prod. xiv. 474, as to Baxter's specimens only.

W. Australia. King George's Sound or to the eastward, *Baxter, Harvey.*

14. **D. serra,** *R. Br. Prot. Nov.* 38. An erect shrub, from 5 to 10 or even 15 ft. high. Leaves 2 to 6 in. long, divided halfway to the midrib into numerous broadly triangular regular lobes, mucronate with short rigid points, flat, reticulate above, tomentose underneath. Flower-heads small, on very short axillary peduncles or branches, surrounded by a few spreading floral leaves. Involucral bracts not very numerous, lanceolate or linear-lanceolate, acute, usually dark-coloured with densely ciliate margins, the inner ones 3 to 4 lines long. Perianths about 7 lines long, slender, silky-villous, the limb small, oblong, obtuse. Style about $\frac{3}{4}$ in. long, with a small but thickened stigmatic end. Capsule falcate, often $\frac{1}{2}$ in. long.—Meissn. in Pl. Preiss. i. 591, and in DC. Prod. xiv. 470.

W. Australia. King George's Sound or neighbouring districts, *Baxter, Drummond, n.* 172, *3rd coll. n.* 296, *Preiss, n.* 513; Wuljenup, *Maxwell.*

15. **D. concinna,** *R. Br. Prot. Nov.* 38, *not of Meissn.* A shrub, probably tall, with tomentose branches. Leaves 2 to 4 in. long, pinnatifid with triangular finely pointed lobes, reticulate above and tomentose underneath as in *D. serra,* but the leaf usually rather broader, with fewer lobes reaching about halfway to the midrib. Flower-heads small, globular, on very short axillary peduncles surrounded by spreading floral leaves as in *D. serra,* but the bracts more numerous, oblong or oblong-linear, very obtuse and tomentose all over, the inner ones 3 to 4 lines long. Perianths more villous than in *D. serra,* otherwise apparently the same but only seen withered. Capsule nearly $\frac{1}{2}$ inch long, oblique but not so falcate as in *D. serra.*

W. Australia. King George's Sound or to the eastward, *Baxter, Drummond, n.*
101.

16. **D. foliolata,** *R. Br. Prot. Nov.* 38. Apparently a tall shrub,
the branches tomentose and hirsute with spreading hairs or nearly
glabrous. Leaves 3 to 6 in. long, ½ to 1 in. broad, divided more than
halfway to the midrib into obliquely ovate-triangular lobes, acute or
mucronate, flat or nearly so, reticulate above, tomentose and transversely veined underneath. Flower-heads small, globular, on very
short axillary peduncles or branches, surrounded by spreading floral
leaves. Involucral bracts not very numerous, linear, softly villous, 3
or 4 lines long, mostly expanded at the end into a small lamina.
Perianths very villous, about ½ in. long. Style ¾ in. long, with a small
but thickened stigmatic end. Capsule obliquely rounded, about 5 lines
broad.—Meissn. in DC. Prod. xiv. 471 ; *D. mutica,* Meissn. l.c.

W. Australia. King George's Sound or neighbouring districts, *Baxter, Drummond, 4th coll. n.* 309; Stirling Range, *Oldfield, F. Mueller.*

SERIES 4. FORMOSÆ.—Flower-heads usually large, broad, terminal
or axillary, surrounded by long floral leaves. Involucral bracts broad,
villous. Styles long with a long narrow stigmatic end. Leaves flat or
nearly so, tomentose underneath, pinnatifid or pinnate, with numerous
contiguous triangular lobes or segments, acute or mucronate but not
pungent-pointed.

The inflorescence and flowers are nearly those of the *Armatæ,* but the foliage gives a
very different aspect to the specimens.

17. **D. stupposa,** *Lindl. Swan Riv. App.* 33. A shrub of about 10
ft., closely resembling *D. formosa,* but the leaves are not divided to the
midrib, the lobes often larger and more acute, and the flower-heads,
either terminal or on short lateral branches, are rather larger. Perianth
nearly 1½ in. long, woolly-villous above the glabrous base, the upper
part of the tube and limb silky-villous. Style longer than the perianth,
with a narrow furrowed stigmatic end.—Meissn. in Pl. Preiss. i. 591,
and in DC. Prod. xiv. 470.

W. Australia. Swan river, *Drummond, 1st coll. n.* 643 ; near Grantham, *Preiss,
n.* 502 (the latter specimen not seen).

18. **D. nobilis,** *Lindl. Swan Riv. App.* 33. A shrub of 4 to 7 ft.
very nearly allied to *D. formosa.* Leaves longer, the lobes broader,
separated by more open sinuses and not always divided to the midrib.
Flower-heads still larger than in *D. formosa,* but the involucre rather
smaller, and all on exceedingly short lateral branches, surrounded by
numerous floral leaves. Perianths 1½ in. long, woolly-villous above
the glabrous base, then silky-villous. Styles nearly 2 in. long.—Meissn.
in Pl. Preiss. i. 592, and in DC. Prod. xiv. 469 ; Bot. Mag. t. 4633,
copied into Lem. Fl. Jard. t. 226, and into Fl. des Serres. vii. t. 728.

W. Australia. Swan river, *Drummond, 1st coll. n.* 646 ; near Wicklow, *Preiss,
n.* 523 (*Meissn.*).

19. **D. mucronulata,** *R. Br. in Trans. Linn. Soc.* x. 213, *Prod.* 398.
A shrub very closely allied to *D. nobilis* and *D. formosa.* Branches tomentose and villous. Leaves very long and narrow, with very numerous
triangular-falcate rigid acute lobes which as in *D. nobilis* do not reach
the midrib, all nearly flat and tomentose underneath. Flower-heads
on very short axillary branches or almost sessile, surrounded by
numerous floral leaves, smaller than in *D. formosa.* Outer involucral
bracts ovate acuminate, the inner ones oblong-linear, obtuse, nearly
1 in. long and 2 lines broad, silky-villous. Perianths 8 to 10 lines
long, woolly-villous above the glabrous base, the remainder silky-
villous but the hairs not so long and fulvous as in *D. formosa.* Style
under 1 in. long. Capsule nearly ¾ in. broad.—Meissn. in DC. Prod.
xiv. 470.

W. Australia. King George's Sound or the neighbouring districts, *R. Brown,
Baxter, Drummond, 4th coll. n.* 311; Gordon plains, *Maxwell;* summits of Stirling
Range, *F. Mueller.*

20. **D. formosa,** *R. Br. in Trans. Linn. Soc.* x. 313, *t.* 3, *Prod.* 397.
An erect shrub attaining 8 to 15 ft., the branches tomentose and often
hirsute with long fine spreading hairs. Leaves 4 to 8 in. long, regularly divided to the midrib into obliquely triangular or broadly falcate
segments, 2 to 3 lines long and broad, mostly acute, flat and not very
thick, tomentose underneath. Flower-heads terminal, broad, surrounded
by floral leaves longer than the flowers, the inner ones dilated at the
base and passing into the involucral bracts. Involucre hemispherical,
1 to 1½ in. diameter, the outer bracts ovate acuminate, the inner ones
narrow and obtuse, all tomentose-villous. Perianths 1¼ to 1½ in. long,
woolly-villous above the short glabrous base, the remainder silky-
villous, the limb narrow acuminate, about 2 lines long, densely villous,
with long often fulvous hairs. Style scarcely longer than the perianth,
with a narrow furrowed stigmatic end. Capsule about 5 lines long
and 3 lines broad.—Meissn. in Pl. Preiss. i. 593, and in DC. Prod. xiv.
471; Sweet, Fl. Austral. t. 53; Bot. Mag. t. 4102.

W. Australia. King George's Sound, *R. Brown, Fraser, Drummond, 3rd coll.
n.* 293, *Preiss, n.* 501, and many others. The flower-heads and flowers vary in size,
even on the same specimens; some specimens from Barker and from Oldfield have them
all smaller than usual. The capsules appear to be always small.

21. **D. Baxteri,** *R. Br. Prot. Nov.* 38. A shrub of 4 to 6 ft., the
branches densely tomentose. Leaves very narrow, often above 1 ft.
long, divided to the midrib into very numerous small triangular-falcate
rigid acute segments, the largest scarcely 2 lines long and broad, all
with recurved margins and white underneath. Flower-heads almost
sessile in the axils, surrounded by long floral leaves. Involucre hemispherical, above 1 in. broad, densely ferruginous-villous, the bracts lanceolate, acuminate, the inner ones ¾ to 1 in. long. Perianths nearly 1
in. long, woolly near the base, then silky-villous, the limb 2 lines
long, narrow, acute, tipped with a tuft of long fine hairs. Style ex-

ceeding the perianth, the slender stigmatic end scarcely distinct.—
Meissn. in Pl. Preiss. i. 593, and in DC. Prod. xiv. 471.

W. Australia. King George's Sound or to the eastward, *Baxter, Drummond, Preiss, n.* 500.

SERIES 5. NIVEÆ.—Low shrubs with a creeping trunk and very
short ascending flowering stems bearing one or few ovoid flower-heads
surrounded by long floral leaves. Leaves pinnate with numerous rigid
segments, the margins usually but not always revolute and white underneath.

The species here enumerated differ in habit from all except some states of *D. vestita*
and two species of the section *Aphragmia*, which require further comparison with *D.
Preissii* as to their carpological characters.

22. **D. nivea,** *R. Br. in Trans. Linn. Soc.* x. 214, *Prod.* 398. A
dwarf shrub, the stems sometimes scarcely any besides the underground
or creeping trunk, rarely ascending to nearly 1 ft. Leaves 4 to 8 in. long,
pinnate, divided almost or quite to the midrib into numerous regular
triangular or falcate segments, obtuse or rarely acute, 1 to 3 lines long,
varying in breadth, those towards the end of the leaf usually separated
by acute sinuses, the lower ones more distant and decurrent, or all different in this respect in different leaves, all rather thick, with revolute
margins, white underneath. Flower-heads terminal, closely surrounded
by long floral leaves. Involucre ovoid, usually about 1 in. long; bracts
numerous, narrow, glabrous or minutely ciliate, or with the ends more
or less woolly, the outer short ones sometimes subulate, the inner ones
obtuse or scarcely acute. Perianths about as long as the involucre,
loosely villous except the undivided base, the limb scarcely 1½ lines long.
Style considerably longer than the perianth, with a small narrow stigmatic end slightly thickened at the base. Capsule obovate-falcate,
about ½ in. broad.—Meissn. in Pl. Preiss. i. 594, and in DC. Prod. xiv.
472; *Banksia nivea,* Labill. Voy. i. 411, t. 24; *Josephia rachidifolia,*
Knight, Prot. 111.

W. Australia. King George's Sound, *R. Brown,* and many others; eastward to
Cape Legrand, *Labillardière;* northward to Vasse, Swan, Moore and Murchison rivers,
Drummond, Preiss, Oldfield, and others.

This species, evidently widely spread in the sandy plains of W. Australia, includes
Drummond's, n. 64, 125, 134, 1st coll. n. 640, 641, 645, 2nd coll. n. 346, 5th coll. n. 419,
and Preiss's n. 506, 510, and (according to Meissner) 504 and 508, besides numerous
specimens from other collectors. Drummond's 4th coll. n. 313, with rather longer
flowers (*D. Brownii,* Meissn. in Pl. Preiss. i. 595, and in DC. Prod. xiv. 472), Preiss's
n. 511, from near Pointwater, with the involucral bracts rather more woolly at the end
(*D. Lindleyana,* Meissn. in Pl. Preiss. i. 598, and in DC. l.c.); and Drummond's 6th
coll. n. 212, from between Moore and Murchison rivers, with the leaf-segments rather
narrower and more distinct than usual (*D. stenoprion,* Meissn. in Hook. Kew Journ. vii.
122, and in DC. Prod. xiv. 473), appear to me to be scarcely distinguishable from specimens of the commoner forms even as marked varieties.

23. **D. arctotidis,** *R. Br. Prot. Nov.* 39. A dwarf shrub with the
habit of *D. nivea.* Leaves much more rigid, 4 to 8 in. long, deeply
divided into numerous linear-falcate rigid acute lobes, 2 to 4 lines long,

separated by broad sinuses, with revolute margins, white underneath. Flower-heads rather large, terminal, surrounded by numerous long floral leaves ciliate at the base with long spreading hairs. Involucre ovoid, above 1 in. long, the bracts numerous, oblong-lanceolate or the inner ones almost linear, nearly glabrous except the densely ciliate margins. Perianths 1¼ in. long, the undivided glabrous base longer than in most species, the remainder loosely villous. Style nearly 2 in. long, with a small narrow dark-coloured stigmatic end.—Meissn in Pl. Preiss. i. 595, and in DC. Prod. xiv. 475; Bot. Mag. t. 4035.

W. Australia. King George's Sound or neighbouring districts, *Baxter, Drummond, 5th coll. n.* 418; Mount Manypeak, *Preiss, n.* 515.

Var. *tortifolia.* Leaf-lobes narrower and more rigid, not so white underneath.—*D. tortifolia,* Kipp. in Hook. Kew Journ. vii. 121 ; Meissn. in DC. Prod. xiv. 475.—Between Moore and Murchison rivers, *Drummond, 6th coll. n.* 211. A specimen of Drummond's 3rd coll. suppl. n. 101, is intermediate as it were between this and the typical form as to foliage, but is not in flower.

24. **D. nana,** *Meissn. in Hook. Kew Journ.* vii. 121, *and in DC. Prod.* xiv. 475. A dwarf or creeping shrub with the habit of *D. nivea.* Leaves 2 to 4 in. long, crowded round the flower-heads on very short ascending stems, divided almost to the midrib into linear-lanceolate acute lobes, all very spreading and often falcate, 2 to 3 lines long, the margins not revolute, scarcely white underneath, with prominent primary veins. Involucre closely sessile within the leaves, ovoid, under ½ in. long, the bracts narrow-lanceolate, silky-villous, the outer ones with subulate points, the inner ones acute. Perianths with the entire base about ½ in. long villous towards the end, the divided portion of the tube about as long, the limb ovoid, reflexed before opening, slightly hairy. Style hairy, very long, doubled down to the limb until released, and then straightening to a length of about 3 in., with a large thick ovoid stigmatic end.

W. Australia. Near Dundagaran, *Drummond, 6th coll. n.* 210. With the habit of the *Niveæ,* this species has a somewhat different foliage, and differs from the whole genus in the remarkable style.

25. **D. Preissii,** *Meissn. in Pl. Preiss.* i. 599, *and in DC. Prod.* xiv. 480. A dwarf shrub with short procumbent stems tomentose and with linear-lanceolate scales on the base of each year's growth. Leaves rarely above 6 in. long, pinnate ; segments numerous, linear, acute or mucronate, rigid, with revolute margins, entire or pinnatifid, the larger ones above 1 in. long, but often all under ½ in., tomentose underneath. Flower-heads terminating the short ascending stems, with a few long floral leaves round them. Involucre ovoid, about 1 line long, the bracts numerous, all narrow, the outer ones with a short broader base and long subulate ends, the others linear or linear-lanceolate, flat and rigid, glabrous or loosely tomentose. Perianths about 1 in. long, loosely hirsute, the tube very slender, the limb broader, about 1 line long. Style 1¼ in. long, the stigmatic end small, narrow-conical. Fruit unknown.

W. Australia, *Drummond, 2nd coll. n.* 301; Gordon river, *Preiss, n.* 528 ; Stirling range and Hay river, *F. Mueller.* This species is placed next to *D. bipinnatifida* by Meissner on account of the foliage; the inflorescence and involucre, usually more indicative of true affinity, are more those of the *Niveæ*. The sectional character and consequently the real place cannot be ascertained until the fruit shall have been examined.

SERIES 6. OBVALLATÆ.—Flower-heads axillary, ovoid or small, enveloped in long floral leaves. Leaves either pinnate with very small rigid segments or more frequently pinnatifid with very rigid pungent-pointed lobes.

26. **D. sclerophylla,** *Meissn. in Hook. Kew Journ.* vii. 123, *and in DC. Prod.* xiv. 474. Apparently a low but erect shrub, not much branched. Leaves under 3 in. long, pinnate; segments numerous, triangular, acute, rarely 2 lines long, rigid, with revolute margins shortly decurrent to the next segments. Flower-heads not numerous, axillary or sometimes terminal, closely surrounded by numerous floral leaves of 2 or 3 in. Involucre 7 to 8 lines long, the bracts lanceolate, tapering into plumose points. Perianth about ¾ in. long, silky-villous, the oblong obtuse limb becoming almost glabrous. Style rather longer than the perianth, with a slightly thickened stigmatic end.

W. Australia. Between Moore and Murchison rivers, *Drummond, 6th coll. n.* 209. The species is very nearly allied to *D. pulchella*.

27. **D. pulchella,** *Meissn. in DC. Prod.* xiv. 473. Apparently a low but stout and erect shrub, the branches scarcely tomentose. Leaves 3 to 6 in. long, crowded, pinnate, with numerous rigid spreading acute or pungent-pointed segments rarely above 1 line long, the margins revolute and decurrent to the next segment. Flower-heads on short axillary branches closely surrounded by floral leaves. Involucre small, ovoid; bracts not numerous, the outer ones more or less leafy, the inner ones narrow, with long points. Perianth about 1 in. long, the tube slightly silky, the limb oblong, covered with rather long silky hairs. Style when set free nearly 1½ in. long, with a narrow but distinct stigmatic cone.

W. Australia, *Drummond, 4th coll. n.* 312.

28. **D. plumosa,** *R. Br. in Trans. Linn. Soc.* x. 214, *Prod.* 398. A shrub of about 2 ft., the branches tomentose and sometimes densely villous with fine spreading hairs. Leaves 6 in. to 1 ft. long, deeply pinnatifid with triangular rigid acute lobes, the larger ones 2 to 3 lines long and broad, the sinuses broad, the margins recurved, tomentose underneath. Flower-heads small, sessile in the axils, surrounded by a few small floral leaves. Involucre 1 in. long, or rather more, the bracts narrow, with long filiform plumose-hairy points. Perianth ½ in. long, densely woolly-villous, the limb oblong, about 1 line long. Style longer than the perianth, with a small slightly furrowed stigmatic end. Capsule about 7 lines broad.—Meissn. in Pl. Preiss. i. 592 and in DC. Prod. xiv. 470.

W. Australia. King George's Sound and neighbouring districts, *R. Brown, Baxter, Drummond, 4th coll. n* 310, *Preiss, n.* 507, *Maxwell*.

29. **D. seneciifolia,** *R. Br. Prot. Nov.* 39.　A shrub of 2 or 3 ft., with stout erect tomentose stems, sometimes nearly simple.　Leaves crowded, 2 to 4 in. long, deeply pinnatifid with rather distant linear or lanceolate pungent-pointed lobes 1 to 2 or rarely 3 lines long, the margins revolute, white underneath.　Flower-heads small, narrow. sessile in the axils and buried in the numerous floral leaves.　Involucral bracts, many of them leafy, the inner ones linear-subulate, with plumose-villous points, about ½ in. long.　Perianth about 5 lines long, woolly-villous above the glabrous base, the limb glabrous or sprinkled with few silky hairs.　Style scarcely exceeding the perianth, the stigmatic end not thickened and smooth.　Capsule ovate, scarcely ¼ in. long. —Meissn. in DC. Prod. xiv. 476 ; *D. cryptocephala*, Meissn. in Pl. Preiss. i. 596, and in DC. Prod. xiv. 479 ; Planch. Hort. Donat. t. 2.

W. Australia. King George's Sound or adjoining districts, *Baxter, Drummond, 3rd coll. n.* 297, *4th coll. n.* 316 ; rocky ridges, Perongerup range, *Maxwell.*

30. **D. vestita,** *Kipp. in Hook. Kew Journ.* vii. 121.　Stems in our specimens erect from a very thick woody trunk, ½ to 1½ ft. high, very rigid, hoary-tomentose or almost woolly.　Leaves linear or linear-cuneate, 3 to 5 in. long, very rigid, bordered by distant teeth or lobes rarely reaching halfway to the midrib, all divaricate acute or pungent-pointed, 1 to 1½ lines long, the entire centre or rhachis 1½ to 3 lines broad, transversely reticulate underneath.　Flower-heads axillary and terminal, closely surrounded by floral leaves.　Involucre ovoid-oblong, softly villous, 1¼ to 1½ in. long ; bracts numerous, narrow linear-lanceolate or linear, acuminate, articulate above the base.　Perianth above 1 in. long, woolly-villous above the glabrous base, the limb glabrous, 3 lines long.　Style about as long as the perianth, the long stigmatic end scarcely distinguishable.　Capsule oblique, above ½ in. long.—Meissn. in DC. Prod. xiv. 477.

W. Australia, *Drummond, n.* 158, *and 5th coll. suppl. n.* 20. This species ap-proaches the *Niveæ* in habit but is much more rigid and erect, with the thistle-like aspect of the *Obvallatæ.*

31. **D. cirsioides,** *Meissn. in DC. Prod.* xiv. 476.　Branches stout, tomentose and villous.　Leaves crowded, 2 to 3 in. long, deeply pin-natifid, but not quite to the midrib ; the lobes lanceolate, ¼ to nearly ½ in. long, very rigid and pungent-pointed, the margins slightly recurved, hoary or whitish underneath.　Flower-heads axillary, enclosed in nu-merous floral leaves.　Involucre ovoid, nearly 1 in. long, villous, the bracts numerous, linear-lanceolate or linear, rigid, appressed.　Perianths (only seen very few in a withered state) above 1 in. long, slender, villous above the glabrous base, the limb very narrow, 3 lines long. Styles all fallen from our specimens, the stigmatic end according to Meissner slender.

W. Australia, *Drummond, 4th coll. n.* 308.

32. **D. Hewardiana,** *Meissn. in DC. Prod.* xiv. 477.　Branches tomentose or nearly glabrous.　Leaves from 3 or 4 in. to nearly 1 ft.

long, pinnatifid, the lobes reaching more than half way to the midrib, obliquely lanceolate or triangular, rigid and pungent-pointed, often distant, 2 to 4 lines long, with recurved margins, white underneath. Flower-heads axillary, distant or crowded, surrounded by a few long floral leaves.. Involucre campanulate, rarely above ½ in. long, the outer bracts rather broad and acute, the inner ones narrow, all tomentose or villous and some or all tapering into fine often recurved points. Perianths nearly 1 in. long, woolly-villous above the short glabrous base, the remainder densely silky-hairy except the glabrous limb. Style not much longer than the perianth, with a small slightly clavate stigmatic end. Capsule about ½ in. long, densely villous.

W. Australia, *Drummond, 4th coll. n.* 315. This and the following species have the involucres and flowers almost of the *Concinnæ*.

33. **D. patens,** *Benth.* A branching shrub, nearly allied to *D. Hewardiana*, but with the fewer more spreading floral leaves of the *Concinnæ* and a different involucre. Leaves 4 to 10 in. long, deeply pinnatifid, the lobes lanceolate or triangular, very rigid and pungent-pointed, mostly distant, with recurved margins, white underneath. Flower-heads axillary, sessile or shortly pedunculate. Involucre campanulate, under ½ in. long as in *D. Hewardiana*, but tomentose not villous, the bracts broader obtuse or rarely mucronate, all appressed. Perianth nearly 1 in. long, woolly-villous above the short glabrous base, then silky-hairy except the glabrous limb. Style scarcely exceeding the perianth, with a small slightly clavate stigmatic end.—*D. concinna*, Meissn. in Pl. Preiss. ii. 266, and in DC. Prod. xiv. 477, not of R. Br.

W. Australia, *Drummond, 2nd coll. n.* 341.

34. **D. conferta,** *Benth.* A shrub apparently low, but with the stout erect stems of *D. cirsioides* and its allies. Leaves crowded, narrow, 3 to 6 in. long, pinnatifid, the lobes rather distant, rigid, pungent-pointed, white underneath with revolute margins as in *D. Hewardiana*. Flower-heads axillary, closely surrounded by long floral leaves. Involucre ovoid, villous and perhaps somewhat viscid, nearly ¾ in. long, the bracts numerous, narrow-lanceolate or linear, obtuse or scarcely acute, closely appressed or inflexed after flowering. Perianths under 1 in. long, densely woolly-villous above the short glabrous base, the limb narrow, 1½ lines long, villous with a few long hairs, as well as the upper part of the tube. Style scarcely exceeding the perianth, with a small slender stigmatic end.

W. Australia, *Drummond, 3rd coll. n.* 295. These specimens referred by Meissner to *D. patens* (*D. concinna*, Meissn., not of R. Br.), with doubt in Pl. Preiss. ii. 266, more positively in DC. Prod. xiv. 477, appear to me to differ too much in the involucres and perianths to be united with that species.

There are other specimens from Drummond, n. 7, with the foliage of this and the preceding species, with glabrous lanceolate involucral bracts approaching those of *D. serratuloides;* the flowers are however all fallen away, and the species, if really distinct, cannot be accurately described.

35. **D. horrida,** *Meissn. in DC. Prod.* xiv. 476. Branches thick, hoary-tomentose. Leaves crowded, narrow, 3˙to 6 in. long, pinnatifid, the lobes distant, rigid, pungent-pointed, 1½ to 3 lines long, divaricate or incurved, the margins of the lobes and of the narrow rhachis closely revolute, the under surface hoary or white where open. Flower-heads axillary, closely surrounded by long floral leaves. Involucre narrow ovoid, the bracts linear or a few of the shorter ones lanceolate with long points, all villous with rather long soft hairs, the inner ones 1½ in. long, but the upper half reflexed when the flowers are open and falling away soon after. Perianths 1¼ in. long, shortly woolly-villous above the glabrous base, the limb narrow, 3 lines long, glabrous as well as the upper part of the tube. Style longer than the perianth, with a long stigmatic end scarcely distinguishable from the remainder.

W. Australia, *Drummond, n.* 156, *4th coll. n.* 314.

36. **D. serratuloides,** *Meissn. in Hook. Kew Journ.* vii. 123, *and in DC. Prod.* xiv. 475. Branches hoary-tomentose. Leaves crowded, 2 to 3 in. long, deeply pinnatifid but not quite to the midrib, the lobes linear-lanceolate, often falcate, rigid and pungent-pointed, nearly flat, pale or scarcely white underneath, 2 to 4 lines long. Flower-heads axillary, closely surrounded by floral leaves. Involucre broadly ovoid or almost globular, about ¾ in. long, the bracts lanceolate or the outer ones ovate, obtuse, appressed, at first minutely ciliate, at length glabrous and smooth. Perianths about 1 in. long, silky-hairy except the glabrous base, and the hairs of the limb fewer and deciduous. Style considerably longer than the perianth, the stigmatic end not thicker but darker coloured and furrowed.

W. Australia. Moore river, *Drummond, 6th coll. n.* 213.

37. **D. comosa,** *Meissn. in DC. Prod.* xiv. 478. Branches slightly tomentose. Leaves 6 in. to above 1 ft. long, very narrow, rigid, flexuose, bordered by small pungent-pointed distant teeth or lobes, 1 to 1½ lines long, the margins of the teeth and rhachis revolute, leaving a narrow white under surface or channel between them and the broad midrib. Flower-heads axillary, with a few long floral leaves round them. Involucre broadly ovoid, 1¼ to 1½ in. long, the outer bracts ovate with short points, the inner ones lanceolate to linear, acute or the innermost almost obtuse, all glabrous or the margins minutely ciliate. Paleæ of the receptacle very slender but longer than in most species. Perianth-tube ¾ in. long, villous in the upper part of the undivided base, the limb narrow, about 2 lines long, silky-hairy. Style exceeding the perianth, the stigmatic end not thicker, but slightly furrowed.

W. Australia, *Drummond, 4th coll. n.* 313.

SERIES 7. GYMNOCEPHALÆ.—Flower-heads lateral, on very short scaly peduncles without floral leaves outside the involucre. Involucral bracts very numerous and narrow, a few of them leaf-like in one species.

38. **D. Shuttleworthiana,** *Meissn. in Hook. Kew Journ.* vii. 122, *and in DC. Prod.* xiv. 474. Apparently a low shrub, the leafy branches hoary-tomentose. Leaves narrow, 2 to 4 in. long, divided almost to the midrib into numerous contiguous obliquely-triangular lobes of 1 to 2 lines, all rather obtuse, rigid, with recurved margins, white underneath. Flower-heads almost sessile on the main stem below the leafy branches and without floral leaves, the very short peduncle covered with small or subulate and recurved scales. Involucre campanulate, the bracts narrow, mostly linear, very numerous, the inner ones 1½ in. long, recurved or reflexed from the middle, the long filiform ends usually ciliate with long fine hairs. Perianths 1 in. long or rather more, woolly-villous above the glabrous base, the limb very narrow, 3 lines long, glabrous. Style not exceeding the perianth, the stigmatic end scarcely distinguishable from the remainder. Capsule obovate, nearly ¾ in. long, densely rufous-villous.

W. Australia. Between Moore and Murchison rivers, *Drummond, 6th coll. n.* 208.

39. **D. speciosa,** *Meissn. in DC. Prod.* xiv. 479. Branches erect, tomentose. Leaves very narrow-linear, mucronate-acute, quite entire, with closely revolute margins, 3 to 5 in. long. Flower-heads very large, nodding, terminating very short leafy branches but not closely surrounded by floral leaves, the very short peduncles covered by small scales. Involucre very broad and above 2 in. long, the bracts very numerous, linear with fine points at first elegantly ciliate with spreading hairs which wear off. Perianth 1¼ to 1½ in. long and therefore shorter than the involucre, woolly-hirsute above the short glabrous base, the upper half glabrous, the limb very narrow, 4 lines long. Style scarcely exceeding the perianth, the stigmatic end not distinguishable from the remainder.

W. Australia, *Drummond, 5th coll. suppl. n.* 19.

40. **D. tridentata,** *Meissn. in Hook. Kew Journ.* vii. 120, *and in DC. Prod.* xiv. 479. Stems 6 in. to 1 ft. high from an underground creeping trunk. Leaves linear or linear-cuneate, mostly 3-toothed at the end, tapering into a very short petiole, 1 to 1½ in. long, flat, prominently reticulate underneath. Flower-heads large, on very short peduncles from below the foliage, leafless except a few narrow scales passing into the outer involucral bracts. Involucres broad, hemispherical, a few of the outer bracts leaf-like and longer than the flowers but mostly entire and dilated at the base, passing into ovate-lanceolate bracts with long narrow points and these again into the inner linear-lanceolate ones, the paleæ within the flowers few and very narrow. Perianths about 1 in. long, loosely villous, the limb narrow and acute. Style much longer, rarely quite straight, the slightly furrowed stigmatic end scarcely distinct. Capsule above ½ in. broad.

W. Australia. Near Dundagaran, *Drummond, 6th coll. n.* 207.

SECT. 2. APHRAGMIA.—Outer integuments of the 2 seeds in each capsule not connate or readily separable from each other, either remaining adnate to the seeds leaving no loose plate between them, or separating from the seeds and forming two parallel plates between them. Involucres large, with numerous broad bracts.

As far as known the carpological differences between the two sections of *Dryandra* appear to be constant, but there are several species of both in which the seed has not been examined, and the characters they furnish are very little available for practical purposes. The involucres, however, give to the species here included in *Aphragmia* a different aspect from all others of the genus. The structure of the seeds is perhaps not so different in the two as would at first appear. In both the nucleus has a double integument, whilst the wing is apparently formed of a prolongation of the outer integument, only with a different venation in the inner and outer layer (the prolongation of the inner and outer faces of the seed) which occasions the ready separation of the two layers when ripe. In *Eudryandra*, as in *Banksia*, this outer integument, wing-like, detaches itself from the inner face of the seed, becomes or remains connate with the corresponding integument of the other seed to the extent of the nucleus, the wing-like prolongations forming the two wings or lobes to the plate thus interposed between the ripe seeds, the wing-like prolongation of the outer integument on the outer face forming the simple wing to the seed. In *Aphragmia* the outer integument either remains adherent to the nucleus on both faces, the wing-like prolongations forming a double wing of which the external layer is deciduous and has been called an appendicular membrane, although the homologue of the wing in *Eudryandra*, or on the inner faces of the two seeds the respective outer integuments separate from the nucleus bearing with them their respective wing-like prolongations and forming two plates between the seeds. The species in which the latter peculiarity has been observed, *D. bipinnatifida*, has been separated on that account into a distinct section, *Diplophragma*, but in the few seeds that I have been able to examine, the separation of the integument from the nucleus when not consolidated with the corresponding integument of the other seed has not appeared to me to be at all constant. The whole question requires further investigation on the part of those who may have a sufficient supply of good fruits of the several species.

41. **D. tenuifolia,** *R. Br. in Trans. Linn. Soc.* x. 215, *Prod.* 398. A robust shrub, sometimes low and procumbent, sometimes erect bushy and attaining 3 or 4 ft., the branches nearly glabrous, with few narrow scales at the base of each year's shoot. Leaves very narrow, often 6 to 8 in. long, with closely revolute margins, tomentose underneath, rarely all entire, frequently toothed towards the end or in the upper half only, or in the typical forms regularly divided for more than half the length or quite to the base into short recurved lobes or teeth. Flower-heads large, lateral without any or with very few small linear floral leaves. Involucres at first ovoid, at length very broad, black and glabrous or when young slightly woolly, 1½ to 2 in. long; outer bracts broad, sometimes with short subulate points, inner ones broadly linear, obtuse. Perianths not exceeding the involucre, villous above the glabrous face, pubescent or glabrous towards the end, the limb very narrow, 3 lines long. Style not exceeding the perianth, with a slightly furrowed but not thickened stigmatic end. Capsule above ½ in. broad. Seeds (in the fruit examined perhaps not quite ripe) entirely separating without leaving any intermediate plate, the wing very thin though formed of two separable layers.—Meissn. in Pl. Preiss. i. 597, and in DC. Prod. xiv. 478; Bot. Mag. t. 3513; *D. uncata,* A. Cunn. Herb.

W. Australia. King George's Sound or to the eastward, *R. Brown, Baxter, Drummond,* 3rd coll. n. 294 ; Beaufort river, *Preiss, n.* 505 ; Tone river, *Oldfield.*— In Drummond's n. 294 the involucres are some of them as large as in *D. proteoides.*

Var. *elegans*. Leaves as in the typical form divided into numerous small segments with revolute margins white underneath; flower-heads and flowers smaller, the perianths more villous.—*D. elegans*, Meissn. in DC. Prod. xiv. 473.—W. Australia, *Drummond, 4th coll. n.* 317, *Maxwell*. In the only capsule I could examine I found only one perfect seed with two equal wings, each formed of 2 plates, probably the outer integuments of both seeds had attained their full growth and become consolidated at the base, the nucleus of one of them having aborted.

42. **D. proteoides,** *Lindl. Swan Riv. App.* 33. Very near *D. tenuifolia*, the stems more scaly, the leaves longer, broader although always under ½ in. broad, more rigid, divided into triangular rigid lobes contiguous or distant, very acute or even pungent-pointed or rarely almost obtuse. Flower-heads larger than in *D. tenuifolia*, on short lateral peduncles covered with imbricated scales without floral leaves. Involucre broadly ovoid, with very numerous broad black glabrous bracts, the innermost rows very much longer than the others and often attaining 3 in. Perianths not exceeding the involucre, glabrous or nearly so, the limb 4 lines long. Style about as long as the perianth, with a faintly sulcate but not thickened stigmatic end.—Meissn. in Pl. Preiss. i. 598, and in DC. Prod. xiv. 478.

W. Australia. Swan river, *Drummond, 1st coll., Preiss, n.* 503.

Var. *ferruginea*. Leaf-lobes rather broader, less acute and more distant, but not always so.—*D. ferruginea*, Kipp. in Hook. Kew Journ. vii. 123; Meissn. in DC. Prod. xiv. 478.—W. Australia, *Drummond, 5th coll. n.* 416.

43. **D. runcinata,** *Meissn. in DC. Prod.* xiv. 469. A dwarf shrub, the stems scarcely any or the leafy branches scarcely above ½ ft. long. Leaves 6 in. to nearly 1 ft. long, deeply divided into numerous triangular lanceolate or falcate acute lobes, the largest ½ in. long, with recurved margins, tomentose several-nerved and reticulate underneath. Flower-heads nearly sessile, terminal or lateral. Involucres ovoid, 2 in. long, entirely like those of *D. obtusa*, as well as the flowers and style.

W. Australia, *Drummond, 4th coll. n.* 318.

44. **D. obtusa,** *R. Br. in Trans. Linn. Soc.* x. 214, *Prod.* 398. Stems short and procumbent, tomentose-villous or concealed by imbricate scales. Leaves 6 in. to 1 ft. long, divided to the midrib or nearly so into numerous small triangular or oblong very obtuse segments, 1½ to 4 lines long, thick, with revolute margins, white underneath. Flower-heads terminal with a few floral leaves rather below them. Involucres ovoid, 2 in. long, the outer bracts short, ovate, passing into the long narrow inner ones, all obtuse, at first loosely tomentose, but soon quite glabrous turning black and finely striate like those of the three preceding species. Perianth nearly as long as the involucre, the tube slightly pubescent, the limb narrow, glabrous or hairy, 3 lines long. Style about as long as the perianth, the stigmatic end long narrow and furrowed.—Meissn. in DC. Prod. xiv. 471; *D. multiserialis*, F. Muell. Fragm. v. 185.

W. Australia. King George's Sound or to the eastward (Lucky Bay?) *R. Brown, Baxter, Drummond, 5th coll. n.* 420; inland from Cape Legrand, *Maxwell*.

45. D. bipinnatifida, *R. Br. Prot. Nov.* 39. Stems very short or procumbent and ½ ft. long or rather more, densely woolly-villous but the base of each year's growth concealed by imbricate scales. Leaves 6 in. to 1 ft. long, pinnate with linear acute segments, entire or again pinnatifid as in *D. Preissii*, 1 to 2 in. long in some specimens, much smaller in others, all with revolute margins, reticulate and tomentose underneath. Flower-heads terminal but not closely surrounded by floral leaves. Involucre ovoid-oblong, 2 to 2½ in. long, the outer bracts ovate, the inner ones narrow-lanceolate, all obtuse, more or less woolly-villous or at length glabrous, but not black as in the preceding species, the paleæ within the head shorter and narrow. Perianth shorter than the involucre, about 1½ in. long, loosely villous or pubescent below the middle, glabrous towards the end, the very narrow limb ½ in. long. Style exceeding the perianth, with a long furrowed stigmatic end. Capsule about ½ in. broad.—Meissn. in Pl. Preiss. i. 599, and in DC. Prod. xiv. 480.

W. Australia. Swan river, *Fraser, Drummond, 1st coll. n.* 644, *Preiss, n.* 522. In the only fruit I could examine the seed was destroyed by insects. According to R. Brown, the outer integuments of the inner faces of the two seeds are free from the seeds and from each other (or separable), forming a double plate between the seeds.

46. D. pteridifolia, *R. Br. in Trans. Linn. Soc.* x. 215, *Prod.* 399. Stems very short and thick, densely tomentose and villous. Leaves often above 1 ft. long, pinnately divided almost or quite to the midrib into numerous linear or lanceolate straight or falcate segments, ¾ to 1½ or even 2 in. long, often distant but usually dilated at the base and frequently confluent, all with recurved or revolute margins, more or less tomentose underneath, 1-nerved in some leaves, 3- to 5-nerved in other leaves on the same stem. Flower-heads large, terminal, closely surrounded by long floral leaves. Involucre hemispherical, the bracts densely villous, the outer ones ovate, the inner ones lanceolate, ¾ to above 1 in. long. Perianths about 1¼ in. long, silky or loosely villous with long hairs, the limb 4 to 5 lines long. Style about as long as the perianth, with a long furrowed stigmatic end. Capsule about ¾ in. broad. Seeds in the two fruits examined quite separate without any intervening plate, each with a double wing, the inner one more transparent with flexuose fibres, the outer one (membranous appendage, *R. Br.*) more opaque.—Meissn. in DC. Prod. xiv. 480 ; Bot. Mag. t. 3500 ; *D. blechnifolia*, R. Br., in Trans. Linn. Soc. x. 215, Prod. 399 ; *D. nervosa*, R. Br., in Sweet, Fl. Austral. 22, Prot. Nov. 39 ; Meissn. in Pl. Preiss. i. 600, and in DC. Prod. xiv. 481 ; Bot. Mag. t. 3063.

W. Australia. King George's Sound and adjoining districts, frequent, *R. Brown, Baxter, A. Cunningham, Drummond, n.* 131, *4th coll. n.* 320, *5th coll. n.* 423, *Preiss, n.* 512, and others ; scrubby plains from Stirling to Phillips Ranges and to Cape Arid, *Maxwell.* The breadth of the leaf-segments and the size of the flower-heads do not appear to be sufficiently constant to establish distinct varieties.

47. D. calophylla, *R. Br. Prot. Nov.* 40. A low shrub, the villous stems either very short and thick or rather longer and prostrate.

Leaves often above 1 ft. long, pinnate with numerous ovate-lanceolate or triangular-acute rigid segments; contiguous at the base and mostly separated by acute sinuses, pale, tomentose and several-nerved underneath, the larger ones 1 to 1½ in. long. Flower-heads terminal, closely surrounded by long floral leaves. Involucre broad, densely villous, a few of the outer bracts long and narrow, sometimes resembling reduced floral leaves, others broad and short, the inner ones linear-lanceolate. Perianths villous, at least 1¼ in. long, the limb 4 to 5 lines long. Style about as long as the perianth, with a long narrow furrowed stigmatic end. Capsule of *D. pteridifolia*, or rather larger.—Meissn. in DC. Prod. xiv. 481; *D. Drummondii*, Meissn. in Pl. Preiss. ii. 267, and in DC. l.c.

W. Australia. King George's Sound or neighbouring districts, *Baxter, Drummond, 2nd coll. n.* 299, 300, 301, *4th coll. n.* 319; Kalgan river, *Oldfield.*

ADDENDUM.

Under Verbenaceæ, after the synopsis of genera, p. 33, add—

Pentaptelion involucratum, Turcz. in Bull. Soc. Imp. Nat. Mosc. 1863, ii. 194, proposed as a new genus of Verbenaceæ, is *Leucopogon plumuliflorus*, described above, vol. iii. p. 205.

INDEX OF GENERA AND SPECIES.

————◆————

The Synonyms and Species incidentally mentioned are printed in Italics.

598 INDEX OF GENERA AND SPECIES.

END OF VOL. V.